Complete Automotive Engine Rebuilding and Parts Machining

by Robert Scharff & the Editors of AUTOMOTIVE REBUILDER

WE ENCOURAGE
PROFESSIONALISM

THROUGH TECHNICIAN
CERTIFICATION

Delmar Publishers Inc.®

NOTICE TO THE READER

Scharff Staff
Production Manager: Marilyn Strouse-Hauptly
Editor: Lois Breiner
Layout Design: Karen Weaver
Cover Design: Eric Schreader
Logo Design: Ed Foulk

For information address Delmar Publishers, Inc.
2 Computer Drive West, Box 15-015,
Albany, New York 12212

Delmar Staff
Editor-in-Chief: Mark W. Huth
Administrative Editor: Joan Gill
Project Editor: Carol Micheli

Printed in the United States of America
Published simultaneously in Canada
by Nelson Canada,
a division of The Thomson Corporation

10 9 8 7 6 5 4 3 2 1

Library of Congress Cataloging-in-Publication Data

Scharff, Robert.
    Complete automotive engine rebuilding and parts machining / by
Robert Scharff & the editors of Automotive rebuilder.
        p.   cm.
    ISBN 0-8273-3612-8. — ISBN 0-8273-3613-6 (instructor's guide). —
ISBN 0-8273-3614-4 (shop manual)
    1. Automobiles—Motors—Maintenance and repair.   I. Automotive
rebuilder.   II. Title.
TL210.S34 1991
629.25'04—dc20   90-3301
                                                  CIP

CONTENTS

Preface   v

Acknowledgments   v

CHAPTER ONE   Shop Practices, Tools, and Safety

Engine Identification 3   Tools and Machines 5 General Shop Safety 24   ASE Certification 30 Review Questions 31

CHAPTER TWO   Modern Vehicle Engines

Engine Classifications 34   The Four-Stroke Gasoline Engine 34   Characteristics of Engine Design 38   Intake and Exhaust Valves 56 Camshaft 60   Crankshaft 67   Piston and Piston Rings 70   Engine Gaskets and Oil Seals 73   Gasoline Two-Stroke Engines 74 Gasoline Engine Systems 75   Diesel Engines 76   Review Questions 79

CHAPTER THREE   Diagnosing Before Engine Teardown

Compression Testing 81   Cylinder Leakage Testing 86   Cylinder Power Balance Testing 92   Evaluating the Engine's Condition 94   Oil Consumption 94   Noise Diagnosis 98   Diagnosing Diesel Engine Problems 100   Review Questions 101

CHAPTER FOUR   Engine Removal and Disassembly

Preparing the Engine for Removal 104   Lifting an Engine 108   Engine Disassembly 110 Ordering Replacement Parts 118   Review Questions 121

CHAPTER FIVE   Cleaning Engine Parts

Removing Soils 123   Abrasive Cleaners 135 Cleaning by Hand 139   Review Questions 142

CHAPTER SIX   Inspecting the Cylinder Head and Related Parts

Valves 144   Valve Inspection 146   Valve Train Inspection 167   Review Questions 179

CHAPTER SEVEN   Inspecting the Cylinder Block and Related Parts

Crankshaft Inspection 181   Valve Timing Drive Assembly 188   Piston Assembly Inspection 191   Cylinder Block Inspection 210 Bearing Inspection 215   Replacement Bearings 225   Review Questions 228

CHAPTER EIGHT   Crack Repair

Detecting Cracks 230   Cold Crack Repair— Pinning Process 234   Repairing Cracks by Welding 240   Review Questions 248

CHAPTER NINE   Reconditioning Cylinder Heads and Blocks

Straightening Aluminum Cylinder Heads 249 Resurfacing Cylinder Heads 253   Review Questions 269

CHAPTER TEN   Reconditioning Valve Train Components

Grinding Valves 271   Valve Guides 276   Valve Stem Seals 300   Valve Train Components 303 Review Questions 312

CHAPTER ELEVEN   Reconditioning Related Cylinder Block Parts

Cylinder Reconditioning 315   Pistons 329 Review Questions 346

CHAPTER TWELVE   Reconditioning Crankshafts, Camshafts, and Engine Balancing

Crankshaft 347   Reconditioning Crankshafts 352   Camshaft 373   Engine Balancing 381   Review Questions 387

CHAPTER THIRTEEN   Lubricating and Cooling Systems

Lubrication System 389   Cooling System 406   Review Questions 410

CHAPTER FOURTEEN Sealing the Engine

Fasteners *411* Gaskets *420* Adhesives, Sealants, and Other Chemical Sealing Materials *434* Oil Seals *437* Sealing Diesel Engines *443* Review Questions *445*

CHAPTER FIFTEEN Engine Reassembly and Installation

Rechecking and Recleaning the Block *447* Installing Core Plugs *449* Installing the Cam Bearings and Camshaft *450* Installing Main Bearings and Crankshaft *456* Installing Pistons and Connecting Rods *459* Installing the Timing Components *463* Installing the Cylinder Head and Valve Train *466* Installing the Oil Pump *471* Installing Valve, Oil Pan, and Timing Front Covers *473* Installing the Engine *475* Review Questions *479*

Appendix A	Glossary	481
Appendix B	Measurement Equivalents	490
Appendix C	Tap and Drill Bit Data	493
Appendix D	Reference Tables	496
Appendix E	Torque Values	497
Index		500

PREFACE

It has been said that the day after the first car was built, the repair industry was started. Maybe it was not until two or three days later, but the truth is that the early cars needed plenty of help because most of them were "working experiments."

By the early 1920s there were already machine shops capable of performing more than just a simple overhaul. By the early 1930s, there were a number of engine rebuilders doing business. Some were automotive jobber machine shops that grew as an adjunct to parts sales, especially on the West Coast. Others were repair shops that would remove, rebuild, and install the same engine. Following World War II, the rebuilding industry grew at a phenomenal rate and has been expanding ever since. But today, the rebuilder or machine shop—whether a small shop or large production remanufacturer—requires skilled automotive technicians or machinists.

This book was prepared to help the rebuilding industry train qualified personnel. To accomplish this task, the subject matter has been broken down into fifteen chapters plus a glossary.

Chapter 1 contains an overview of the rebuilding profession and shop practices, including commonly used tools. (The use and operation of the large, special-purpose tools are described in later chapters.) The important subject of measurement is fully detailed as well as shop safety. The up-to-date right-to-know laws are described, along with the importance of ASE certification and the aid given by the Automotive Engine Rebuilder's Association (AERA).

Chapter 2 gives a complete description of modern vehicle engines, including such up-to-date information as aluminum heads, dual camshafts, and multivalve engines.

The information given in Chapter 3 will help the automotive technician to determine—through the text described—if an engine requires teardown (or disassembly) and rebuilding.

Chapter 4 describes how an engine should be removed from a vehicle and then how the various components should be removed from the engine.

Once the engine is disassembled, all the components must be cleaned. Chapter 5 describes the various methods of cleaning an engine and its components.

Chapters 6 and 7 cover the inspection and analysis of upper and lower engine parts, while Chapters 9, 10, 11, and 12 cover the reconditioning of these components. These chapters also contain procedures for various machining operations. Chapter 8 describes the methods of repairing engine cracks, including the welding of aluminum heads.

Proper operation of the engine's lubricating and cooling systems as described in Chapter 13 is important to the well-being of any engine rebuilding job. Chapter 14 describes the method of sealing an engine.

Chapter 15 covers methods of reassembling an engine and putting it back into service, thus completing the rebuilding procedure.

Each chapter opens with a list of objectives so that the student knows what he or she can expect to learn. At the end of each chapter, there are review questions to check students' understanding of the text. The glossary gives the most commonly used terms in a rebuilder shop.

Complete Engine Rebuilding and Parts Machining contains cautions and warnings that should be carefully studied to avoid personal injury or vehicle damage when rebuilding an engine. However, it is important to understand that these cautions and warnings could not possibly cover all rebuilding procedure variations and their particular hazards, nor can the contributors, the publisher, or the editors and their staff possibly know or investigate all such variations. It is, therefore, the responsibility of anyone using the rebuilding procedures or tools (whether or not recommended in this book) to ensure to their own satisfaction that neither personal safety nor vehicle safety will be jeopardized.

ACKNOWLEDGMENTS

To organize a book requires the help of a great number of sources of information and the aid of many people and organizations. The cooperation of the staff and contributing editors of *Automotive Rebuilder* magazine plus Becky Babcox and David Wooldridge played a major role in the preparation of this text and in furnishing illustrations. In addition we would like to thank the "big three" automotive manufacturers—Chrysler, Ford, and General Motors—for permission to use information and illustrations from their training programs and service manuals. Another important thanks to *Motor* (a Hearst Corporation publication) for the use of the timing data that appears in Chapter 12.

Grateful acknowledgment is also made to the following companies for reference materials and illustrations used in this book.

Am/Pro Machinery, Inc.
Atlas Engineering and Manufacturing, Inc.
Bayco Inc.
Bear Automotive Equipment Company
Black & Decker, Inc.
Botts Auto Parts Company
Bowman Distribution Company
Brush Research Manufacturing Company, Inc.
Central Tools, Inc.
Clayton Dynamometer, Inc.
C&M Cleaning Systems, Inc.
Crane Cams, Inc.
CR Industries, Inc.
Dana Corp.
Detroit Gasket, Inc.
Dresser Industries, Inc.
Dyer Company
Economy Garage and Machine Shop, Inc.
Enginetech, Inc.
Everco Industries, Inc.
Federal Mogul Corp.
Fel-Pro Inc.
Fred V. Fowler Company, Inc.
Goff Corp.
Goodson Shop Supplies
Great Lakes Energy Systems, Inc.
Hall-Toledo, Inc.
Hamilton Test Products, Inc.
Hartridge Equipment Corp.
Hastings Manufacturing Company
Heli-Coil Products, Inc.
Hines Industries, Inc.
Howard Supply Co.
Irontite Products Company, Inc.
Jasper Engine & Transmission Exchange, Inc.
J P Industries, Inc.
Kansas Instruments Inc.
Kent Moore Corp.
K-Line Industries, Inc.
K. O. Lee Company
Kolene Corporation
Krautkramer Branson, Inc.

Kwik-Way Manufacturing Company
Lenox Instrument Company
Lincoln Inc.
Lisle Corp.
Loctite Corp.
Los Angeles Sleeve Company
L. S. Industries, Inc.
L. S. Starrett Company
Lubriplate Corp.
Mac Tools, Inc.
Magnaflux Corp.
Man-Gil Chemical Corp.
Melling Automotive Products, Inc.
Meridian Parts Corp.
Miller Electric Manufacturing Company
NAPA Engine & Chassis Parts Division
Neway Manufacturing, Inc.
Owatonna Tool Company
Peterson Machine Tool, Inc.
Ramco Corporation
Rogers Machine Company
Rottler Manufacturing Company
Safety-Kleen Corp.
Sealed Power Corp.
Silver Seal Products Company, Inc.
Sioux Tools, Inc.
Snap-on Tools Corp.
Stewart-Warner, Inc.
Storm Vulcan, Inc.
Sun Electric Corp.
Sunnen Products Company
T. Hoff Manufacturing, Inc.
Tool Division—SPX Corp.
TRW, Inc.
Unocal
Viking Corp.
Vulcan Material Company
Winona Van Norman Machine Company
Wolverine Gear & Parts Company
Zero Manufacturing Company
Zollner Corp.

We would like to thank the following for reviewing the manuscript and for their helpful comments:

George Benda
S. A. I. T.
Calgary
Alberta, Canada

I. Andrew Norman
Trident Technical College
Charleston, South Carolina

Max Morley
Anchorage Community College
Anchorage, Alaska

Dan Claus
Macomb Community College
Warren, Michigan

Joe Polich
AERA
Buffalo Grove, Illinois

CHAPTER ONE

SHOP PRACTICES, TOOLS, AND SAFETY

Objectives

After reading this chapter, you should be able to
- Define what each digit of the VIN identifies.
- Explain what advantages pneumatic tools have over electrically powered equipment.
- List and describe the various engine measuring tools.
- List and describe the publications that should be available to engine rebuilder technicians and shop machinists.
- Identify and describe the various forms used in the rebuilding shop.
- Explain the rules regarding personal safety, work area safety, tool and machine safety, and hazardous wastes.
- Explain the requirements for ASE/AERA certification.

When an engine reaches a point where it cannot attain its rated power output, performance must be restored through maintenance. Depending on the engine conditions, performance can be restored through a tune-up, a minor overhaul, or a major overhaul.

Broadly defined, a tune-up (Figure 1-1) is a series of adjustments, replacement of parts, and reconditioning operations that can be performed without removing the engine cylinder head or oil pan. This includes service to the fuel system, ignition system, starter system, and possibly the charging system. Engine tuning is a periodic maintenance requirement that is necessary to keep the engine operating at peak efficiency and to prevent premature damage to internal engine components.

A minor overhaul consists of operations performed with the cylinder head or oil pan removed but with the engine still mounted in the vehicle. A minor overhaul will put the engine back into reasonable running order without excessive downtime. A minor

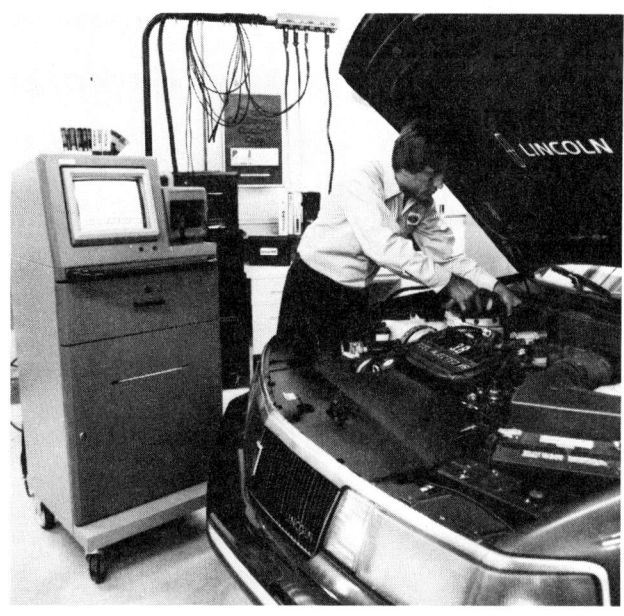

FIGURE 1-1 Engine tune-up

overhaul could involve replacing valves, valve guides, valve springs, pistons, piston rings, main bearings, and connecting rod bearings (Figure 1-2).

A major overhaul or rebuilding job requires the removal of the engine from the vehicle and its complete disassembly (Figure 1-3). A major overhaul involves the cleaning, inspection, replacement, and/or reconditioning of all engine parts. A major overhaul will result in a rebuilt engine that has nearly the same performance characteristics and life expectancy of a new engine.

FIGURE 1-2 Minor overhaul

Over a number of years, an engine can be rebuilt several times to extend its life, provided that each overhaul or rebuild job is performed properly. The useful life of any engine, therefore, can be extended indefinitely.

Overhaul is a word seldom found in an engine rebuilder's vocabulary. It is the term used by full-service shops to describe the engine work that they do, which usually involves a tune-up, simple overhaul, or a thorough examination to determine if the engine requires rebuilding. It must be kept in mind that only a few full-service repair shops have the equipment and/or the skills to completely rebuild an engine. Consequently, if their inspection reveals that the engine must be rebuilt, it is taken from the vehicle and sent to a rebuilder machine shop to have the work done. After the engine has been rebuilt, it is returned to the full-service shop to be reinstalled into the vehicle. However, in some cases, the rebuilder must also determine if the engine requires rebuilding, and the rebuilder shop does the actual removing and reinstallation of the engine.

Engine rebuilder machine shops vary in size. Many small shops (Figure 1-4) have a few machinists who must do all phases of rebuilding. And there are large organizations, often called production shops (Figure 1-5), in which specialists do the various remanufacturing jobs. But regardless of size,

FIGURE 1-3 Rebuilding includes complete disassembly.

FIGURE 1-4   Operations in a small rebuilder shop

FIGURE 1-5   Inside a larger production shop

FIGURE 1-6   Small, 4-cylinder engine

FIGURE 1-7   Large diesel engine

the main function of all rebuilder machine shops is to produce quality remanufactured engines. The primary emphasis of this book is on the procedures involved in rebuilding gasoline and diesel engines. These engines can be small (Figure 1-6) or large (Figure 1-7).

ENGINE IDENTIFICATION

Both the rebuilder technician and machinist must know the Vehicle Identification Number (VIN). The VIN is a code of seventeen numbers and letters stamped on a metal tab that is riveted to the instrument panel close to the windshield. The VIN can be seen through the front windshield on the driver's side. VIN identification is also found on the Vehicle Safety Compliance Certification Label of trucks (Figure 1-8).

The adoption of the seventeen number and letter code became mandatory beginning with 1981 vehicles. The standard VIN of the United States National Highway Transportation and Safety Administration, Department of Transportation is being used by all manufacturers of vehicles both domestic and foreign.

By looking at the VIN a variety of information about the vehicle can be determined (Figure 1-9). The engine calibration number is generally stamped on the valve rocker cover or on a machined pad somewhere on the engine block (Figure 1-10).

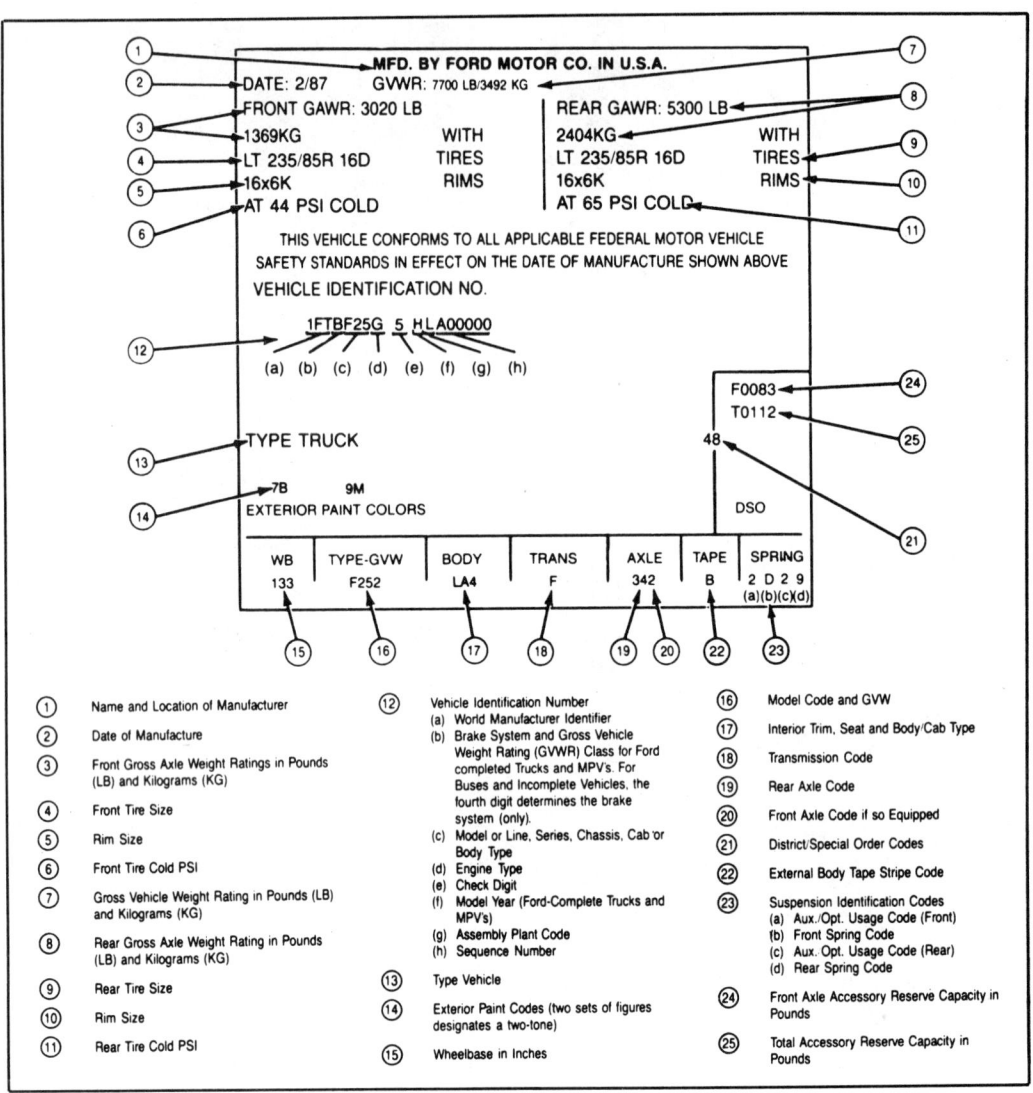

FIGURE 1-8 Typical truck safety compliance certification label *(Courtesy of Ford Motor Co.)*

SAMPLE VIN NUMBER

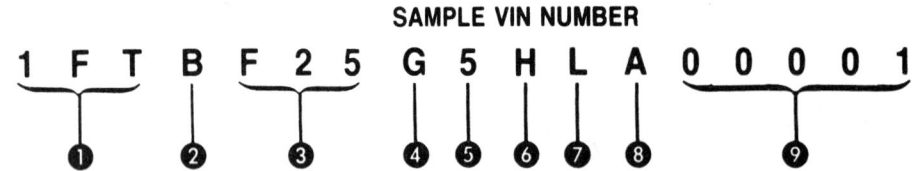

❶ Position 1, 2 and 3 — Manufacturer, Make and Type (World Manufacturer Identifier)

❷ Position 4 — Brakes System GVWR Class for Ford-completed Trucks and MPV's. For Buses and Incomplete Vehicles, Brake System (only).

❸ Position 5, 6 and 7 — Model or Line, Series, Chassis, Cab or Body Type

❹ Position 8 — Engine Type

❺ Position 9 — Check Digit

❻ Position 10 — Model Year (Ford-completed vehicles)

❼ Position 11 — Assembly Plant

❽ Position 12 — Constant "A" until sequence number of 99,999 is reached, then changes to a constant "B" and so on

❾ Position 13 through 17 — Sequence number — begins at 00001

FIGURE 1-9 Vehicle identification number

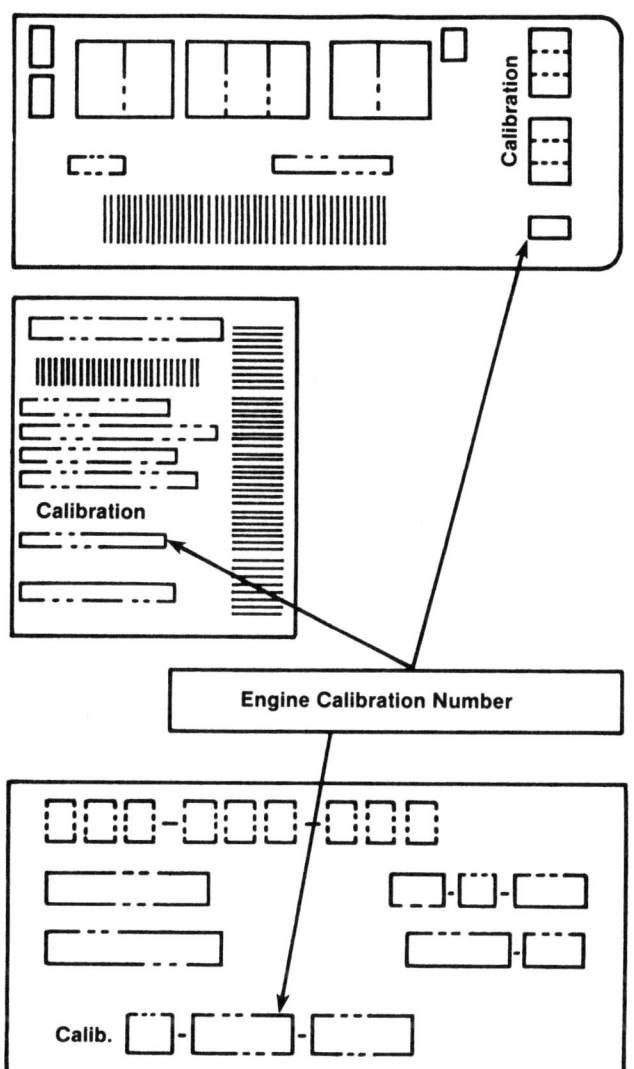

FIGURE 1-10   Typical engine calibration tabs

TOOLS AND MACHINES

There is no substitute for getting the job done right the first time. For parts personnel, that means being able to provide both the correct part(s) and helpful information. For the rebuilder technician or machinist, it means doing the job as quickly and safely as possible.

Many of the tools an engine rebuilder technician or machinist uses every day are general-purpose hand tools (Figure 1-11). For instance, a complete collection of wrenches is indispensable. A variety of auto engine parts, accessories, and related parts, not to mention shop equipment, use common bolt and nut fasteners as well as special hex screws and fasteners. Depending on the make and model of the

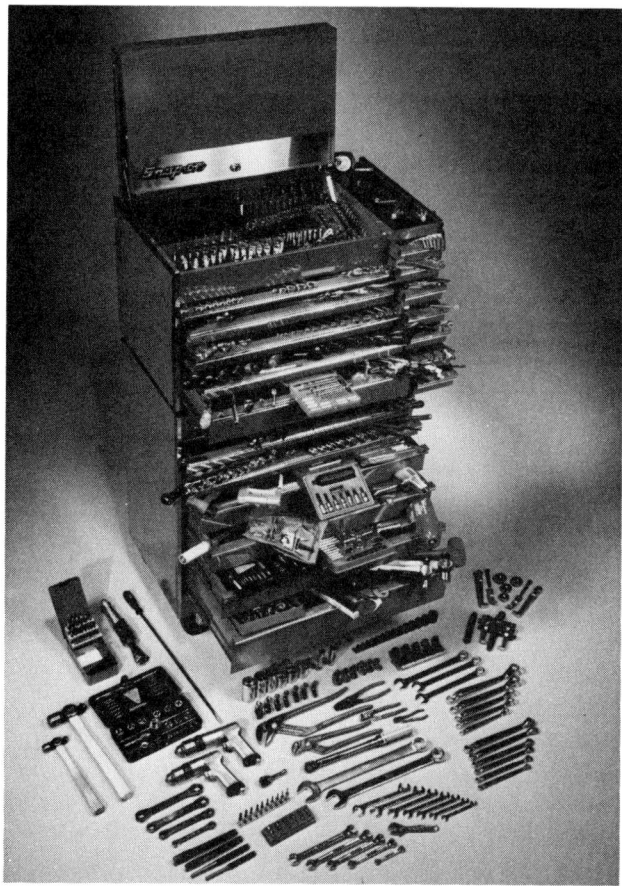

FIGURE 1-11   Professional set of automotive hand tools *(Courtesy of Snap-on Tools Corp.)*

vehicle, the fasteners can be standard SAE or metric size fasteners. So a well-equipped auto engine rebuilder and/or machinist will have both metric and SAE wrenches in a variety of sizes and styles. The proper use of the appropriate hand tools by the technician and machinist is very important for performing quality auto service.

This is also true of power tools, which make the job a great deal easier. Although electric drills, wrenches, grinders, chisels, drill presses, and various other tools are found in shops, pneumatic (air) tools are used more frequently. Pneumatic tools have four major advantages over electrically powered equipment in an engine rebuilding shop:

- *Flexibility.* Air tools run cooler and have the advantage of variable speed and torque; damage from overload or stalling is eliminated. They can fit in tight spaces.
- *Lightweight.* The air tool is lighter in weight and lends itself to a higher rate of production with less fatigue (Figure 1-12).

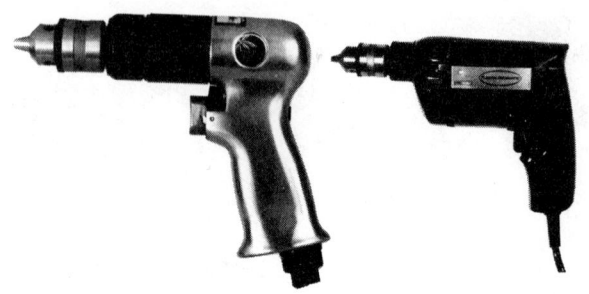

FIGURE 1-12 (Left) 3/8-inch air drill and (right) 3/8-inch electric drill. The air drill weighs 2-1/2 pounds; the electric drill weighs 4-1/2 pounds.

- *Safety.* Air equipment reduces the danger of fire and shock hazards in some environments where the sparking of electric power tools can be a problem.
- *Low-Cost Operation and Maintenance.* Due to fewer parts, air tools require fewer repairs and less preventive maintenance. Also, the original cost of air-driven tools is usually less than the equivalent electric type.

The automotive industry was one of the first industries to recognize the advantages of air-powered tools. Today they are known as the tools of the professional auto engine rebuilder. However, one major disadvantage of air tools is noise.

Some special tools are used by rebuilders (Figure 1-13). The use of these and other specialty tools is covered in later chapters.

The automotive machinist must know how to operate many special machines. Figure 1-14 illustrates a setup of a medium-size full-service machine shop. The key lists the typical equipment found in a rebuilder's machine shop. The operation of this equipment is given in Chapters 9 through 12.

The layout of the machine shop will, of course, have a direct impact on how efficiently work proceeds through engine disassembly, cleaning, machining, and reassembly. In any good setup, it is important to keep various machining equipment for similar products in close proximity. Note in the illustrated shop layout how the crankshaft grinder, crankshaft polisher, crankshaft straightener, and crankshaft storage rack are closely positioned to each other. The arrangement is in essence a crankshaft work cell environment with a workbench in the center. Such a setup will reduce operator fatigue and improve productivity.

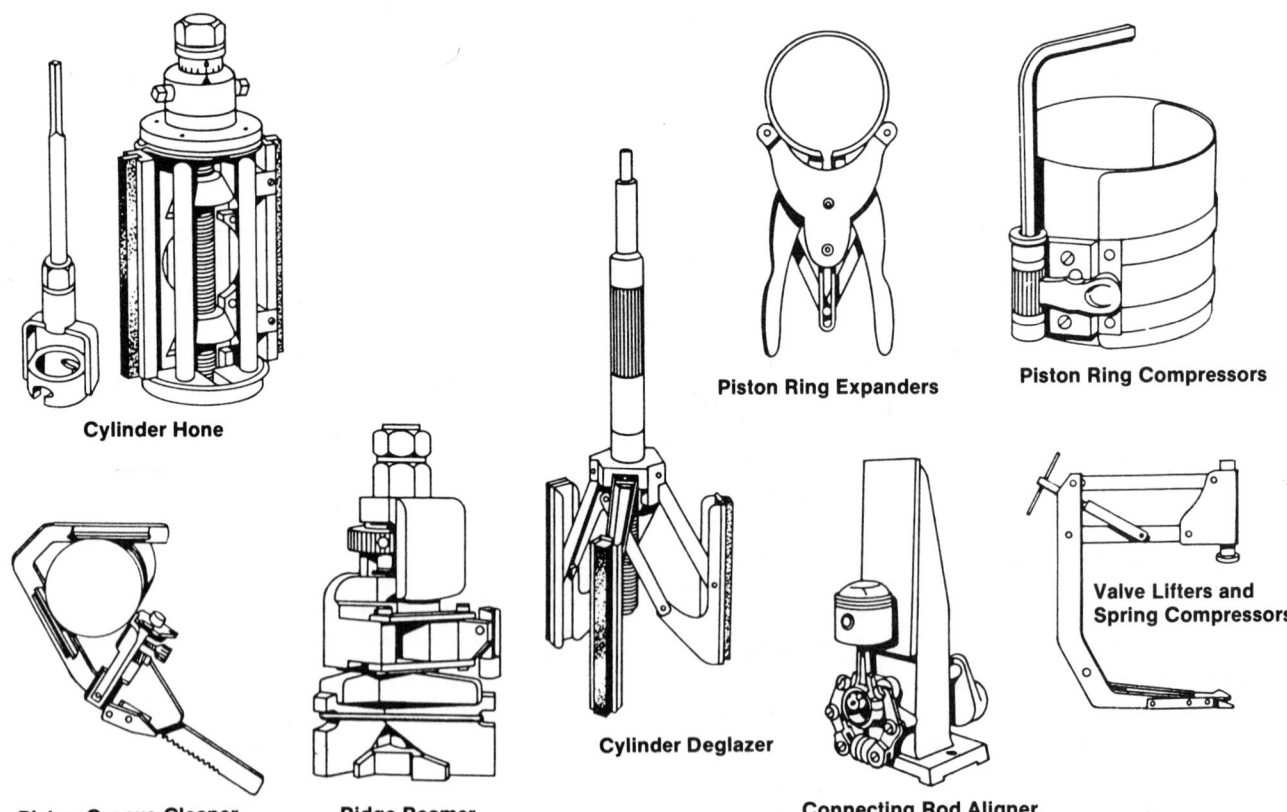

FIGURE 1-13 Special tools used by rebuilders

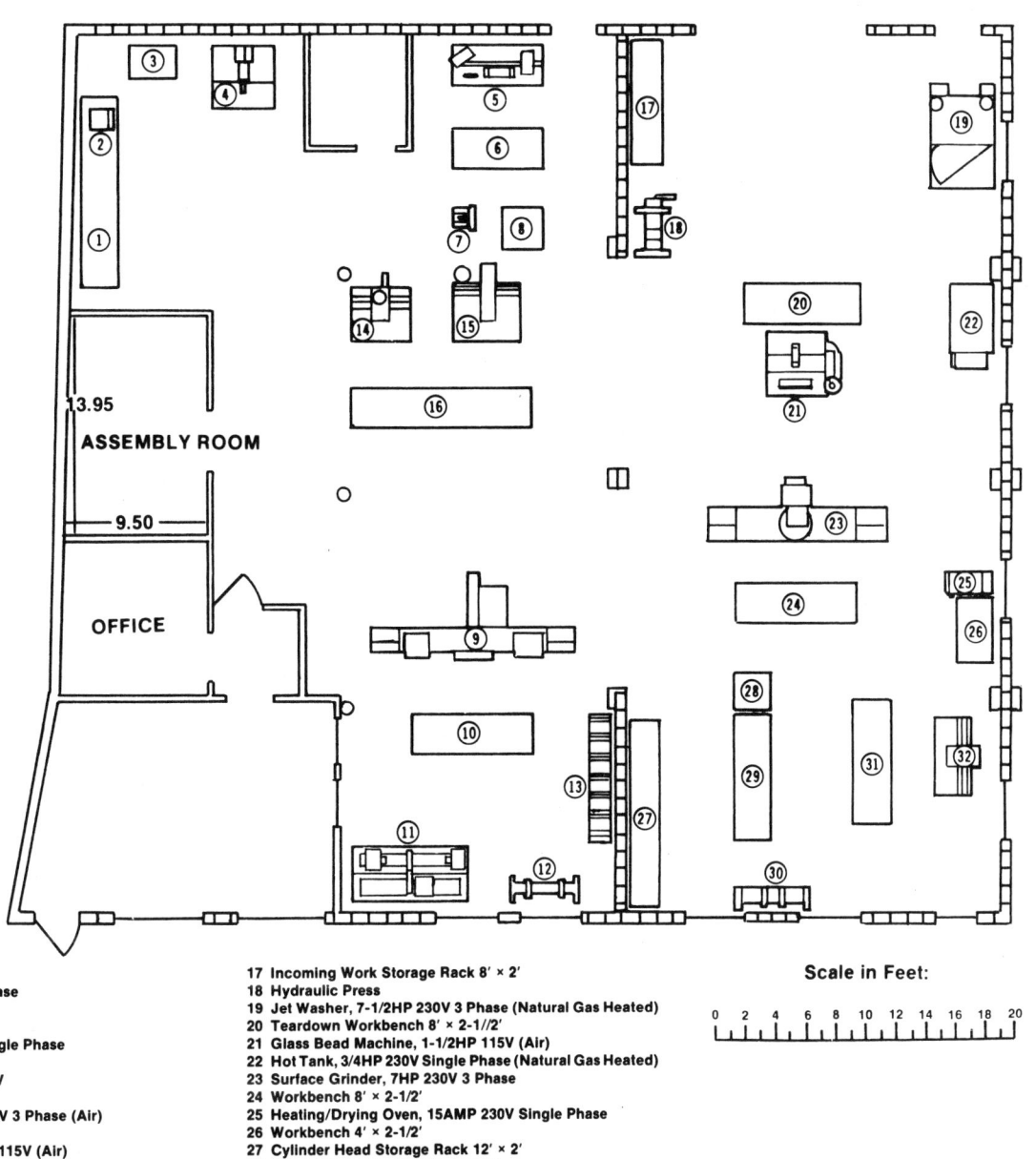

1 Workbench 12' × 2-1/2'
2 Cap Grinder, 1HP 230V 3 Phase
3 Parts Washer, 115V
4 Rod/Pin Hone, 115V
5 Balancer, 115V and 230V Single Phase
6 Workbench 6' × 2-1/2'
7 Belt/Disc Sander, 1.3HP 115V
8 Vertical Milling Machine
9 Crankshaft Grinder, 6HP 230V 3 Phase (Air)
10 Workbench 8' × 2-1//2'
11 Crankshaft Polisher, 1.25HP 115V (Air)
12 Crankshaft Straightening Press
13 Crankshaft Storage Rack
14 Cylinder Boring Machine, 1HP 230V 3 Phase (Air)
15 Cylinder Honing Machine, 1HP 230V 3 Phase (Air)
16 Workbench 12' × 2-1/2'

17 Incoming Work Storage Rack 8' × 2'
18 Hydraulic Press
19 Jet Washer, 7-1/2HP 230V 3 Phase (Natural Gas Heated)
20 Teardown Workbench 8' × 2-1//2'
21 Glass Bead Machine, 1-1/2HP 115V (Air)
22 Hot Tank, 3/4HP 230V Single Phase (Natural Gas Heated)
23 Surface Grinder, 7HP 230V 3 Phase
24 Workbench 8' × 2-1/2'
25 Heating/Drying Oven, 15AMP 230V Single Phase
26 Workbench 4' × 2-1/2'
27 Cylinder Head Storage Rack 12' × 2'
28 Valve Grinder And Cabinet, 115V
29 Workbench 8' × 2-1/2'
30 Cylinder Head Pressure Tester (Air)
31 Workbench 8' × 2-1/2'
32 Guide/Seat Machine, 3/4HP 230V Single Phase

Scale in Feet:

0 2 4 6 8 10 12 14 16 18 20

Note: All work benches should have air
and 115V electricity if possible.

FIGURE 1-14   Setup for a medium-size rebuilding shop

ENGINE MEASURING TOOLS

In the engine rebuilding field, both the rebuilder technician and machinist are frequently required to work to very close measurements, often in ten thousandths (0.0001)of an inch. One thousandth of an inch is approximately the thickness of the cellophane used to enclose a pack of cigarettes. Accurate measurements to this high standard of precision can only be made through the use of measuring devices especially designed to show these very small differences in size.

Measuring tools are precise and delicate instruments. In fact, the more precise they are, the more delicate they are. They should be handled with great care. When using a measuring device, always place it so that it cannot fall or strike other tools. Never pry, strike, drop, or force these instruments; they might be damaged beyond usability.

Precision measuring instruments, especially micrometers, are extremely sensitive to rough handling. Clean them before and after every use. Never touch the measuring surfaces of the micrometer. Fingerprints can help rust to form, and body heat can affect accuracy. All measuring operations must be performed on parts that are room temperature. Never measure parts that are still warm from machining operations.

 SHOP TALK _____

Check measuring instruments regularly against known good equipment to ensure that they are operating properly and capable of accurate measurement. Always refer to the appropriate shop manual or other reference material for the correct specifications before performing service or diagnosis procedures. The close tolerances in current vehicles make using the correct specifications and taking accurate measurements more important than ever.

Machinist's Rule

The machinist's rule looks like an ordinary ruler (Figure 1–15). However, unlike the common ruler, it is precisely divided into small (1/64-inch) increments. A typical machinist's rule is marked on both sides. One side is marked off at 1/16-, 1/8-, 1/4-, 1/2- and 1-inch intervals. The other side is marked at 1/32- and 1/64-inch intervals.

Machinist's rules are also available with metric or decimal graduations. Metric rules are usually divided into 0.5-mm and 1-mm increments. Decimal rules are typically divided into 1/10-, 1/50-, and 1/100-inch (0.10-, 0.50-, and 0.01-inch) increments. Decimal machinist's rules are convenient when measuring component dimensions that are specified in decimals.

Dial Caliper

The dial caliper (Figure 1–16) is a versatile measuring instrument. It is capable of taking inside, outside, depth, and step measurements. It can measure these dimensions from zero to 6 inches. Metric dial calipers typically measure from zero to 150 mm in increments of 0.02 mm.

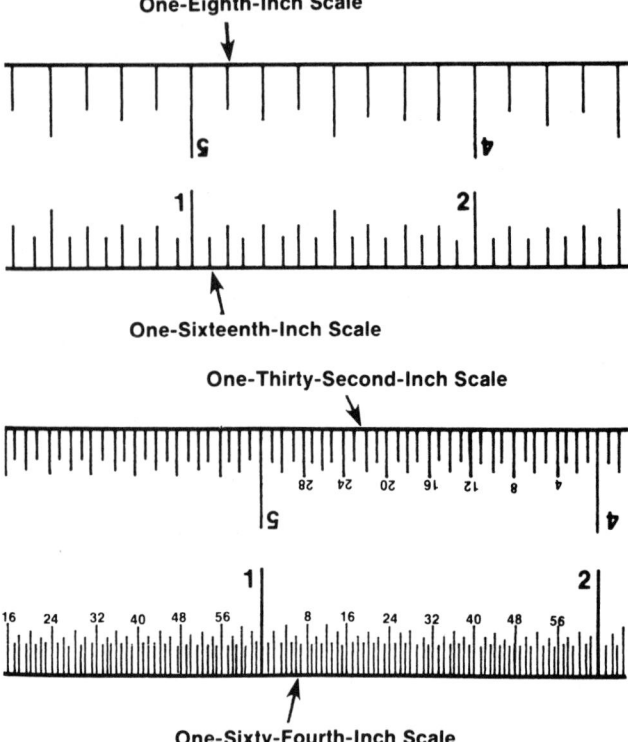

FIGURE 1–15 Typical machinist's rule graduations

The dial caliper features a depth scale, bar scale, dial indicator, inside measurement jaws, and outside measurement jaws. The bar scale is divided into one-tenth (0.10) of an inch graduations. The dial indicator is divided into one-thousandth (0.001) of an inch graduations. Therefore, one revolution of the dial indicator needle equals one-tenth of an inch

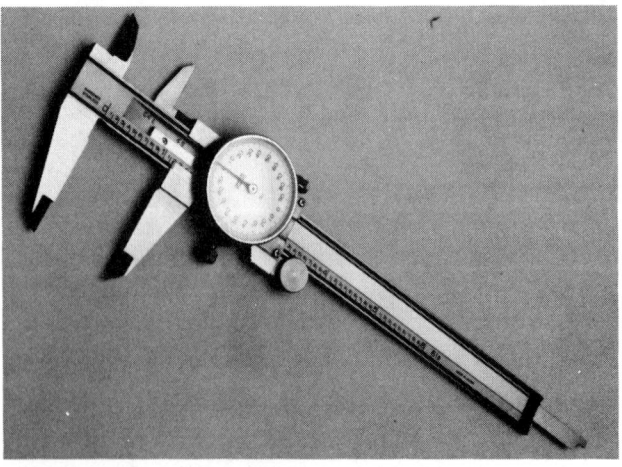

FIGURE 1–16 Dial caliper *(Courtesy of Central Tools, Inc.)*

on the bar scale (one hundred-thousandth of an inch equals one-tenth of an inch).

The metric dial caliper is similar in appearance to the United States (inch reading) model. However, the bar scale is divided into 2-mm increments. Additionally, on the metric dial caliper, one revolution of the dial indicator needle equals 2 mm.

Both inch reading and metric dial calipers use a thumb-operated roll knob for fine adjustment. When you use a dial caliper, always move the measuring jaws backward and forward to center the jaws in the work. Always make sure that the caliper jaws lay flat on the work. If the jaws or the work are tilted in any way, you will not obtain an accurate measurement. However, although dial calipers are precision measuring instruments, they are only accurate to within ± 0.002 inch. The factors that limit dial caliper accuracy include jaw flatness and "feel." Micrometers are better suited to measuring tasks that require extreme precision.

Micrometers

Measurements required in an engine rebuilding job are either the size of the outside or inside diameter or the diameter of a shaft and the bore of a hole. The micrometer is the common instrument for taking these measurements. Both outside and inside micrometers are calibrated and read in the same manner and are both operated so that the measuring points exactly contact the surfaces being measured.

The major components and markings of the micrometer include the frame, anvil, spindle, locknut, sleeve, sleeve numbers, sleeve long line, thimble marks, thimble, and ratchet (Figure 1–17). Micrometers are calibrated in either inch- or metric-graduations.

On both the outside and inside micrometers, the thimble is revolved between the thumb and forefinger (Figure 1–18). Very light pressure is required when bringing the measuring points into contact with the surfaces being measured. It is *important* to remember that the micrometer is a delicate instrument and that even slight excessive pressure will result in an incorrect reading.

Reading an Inch-Graduated Outside Micrometer. These micrometers are made so that each turn of the thimble moves the spindle 0.025 inch (twenty-five thousandths of an inch). This is accomplished by using forty threads per inch on the thimble. The sleeve long line is marked with sleeve numbers 1, 2, 3, and so on. These sleeve numbers represent 0.100 inch, 0.200 inch, 0.300 inch, and so on. The sleeve of the micrometer

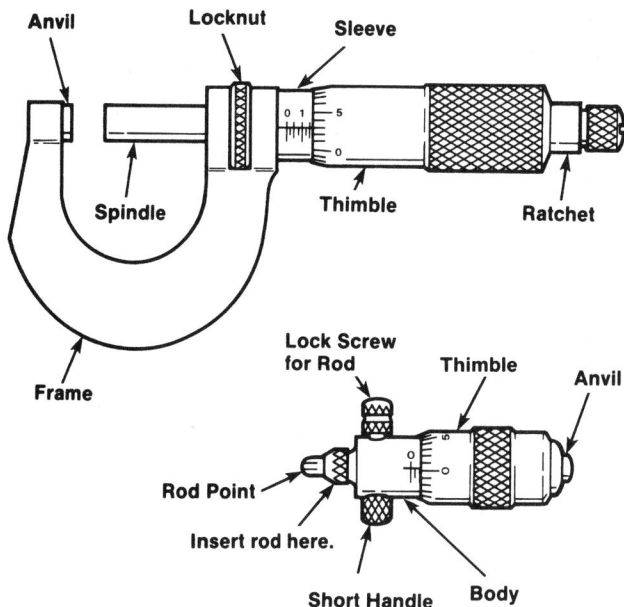

FIGURE 1-17 The main components of an outside and an inside micrometer

FIGURE 1-18 The proper way to hold a micrometer and control the thimble.

contains sleeve marks that represent 1 inch in 0.025-inch (twenty-five thousandths of an inch) increments. Each of the thimble marks represents 0.001 inch (one-thousandth of an inch). In one complete turn, the spindle will move 25 marks or 0.025 inch (twenty-five thousandths of an inch). Inch-graduated micrometers come in a range of sizes—zero to 1 inch, 1 inch to 2, 2 inches to 3, 3 inches to 4, etc. The most commonly used micrometers are calibrated in one-thousandths of an inch increments.

To read a micrometer, first read the last whole sleeve number visible on the sleeve long line. Next, count the number of full sleeve marks past the number. Finally, count the number of thimble marks past the sleeve marks. Add these measurements together for the measurement. These three readings indicate tenths, hundredths, and thousandths of an inch respectively. For example, a 2- to 3-inch micrometer that has taken a measurement is described as follows (Figure 1–19):

1. The largest sleeve number visible is 4, indicating 0.400 inch (four-tenths of an inch).
2. The thimble is three full sleeve marks past the sleeve number. Each sleeve mark indicates 0.025 inch, so this indicates 0.075 inch (seventy-five hundredths of an inch).
3. The number 12 thimble mark is lined up with the sleeve long line. This indicates 0.012 inch (twelve-thousandths of an inch).
4. Add the readings from steps 1, 2, and 3. The total of the three is the correct reading. In our example:

Sleeve	0.400 inch
Sleeve marks	0.075 inch
Thimble marks	0.012 inch
Total =	0.487 inch

5. Now add 2 inches to the measurement, since this is a 2- to 3-inch micrometer. The final reading is 2.487 inches.

Reading an Outside Micrometer with a Vernier Scale. In cases where a measurement must be within 0.0001 inch (one ten-thousandth of an inch), a micrometer with a vernier scale should be used. This micrometer is read in the same way as a standard micrometer. However, in addition to the three scales found on the typical micrometer, this type has a vernier scale on the sleeve. When taking measurements with this mike, read it in the same way as you would a standard mike. Then, find the thimble mark that lines up precisely with one of the vernier scale lines (Figure 1–20). Only one of these lines will match up correctly. All other lines will be mismatched. The vernier scale number that matches up with a thimble mark is the 0.0001-inch (one ten-thousandth of an inch) measurement.

Reading a Metric Outside Micrometer. The metric micrometer is read in the same manner as the inch-graduated micrometer,

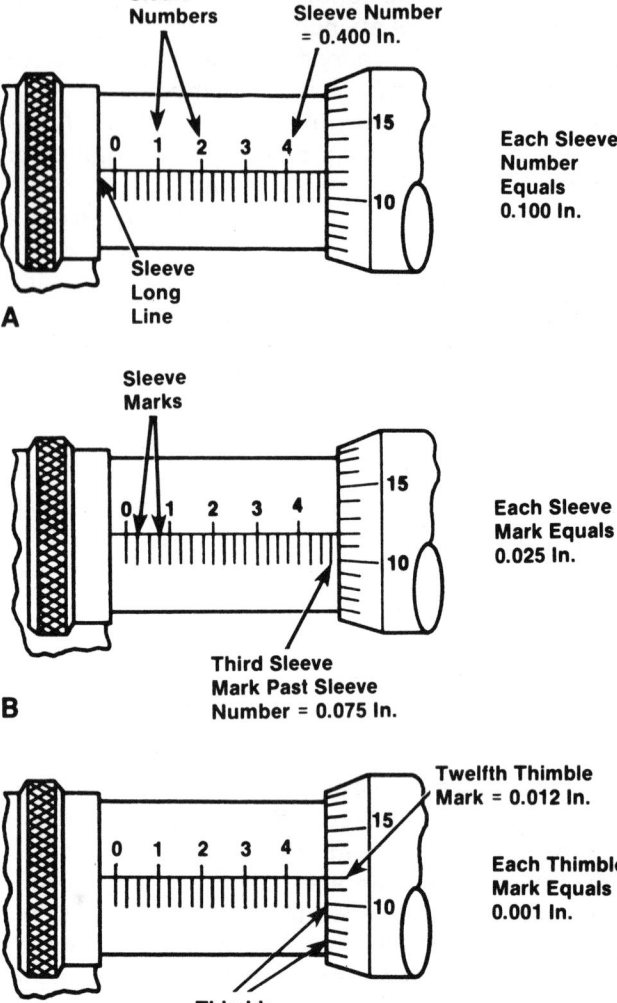

FIGURE 1-19 The three steps in reading a micrometer: (A) measuring tenths of an inch; (B) measuring hundredths of an inch; (C) measuring thousandths of an inch

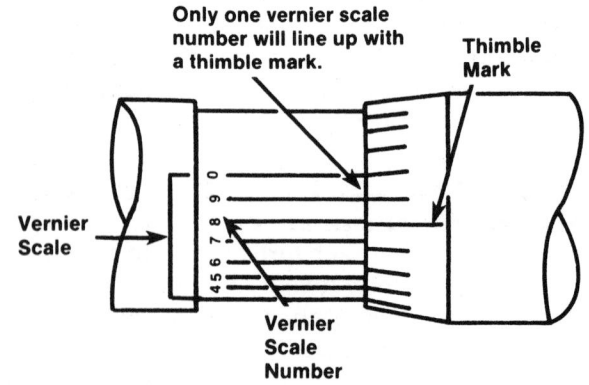

FIGURE 1-20 Measuring ten thousandths of an inch using a micrometer with vernier scale

except that the graduations are in the metric system of measurement. Readings are obtained as follows:

- Each number on the sleeve of the micrometer represents 5 millimeters or five thousandths of a meter (Figure 1-21A).
- Each of the ten equal spaces between each number, with index lines alternating above and below the horizontal line, represents 0.5 millimeter or five tenths of a millimeter. One revolution of the thimble changes the reading one space on the sleeve scale or 0.5 mm (Figure 1-21B).
- The beveled edge of the thimble is divided into 50 equal divisions with every fifth line numbered 0, 5, 10 . . . 45. Since one complete revolution of the thimble advances the spindle 0.5 mm, each graduation on the thimble is equal to 1/50 of 0.5 mm or one hundredth of a millimeter (Figure 1-21C).

As with the inch-graduated micrometer, the three separate readings are added together to obtain the total reading (Figure 1-22). That is:

1. Read the largest number on the sleeve that has been uncovered by the thimble. In the illustration it is 5, which means the first number in the series is 5 mm.
2. Count the number of lines past the number 5 that the thimble has uncovered. In the example, this is 4, and since each space is equal to 0.5 mm, 4 spaces equal 4 × 0.5 or 2 mm. This added to the figure obtained in step 1 gives 7 mm.
3. Read the graduation line on the thimble that coincides with the horizontal line of the sleeve scale and add this to the total obtained in step 2. In the example, the thimble scale reads 28 or 0.28 mm. This added to the 7 mm from step two gives a total reading of 7.28 mm.

Using an Outside Micrometer.

To measure small objects using an outside micrometer, grasp the micrometer with the right hand and slip the object to be measured between the spindle and anvil. While holding the object against the anvil, turn the thimble using the thumb and forefinger until the spindle contacts the object. Never clamp the micrometer tightly. Use only enough pressure on the thimble to allow the work to just fit between the anvil and spindle. If the micrometer is equipped with a ratchet screw, use it to tighten the micrometer around the object for final adjustment. For a correct measurement, the object must slip

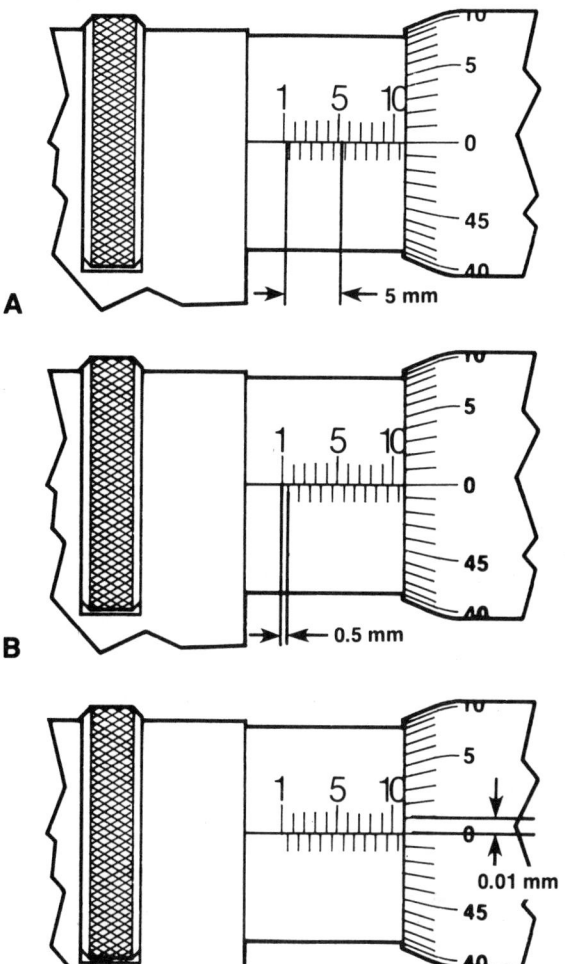

FIGURE 1-21   Reading a metric micrometer: (A) 5 mm; (B) 0.5 mm; (C) 0.01 mm

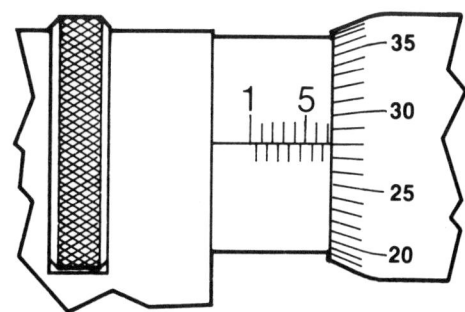

FIGURE 1-22   Micrometer set at 7.28 mm

through the micrometer with only a very light resistance. When a satisfactory adjustment has been made, read the micrometer. Be careful not to change the setting.

To measure larger objects such as a piston, hold the frame of the micrometer and slip it over the work while adjusting the thimble (Figure 1–23). It is important to slip the mike back and forth over the work until you feel a very light resistance, while at the same time rocking the mike from side to side to make certain the spindle cannot be closed any further (Figure 1–24). These steps should be taken with any precision measuring device to ensure correct measurements.

Measurements will be reliable if the mike is calibrated correctly. To calibrate a micrometer, close the mike over a micrometer standard. If the reading differs from that of the known micrometer standard, then the mike will need adjustment.

Reading an Inside Micrometer.

Inside mikes (Figure 1–25) are used to measure bore

FIGURE 1–23 Using outside micrometer for measuring small objects

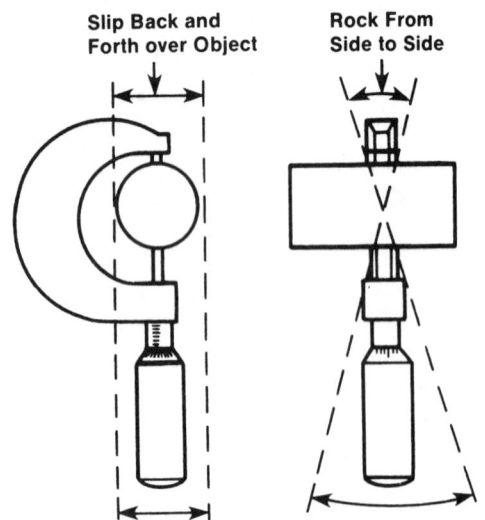

FIGURE 1–24 Slip the mike

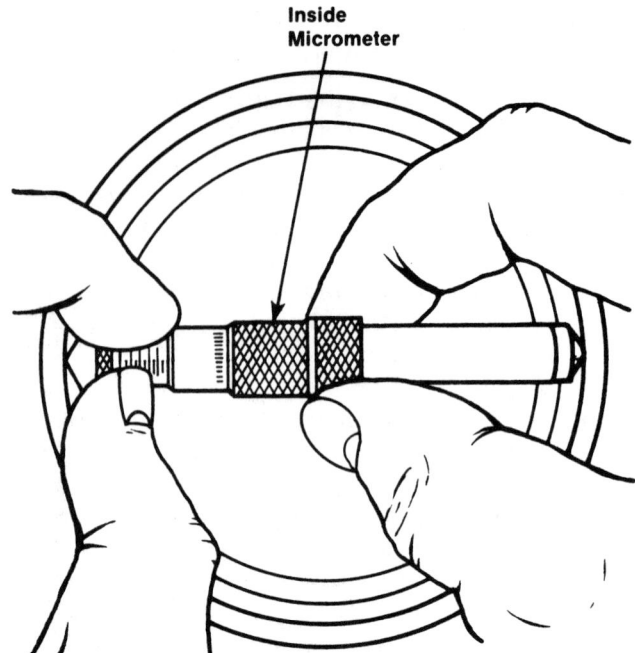

FIGURE 1–25 Obtaining a precise measurement with an inside micrometer

or hole sizes. They are frequently used with outside mikes to reduce the chance of error.

To use an inside mike, place it inside the bore or hole and extend the measuring surfaces until each end touches the bore's surface. If the bore is large, it might be necessary to use an extension rod to increase the mike's range. These extension rods come in various lengths. The inside micrometer is read in the same manner as an outside micrometer.

To obtain a precise measurement in either inch or metric graduations, keep the anvil firmly against one side of the bore and rock the inside mike back and forth and side to side. This ensures that the mike fits in the center of the work with the correct amount of resistance. As with the outside micrometer, this procedure will require a little practice until you get the feel for the correct resistance and fit of the mike.

After taking the measurement with the inside mike, use an outside mike to take a comparison measurement. This reduces the chance of errors and ensures an accurate measurement.

Reading a Depth Micrometer.

The depth micrometer (Figure 1–26) is used to measure the distance between two parallel surfaces. Depth micrometers are available in inch-graduated and metric models. The sleeves, thimbles, and ratchet screws operate in the same manner as the previously described inside and outside micrometers. Depth mikes are read in the same way as other micrometers.

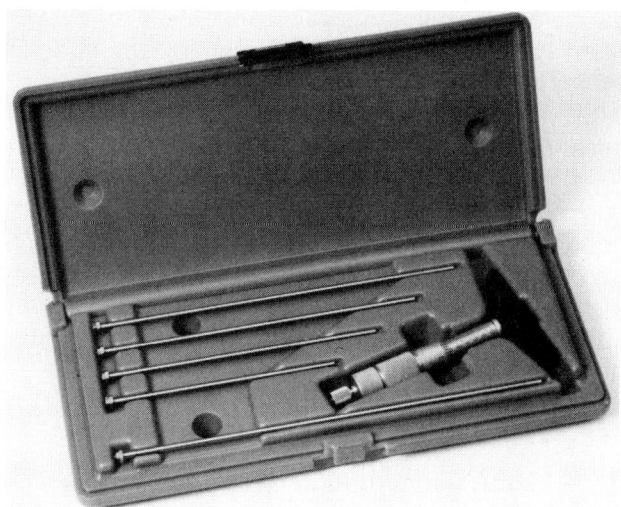

FIGURE 1-26   Depth micrometer *(Courtesy of Central Tools, Inc.)*

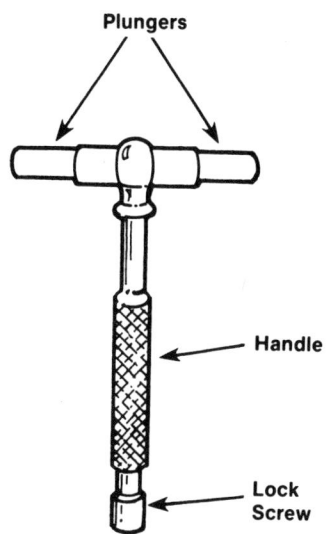

FIGURE 1-27   Typical telescoping gauge

If the depth micrometer is used with a gauge bar, it is essential to keep both the bar and the micrometer from rocking. Any movement of either part will result in an inaccurate measurement.

 SHOP TALK ————

Follow these tips for taking care of a mike:

* *Always clean the mike before using it.*
* *Do not touch measuring surfaces.*
* *Store the mike properly. The spindle face should not touch the anvil face, or a change in temperature might spring the mike.*
* *Clean the mike after use. Wipe it clean of any oil, dirt, or dust using a lint-free cloth.*
* *Do not drop the mike. It is a sensitive instrument and must be handled with care.*
* *Check the calibration weekly. If it drops at any time, check it immediately.*

Telescoping Gauge

Telescoping or snap gauges (Figure 1-27) are used for measuring bore diameters and other clearances. They are normally offered in sizes ranging from fractions of an inch through 6 inches. Each gauge consists of two telescoping plungers, a handle, and a lock screw. Telescoping gauges are usually used with an outside micrometer.

To use the telescoping gauge, insert it in the bore and loosen the lock screw. This will allow the plungers to "snap" against the bore. Lock the screw in place before removing. Then, the telescoping gauge is measured using the micrometer (Figure 1-28).

Small Hole Gauge

The small hole gauge functions in the same manner as the telescoping gauge. It is expanded in a bore, then measured with an outside micrometer (Figure 1-29). Like the telescoping gauge, the small hole gauge consists of a lock, handle, and an ex-

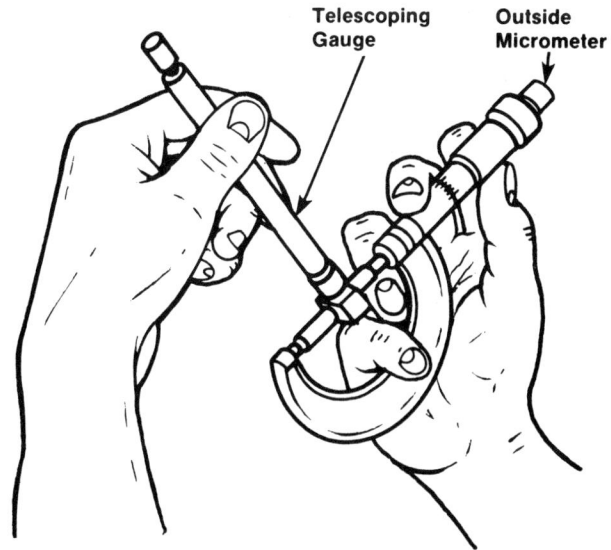

FIGURE 1-28   Measuring a telescoping gauge using an outside micrometer

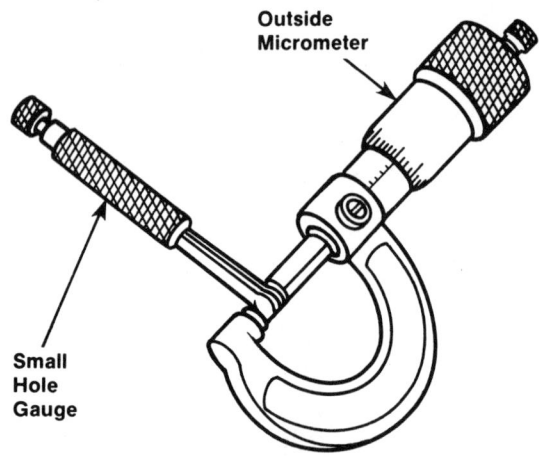

FIGURE 1-29 Measuring a small hole gauge using an outside micrometer

panding end. It expands or retracts by turning the gauge handle. The small hole gauge is commonly used to determine the diameter of valve guides.

Feeler Gauge

The feeler gauge is a strip of metal of known and closely controlled thickness. Several of these metal strips are often combined into a multiple measuring instrument that pivots in a manner similar to a pocket knife (Figure 1–30). The desired thickness

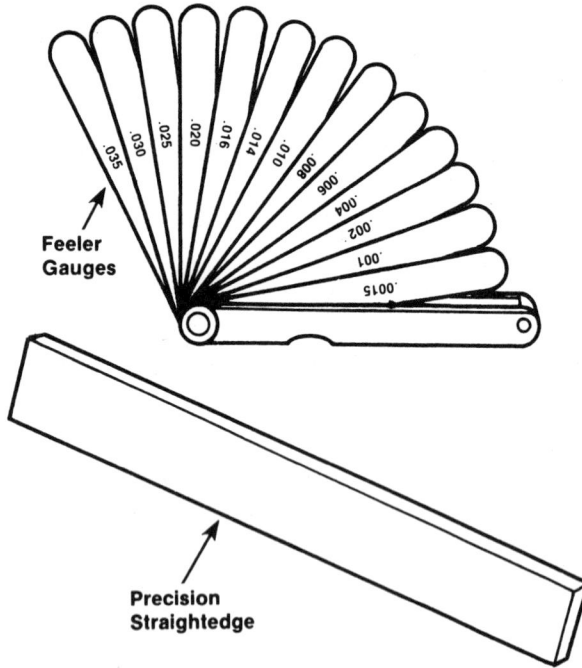

FIGURE 1-30 Typical feeler gauge and precision straightedge

gauge can be pivoted away from others for convenient use. A steel feeler pack usually will contain leaves of 0.002 to 0.010-inch thickness (in steps of 0.001 inch) and leaves of 0.012 to 0.024-inch thickness (in steps of 0.002 inch).

The feeler gauge can be used by itself to measure piston ring side clearance, piston ring end gap, connecting rod side clearance, crankshaft end play, and other similar procedures. The feeler gauge can also be used with a precision straightedge to measure main bearing bore alignment and cylinder head/block warpage (Figure 1–31).

Radius Gauge

A radius gauge (Figure 1–32) is used to check the crankshaft fillet radii at the edge or side of the

FIGURE 1-31 Using a precision straightedge and feeler gauge to check for engine warp *(Courtesy of Fel-Pro, Inc.)*

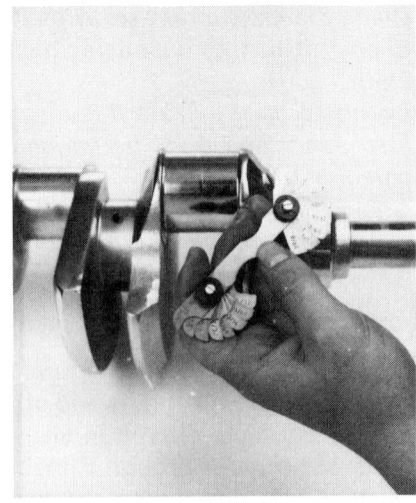

FIGURE 1-32 Set of radius gauges *(Courtesy of Goodson Shop Supplies)*

rod and main bearing journals to make sure they are the same size as specified in the service manual. They can be used to measure the radii of the grooves and the external or internal fillets (round corners).

Screw Pitch Gauge

Sometimes it is necessary to determine the pitch (threads per inch) of a fastener. The use of a screw pitch gauge (Figure 1–33) provides a quick and accurate method of checking the threads per inch. The leaves of this measuring tool are marked with the various pitches. Just match the teeth of the gauge with the threads of the fastener and the correct pitch can be read directly from the leaf.

Screw pitch gauges are available for the various types of fastener threads used in the automotive industry: American National coarse and fine threads, metric threads, International Standard threads, and Whitworth threads.

Dial Indicator

The dial indicator (Figure 1–34) is calibrated in 0.001-inch (one-thousandth of an inch) increments. Metric dial indicators are also available. Both types are used to measure movement. Common uses of the dial indicator include measuring valve lift, journal concentricity, flywheel or brake rotor runout, gear backlash, and crankshaft end play. Dial indicators are available with various face markings and measurement ranges to accommodate many measuring tasks.

To use a dial indicator, position the indicator rod against the object to be measured. Then, push the indicator toward the work until the indicator needle travels far enough around the gauge face to permit movement to be read in either direction (Figure 1–35). Zero the indicator needle on the gauge.

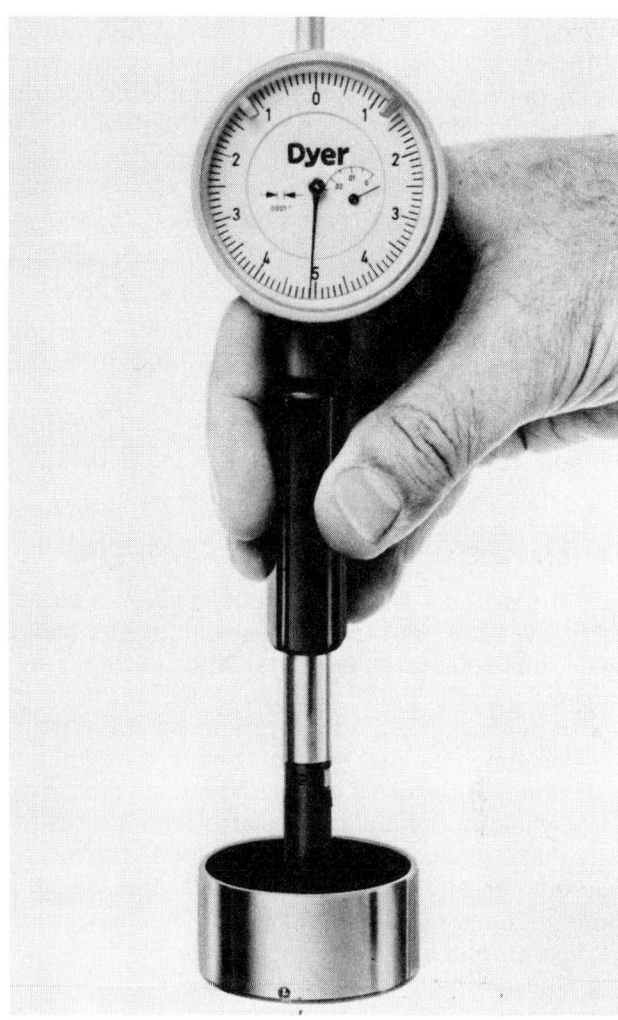

FIGURE 1-34 Dial indicator *(Courtesy of The Dyer Company)*

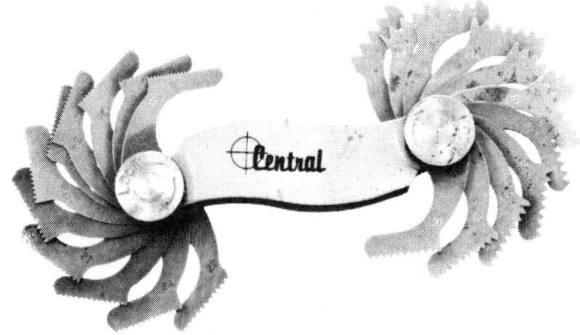

FIGURE 1-33 Screw pitch gauge *(Courtesy of Central Tools, Inc.)*

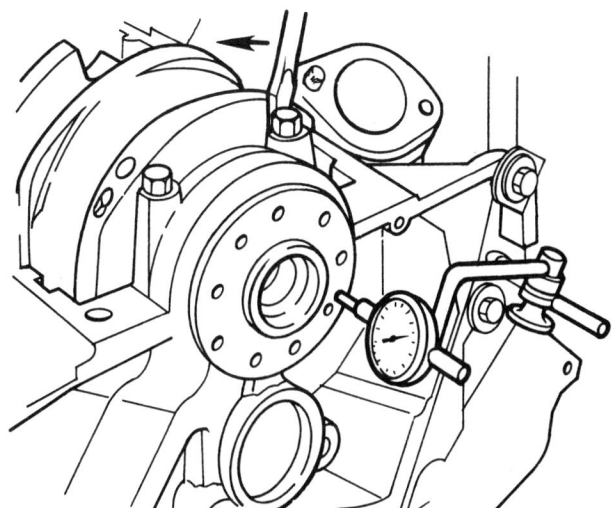

FIGURE 1-35 Measuring crankshaft end play using a dial indicator and bracketry

Always be sure that the range of the dial indicator is sufficient to allow the amount of movement required by the measuring operation. For example, do not use a 1-inch indicator on a component that will move 2 inches.

Valve Seat Runout Gauge

The valve seat runout gauge is similar to the dial indicator. It features a gauge face divided into 0.001-inch (one-thousandth of an inch) increments, an arbor that centers the instrument in the valve guide bore, and an indicator bar that can be adjusted so that it bears on the valve seat (Figure 1–36). The tool is then slowly rotated around the circumference of the valve seat to check its concentricity (runout).

Cylinder Bore Dial Gauge

The cylinder bore dial gauge is used to determine cylinder bore size out-of-round and taper. These three measurements, in addition to main bearing saddle alignment, provide basic information about the condition of the cylinder block. A block outside specification in any of these areas requires machining to restore it to specified tolerances.

Cylinder bore dial gauges are read in the same way that a dial indicator is read. As with the inside micrometer and telescoping gauge, the cylinder bore micrometer must be rocked to obtain the correct measurement.

The parts of the cylinder bore dial gauge include the handle, guide blocks, lock, indicator contact, and dial indicator (Figure 1–37). Various indicator contacts can be installed to measure several bore sizes with one gauge.

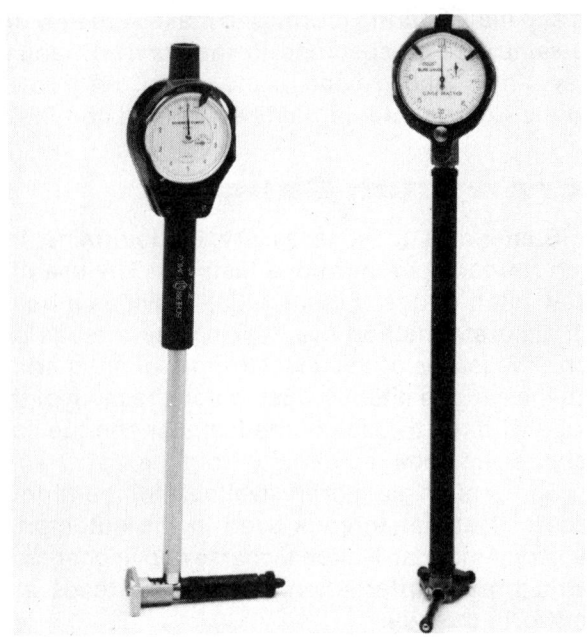

FIGURE 1-37 Typical cylinder bore dial gauge *(Courtesy of Goodson Shop Supplies)*

Cylinder bores do not always wear in a perfectly round pattern; one side of the cylinder wall usually wears more than the other. For this reason, cylinder roundness must be determined (Figure 1–38). To measure out-of-round, measure the cylinder bore at uniform depth in three directions: front/rear, right/left, and diagonally (Figure 1–39). Then, the smallest reading obtained from the cylinder is subtracted

FIGURE 1-36 Valve seat runout gauge *(Courtesy of Central Tools, Inc.)*

FIGURE 1-38 Determining cylinder roundness with a cylinder bore dial gauge *(Courtesy of Central Tools, Inc.)*

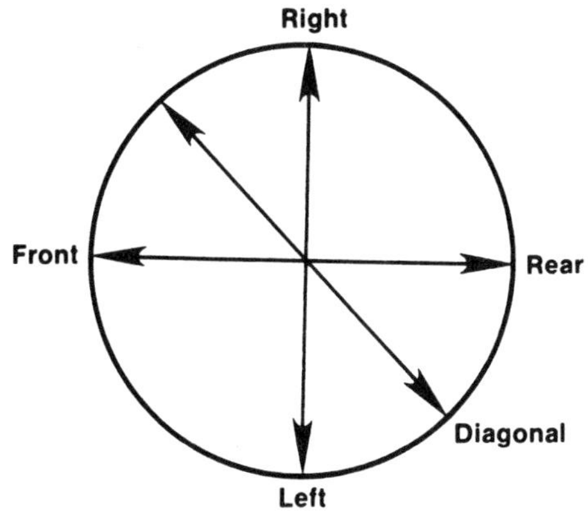

Right

Front ←————→ Rear

Diagonal

Left

**The difference
between the largest
and smallest diameters
equals out-of-round.**

FIGURE 1-39 Recommended directions for measuring cylinder out-of-roundness

from the largest reading. The resulting figure is cylinder bore out-of-round. By measuring the bore at the top of the ring travel, it is possible to determine the maximum cylinder wear.

Another cylinder bore gauge that is used to check cylinder taper and out-of-roundness is shown in Figure 1-40. The cylinder diameter can also be determined when used with a micrometer.

Out-of-Roundness Gauge

The present-day thin-wall bearing insert is designed to assure complete contact between the bearing back and saddle bore. A properly assembled bearing insert will, therefore, conform exactly to the contour of the saddle bore, making it imperative that the saddle bore is a perfect circle. Special gauges are available for checking connecting rod out-of-roundness.

The gauge shown in Figure 1-41 consists of a base on which is mounted a thousandths reading dial indicator and on the bottom a positive locking adjustable slide, which is readily set to the approximate hole size. The sliding head has two line contact points under spring tension insuring alignment at all times. All contacting surfaces and points are hardened. When checking a rod for out-of-roundness, move the locking adjustable slide to suit the diameter of the bore and then set the dial indicator at zero. The dial will then give a plus or minus reading in thousandths for any variation in the rod diameter.

Out-of-roundness can be checked by using an inside micrometer. However, the out-of-roundness gauge is a much quicker and easier method.

Aligning Bar

An aligning bar is an excellent tool for checking the alignment of the crankcase saddle bores (Figure 1-42). With the bearings removed, insert the aligning

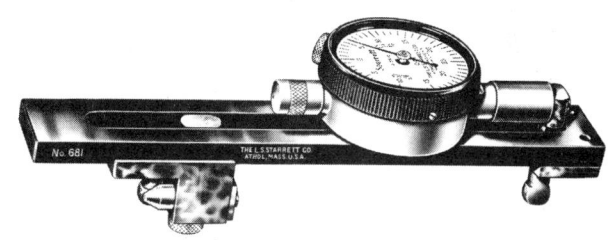

FIGURE 1-41 Typical out-of-roundness gauge *(Courtesy of L. S. Starrett Company)*

FIGURE 1-42 Checking crankcase saddle bores' alignment *(Courtesy of Sunnen Products Co.)*

FIGURE 1-40 Another type of cylinder bore *(Courtesy of L. S. Starrett Company)*

bar into the full length of the crankcase and tighten the bearing caps. If the bar can then be turned by hand using a short bar (12 inches) for leverage, the saddle bores are in alignment. If the aligning bar will not turn, the saddle bores are out of alignment.

V-Blocks

The use of V-blocks and a dial indicator is the recommended method for checking crankshaft alignment (Figure 1-43). With the front and rear main journals in the V-blocks, the dial indicator is used to check the center main journal.

On short crankshafts with three or four main journals, the shaft can be supported and rotated on the front and rear journals, with the dial indicator registering the alignment or concentricity at the vertical centerline on the other main journals. On longer shafts having five or seven main journals, it is recommended that the shaft be supported at the intermediate mains to prevent possible sag. The alignment from journal to journal should be within 0.001 inch, with overall alignment within 0.002 inch.

Each of the end journals is checked by moving one of the V-blocks to the center main journal and taking the dial reading on the unsupported journal. A strip of paper should be placed in each V to prevent the journals from becoming scratched. It is also important to check the nose of the crankshaft for runout.

Torque Indicating Wrench

The fact that practically every vehicle and engine manufacturer publishes a list of torque recommendations is ample proof of the importance of using the proper amounts of torque when drawing up nuts or cap screws on bearing caps and other engine parts.

Several types of torque indicating wrenches are available (Figure 1-44). Any one of these will enable the rebuilder technician to duplicate the pressure used when the engine was assembled originally, thereby duplicating conditions of tightness and stress that the manufacturer has found to be the most desirable (Figure 1-45).

Bearing caps that are too tight will distort the bearings, causing excessive wear and incorrect oil clearance, which often results in rapid wear of other

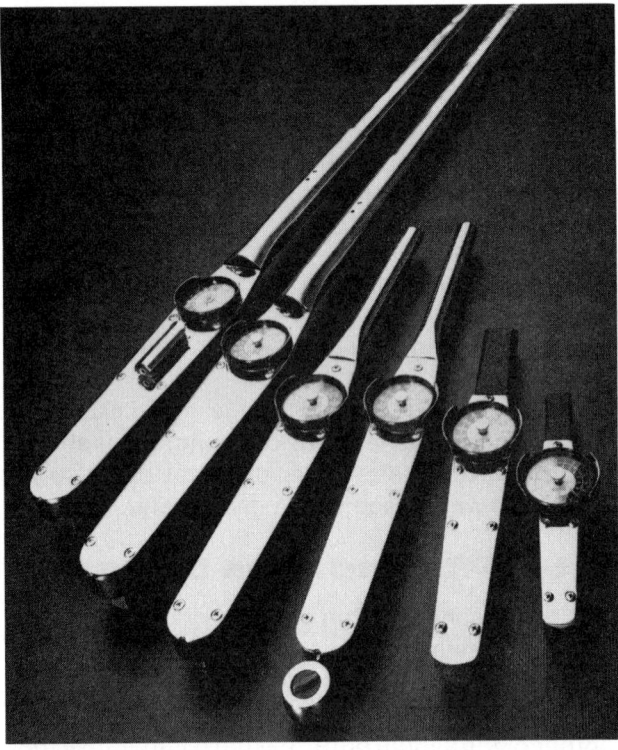

FIGURE 1-44 Dial-type torque wrenches *(Courtesy of Central Tools, Inc.)*

FIGURE 1-43 Checking crankshaft alignment in V-blocks *(Courtesy of Perfect Circle/Dana)*

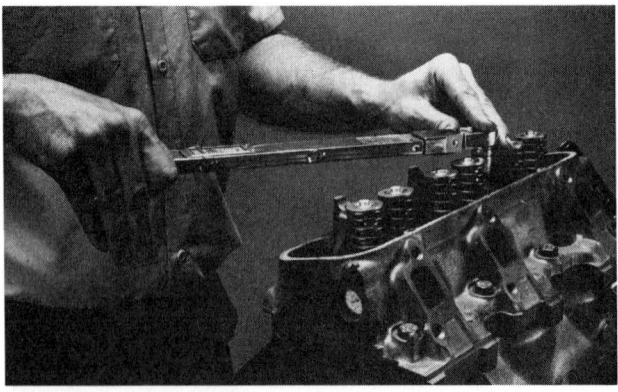

FIGURE 1-45 Typical torque indicating wrench in use *(Courtesy of Perfect Circle/Dana)*

engine parts due to the effect of oil clearance on the engine's lubricating system.

When the backs of the bearings and the bores into which they seat do not make perfect contact due to insufficient torquing, trouble can occur.

Insufficient torque can result in out-of-round bores with less horizontal than vertical clearance and subsequent failure due to side contact. In extreme cases of inadequate bolt torque, bolt and rod or cap breakage result, causing drastic engine failure.

 SHOP TALK _____

The torque wrench, as already noted, has been the standard for years for tightening bolts and nuts that clamp parts together. Torque is actually the force used to turn a fastener (bolt or nut) against friction (between threads and between fastener head and component surface). It is a fact that up to 90 percent of torque is used up by friction. The remaining 10 percent is used to clamp parts together. So, by using fastener torque tightening methods, clamping forces can vary significantly over a single cylinder head or other engine component. Because new high tech engines are designed with extremely close tolerances, it is not surprising that torque alone is no longer adequate for proper service procedures on this type of product. Torque/angle measurement of fastener installations is a method of clamping parts together that insures like fasteners will exert the same clamp force without deviation from one fastener to the next. Without proper clamping forces, distortion of cylinder bores and bearing bores can occur as well as leaking head gaskets. The torque/angle method first specifies a low threshold torque on the fastener to get components in close touch. Torque is least affected by fastener friction at these low values. The fasteners are then turned through a specified angle. The clamping force of each fastener will then be consistent. The unit shown in Figure 1-46 is a precision, microprocessor-controlled torque/angle meter that makes sure fasteners are tightened to true clamp load specifications consistently and easily.

Electronic Gauging Equipment

Electronic gauging, or statistical process control (SPC), is a way of checking and tracking toler-

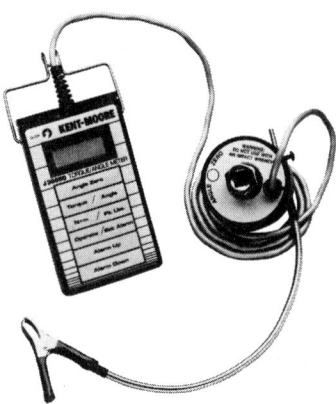

FIGURE 1-46   Precision torque/angle meter *(Courtesy of Kent Moore Corp.)*

ances to reduce the number of rejects in a batch of manufactured products. By carefully measuring and recording tolerances at each step in the manufacturing or remanufacturing process and correcting tolerances when they begin to deviate from acceptable limits, the likelihood of bad parts reaching the end of the line or going out the door is greatly reduced, at least in theory.

SPC can also be used to estimate the probable number of good or bad parts in a batch without having to test each one. The confidence factor or accuracy of the estimate goes up as the number of samples tested increases to a certain point, beyond which there are diminishing returns.

One of the most obvious advantages is the ease with which such instruments can be read. Take a micrometer, for example. With an ordinary mechanical micrometer, thickness is determined by reading and counting the calibrated notches on the handle. It is a highly accurate tool when used properly, but some people have difficulty setting the mike and reading the numbers correctly. In this situation, electronics can reduce the chance of error by eliminating the need to interpret graduated markings. The instrument calculates the number, not the user. By removing the "fear factor" with an easy-to-read digital display, people who were previously reluctant to use a precision instrument gain confidence and are more apt to make dimensional checks, according to the manufacturers. Such an instrument can also make it much easier for an inexperienced person to take accurate measurements, which means quality control can be integrated into more phases of the rebuilding operation.

Another feature electronic gauging offers is the ability to switch scales from inch to metric units with the push of a button. The usual alternative is to have two instruments on hand—one for reading inches, the other for millimeters—a cumbersome process that frequently results in mistakes.

Electronics can also provide a higher degree of resolution in some instances. For example, the standard vernier calipers and dial calipers could both be calibrated to within 0.001 inch, while the electronic caliper could read to the nearest 0.0005 inch (Figure 1-47). A standard inside micrometer offered a resolution of 0.001 inch compared to 0.0001 inch for an electronic version. And for outside micrometers, the mechanical ones were rated at 0.0001 inch while the electronic version could read to 0.00001 inch—a tenfold difference.

Another advantage that can be found on some of the electronic gauging equipment is the ability to set tolerances. By pressing a zero button, the instrument can be reset to zero at any point along its scale (Figure 1-48). Comparative measurements can then be taken to determine the amount of deviation from the norm. Some can even recall a preset value with the press of a button. This feature can be used to compare dimensions to specs and to help people who have trouble remembering numbers.

One of the most useful features, however, is the ability to tell when a measurement exceeds a predetermined dimension. Some electronic micrometers, for example, can be preset to a particular dimension and will then flash a green (pass) or red (fail) warning light when subsequent measurements are taken. One of the major advantages of an electronic micrometer over a conventional micrometer is its digital display. As mentioned earlier, an electronic micrometer can offer a higher degree of resolution, change from inch to metric scales with the press of a button, and have a zero reset capability for comparing dimensions. It can also recall a previously set value and alert the user via a flashing light when a measurement exceeds a preset tolerance. Thus, an

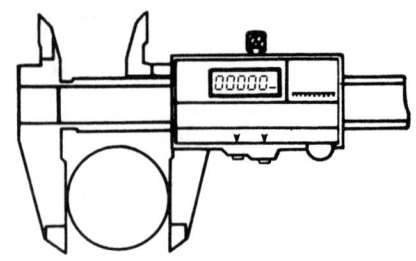

A

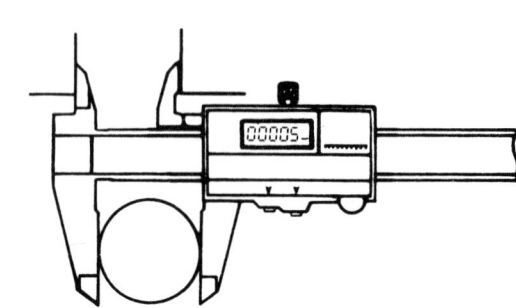

B

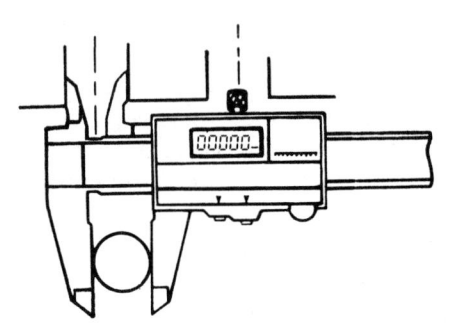

C

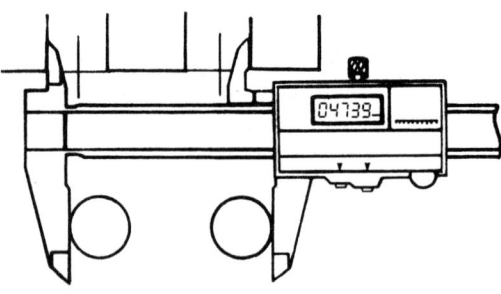

D

FIGURE 1-48 Range of applications due to zero function: (A) zero reset on nominal; measure of deviation ± from a nominal; (B) deviation/clearance on display; measure of clearance between shaft OD and hole ID; (C) zero reset on datum; measure of pitch of holes of same diameter; (D) pitch on display; measured of pitch of pins of same diameter.

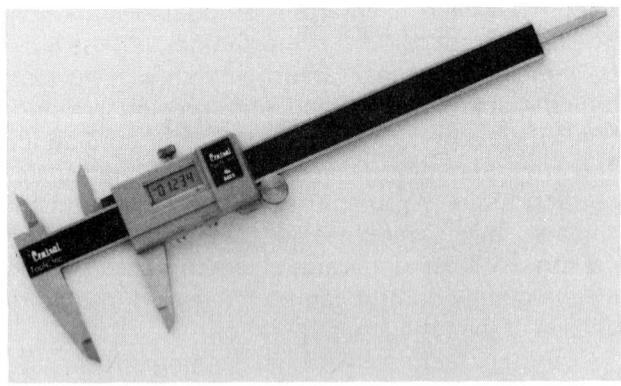

FIGURE 1-47 Electronic digital caliper *(Courtesy of Central Tools, Inc.)*

electronic micrometer gives the user a lot more utility.

As for reliability, electronic instruments require the same care and respect as any other precision instrument. Most such instruments are built rugged enough to withstand ordinary use, but repeated abuse will take its toll. Electronics are vulnerable, however, to dirt and moisture, so they must be kept clean. Some calipers, for example, read a calibrated glass scale. Any dirt on the glass scale could interfere with the accuracy of the measurement. Others determine position by reading capacitance across an internal air gap. This type of instrument is less sensitive to dirt contamination.

Battery replacement is another thing to consider. Maintenance is minimal, because the LCD displays used on almost all of the equipment draw little power. Consequently, the typical battery usually lasts a year or more. The only disadvantage to having battery power is that if the battery goes dead, the digital display is useless until a new battery is installed. Most units have some warning of when the battery needs to be replaced because the LCD display will flicker or only partially illuminate.

The real power of electronic gauging comes into play when it is coupled with other recording devices and microprocessing. This can become the basis for installing a statistical process control plan in your rebuilding operations to better control quality and to reduce rejects and warranty returns. Many electronic measuring instruments have a plug-in connector so the instrument can be linked to a printer or computer. If a printer is being used, a press of a button prints out a hard copy of the measurement that has just been taken. The number can also be combined with other information on the printer tape such as scale limits, an indentification code, or other data. The equipment available to the rebuilder and engine machinist ranges from hand-held data analyzers (Figure 1–49) to sophisticated programs for use with a personal computer (Figure 1–50).

SERVICE MANUALS

Some of the most important tools of the trade are repair or service manuals. Publications that should be available to engine rebuilder technicians and shop machinists include the following:

- *Auto Manufacturer's Service Manuals.* The main source of repair and specification information for any car, van, or truck is the manufacturer. Service manuals are published each year for every vehicle built (Figure 1–51). The manuals are written in technical language for professional technicians

FIGURE 1-49 A hand-held ultrasonic thickness gauge that can be used to inspect aluminum or cast iron for sufficient wall material and uniformity *(Courtesy of Krautkramer Branson Inc.)*

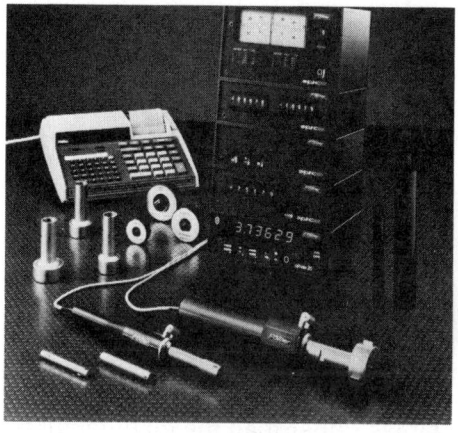

FIGURE 1-50 Electronic bore gauging system, complete with a remote display unit, analyzer, and printout *(Courtesy of Fred V. Fowler Company, Inc.)*

FIGURE 1-51 Typical manufacturer's service manual

working in the dealerships. Technicians must be able to understand the language, procedures, and specifications. Because of the enormous amount of information, some manufacturers publish more than one manual per year per car model. They can be separated into sections such as chassis, suspension, steering, emission controls, fuel systems, brakes, basic maintenance and tune-up, engine, transmission, body, and so on.

When complete information with step-by-step testing, repair, and assembly procedures is desired, nothing can match the auto manufacturers' repair manuals. They are the most reliable because they cover all repairs, adjustments, specs, detailed diagnostic procedures, and special tools required. They can also be purchased directly from the automobile manufacturer.

Automotive manufacturers also publish a series of so-called "technician reference books." The publications provide general instructions on all their current vehicles for accomplishing service and repair with their tested, effective techniques.

- *Aftermarket Suppliers' Guides and Catalogs.* Many of the larger manufacturers have excellent guides on the various parts that they manufacture or supply. They also provide updated service bulletins on their products.
- *General and Specialty Repair Manuals.* These are published by independent companies rather than the manufacturers. However, they pay for and get most of their information from the carmaker. They contain component information, diagnostic steps, repair procedures, and specs for several car makes in one book. Information is usually condensed and is more general in nature, depending on which manual is used. The condensed format allows for more coverage areas in less space and therefore is not always specific. They also can contain several years of models as well as several car makes in one book.
- *Flat-Rate Manuals.* These contain figures dealing with the length of time a rebuilding job is supposed to require. Usually they contain a parts list with approximate or exact parts prices. They are excellent for making rebuild estimates of costs and are published by manufacturers and independents.

One of the best sources of up-to-date engine rebuilding information is trade magazines. Most are published monthly.

Trade associations, such as the Automotive Engine Rebuilder's Association (AERA), are other good sources of updated information. The AERA publishes a monthly bulletin, which would be of interest to all rebuilders: Technical, Service, and Shop Procedures. They also put out an annual publication plus several regional ones. For more information on the AERA, write to the Automotive Engine Rebuilder's Association, 330 Lexington Drive, Buffalo Grove, Illinois 60089.

SHOP RECORDS

As with any business, the automotive engine rebuilding and/or machine shop operation runs more smoothly when the paperwork is done accurately. The most frequently used shop form is the work order. In every shop, a work order form is completed after performing the vehicle diagnosis test before engine disassembly (see Chapter 3). Even in a small shop where the rebuilder deals directly with the customer problems can still arise because information is not written down; for example, parts that are used in the job are not recorded or requested, operations are not performed, or too many operations are performed, which destroys the quote the customer was given. No matter how careful anyone is, problems will occur unless the work order system is complete and information is recorded properly. This problem is compounded as the shop grows in size.

There are various work orders available from different associations. With the inexpensive printing that is available today, it takes little effort to create a work order that addresses the shop's needs. The work order shown in Figure 1–52 was designed for a medium-size full-service shop. It not only allows the machinist to ask the needed questions, it also has boxes to be checked when the work is completed or when parts are ordered or in stock. Prices can be written beside each operation or part so billing will be accurate.

Keep in mind that many states have information disclosure laws to protect consumers from fraud. Written job estimates must be furnished to the customer by the rebuilding shop. If the cost will be higher than the original written estimate, additional approval must be received from the customer. Also, any replaced part is returned to the customer.

Other forms that are used in the rebuilding shop include the following:

- *Parts Requisition.* To order new parts, the mechanic writes the names of what is needed in the parts requisition section of the repair order. Then the rebuilder technician/machin-

NAME Midland Nursery	HOME PHONE	BUS. PHONE 555-2711	REPAIR ORDER NO. 11402	B&J ENGINE SUPPLY
ADDRESS	CITY	CUSTOMER P. O. NO. F12R4		20 WEST BROAD ST. AUBURN, PA 17922
MAKE FORD / C.I.D. 390 / YEAR 65 / BORE / STROKE	MAIN	THROW	DATE 7/4	TELEPHONE: (717) 555-7764

SERVICES REQUIRED

DESCRIPTION	CHECK	PER-FORM	COM-PLETE	PRICE	DESCRIPTION	CHECK	PER-FORM	COM-PLETE	PRICE
Grind crankshaft	✓	X			Cut for big springs				
Polish crankshaft					C-c cylinder heads				
Comp. prep. crankshaft					Balance chambers				
Shot-peen crankshaft					Mill heads			X	
Weld ( ) journal(s)					Repair cracked block or head(s)				
Cross-drill crankshaft					Ceramic coat & pressure check head(s)				
					Mill ( ) manifold side(s)				
					Surface flywheel				
					R & R press fit pistons				
					Install bushings & fit pins				
					Recondition ( ) con rods				
Balance complete assy.					Side clearance con rods				
Balance crank only					Full float con rods				
Balance flywheel and clutch					Fit pins for clearance				
					Shot-peen con rods				
Hot tank block		X			Magnaflux crankshaft				
R & R camshaft bearings		X			Magnaflux rod & bolts				
Hot tank misc. parts					Magnaflux main caps & bolts				
Line-hone block					Magnaflux heads				
Deburr block					Magnaflux block				
Mill block									
Bore & wet hone block									
Hone cyl w/o plate									
Hone cyl w/ plate									
Bore & hone ( ) cyl.(s)									
Install ( ) cylinder sleeve(s)					Knurl pistons				
					Install ( ) top groove spacers				
					Clean ( ) pistons				
Stock valve grind		X							
Legal comp. valve grind					Tear down & inspect engine				
Super comp. valve grind					Assemble short block		X		
Knurl ( ) guides					Assemble engine				
Bore & install ( ) guides		X			Competition assembly of engine				
Install ( ) seats		X							
R & R replaceable guides									
Install screw-in studs									
Machine stud bosses									
Cut for pc seals									
Install ( ) oversize valves									

PARTS REQUIRED

QUAN	DESCRIPTION	OR-DERED	IN STOCK	DATE P.U.	BY	PRICE
	Pistons					
	Piston pin bushings		✓			37.38
	Rings KO 371		✓			15.04
	Rod bearings STD 323		✓			23.06
	Main bearings STD 442					
	Cam bearings					
	Camshaft					
	Lifters					
	Gaskets CS 8016	1660	✓			11.61
	F/S 8016	825	✓			5.95
	Timing chain					
	Timing gear					
	Crank gear					
	Oil pump					
	Oil pump drivershaft					
	Valves - intake 2383	6.22	✓			4.19
	Valves - Exhaust 2345	2.66	✓			1.70
	Valve springs					
	Valve seats					
	Valve guides					
	Aluminum retainers					
	Hardened valve locks					
	Clutch pressure plate					
	Clutch disc					
	Release assembly					
	Crank core					
	Connecting rod cores					
	Rod bolts & nuts					

I understand that there is no guarantee stated or implied on any engine used for racing. Completely stock engines used for normal passenger car service will be warranteed for 90 days or 4000 miles, whichever occurs first. Truck engines are guaranteed for 30 days after installation. B&J Engine Supply cannot be held responsible for improper installation or operation of said engine. All warrantee work will be performed by B&J Engine Supply.

A 50% deposit is required on all C.O.D. transactions. All work left over 30 days after notification of completion will be sold for labor costs. Prices valid for 14 days. Absolutely no guarantee given on products used for racing. We are not responsible for bad castings due to factory flaws, mishandling beyond our control or conditions of wear beyond initial appearance. B&J Engine Supply reserves the right to require the final payment to be made by cash or certified check.

Authorization _____

FIGURE 1-52 A typical work order sheet

ist takes the form to the parts department where it is fulfilled.

Parts managers must fill out forms to keep an adequate number of parts in stock. When the stock gets low, new parts are ordered. Occasionally it is necessary to order special parts. To ensure customer and shop satisfaction, these special orders must always be coordinated between the service and parts department. Parts can be ordered by using standard or computerized catalogs.

In some small shops, the technician or machinist might also have the job of ordering parts (see Chapter 4).

- *Dispatch Sheet.* A dispatch sheet, or work schedule, keeps track of dates when the work is to be completed. Some dispatch sheets follow the job through each step of the rebuilding process. Whenever a customer calls, the dispatch sheet is consulted. This sheet is usually posted in the office of the shop or wherever it is convenient.

- *Labor Charges.* Once the engine is rebuilt, the technician turns in the work order to the service manager, or service writer, who adds the labor charges.
- *Billing.* Billing is the total cost of servicing. It includes all labor charges, parts charges, sales tax, and any other charges. The billing department totals the amount of and receives payment from the customer.

GENERAL SHOP SAFETY

The most important considerations in any automotive repair shop should be accident prevention and safety. Carelessness and the lack of safety habits cause accidents. Accidents have a far-reaching effect, not only on the victim, but also on the victim's family and society in general. More important, accidents can cause serious injury, temporary or permanent, or even death. Therefore, it is the obligation of all shop employees and the employer to develop a safety program to protect the health and welfare of those involved.

In the following chapters of this book, the text contains special notations labeled **SHOP TALK,** **CAUTION,** and **WARNING.** Each one is there for a specific purpose. **SHOP TALK** gives added information that will help the rebuilder or machinist to complete a particular procedure or make a task easier. **CAUTION** is given to prevent the rebuilder or machinist from making an error that could damage the vehicle. **WARNING** reminds the rebuilder or machinist to be especially careful of those areas where carelessness can cause personal injury. The following text contains some general **WARNINGs** that should be followed when working in an automotive rebuilding shop.

In addition to the safety procedures given in future chapters, there are certain rules regarding personal safety, work area safety, and tool and machine safety that must be followed at all times.

PERSONAL SAFETY

Personal appearance and conduct can help prevent accidents. Guidelines for personal safety are as follows:

- Always wear safety glasses, goggles, or a safety face shield (Figure 1–53), protective clothing, or protective equipment whenever required (Figure 1–54).
- Keep clothing away from moving parts when an engine or a machine is running. Any loose

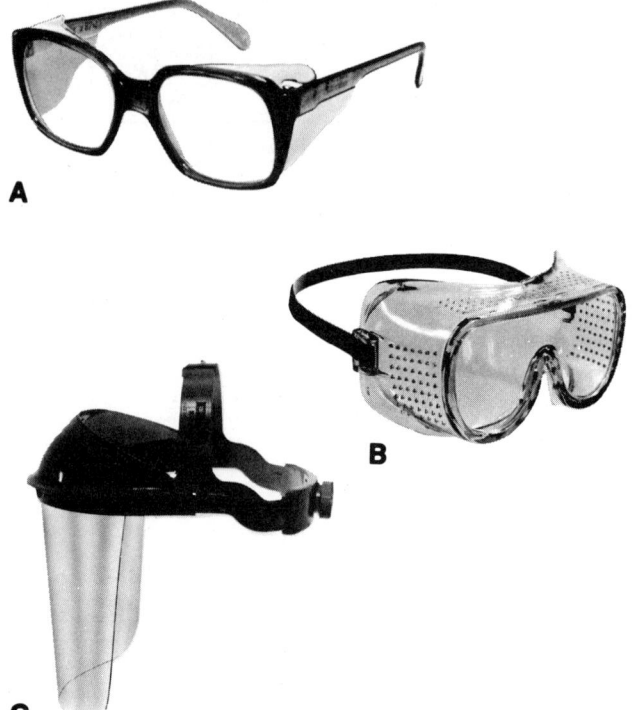

FIGURE 1–53 (A) Safety glasses; (B) goggles; (C) face shield *(Courtesy of Goodson Shop Supplies)*

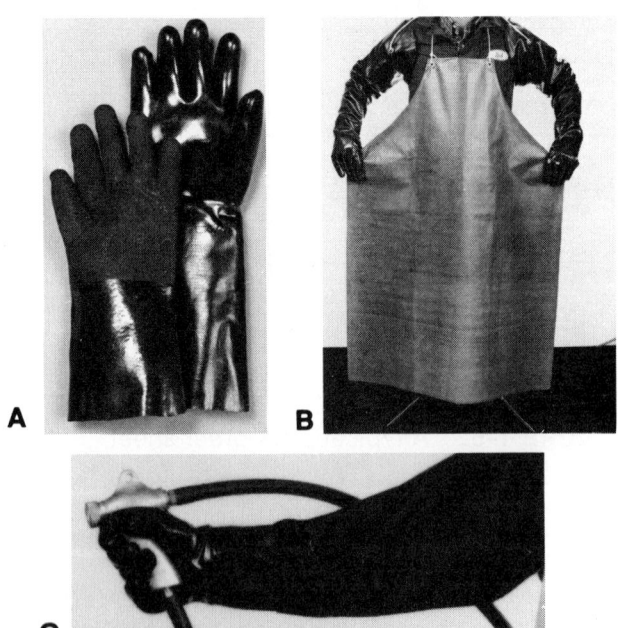

FIGURE 1–54 Engine cleaning protective gear: (A) hot tank gloves, (B) full-length bib-style apron, and (C) glass bead gloves. The latter are often fitted in the blasting equipment. *(Courtesy of Goodson Shop Supplies)*

or hanging clothing, such as shirttails, ties, lapels, cuffs, or scarves, creates a risk of getting wrapped up in moving parts of the vehicle or machinery, causing serious bodily injury. No type of jewelry should be worn.

- Long hair should be tied back or tucked into a cap or bandana to prevent it from falling into the eyes, work, or machine.
- When lifting and carrying objects, bend with the knees, not the back. Also, do not bend the waist when lifting. Remember, heavy objects should be lifted and moved with the proper equipment for the job.
- Never smoke while working on any vehicle or machine.
- Proper conduct can also help prevent accidents. Horseplay is not fun when it sends someone to the hospital. Such things as air nozzle fights, creeper races, or practical jokes do not have any place in the shop.
- Do not risk injury through the lack of knowledge; use shop machinery or perform rebuilding operations only after receiving proper instruction.
- To prevent serious burns, avoid contact with hot metal parts such as the radiator, exhaust manifold, tailpipe, catalytic converter, and muffler.
- Use safety stands whenever a procedure requires getting under the vehicle. Never trust jacks alone.
- When it is necessary to drive a car in a shop, watch out for other cars and people. It is best to have someone act as a guide. Leave the window open too, so that the guide's directions can be heard.

WORK AREA SAFETY

It is very important that the work area be kept safe.

- All surfaces should be kept clean, dry, and orderly. Any oil, coolant, or grease on the floor can cause slips that could result in serious injuries. To clean up oil, be sure to use a commercial oil absorbent.
- Keep all water off the floor; remember that water is a conductor of electricity. A serious shock hazard will result if a live wire happens to fall into a puddle in which a person is standing.
- Make sure that aisles and walkways are kept clean and wide enough for a safe clearance.

Cluttered walking areas contain items waiting to cause accidents.

- Use antiskid floor strips on the floor area where the operator normally stands and mark off each machine's work area. Make certain the work area is well lighted and ventilated. Provide for adequate work space around the machine. The work area should not be readily accessible to anyone except the operator.
- Operate the engine only in a well-ventilated area to avoid the danger of carbon monoxide. Most shop exhaust systems carry the exhaust away from a vehicle directly to the outside of the shop.
- There should be a list of emergency telephone numbers clearly posted next to the telephone. These numbers should include a doctor, hospital, and fire and police departments. Also the work area should have a first-aid kit for treating minor injuries. This kit should include some sterile gauze, bandages, scissors, and other related items. Facilities for flushing the eyes should also be in or near the shop area.
- When diagnosing an engine, set the parking brake when working on the vehicle. If it is an automatic transmission, set it in PARK unless instructed otherwise for a specific service operation. If it is a manual transmission, it should be in REVERSE (engine off) or NEUTRAL (engine on) unless instructed otherwise for a specific service operation.
- All bolts, nuts, lock rings, and other fastening components mentioned in the manufacturer's service manual are crucial to the safe operation of the car. Failure to use those specific items could cause extensive damage. Manufacturer's torque specifications must be followed.
- Gasoline is a highly flammable liquid that vaporizes rapidly. For this reason, always keep gasoline or diesel fuel in an approved safety can (Figure 1–55) and never use it to wash hands or tools. Other flammables, such as paints, thinners, and pressurized containers, should be stored in approved metal cabinets away from any source of heat.
- Oily rags should also be stored in an approved metal container. When these oily, greasy, or paint-soaked rags are left lying about or are stored improperly, they are prime candidates for spontaneous combustion, that is, fires that start by themselves.

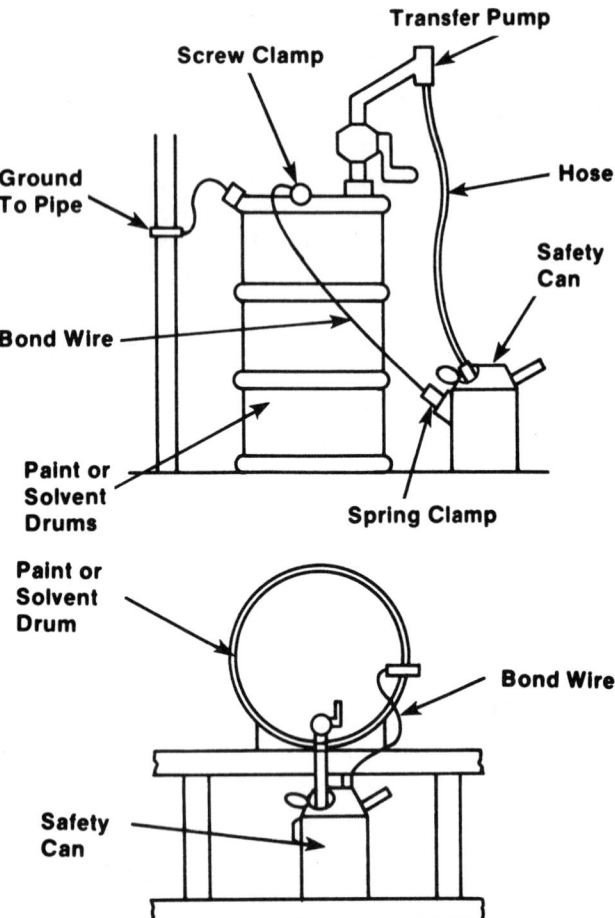

FIGURE 1-55 Two safe methods of moving flammable liquids from a drum to a portable safety can

- Check to be sure that all drain covers are snugly in place. Open drains have caused many toe, ankle, and leg injuries.
- Keep workbenches clean and orderly. When things are stacked on top of benches and tables, they can easily fall. Generally, they cause a cluttered work area, which invites injury.
- Floor jacks, bumper jacks, jack stands, and creepers should be kept in one area, out of aisles and walkways.
- Never use gasoline to clean parts, clothing, or hands.
- Know where the fire extinguishers are and what types of fires they put out (Table 1-1). A multipurpose dry chemical fire extinguisher will put out ordinary combustibles, flammable liquids, and electrical fires. In case of a gasoline fire, do not put water on it; the water will just spread the fire. Use a fire extinguisher to smother the flames. Remember, during a fire, never open doors or windows

unless it is absolutely necessary; the extra draft will only make the fire worse. Keep doors and windows closed. A good rule is to call the fire department first and then attempt to extinguish the fire. But if it gets too hot or too smoky, get out. Remember, never go back into a burning building for anything.

TOOL AND MACHINE SAFETY

The rebuilder technician must observe the following hand and power tool safety guidelines:

- Hand tools should always be clean and in workable condition. Greasy, oily, or chipped hand tools can easily slip out of your grasp, causing skinned knuckles or broken fingers.
- Check all hand tools for cracks, chips, burrs, broken teeth, or other dangerous conditions before using them. If any tools are defective, do not use them.
- Be careful when using sharp or pointed tools that can slip and cause injury. If a tool is supposed to be sharp, make sure it is sharp.
- Do not use hand tools for any job other than that for which they were specifically designed.
- Do not carry screwdrivers, punches, or other sharp hand tools in your pockets. It is possible to injure yourself or damage the vehicle being worked on.
- When using an electric power tool or rebuilding machine, make sure that it is properly grounded and check the wiring for cracks in the insulation, as well as for bare wires. Also, when using electrical power tools, never stand on a wet or damp floor.
- Do not operate a power tool or machine without its guard(s).
- Disconnect electrical power before performing any service or maintenance on the machine or tool.
- When doing any power grinding, chipping, sanding, or similar operation, always wear safety glasses. When using power equipment on small parts, never hold the part in your hand. It could slip. Always use vise grips instead.
- Before plugging in any electric tool, make sure the switch is off to prevent serious injury. When you are through using it, turn it off for the next person.
- Do not attempt to use the machine or tool beyond its stated capacity or for operations

TABLE 1-1: GUIDE TO EXTINGUISHER SELECTION

	Class of Fire	Typical Fuel Involved	Type of Extinguisher
Class **A** Fires (green)	**For Ordinary Combustibles** Put out a class A fire by lowering its temperature or by coating the burning combustibles.	Wood Paper Cloth Rubber Plastics Rubbish Upholstery	Water*[1] Foam* Multipurpose dry chemical[4]
Class **B** Fires (red)	**For Flammable Liquids** Put out a class B fire by smothering it. Use an extinguisher that gives a blanketing, flame-interrupting effect; cover whole flaming liquid surface.	Gasoline Oil Grease Paint Lighter fluid	Foam* Carbon dixoide[5] Halogenated agent[6] Standard dry chemical[2] Purple K dry chemical[3] Multipurpose dry chemical[4]
Class **C** Fires (blue)	**For Electrical Equipment** Put out a class C fire by shutting off power as quickly as possible and by always using a nonconducting extinguishing agent to prevent electric shock.	Motors Appliances Wiring Fuse boxes Switchboards	Carbon dioxide[5] Halogenated agent[6] Standard dry chemical[2] Purple K dry chemical[3] Multipurpose dry chemical[4]
Class **D** Fires (yellow)	**For Combustible Metals** Put out a class D fire of metal chips, turnings, or shavings by smothering or coating with a specially designed extinguishing agent.	Aluminum Magnesium Potassium Sodium Titanium Zirconium	Dry power extinguishers and agents only

*Cartridge-operated water, foam, and soda-acid types of extinguishers are no longer manufactured. These extinguishers should be removed from service when they become due for their next hydrostatic pressure test.

Notes:
(1) Freeze in low temperatures unless treated with antifreeze solution, usually weighs over 20 pounds, and is heavier than any other extinguisher mentioned.
(2) Also called ordinary or regular dry chemical. (sodium bicarbonate)
(3) Has the greatest initial fire-stopping power of the extinguishers mentioned for class B fires. Be sure to clean residue immediately after using the extinguisher so sprayed surfaces will not be damaged. (potassium bicarbonate)
(4) The only extinguishers that fight A, B, and C classes of fires. However, they should not be used on fires in liquefied fat or oil of appreciable depth. Be sure to clean residue immediately after using the extinguisher so sprayed surfaces will not be damaged. (ammonium phosphates)
(5) Use with caution in unventilated, confined spaces.
(6) May cause injury to the operator if the extinguishing agent (a gas) or the gases produced when the agent is applied to a fire is inhaled.

requiring more than the rated horsepower of the motor. Never use a tool or machine for operations it was not designed for.
- Keep hands away from moving parts when the machine or tool is under power. Never clear chips or debris when the machine is under power and never use your hands to clear chips; use a brush or chip rake.
- Never overreach. Maintain a balanced stance and avoid slipping.
- Utmost caution should accompany the use of compressed air. Pneumatic tools must be operated at the pressure recommended by their manufacturer. The downstream pressure of compressed air used for cleaning purposes must remain at a pressure level below 30 psi whenever the nozzle is dead-ended (Figure 1-56). Do not use compressed air to clean clothes. Even at low cleaning pressure, compressed air can cause dirt particles to become embedded in the skin, which can result in infection.
- Store all parts and tools properly by putting them away neatly where people will not trip

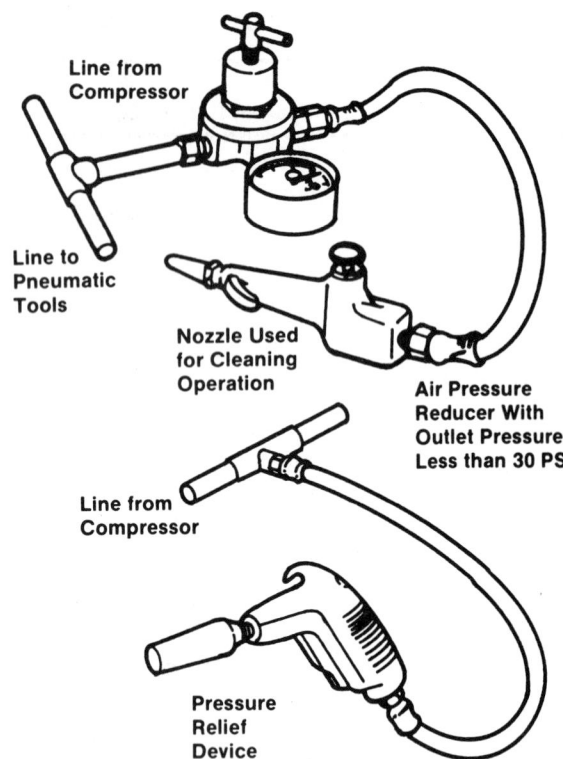

Line from Compressor

Line to Pneumatic Tools

Nozzle Used for Cleaning Operation

Air Pressure Reducer With Outlet Pressure Less than 30 PSI

Line from Compressor

Pressure Relief Device

FIGURE 1-56 Air pressure reducing devices

over them. This practice not only cuts down on injuries, it also reduces time wasted looking for a misplaced part or tool.

- When working with a hydraulic press, make sure that hydraulic pressure is applied in a safe manner. It is generally wise to stand to the side when operating the press. Always wear safety glasses.
- If the shop has a hydraulic lift, be sure to read the instruction manual before using it. Check the pads to see that they are making proper contact with the frame. Then raise the vehicle about 6 inches and shake it to make sure it is well balanced on the lift. If there are any rattling or scraping sounds, it means that the vehicle is not locked in place properly. If this happens, lower the lift and realign the pads to the vehicle. Test it again as previously described. Then, after lifting the vehicle to full height, put the safety catch on before working underneath the vehicle. Never permit anyone, either technician or customer, to remain in the car while it is being lifted.

HAZARDOUS WASTES

In rebuilding and machining shops, hazardous wastes are generated. Every employee in these

shops is protected by Right-to-Know laws. These laws started with the Occupational Safety and Health Administration (OSHA) Hazard Communication Standard published in 1983. This document was originally intended for chemical companies and manufacturers that require employees to handle potentially hazardous materials in the workplace. Since then, the majority of states have enacted their own Right-to-Know laws and the federal courts have decided that these regulations should apply to all companies, including the engine rebuilding profession.

The general intent of the law is for employers to provide their employees with a safe working place as it relates to hazardous materials. Specifically, there are three areas of employer responsibility:

1. *Training/Educating Employees.* All employees must be trained about their rights under the legislation, the nature of the hazardous chemicals in their workplace, the labeling of chemicals, and the information about each chemical contained in the Material Safety Data Sheet (MSDS) (Figure 1-57). Employees must be familiar with the general uses, characteristics, protective equipment, accident or spill procedures, and so on associated with major groups of chemicals. This training must be given to employees annually and provided to new employees as part of their job orientation.

2. *Labeling/Information about Potentially Hazardous Chemicals.* All hazardous materials must be properly labeled, indicating what health, fire, or reactivity hazard it poses and what protective equipment is necessary when handling each chemical. A list of all hazardous materials used in the shop must be posted for the employees to see.

3. *Record Keeping.* Companies must maintain documentation on the hazardous chemicals in the workplace, proof of training programs, records of accidents and/or spill incidents, satisfaction of employee requests for specific chemical information via the MSDSs, and a general written Right-to-Know compliance procedure manual utilized within the company.

Hazardous waste as determined by the EPA must be in the form of "solid" material, but the EPA includes many liquids in this definition. If the waste is on the EPA list of known harmful materials or has

Vulcan CHEMICALS
A Division of Vulcan Materials Company

CAUSTIC SODA

MATERIAL SAFETY DATA SHEET
(ESSENTIALLY SIMILAR TO FORM OSHA-20)
SEE IMPORTANT NOTICE ON BOTTOM OF OTHER SIDE
24 Hour Emergency Phone (316) 524-5751

I - PRODUCT IDENTIFICATION

MANUFACTURER'S NAME AND ADDRESS
Vulcan Materials Company, Chemicals Division, P. O. Box 7689, Birmingham, AL 35253-0689

CHEMICAL NAME	CHEMICAL FORMULA
Sodium Hydroxide Anhydrous	NaOH

TRADE NAME AND SYNONYMS	CHEMICAL FAMILY
Caustic, Caustic Soda, Lye	Alkali

CAS REGISTRY NO.	DOT IDENTIFICATION NO.
1310-73-2	UN 1823

II - HAZARDOUS INGREDIENTS

MATERIAL OR COMPONENT	% (wt)	PEL (Units)

III - PHYSICAL DATA

BOILING POINT (°F)	N/A	SPECIFIC GRAVITY (H₂O = 1)	2.13
VAPOR PRESSURE (mm Hg.)	N/A	PERCENT, VOLATILE BY VOLUME (%)	N/A
VAPOR DENSITY (AIR = 1)	N/A	EVAPORATION RATE	N/A
SOLUBILITY IN WATER	Appreciable	APPEARANCE AND ODOR	white odorless solid or bead

IV - FIRE AND EXPLOSION HAZARD DATA

FLASH POINT (Method used)	Nonflammable	FLAMMABLE LIMITS	Lower N/A	Upper N/A

EXTINGUISHING MEDIA
N/A

SPECIAL FIRE FIGHTING PROCEDURES
N/A

UNUSUAL FIRE AND EXPLOSION HAZARDS Will react with metals, i.e. aluminum, tin and zinc, to release flammable hydrogen gas.

V - REACTIVITY DATA

STABILITY	UNSTABLE		CONDITIONS TO AVOID
	STABLE	X	

INCOMPATIBILITY (Materials to avoid) Acids, flammable liquids, organic halogenated compounds. Contact with nitro compounds may form shock sensitive salts.

HAZARDOUS DECOMPOSITION PRODUCTS None

HAZARDOUS POLYMERIZATION	MAY OCCUR		CONDITIONS TO AVOID
	WILL NOT OCCUR	X	

VI - HEALTH HAZARD DATA

OSHA PERMISSIBLE EXPOSURE LIMIT   2 mg/m³ 8 hour TWA (29 CFR Part 1910.1000)

ACGIH: 2 mg/m³ ceiling.

EFFECTS OF OVEREXPOSURE

INHALATION:
Caustic dust can cause destructive burns of the mucous membranes. Severe pneumonitis may occur.

SKIN CONTACT/ABSORPTION:
Caustic solids or dust will cause skin irritation. Prolonged exposure will cause severe burns with scarring.

INGESTION:
Swallowing will cause severe burns of the mouth, throat and stomach. May result in death in extreme cases.

EYES:
Contact with eyes causes rapid tissue destruction leading to permanent eye damage and possible blindness.

EMERGENCY AND FIRST AID PROCEDURES

EYES AND SKIN   Wash eyes immediately with large amounts of water, lifting the lower and upper eyelids occasionally. Get medical attention immediately.

INHALATION   Move person to fresh air. If breathing stops, administer artificial respiration. Get medical attention immediately.

INGESTION   If person is conscious, give large quantities of water to dilute caustic. Do not induce vomiting. Get medical attention immediately.

VII - SPILL OR LEAK PROCEDURES

STEPS TO BE TAKEN IN CASE MATERIAL IS RELEASED OR SPILLED
Cleanup personnel must have proper protective equipment. Reclaim into closed containers for possible normal use or proper disposal. Can be flushed and dissolved with water and neutralized.

WASTE DISPOSAL METHOD
Dispose of in approved landfill, or contact approved waste disposal firm. Comply with federal, state, local regulations.

VIII - SPECIAL PROTECTION INFORMATION

SPECIFIC PERSONAL PROTECTIVE EQUIPMENT
RESPIRATORY   NIOSH-approved self-contained breathing apparatus for exposures above OSHA PEL.
EYE   Chemical goggles which are dust and splashproof.
SKIN   Impervious clothing, rubber gloves and shoes.
OTHER   Safety shower and eyewash fountain should be located in the immediate work area.
VENTILATION REQUIREMENTS
Use ventilation or exhaust to maintain exposure below OSHA PEL.

IX - SPECIAL PRECAUTIONS

PRECAUTIONS TO BE TAKEN IN HANDLING AND STORING   When dissolving in water, add caustic slowly to surface of water with constant stirring, to avoid violent spattering. Store in dry area in closed containers.

OTHER PRECAUTIONS   Contact of caustic soda cleaning solutions with food and beverage products may produce lethal concentrations of carbon monoxide gas in enclosed vessels or spaces.

DATE   July 1983

NOTICE: Vulcan Chemicals believes that the information contained on this Material Safety Data Sheet is accurate. The suggested procedures are based on experience as of the date of publication. They are not necessarily all-inclusive nor fully adequate in every circumstance. Also, the suggestions should not be confused with nor followed in violation of applicable laws, regulations, rules or insurance requirements. NO WARRANTY, EXPRESS OR IMPLIED, OF MERCHANTABILITY, FITNESS OR OTHERWISE IS MADE.

Form 3239-230

FIGURE 1-57 Typical Material Safety Data Sheet *(Courtesy of Vulcan Material Company)*

one or more of the following four known dangerous characteristics, it is considered hazardous.

- *Ignitability.* The waste fails the ignitability test if it is a liquid with a flash point below 140 degrees Fahrenheit or a solid that can spontaneously ignite.
- *Corrosivity.* A material or waste is considered corrosive if it dissolves metals and other materials or burns the skin. It is an aqueous solution with a pH of 2 and below or 12.5 and above. Acids have the lower value and alkalis have the higher value.
- *Reactivity.* Any material that reacts violently with water or other materials or releases cyanide gas, hydrogen sulfide gas, or similar gases when exposed to low pH solutions (acid) is considered hazardous. This also includes material that generates toxic mists, fumes, vapors, and flammable gases.

- *EP Toxicity.* Materials that leach one or more of eight heavy metals in concentrations greater than 100 times primary drinking water standard concentrations are considered hazardous. These heavy metals include lead, cadmium, chromium, and arsenic, the most common heavy metals found in a typical machine shop operation.

Complete EPA lists of hazardous wastes can be found in the Code of Federal Regulations. Materials and wastes of most concern to the engine rebuilder are organic solvents: ignitable, corrosive, and/or toxic materials, and wastes that contain heavy metals, especially lead. It should be noted that no material is considered hazardous waste until the shop is finished using it and ready to dispose of it. For instance, a caustic cleaning solution with a heavy concentration of lead in the cleaning tank is not considered hazardous waste until it is ready to be replaced. Now that the shop is ready to dispose of it, it is hazardous waste and must be handled accordingly (Figure 1-58). The EPA says it is the owner's responsibility to determine whether the waste is hazardous, but the owner must have adequate test results to support the shop's claim.

Testing for hazardous wastes can be done by any qualified laboratory that performs tests on drinking water. They should be capable of performing tests for hazardous materials. The shop's owner(s) should contact them to see how they would like the samples taken and shipped to give accurate results. The area of most concern to the engine rebuilder will be in the cleaning operation (see Chapter 5). Spray cabinets and dip tanks that use caustic chemicals produce high alkaline solutions and contain heavy metals. Baking ovens generate ash containing heavy metals. Small parts washers generally use solvents that are classified as hazardous materials. Other areas to be concerned about are the coolants used in flywheel, crankshaft, camshaft, valve, and surface grinders.

WARNING: When handling any hazardous waste material be sure to wear the safety equipment (Figure 1-59) covered under the right-to-know law and follow all required procedures correctly.

FIGURE 1-58 Many automotive service operations, including engine rebuilders, are hiring outside contractors to handle their hazardous waste materials.

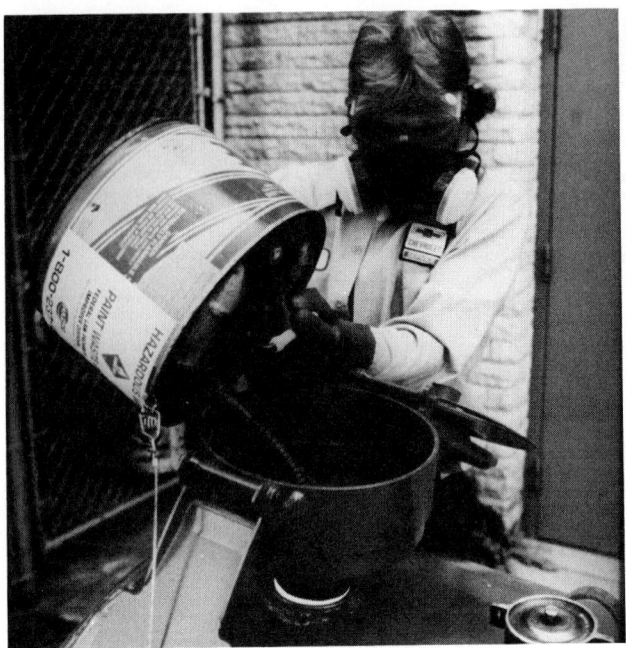

FIGURE 1-59 Handle all hazardous wastes as per EPA regulations.

ASE/AERA CERTIFICATION

Just as doctors, nurses, accountants, electricians, and other professionals are licensed or certified in order to practice their profession, so the technician can be certified. Certification protects the general public and the practitioner or professional. It assures the general public and the prospective employer that certain minimum standards of performance have been met. Many employers now expect their rebuilder technicians to be certified. The certified machinist/technician is recognized as a professional by the public, employers, and peers. For this reason, the certified machinist/technician usually receives higher pay than one who is not certified.

A testing and certification program for automotive machinists is called the Engine Machinist test series. It was created by the National Institute for Automotive Service Excellence (ASE) and the Automotive Engine Rebuilders Association (AERA).

The series consists of three tests: the Cylinder Head Specialist, Engine Specialist, and Engine Assembly Specialist. The tests cover cylinder head rebuilding, block reconditioning, and general machine and assembly (Figure 1-60). To become certified by ASE, a machinist/technician must pass the appropriate test(s) and show two years hands-on work

FIGURE 1-60 Machinist/technician must be able to perform general machining and rebuilding.

experience or a combination of work experience and appropriate vocational training.

To help prepare for these programs, the test questions at the end of each chapter are similar in design and content to those used by the ASE. For further information on the ASE certification program, write: National Institute for Auto Service Excellence, 13505 Dulles Technology Drive, Herndon, Virginia 22071. Remember that a certified rebuilder (or specialist) is usually better paid than the general line mechanic.

REVIEW QUESTIONS

1. In which of the following operations is the cylinder head or oil pan removed with the engine still mounted in the vehicle?
   a. tune-up
   b. minor overhaul
   c. major overhaul
   d. none of the above

2. Which position of the VIN identifies the engine type?
   a. position 6
   b. position 8
   c. position 10
   d. position 12

3. Which of the following is an advantage of pneumatic tools over electrically powered equipment?
   a. flexibility
   b. lightweight
   c. safety
   d. both a and b
   e. all of the above

4. Which of the following measurements is the dial caliper capable of taking?
   a. inside
   b. outside
   c. depth
   d. step
   e. all of the above

5. What increments are indicated by a vernier scale?
   a. 0.01 inch
   b. 0.025 inch
   c. 0.001 inch
   d. 0.0001 inch

6. A telescoping gauge is usually used with a(n) _____.
   a. machinist's rule
   b. dial caliper
   c. inside micrometer
   d. none of the above

7. A small hole gauge is used to measure the _____.
   a. depth of a bore
   b. outside diameter of a cylinder
   c. diameter of a bore
   d. all of the above
   e. none of the above

8. Which of the following measures motion?
   a. dial caliper
   b. telescoping gauge
   c. dial indicator
   d. none of the above

9. Which of the following must be rocked to obtain the correct measurement?
   a. inside micrometer
   b. telescoping gauge
   c. cylinder bore dial gauge
   d. all of the above
   e. none of the above

10. Which of the following is an advantage of electronic gauging equipment?
    a. easier to read
    b. higher resolution
    c. ability to tell when a measurement exceeds a predetermined dimension
    d. all of the above
    e. none of the above

11. Which of the following contains figures dealing with the length of time a rebuilding job is supposed to require?
    a. auto manufacturer's service manual
    b. aftermarket supplier's guide
    c. specialty repair manual
    d. all of the above
    e. none of the above

12. Which form is completed after performing the vehicle diagnosis test and before engine teardown?
    a. work order
    b. parts requisition
    c. billing
    d. none of the above

13. A fire breaks out in the shop. Technician A opens the windows to enhance ventilation. In the same situation, Technician B calls the fire department. Who is right?
    a. Technician A
    b. Technician B
    c. Both A and B
    d. Neither A nor B

14. Technician A will operate a tool beyond its stated capacity only for a limited time. Technician B pays no attention to the limits of the machine. Who is right?
    a. Technician A
    b. Technician B
    c. Both A and B
    d. Neither A nor B

15. How many years of experience are required to become ASE certified?
    a. one
    b. two
    c. three
    d. five

CHAPTER TWO

MODERN VEHICLE ENGINES

Objectives

After reading this chapter, you should be able to
- Explain the four-stroke cycle in an automotive gasoline engine.
- Describe the characteristics of various gasoline engine designs.
- Explain the design and function of various engine parts.
- Describe the three basic valve and camshaft placement configurations.
- Explain the function of the main engine systems.
- Identify the classifications of four-stroke engines.
- Describe alternative engine designs, including diesel, two-stroke, and rotary.

Modern engines are highly engineered state-of-the-art power plants designed to meet the special demands of the automobile buying public for achieving greater performance and fuel efficiency. The days of the "gas hog," heavy, cast-iron V-8 engine are quickly drawing to a close. Today, these engines have been replaced by compact, high-tech, four-cylinder and V-6 power plants (Figure 2–1). Modern engine technology utilizes many extremely light-weight engine castings and stampings; nontraditional materials for power plant applications (for example, aluminum, magnesium, fiber-reinforced plastics); and fewer and smaller fasteners to hold things together, due to computerized joint designs that optimize loading patterns. Each of these newer engine designs has its own distinct personality, based on

FIGURE 2–1 (Left) High-tech 4-cylinder engine; (right) a typical V-6 power plant. *(Courtesy of General Motors Corporation)*

construction materials, casting configurations, and design optimizations that concern weight, power, manufacturing costs, and serviceability.

As will be covered in future chapters of this book, these modern engine technologies have created problems for the engine rebuilding profession. But before examining how these problems might affect the rebuilder, it is important for the technician to have a basic knowledge of engine design and operation (Figure 2–2).

ENGINE CLASSIFICATIONS

Modern automotive engines can be classified in several ways depending on the following design features:

- *Operational Cycles.* Most automotive rebuilders will generally come in contact with only two- or four-cycle engines. The most popular is the four-cycle engine (Figure 2–3).
- *Number of Cylinders.* Current engine designs include 3-, 4-, 5-, 6-, 8-, and 12-cylinder engines.
- *Cylinder Arrangement.* An engine can be flat (opposed), in-line, or V-type. Other more complicated designs have also been used.
- *Valve Train Type.* Engine valve trains can be either the overhead camshaft (OHC) type or the in-block overhead valve (OHV) type. It is

also possible to use separate camshafts for intake and exhaust valves. Some high-performance sports cars and racing cars use dual overhead camshafts (DOHC).

- *Valve Arrangement.* There are several types of valve arrangements as described later in this chapter.
- *Ignition Type.* There are two types of ignition systems: spark and compression. In a spark ignition system, the air/fuel mixture is ignited by an electrical spark. Diesel engines, or compression ignition engines, have no spark plugs. An automotive diesel engine has a higher compression ratio than a spark ignition engine. The higher compression generates enough heat to ignite the air/fuel mixture for the power stroke.
- *Cooling Systems.* There are both air-cooled and liquid-cooled engines in use. Most engines have liquid cooling systems.
- *Fuel Systems.* Fuel systems currently used in automobile engines include gasoline, diesel, and propane.

THE FOUR-STROKE GASOLINE ENGINE

In a passenger car or truck, the engine provides the rotating power to drive the wheels through the transmission—usually by a clutch or torque converter—and driving axle. All automobile engines,

FIGURE 2–2 (Left) V-6 gasoline engine; (right) a cutaway of the same engine *(Courtesy of General Motors Corporation)*

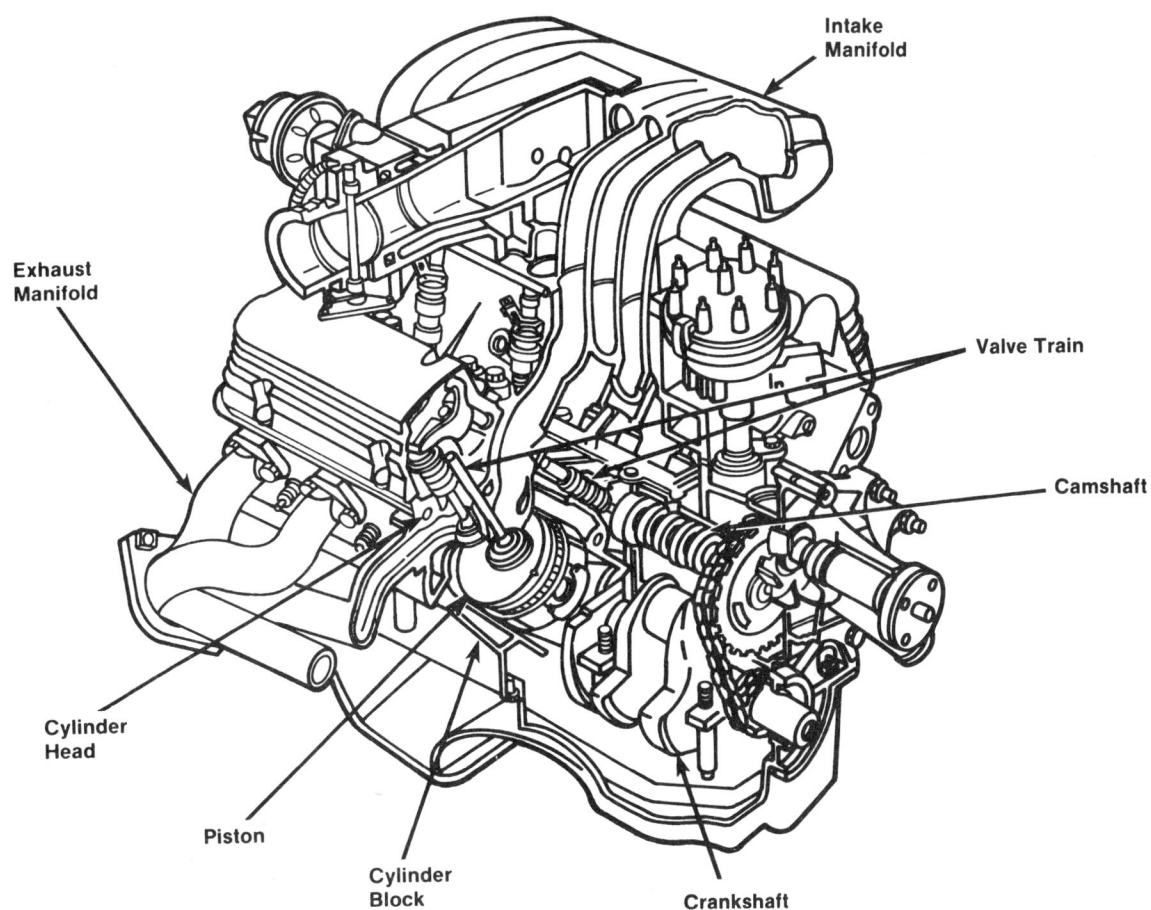

FIGURE 2-3   Components of a conventional four-stroke engine

both gasoline and diesel, are classified as "internal combustion engines" because the combustion or burning that creates heat energy takes place inside the engine. These systems require an air/fuel mixture that arrives in the combustion chamber with exact timing and an engine constructed to withstand the temperatures and pressures created by thousands of fuel droplets burning.

The combustion chamber is the space between the top of the piston and cylinder head. It is an enclosed area in which the gasoline and air mixture is burnt. The piston is a hollow metal tube with one end closed that moves up and down in the cylinder. This reciprocating motion is produced by the burning of fuel in the cylinder.

The reciprocating motion must be converted to rotary motion before it can drive the wheels of a vehicle. This conversion is achieved by linking the piston to a crankshaft with a connecting rod. The upper end of the connecting rod moves with the piston as it moves up and down in the cylinder. The lower end of the connecting rod is attached to the crankshaft and moves in a circle. The end of the

crankshaft is connected to the transmission to continue the power flow through the drivetrain and to the wheels.

For the combustion action in the cylinder to take place completely and efficiently, precisely measured amounts of air and fuel must be combined in the right proportions. The carburetor (or in most cases, a fuel injection system) makes sure that the engine gets exactly as much fuel and air as it needs for the many different conditions under which the vehicle must operate: starting, idling, power accelerating, or cruising.

There are at least two valves at the top of each cylinder. The air/fuel mixture enters the combustion chamber through an intake valve and leaves (after having been burned) through an exhaust valve. The valves are accurately machined plugs that fit into machined openings. A valve is said to be seated or closed when it rests in its opening. When the valve is pushed off its seat, it opens.

A rotating camshaft, connected to the crankshaft, opens and closes the intake and exhaust valves (Figure 2-4). Cams are raised sections of the

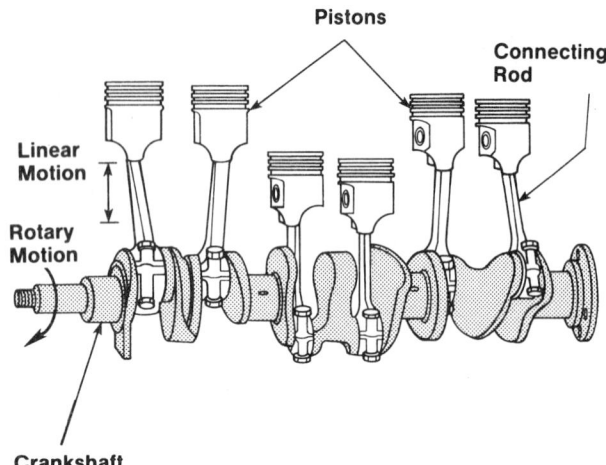

FIGURE 2-4 Linear (reciprocating) motion must be converted to rotary motion by the crankshaft.

shaft, or collars, with high spots called *lobes.* As the camshaft rotates, the lobes rotate and push away a spring-loaded valve. The tappet transfers the motion to a pushrod and perhaps a rocker arm to open the valve by lifting it off its seat. Once the lobe on the cam rotates out of the way, the valve, forced by a spring, moves down and reseats. The camshaft can be located either in the cylinder block or in the cylinder head.

In summary, the essentials for the complete combustion process include the following:

- Admit a proper mixture of air and fuel into the cylinder.
- Compress (squeeze) the mixture so it will burn better and deliver more power.
- Ignite and burn the mixture.
- Remove the burned gases from the cylinder so that the process can be completed and repeated.

With the proper timing of the action of the valves and spark plug to the movement of the piston, the combustion cycle takes place in four strokes of the piston. The basis of automotive gasoline engine operation is the four-stroke cycle. A stroke is the full travel of the piston up or down. There are four strokes in this cycle: the intake stroke, the compression stroke, the power stroke, and the exhaust stroke.

- *Intake Stroke.* As the piston moves away from top dead center (TDC), the intake valve opens (Figure 2–5A). The downward movement of the piston increases the volume of

the cylinder above it. This, in turn, reduces the pressure in the cylinder below atmospheric pressure. The reduced pressure causes atmospheric pressure to push a mixture of air and fuel through the open intake valve. As the piston reaches the bottom of its stroke, the reduction in pressure stops and the intake of air/fuel mixture nearly ceases. But due to the weight and movement of the air/fuel mixture, it will continue to enter the cylinder until the intake valve closes. The delayed closing of the intake valve increases the volumetric efficiency of the cylinder by packing as much air and fuel into it as possible.

- *Compression Stroke.* The compression stroke begins as the piston starts to move from bottom dead center (BDC). The intake valve closes, trapping the air/fuel mixture in the cylinder (Figure 2–5B). Upward movement of the piston compresses the air/fuel mixture, thus heating it up. At TDC, the piston and cylinder walls form a combustion chamber in which the fuel will be burned. The volume of the cylinder with the piston at BDC compared to the volume of the cylinder with the piston at TDC determines the compression ratio of the engine.

- *Power Stroke.* The power stroke begins as the compressed fuel mixture is ignited in the combustion chamber (Figure 2–5C). An electrical spark across the electrode of a spark plug ignites the air/fuel mixture. The burning fuel rapidly expands, creating a very high pressure against the top of the piston. This drives the piston down toward BDC. The downward movement of the piston is transmitted through the connecting rod to the crankshaft.

- *Exhaust Stroke.* The exhaust valve opens just before the piston reaches BDC on the power stroke (Figure 2–5D). Pressure within the cylinder when the valve opens causes the exhaust gas to rush past the valve and into the exhaust system. Movement of the piston from BDC pushes most of the remaining exhaust gas from the cylinder. As the piston nears TDC, the exhaust valve begins to close as the intake valve starts to open. The exhaust stroke completes the four-stroke cycle. The opening of the intake valve begins the cycle again. This cycle occurs in each cylinder and is repeated over and over, as long as the engine is running. The up-and-

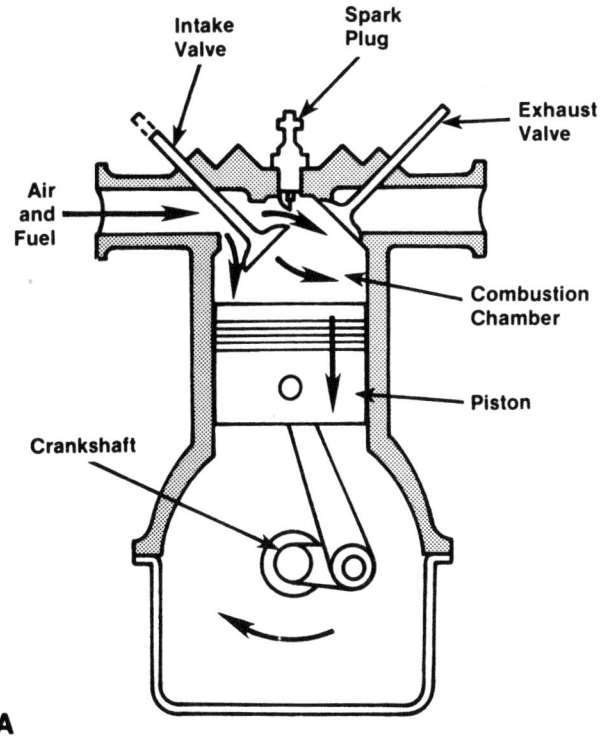

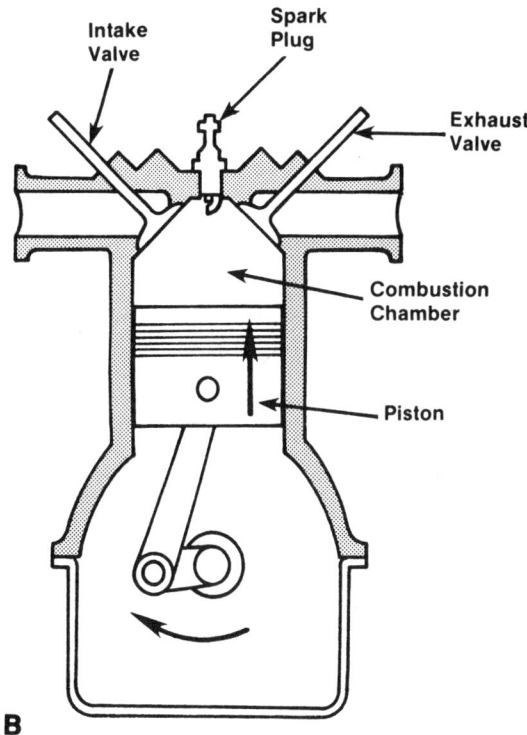

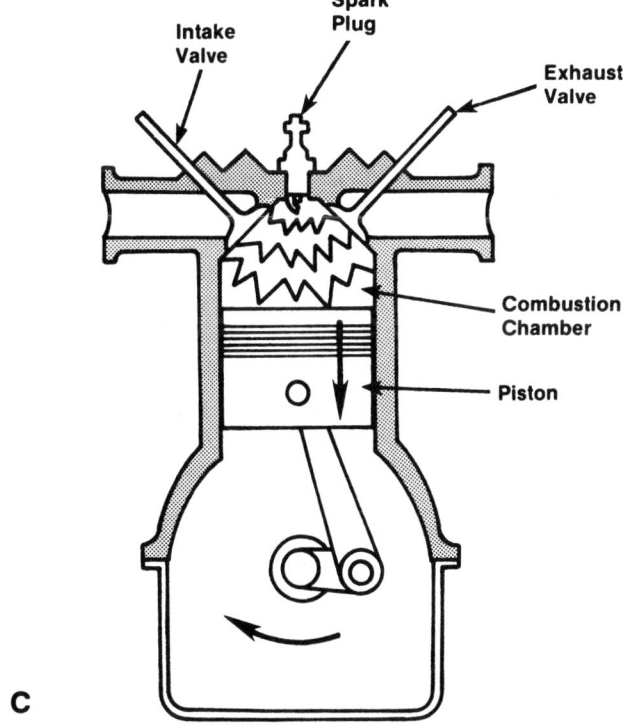

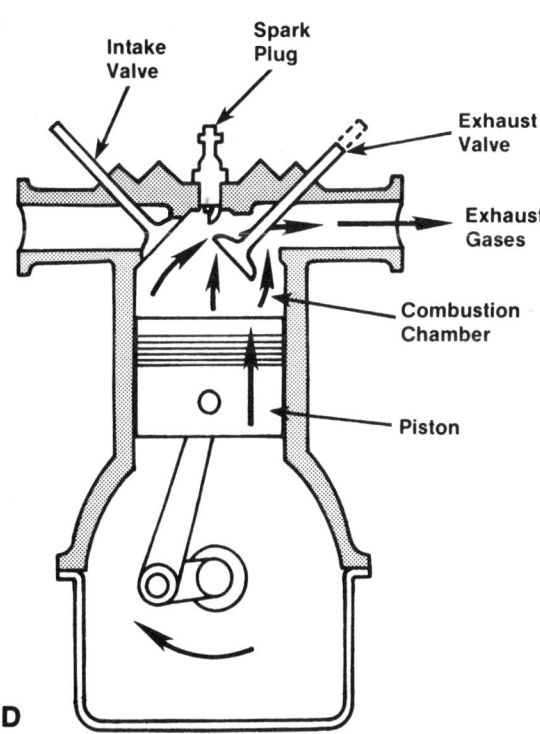

FIGURE 2-5 Strokes in a four-stroke cycle

down movement of the piston on all four strokes is converted to rotary motion of the crankshaft.

The four strokes of the cycle require two full revolutions of the crankshaft. Also, the piston is being acted on by combustion pressure during only about half of one stroke, or about one-quarter of one revolution. This makes it easier to understand the function of the flywheel. Even if the engine has multiple cylinders, a certain amount of power it produces has to be stored momentarily in the flywheel. From there, it is used to keep the piston in motion during about seven-eighths of the total cycle and to compress the fuel mixture just before combustion.

CHARACTERISTICS OF ENGINE DESIGN

Depending on the vehicle, either an in-line, V-type, slant, or opposed cylinder design can be used. The most popular designs are in-line and V types.

In-line Engines. In this design (Figure 2–6), the cylinders are all placed in a single row. There is one crankshaft and one cylinder head for all of the cylinders. The block is a single cast piece with all cylinders located in an upright position.

In-line engine designs have certain advantages and disadvantages. They are easy to manufacture, which brings the cost down somewhat. They are very easy to work on and to perform maintenance on. In-line engines have adequate room under the hood to work on other vehicle parts. However, because the cylinders are positioned vertically, the front of the vehicle must be higher. This affects the aerodynamic design of the car. Aerodynamic design refers to the ease at which the car can move through the air. The front of the vehicle cannot be made lower as with other engines. This means that the aerodynamic design of the car cannot be improved easily.

V-Type Engines. The V- or Y-configuration cylinder design has two rows of cylinders (Figure 2–7). These cylinder rows are approximately 90 degrees from each other. This is the angle in most V- or Y-configurations. However, other angles ranging from 60 to 90 degrees are used. This design utilizes one crankshaft that operates the cylinders on both sides. Two connecting rods are attached to each journal on the crankshaft. However, there must be two cylinder heads for this type of engine.

One advantage of using a V-configuration is that the engine is not as vertically high as with the in-line configuration. The front of a vehicle can now be made lower. This design improves the outside aerodynamics of the vehicle. If eight cylinders are needed for power, a V-configuration makes the engine much shorter and more compact. Manufacturers used to produce in-line 8-cylinder engines, which made the engine rather long. The vehicle was hard to design around this long engine. The long

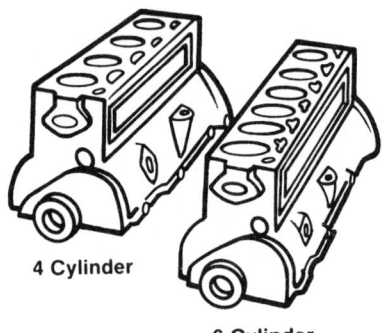

FIGURE 2–6 In-line arrangement of four- and six-cylinder engines

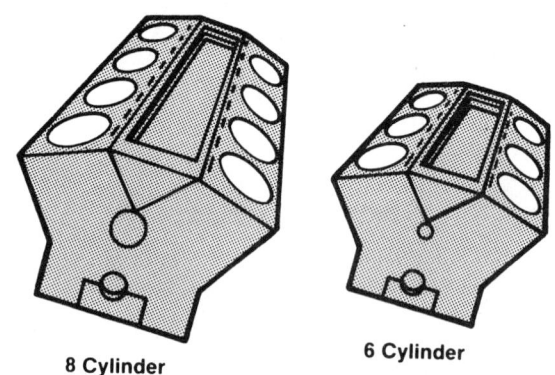

FIGURE 2–7 V-8 and V-6 engines

crankshaft also caused increased torsional vibrations in the engine.

Slant Cylinder Engines. Another way of arranging the cylinders is in a slant configuration (Figure 2–8A). This is much like an in-line engine except that the entire block has been placed at a slant. The slant engine was designed to reduce the distance from the top to the bottom of the engine. Vehicles using the slant engine can be designed more aerodynamically.

Opposed Cylinder Engines. In this design, two rows of cylinders are located opposite the crankshaft (Figure 2–8B). Opposed cylinder engines are used in applications where there is very little vertical room for the engine. For this reason, opposed cylinder designs are commonly used on vehicles that have the engine in the rear. The angle between the two cylinders is typically 180 degrees. One crankshaft is used with two cylinder heads. There are two connecting rods attached to each journal on the crankshaft.

Following two basic valve and camshaft placement configurations of the four-stroke gasoline engines that are used in automobiles (Figure 2–9):

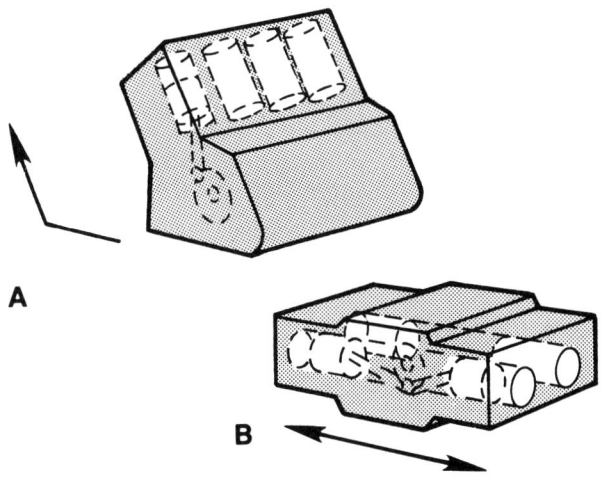

1. *Overhead Valve (OHV).* As the name implies, the intake and exhaust valves are mounted in the cylinder head and are operated by a camshaft located in the cylinder block. This arrangement requires the use of valve lifters, pushrods, and rocker arms to transfer camshaft rotation to valve movement. The intake and exhaust manifolds are attached to the cylinder head.

2. *Overhead Cam (OHC).* This engine also has the intake and exhaust valves located in the cylinder head, but as the name implies, the cam has now been relocated to the cylinder head as well. In an overhead cam engine, the valves are operated directly by the camshaft and cam follower or tappets.

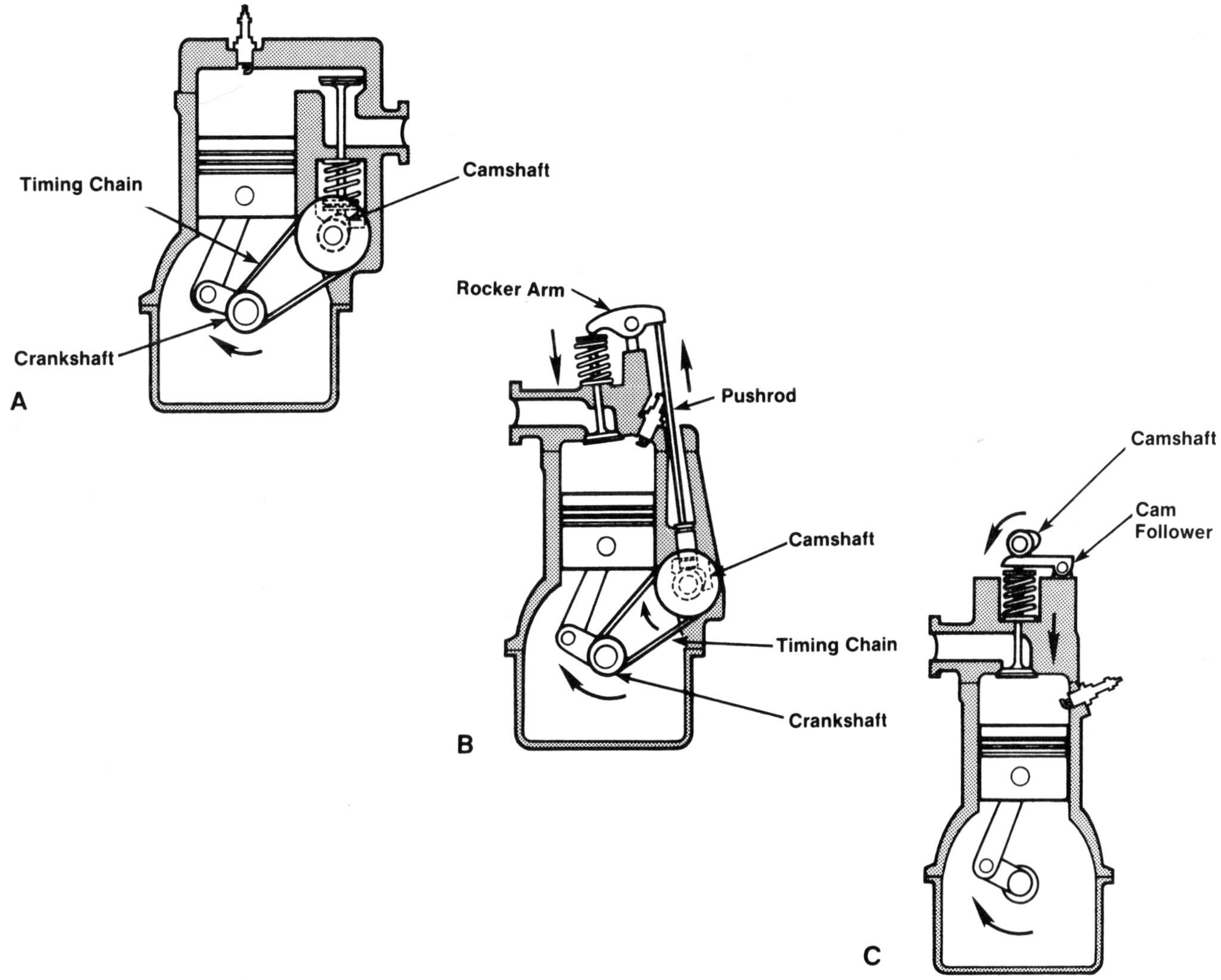

FIGURE 2-8   (A) Slant cylinder engine and (B) opposed cylinder engine

Timing Chain

Camshaft

Crankshaft

A

Rocker Arm

Pushrod

Camshaft

Timing Chain

Crankshaft

B

Camshaft

Cam Follower

C

FIGURE 2-9   Popular engine valve arrangement: (A) flathead or side valve; (B) overhead valve; and (C) overhead cam

As shown in Figure 2-10, there are several valve arrangements including the following designs:

- *L-head Engine Design.* This engine is also known as the *flathead* or *side valve engine.* No moving parts are contained in the cylinder head. The head serves only as the combustion chamber. The valves are located in the cylinder block alongside each cylinder, hence the name side valve. The intake and exhaust manifolds mount to the cylinder block. In automobiles this type of engine has been lost to progress.

- *I-head Engine Design.* I-head means the valves are directly above the piston (overhead valves). The valves are located in the cylinder head. The design allows easy breathing of the engine. Air and fuel can move easily into and out of the cylinder with little restriction. This process improves the volumetric efficiency of the engine. The I-head is also easy to maintain. For example, if a valve is damaged, it can be replaced easily. Adjusting the valves is easier, too.

- *T-head Engine Design.* The T-head design has the valves located within the block. The difference between this design and the L-head design is that two camshafts are needed. Because of this extra expense, T-head designs are not commonly used in the automotive industry today.

- *F-head Engine Design.* This design is a combination of the I-head and the L-head designs. There are valves located in the head as well as in the block. It has some of the advantages of the L-head and I-head designs. However, the increased cost of parts is a disadvantage.

In overhead valve engines (Figure 2-11), the valves are operated by valve lifters and pushrods that are actuated by the camshaft. On overhead cam engines, the cam lobes operate the valves directly and there is no need for pushrods or lifters. The cam lobes are oval shaped. The placement of the lobe determines when the valve will open. Design of the lobe determines how far the valve will open and how long it will remain open in relation to piston movement.

The camshaft is driven by the crankshaft through gears, or sprockets and a cogged belt or timing chain. The camshaft turns at half the crankshaft speed and rotates one complete turn during each complete four-stroke cycle.

Some design characteristics that the rebuilder technician should be aware of are as follows:

1. *Bore and Stroke.* The bore of a cylinder is simply its diameter. The stroke is the length of the piston travel between TDC and BDC. Between them, bore and stroke determine the displacement of the cylinders. The bore and stroke ratio (Figure

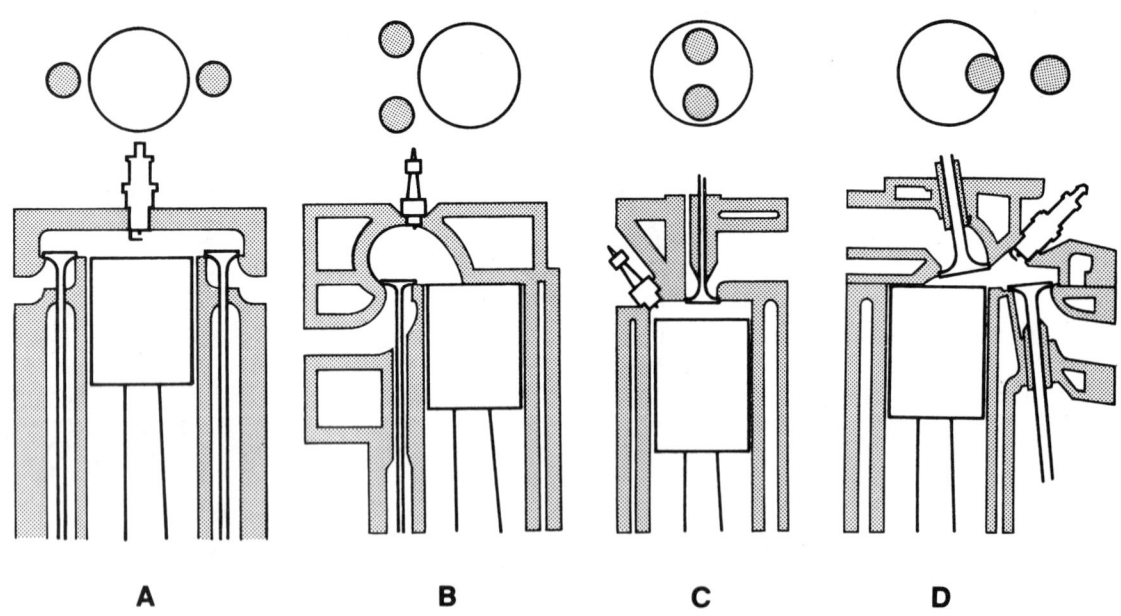

      A   **B**   **C**   **D**

FIGURE 2-10 Other engine valve arrangments; (A) T-head; (B) L-head; (C) I-head; and (D) F-head

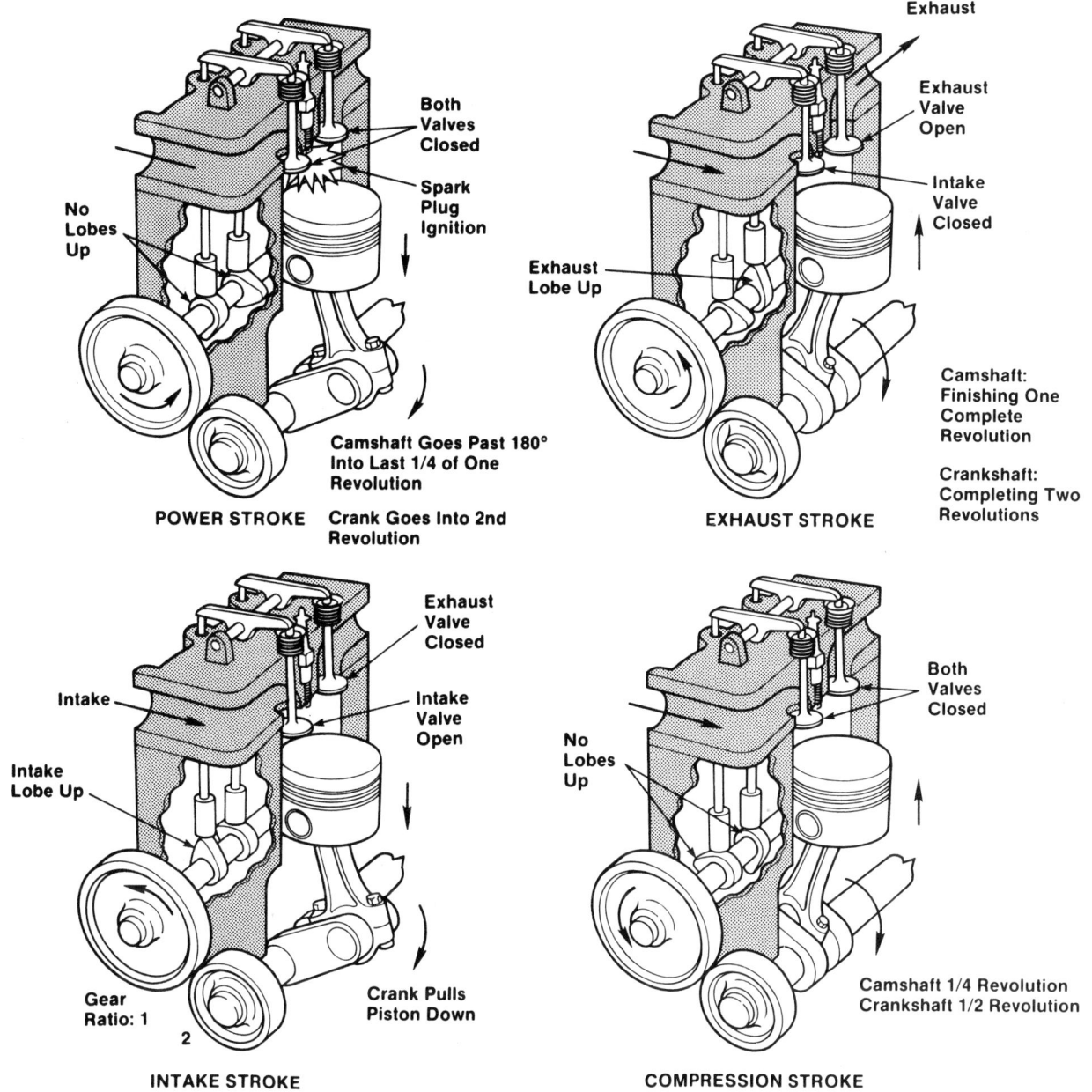

FIGURE 2-11 Operation of the valve in an overhead engine

2-12) is the relationship between the bore and stroke of the engine. When the bore of the engine is larger than its stroke, it is said to be oversquare. For example, an engine that has a bore of 4.00 inches and a stroke of 2.87 inches is considered to be oversquare. The bore and stroke ratio for this engine is 1.39. To find the bore and stroke ratio, take the measurement of the bore and stroke and divide the bore by the stroke. When the answer is 1.00, the bore and stroke are the same. When the answer is more than 1.00, the bore is larger than the stroke, or oversquare. When the answer is less than one, the stroke is larger than the bore, or undersquare. Generally, an oversquare engine will deliver high rpm, such as for automobile use, and an undersquare or long stroke engine will deliver good low end torque, such as an engine for a truck or tractor. The crank throw is the distance from the crankshaft's main bearing centerline to the crankshaft throw centerline.

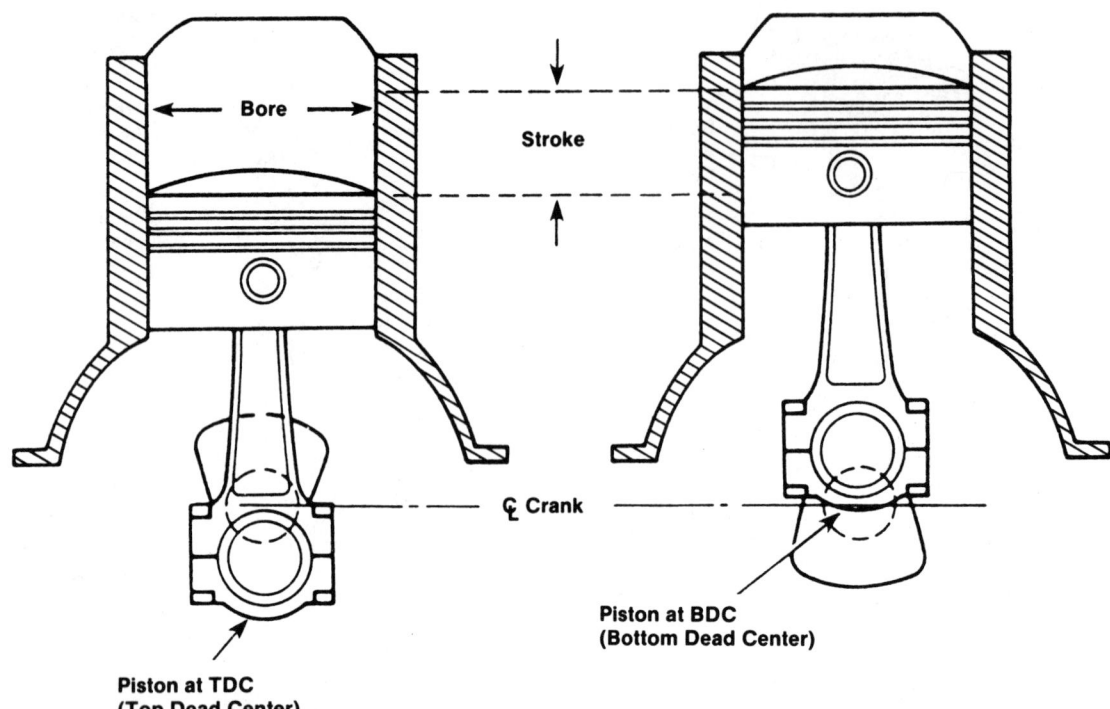

FIGURE 2-12 Bore/stroke ratio

 **SHOP TALK** ───────

The stroke of any engine is twice the crank throw (Figure 2-13).

2. *Displacement*. Displacement is the volume the cylinder holds between the TDC and

BDC positions of the piston. It is usually measured in cubic inches, cubic centimeters, or liters (Figure 2-14). The total displacement of an engine (including all cylinders) is a rough indicator of its power output. Displacement can be increased by opening the bore to a larger diameter or by increasing the length of the stroke. Total

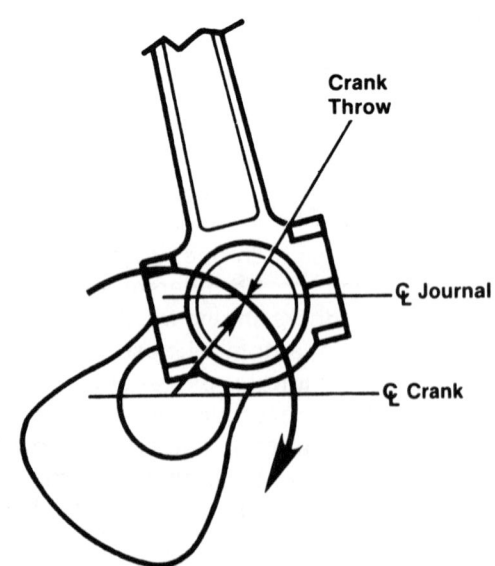

FIGURE 2-13 Crankshaft throw

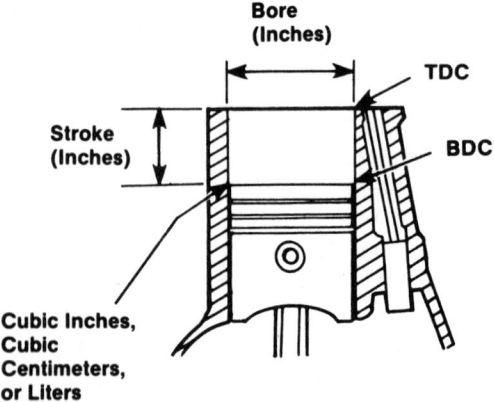

FIGURE 2-14 Cubic inch displacement is the volume of the cylinder from BDC to TDC. It is also stated in cubic centimeters and liters.

displacement is the sum of displacements for all cylinders in an engine. Displacement may be calculated as follows:

$$\pi \times D^2 \times L \times N = \text{displacement (CID)}$$

where $\pi = 3.1416$
D = bore diameter
L = length of stroke
N = number of cylinders

Example: Calculate the cubic inch displacement (CID) of a six-cylinder engine with a 3.7-inch bore and 3.4-inch stroke.

$$3.1416 \times 3.7^2 \times 3.4 \times 6 =$$
219.66 cubic inches of displacement

3. *Compression Ratio.* The compression ratio is a measure of how much the air and fuel have been compressed. The compression ratio is defined as the ratio of the volume in the cylinder above the piston when the piston is at BDC to the volume in the cylinder above the piston when the piston is at TDC. The compression ratio is shown in Figure 2–15. The formula for calculating the compression ratio is:

$$\frac{\text{Volume above the piston at BDC}}{\text{Volume above the piston at TDC}}$$

or

$$\frac{\text{Total cylinder volume}}{\text{Total combustion chamber volume}}$$

In many engines, at TDC the top of the piston is even or level with the top of the cylinder block. The combustion chamber volume is in the cavity in the cylinder head above the piston. This is modified slightly by the shape of the top of the piston. The combustion chamber volume must be added to each volume stated in the formula to give accurate results.

Example: Calculate the compression ratio if the total piston displacement is 45 cubic inches and the combustion chamber volume is 5.5 cubic inches.

$$\frac{45 + 5.5}{5.5} = \text{Compression ratio}$$

9.1 to 1 = Compression ratio

(Be sure to add the combustion chamber volume to the piston displacement to get the total cylinder volume.)

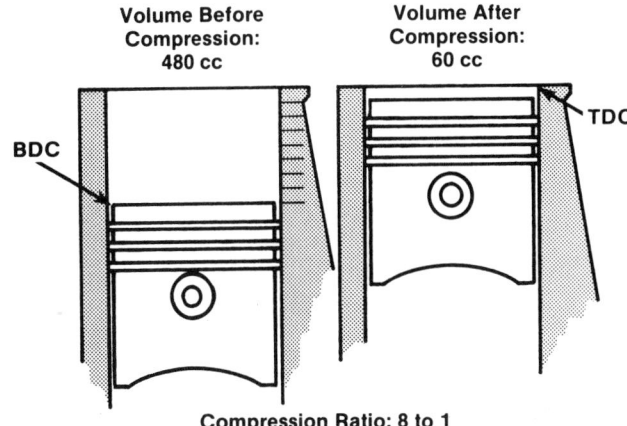

Volume Before Compression: 480 cc   **Volume After Compression: 60 cc**

BDC   TDC

Compression Ratio: 8 to 1

FIGURE 2-15 Compression ratio is the ratio between the volume above the piston at BDC to the volume above the piston at TDC.

Common compression ratios are anywhere from 8 to 1 on low-compression gasoline engines to 25 to 1 on diesel engines.

4. *Engine Efficiency.* Engine efficiency is a measure of the relationship between the amount of energy put into the engine and the amount of energy available out of the engine. For understanding basic engine principles, efficiency is defined as:

$$\text{Efficiency} = \frac{\text{Output energy}}{\text{Input energy}} \times 100$$

Other types of engine efficiencies of interest to the automotive engine technician include mechanical efficiency, volumetric efficiency, and thermal efficiency. They are expressed as a ratio of input (actual) to output (maximum or theoretical). Efficiencies are expressed as percentages. They are always less than 100 percent. The difference between the efficiency and 100 percent is the percentage lost during the process.

Example: If there were 100 units of energy put into the engine and 28 units came back, the efficiency would be equal to 28 percent.

 SHOP TALK _____

Gasoline engines are approximately 25 to 38 percent efficient. Diesel engines are about 10 percent more efficient than gasoline automobile engines or have an overall efficiency of 35 to 38 percent.

5. *Brake Mean Effective Pressure (BMEP).*
This is a theoretical term used to indicate how much pressure is applied to the top of the piston from TDC to BDC. It is measured in pounds per square inch. This term becomes very useful when analyzing the results of different fuels used in engines. For example, if diesel fuel is used in an engine, more BMEP will be produced. This will produce more output power than if gasoline fuel were used. Also, as different injection systems, combustion designs, and new ignition systems are added, the BMEP of the engine is affected.

Complete engine specifications can usually be found in the manufacturer's service manual as shown in Figure 2-16.

Gasoline engines are made of many different parts and components. Those that the rebuilder technician is concerned with include the following:

- Intake and exhaust manifolds
- Cylinder block
- Cylinder head
- Intake and exhaust valves
- Valve seats
- Valve guides
- Camshaft
- Lifters
- Timing set
- Crankshaft
- Crankshaft bearings
- Vibration damper
- Flywheel
- Piston and piston rings

1.9L Escort Engine Specification

GENERAL SPECIFICATIONS

DISPLACEMENT	1.9L
NUMBER OF CYLINDERS	1-4
BORE AND STROKE	
1.9L	82 x 88 (3.23 x 3.46)
FIRING ORDER	1-3-4-2
OIL PRESSURE (HOT 2000 RPM)	240-450 kPa (35-65 PSI)
DRIVE BELT TENSION	178-311 (40-70 Lb-Ft)

CYLINDER HEAD AND VALVE TRAIN ① ②

COMBUSTION CHAMBER VOLUME (cc)	EFI-HO 55 ± 1.6
VALVE GUIDE BORE DIAMETER	EFI 39.9 ± 0.8
Intake	13.481-13.519mm (0.531-0.5324 in.)
Exhaust	13.481-13.519mm (0.531-0.532 in.)
VALVE GUIDE I.D.	
Intake and Exhaust	8.063-8.094 N·m (.3174-.3187 in.)
VALVE SEATS	
Width — Intake & Exhaust	1.75-2.32mm (0.069-0.091 in.)
Angle	45°
Runout (T.I.R.)	0.076mm (0.003 in.) MAX.
Bore Diameter (Insert Counterbore Diameter)	
Intake	(EFI-HO) 43.763mm (1.723 in.) MIN.
	43.788mm (1.724 in.) MAX.
	(EFI) 39.940 mm (1.572 in.) MIN.
	39.965 mm (1.573 in.) MAX.
Exhaust	(EFI-HO) 38.263mm (1.506 in.) MIN.
	38.288mm (1.507 in.) MAX.
	(EFI) 34.940 mm (1.375 in.) MIN.
	39.965 mm (1.573 in.) MAX.
GASKETS SURFACE FLATNESS	0.04mm (0.0016 in.)/26mm (1 in.)
	0.08mm (0.003 in.)/156mm (6 in.)
	0.15mm (0.006 in.) Total
HEAD FACE SURFACE FINISH	0.7/2.5 0.8 (28/100 .030)
VALVE STEM TO GUIDE CLEARANCE	
Intake	0.020-0.069mm (0.0008-0.0027 in.)
Exhaust	0.046-0.095mm (0.0018-0.0037 in.)
VALVE HEAD DIAMETER	
Intake	42.1-41.9mm (1.66-1.65 in.)
Exhaust	37.1-36.9mm (1.50-1.42 in.)
VALVE FACE RUNOUT	
LIMIT	Intake & Exhaust .05mm (0.002 in.)
VALVE FACE ANGLE	45.6°
VALVE STEM DIAMETER (Std.)	
Intake	8.043-8.025mm (0.3167-0.3159 in.)
Exhaust	8.017-7.996mm (0.3156-0.3149 in.)
Oversize	
Intake	8.423-8.405mm (0.3316-0.3309 in.)
Exhaust	8.397-8.378mm (0.3306-0.3298 in.)
Oversize	
Intake	8.843-8.825mm (0.3481-0.3474 in.)
Exhaust	8.777-8.759mm (0.3479-0.347 in.)
VALVE SPRINGS (EFI)	
Compression Pressure (N [Lb] @ Spec. Length)	
Loaded	892.7N (200 Lb.) @ 27.71mm (1.09 in.)
Unloaded	422N (95 lb.) @ 37.1mm (1.461 in.)
Free Length (Approximate)	47.2mm (1.86 in.)
VALVE SPRINGS (EFI-HO)	
Compression Pressure (N [Lb] @ Spec. Length)	
Loaded	960N (216 lb.) @ 25.8mm (1.016 in.)
Unloaded	417N (94 lb.) @ 37.1mm (1.461 in.)

CYLINDER HEAD AND VALVE TRAIN ① ② (Continued)

VALVE SPRINGS (EFI-HO) (Continued)	
Free Length (Approximate)	48.3mm (1.90 in.)
Assembled Height	37.5-36.9mm (1.48-1.44 in.)
Service Limit	5% Pressure Loss @ Specified Height
Out Of Square Limit	1.53mm (0.060 in.)
ROCKER ARM	
Ratio	EFI-HO 1.68, EFI 1.65
VALVE TAPPET, HYDRAULIC	
Diameter (Std.)	22.212-22.200mm (0.8745-0.8740 in.)
Clearance to Bore	0.023-0.065mm (0.0009-0.0026 in.)
Roundness	0.013mm (0.0005 in.)
Runout	0.1
Finish	2 micromet. (8 micro in.)
Service Limit	0.127mm (0.0005 in.)
Collapsed Tappet Gap (Nominal)	
EFI	1.2-3.5mm
EFI-HO	1.5-3.8mm
DISTRIBUTOR SHAFT BEARING BORE DIAMETER	47.05-47.10mm
	(1.852-1.854 in.)
TAPPET BORE DIAMETER	22.25mm ± .015 (.876 ± 0.0006 in.)
CAMSHAFT BORE INSIDE DIAMETER	
No. 1, 2, 3, 4, 5	45.796-45.821mm (1.8030-1.8040 in.)
CAMSHAFT BORE INSIDE DIAMETER — OVERSIZE	
No. 1	45.201-45.176mm (1.7796-1.7786 in.)
No. 2	45.451-45.426mm)1.7894-1.7884 in.)
No. 3	45.701-45.676mm (1.7993-1.7983 in.)
No. 4	45.951-45.926mm (1.8091-1.8081 in.)
No. 5	46.201-46.176mm (1.8189-1.8179 in.)

CAMSHAFT (EFI)

LOBE LIFT	Intake and Exhaust 6.096mm (.240 in.)
Allowable Lobe Lift Loss	0.127mm (.005 in.)
THEORETICAL VALVE MAXIMUM LIFT	
Intake	10.06mm (0.468 in.)
Exhaust	10.06mm (0.468 in.)
CAMSHAFT (EFI-HO)	
Lobe Lift	Intake and Exhaust 6.731mm (0.240 in.)
THEORETICAL VALVE MAXIMUM LIFT	
Intake	11.31mm (0.396 in.)
Exhaust	11.31mm (0.396 in.)
END PLAY	0.152-0.046mm (0.006-0.0018 in.)
Service Limit	0.20mm (0.0078 in.)
JOURNAL TO BEARING CLEARANCE	0.0335-0.0835mm
	(0.0013-0.0033 in.)
JOURNAL DIAMETER	
STANDARD	45.7625-45.7375mm (1.8017-1.8007 in.)
OVERSIZE (NOT APPLICABLE TO EFI)	46.1425-46.1175mm
	(1.8166-1.8156 in.)
Runout Limit	0.127mm (0.005 in.) (Runout of center bearing relative to bearing Nos. 1 & 5)
Out-of-Round Limit	0.008mm (0.003 in.)

CAMSHAFT

Assembled Gear Face Runout	
Crankshaft	0.65mm (0.026 in.)
Camshaft	0.275mm (0.011 in.)

FIGURE 2-16 Typical engine specifications given in the service manual.

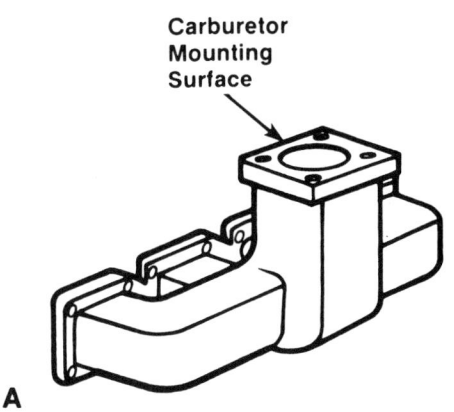

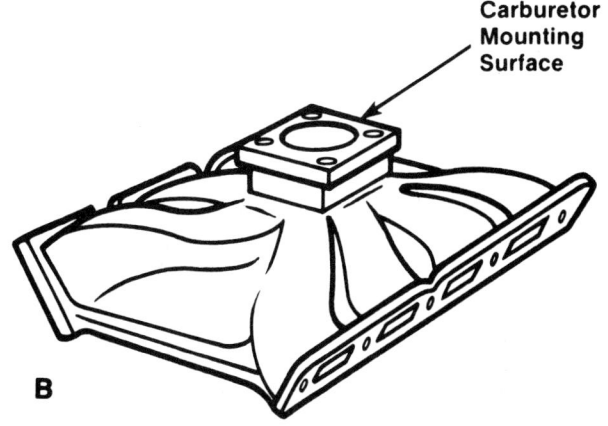

FIGURE 2-17   (A) Typical four-cylinder and six-cylinder in-line intake manifold; (B) typical V-6 and V-8 intake manifold

- Connecting rod
- Gaskets

MANIFOLDS

Two types of manifolds are used with a conventional four-stroke engine.

- Intake
- Exhaust

Intake Manifold

The intake manifold distributes the air or air/fuel mixture as evenly as possible to each cylinder, helps to prevent condensation, and assists in vaporization of the air/fuel mixture (Figure 2-17). Smooth and efficient engine performance depends on mixtures entering each cylinder that are uniform in strength, quality, and degree of vaporization. This is partly the job of the intake manifold. The ideal air/fuel mixture is completely vaporized when it goes into the combustion chamber. Complete vaporization requires high temperature, but high temperature increases volume and decreases the volumetric efficiency of the engine. Therefore, the best alternative is to introduce an air/fuel mixture into the manifold that is vaporized above the point where fuel particles will be deposited on the manifold and below the point where excess heat results in power losses. Intake manifolds are carefully designed to meet these requirements. In some designs the manifold may also have an air preheater, exhaust heater, or coolant heater; an electric grid under the carburetor; or an electric heater in the manifold under the carburetor (Figure 2-18). These heating devices improve the vaporization of fuel during engine warm-up and reduce carburetor icing.

 SHOP TALK _____

On diesel engines and port injected gasoline engines, the manifold delivers air only. On carbureted systems and throttle body injection systems, the manifold delivers the air/fuel mixture. There are many new intake manifold designs for ported gasoline fuel injection systems. They are often referred to as turned intake manifolds because they have been redesigned to give the best possible equal amount of airflow to each cylinder (Figure 2-19).

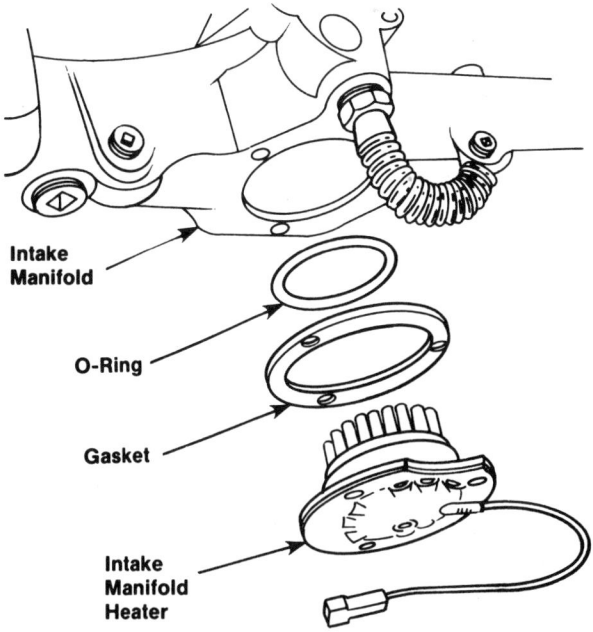

FIGURE 2-18   Intake manifold for an in-line six-cylinder engine with electric heater

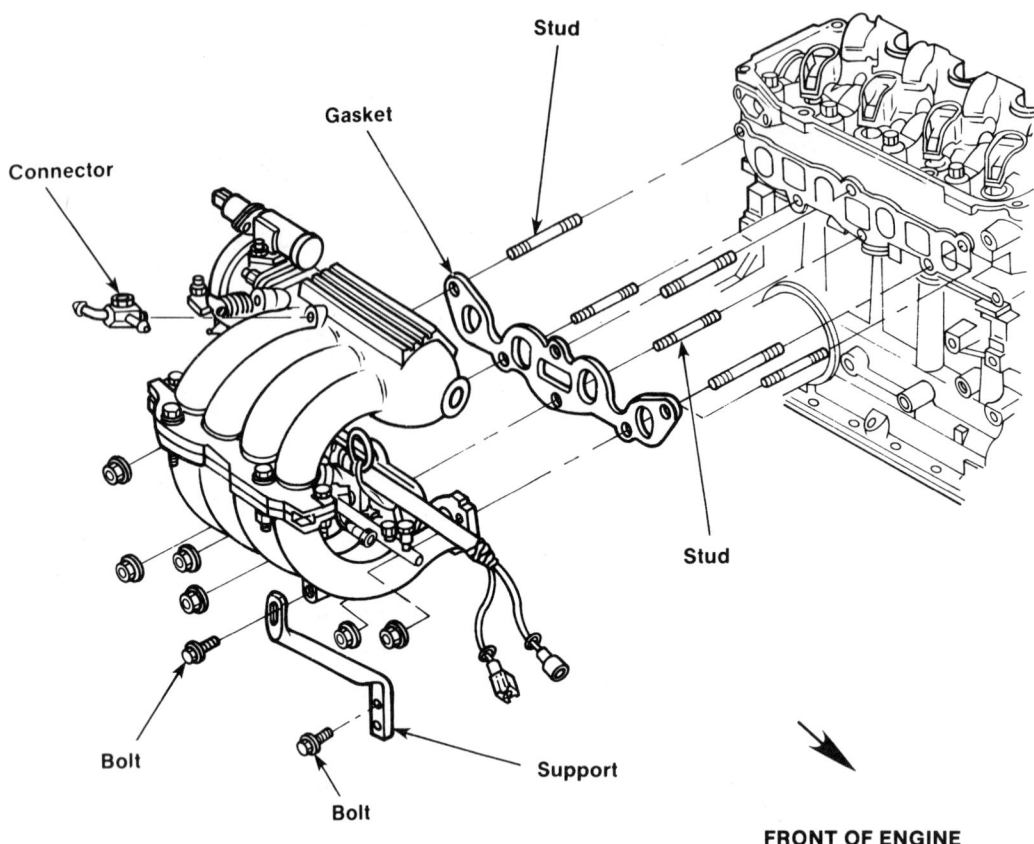

FIGURE 2-19 This type of intake manifold, used with a ported fuel injection, gives a more equal amount of air for each cylinder.

Intake manifolds are of cast-iron or die-cast aluminum construction. Aluminum manifolds reduce engine weight. Some are cast integrally with the cylinder head. The interior of the manifold must be smooth and offer no obstruction to the flow of the air/fuel mixture. Design must also prevent collecting of fuel at the bends in the manifold.

Manifolds can also be either wet or dry. Wet manifolds have coolant passages cast directly into the manifold. Dry manifolds do not have cooling passages. Some naturally aspirated engine intake manifolds have an integrally cast water crossover passage. This allows coolant to flow through a passage below the carburetor and warm the incoming air/fuel mixture.

Manifold design varies greatly depending on engine type. For example, on four-cylinder engines, the intake manifold has either four runners or two runners that break into four near the intake manifold (Figure 2-20). On in-line six-cylinder engines, there are either six runners or three that branch off into six near the intake manifold. On V-configuration en-

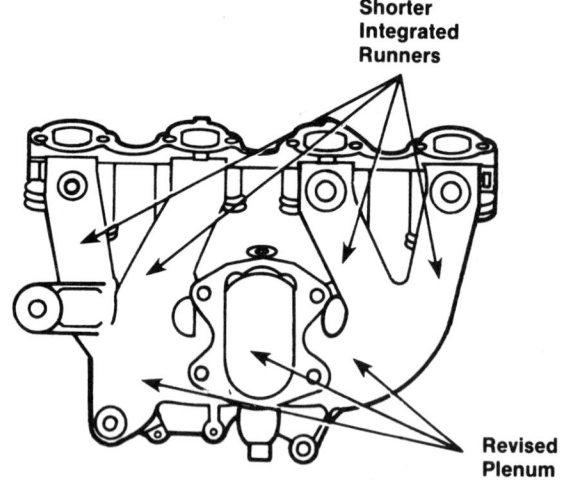

FIGURE 2-20 The intake manifold on an in-line four-cylinder engine

gines (V-6 and V-8), both open and closed intake manifolds are made. Open intake manifolds have an open space between the bottom of the manifold and the valve lifter valley. Closed intake manifolds act as the cover to the intake lifter valley.

On some manifolds, there is an exhaust crossover passage (Figure 2–21). This passage allows the exhaust from one side of the engine to cross over through the intake manifold to the other side to be exhausted. The exhaust crossover provides heat to the base of the carburetor to improve the vaporization of the fuel while the engine is warming up. On many engines, the exhaust crossover is also used to provide automatic choke heat. On four- and six-cylinder in-line engines the heat riser valve also directs heat to the base of the carburetor for the same reasons.

Intake manifolds for V-type engines may include coolant connecting passages between cylinder heads. Some include a provision for mounting the thermostat and thermostat housing. In addition, connections to the intake manifold in some vehicles provide the necessary vacuum source exhaust gas recirculation valves (EGR), automatic transmission vacuum modulators, distributor vacuum advance units, thermostatic air cleaners, power brakes, and heater and air-conditioning airflow control doors. Other devices include manifold absolute pressure

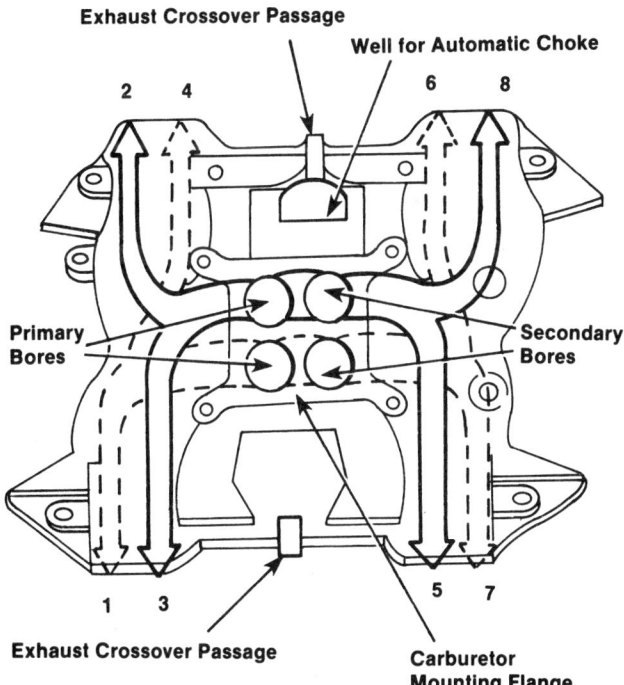

FIGURE 2-21 V-8 engine intake manifold. Note how cylinders are fed from carburetor boxes. Exhaust crossover passage provides heat for improved fuel vaporization and prevents carburetor icing.

sensors, manifold air temperature sensors, knock sensors, and exhaust gas recirculation passages.

Exhaust Manifold

The exhaust manifold collects engine exhaust gases from cylinder ports and carries them to an exhaust pipe. These manifolds are usually of one-piece, cast-iron, or steel construction. Extreme temperature variations are encountered by exhaust manifolds. A hot exhaust manifold is often splashed with cold water when driving through puddles on the road. Manifolds expand and contract considerably owing to temperature change, and cracking can result. Some manifolds use reinforcing ribs to reduce distortion and the possibility of cracking.

Like the intake manifold, the exhaust design depends on the engine type. Four-cylinder in-line engines have either three- or four-runner exhaust manifolds. On the four-runner manifold the center two runners collect exhaust gases from the two center cylinders (Figure 2–22A). Six-cylinder in-line engines have either a six- or a four-runner manifold. On the four-runner exhaust manifold the two middle runners each collect exhaust gases from two cylinders (Figure 2–22B). The V-6 engine has two exhaust manifolds each with three runners. The V-8 engine has two exhaust manifolds. Each manifold has either three or four runners. The three-runner manifold has the center runner collecting exhaust gases from the center two cylinders on each side.

CYLINDER BLOCKS AND HEADS

The automotive industry's requirements for lighter weight engines have caused radical changes in engine rebuilding procedures. The thin-walled, bimetal engine type (block of cast iron and head of aluminum) is becoming very popular with automakers, but it has caused problems for auto machinists and rebuilders. This section covers the various problems created by these new engines. The solutions to the problems are covered in later chapters.

The modern four-stroke engine has two major areas (Figure 2–23). They are:

- *Cylinder Block.* Located within the lower engine are enclosed areas where compression, ignition, and combustion take place. A piston slides up and down in each cylinder to compress the air/fuel mixture.
- *Cylinder Head.* The cylinder head is the metal section bolted on top of the cylinder block. In this position, it covers the tops of cylinders

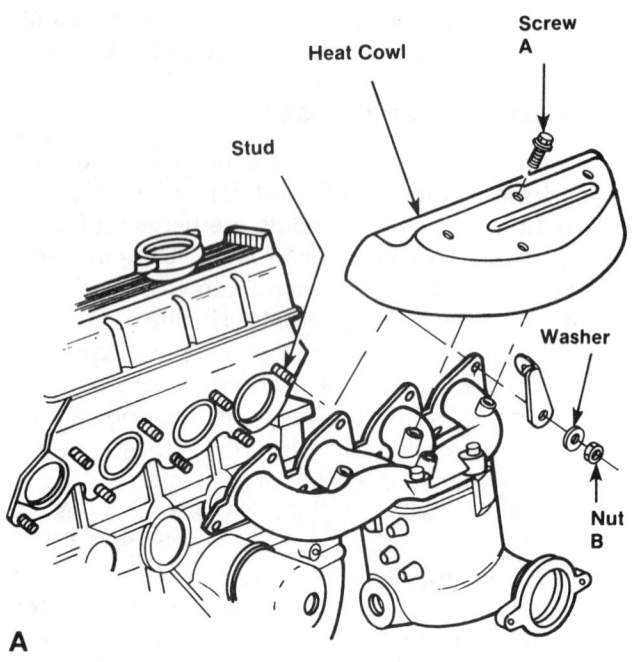

A

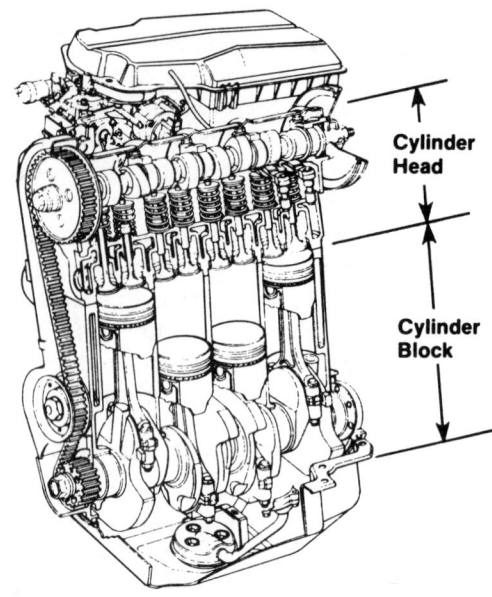

FIGURE 2-23 Typical upper and lower engine areas

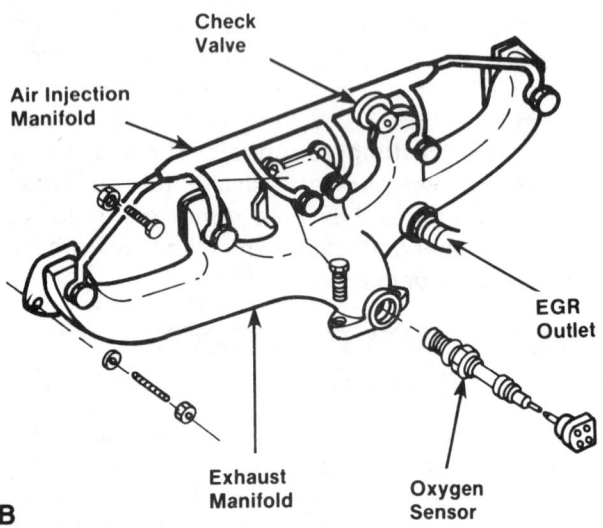

B

FIGURE 2-22 (A) Four-cylinder exhaust manifold with a closed coupled catalytic converter and heat cowl; (B) six-cylinder exhaust manifold with air injection as well as oxygen sensor

FIGURE 2-24 The V-6 cylinder block. The area between the heads is known as the valley.

and creates an enclosed upper engine area where the air/fuel mixture is burned. Thus, the cylinder head may form part of the combustion chamber. In addition to covering the tops of the cylinders and forming part of the combustion chambers, the cylinder head often contains the valve train.

Cylinder Blocks

The cylinder block (Figure 2-24) is normally one piece, cast, and machined so that all the parts con-

tained in it fit properly. They may be cast from several different materials: some are cast iron, some cast aluminum. Today, even plastic is being tested as a material for block construction.

The word *cast*, with regard to the engine block, refers to how it is made. To cast is to form molten metal into a particular shape by pouring or pressing it into a mold. This molded piece must then undergo a number of machining operations to make sure all the working surfaces are smooth and true. The top of the block must be perfectly smooth because the cylinder head will later be attached at this point. The base or bottom of the block must also be machined because the oil pan attaches here. All block sealing areas are also machined. The cylinder bores must be smooth and the correct diameter to accept the pistons.

The main bearing area of the block must be align bored to a diameter that will accept the crankshaft. Camshaft bearing surfaces must also be bored. The word *bore* means to drill or machine a hole; align boring is a series of holes in a straight line.

As already mentioned, blocks can be made from either cast iron or aluminum. In the past, most blocks were made of cast iron. This improved strength and controlled warpage from heat. With the increased concern for improved gasoline mileage however, car manufacturers are trying to make the vehicle lighter. One way is to reduce the weight of the block. Aluminum is used for this purpose because it is a very light metal. Certain materials are added to the metal before it is poured into the mold. These metals are used to make the aluminum stronger and less likely to warp when heat is applied from combustion. Aluminum blocks must also have a sleeve or steel liner placed in them. Steel liners are placed in the mold before the metal is poured. After the metal is poured, the steel liner cannot be removed.

Silicon is also added to the aluminum. Through a special process, called *silicon-impregnated cylinder walls,* the silicon is concentrated on the cylinder walls, eliminating the need for a steel liner. One problem with this design is that it requires the use of very high-quality engine oils. Because of owner neglect, this engine does not usually survive its intended service life.

Cylinder Heads

The cylinder head (Figure 2-25) is made of cast iron or aluminum. It contains, on overhead valve engines, the valves, valve seats, valve guides, valve springs, rocker arm supports, and a recessed area called the *combustion chamber.* (In overhead camshaft engines can be found the supports for the camshaft and the camshaft bearings.) Both types of cylinder heads contain passages that match passages in the cylinder block to allow coolant to circulate in the head. The cylinder head also contains tapped holes in the combustion chamber to accept the spark plugs.

The surface of the head that contacts the block must be perfectly smooth because this area must contain the force of the burning fuel mixture. To aid in the sealing, a gasket is placed between the head and block. This gasket is called the *head gasket.* It is

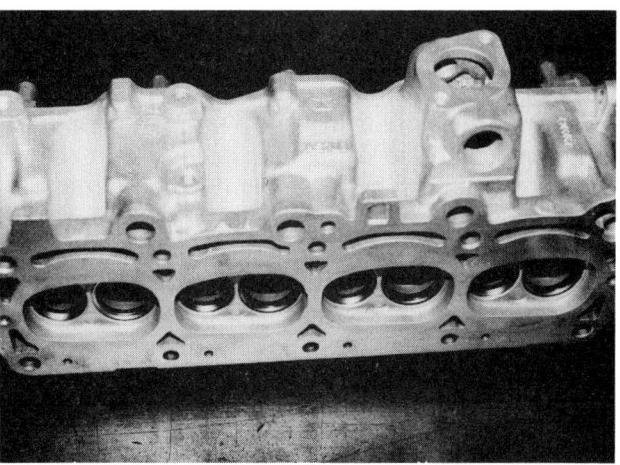

FIGURE 2-25   The cylinder head is the main structure of the engine. Most of the other engine parts are attached to the cylinder block

made of special material that can withstand high temperatures, high pressures, and the expansion of the metals around it. The head also serves as the mounting point for the intake and exhaust manifolds and contains the intake and exhaust ports.

The primary difference between lightweight and heavy heads is changes in the internal coring of the water jackets. The design contour of the heavy head (Figure 2-26) shows a straight-edged structure at the bottom. The design of the lower structure of the lightweight head reveals a scalloped contour.

Additionally, the factory has narrowed some of the interior casting structure around the exhaust ports and left large water hole voids in the head for ease of casting and to recover the sand from the molds. It also saves the factory cast iron.

Cross-sectional samples of early (heavy) and late (lightweight) heads show that there is considerable difference in metal thickness in certain areas.

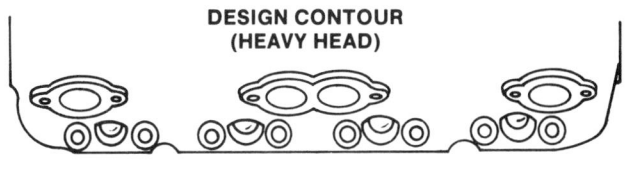

FIGURE 2-26   Design contour of heavy head (top) and lightweight head (bottom)

On the late heads there is much less stock in the exhaust seat area to work with. Also, valve guide areas are thin on the backside. Sometimes when installing a false guide in a late head, the press-fit is sufficient to create a fracture in the iron casting behind it. This can cause coolant to seep from behind the valve guide and enter the combustion chamber or enter the area under the valve cover and contaminate the oil. Due to this problem, some shops trim the guide to a smaller outside diameter. This eliminates core drilling (boring) as far into the head to use the false guide.

One major problem with thin-walled, lightweight heads is cracking (Figure 2–27). These heads have also created head gasket sealing problems in the spark plug cooling port area (Figure 2–28). Head resurfacing when there is an existing core shift causes the narrow land area to move inward toward the combustion chamber and reduce gasket contact. Coolant can then leak into the cylinder. Before installing a lightweight head, carefully check this area for proper head gasket fit and possible hairline cracks.

Aluminum Heads. Aluminum heads have become popular with car manufacturers in recent years primarily because of the weight saving they offer. A typical aluminum head weighs roughly half as much as an iron head. Eliminating anywhere from 20 to 40 pounds of weight is a plus for fuel economy, but it has its drawbacks. In the construction of bimetal engines, the most troublesome problem is aluminum's thermal expansion characteristics. Aluminum expands and contracts almost twice as much as cast iron in response to temperature changes (Figure 2–29), and this creates a number of

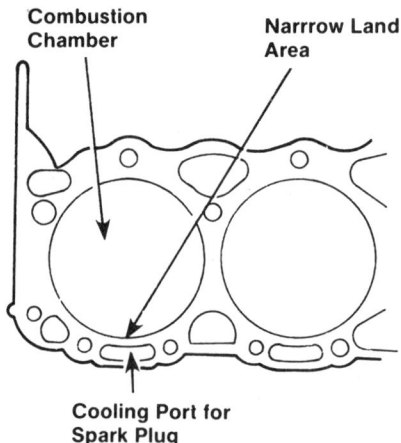

FIGURE 2-28 Before installing a rebuilt lightweight cylinder head, it is very important to check the head gasket-to-casting fit around the cooling port areas.

problems. When an aluminum head is mated to an iron block (which most are), the difference in thermal expansion between head and block creates a great deal of scrubbing stress on the head gasket. Unless the gasket is engineered to take such punishment, the result can be leakage and premature gasket failure.

Increased thermal expansion and stress also lead to cracking. The most crack-prone areas in the head are usually the areas around the valve seats. The interference fit of the seat in the head combined with different rates of expansion between seat and head, high combustion temperatures, and the constant pounding of the valve against the seat often cause cracking between the intake and exhaust seats or just under the exhaust seat. If the seats are not machined to very close tolerances and installed properly, cracking is virtually guaranteed.

The difference in thermal expansion between aluminum and iron creates a lot of stress throughout the head. The head wants to expand in all directions at once as it heats up, but the head bolts keep it from going sideways or lengthwise. The only place left to go is up so the head tends to bow up in the middle.

Lack of rigidity in the head itself is another factor to consider and one that can contribute to other problems in an engine. Aluminum is not as strong as steel, so consequently, the head provides less top end support for the block. This can allow more distortion in the upper cylinder bore area, affecting combustion sealing, blowby, and ring life. Using deck plates when boring the block can help minimize some of the distortion that will occur after the head is torqued down.

In engines with overhead cams and aluminum heads, the cam journals often run in machined bores

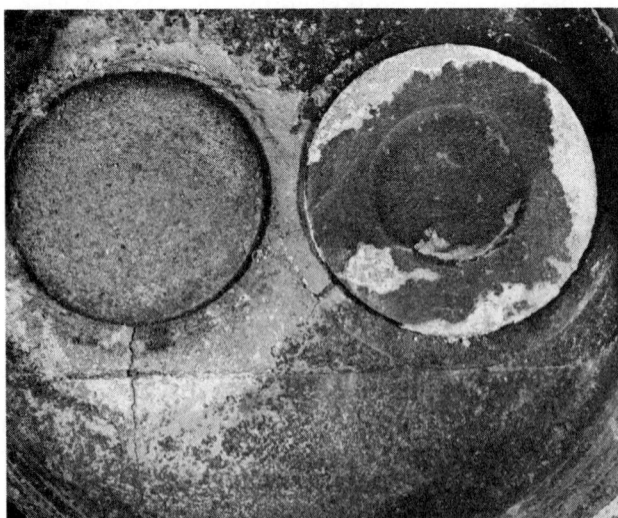

FIGURE 2-27 Typical crack in a lightweight head. The exhaust valve seat has started to recess.

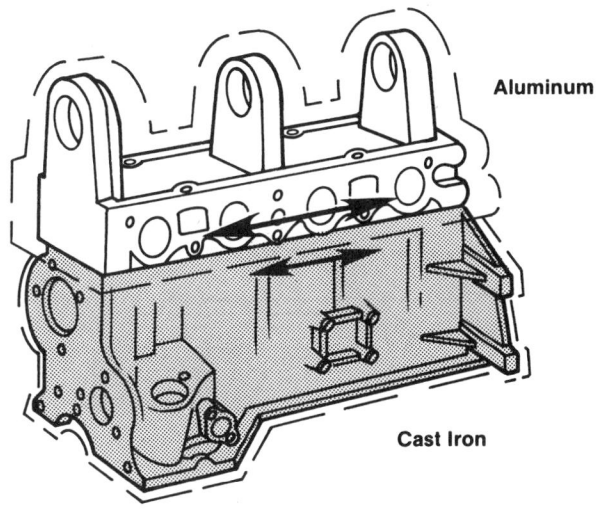

FIGURE 2-29 Bimetal engine expansion

in the head itself. Aluminum makes a fairly good bearing material. It is soft and provides good imbedability to foreign particles. But it lacks the support and rigidity of a conventional steel-backed bearing in an iron saddle. Because of this, overhead cam bores in such heads typically experience more flex than their cast-iron counterparts. The result is usually accelerated wear and egg-shaped bores. If the head overheats and warps, alignment through the cam bores is destroyed. In some instances, the misalignment can be so bad the cam will not turn once the head is unbolted from the engine. Since many aluminum heads lack serviceable cam bearings and, in many instances, enough metal for boring and sleeving (repair sleeves are generally unavailable anyway), the only cure short of replacing the head is to machine out the cam bores and install a cam with oversized journals.

Aluminum has another drawback—porosity. The casting process sometimes leaves microscopic pores in the metal, which can weep oil or coolant. In most instances, the problem is cosmetic only in that it does no real harm. But the customer may not agree. To him or her a wet spot on the outside of a cylinder head looks like a leak.

If these shortcomings are not enough, aluminum is also highly vulnerable to electrolysis corrosion within the cooling system. Whenever two different metals such as aluminum and iron are in contact with one another it creates a battery-like condition in which the lesser of the two metals is eaten away. The lesser metal in this case is aluminum, and the only way to prevent such corrosion is to use an antifreeze with the right kind of corrosion inhibitors and to change it regularly. Unfortunately,

many motorists do not follow this bit of advice, so by the time a rebuilder sees the head it can be severely corroded.

With so many shortcomings it makes one wonder why the auto manufacturers use aluminum at all. The weight savings apparently exceed all other trade-offs, so it looks as if aluminum is here to stay. This is good for the aftermarket because it means more aluminum head work for the rebuilder who can do it correctly.

Combustion Chamber

The performance of a vehicle's engine, its fuel efficiency, and the levels of pollutants in the exhaust gases all depend to a large extent on the shape of its combustion chambers—the space between the piston when it is at the top of its stroke and the top of the cylinder where combustion takes place.

An efficient combustion chamber must be compact to minimize the surface area of the walls through which heat is lost to the engine's cooling system. This makes the ideal shape a hemisphere with the point of ignition (the nose of the spark plug) at its center. This also has the advantage of minimizing the flame path, or the distance from the spark to the furthermost point in the chamber. The shorter the flame path, the more progressive and even the combustion of the air/fuel mixture, and there is less likelihood of knocking.

Manufacturers have designed several different shapes of combustion chambers in attempts to approach this ideal. Before looking at the popular combustion chamber designs, there are two terms that should be defined:

1. *Turbulence.* This is a very rapid movement of gases. When gases move, they make contact with the combustion chamber walls and pistons. Turbulence causes better combustion because the air and fuel are mixed better.
2. *Quenching.* This is the cooling of gases by pressing them into a thin area. The area in which gases are thinned is called the *quench area.*

Wedge Chamber. In the wedge-type combustion chamber, the spark plug is located at the wide part of the wedge (Figure 2-30). As the piston comes up on the compression stroke, the air/fuel mixture is squashed in the quench area. The quench area causes the air and fuel to be mixed thoroughly before combustion. This helps to improve the combustion efficiency of the engine. Spark plugs are positioned to get the greatest advan-

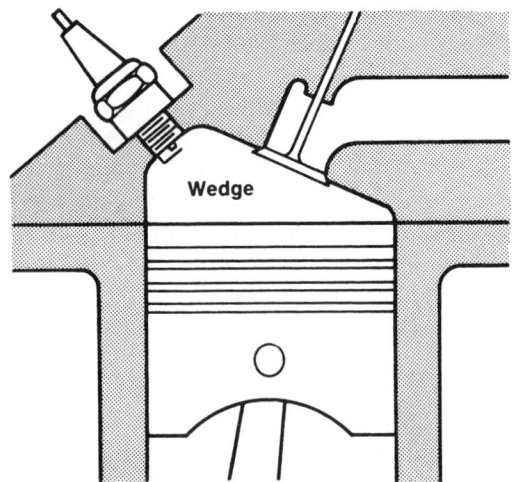

FIGURE 2-30 The wedge combustion chamber is shaped like a wedge to improve the turbulence within the chamber.

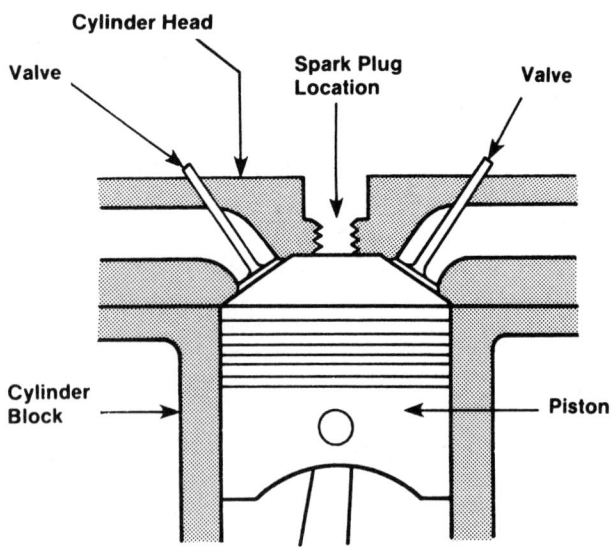

FIGURE 2-31 The hemispherical combustion chamber is shaped like a half circle. The valves are placed on both sides of the spark plug.

tage for combustion. When the spark occurs, smooth and rapid burning moves from the spark plug outward. The wedge-shaped combustion chamber is also called a *turbulence-type combustion chamber.* On newer model cars, the quench area has been reduced, which helps reduce exhaust emissions.

Hemispherical Chamber. The hemispherical combustion chamber gets its name from the chamber shape. *Hemi* is defined as half, and *spherical* means circle. The combustion chamber is shaped like a half circle. This type of chamber is also called the *hemi-head.* The piston top forms the base of the hemisphere, and the valves are inclined at an angle of 60 to 90 degrees to each other with the spark plug positioned centrally between them (Figure 2-31).

This design has several advantages. The flame path from the spark plug to the piston head is short, which gives efficient burning. The crossflow arrangement of the inlet and exhaust valves allows for a relatively free flow of gases into and out of the chamber, yet the shape of the chamber also creates some turbulence to ensure a thorough mixing of the fuel and air. The result is that the engine can "breathe deeply," meaning that it can draw in a large volume of gas for the space available and so give a high power output.

The hemispherical combustion chamber is considered a nonturbulence-type combustion chamber. Little or no turbulence is produced in this chamber. The air/fuel mixture is compressed evenly on the compression stroke. When flat-top pistons are used, little turbulence can be created. The spark plug is located directly in the center of the valves. Combus-

tion radiates evenly from the spark plug, completely burning the air/fuel mixture.

One of the more important advantages of the hemispherical combustion chamber is that air and fuel can enter the chamber very easily. The wedge combustion chamber restricts the flow of air and fuel to a certain extent. This is called *shrouding.* Figure 2-32 shows the valve very close to the side of the combustion chamber, which causes the air and fuel to be restricted. Volumetric efficiency is reduced. Hemispherical combustion chambers do not have this restriction. Hemispherical combustion chambers are used on many high-performance applications. This is especially true when large quantities of air and fuel are needed in the cylinder.

Some high-performance engines use a domed piston. This type of piston has a quench area to

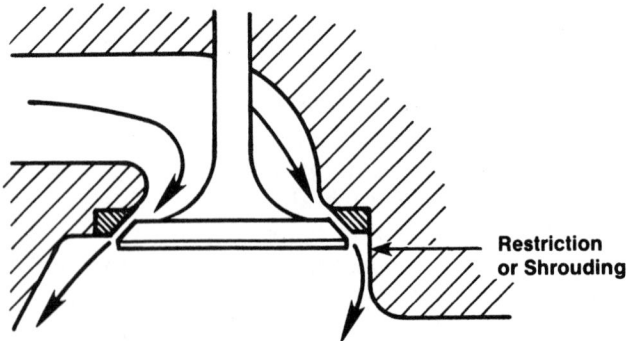

FIGURE 2-32 Shrouding is defined as a restriction in the flow of intake gases caused by the shape of the combustion chamber.

improve turbulence (Figure 2-33). Several variations of this design are used by different engine manufacturers.

Swirl Chamber. Swirl combustion uses compound curve port design to cause the air/

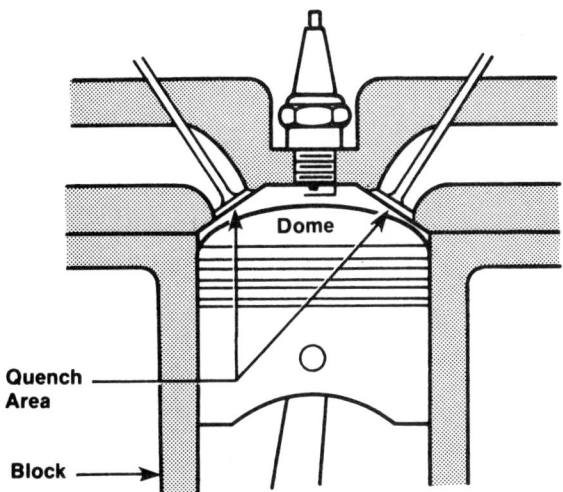

FIGURE 2-33 Domed-shaped pistons improve the eficiency of the hemispherical combustion chamber by producing a quench area.

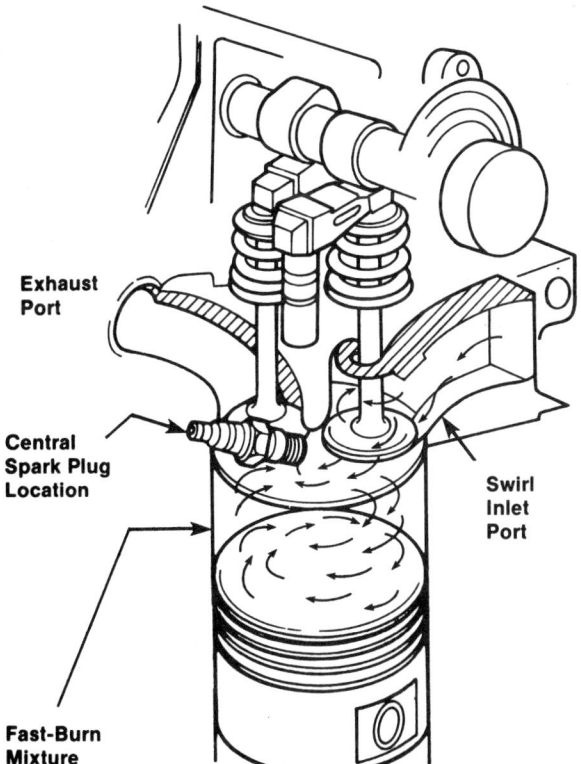

FIGURE 2-34 Swirl combustion chamber uses a compound curve port design to cause the air/fuel mixture to swirl in a corkscrew pattern to improve combustion.

fuel mixture to swirl in a corkscrew pattern to improve combustion. That is, the swirl effect is accomplished by intake port design, port location in the combustion chamber, and the shape of the chamber itself. As the piston comes up on the compression stroke, this agitation of the air/fuel mixture continues and is compounded by compression (Figure 2-34). This swirl action provides better fuel economy and lower exhaust emissions since fewer unburned hydrocarbons enter the exhaust system. A comparison of the swirl and hemispherical combustion chambers is shown in Figure 2-35.

Chamber-in-Piston. The valves are positioned vertically in the top of the cylinder head

A

B

FIGURE 2-35 (A) Swirl combustion chamber and (B) hemispherical combustion chamber

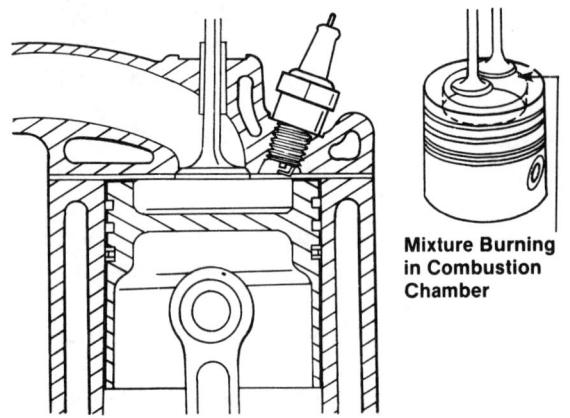

FIGURE 2-36 Typical chamber-in-piston combustion design

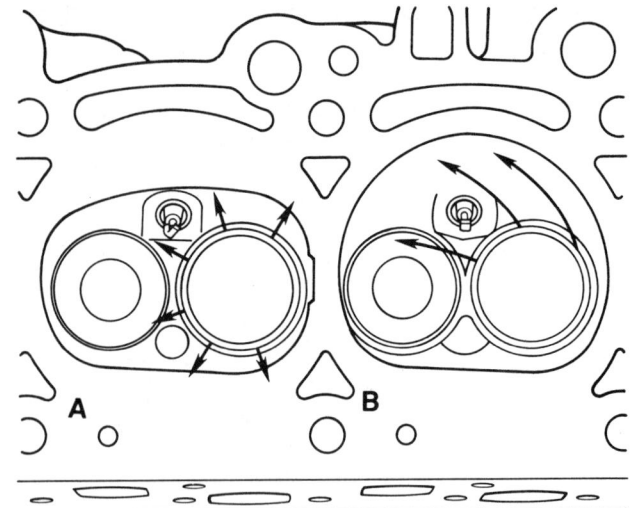

FIGURE 2-37 (A) Standard uniform flow combustion chamber and (B) fast burn tangential flow combustion chamber.

and are generally set flush with the face of the head (Figure 2-36). The spark plug is set to one side, midway between the inlet and exhaust valves, and is sometimes inclined a little away from the vertical.

In some designs, the inlet valve is well recessed into the inlet port and may even have a shroud on the head. Opening the inlet valve therefore ensures a strong swirl and thorough mixing of the incoming charge of gases. The rim of the bowl in the piston head creates some squish because, as the piston reaches the top of its stroke, some of the combustion gases around the rim are forced into the bowl itself, producing enough turbulence to minimize knocking.

An advantage of this shape is that because the combustion chamber lies in the piston itself, it remains hot. This helps to ensure the complete vaporization of combustion gases under light load, which, in turn, ensures smooth combustion.

Fast Burn Combustion Chamber. A fast burn combustion chamber is used on certain four-cylinder engines to improve their efficiency (Figure 2-37). Faster combustion is achieved by directing airflow tangentially through the intake valves to create turbulence. Fast burn combustion chambers also decrease potential engine knock. With less potential for knock, compression ratios can be increased without increasing fuel octane requirements.

Intake and Exhaust Ports

Intake and exhaust ports must be cast into the cylinder head. These ports are made so the air and fuel can pass through the cylinder head into the combustion chamber. It would be ideal if one port could be used for each valve. Because of space, however, ports are sometimes combined. These ports are called *siamese ports* (Figure 2-38). Sia-

mese ports can be used because each cylinder uses the port at a different time.

Cross flow ports are used on some engines. Cross flow heads have the intake and exhaust ports on the opposite sides.

Coolant Passages

As mentioned previously in this chapter, the large openings that allow coolant to pass through the head are cast into the cylinder head (Figure 2-39). Coolant must circulate throughout the cylinder head so excess heat can be removed. The coolant flows from passages in the cylinder block through the head gasket and into the cylinder head. Depending upon the engine configuration, the coolant then passes back to other parts of the cooling system.

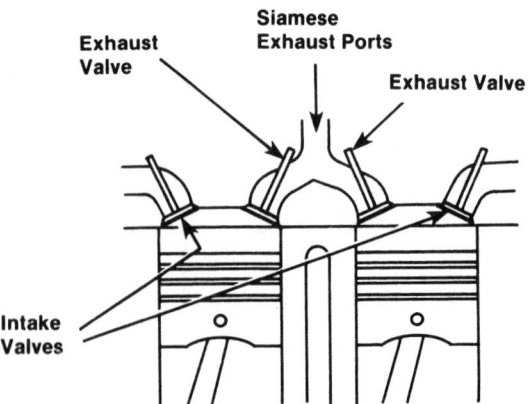

FIGURE 2-38 With siamese ports, two cylinders feed the same exhaust port.

FIGURE 2-39 Coolant passage in an aluminum cylinder head

Core Plugs

All cylinder blocks use core plugs. These are also called *freeze* or *expansion plugs.* During the manufacturing process, sand cores are used. These cores are partly broken and dissolved when the hot metal is poured into the mold. However, holes have to be placed in the block to get the sand out of the internal passageways. These are called *core holes.* The holes are machined and core plugs are placed into these holes (Figure 2–40).

Core plugs are made of soft metal. They can also protect the block from cracking. If the coolant in the block freezes near the core plugs, the coolant will expand. This may cause the block to expand and crack. Rather than having the block crack, the core plugs may pop out and possibly save the block.

Cylinder Sleeves

Some automobile manufacturers use cylinder sleeves (Figure 2–41). Rather than casting the cylinder bores directly into the block, they insert a machined sleeve after the block has been machined. The purpose of using a sleeve is that if the cylinder is damaged, the sleeve can be removed and replaced rather easily. Blocks that do not have sleeves have to be bored out to remove any damage. After boring, larger pistons will be needed.

As described in Chapter 11, there are two types of sleeves: wet and dry. The dry sleeve is pressed into a hole in the block. It can be machined quite thin because the sleeve is supported from top to bottom by the cast-iron block.

The wet sleeve is also pressed into the block. The cooling water touches the center part of the sleeve. This is why it is called a wet sleeve. It must be machined thicker than the dry sleeve because it is supported only on the top and bottom. Seals must be

FIGURE 2-40 Typical core plug locations

used on the top and bottom of the wet sleeve. Seals are used to keep the cooling water from leaking out

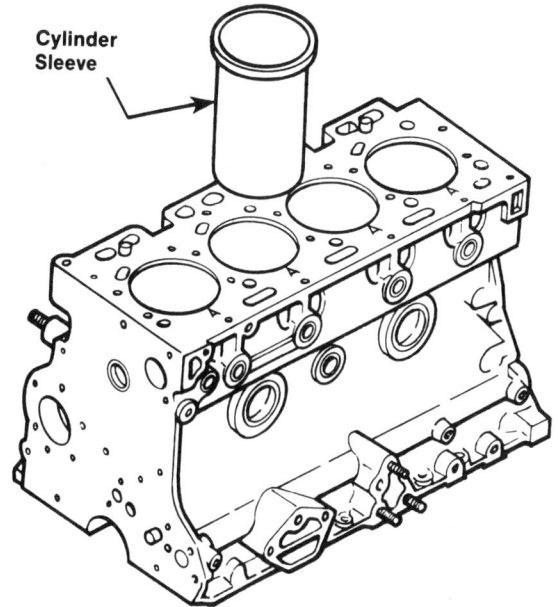

FIGURE 2-41 Cylinder sleeves

55

of the cooling system. Wet sleeves are used on some larger diesel engines.

INTAKE AND EXHAUST VALVES

Every cylinder of a four-stroke cycle engine contains at least one intake valve to permit the air/ fuel mixture to enter the cylinder and one exhaust valve to allow the burned exhaust gases to escape. The intake and exhaust valves, along with the cylinder head gasket, must also seal the combustion chamber (Figure 2-42).

The type of valve used in automotive engines is called a *poppet*. The word *poppet* is derived from the popping action of the valve as it opens and closes. A poppet-type valve has a round head with a tapered face, a stem that is used to guide the valve, and a slot that is machined at the top of the stem for the valve spring retainers.

The head of the valve is the large diameter end and is used to seal the intake or exhaust port. This seal is made by the valve face contacting the valve seat. The valve face is the tapered area machined on the head of the valve. The angle of this taper is determined by design and manufacture of the engine. The taper will vary from one engine family to another and may vary between intake and exhaust valves in the same engine. The area between the valve face and the head of the valve is called the *margin*. The margin allows for some machining of the valve face, which is sometimes necessary to restore its finish, and allows the valve an extra capacity to hold heat.

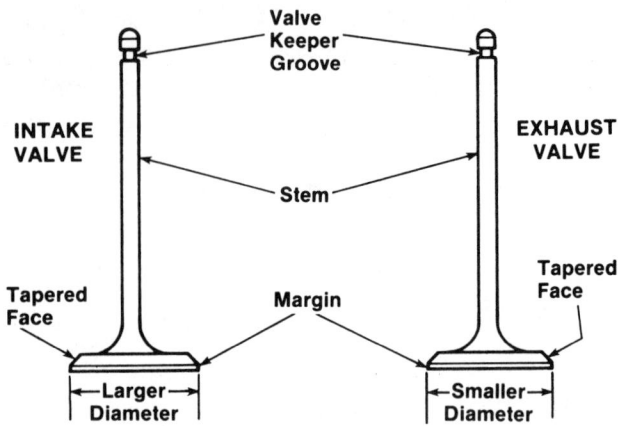

FIGURE 2-43 Valve nomenclature

The intake and exhaust valve heads are different diameters, the intake being the larger of the two (Figure 2-43). The size or diameter of the valves is determined by the engine design and its uses. As mentioned, the stem guides the valve during its up-and-down movement and serves to connect the valve to its spring through its valve spring retainers and keepers. The stem rides in a guide that is either machined into the head or pressed into the head as a separate replaceable part.

The valve seat is the area of the head contacted by the face of the valve. The seat may be machined in the head or it may be pressed in like the valve guide.

The valves found in today's modern engines are made from special high-strength steel or ceramic that is highly heat resistant. Heat resistance is very critical in exhaust valves because they must withstand working temperatures of between 1500 and 4000 degrees Fahrenheit. Heat resistance is much less of a problem for intake valves because they receive extra cooling from the fuel mixture during the intake stroke.

There are two ways for the exhaust valve to rid itself of its heat. First, when the valve face is in contact with its seat, the heat from the valve will be transferred to the cylinder head, which is liquid cooled. The second is through the valve stem to the valve guide and again to the cylinder head. To aid in this second method of heat transfer, some exhaust valve stems are hollow. This hollow section is filled with sodium (Figure 2-44). Sodium is a silver-white alkaline metallic chemical element that transfers heat much better than steel.

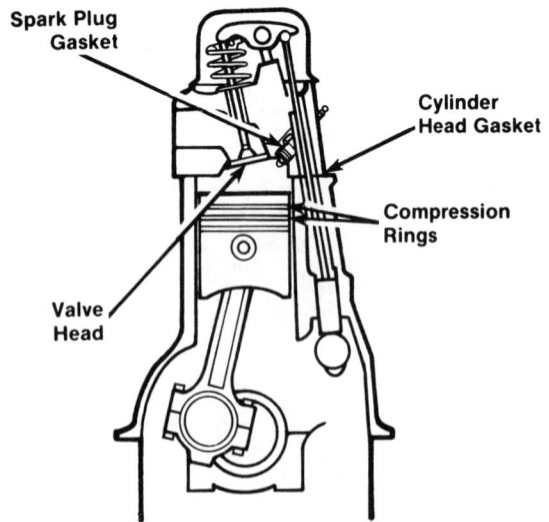

FIGURE 2-42 Sealing points of typical engine

WARNING: Never cut open any sodium-filled valves. Sodium will burn violently when it contacts water.

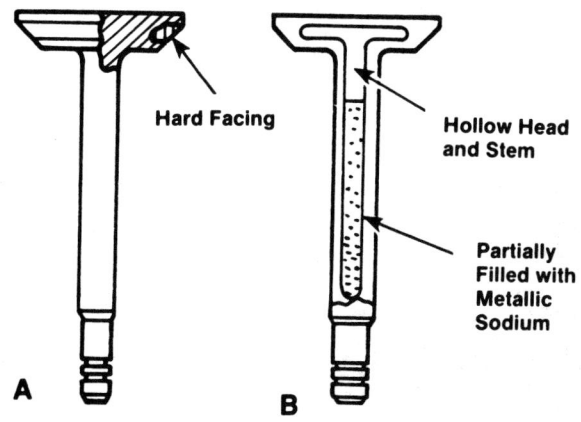

FIGURE 2-44   (A) Hard-faced valve and (B) sodium-filled valve

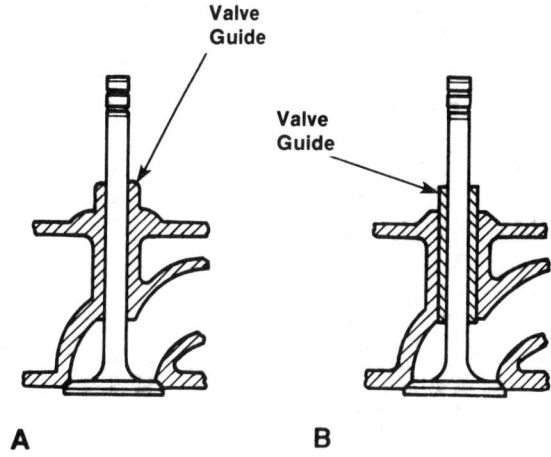

FIGURE 2-45   Two types of valve guides: (A) integral and (B) removable

The valve seat is the area contacted by the valve face when the valve is in the *closed* position. Both intake and exhaust valves have seats. The valve seat area must be hard enough to withstand the constant closing of the valve and supply good heat transfer. Due to corrosive products found in the exhaust gas, the seats must be highly resistant to corrosion. When the cylinder head material meets these requirements the seats are machined directly into it. When it does not, the seats are then made of material that will meet the requirements and the seats are pressed into the head.

The following are the important valve components found in a four-stroke engine.

Valve Guides.   The valve guides are the parts that support the valves in the head. They are machined to a fit of a few thousandths of an inch clearance with the valve stem. This close clearance is important for the following reasons:

- It keeps the lubricating oil from getting sucked into the combustion chamber past the intake valve stem during the intake stroke.
- It keeps exhaust gases from getting into the crankcase area past the exhaust valve stems during the exhaust stroke.
- It keeps the valve face in perfect alignment with the valve seat.

Valve guides can be cast integrally with the head, or they can be removable (Figure 2-45). Removable valve guides usually are press-fit into the head.

Valve Springs, Retainers, and Seals.   The valve assembly is completed by the spring, retainer, and seal. Before the spring and the retainer fit into place, a seal is placed over the valve stem. The seal acts like an umbrella to keep the valve operating mechanism oil from running down the valve stem and into the combustion chamber. The spring, which keeps the valve in a normally closed position, is held in place by the retainer. The retainer locks onto the valve stem with two wedged-shaped parts that are called *valve keepers*. Some engines utilize a single valve spring per valve (Figure 2-46A); others use two or three springs (Figure 2-46B). In a three-spring arrangement, two coil springs are usually separated by a flat spring called a *damper*.

Valve Rotators.   It is common in heavy-duty applications to use mechanisms that make the exhaust valves rotate. The purpose is to keep carbon from building up between the valve face and seat, which could hold the valve partially open, causing it to burn. The release-type rotator (Figure 2-47A) releases the spring tension from the valve while open. The valve will then rotate from engine vibration. The positive rotator (Figure 2-47B) is a two-piece valve retainer that has a flexible washer between the two pieces. A series of balls between the retainer pieces roll on machined ramps as pressure is applied and released from the opening and the closing of the valve. The movement of the balls up and down the ramps translates into rotation of the valve.

MULTIVALVE ENGINES

As mentioned earlier in this chapter, some newer engines, especially smaller ones, employ multivalve arrangements. Automotive engineers have long been obsessed with the idea of additional valves in the cylinder head. It all started in 1918 with the "dual valve" Pierce Arrow, one of the first cars to use four valves per cylinder as a way to enhance gas flow and increase horsepower.

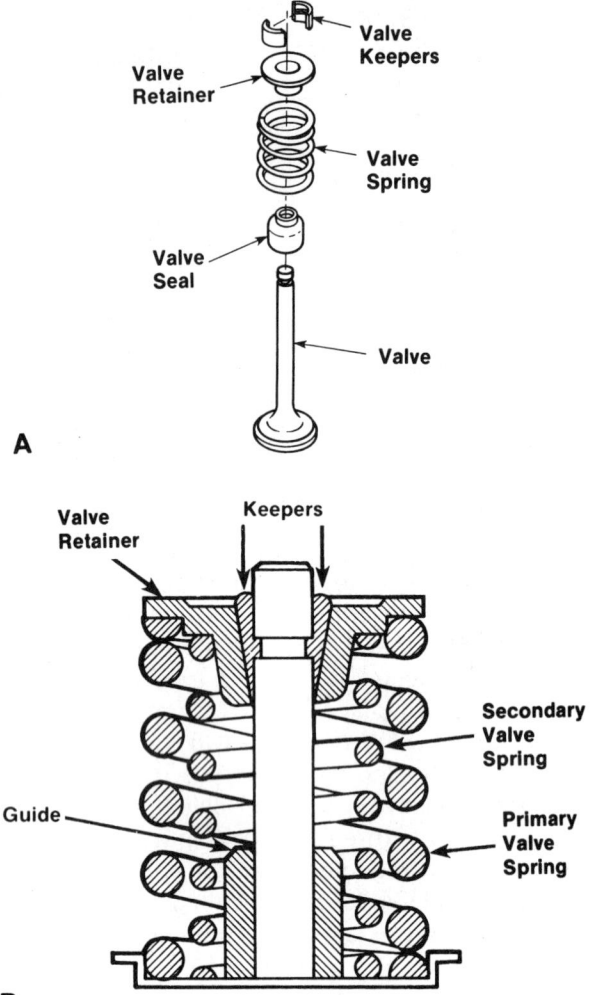

FIGURE 2-46 Typical valve spring assemblies: (A) single spring setup and (B) reverse wound secondary spring, which dampens spring vibrations and increases total spring pressure.

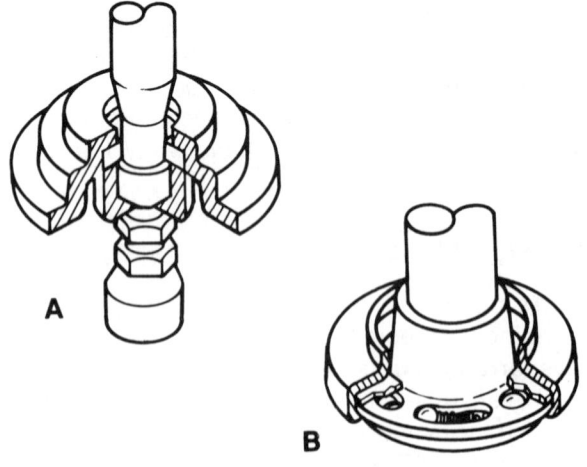

FIGURE 2-47 Two types of valve rotators: (A) release and (B) positive

Multivalve can be three valves per cylinder (Figure 2–48) or four valves (Figure 2–49) per cylinder. This means that a four-cylinder car has either twelve valves (two intakes and one exhaust) or sixteen valves (two intakes and two exhausts); a six-cylinder engine has either eighteen or twenty-four valves; and an eight-cylinder engine has twenty-four or thirty-two valves. To aid in combustion, some multivalve arrangements feature a jet valve (Figure 2–50). Air from the carburetor throttle plate is directed to the jet valve (a smaller intake valve) at lower engine speeds. This results in a swirling action in the combustion chamber and increased turbulence. The location of the air inlet above the throttle plate is located so that very little air enters the passage to the jet valve at higher engine speeds and greater throttle opening. In all multivalve engines, the heads are of the cross flow design (Figure 2–51).

Replacing the conventional single intake and single exhaust valve with two intake and one or two

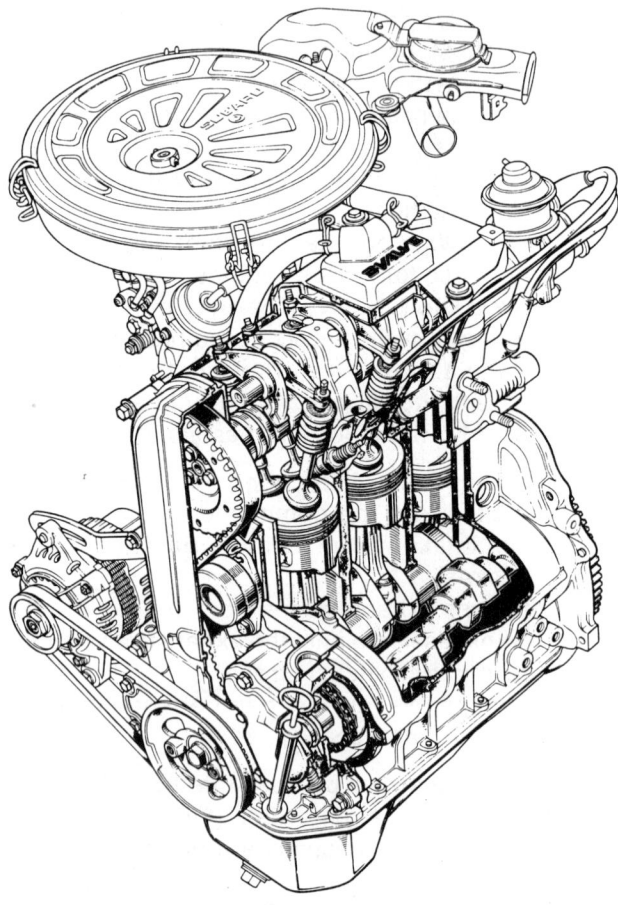

FIGURE 2-48 Single overhead camshaft (SOHC) engine that has two intake valves and one exhaust valve for each of its three cylinders

FIGURE 2-49 Cutaway view of four valves per cylinder multivalve arrangement

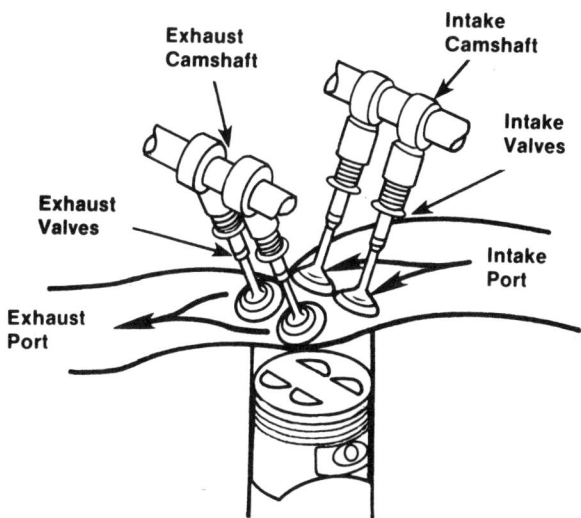

FIGURE 2-51 Typical cross flow pattern found in multivalve engines

exhaust valves increases the area of the passages that let the air/fuel mixture into the combustion chamber and let burned fuel and exhaust gases exit. Thus, multivalve engines have a more complete combustion, which reduces the chances of misfire and detonation. This results in enhanced fuel efficiency, cleaner exhaust, and increased power out-

put. A related point is intake velocity, which is easier to keep high with small multiple ports than a large single passage, and a fast-moving charge promotes good torque production during low- and mid-range rpm operation. The smaller valves naturally have less mass than big ones, so mechanical inertia is reduced, making a higher tachometer red line read-

INDUCTION AND COMPRESSION

IGNITION AND COMBUSTION

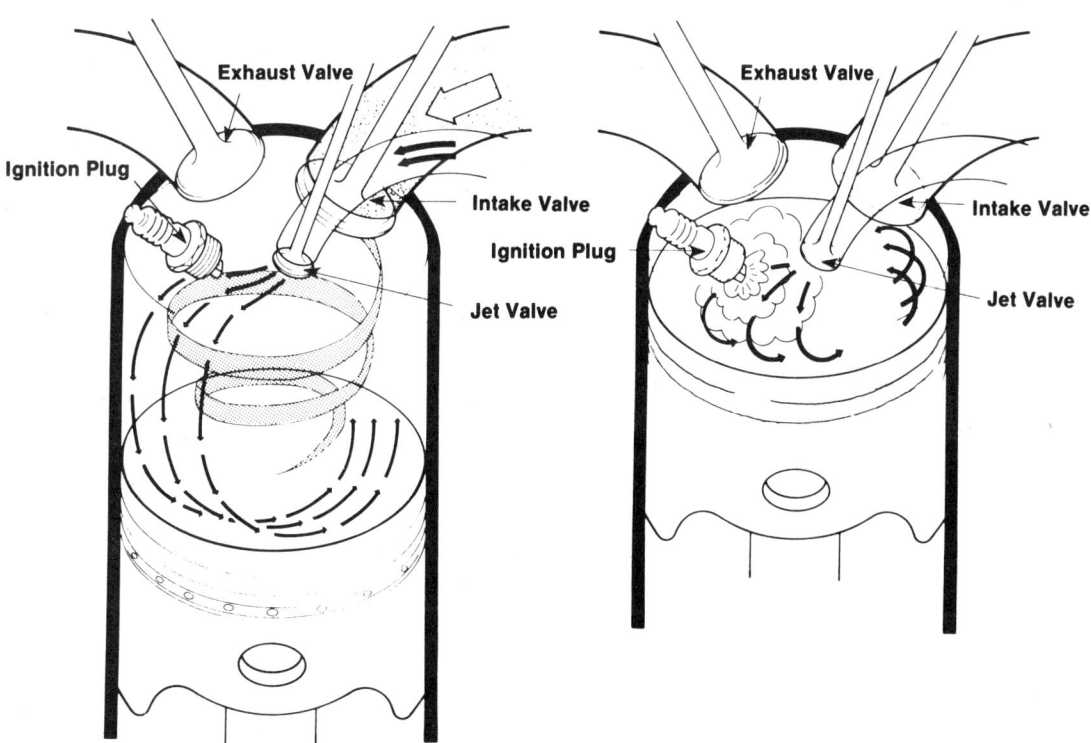

FIGURE 2-50 How a jet valve operates in multivalve arrangement

ing possible before float occurs. And the more times the cylinder can be filled and evacuated per second, the more horsepower can be obtained.

The benefits of multivalves are offset to some extent, however, by a more complicated camshaft arrangement. The easiest way to actuate four valves per cylinder is with dual overhead camshafts (Figure 2-52), which are sometimes difficult to lubricate. Because increased gas velocity is the main benefit of a four-valve cylinder head, the technology works best at high engine speeds. Unfortunately, multiple-valve technology is also inhibited by manifold con-

strictions and exhaust back pressure. This is the reason some manufacturers are researching new engine blocks that feature a one-piece cast cylinder head and block. The lack of cylinder head bolts allows a highly efficient manifold that maximizes the effect of multivalve design.

The cam drive is even more complicated with V-power plants, and most today are using a single overhead cam per cylinder bank, with some kind of lever arm actuating the opposite bank of valves.

The high revolution rate inherent in multivalve engine design puts a premium on balancing techniques. Four-cylinder engines without balance shafts tend to be prone to high shaking forces. To overcome this problem, some multivalve engines use relatively long connecting rods to reduce angular changes during a piston stroke, a solution that has only been partially successful in the past. Using lightweight materials in the manufacturing of the piston rod assembly also contributes to the solution.

Servicing and rebuilding multivalve engines present no major problem for the builder. That is, as far as the rebuilder is concerned, there simply are more seats, faces, guides, seals, and springs to take care of; there is no new technology to be learned.

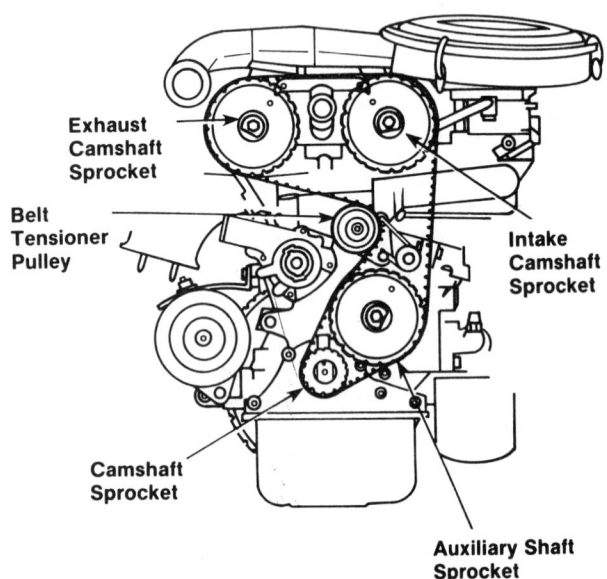

FIGURE 2-52 Some engines are equipped with two camshafts in the cylinder head, one for the intake valves and one for the exhaust valves. The primary advantage of dual overhead camshaft (DOHC) configuration is increased control over the opening and closing of the valves which is very important in the operation of multivalve.

CAMSHAFT

A camshaft is a shaft (Figure 2-53) with a cam for each exhaust and intake valve, each one placed to allow for the proper timing of each valve. A cam is a device that changes rotary motion into reciprocating motion. Each cam has a high spot or cam lobe that controls the opening of the valves. The height of the lobe is proportional to the amount the valve will open. Some camshafts may also be equipped with an eccentric to operate the fuel pump and a gear to drive the distributor and oil pump. Heavy-duty en-

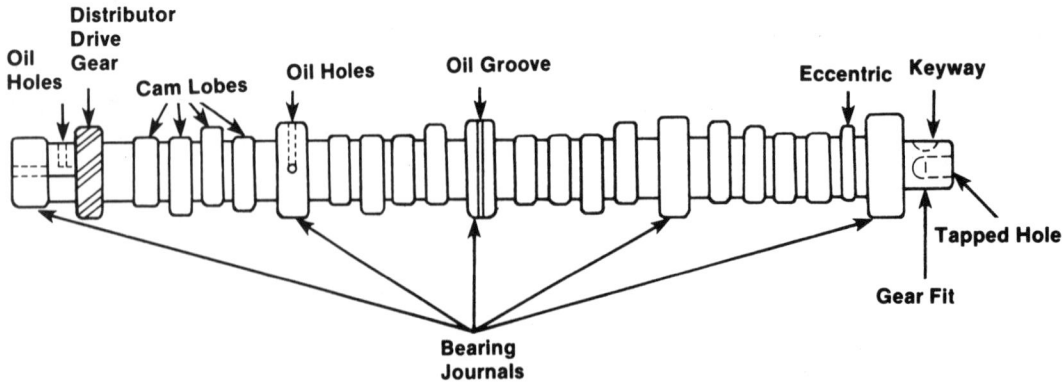

FIGURE 2-53 Camshaft and lobe nomenclature

gines utilize lobes for fuel spray valves, fuel injection pumps, or air starting valves.

As mentioned earlier in the chapter, the camshaft can be located in either the cylinder block or in the cylinder head. If the camshaft is in the block (Figure 2-54), it is positioned above the crankshaft, and the valves are opened through lifters, pushrods, and rocker arms. As the cam lobe rotates, it pushes up on the lifter, which lifts up the pushrod, moving one end of the rocker arm up while the other end pushes the valve down to open it (Figure 2-55). As the cam rotates, the valve spring forces the valve to close and maintains the contact between the valve and the rocker arm, thereby keeping the pushrod and the lifter in contact with the rotating cam. Engines with the camshaft in the engine block and the valves in the cylinder head are referred to as overhead valve (OHV) engines. Most diesel engines are of the OHV type.

Overhead camshaft (OHC) engines have the camshaft mounted in or on the cylinder head (Figure 2-56). The OHC engines have no need for pushrods. As the camshaft rotates, the cams ride directly above the valves. The lobes open the valves by either directly depressing the valve or by depressing the valve through the use of a cam follower or bucket-type tappet. The closing of the valves is still the responsibility of the valve springs.

An advantage of the overhead camshaft engine is that through the elimination of the pushrod and, on some engines, the valve lifter, the inertia of the valve train is lower and there is less deflection in the

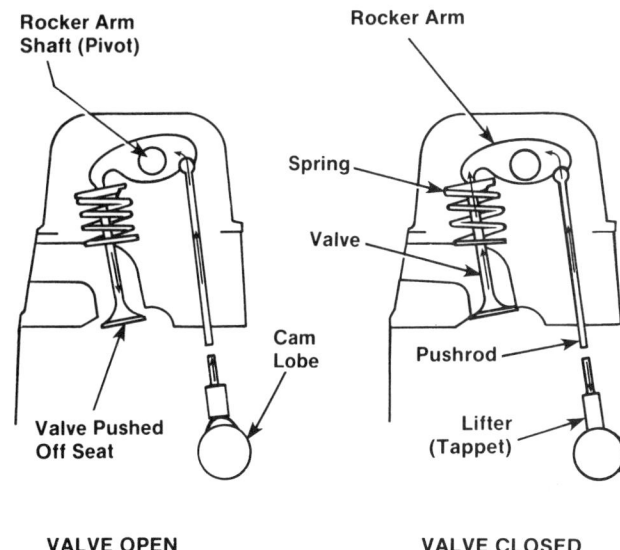

VALVE OPEN   VALVE CLOSED

FIGURE 2-55 How the camshaft, pushrod, rocker arm, and valve work together in a valve train.

system. With lower inertia, the chances of valve float at higher engine speeds are reduced. Valve float occurs when the valve is momentarily thrown free from the direct influence of the cam's shape. This can occur at high engine speeds when the inertia of the lifter and pushrod momentarily overcomes the force of the valve spring, allowing gaps to form in the valve train. When the spring pressure finally resists more movement, the parts come together strongly and quickly. This can cause damage and definitely cause a loss of valve control, which affects the performance of the engine. Valve float is less likely to occur in OHC engines because they have fewer valve train parts and, therefore, less inertia. *Inertia* is best defined as the tendency of an object to continue moving in the same direction without a direct force applied to it. The heavier the object is and the greater its speed, the higher the inertia force will be.

Without the worry of keeping the lifter and pushrod in constant contact with the cam lobes, OHC camshafts can be machined to provide more rapid valve opening and closing. This can result in improved engine efficiency through increased volumetric efficiency. This is another advantage of OHC engines.

VALVE LIFTERS

Valve lifters, sometimes called *cam followers* or *tappets*, follow the contour or shape of the cam lobe. Lifters are either mechanical (solid) or hydraulic. Solid lifters provide for a rigid connection between the camshaft and the valves; hydraulic lifters provide for the same connection but use oil to absorb the

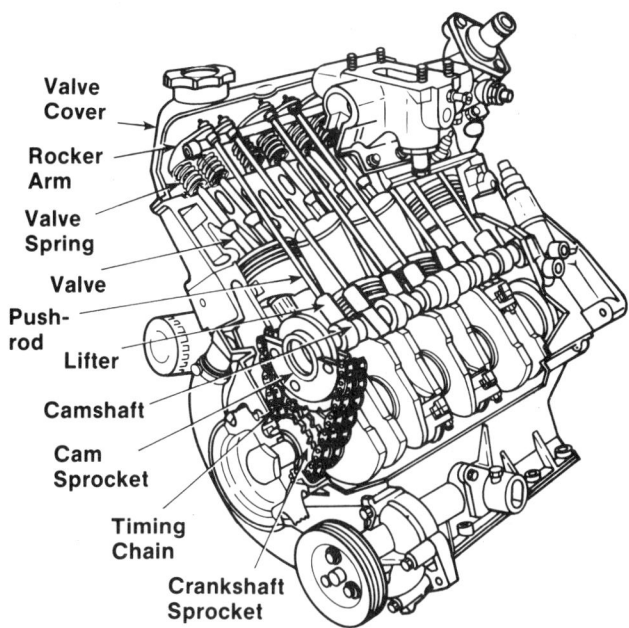

FIGURE 2-54 Camshaft in the block

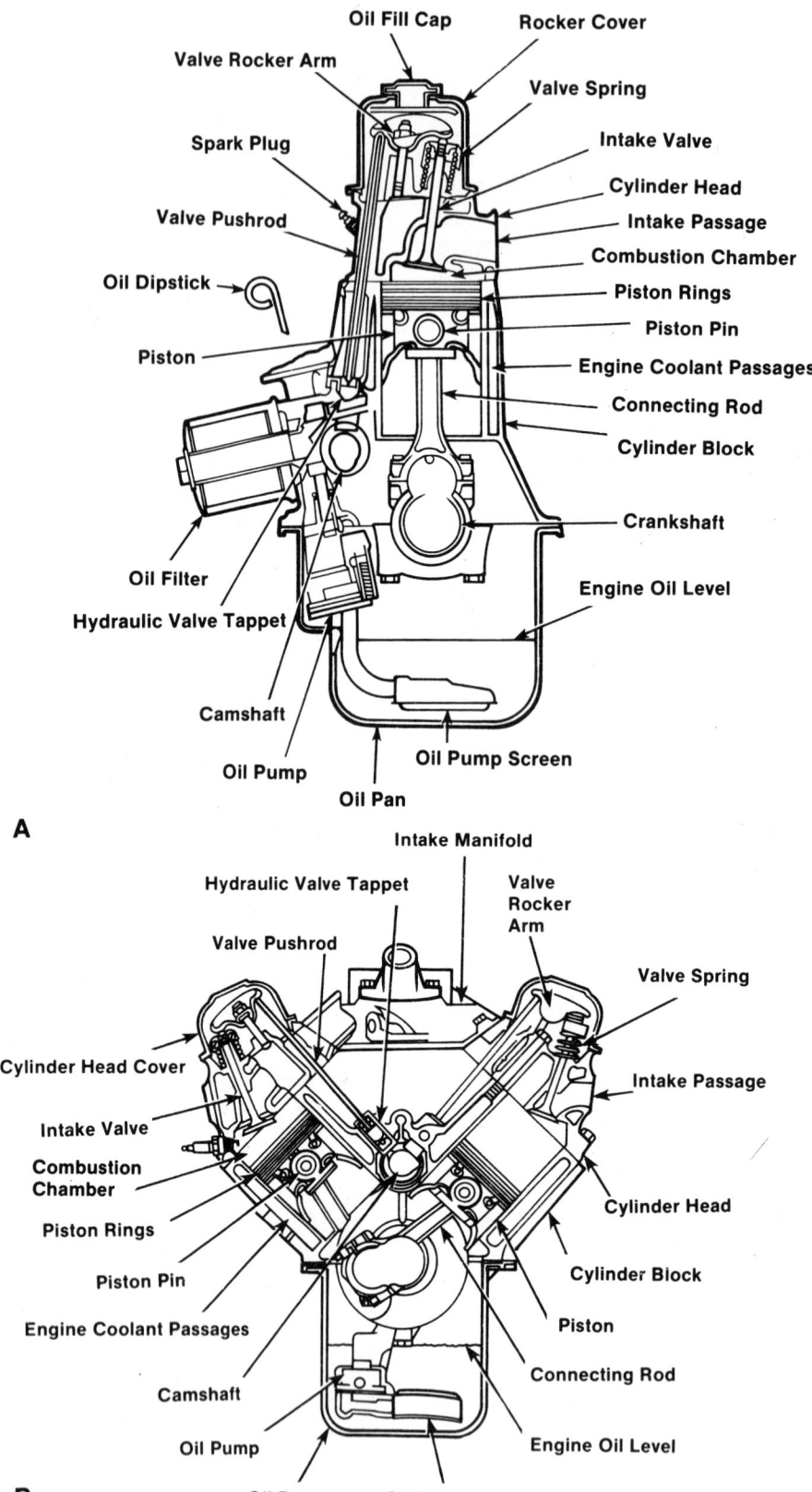

FIGURE 2-56 Typical OHC components (A) Typical in-line six-cylinder engine (front view) and (B) typical V-8 engine (front view)

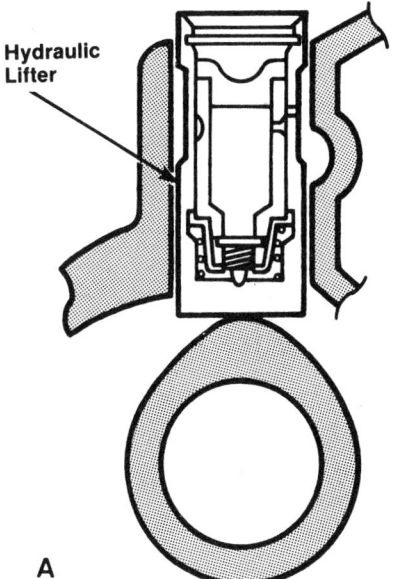

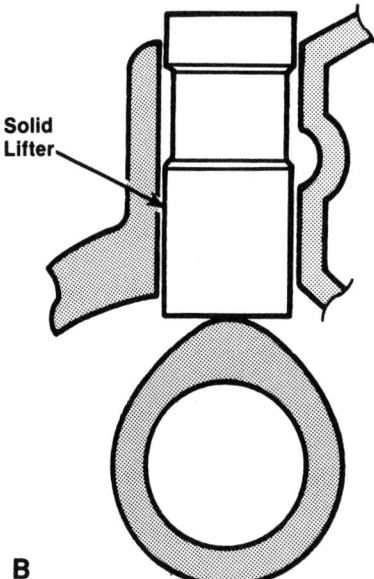

FIGURE 2-57   Two types of lifter designs: (A) hydraulic and (B) solid

shock that results from the movement of the valve train. Although both types function the same, solid lifters have certain disadvantages when compared to hydraulic lifters.

Hydraulic lifters (Figure 2-57A) are designed to automatically compensate for the effects of engine temperature. Changes in temperature cause valve train components to expand and contract. Hydraulic lifters are designed to automatically maintain a direct connection between valve train parts.

Solid lifters (Figure 2-57B) do not have this built-in feature and require a clearance between the parts of the valve train. This clearance allows for expansion of the components as the engine gets hot. Periodic adjustment of this clearance must be made, and excessive clearances might cause a clicking noise. This clicking noise is also an indication of the hammering of valve train parts against one another, which will result in reduced camshaft and lifter life.

In an effort to reduce the friction—and the resulting power loss—from the lifter rubbing against the cam lobes, engine manufacturers often use roller-type hydraulic lifters. Roller lifters (Figure 2-58) are manufactured with a large roller on the camshaft end of the lifter. The roller acts like a wheel and allows the lifter to follow the cam lobe contour better than a flat-type lifter with reduced friction between the contacting surfaces. Friction is reduced by the lifter rolling along the surface of the cam lobe as opposed to rubbing along it. Currently, nearly all heavy-duty diesels use roller lifters that are the solid type.

Operation of Hydraulic Lifters

When the lifter is resting on the base circle of the cam (not on the lobe), the valve is closed and the plunger spring inside the lifter maintains a zero clearance in the valve train. Oil is pumped by the oil pump from the oil pan, to the oil filter, main bearings, camshaft bearings, and from there to the oil feed holes of the valve lifter bore. That is, pressurized oil enters the lifter and flows through the plunger. Oil flow continues through the lifter until it is filled. The

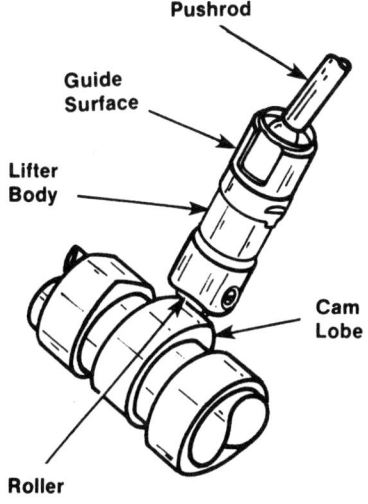

FIGURE 2-58   Roller-type hydraulic lifters

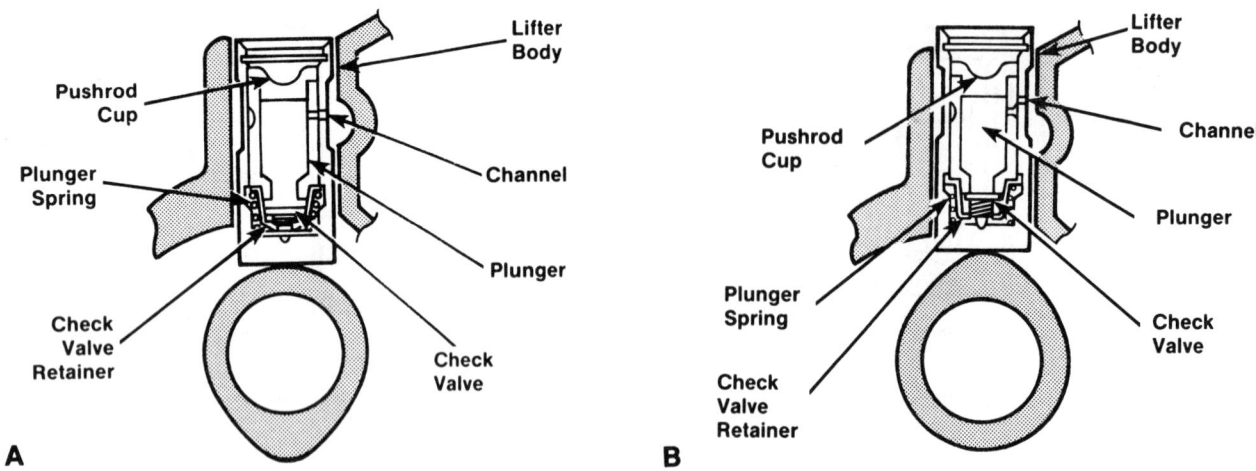

FIGURE 2-59 Operation of a hydraulic lifter: (A) valve closed and (B) valve open

pressure of the oil seals the oil in the lifter by forcing down the check valve inside the lifter. The oil between the plunger and the check valve forms a rigid connection between the lifter and the pushrod. The position of the plunger inside the lifter is determined by the force of the pushrod on the plunger.

As the valve train grows from heat, clearance is reduced so the plunger is held lower in the lifter (Figure 2-59A). As the cam lobe turns, the lifter's oil feed hole is no longer lined up with the feed in the block and new supplies of oil are not available until the lifter returns to the lower part of its bore (Figure 2-59B). Also, the amount of force on the plunger increases as new forces result from the opening of the valve against the pressure of the valve spring. This new pressure causes the plunger to depress slightly which allows a small amount of oil to leak out. This leaking out of oil is called *leakdown*.

Leakdown is defined as the relative movement of the plunger with respect to the lifter body after the check valve is seated by the pressurized oil. When the cam rotates and the lifter returns to the base of the cam, oil again can enter the lifter. At this point, the oil fills the lifter, and the lifter is ready to open the valve again.

If the lifter is unable to fill with oil or if it does not leak down the proper amount, a noise will be heard when the engine is running. Quite often light tapping noises are referred to as *lifter noises*. Not always is the lifter at fault; there are other causes for similar sounding engine noises, such as worn cam lobes, stuck valve, broken valve spring, insufficient lubrication between the rocker arm and the pushrod or valve, or a loose rocker arm assembly. All of these problems relate to the opening and closing of the valves and will take on the rhythm of valve operation.

Careful listening and troubleshooting will help determine if the cause of the noise is a lifter or one of the other possible problems.

The valves in overhead valve and overhead camshaft engines use additional components to link the camshaft to the valves. Overhead valve engines use pushrods and rocker arms. Overhead camshaft engines use various configurations of rocker arms.

Pushrods. Pushrods transfer motion between the lifters and the rocker arms. They are needed when the camshaft is located in the cylinder block. Pushrods are hollow metal tubes with specially formed ends (Figure 2-60). Some pushrods have a ball on both ends. Others have a ball on one end and a female socket on the other end. Hollow pushrods with holes in the ends can be used to feed oil from the lifters to the rocker arms. This prevents wear on the tip of the pushrod and on the rocker arm.

Pushrod Guide Plates. In some engines, pushrod guide plates are used to limit side movement on the pushrods. The guides hold the pushrods in alignment with the rocker arms. When the pushrods pass through holes in the cylinder head or intake manifold guide plates are *not* needed.

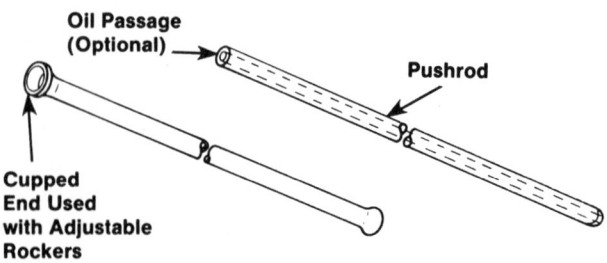

FIGURE 2-60 Two common pushrod types

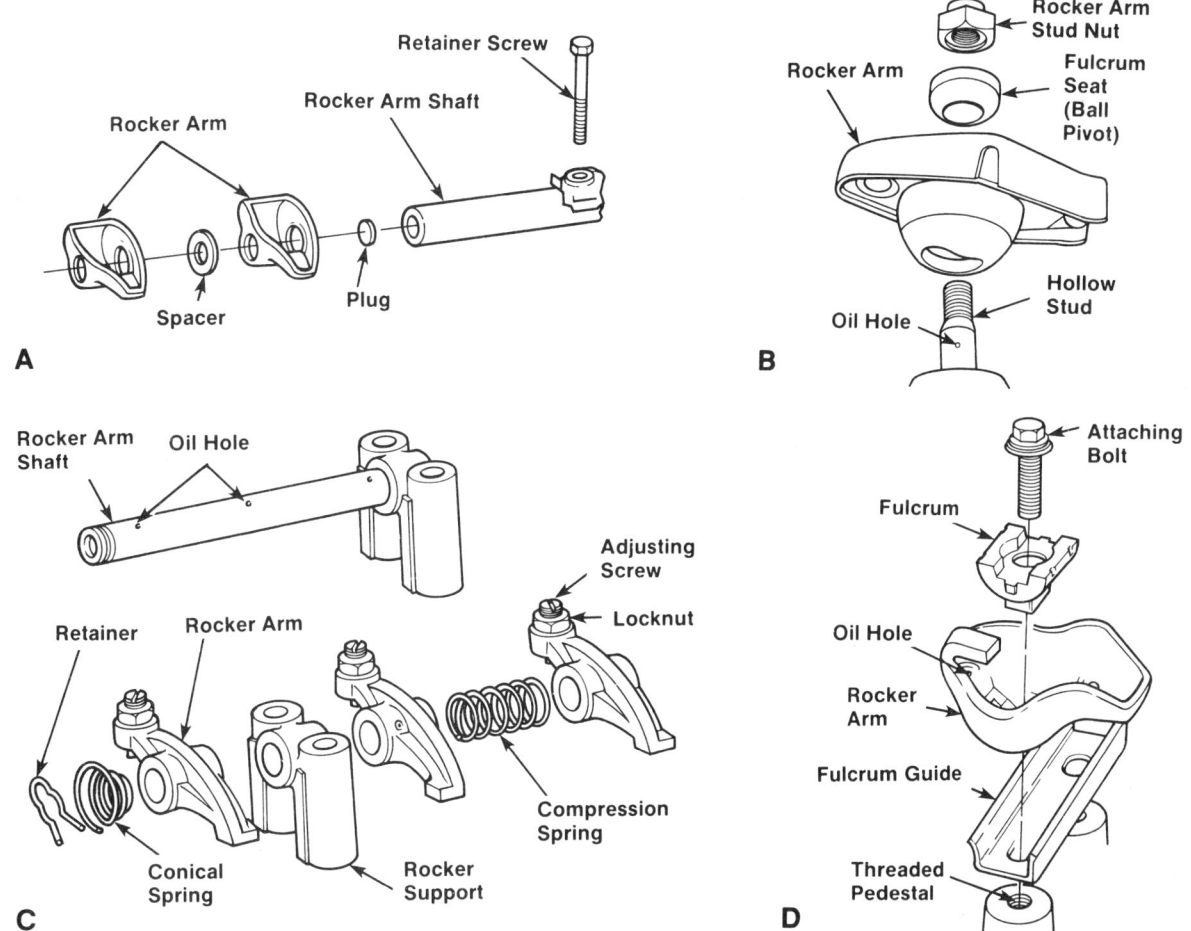

FIGURE 2-61 Various rocker mountings: (A) shaft-mounted nonadjustable rocker arm; (B) stud-mounted rocker arm; (C) shaft-mounted adjustable rocker arm; and (D) pedestal-mounted rocker arm

Rocker Arms. Rocker arms (Figure 2-61) are manufactured of steel, aluminum, or cast iron. The most common for current use are the stamped steel variety (Figure 2-62). They are lightweight, strong, and inexpensive to manufacture. They usually pivot on a stud and ball that is pushed up at one end by the lifter and pushrod, which causes the opposite end to depress the valve open.

Cast-iron rockers are used in larger, low-speed engines. They almost always pivot on a common shaft. Aluminum rockers, on the other hand, are generally used on high-performance applications and are frequently pivoted in needle bearings.

The difference in length from the valve end of the rocker arm and the center of the pivot point (shaft or stud) compared to the pushrod or cam end of the rocker arm and the pivot point (shaft or stud) is expressed as a ratio. Usually, rocker arm ratios range from 1:1 to 1:1.75. A ratio larger than 1:1 re-

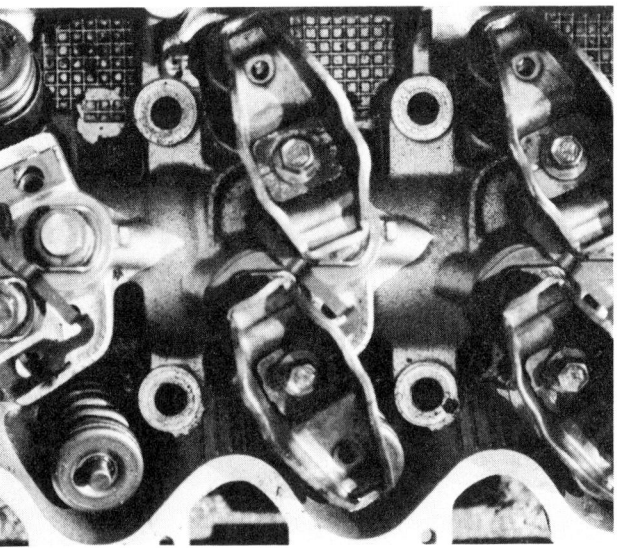

FIGURE 2-62 Typical stamped steel rocker arms

sults in the valve opening farther than the actual lift of the cam lobe (Figure 2–63).

TIMING MECHANISMS

On a four-stroke cycle engine, the camshaft is driven by the crankshaft at half the speed of the engine. This is accomplished through the use of a camshaft drive gear or drive sprocket that is twice as large as the crankshaft sprocket. For every two complete turns of the crankshaft, the camshaft turns once. Because the camshaft rotates at half the speed of the crankshaft and because the exhaust and intake valves are each opened only once each during the four strokes, there is a cam lobe for each exhaust valve and a lobe for each intake valve on the cam-

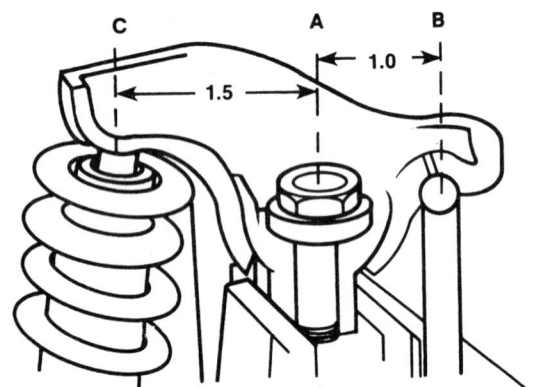

FIGURE 2-63 Rocker arm ratio is determined by the distance from the (A) rocker pivot to the (B) contact points of the pushrod and (C) the valve stem tip.

shaft. As the lobe of the cam rotates, it allows the valves to open and close. In a two-stroke cycle engine with valves, the camshaft is rotated at engine speed.

 SHOP TALK _____

The cam sprocket or gear has twice as many teeth as the crankshaft sprocket or gear.

The following are the basic configurations for driving the camshaft:

- *Gear Drive.* A gear on the crankshaft meshes directly with another gear on the camshaft (Figure 2–64A). The gear on the crankshaft is usually made of steel; the gear on the camshaft may be steel for heavy-duty applications, or it may be made of aluminum or pressed fiber when quiet operation is a major consideration. The gears are helical in design. Helical gears are used because they are stronger and also tend to push the camshaft backward during operation to help control thrust.

- *Chain Drive.* Sprockets on the camshaft and the crankshaft are linked by a continuous chain (Figure 2–64B). The sprocket on the crankshaft is usually made of steel; the sprocket on the camshaft may be steel for heavy-duty applications. When quiet operation is a major consideration, an aluminum sprocket with nylon covering on the teeth is used. There are two common types of timing

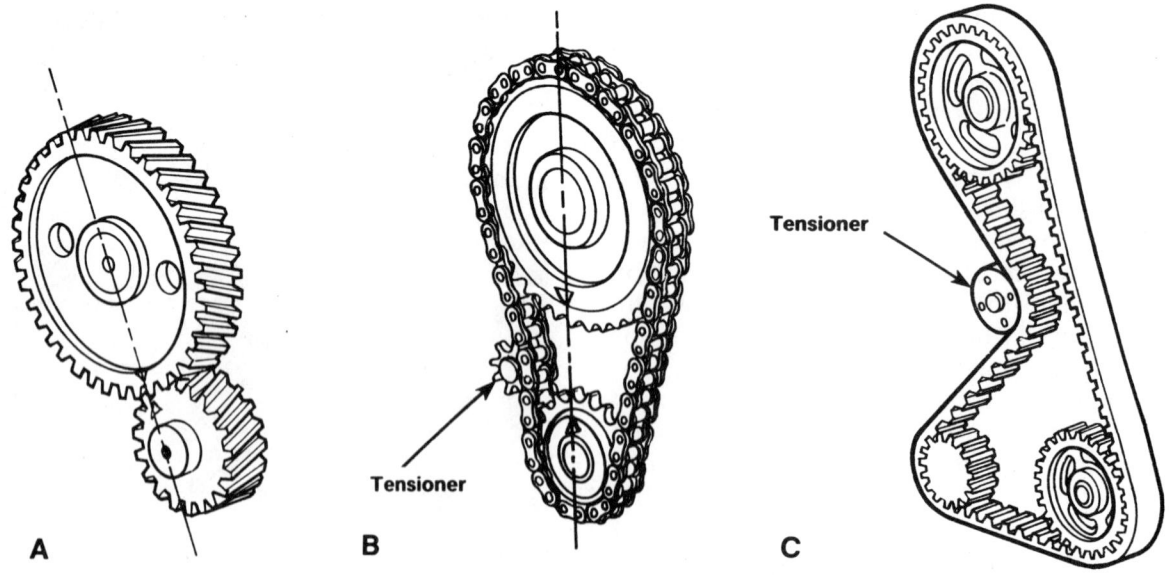

FIGURE 2-64 Timing drives: (A) gear drive; (B) chain drive; and (C) belt drive

chains. One is a silent link-type chain that is used in standard and light-duty applications. The other is the roller link chain, which is used in heavy-duty applications. The roller link chain may have a single or double row of links.

• *Belt Drive.* Sprockets on the crankshaft and the camshaft are linked by a continuous neoprene belt (Figure 2–64C). The belt has square-shaped internal teeth that mesh with teeth on the sprockets. The timing belt is reinforced with nylon or fiberglass to give it strength and prevent stretching. This drive configuration is limited to overhead camshaft engines.

• *Timing Belt and Chain Tensioners.* Many engines with chain-driven and all engines with belt-driven camshafts employ a tensioner. The tensioner pushes against the belt or chain to keep it tight. This serves to keep it from slipping on the sprockets, provide more precise valve timing, and compensate for component stretch and wear. Engines with belt-drive configurations usually use a spring-loaded idler wheel. Chain-driven configurations usually use a fiber rubbing block that is either spring loaded or hydraulic. The hydraulic tensioner is a device that works on the same principle as a hydraulic lifter. The hydraulic tensioner is much more desirable for use with a rubbing block because it takes up the slack in the chain without exerting excessive pressure, resulting in longer component life.

• *Timing Marks.* The camshaft and the crankshaft must always remain in the same relative position to each other. Because the crankshaft must rotate twice as fast as the camshaft, the drive member on the crankshaft must be exactly half as large as the driven member on the camshaft. For the camshaft and crankshaft to work together, they must be in the proper initial relation to each other. This initial position between the two shafts is designated by marks that are called *timing marks.* To obtain the correct initial relationship of the components, the corresponding marks are aligned at the time of assembly (Figure 2–65).

• *Auxiliary Camshaft Functions.* The camshaft, after being driven by the crankshaft, in turn drives other engine components. On gasoline engines, the oil pump and the distributor are usually driven from a common gear that is machined into the camshaft. The

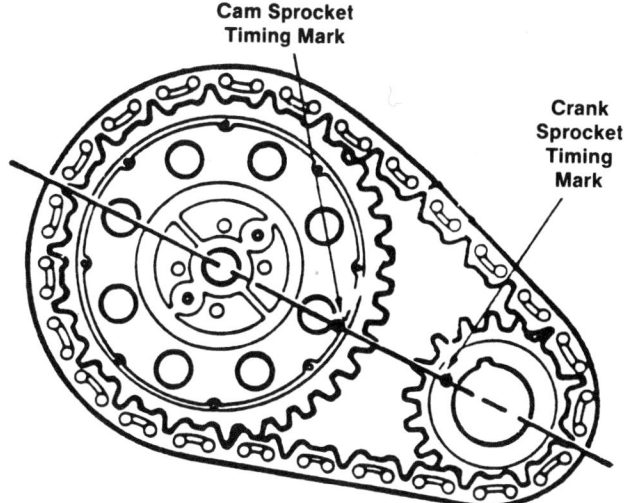

FIGURE 2–65   Timing mark alignment

fuel pump can also be driven by the camshaft. This is usually accomplished by machining an extra lobe on the camshaft to operate the pump. On some diesel engines, the camshaft is often utilized to drive the fuel injection pump and on some other systems to actuate the injector at the top of each cylinder.

CRANKSHAFT

The crankshaft is used to convert the up-and-down movement of the pistons to rotary movement, which can be used to supply power (Figure 2–66). Crankshafts are generally made of cast iron, forged cast steel, nodular iron, and then machined. At the centerline of the crankshaft are the main bearing journals (Figure 2–67). These journals must be machined to a very close tolerance because the weight and movement of the crankshaft will be supported at

FIGURE 2–66   Typical crankshaft

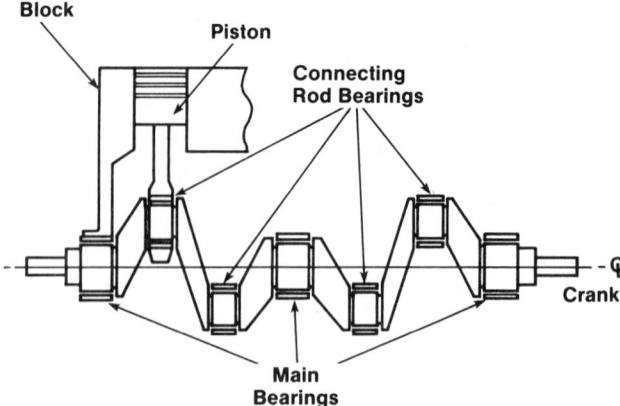

FIGURE 2-67 Crankshaft nomenclature

these points. The number of main bearings is determined by the design of the engine. V-block engines generally have fewer main bearings than an in-line engine with the same number of cylinders because the V-block engine will use a shorter crankshaft. Offset from the crankshaft centerline are the connecting rod bearing journals. The degree of offset and the number of journals are determined by the engine design. An engine having six cylinders in-line will have six connecting rod journals; a V-8 engine will have only four because each journal will be connected to two connecting rods, one from each side of the V. The connecting rod journal is also called the *crank pin*. This area is machined to a very close tolerance just like the main bearing journal. The machining of the main and rod bearing journals must be done to have a very smooth surface at the bearing area. The bearings must fit tightly enough to eliminate noise but must also have a clearance be-

tween them and their bearings for an oil film of 0.0015 inch and 0.002 inch to form. This is usually attained by the proper use of torque wrench specifications and checked by the use of plastigage that is available at most jobber stores.

The crankshaft does not turn directly on the bearings. It turns on a film of oil trapped between the bearing surface and the journal surface (Figure 2–68). This oil is supplied by the engine's oil pump. Should the crankshaft journals become out of round, tapered, or scored, the oil film will not form properly and the journal will contact the bearing surface. This causes early bearing and/or crankshaft failure. The main and rod bearings are generally made of lead-coated copper or tin aluminum. Both of these are softer material than that used to form the crankshaft. By using the soft material, any wear at the crankshaft journals will appear first on the bearings. Early diagnosis of bearing failure most often will spare the crankshaft and only the bearings will have to be replaced (see Chapter 7). Common types of engine bearings are shown in Figure 2–69.

As mentioned in this chapter, the connecting rod journals are offset from the centerline of the crankshaft. This puts weight and piston pressure off center on the crankshaft. To balance this for smooth engine operation, counterweights must be added to the crankshaft. These weights can be cast as part of the crankshaft. They will be found opposite the connecting rod journals.

The journals of the crankshaft must be smooth and highly polished. Bearings surround each journal and are fed oil under pressure (Figure 2–70). In order for the oil to reach these bearings oil passages must be drilled into the crankshaft. Each main bearing will receive oil under pressure from the pump. Each main bearing journal will have a hole drilled into it with a connecting hole or holes leading to one or more rod bearing journals. In this way all bearing journals receive oil under pressure to protect both the bearing and the journal. The crankshaft configuration determines the engine block design, or the

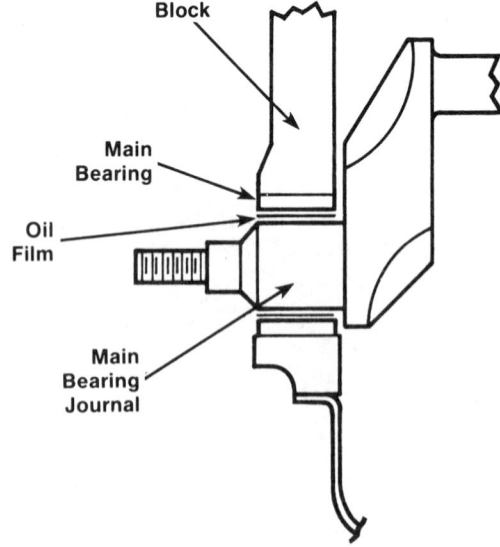

FIGURE 2-68 Crankshaft turns on oil film

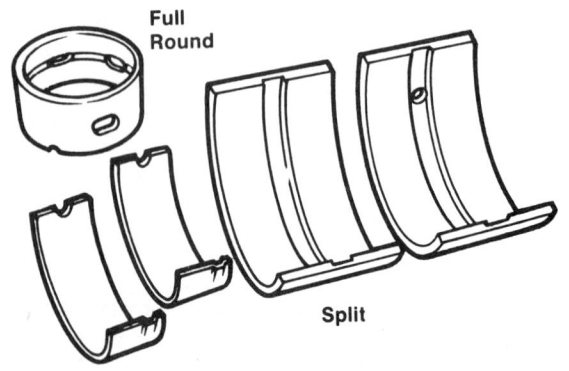

FIGURE 2-69 Insert or sleeve bearings

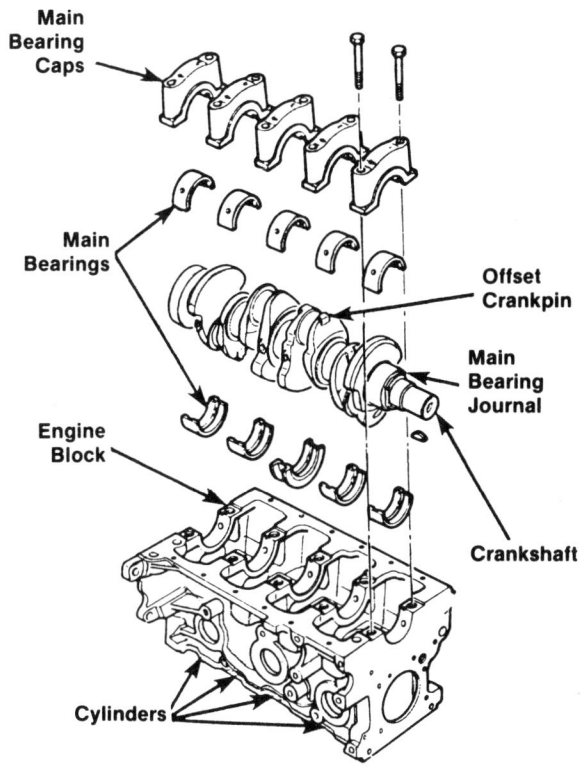

FIGURE 2-70 Crankshaft assembly showing main engine bearings

positioning of the connecting rod journals around the centerline (C) of the crankshaft (Figure 2-71).

The crankshaft has two distinct ends (Figure 2-72). One is called the flywheel end and, as its name implies, this is where the flywheel is connected to it. The front end or belt drive end of the crankshaft contains a threaded snout or is drilled and tapped. This is for attaching a vibration damper.

CAUTION: Never lay down a crankshaft or stand it on end because it can warp out of shape or become bent. It should not be stored without proper support in the middle. Remember that all engine parts must be carefully stored.

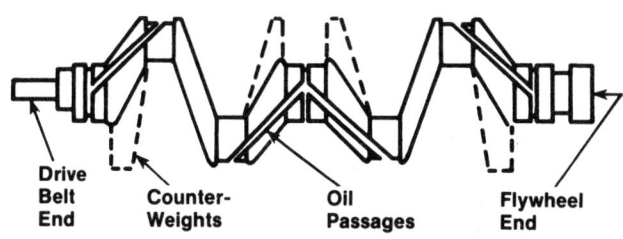

FIGURE 2-72 Flywheel and drive belt ends

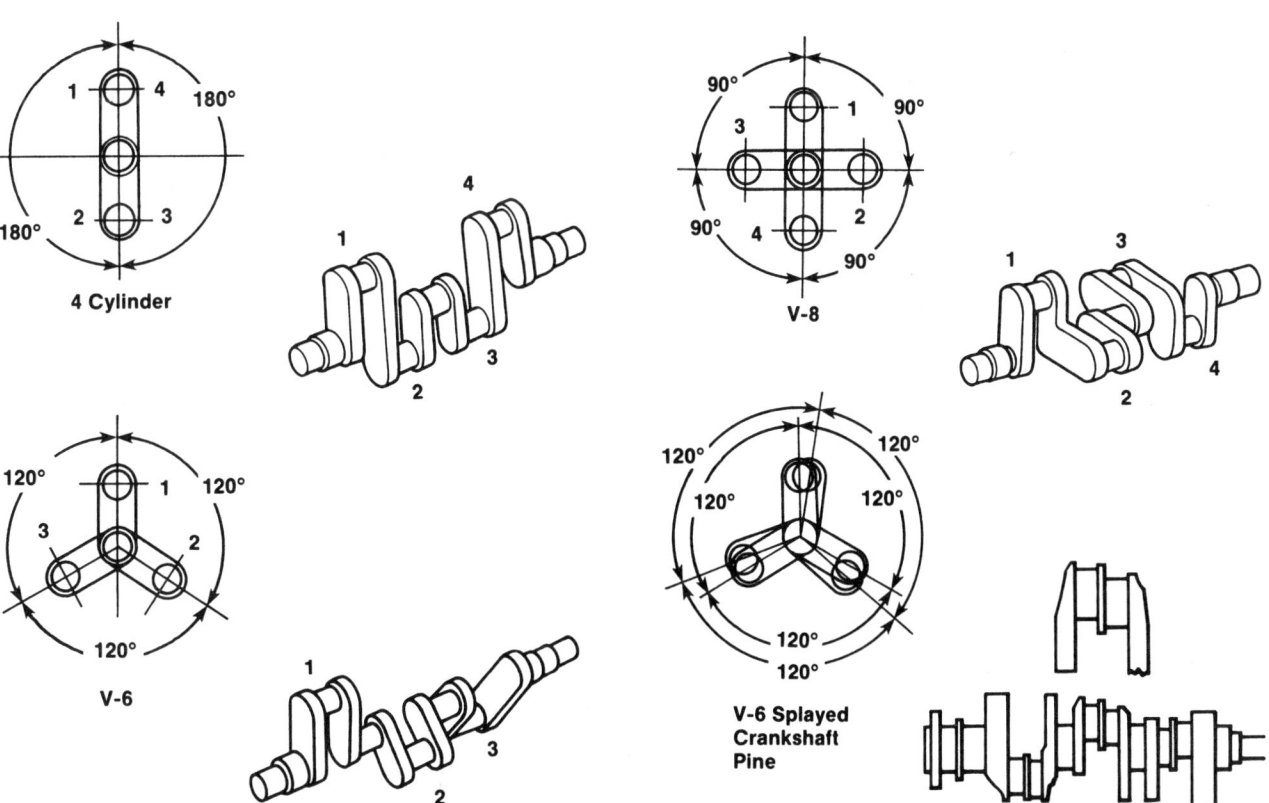

FIGURE 2-71 Crankshaft configurations for four-cylinder, V-6, and V-8 engines

VIBRATION DAMPER

The purpose of the vibration damper is to dampen crankshaft vibration. Vibration is a back-and-forth motion. The crankshaft of a running engine will have a back-and-forth or twisting motion each time a cylinder fires. The force applied to the crankshaft can be more than two tons. This causes the crank to momentarily twist and snap back. The vibration damper (Figure 2-73) consists of a center section, which is attached to the crankshaft. Surrounding the center section is a strip of rubber-like material. Attached to the material is a grooved counterweight. As the crankshaft twists, the center section "A" applies a force to the material. The material must then apply this force to the counterweight "C." The weight is snapped in the direction of the crankshaft rotation to counterbalance the crankshaft connecting rod journal snapping back against the force due to ignition "B." The back-and-forth movement of the crankshaft is counterbalanced by the back-and-forth movement of the vibration damper, and the engine runs smoothly.

FLYWHEEL

The flywheel also adds to the smooth running of the engine by applying a constant moving force to carry the crankshaft from one firing stroke to the next. Because of its large diameter, the flywheel makes a convenient point for the starter to connect to the engine. The large diameter supplies good gear reduction for the starter, making it easy for the start-

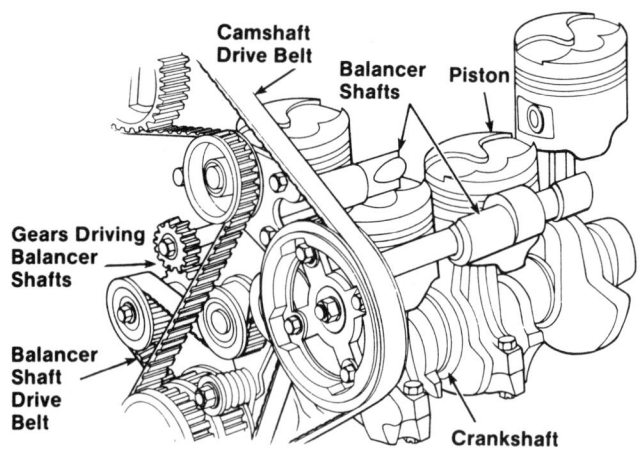

FIGURE 2-74 A balancer shaft

er to turn the engine against its compression. The surface of a flywheel may be used as part of the clutch. On an engine that drives an automatic transmission, a flexplate is used and the automatic transmission converter provides the weight required to attain flywheel functions.

BALANCER SHAFTS

The crankshaft is one of the main sources of engine vibration because its shape makes it inherently out of balance as it rotates. On some engines a balancer shaft (Figure 2-74) is employed to reduce vibration.

In its basic form, the balancer shaft is fitted with counterweights designed to mirror the throws on the crankshaft and is driven directly from the crankshaft but in the opposite direction. As the engine turns, the two out-of-balance shafts mutually cancel out any vibrations.

PISTON AND PISTON RINGS

The piston forms the lower portion of the combustion chamber. The force of the expanding air/fuel mixture at the time of ignition is exerted against the head or dome of the piston. This force then pushes the piston down in the cylinder. The force applied to the piston can be as high as 2-1/2 tons in a gasoline engine. For this reason the piston must be very strong. In the past, pistons were made primarily of cast iron or a mixture of iron and steel. This type of piston is strong but is also heavy. Due to advances in aluminum technology, the pistons in most modern automobile engines are made of aluminum or aluminum alloy. Aluminum can be made strong enough to

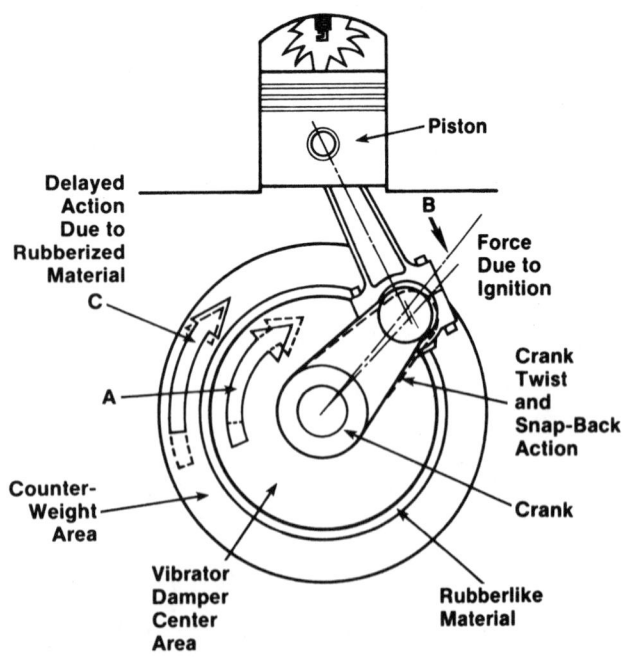

FIGURE 2-73 Operation of vibration damper

withstand combustion pressure and has the advantage of light weight. The aluminum piston's disadvantage is that it will expand more than the iron or iron and steel pistons. To counteract this, aluminum pistons usually have steel struts or inserts cast into them to help hold heat and contain the piston expansion.

The top of the piston is called the *head* or *dome*. Just below the dome on the side of the piston is a series of grooves. The grooves are used to contain the piston rings; the high parts between the grooves are called *ring lands*. Below the grooves, as shown in Figure 2–75, there is a bore, or hole, which is used for the piston pin, sometimes called the wrist pin. The pin is used to connect the piston to the connecting rod. This hole is not always centered in the piston; it can be offset to one side. The piston pin offset is toward the major thrust side of the piston, the side that will contact the cylinder wall during the power stroke. By offsetting the pin, piston slap caused by the piston changing direction in the cylinder is eliminated.

 SHOP TALK _____

The term piston slap or bang is used to describe the noise made by the piston when it contacts the cylinder wall. This noise is usually heard only in older, high-mileage engines that have worn pistons or cylinder walls.

To ensure that the piston is installed correctly so the offset is on the proper side, the top of the piston will have a mark. The most common mark is a notch, machined into the top edge of the piston. This mark must always face the timing chain end of the engine when the piston is installed. The base of the piston, the area below the piston pin, is called the *piston skirt*. The area from just below the bottom ring groove to the tip of the skirt is the piston thrust surface. There are two basic types of piston skirts: the slipper type and the full skirt. The full skirt is used primarily in truck and commercial engines; the slipper type is used for automobile engines. The slipper-type skirt allows the piston enough thrust surface for normal operation and has the advantage of allowing the piston to be lighter. This design also cuts down on piston expansion because there is less material to hold heat. When an engine is designed, the piston expansion determines how much piston clearance will be needed in the cylinder bore. Too little clearance will cause the piston to bind at operating temperature; too much will cause piston slap. The normal piston clearance for an automobile en-

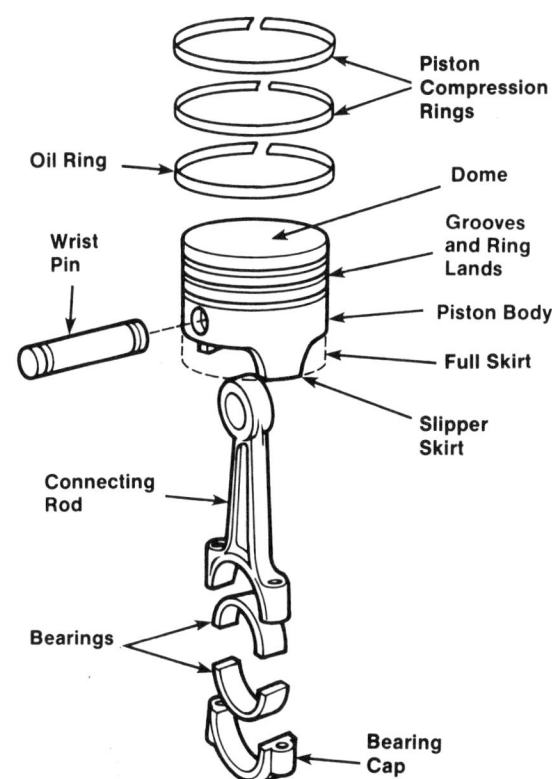

FIGURE 2–75 Components of a piston

gine, using slipper skirted pistons, is about 0.001 to 0.002 inch. This clearance is measured cold between the piston skirt and the cylinder wall. Because of this necessary expansion clearance, the piston does not seal the combustion area. This is the job of the piston rings. The piston clearance also supplies a space for piston lubrication.

The fitting or matching of the piston to its cylinder bore is a very important step in engine servicing. If the clearance between the piston and cylinder is too large, the piston will "slap"; if the clearance is too small, the piston will bind in the bore and cause early piston failure.

Methods of measuring piston bores and installing the proper size are covered in Chapters 7 and 12.

PISTON RINGS

Piston rings are used to fill the expansion gap between the piston and cylinder wall. It is the piston rings that seal the combustion chamber at the piston. Piston rings, which are available in several coatings, must serve three functions:

1. Seal the combustion chamber at the piston
2. Remove oil from the cylinder walls. This is necessary to keep oil from reaching the upper cylinder where it will be burned.

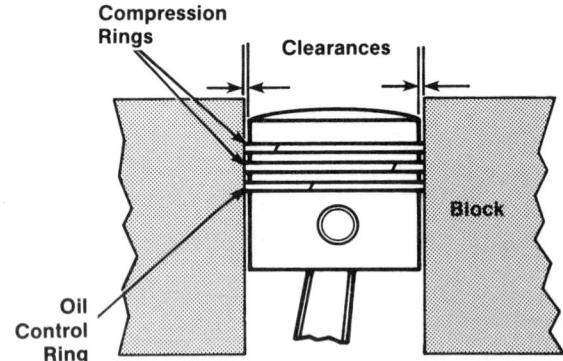

FIGURE 2-76 Piston is filled with one oil control ring and two compression rings.

3. Carry heat from the piston to the cylinder walls to help cool the piston

There are two basic ring families: compression rings and oil control rings. In most modern automobile engines the pistons will be fitted with two compression rings and one oil control ring (Figure 2-76). The compression rings are found in the two upper grooves closest to the piston head. The oil ring is fitted to the groove just above the wrist pin.

Compression Rings

The compression rings form the seal between the piston and cylinder walls (Figure 2-77). They are designed to use combustion pressure to force the ring against the cylinder wall and against the bottom edge of the ring groove. The top ring is the primary seal with the second ring used to seal any small amount of pressure which may reach it. The compression rings are also used to remove excess oil from the cylinder walls and aid in cooling the piston carrying its heat to the cylinder walls.

During the power stroke, the pressure caused by the expanding air/fuel mixture is applied between the inside of the ring and the piston groove. This forces the ring into full contact with the cylinder walls. The same force is applied to the top of the ring forcing it against the bottom of the ring groove. The combination of combustion pressure and the compression ring join together to form a tight ring seal.

Compression rings are generally made of cast iron. They come in a number of variations in cross section design. Most compression rings have a coating on their face, which will aid in the wear in process. Wear in is the time necessary for the rings to conform to the cylinder wall. Typical soft coatings are graphite, phosphate, iron oxide, and molybdenum. Some compression rings have a hard coating, such as chromium.

Oil Control Rings

Oil is constantly being applied to the cylinder walls. The oil is used for lubrication as well as to clean the cylinder wall of carbon and dirt particles. This oil bath will also aid somewhat in cooling the piston. Controlling this oil is the primary function of the oil ring. The two most common types of oil rings are the segmented oil ring and the cast-iron oil ring. Both types of rings are slotted so that excess oil from the cylinder wall can pass through the ring. The oil ring groove of the piston is also slotted. After the oil passes through the ring it can then pass through the slots in the piston and return to the oil sump through the open section of the piston.

Segmented oil rings are made up of three pieces: an upper and lower scraper and an expander. The scrapers and expanders are made of spring steel. Segmented oil rings are a free fit design. This means that the expander is larger than the diameter of the cylinder. They must be carefully compressed when installed. The expander is used to form a tight seal of the scrapers at the cylinder walls.

CONNECTING ROD

The connecting rod is used to transmit the pressure applied on the piston to the crankshaft (Figure 2-78). The rod must be very strong and at the same time be kept as light as possible. Connecting rods are generally forged from high-strength steel. The center section is made in the form of an "I" for maximum strength with minimum weight. The small end or piston pin end is made to accept the piston pin. The pin is used to connect the piston and connecting rod. The rod must be free at the piston to move back and forth as the crankshaft rotates. The piston pin can be a pressed fit in the piston and free fit in the rod. When this is the case, the small end of the rod will be fitted with a bushing. The pin can also be a free fit in the piston and pressed fit in the rod; in this case no bushings are used. The pin simply moves in the piston using the piston hole as a bearing surface.

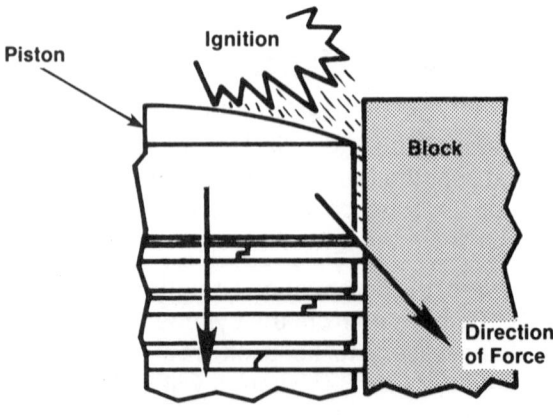

FIGURE 2-77 Compression ring

A third mounting allows the pin to move freely in both the piston and the rod.

The larger end of the rod is used to attach the connecting rod to the crankshaft. This end is made in two pieces. The upper half is part of the rod, the lower half is called the *rod cap* and is bolted to the rod. The connecting rod and its cap are manufactured as a unit and must always be kept together. The large crankshaft end of the rod is fitted with bearing half shells made of the same material as the main bearings (Figure 2–79). As mentioned earlier in the chapter, the crankshaft has oil passages that lead to the crank pins for lubricating the rod bearings. Some connecting rods have a hole drilled through the big end to the bearing area. The bearing insert might have a hole, which will align with this drilling. This hole is used to supply oil for lubricating and cooling the piston skirt. When the rod is properly installed, the oil hole should be pointing to the major thrust area of the cylinder wall.

ENGINE GASKETS AND OIL SEALS

Figure 2–80 shows the various gaskets used on a typical engine. To the uneducated eye, a gasket is just a piece of die-cut metal or other material. Its sole purpose is to be sandwiched between two pieces of metal to seal the mating surfaces. However, to the engine technician the selection of gaskets for a high-tech engine is a very important decision.

Oil seals are used for the rear of the crankshaft and timing gear cover to prevent oil from escaping around a moving shaft. They are passed into a stationary bore in the block or cover. Complete information on the sealing of engines with gaskets and oil seals can be found in Chapter 14.

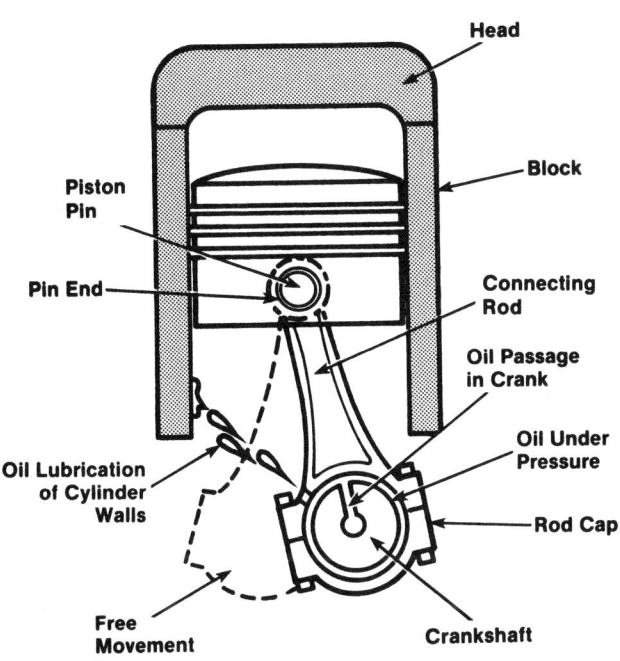

FIGURE 2-78 Connecting rod operation

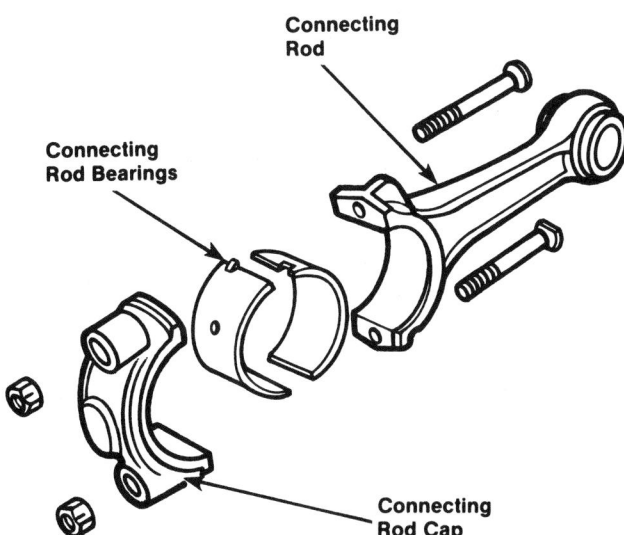

FIGURE 2-79 Connecting rod assembly showing bearings

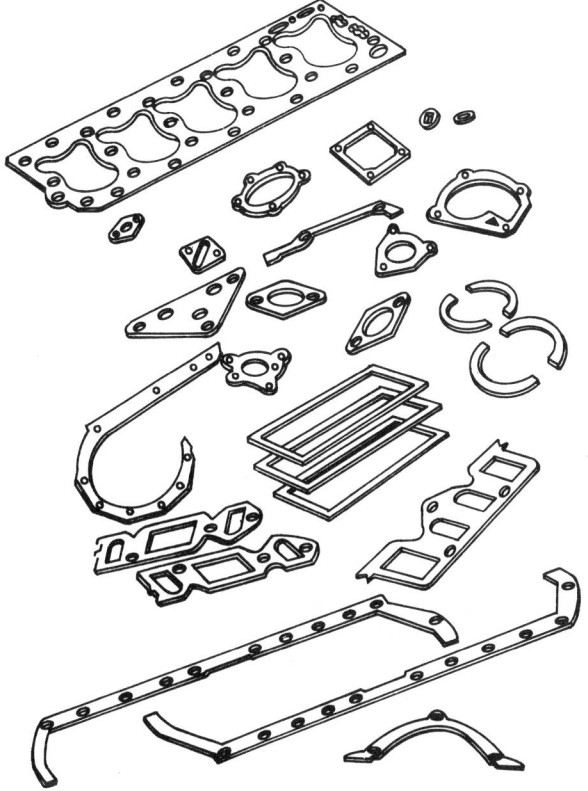

FIGURE 2-80 Engine gaskets

GASOLINE TWO-STROKE ENGINES

In the past, several imported vehicles have used two-stroke engines. As the name implies, this engine requires only two strokes of the piston to complete all four operations: intake, compression, power, and exhaust. As shown in Figure 2-81, this is accomplished as follows:

1. Movement of the piston from BDC to TDC completes both intake and compression (Figure 2-81A).
2. When the piston nears TDC, the compressed air/fuel mixture is ignited, causing expansion of the gases. Note that the reed valve is closed (Figure 2-81B).
3. Expanding gases in the cylinder force the piston down, rotating the crankshaft. Downward movement of the piston compresses the air/fuel mixture in the crankcase (Figure 2-81C).
4. With the piston at BDC, the intake and exhaust ports are both open, allowing exhaust gases to leave the cylinder and air/fuel mixture to enter (Figure 2-81D).

Although the two-stroke-cycle engine is simple in design and lightweight because it lacks a valve train, it has never been seriously considered for automotive applications for several reasons. Poor cylinder scavenging and blowback prevent a two-stroke from breathing as efficiently as a four-stroke; thus, fuel consumption is higher.

The difference in breathing characteristics also means a four-stroke develops more torque at lower rpm than a two-stroke, making a four-stroke better suited for automotive use. Two-strokes tend to be high revving engines that develop peak power too far up the rpm scale for everyday driving. A two-stroke also leaks a fair amount of its intake charge into the exhaust port, creating high hydrocarbon emissions in the exhaust.

In recent years, however, thanks to a revolutionary pneumatic fuel injection system, there has been increased interest in the two-stroke engine. The injection system, which works something like a spray paint gun, uses compressed air to blow highly atomized fuel directly into the top of the combustion chamber. The system was designed to improve the fuel efficiency of the original orbital engine, but soon proved to be the long sought-after answer to overcoming the fuel economy and emissions problems of the conventional two-stroke engine. Thus

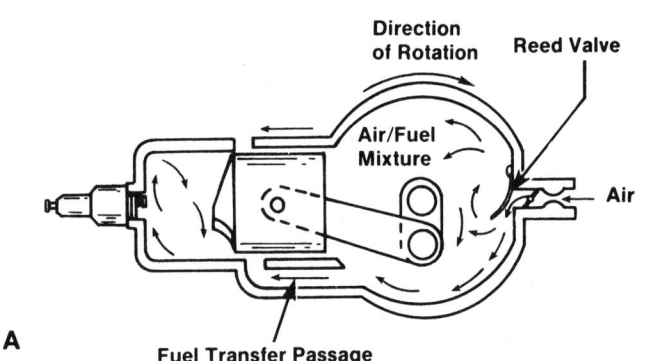

A

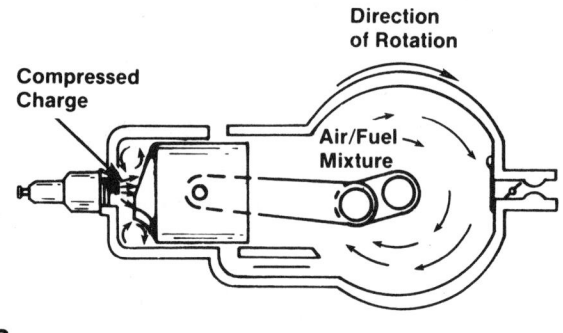

B

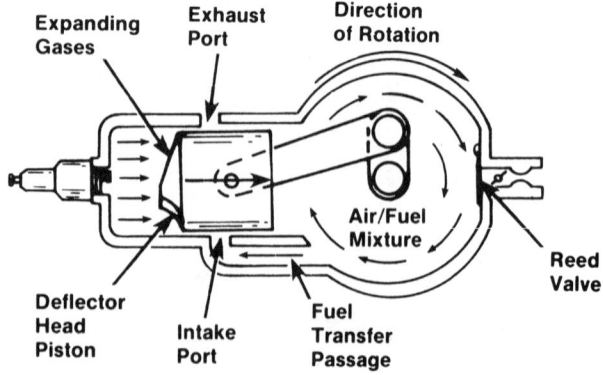

C

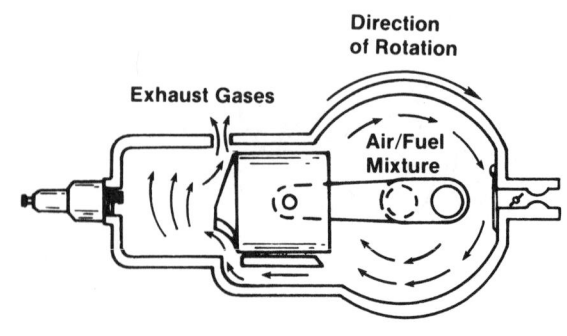

D

FIGURE 2-81 Operation of two-stroke engine

evolved the orbital two-stroke direct injection piston engine.

A two-stroke can deliver as much horsepower as a larger displacement four-stroke engine because a two-stroke fires every crankshaft revolution rather than every other revolution (Figure 2–82). That means a smaller displacement two-stroke engine can be used in the same vehicle application. The improvement in fuel economy the orbital has achieved, therefore, is due in part to a number of mechanical design features in addition to its pneumatic direct injection fuel system.

One of these is the use of a three-cylinder engine block (Figure 2–83). Though the same direct injection technology can be used with a four-cylinder (or a V-6, both of which orbital has already done), a three-cylinder block saves the weight, bulk, and manufacturing expense of an extra cylinder. The reduction in internal engine friction by eliminating the fourth piston and the valve train, according to the orbital, is in itself worth a significant improvement in fuel economy (Figure 2–84).

Other design features found in the orbital two-stroke to reduce frictional losses include the absence of oil scraper, or piston, rings on the pistons and the use of roller crankshaft bearings rather than plain bearings. Lubrication is controlled electronically to reduce oil consumption.

The oiling system uses a separate oil reservoir, which meters oil into the engine as required. An indicator light mounted in the vehicle's dash alerts the driver when to add oil. The three-cylinder two-stroke oil reservoir has a 4-quart capacity and can log 8000 miles before additional oil is required.

Additional performance improvements are obtained in the orbital two-stroke design by using a high-turbulence combustion chamber that pro-

FIGURE 2–83 A three-cylinder two-stroke orbital engine (left) puts out as much power as a four-cylinder four-stroke engine (right).

FIGURE 2–84 A three-cylinder two-stroke engine leaves plenty of room in the engine compartment of a small car.

motes rapid mixing at high and low loads. The engine also has an exhaust port scavenge flow control valve, which is controlled by the engine computer to increase low-speed torque and assist in emissions control. The valve can be partially closed to restrict the flow of exhaust out of the cylinder under certain operating conditions, creating an exhaust gas recirculation effect to reduce NO_x.

GASOLINE ENGINE SYSTEMS

Besides the major power-generating system in the gasoline engine, there are, of course, several other systems essential to the engine operation.

FIGURE 2–82 Cutaway of two-stroke engine

- *Air/Fuel System.* This system ensures that the engine gets the right amount of both air and fuel needed for efficient operation. These two systems join in the carburetor, which blends the fuel and air and supplies the resulting mixture to the cylinder. Some automobiles have a fuel injector system that replaces the carburetor.
- *Ignition System.* This system supplies a precisely timed spark to ignite the compressed air/fuel mixture in the cylinder at the end of the compression stroke. The ignition cylinder firing order is determined by the engine manufacturer and can be found in the vehicle's service manual. Typical firing orders given in a manual are illustrated in Figure 2–85.
- *Lubrication System.* The system supplies oil to the various moving parts in the engine. The oil lubricates all parts that slide in or on other parts, such as the piston, bearings, crankshaft, and valve stems. The oil enables the parts to move easily so that little power is lost and wear is kept to a minimum. The lubrication system also helps transfer heat from one part to another for cooling.

- *Cooling System.* This system is also extremely important. Coolant circulates in jackets around the cylinder and in the cylinder head. This removes part of the heat produced by the combustion of the air/fuel mixture and prevents the engine from being damaged by overheating.
- *Exhaust System.* This system efficiently removes the burned gases and limits noise produced by the engine. Important parts of the exhaust system in many new engines are the superchargers and turbochargers.
- *Emission Control System.* Several control devices, which are designed to reduce emission levels of combusted fuel, have been added to the engine. Engine design changes, such as reshaped combustion chambers and altered tune-up specs, have also been implemented to help control the auto's smog-producing by-products. These devices and adjustments have reduced emissions considerably but have changed automotive engine servicing and rebuilding to a great extent.

In this book, these systems are described only when used in conjunction with the engine and its rebuilding.

DIESEL ENGINES

Diesel engines (Figure 2–86) and gasoline-powered engines share several similarities. They have a number of components in common, such as the crankshaft, pistons, valves, camshaft, and water and oil pumps. They both are available in four-stroke combustion cycle models. However, the diesel engine, and four-stroke or compression-ignition engine, is easily recognized by the absence of the conventional ignition components found on gasoline engines. Instead of relying on a spark for ignition, the diesel engine uses the heat produced by compressing air in the combustion chamber to ignite the fuel. The engine systems used in diesel-driven vehicles are essentially the same as those used in gasoline types.

Figure 2–87 shows one cycle of a four-stroke diesel engine operation. On the intake stroke of a diesel engine, the piston is pulled down in the cylinder by the crankshaft and connecting rod. During this time, the camshaft holds open the intake valve and air is drawn into the cylinder due to the vacuum caused by the downward-moving piston. The compression stroke begins when the piston starts to

Firing Order 1-5-3-6-2-4

6 CYLINDER

Right Bank
Left Bank
Firing Order 1-5-4-8-6-3-7-2
V-8

Right Bank
Left Bank
Firing Order 1-8-4-3-6-5-7-2
V-8

Firing Order 1-3-4-2
1-2-4-3
4 CYLINDER

Right Bank
Left Bank
Firing Order 1-4-5-2-3-6
V-6

Right Bank
Left Bank
Firing Order 1-6-5-4-3-2
V-6

Right Bank
Left Bank
Firing Order 1-8-7-2-6-5-4-3
V-8

Right Bank
Left Bank
Firing Order 1-5-4-2-6-3-7-8
V-8

FIGURE 2–85 Engine manufacturers use a variety of cylinder numbering systems and firing orders. Some of the most common type are shown here.

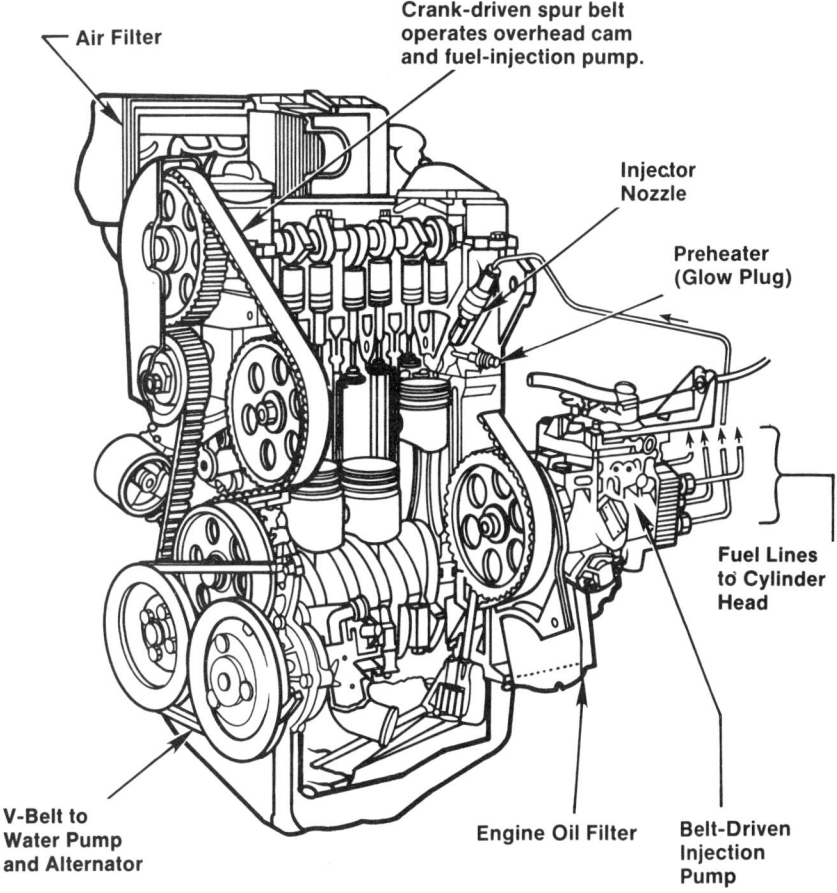

Air Filter

Crank-driven spur belt
operates overhead cam
and fuel-injection pump.

Injector
Nozzle

Preheater
(Glow Plug)

Fuel Lines
to Cylinder
Head

V-Belt to
Water Pump
and Alternator

Engine Oil Filter

Belt-Driven
Injection
Pump

FIGURE 2-86   Typical diesel engine

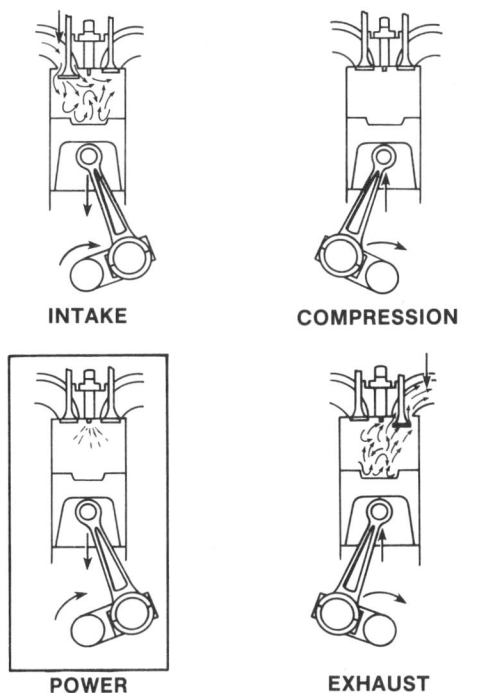

INTAKE

COMPRESSION

POWER

EXHAUST

FIGURE 2-87   Four-stroke diesel engine in operation

move upward. The intake valve closes and air is trapped in the cylinder above the piston. By the time the piston reaches top dead center, the air is compressed to approximately 1/20 of its original volume, a compression ratio of 20:1. Compression ratios can vary from 13:1 to about 25:1. At the end of the compression stroke, fuel is sprayed into the combustion chamber. Because of the amount of friction between the air molecules in a high-compression situation, sufficient heat is generated to ignite the fuel when it is injected. Thus, a diesel engine does not require a spark ignition system. On the power stroke, the piston is forced downward by the combustion of the air/fuel mixture. As the mixture burns, it expands, exerting pressure against the piston. Combustion occurs at a controlled rate. The exhaust stroke begins when the camshaft opens the exhaust valve and the piston starts to move upward, forcing the exhaust gases out of the cylinder. At the end of the intake stroke, the exhaust valve closes and the entire cycle begins again.

Because fuel timing, rather than ignition timing, is essential for proper diesel operation, fuel injection is used with all diesel engines. The amount of fuel sent to the cylinders determines the power and speed developed by the engine.

A diesel engine does not use spark plugs; instead, it uses glow plugs. These plugs serve only to warm the combustion chamber when the engine is cold. Cold starting is impossible without these plugs because even the high-compression ratios cannot heat the air sufficiently for combustion. To start the engine, a starting motor is used to turn the crankshaft and to begin piston movement. Once the power strokes generate enough energy to continue crankshaft rotation, the starter can be disengaged by releasing the key that activates it.

Diesel combustion chambers are different from gasoline combustion chambers. Diesel fuel burns differently so the combustion chamber must be different. Three types of combustion chambers are used in diesel engines: open combustion chamber, precombustion chamber, and turbulence combustion chamber. The open combustion chamber has the combustion chamber located directly inside the piston. Figure 2-88 shows the open combustion chamber with diesel fuel being injected directly into the center of the chamber. The shape of the chamber and the quench area produces turbulence. The precombustion chamber shown in Figure 2-89 is used on both the gas and diesel engines. A smaller, second chamber is connected to the main combustion chamber. On the power stroke, fuel is injected into the small chamber. Combustion is started and then spreads to the main chamber. This design allows lower fuel injection pressures and simpler injection

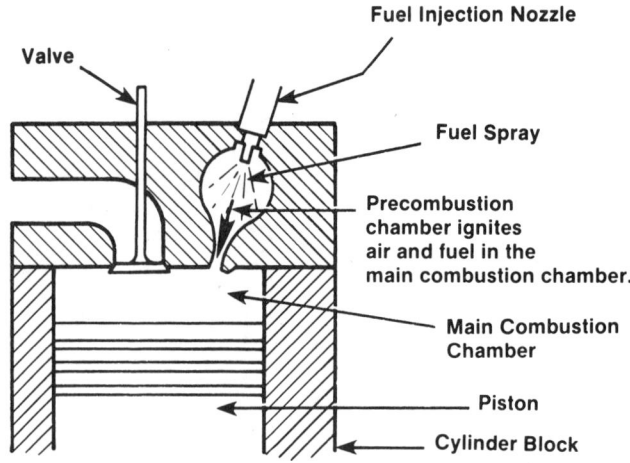

FIGURE 2-89 Precombustion chambers are used to ignite air and fuel in a small prechamber. The combustion in this chamber ignites the air and fuel in the main combustion chamber.

systems on the diesel engines. On gas engines, the precombustion chamber has a very rich mixture, but the main chamber can be very lean. The overall effect is a leaner engine, producing better fuel economy. The turbulence combustion chamber is shown in Figure 2-90. The chamber is designed to create an increase in air velocity or turbulence in the combustion chamber. The fuel is injected into the turbulent air and burns more completely.

The prechambers on a diesel engine head must be properly indexed with the head and correctly installed (Figure 2-91). They must be perfectly flush (not above or below) with the cylinder head. Failure to follow this will cause the head gasket to fail.

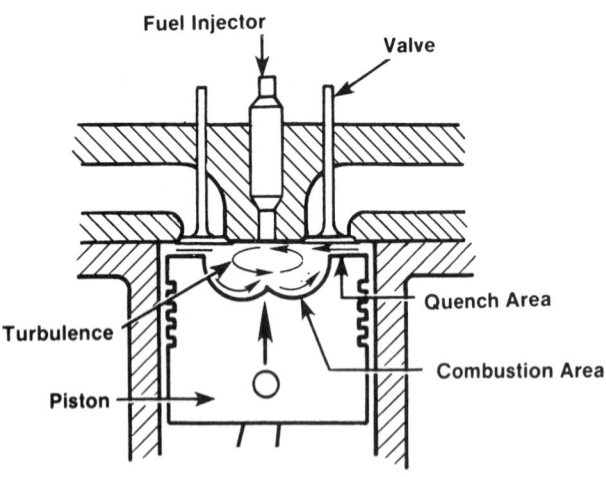

FIGURE 2-88 On certain diesel applications, the open combustion chamber is located directly within the top of the piston. This design has both a quench and turbulence area.

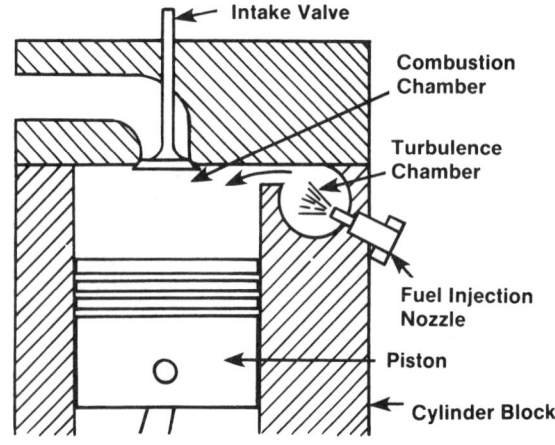

FIGURE 2-90 The turbulence combustion chamber is used on certain diesel applications to increase turbulence of air and fuel.

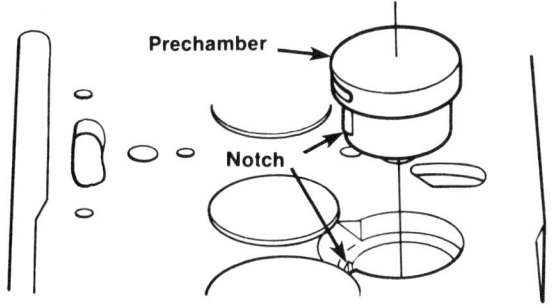

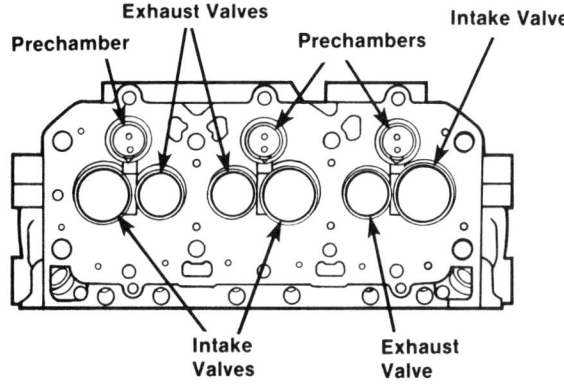

FIGURE 2-91 Prechambers correctly installed

Table 2-1 compares the gasoline and diesel four-stroke-cycle engines. Remember that diesel engines are also available in two-stroke-cycle models (Figure 2-92).

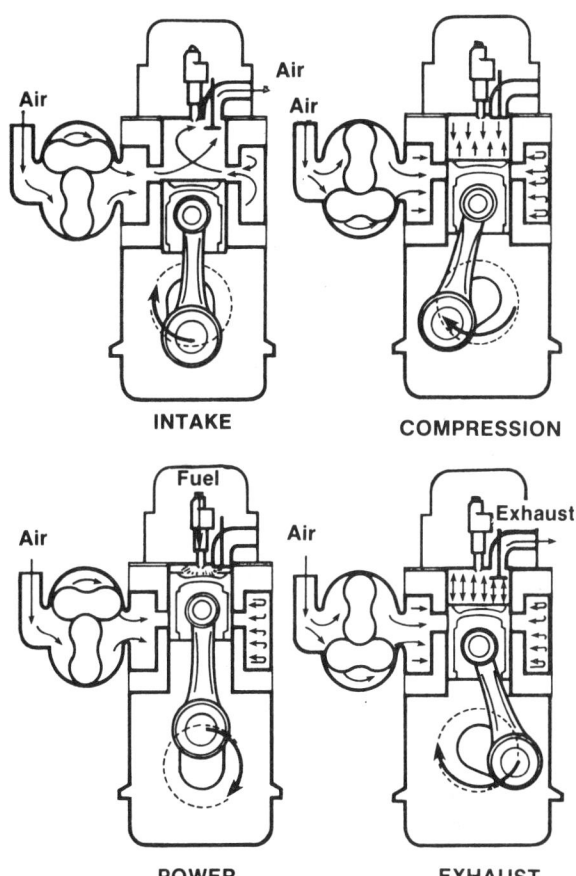

FIGURE 2-92 Two-stroke diesel engine in operation

TABLE 2-1: COMPARISON BETWEEN GASOLINE AND DIESEL ENGINES		
	Gasoline	**Diesel**
Intake	Air/fuel	Air
Compression	8–10 to 1 130 psi 545° F	13–25 to 1 400–600 psi 1,000° F
Air/fuel mixing point	Carburetor or before intake valve with fuel injection	Near TDC by injection
Combustion	Spark ignition	Compression ignition
Power	464 psi	1,200 psi
Exhaust	1,300°–1,800° F CO = 3%	700°–900° F CO = 0.5%
Efficiency	22–28%	32–38%

REVIEW QUESTIONS

1. How many revolutions does the crankshaft make during one cycle in a four-stroke engine?
   a. one
   b. two
   c. four
   d. eight

2. Which of the following configurations has no moving parts in the cylinder head?
   a. flat-head engine
   b. overhead valve
   c. overhead cam
   d. none of the above

3. How many revolutions does the camshaft make with each complete four-stroke cycle?
   a. one
   b. two

c. four
d. eight

4. Which has the larger diameter?
   a. intake valve head
   b. exhaust valve head
   c. poppet valve head
   d. all are the same diameter

5. Why is a close clearance between the valve guides and stems so important?
   a. It keeps oil from getting into the combustion chamber.
   b. It keeps exhaust from getting into the crankcase.
   c. It keeps the valve face in perfect alignment with the valve seat.
   d. All of the above

6. How many rod bearing journals does a V-8 engine have?
   a. one
   b. two
   c. four
   d. eight

7. What is the top of the piston called?
   a. head
   b. dome
   c. both a and b
   d. neither a nor b

8. The noise made by the piston when it contacts the cylinder wall is called

   _____ .

   a. pinging
   b. knocking
   c. piston slap
   d. none of the above

9. Which of the piston rings is the primary seal?
   a. top compression ring
   b. bottom compression ring
   c. oil control ring
   d. none of the above

10. Which of the following is part of a segmented oil ring?
    a. upper scraper
    b. lower scraper
    c. expander
    d. all of the above

11. Which is the most common type of valve arrangement?
    a. L
    b. I
    c. T
    d. F

12. Technician A says that displacement is the amount of air displaced when the piston moves from TDC to BDC. Technician B says it is the amount of air displaced when the piston moves from BDC to TDC. Who is right?
    a. Technician A
    b. Technician B
    c. Both A and B
    d. Neither A nor B

13. Technician A says proper diesel operation depends on ignition timing. Technician B says proper diesel operation depends on fuel timing. Who is right?
    a. Technician A
    b. Technician B
    c. Both A and B
    d. Neither A nor B

14. In recent years, there has been increased interest in the two-stroke engine because of

    _____ .

    a. stricter pollution regulations
    b. the demand for more fuel-efficient vehicles
    c. a revolutionary pneumatic fuel injection system
    d. none of the above

15. Which of the following is used in a diesel engine?
    a. open combustion chamber
    b. precombustion chamber
    c. turbulence combustion chamber
    d. all of the above

CHAPTER THREE

DIAGNOSING BEFORE ENGINE TEARDOWN

Objectives

After reading this chapter, you should be able to
- Prepare for and perform both a dry and wet compression test.
- Prepare for and perform a cylinder leakage test.
- Describe the necessary test precautions for a cylinder power balancing test and perform the test.
- Perform a vacuum test and a cranking vacuum test.
- Evaluate the engine's condition.
- Determine the cause of excessive oil consumption.
- List and describe nine abnormal engine noises.
- List and perform the maintenance methods used to restore proper engine performance.

Correct analysis and diagnosis of engine performance problems are very important to the rebuilder. The results of the analyzing and diagnosing testing will determine the extent and degree of rebuilding necessary. Each year, countless engines are needlessly overhauled or rebuilt as a result of a hasty decision made off the top of the head of the rebuilder. If the same rebuilder had only taken the time to verify and prediagnose the customer's complaint, the cost to the owner might have been considerably reduced. It is very important to keep in mind that the success of any rebuilding business depends to a great degree on customer relations and the quality of the service.

A gasoline engine performance problem can usually be related to any problem or adverse condition that affects the power, fuel economy, emission output levels, and engine dependability. Use a systematic approach rather than hit-or-miss when locating performance problems. Most legitimate engine rebuilding operations stem from a complaint in one or more of the following areas:

- Power or performance
- Oil consumption
- Engine noise

With modern test equipment it is possible to determine fairly accurately which mechanical repairs the engine requires before it is dismantled. If available, an analyzer (Figure 3–1A), chassis dynamometer (Figure 3–1B), gas analyzer (Figure 3–1C), or borescope (Figure 3–1D) are very useful in diagnosing poor performance. However, testing with a compression gauge and a vacuum gauge plus careful visual and audible observation will provide to be valuable information, and removal of the oil pan will reveal other facts. An experienced and educated ear is one of the best diagnostic tools anyone can have (Figure 3–2).

WARNING: A hot exhaust or cooling system and other engine components can cause severe burns. Avoid touching hot engine surfaces during test procedures. Also be sure to avoid moving engine parts.

COMPRESSION TESTING

The amount of compression generated in each cylinder has a great effect on the total power output

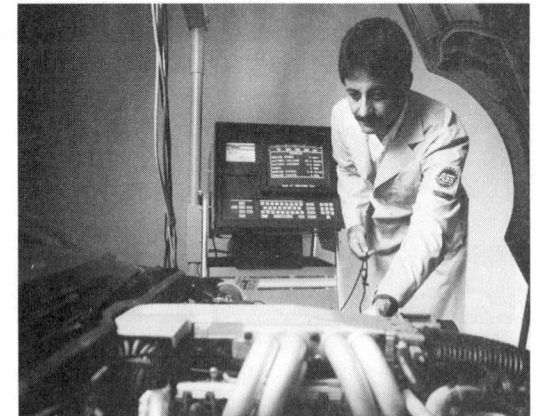

A

B

C

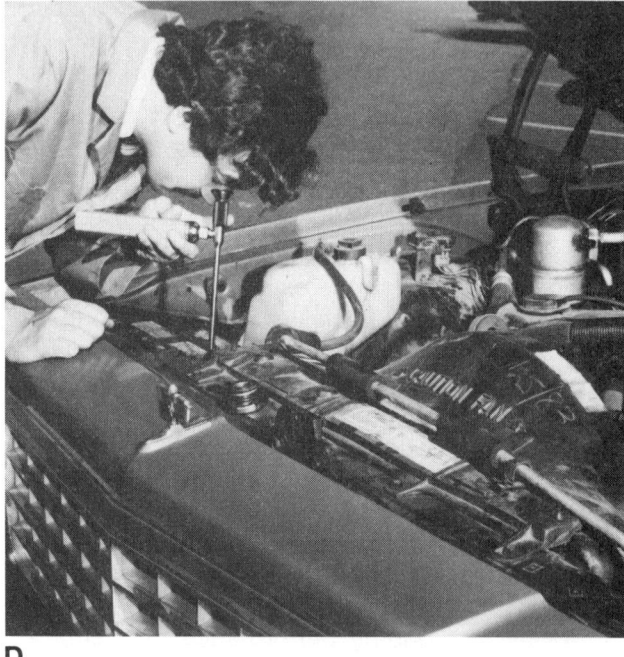

D

FIGURE 3-1 (A) Making an engine diagnosis test with an engine analyzer *(Courtesy of Hamilton Test Products, Inc.);* (B) typical dynamometer in use *(Courtesy of Clayton Industries, Inc.);* (C) typical four-gas analyzer *(Courtesy of Bear Automotive Service Equipment Company);* and (D) checking inside an engine with a fiber-optic borescope *(Courtesy of Lenox Instrument Company)*

of an engine. Likewise, the amount of compression depends on how well the cylinder is sealed by the cylinder head gasket, piston rings, and valves (Figure 3-3). A leak at any of these points will result in a loss of compression and poor driveability. A com-

pression test measures the compression in each individual cylinder. It also points out any significant pressure variations between all of the cylinders.

Every car manufacturer provides compression-pressure specifications for its engines. Always com-

FIGURE 3-2 An experienced, well-trained ear is an invaluable diagnostic tool. *(Courtesy of Perfect Circle/Dana)*

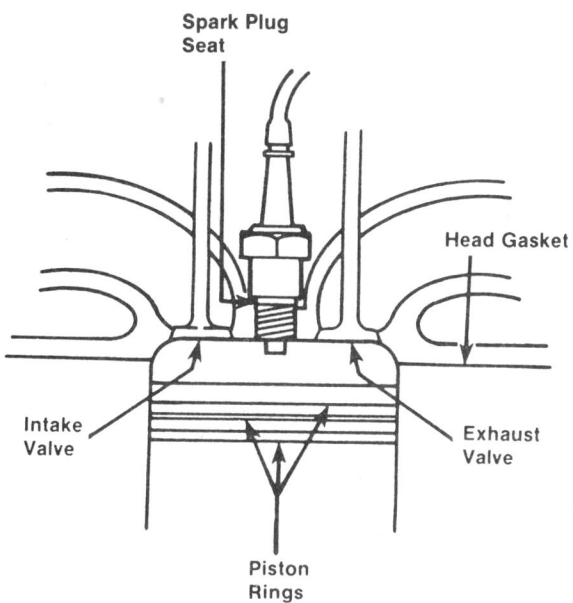

FIGURE 3-3 The amount of compression depends on how well the cylinder is sealed at various points.

pare the compression test results with these figures. The specifications are usually given one of two ways: percentage or minimum.

- A typical percentage specification might state that the compression pressure be 150 psi plus or minus 10 percent. This plus or minus tolerance is crucial because it limits the pressure variation between cylinders. Another percentage specification might specify that the lowest compression reading of any cylinder can be no less than 75 percent of the highest reading. For example, if the

highest reading is 200 psi, the lowest must be at least 150 psi.

- A minimum figure specification generally also includes an allowable pressure variation between cylinders. For example, the minimum compression pressure for an engine might be 150 psi, with a variation of no more than 30 psi. In this case, each cylinder must have a compression pressure of at least 150 psi, but no higher than 180 psi.

PREPARING FOR THE TEST

The two special tools needed to do a compression test are the compression gauge and remote starter switch. Always be sure that the tools used are high quality and in good condition; otherwise, the test results will be invalid.

Compression Gauge

For gasoline engines, the typical compression gauge can measure pressures up to 300 psi. Diesel engines require specialized gauges that measure higher pressures. A popular compression gauge design features a threaded adapter that fits into the spark plug hole; a hose connects the adapter to the gauge (Figure 3-4). The vent valve traps the pressure in the gauge until the reading is taken.

Another type of compression gauge utilizes a rubber tip that fits on the end of a metal stem; the stem threads into the gauge (Figure 3-5). The tips fit any size spark plug hole, making this gauge very adaptable. The stems come in various lengths and shapes, a very useful feature for working around obstructions. A vent valve is also used on this gauge.

FIGURE 3-4 Compression gauge

FIGURE 3-5 Holding the rubber tip on the spark plug opening is necessary with some compression gauges.

To take a reading, the rubber tip must be manually held in the spark plug opening, which can be rather difficult when working with a high-compression engine.

The procedure for testing compression on a diesel engine is basically the same as that for a gasoline engine. The major exception is that the compression gauge for making the test must register at least 500 to 600 psi. A normal automotive gauge will not do the job because it does not register high enough. The diesel gauge is inserted into the glow plug hole after the glow plug is removed. Some glow plugs might require a special tool to remove them. Because of the unique design of the rotary engine, special equipment is necessary to measure the rotary engine's compression.

Remote Starter Switch

A remote starter switch can be an individual tool or part of a compression gauge. Since it permits the engine to be cranked from under the hood, it is an indispensable tool when working alone. In addition to checking the compression, tasks such as adjusting valves and setting ignition points can be done without the aid of a second person.

WARNING: When working under the hood with the engine running, use extreme care when working near moving parts.

READYING THE ENGINE

Use the following procedure to prepare an engine for a compression test:

1. Run the engine until it reaches its normal operating temperature, then turn it off.
2. Disconnect the spark plug cables from the spark plugs.
3. Use compressed air to clean all dirt and other foreign matter out of the spark plug wells (Figure 3-6).

WARNING: Be sure to wear eye protection.

4. Remove all of the spark plugs.
5. Be sure to remove all spark plug gaskets from the cylinder head.
6. Remove the air cleaner from the carburetor and block the choke and throttle plate in the wide-open position.
7. Remove the ignition coil's high-tension wire from the distributor cap and ground it (Figure 3-7). By disabling the ignition system in this manner, the chance of electrical shock or fire is greatly reduced. However, do not leave the plug disconnected too long or you risk overheating the catalytic converter. The rebuilder can also damage certain electronic and HEI systems; consult the service manual for information.

 SHOP TALK _____

Be sure to check the service manual before removing or making any ignition system inoperative.

8. Connect a remote starter switch to the starter relay or solenoid (Figure 3-8). Attach one of its leads to the positive battery terminal and the other to the S-terminal on the relay or solenoid.

DRY COMPRESSION TEST

Use the following procedure to perform a dry compression test:

1. Depending on the type of gauge being used, either thread the adapter into the

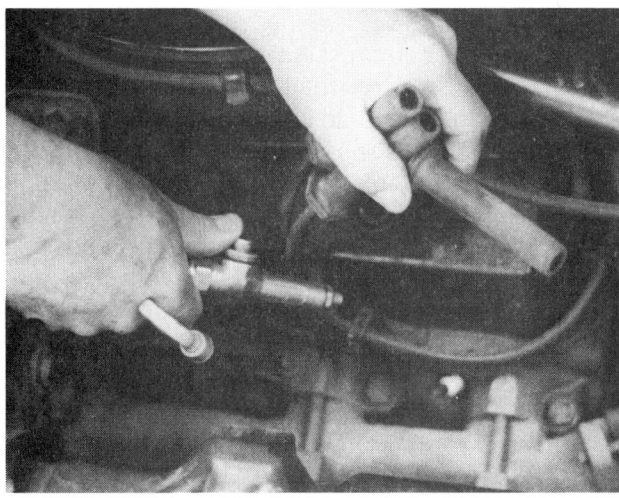

FIGURE 3-6 Using compressed air to clean out the spark plug walls

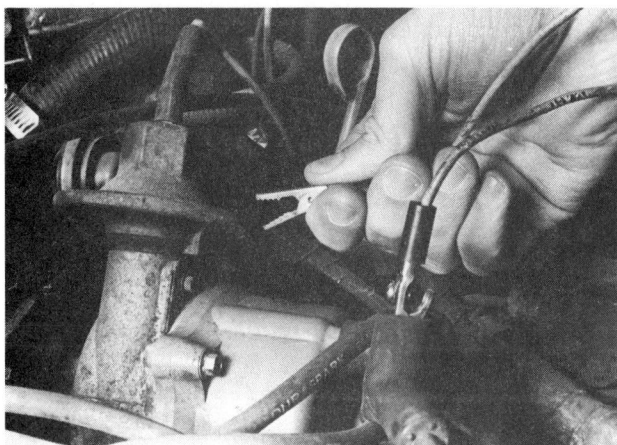

FIGURE 3-7 Remove the ignition coil's wire and ground it.

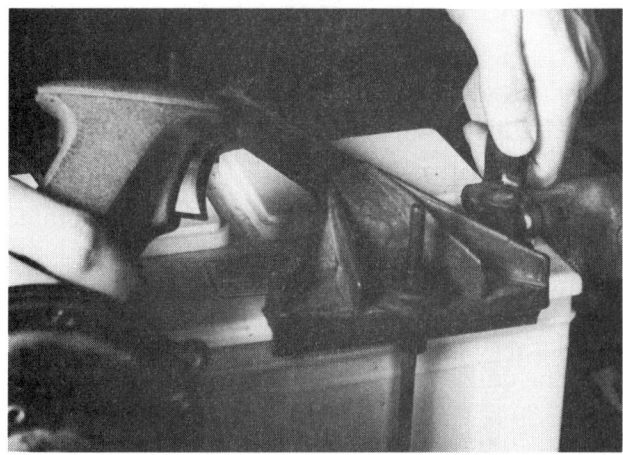

FIGURE 3-8 Connecting a remote starter switch

spark plug hole, fingertighten, and connect the gauge to the adapter, or insert the rubber tip into the hole and hold it firmly in place by hand.

2. Turn the engine over four full compression strokes, keeping an eye on the gauge needle at all times.
3. Record the first and fourth gauge readings. The needle should rise steadily with each subsequent stroke. The fourth and final reading should be well within specifications.
4. Open the vent valve to release the compression pressure.
5. Disconnect and remove the gauge.
6. Repeat the procedure on the remaining cylinders.

To interpret the results of the dry compression test, use the following guidelines:

- If any reading is low on the first stroke but gradually gets higher without reaching the specified pressure, the piston rings are probably worn badly.
- If any reading is low on the first stroke and gets only a little higher during subsequent strokes, the cause may be a sticking or burned valve.
- If the reading is equally low on two adjacent cylinders, the probable cause is a leaking head gasket.
- If the total gauge reading is higher than specified, excessive carbon deposits in the combustion chamber are the likely cause.

Any cylinders that record a low pressure reading should be given a wet compression test.

WET COMPRESSION TEST

A wet compression test should be performed if a dry test is inconclusive or shows a problem. However, if a cylinder has zero psi, there is no need to conduct a wet test.

A wet compression test is performed in the following manner:

1. Squirt one or two ounces of medium-viscosity oil into the cylinder through the spark plug hole (Figure 3-9).
2. Turn the engine over several times; this will allow the oil to work its way down around the rings.

FIGURE 3-9 Oil must be added to the cylinder before performing the wet compression test.

3. Connect the compression gauge to the cylinder and conduct the compression test as described earlier for the dry test procedure.

Guidelines for the wet compression test are as follows:

- If the readings are now normal or very close to normal, either the rings, pistons, or cylinders need servicing.
- If the readings do not improve at all, the leakage is probably at the valves or head gasket.
- If the readings improve only slightly, the rings and valves should both be looked at carefully.

 SHOP TALK —————

The wet compression test should not be performed on horizontally opposed cylinder engines such as those found on Subarus or older Volkswagens. For this engine, a leakage test is the most accurate method of discovering the source of low compression.

From the information gathered from the dry and wet compression tests a great deal can be learned about the general condition of the engine. However, since it is almost impossible to give any hard and fast answers, use the following interpretations as a guide to help support and confirm any conclusions that you might have about compression:

1. If the gauge reading rises steadily during each compression stroke and the final compression readings are all within the range specified by the manufacturer, then compression is good.
2. After completing a minimum of four compression strokes, if the compression is low on the first stroke but gradually builds and improves to nearly normal with the addition of oil, the problem is a piston, piston ring, or cylinder wear.
3. A compression reading that stays low throughout all strokes and does not improve when you add oil indicates valve trouble.
4. Low compression readings in two adjacent cylinders accompanied by water or oil-fouled spark plugs is a sure sign of a bad head gasket.
5. Higher than normal compression readings are generally caused by excessive combustion chamber deposits.

CYLINDER LEAKAGE TESTING

Although the compression test gives a good indication of the amount of pressure in the cylinders, the leakage test provides a more accurate method of testing engine condition. Even minute leaks in the valves, rings, or head gasket can be detected and measured. The leakage test also points out leaks around the exhaust and intake valves, leaks between cylinders, leaks into the water jacket, and any other cause of compression loss.

LEAKAGE TESTER

The leakage tester contains a precision gauge for extremely accurate readings (Figure 3-10). Its scale ranges from zero to 100 percent; zero means that the cylinder is perfectly sealed, 100 indicates that it is holding no air at all. To receive as accurate a reading as possible, the gauge should be calibrated before every leakage test. To do this, first turn the control regulator knob counterclockwise until it rotates freely. Connect a 70 to 200 psi air supply to the tester's air input fitting. Turn the knob clockwise until the gauge reads zero. Connect and disconnect a test adapter to the tester's cylinder connection fitting. The gauge should rise to 100 percent, then return to zero; if it does not, the control regulator knob must be readjusted.

FIGURE 3-10 Leakage tester *(Courtesy of Snap-on Tools Corp.)*

PREPARING FOR THE TEST

To prepare an engine for a cylinder leakage test, do the following:

1. Check the coolant level; fill if needed.
2. Run the engine until it reaches its normal operating temperature, then shut it off.
3. Disconnect the spark plug cables from the spark plugs.
4. Use compressed air to clean all dirt and other foreign matter out of the spark plug wells.
5. Remove all the spark plugs as well as any gaskets or tubes that may have been used.
6. Remove the air cleaner from the carburetor.
7. Disconnect the positive crankcase ventilation (PCV) hose from the crankcase.

PERFORMING THE TEST

In addition to the tester, a top dead center (TDC) indicator and indicator light are also needed to conduct a cylinder leakage test. Always begin at the number 1 cylinder. Use the following procedure:

1. Install the proper test adapter hose in the number 1 cylinder spark plug hole. Connect the tester whistle to the adapter hose (Figure 3-11).
2. Use a wrench on the crankshaft pulley nut or bolt to slowly rotate the engine in the normal direction until the whistle blows, which indicates the beginning of a compression stroke.

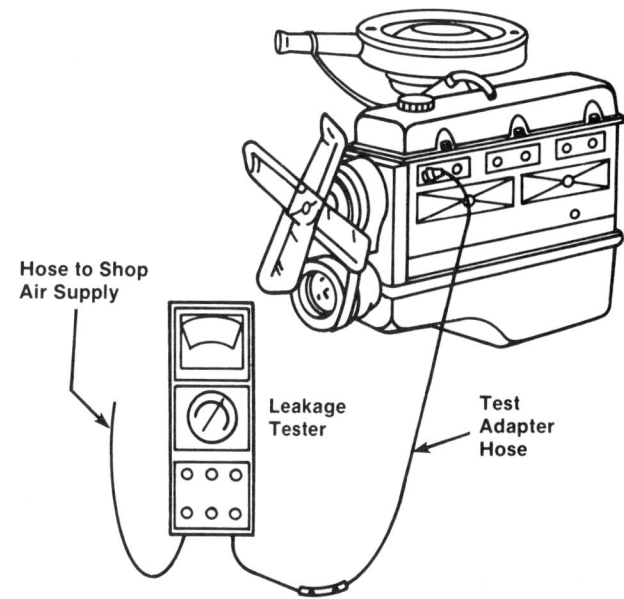

FIGURE 3-11 Leakage tester and whistle connected to the number 1 spark plug hole via test adapter hose

3. Continue the rotation until the timing mark on the crankshaft pulley lines up with the engine-timing pointer on the timing chain cover. Remove the whistle from the adapter.
4. Use a jumper lead to connect the coil-to-distributor secondary cable to a good ground.
5. Remove the distributor cap and rotor, then mount a TDC indicator (Figure 3-12) on the distributor shaft. Mark a chalk reference point on the engine that lines up with the appropriate cylinder marking on the TDC indicator.
6. Connect the tester to the adapter hose. If the gauge shows more than 20 percent leakage, air might be escaping through the carburetor, exhaust pipe, and oil filler opening or crankcase breather cap.
7. Disconnect the tester from the adapter hose. Resume rotating the engine until the next appropriate cylinder mark on the TDC indicator lines up with the chalk mark on the engine. The indicator light should glow, meaning that the piston is in firing position.
8. Remove the adapter from the previously tested cylinder. Install it in the spark plug hole of the next cylinder in the engine's

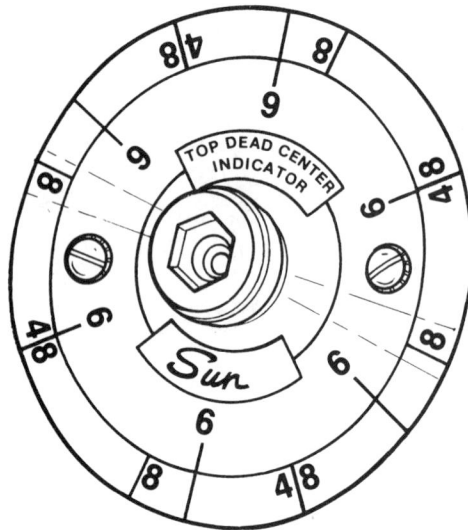

FIGURE 3-12 TDC indicator identifies the compression stroke of the next cylinder

FIGURE 3-13 Air bubbles in the radiator can indicate a crack in the engine block or head or a leaking head gasket.

firing order. (The piston in this cylinder is at top dead center.)

9. Repeat steps 7, 8, and 9 on all of the cylinders.

WARNING: Remember that the engine could rotate suddenly while a cylinder is under pressure.

All readings should be relatively even and less than 20 percent. Any reading of 30 percent or higher indicates a definite problem. Use a hydrosonic leak detector and the following guidelines to determine where the leak is and its cause:

- Air escaping from the exhaust pipe means that an exhaust valve is leaking.
- Air escaping through the carburetor means that an intake valve is leaking.
- Air bubbles in the radiator indicate a crack in the engine block or head, or a leaking head gasket (Figure 3-13). When removing a radiator cap, do so slowly and carefully.
- Leakage from two adjacent cylinders might also indicate a cracked block or head or a leaking blown head gasket.
- Air escaping through the oil filler cap or crankcase breather cap means the piston rings are worn or damaged, or the cylinder wall or piston is damaged.
- Crankcase leakage could also mean worn piston rings or cylinder walls or a cracked piston (Figure 3-14). However, keep in mind

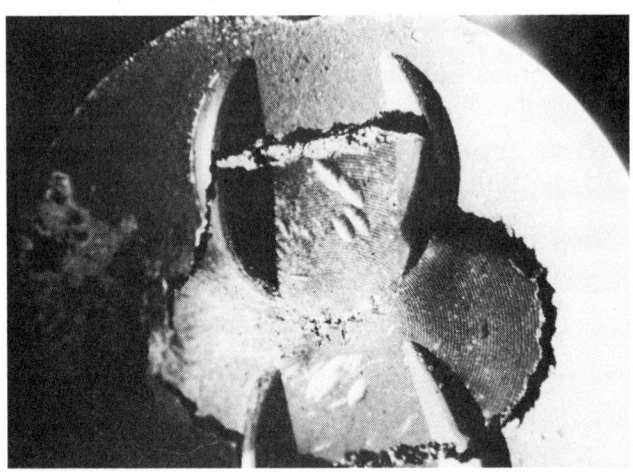

FIGURE 3-14 A cracked piston can cause crankcase leakage.

the age of the engine; if it is relatively new, the problem could just be that the rings are stuck or not yet seated.

In many cases, it is useful to compare results of compression and cylinder leakage tests in order to best analyze a difficult or unusual mechanical engine problem. For example, an older engine whose compression reading is within specifications but has air escaping from the crankcase opening is probably just showing signs of old age. High-mileage engines typically suffer from excessive blowby and poor gas mileage, as well as heavy carbon deposits in the combustion chambers. These deposits can raise the compression readings, thus creating the impression that the engine is more mechanically sound than it really is.

Low compression combined with minimal cylinder leakage usually means a valve-related problem. An improperly installed timing chain or gears, a loose timing chain, the wrong type of camshaft, or worn camshaft lobes can also be the cause of this situation. In all of these cases, the valves are not opening and closing at the correct point in the engine cycle.

If the compression test and leakdown test prove to be acceptable, then the problem is not mechanical. A problem could exist with the ignition, fuel, or computer system of the engine.

VACUUM TESTING

An engine vacuum test is one of the quickest and easiest ways to test an engine. Since any piston-driven engine is basically a combination vacuum pump and heat exchanger, it is the higher pressure on the outside of the cylinders that causes air to be pushed into the cylinders. When an engine loses the ability to create this pressure differential, its performance suffers.

CAUTION: Like compression and leakage tests, vacuum testing alone should not be used to locate the exact source of a problem. Always perform it as part of a series of diagnostic tests.

The vacuum gauge itself may be a separate tool or it can be part of an engine analyzer. Increments are in inches of mercury (Hg), and range from zero to 30. With the engine running, the "normal" range is between 18 and 20 inches Hg, with most of today's emission-controlled cars falling on the low end of this scale (Figure 3-15). It is important to keep in mind that all vacuum gauge readings are dependent on altitude: for every 1,000 feet above sea level, the reading will be low by 1 inch. Therefore, add 1 inch Hg to the reading for every 1,000 feet above sea level.

PERFORMING THE TEST

Use the following procedure to perform a vacuum test:

1. Use a length of hose to connect the vacuum gauge to a nonrestricted port on the intake manifold. The hose should be at least 3 feet long in order to dampen vibrations from the needle.

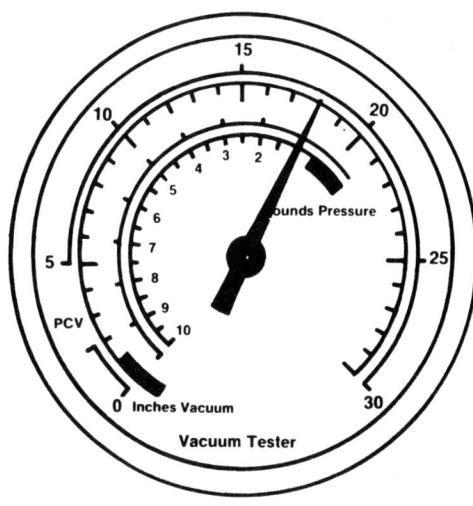

FIGURE 3-15 A normal vacuum gauge reading

2. In some cases, it may be necessary to further dampen the needle by clamping the hose; this will slightly restrict its passageway.
3. Run the engine until it reaches its normal operating temperature. With a few exceptions, vacuum tests must always be performed with the engine at idle rpm; consult a service manual for details.
4. If the needle remains constant between 15 and 20 inches Hg, the engine vacuum is good. In addition to engines with emission control devices, new or recently overhauled engines also have a tendency toward lower vacuum readings.

Because the results can be so varied, it is very important to interpret the vacuum test correctly. Use Table 3-1 to arrive at a diagnosis. Figure 3-16 shows typical vacuum diagnosis readings. The white pointer indicates a steady reading; the black pointer indicates a fluctuating movement.

If the valve guides are suspect, check them by warming up the engine to its normal operating temperature. Turn off the engine, then remove the valve covers and squirt oil over the tops of the guides. Turn the engine back on; if blue smoke exits the exhaust pipe and the gauge needle steadies, the guides are definitely worn. As for intake system leaks, they are usually caused by defective intake manifold or carburetor mounting gaskets, or defective vacuum hoses. To test for vacuum leaks at the gaskets, run the engine at idle rpm and squirt cleaning solvent along the gasket joints. If the vacuum increases and the idle smoothes out, the leak has been found.

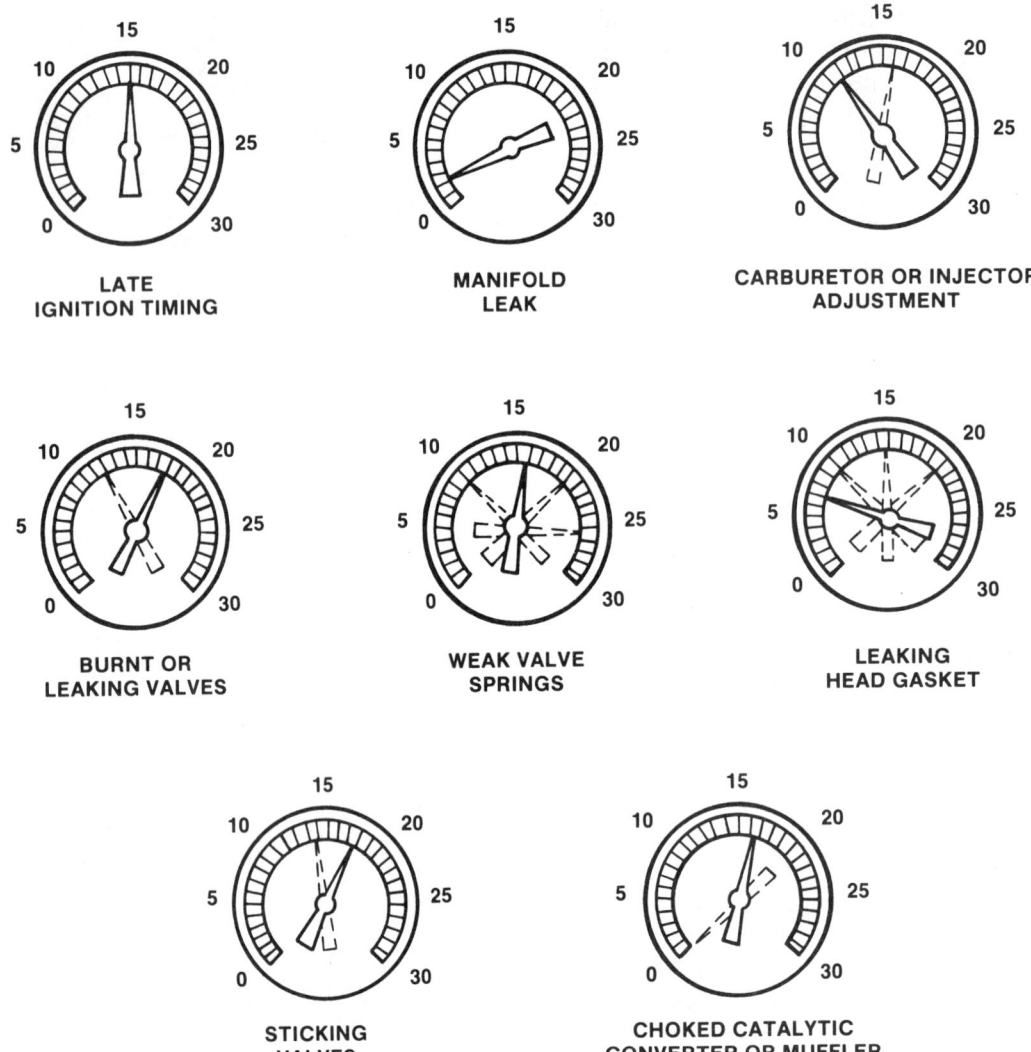

LATE
IGNITION TIMING

MANIFOLD
LEAK

CARBURETOR OR INJECTOR
ADJUSTMENT

BURNT OR
LEAKING VALVES

WEAK VALVE
SPRINGS

LEAKING
HEAD GASKET

STICKING
VALVES

CHOKED CATALYTIC
CONVERTER OR MUFFLER

FIGURE 3-16 Typical vacuum diagnosis readings

CAUTION: Be sure the cleaning solvent used on the gasket joints is noncombustible. Otherwise, the risk of an engine fire is great.

If the lower edges of the intake manifold gasket are not accessible, this method of finding intake leaks might not work; this is often the case with V-type and in-line engines (Figure 3-17).

CRANKING VACUUM TESTING

To perform a cranking vacuum test, follow this procedure:

1. Run the engine until it reaches normal operating temperature, then turn it off.

FIGURE 3-17 On in-line engines like this, the normal method of finding intake leaks often cannot be used.

TABLE 3-1: VACUUM TEST DIAGNOSIS

Reading	Possible Cause	Remedy
Low but steady, between 12 and 15 inches Hg	Leakage around piston rings, late ignition timing, or late valve timing	Replace piston rings or reset timing.
Needle oscillates slowly, then rapidly, between 12 and 18 inches Hg	Ignition timing too far advanced or carburetor idle mixture too lean	Reset timing or carburetor idle mixture.
Regular needle drop between 1 and 2 inches Hg	Burned or leaking valve or spark plug in one of the cylinders is not firing	Replace valve or spark plug.
Irregular needle drop between 1 and 2 inches Hg	Sticking valve, carburetor out of adjustment, or intermittent spark plug misfire	Replace valve, adjust carburetor, or replace spark plug.
Normal at idle speed, but excessive vibrations at higher rpm	Weak valve springs	Replace valve springs.
Excessive vibrations at idle speed, but steadies at higher rpm	Worn valve guides	Replace valve guides.
Excessive vibration at all rpm	Leaky head gasket	Replace head gasket.
Needle oscillates between 3 and 9 inches Hg lower than normal	Intake system leak	Replace faulty component.
Normal at idle speed, but drops to near zero and rises to lower than normal	Restriction in exhaust system	Repair or replace exhaust system.

2. Connect the vacuum gauge hose to a source of engine manifold vacuum (Figure 3-18).
3. Disable the ignition system. On a vehicle with a separate ignition coil, do this by removing the coil wire from the distributor cap and connecting a jumper lead from the coil wire to an engine ground. On a vehicle without a separate ignition coil, disconnect the electrical connector that supplies battery voltage to the system.

CAUTION: Make sure the vacuum gauge hose is well away from the belts, pulley, and fan before cranking the engine.

4. Use a remote starter switch to crank the engine, but do not depress the accelerator pedal.
5. Note the vacuum reading. An engine in good condition should produce a reading of at least 5 inches.
6. If the reading is less than 5 inches, check for external leakage. Inspect the intake manifold gasket and carburetor gasket, and look for broken or disconnected vacuum lines.

EXHAUST SYSTEM TESTING

A vacuum gauge can also be used to test the exhaust system for restriction, as follows:

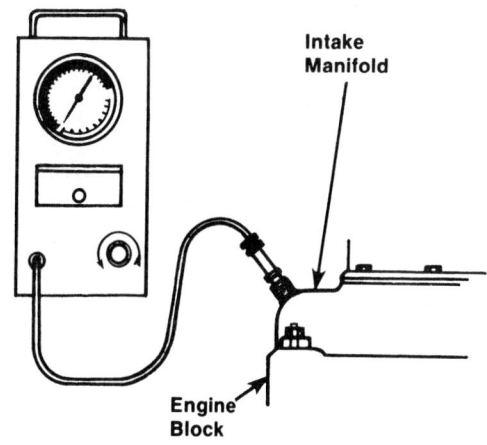

FIGURE 3-18 Making the vacuum gauge connection

1. Attach the gauge to the intake manifold as described earlier, then connect a tachometer to the engine.
2. Run the engine until it reaches its normal operating temperature.
3. Accelerate the engine slowly. After it reaches 2,000 rpm, note the reading on the gauge; it should drop a little, then rise sharply.
4. Close the throttle quickly. The needle should return to the normal idle reading as quickly as it rose.
5. If the needle gives a normal reading at idle speed and at 2,000 rpm, drops to near zero, and rises to a below normal reading, there is a restriction in the exhaust system. Check the muffler and tailpipe for damage, and inspect the manifold heat control valve to see if it is stuck or frozen.

If working with a large displacement engine, it will probably be necessary to do this test while driving the vehicle. Proceed as follows:

1. Connect the gauge to the intake manifold. Use a hose long enough so that the gauge is inside the vehicle.
2. Connect a tachometer to the engine; route its wires so that it is also entirely inside the vehicle (Figure 3–19).
3. Drive the vehicle until it reaches the speed where the engine loses power. The gauge reading should be approximately the same each time a power loss occurs, with the needle dropping toward zero. The greater

FIGURE 3–19 Connecting a tachometer to the engine

the restriction, the closer to zero the needle will drop.

VACUUM TESTING FOR LOSS OF COMPRESSION

A vacuum gauge can even be used to test for compression loss due to leakage around the pistons. However, this test should not be performed unless normal readings were produced on all previous vacuum tests. Conduct the test as follows:

1. Make sure the engine oil level is full and the oil is not too old. Dirty oil can cause an incorrect vacuum reading.
2. Connect the vacuum gauge to the intake manifold, and attach a tachometer to the engine.
3. Quickly accelerate the engine to 2,000 rpm, then close the throttle fast.
4. As the throttle closes, the needle should rise 5 inches Hg or more above the normal reading. An increase of less than 5 inches Hg means there is a compression loss around the pistons, rings, or cylinder walls.

CYLINDER POWER BALANCE TESTING

The cylinder power balance test is useful in determining if a cylinder or bank of cylinders is producing its share of engine power. Ideally, all the cylinders should be doing the same amount of work, and changes in engine rpm should be about equal as each cylinder is shorted out. Unequal cylinder power can mean a problem in the cylinders themselves, as well as the rings, valves, intake manifold, head gasket, fuel system, or ignition system.

The power balance test is performed quickest and easiest using an engine analyzer, because the spark plugs can be controlled with push buttons. Changes are measured in "rpm drop." Keep in mind that the push-button numbers refer to the cylinder firing order, not the cylinder number designation. For example, when testing an engine with a firing order of 1-3-4-2, pushing the first button shorts out the number 1 cylinder, pushing the second button shorts out the number 3 cylinder, and so on.

TEST PRECAUTIONS

On some computer-controlled or fuel-injected engines, certain components must be disconnected before attempting the power balance test. Because of the wide variation from manufacturer to manufacturer, consult the vehicle's service manual for specific instructions.

If the engine being tested has an exhaust gas recirculation (EGR) system, added precautions must be taken. If the system is valve controlled, disconnect the vacuum or electrical connection to the EGR valve. This will prevent the valve from cycling due to vacuum changes when the cylinders are shorted out. For engines with a floor jet EGR system, the power balance test cannot be performed accurately because of the possibility of the unburned fuel mixture being sent back into the cylinders. The compression test is the recommended alternative in such cases.

Care must be taken when performing the power balance test on vehicles with catalytic converters. To prevent unburned fuel from building up in the converter, short each spark plug for less than 15 seconds, then wait another 30 seconds before shorting another one.

 SHOP TALK —————

The old style power balance testers, on which a single knob controls individual spark plugs, might not be safe for use on vehicles with electronic ignitions. The ignition system or solid state components could be damaged, not to mention the possibility of inaccurate test results. Make certain the tester being used is compatible with electronic ignition systems.

PERFORMING THE TEST

The standard power balance test is fairly simple. Use the following procedure:

1. If the engine has an air/fuel mixture feedback control or O₂ sensor (Figure 3-20A), disconnect and plug either the air pump hose going to the catalytic converter or the downstream hose between the air switching valve and the check valve. On some Ford models, the air switching valve can routinely have up to 10 percent leakage, in which case both hoses must be disconnected and plugged.

2. If the engine does not have an air/fuel mixture feedback control or O₂ sensor (Figure 3-20B), disconnect and block the air pump on the valve side.

3. Override the controls of the electric cooling fan by jumper wiring the controls so that the fan runs constantly. If the fan cannot be bypassed, disconnect it, but be careful that the engine does not overheat during the test.

4. Connect the engine analyzer's leads, referring to the vehicle's service manual for specific instructions.

5. Turn on the engine and let it reach its normal operating temperature before beginning the test. Engine speed should be stabilized at approximately 1,000 rpm. When a cylinder is shorted, note any drop in rpm or manifold vacuum.

As each cylinder is shorted out, a noticeable drop in engine speed should occur. Little or no decrease in rpm indicates a weak cylinder. If all the readings are fairly close to each other, the engine is in sound mechanical condition; if a reading in one or

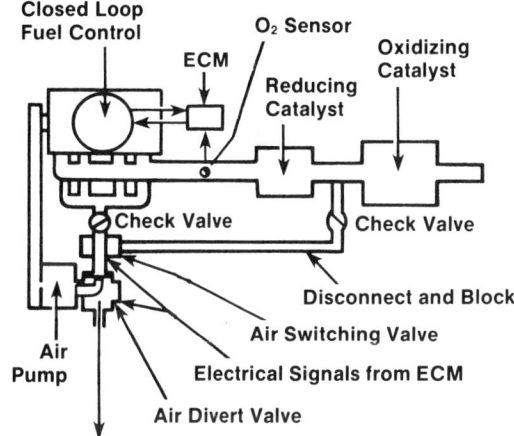

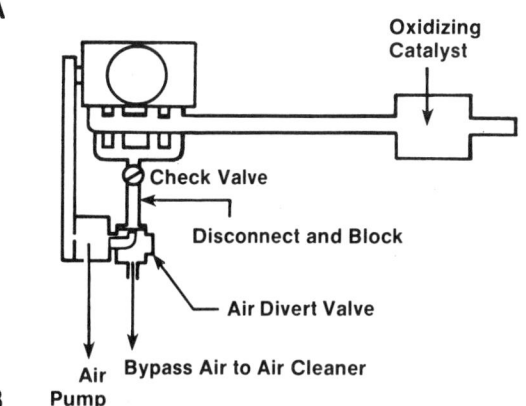

FIGURE 3-20 A cylinder power balance test differs, depending on whether the engine (A) has an air/fuel mixture feedback control or (B) does not.

more cylinders differs greatly from the rest, there is a problem. Further testing should be done to determine if the problem is purely mechanical, or if it is in the ignition or fuel system.

EVALUATING THE ENGINE'S CONDITION

Once the compression tests (including the leakage test), vacuum tests, and power balance tests are performed the rebuilder should be ready to start evaluating the engine's condition. For example, an engine with good relative compression but heavy cylinder leakage past the rings is typical of a high-mileage engine that is worn out. Other symptoms include excessive blowby, lack of power, poor performance, and reduced fuel economy.

If these same compression and leakage conditions are noted on an engine with comparatively low mileage, the problem is most likely that the piston rings are stuck and not expanding properly. If such is the case, try treating the engine with a combustion cleaner, oil treatment, or engine flush. If this effort proves futile, the only other remedy requires a complete engine disassembly.

A cylinder that has poor compression but minimal leakage indicates a valve train problem. Under these circumstances, a valve might not be opening at the right time, might not be opening enough, or might not be opening at all. This condition can be confirmed on engines with a pushrod-type valve train by pulling the rocker covers and watching the valves operate while the engine is cycled. If one or more valves fail to move, either the lifters are collapsed or the cam lobes are worn. If all of the cylinders have low compression with minimal leakage, the most likely cause is incorrect valve timing.

If compression and leakage are both good, but the power balance test revealed weak cylinders, the cause of the problem is outside the combustion chamber. Assuming there are no ignition or fuel problems, check for broken, bent, or worn valve train components, collapsed lifters, leaking intake manifold, and/or excessively leaking valve guides. If the latter is suspected, squirt some oil on the guides. If they are leaking, in some of the older systems, blue smoke will be seen in the exhaust.

OIL CONSUMPTION

Excessive oil consumption can be a result of external and internal leaks, faulty accessories, piston rings, and valve guides. Internal leaks (Figure 3-21), which usually result in oil burning, are usually more difficult to diagnose and the problems are solved by a teardown of the engine.

To start an oil consumption diagnosis, examine the engine thoroughly for external leaks. These leaks can occur externally, at the valve cover gasket, camshaft expansion plug, oil filter, front and rear oil seals, oil pan gasket, fuel pump gasket, and timing gear cover.

Even the smallest oil leak can cause major oil consumption. Three drops of oil lost externally every 100 feet amounts to 3 quarts every thousand miles. External leaks occur under two different conditions:

1. Normal crankcase pressure
2. Abnormal crankcase pressure

Normal crankcase pressure will cause oil leaks at gaskets or past metal-to-metal joints which are in direct contact with oil. Worn seals, faulty gaskets, and loose cover or housing bolts could be the problem. Fresh oil on the clutch housing, oil pan (Figure 3-22), fuel pump, edges of valve covers, external oil lines, distributor shaft housing, base or crankcase filler tube, or at the bottom of the timing gear or chain cover usually indicates a leak close to that point.

Check for proper oil pressure at the sending unit passage with an externally mounted mechanical oil pressure gauge (Figure 3-23) (as opposed to relying on an OE installed dash-mounted gauge). To insure accurate test results, follow the car manufacturer's pretest recommendations and compare your readings to the minimum/maximum oil pressure specs listed in the manual. Low oil pressure readings can be attributed to internal component wear, pump-related problems, a low oil level, or oil viscosity that

FIGURE 3-21 Indication of internal oil leak condition *(Courtesy of Perfect Circle/Dana)*

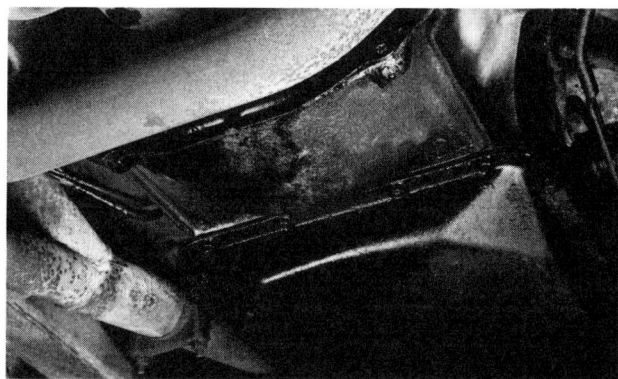

FIGURE 3-22 Fresh oil on the clutch housing or oil pan usually indicates a leak *(Courtesy of Perfect Circle/Dana)*

FIGURE 3-23 Typical mechanical oil pressure gauge. *(Courtesy of Snap-On Tools Corp.)*

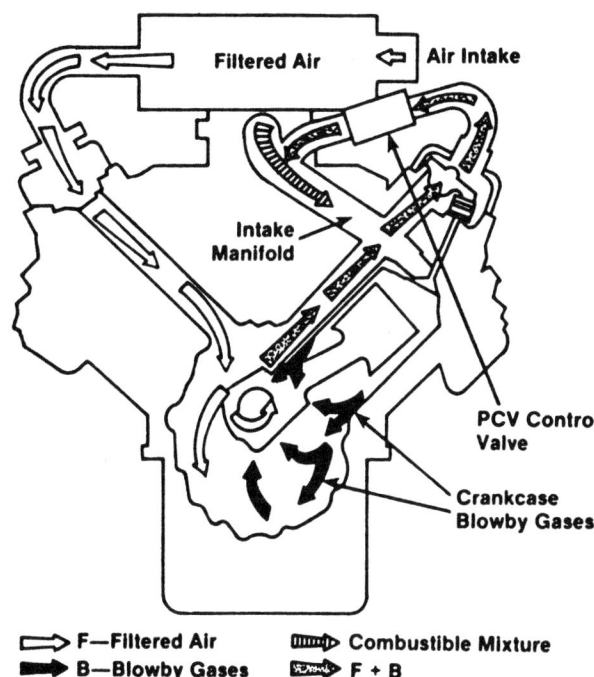

F—Filtered Air   Combustible Mixture
B—Blowby Gases   F + B

FIGURE 3-24 Operation of a PCV system

 SHOP TALK _____

Internal leaks are frequently the result of aluminum intake manifolds on V-6 and V-8 engines because aluminum often does not seal as well as cast iron. If an unacceptable amount of warpage is found, removing and milling the manifold will be necessary. Sometimes properly installing a new intake manifold will cure the problem.

is too low. Conversely, a pressure reading that is too high could be caused by an overfilled crankcase, too high of an oil viscosity, the wrong pump (remote but possible, especially if the pump has been replaced), or a faulty pressure relief valve.

When crankcase pressure is abnormal, oil is forced out through joints that normally would not leak. Pressure develops when the crankcase ventilator inlet becomes clogged, when blowby becomes excessive, or when a positive crankcase ventilation (PCV) valve is malfunctioning. The latter system provides a continuous flow of fresh air through the crankcase to inhibit formation of corrosive contaminants (Figure 3-24). It is important that the correct replacement valve is installed because each is designed for a particular engine's operating characteristics. Use of the wrong valve can cause oil consumption. If the PCV valve or connecting hoses become clogged, excessive pressure will develop in the crankcase, which might force oil into the air cleaner or cause it to be sucked into the intake manifold. This problem can be prevented by maintenance

of the system and replacement of the PCV valve as recommended by the vehicle manufacturer.

Most oil consumption problems, as shown in Figure 3-25, can be corrected by proper rebuilding of the engine. But two common causes of oil consumption often overlooked are thinning of the oil resulting from higher than normal operating temperature and oil dilution by unburned fuel, which mixes with the oil.

Oil viscosity, or thickness, decreases as the engine temperature increases. Thinner oil works its way more easily into the combustion chambers where it is burned. Abnormally high oil temperatures can develop under some circumstances by such conditions as towing a large trailer with a vehicle that is not adequate for the job. Assuming that the vehicle is otherwise capable of handling the load, proper installation of an oil cooler will help control oil temperature (see Chapter 13).

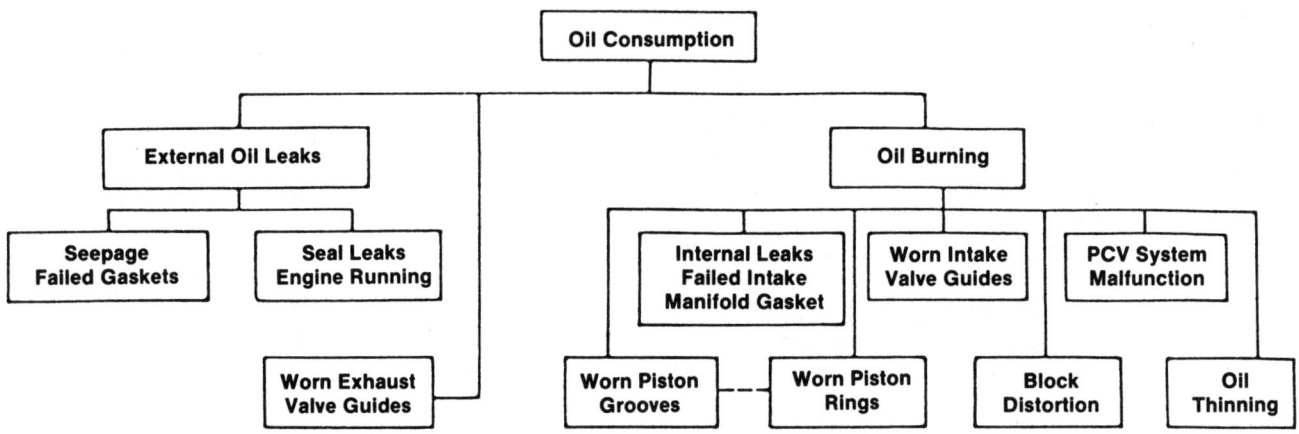

FIGURE 3-25 Common oil consumption problems and their causes

Oil dilution usually occurs during low-temperature operating conditions when a vehicle is used regularly for short trips that do not permit the engine to warm up sufficiently. The choke is on while the engine is cold, which allows a greater than normal amount of fuel to enter the engine. Some of this fuel is not burned. The excess can dilute the oil. Dilution can also occur if the carburetor is malfunctioning or if the choke is incorrectly adjusted so that it remains on for an abnormally long time. The effects of oil dilution show up when the engine is operated at normal temperatures for a period of time, such as a highway trip. Diluted oil not only is consumed at a higher rate, but also is less able to adequately lubricate the engine's moving parts. Oil dilution can be reduced by proper maintenance or alteration of operating habits so that the engine is allowed to warm up.

The black light and fluorescent tracer (or oil) dye is one way of testing for oil leaks. The dye is actually mixed with the engine's oil supply; it appears at the source of the oil leaks. The black light makes it easier to detect the leaks (see Chapter 8). To start the black light procedure, pour the ultraviolet sensitive dye into the crankcase (Figure 3-26), start the engine, and allow sufficient time for the dye to circulate throughout the entire engine before looking for leaks. (Full details on the black light and other dye detection methods can be found in Chapters 8 and 12.) Typical leakage points are shown in Figure 3-27.

Low- and high-pressure air testing are other methods of locating oil leaks. For the former, lower the air compressor regulator down to 4 or 5 psi and rig up a tight connection between the blowgun nozzle and the dipstick tube (a length of fuel hose and two hose clamps work well). Next, pull the PCV valve

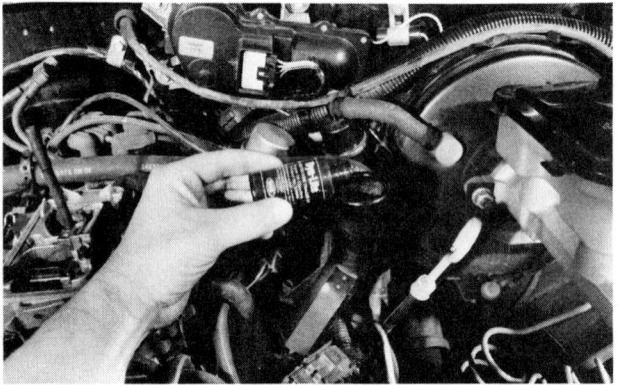

FIGURE 3-26 Pouring ultraviolet sensitive dye into the crankcase

and breather tubes and plug their grommets to prevent pressure loss. Once the engine is set, tape the blowgun's trigger in the *on* position (Figure 3-28), then mix some detergent and water and brush the sudsy mixture over likely seams and around suspect parts. If there is a leak, it will be indicated by the bubbles created by the escaping air. If the leak is large enough, it will be easy to hear. To help zero in on its exact location, a rubber hose or stethoscope held to the ear makes an effective sound amplifier.

For high-pressure testing, set the compressor's psi at 80 to 100, remove the oil pressure sending unit, and screw in whatever adapters are needed to tap into the oiling system. Again, apply soapy water and listen as above. This method is especially useful for finding leaky casting plugs.

Oil consumption problems in some engine arrangements can sometimes be detected in the color of the exhaust pipe smoke using a smoke check. Blue-gray smoke indicates oil escaping to and burn-

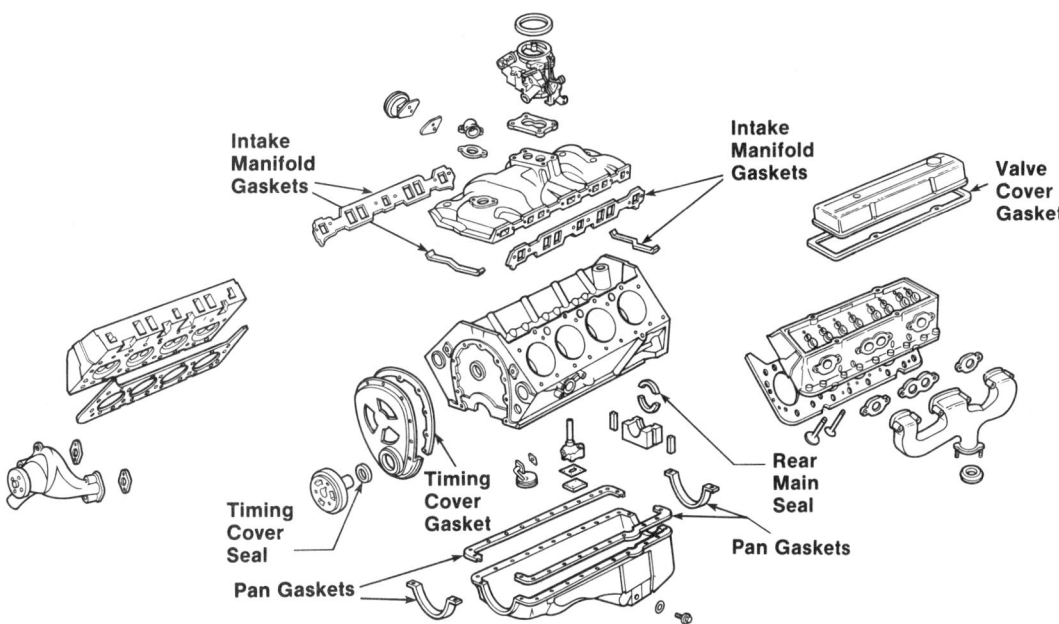

FIGURE 3-27 Typical leakage points

ing in the combustion chamber or exhaust manifold. Black smoke indicates a sign of rich carburetion, and white smoke indicates the condensation of steam.

Fouled spark plugs are also related to oil consumption. They are usually caused by oil leaking through worn piston rings or valve guides and can be identified by oil deposits around the insulator tips (Figure 3-29).

Check for oil loss through the valve guides. Oil passing the guides usually enters the combustion chamber, burns, then shows up as blue-gray smoke at the tailpipe. Improper installation of intake manifold-to-head gaskets can result in external oil leaks which are easily detected by inspection. The vacu-

um test can identify and pinpoint oil leaks as well as vacuum leaks. When the faulty installation results in internal leaks, diagnosis is sometimes more difficult. Oil leakage into the intake ports causes coking, which causes deposits similar in appearance to those caused by oil passing the valve guides. These deposits can be found by removing the manifold.

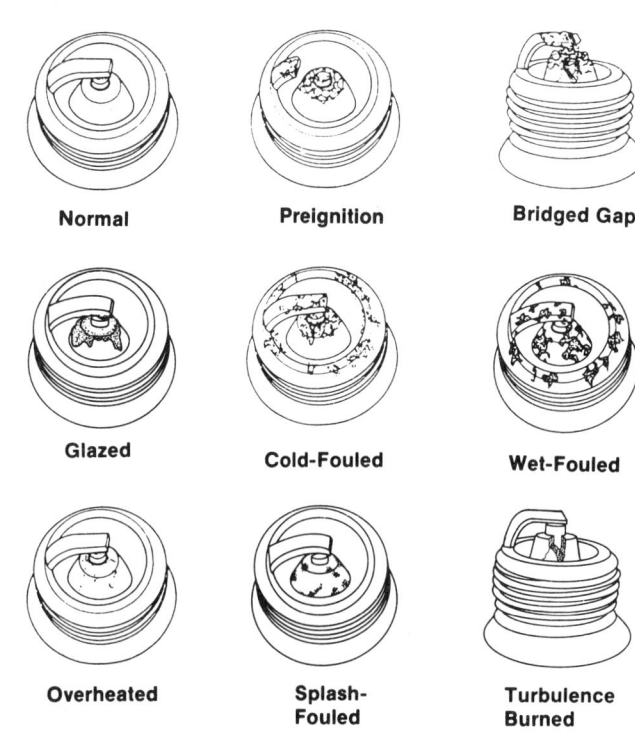

FIGURE 3-28 Blowgun trigger taped in the *on* position

Normal	Preignition	Bridged Gap
Glazed	Cold-Fouled	Wet-Fouled
Overheated	Splash-Fouled	Turbulence Burned

FIGURE 3-29 Common faulty spark plugs

Internal leaks can also be related to rough engine idle.

Further details on lubrication systems and their problems can be found in Chapter 13.

NOISE DIAGNOSIS

More often than not, a malfunction in the engine will reveal itself first as an unusual noise. This can happen before the problem affects the driveability of the vehicle. Problems such as loose pistons, badly worn rings or ring lands, loose piston pins, worn main bearings and connecting rod bearings, loose vibration damper or flywheel, and worn or loose valve train components all produce telltale sounds. Of course, unless the technician has experience in listening to and interpreting engine noises, it can be very hard to distinguish one from the other.

When correctly interpreted, engine noise can be a very valuable diagnostic aid. For one thing, a costly and time-consuming engine teardown might be avoided. *Always* make a noise analysis before doing any repair work; this way, there is a much greater likelihood that only the necessary repair procedures will be done. Careful noise diagnosis also reduces the chances of ruining the engine by continuing to use the vehicle despite the problem.

WARNING: Be very careful when listening for noises around moving belts and pulleys at the front of the engine. Keep the end of the hose or stethoscope probe away from moving parts. Physical injury can result if the hose or stethoscope is pulled inward or flung outward by moving parts.

USING A STETHOSCOPE

Some engine sounds can be easily heard without using a listening device, but others are impossible to hear unless amplified. A stethoscope or rubber hose (as mentioned earlier) is very helpful in locating engine noise by amplifying the sound waves; it can also distinguish between normal and abnormal noise (Figure 3–30). The procedure for using a stethoscope is simple: Use the metal prod to trace the sound until it reaches its maximum intensity. Once the precise location has been discovered, the sound can be better evaluated. A sounding stick, which is nothing more than a long, hollow tube, works on the same principle, though a stethoscope gives much clearer results.

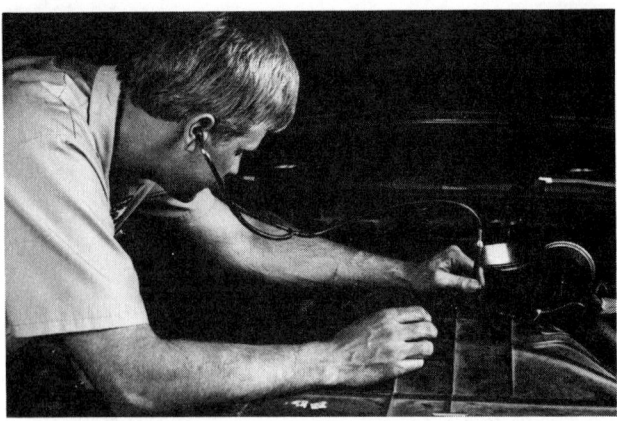

FIGURE 3–30 A stethoscope can help locate engine noise and distinguish between normal and abnormal noise. *(Courtesy of Perfect Circle/Dana)*

COMMON NOISES

Following are examples of abnormal engine noises, including a description of the sound, its likely cause, and ways of eliminating it. An important point to keep in mind is that everyone has his or her own way of describing a particular noise; one person's "rattle" can be another person's "thump." Although the owner's descriptions can be helpful, the final diagnosis is in the hands of the technician. It should also be noted that insufficient lubrication is the most common cause of engine noise. For this reason, always check the oil level first before moving on to other areas of the vehicle (Figure 3–31). Some noises are more pronounced on a cold engine because clearances are greater when parts are not

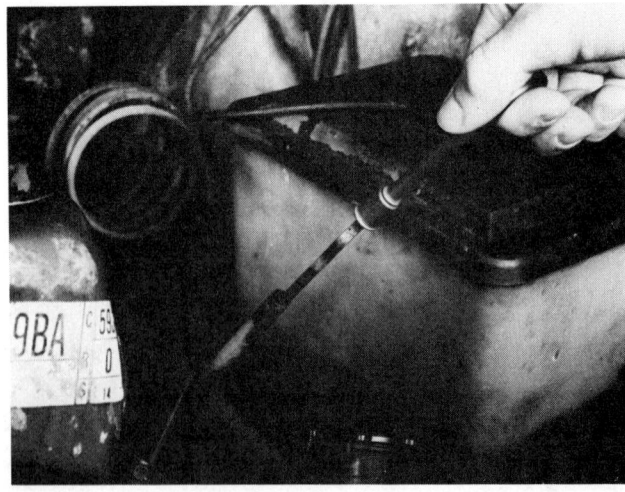

FIGURE 3–31 Always check the oil level first when trying to locate the source of engine noise.

expanded by heat. Remember that aluminum and iron expand at different rates as temperatures rise. For example, a cold knock that disappears as the engine warms up probably is piston slap or knock. An aluminum piston expands more than the iron block, allowing the piston to fit more closely as engine temperature rises.

Ring Noise

This sound can be heard during acceleration as a high-pitched rattling or clicking in the upper part of a cylinder. It can be caused by worn rings or cylinders, broken piston ring lands, or insufficient ring tension against the cylinder walls. Ring noise is corrected by replacing the rings, pistons, or sleeves or reboring the cylinders. Shorting out the spark plug of the affected cylinder usually will not help to eliminate ring noise.

Piston Slap

This is a common sound when the engine is cold, and often intensifies when the vehicle accelerates. When a piston slaps against the cylinder wall, the result is a hollow, bell-like sound. Piston slap is caused by worn pistons or cylinders, collapsed piston skirts, misaligned connecting rods, excessive piston-to-cylinder wall clearance, or lack of lubrication, resulting in worn bearings. Correction requires either replacing the pistons, reboring the cylinder, replacing or realigning the rods, or adding oil to the engine and replacing the bearings. Shorting out the spark plug of the affected cylinder might quiet the noise.

Piston Pin Knock

Piston pin knock is a sharp, metallic rap that can sound more like a rattle if all the pins are loose. It originates in the upper portion of the engine and is most noticeable when the engine is idling. Piston pin knock is caused by a worn piston pin, piston pin boss, or piston pin bushing or lack of lubrication, resulting in worn bearings. To correct it, either install oversize pins, replace the boss or bushings, or replace the piston.

Ridge Noise

This noise is less common, but very distinct. As a piston ring strikes the ridge at the top of the cylinder, the result is a high-pitched rapping or clicking that becomes louder during deceleration (Figure 3–32).

There can be more than one reason for the ridge interfering with the ring's travel. For one thing, if new rings are installed without removing the old ridge, the new rings will contact the ridge and make a noise. Also, if the piston pin is very loose or the connecting rod has a loose or burned-out bearing, the piston will go high enough in the cylinder for the top ring to contact the ridge. Thus, in order to eliminate ridge noise, remove the old ring ridge and replace the piston pin or piston.

Rod Bearing Noise

The result of worn or loose connecting rod bearings, this noise is heard at idle as well as at speeds over 35 mph. Depending on how badly the bearings are worn, the noise can range from a light tap to a heavy knock or pound. Shorting out the spark plug of the affected cylinder can lessen the noise, unless the bearing is totally burned out; in this case, shorting out the plug will have no effect. Rod-bearing noise is caused by a worn bearing or crankpin, a misaligned rod, or lack of lubrication, resulting in worn bearings. To correct it, service or replace the crankshaft, realign or replace the connecting rods, and replace the bearings.

Main or Thrust Bearing Noise

A loose crankshaft main bearing produces a dull, steady knock, while a loose crankshaft thrust bearing produces a heavy thump at irregular intervals. The thrust bearing noise might only be audible on very hard acceleration. Both of these bearing

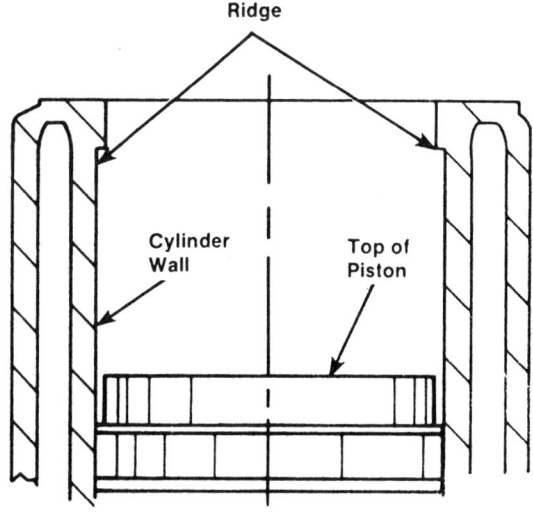

FIGURE 3–32   As the piston strikes the ridge at the top of the cylinder, a high-pitched rapping or clicking sound is made.

noises are usually caused by worn bearings or crankshaft journals. To correct the problem, replace the bearings and/or crankshaft.

Tappet Noise

Tappet noise is characterized by a light, regular clicking sound that is more noticeable when the engine is idling. It is the result of excessive clearance in the valve train. The clearance problem area is located by inserting a feeler gauge between each lifter and valve, or between each rocker arm and valve tip, until the noise subsides. Tappet noise can be caused by improper valve adjustment, worn or damaged parts, dirty hydraulic lifters, or lack of lubrication. To correct the noise, adjust the valves, replace any worn or damaged parts, or clean or replace the lifters.

Abnormal Combustion Noises

Preignition and detonation noises are caused by abnormal engine combustion. Their causes are fully discussed in Chapter 7. For instance, detonation knock or ping is a noise most noticeable during acceleration with the engine under load and running at normal temperature. Excessive detonation knock can be very harmful to the engine. It is often caused by advanced ignition timing or substantial carbon buildup in the combustion chambers that increases the combustion pressure. Carbon deposits that get so hot they glow will also preignite the air/fuel mixture, causing detonation. Another possible cause is fuel whose octane is too low. Detonation knock can usually be cured by removing carbon deposits from the combustion chambers with a rotary wire brush as well as recommending the use of a higher octane gasoline. A malfunctioning EGR valve can also cause detonation and even rod knock.

Sometimes abnormal combustion combines with other engine parts to cause noise. For example, rumble is a term that is used to describe the knock or noise resulting from another form of abnormal ignition. Rumble is a bending vibration of the crankshaft and connecting rods that is caused by multisurface ignition. Rumble is a form of preignition in which several flame fronts occur simultaneously from overheated deposit particles. Multisurface ignition causes a tremendous sudden pressure rise near top dead center. It has been reported that the rate of pressure rise during rumble is five times the rate of normal combustion.

Damper or Flywheel Noise

A loose vibration damper causes a heavy rumble or thump in the front of the engine that is more apparent when the vehicle is accelerating from idle under load or is idling unevenly. A loose flywheel causes a heavy thump or light knock at the back of the engine, depending on the amount of play and the type of engine. Both of these problems are corrected either by tightening or replacing the damper or flywheel.

DIAGNOSING DIESEL ENGINE PROBLEMS

Like the gasoline engine, the diesel engine should be carefully diagnosed before it is disassembled. Although there are differences, as stated in Chapter 2, the basic diagnostic procedure is the same. Typical diesel engine problems that must be diagnosed include:

- Overheating
- Abnormal engine knock
- Loss of power
- Smoky exhaust
- High fuel consumption

It is possible for these problems to be caused by the fuel injection system, but they can also be caused by the other engine systems.

Overheating problems are caused by the same conditions that cause a gasoline engine to overheat—low coolant level, slipping fan belt, defective head gasket, or a thermostat stuck closed.

When too much fuel is injected into the cylinder too early in the combustion cycle, engine knock will result. If engine knock is noted on all cylinders, it could indicate incorrect injection pump timing. Normally, engine knock will occur only on one or two cylinders and signals that the applicable injector is at fault. Incorrect operation of an injector will result in a much larger uncontrolled burning period and produce noticeable engine knock.

A restricted air inlet system or fuel system can be the cause of a loss of power. A check should be made to ensure that injector nozzle tips are clean and opened, that the fuel lines are not plugged, that the car filter is clean, and that the engine compression is correct.

A common problem with diesel engines is a smoky exhaust. This is caused by incomplete combustion of the fuel. At idle speeds this is more pronounced because the combustion chamber might still be cool and will tend to quench combustion before all the fuel is consumed. At higher engine speeds, if the smoky exhaust condition is still noted, the air/fuel mixtue is too rich. Possible causes could be a faulty EGR valve, faulty injection pump, or an incorrect injection pump timing.

The injection of too much fuel results in poor fuel economy. A fuel leak in the system also produces the same symptom. High fuel consumption is always caused by low compression and incorrect injection pump timing.

For a specific diagnosis, such as a gasoline engine, always check the proper procedure in the engine's service manual and follow it to the letter.

REVIEW QUESTIONS

1. A leak at which of the following points will result in a loss of compression and poor driveability?
   a. cylinder head gasket
   b. piston rings
   c. valves
   d. all of the above

2. When a dry compression test is performed, a reading is low on the first stroke and gets only a little higher during subsequent strokes. Technician A says the cause could be badly worn piston rings. Technician B says the cause could be a sticking valve. Who is right?
   a. Technician A
   b. Technician B
   c. Both A and B
   d. Neither A nor B

3. A wet compression test should not be performed on a _____ .
   a. horizontally opposed cylinder engine
   b. four-cylinder engine
   c. V-6 engine
   d. both a and b

4. Technician A says a compression test is more accurate than a cylinder leakage test. Technician B says a cylinder leakage test is more accurate than a compression test. Who is right?
   a. Technician A
   b. Technician B

c. Both A and B
d. Neither A nor B

5. Which of the following is needed to perform a cylinder leakage test?
   a. leakage tester
   b. top dead center indicator
   c. indicator light
   d. all of the above

6. Which of the following determines where a leak is and what is causing it?
   a. compression tester
   b. leakage tester
   c. hydrosonic leak detector
   d. all of the above

7. If the compression test and leakdown test prove to be acceptable, then the problem is not _____ .
   a. mechanical
   b. with the ignition system
   c. with the computer system
   d. all of the above

8. When performing a power balance test, the spark plugs are controlled with push buttons. Technician A says the push-button numbers refer to the cylinder firing order. Technician B says the numbers refer to the cylinder number designation. Who is right?
   a. Technician A
   b. Technician B
   c. Both A and B
   d. Neither A nor B

9. When a vacuum reading is normal at idle speed but excessive vibrations are present at higher rpm, which of the following could be the cause?
   a. valve springs
   b. worn valve guides
   c. leaky head gasket
   d. intake system leak

10. What condition is likely to show good compression but heavy cylinder leakage past the rings in a low-mileage car?
    a. a valve train problem
    b. incorrect valve timing
    c. stuck piston rings
    d. none of the above

11. A cylinder that has poor compression but minimal leakage indicates _____ .
    a. a valve train problem
    b. incorrect valve timing
    c. the piston rings are stuck
    d. none of the above

12. Which method is especially useful for finding leaky casting plugs?
    a. black light and fluorescent tracer dye method
    b. low-pressure air testing method
    c. high-pressure air testing method
    d. none of the above

13. Which of the following produces a heavy thump at irregular intervals?
    a. loose crankshaft main bearing
    b. loose crankshaft thrust bearing
    c. piston slap
    d. ring noise

14. Which is the most common source of engine noise?
    a. piston pin knock
    b. ring noise
    c. piston slap
    d. insufficient lubrication

15. Which of the following is a maintenance method used to restore proper engine performance?
    a. minor overhaul
    b. major overhaul
    c. tune-up
    d. all of the above
    e. both a and b

CHAPTER FOUR

ENGINE REMOVAL AND DISASSEMBLY

Objectives

After reading this chapter, you should be able to
- List general guidelines for preparing the engine for removal.
- Explain how to lift an engine out of its compartment.
- Describe the disassembly and visual inspection of the upper engine.
- List the recommended steps for disassembling the short block.
- Explain how to remove freeze plugs, oil gallery plugs, timing components, pistons, rods, main bearings, crankshaft, camshaft, and lifter.
- List the components of the various engine kits.
- Explain the usual care policies of engine remanufacturers.

If engine diagnosis indicates that a major problem exists, the best solution might be to remove the engine from the vehicle, disassemble it (Figure 4-1), and rebuild it.

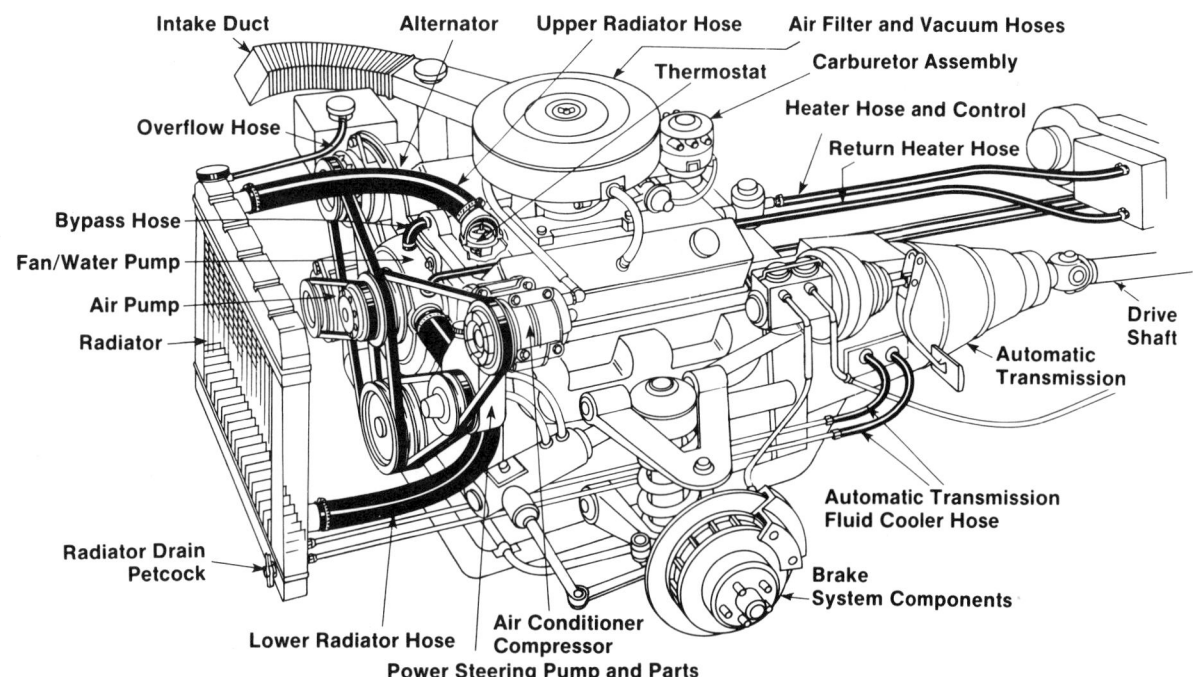

FIGURE 4-1 Some parts that must be removed before the engine can be completely disassembled and rebuilt. Follow the removal procedures given in the vehicle's service manual.

PREPARING THE ENGINE FOR REMOVAL

When preparing an engine for rebuilding, the rebuilder should always follow the specific engine removal and disassembly procedures for the particular vehicle being worked on. The procedures given here are intended to be used only as guidelines.

1. Open the hood and install fender covers on the fenders. Also mask any areas where there is any possibility of scratching the paint.
2. Scribe a line to mark the location of the hood hinges for reference in reassembly. Unbolt the hood hinges (Figure 4–2) and remove the hood.
3. Disconnect the battery cables from the battery and the body ground from the battery tray (Figure 4–3). The battery should

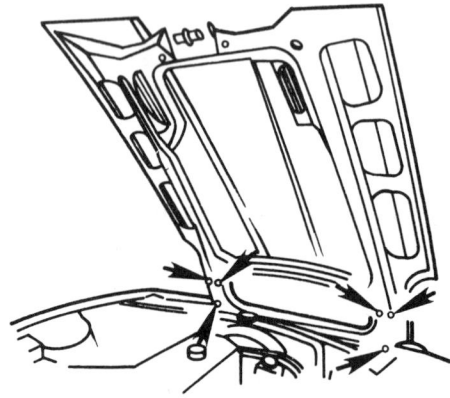

FIGURE 4-2 Hood hinges are marked then disconnected.

FIGURE 4-3 Disconnect the battery before starting any work on an engine.

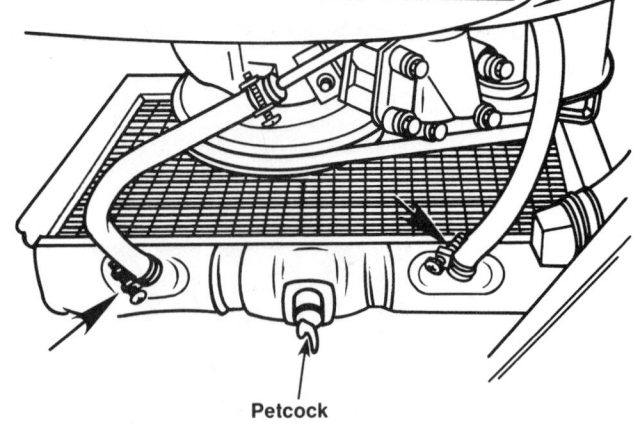

Petcock

FIGURE 4-4 Fluid cooler lines are disconnected from the bottom of the radiator.

be removed from the vehicle and stored on a charger to prevent it from discharging. On some vehicles it might be necessary to disconnect the ground cable at the cylinder head before the engine can be lifted off the chassis.

4. Remove the air cleaner, intake air duct, and heat tube to aid visibility and to increase the size of the working area.
5. Drain the cooling system by disconnecting the engine coolant hoses at the top and bottom of the radiator (Figure 4–4).
6. On vehicles with an automatic transmission, disconnect the oil cooler lines at the radiator.
7. Disconnect the heater hoses (Figure 4–5) from the water pump and heater tube.

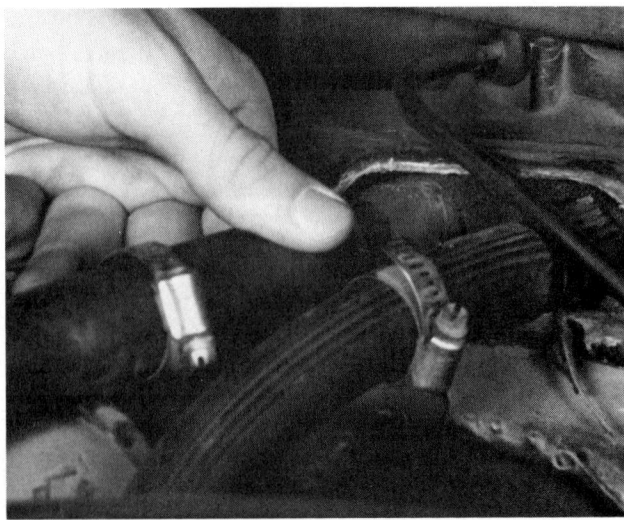

FIGURE 4-5 Disconnecting the heater hose

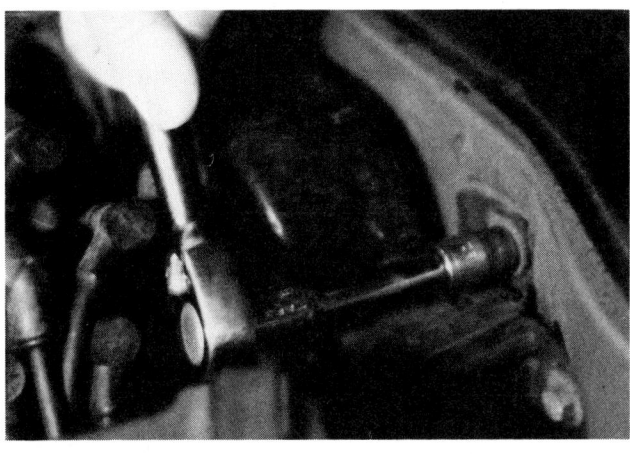

FIGURE 4-6   Removing the radiator mounts

8. Remove the fan shroud and fan assembly, then loosen the accessory drive belt and the water pump pulley. Remove the radiator (Figure 4-6).
9. Disconnect and plug all fuel lines to prevent fuel loss (Figure 4-7) and to keep dirt from entering the system. Remove the carbon canister purge hose from the PCV valve and disconnect the throttle linkage from the carburetor or fuel injection line.

CAUTION: Some fuel injection lines are under pressure even when the engine is off. Check the bleed-down procedure in the service manual when disconnecting a fuel injection line.

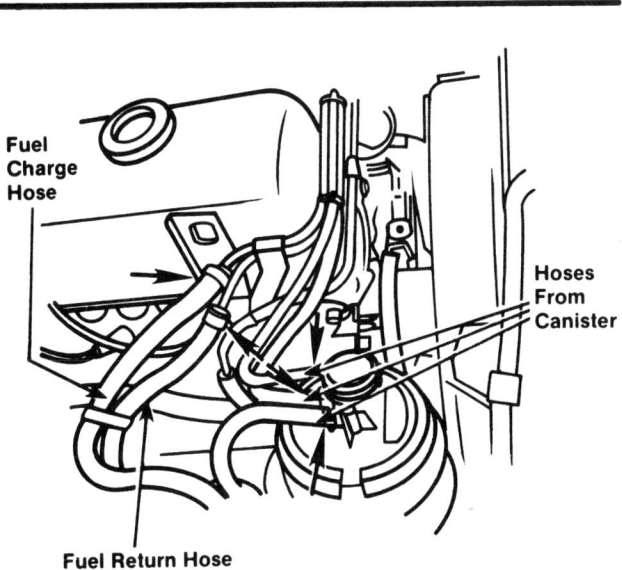

Fuel Charge Hose

Hoses From Canister

Fuel Return Hose

FIGURE 4-7   Typical fuel lines that must be disconnected

10. Remove the carburetor assembly from the intake manifold. Also remove the carbon canister, fuel injector controls or fuel injector body, the managed air injector system, and any other emission control devices that could pose engine removal problems.
11. Disconnect all vacuum hoses, wiring connections, and harnesses (Figure 4-8). But before starting to disconnect the vacuum lines or electrical lines, be sure to attach a piece of masking tape to both sides of the parts that are disconnected (Figure 4-9). Mark the same code letter or number on both sides of what has been disconnected.

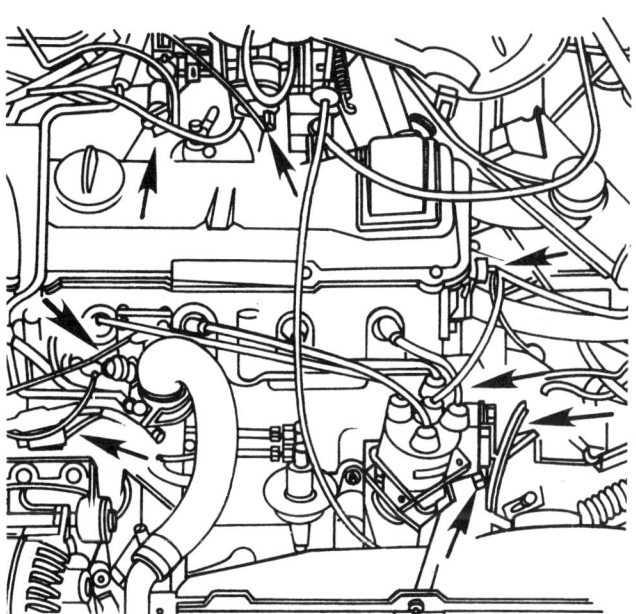

FIGURE 4-8   Vacuum and electrical lines that must be disconnected

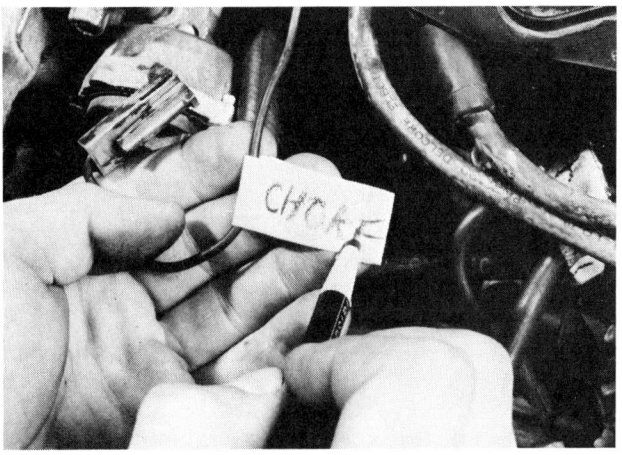

FIGURE 4-9   Label parts as they are disconnected.

The rebuilder can feel assured that when putting everything back together it will be possible to fit it right the first time.

12. On models equipped with electronic or computer controls, disconnect the air charge temperature, the engine coolant temperature sensor, and the exhaust gas oxygen sensor electrical connectors.

13. Disconnect the ignition coil, water temperature sending unit, and oil pressure sending unit. Remove the distributor, distributor cap, and spark plug wires from the engine and alternator.

FIGURE 4-10 Removing the air-conditioning compressor

14. Disconnect the speedometer cable where it attaches to the transmission.

15. If so equipped, remove the bolts attaching the power steering pump mounting bracket, then move the pump and bracket assembly and set aside in the chassis engine compartment with the hoses attached. By removing the unit in this manner, the hydraulic lines do not have to be disconnected.

16. On models with air conditioners, remove the compressor mounting bracket attaching bolts (Figure 4-10), then remove the compressor and the mounting bracket assembly and set aside without disconnecting the refrigerant lines. If the lines are disconnected, the system will require recharging during reassembly.

17. Raise the vehicle following the safety procedures recommended in Chapter 1. Then drain the engine oil.

18. Disconnect the exhaust system from the manifold. Loosen the exhaust pipe clamp and slide off the support bracket on the engine. To help in removing bolts and studs soak them with penetrating oil. Remove the turbocharger if so equipped.

19. Remove the flywheel or converter housing cover. If equipped with manual transmission, remove the flywheel housing lower attaching bolts. If equipped with automatic transmission, remove the converter-to-flywheel bolts, then remove the converter housing attaching bolts.

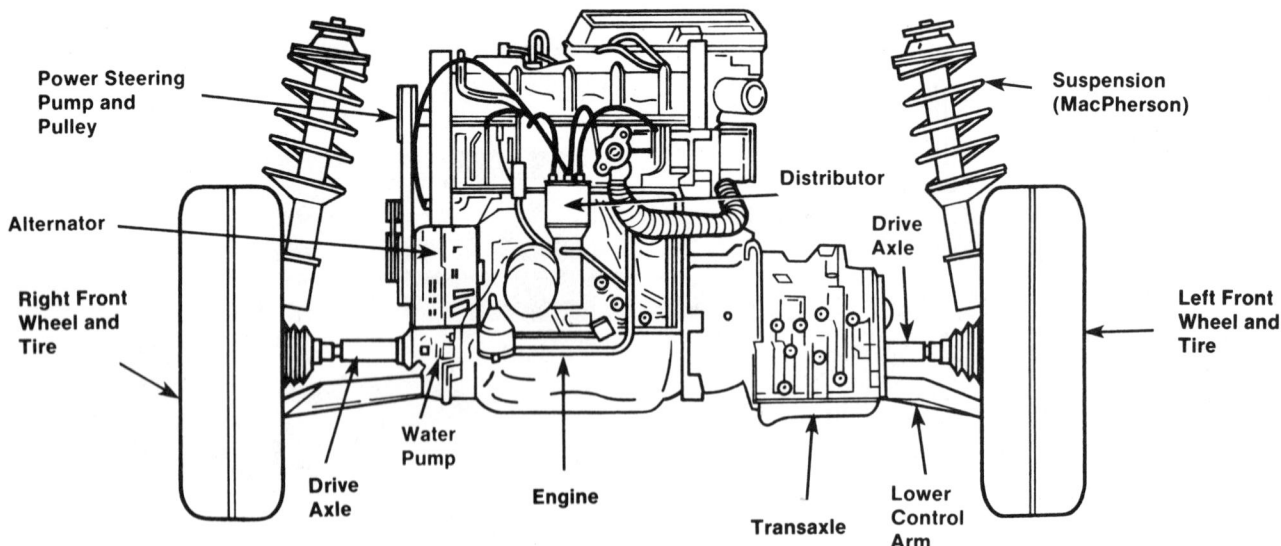

FIGURE 4-11 Basic transverse-mounted front-wheel-drive engine

FIGURE 4-12 Typical transverse engine support bar provides the necessary support when removing the lower driveline cradle on today's front-wheel drive vehicles. *(Courtesy of Tool Division—SPX Corp.)*

 SHOP TALK _____

When lifting a transverse-mounted engine with front-wheel drive (Figure 4-11), most service manuals recommend removing the engine and transaxle assembly as a unit. The automatic transmission can be separated from the engine once it has been lifted out of the vehicle. In this case, the drive axles must be disconnected from the transaxle. It is wise to employ a transverse engine support bar before attempting to remove this type of engine (Figure 4-12).

20. Remove the starter motor and all necessary brake system components.
21. Mark the drive shaft for reassembly reference, then unbolt and remove it. Never let the drive shaft hang unsupported.
22. Disconnect the engine mounts from the brackets on the frame (Figure 4-13).
23. Remove all transmission bolts. Then lower the vehicle and position a suitable transmission jack under the transmission (Figure 4-14). Raise the jack just enough to support the weight of the transmission.

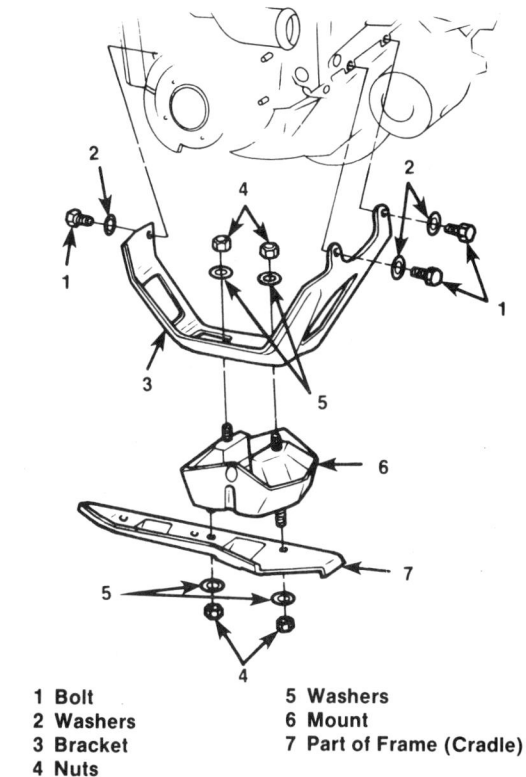

1 Bolt
2 Washers
3 Bracket
4 Nuts
5 Washers
6 Mount
7 Part of Frame (Cradle)

A

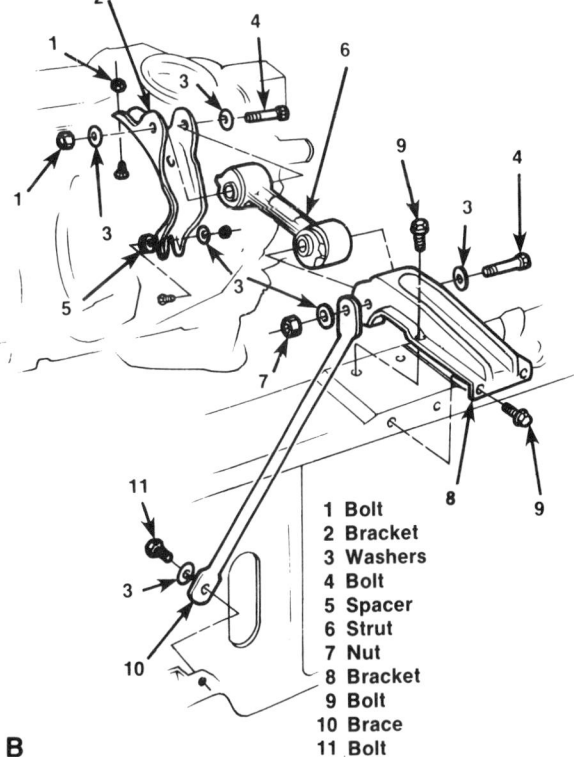

1 Bolt
2 Bracket
3 Washers
4 Bolt
5 Spacer
6 Strut
7 Nut
8 Bracket
9 Bolt
10 Brace
11 Bolt

B

FIGURE 4-13 (A) Typical front engine mount removal; (B) typical mount strut and bracket removal procedure

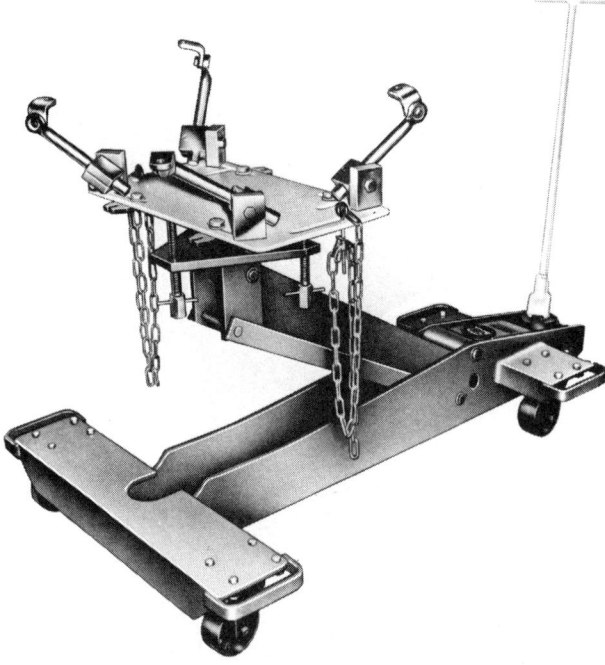

FIGURE 4-14 Typical transmission jack *(Courtesy of Tool Division—SPX Corp.)*

24. Carefully raise the engine slightly, then pull away the transmission and lift from the vehicle.
25. Make a final check to be sure that everything has been disconnected from the engine. The engine is now ready to be lifted from its compartment.

FIGURE 4-15 Canvas slings are also in common use.

LIFTING AN ENGINE

To lift an engine out of its compartment use either a canvas hoist (Figure 4-15) and chain (Figure 4-16), which consists of two gears to provide the necessary mechanical advantage to do the lifting, or a mobile crane, frequently called a *cherry picker,* that uses hydraulic power to do the lifting (Figure 4-17).

FIGURE 4-16 Engines can be lifted with a chain hoist. *(Courtesy of Tool Division—SPX Corp.)*

FIGURE 4-17 A mobile crane, or cherry picker, can be used to lift out the engine. *(Courtesy of Tool Division—SPX Corp.)*

To lift an engine, proceed as follows:

1. Attach a pulling sling (Figure 4–18) or chain (Figure 4–19) to the engine. Some engines have eye plates for use in lifting. If they are not available the sling must be bolted to the engine. The sling attaching bolts must be large enough to support the engine and must thread into the block a *minimum* of 1-1/2 times the bolt diameter.

FIGURE 4-18 A pulling sling helps to lift an engine out of a vehicle. *(Courtesy of Tool Division—SPX Corp.)*

FIGURE 4-19 Lifting the engine out of its compartment *(Courtesy of Tool Division—SPX Corp.)*

2. Connect the chain hoist or mobile crane to the pulling cable. Double-check the engine to be sure that everything is disconnected.
3. Raise the engine slightly and make certain that the sling attachments are secure. Then carefully lift the engine out of its compartment. Be sure that the engine does not bind or damage any compartment components during this procedure.
4. Lower the engine close to the floor so it can be transported to the desired location.
5. If you have not already disconnected the transmission and torque converter or clutch from the engine, do so now. The torque converter must remain with the transmission when the engine is removed. A C-clamp prevents the converter from dropping out of the bell housing.
6. Raise the engine and position it next to an engine stand. Mount the engine to the engine stand with bolts (Figure 4–20). Most stands (Figure 4–21) use a plate with sev-

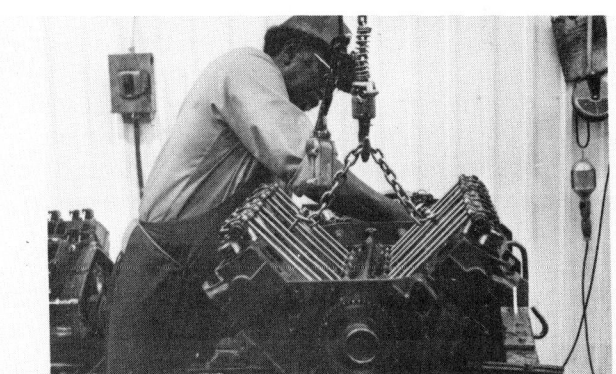

FIGURE 4-20 Engine on a specially designed worktable

FIGURE 4-21 Engine on an engine stand

eral holes or adjustable arms. The engine must be supported by at least four bolts that fit solidly into the engine.

7. Set the engine at a comfortable working height (Figure 4-22) and make certain that it is held securely.

8. Remove the pulling cable or chain from the engine. The engine can now be disassembled.

ENGINE DISASSEMBLY

Before engine disassembly, be sure it is securely bolted to an engine stand or sitting on blocks. Go slowly and visually inspect each part for any signs of damage. Look for excessive wear on the valves, stems, guides, and seats; check rocker arms, pushrods, cam lobes, and lifters for signs of overheating, unusual wear, and chips. Also look for signs of gasket and seal leakage.

A disassembled engine is shown in Figure 4-23.

The following engine teardown of both cylinder head and block can be considered typical. Exact details will vary slightly depending on the style and type of engine. For instance, in some engines, the overhead camshaft is mounted directly in the cylinder head; in other engines, it is located in a separate housing that is mounted on the cylinder head. The camshaft housing is unbolted from the cylinder head. The bearing caps must be removed in order to remove a camshaft mounted to the cylinder head.

Although general methods of parts removal and procedures to inspect the various engine components are given in later chapters of this book, the vehicle's OEM service manual is always the final word.

CYLINDER HEAD DISASSEMBLY

The first step in disassembly of an engine is usually the removal of the intake and exhaust manifolds (Figure 4-24). On some in-line engines, the intake and exhaust manifolds are often removed as an assembly and are not disconnected from each other unless a problem exists that would require disassembly.

 SHOP TALK

It is important to let an aluminum cylinder head cool completely before removing it.

To start the cylinder head disassembly, remove the valve cover or covers. The cylinder head bolts are first loosened one or two turns each, working from the center of the cylinder head outward (Figure 4-25). This procedure prevents the distortion that can occur if bolts are all loosened at once. The bolts are then removed, again following the center-outward sequence. With the bolts removed, the cylinder head can be lifted off. The cylinder head gasket should be saved to compare with the new head gasket during reassembly.

After the valve cover has been removed, disassemble the rocker arm components (Figure 4-26). When removing the rocker assembly, remember that each has a different disassembly procedure. It is best to check the manufacturer's manual for specific procedures.

After removing the rocker arm and pushrods, check the rocker area for sludge (Figure 4-27). Excessive buildup can indicate a poor oil change schedule and is a signal to look for similar wear patterns on other components.

When disassembling the cylinder head, keep the pushrods and rocker arms or rocker arm assemblies in exact order if they will be reused (Figure 4-28). These parts are wear mated to each other and should be reassembled in the same position on the camshaft. Use an organizing tray or label the parts with a felt-tipped marker to keep them together and labeled accurately. Check lifters for a "dished" bottom or scratches, which indicate poor rotation.

FIGURE 4-22 Engine in an overhaul stand that allows it to be turned over for easy access. *(Courtesy of Tool Division—SPX Corp.)*

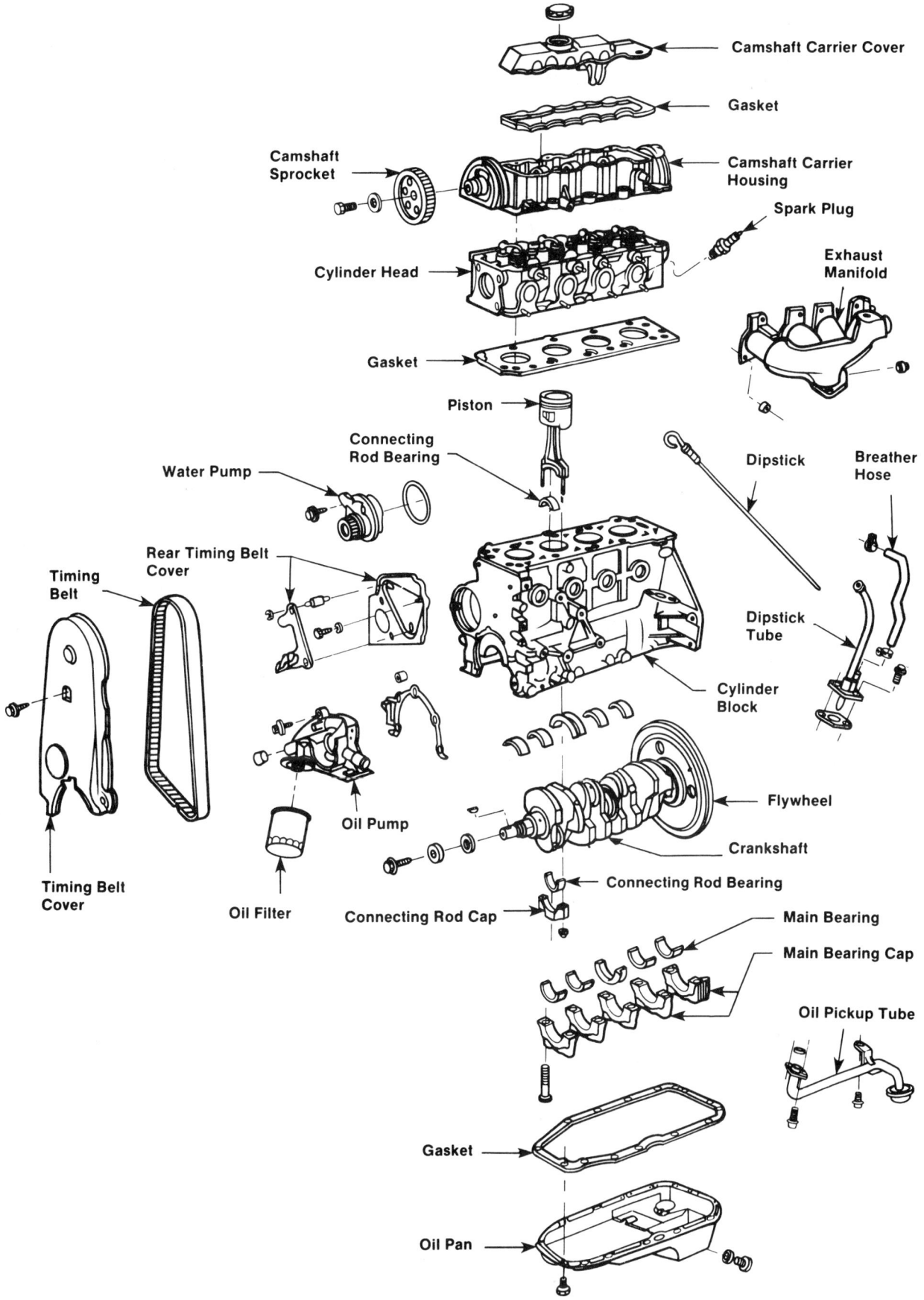

FIGURE 4-23 Disassembled engine

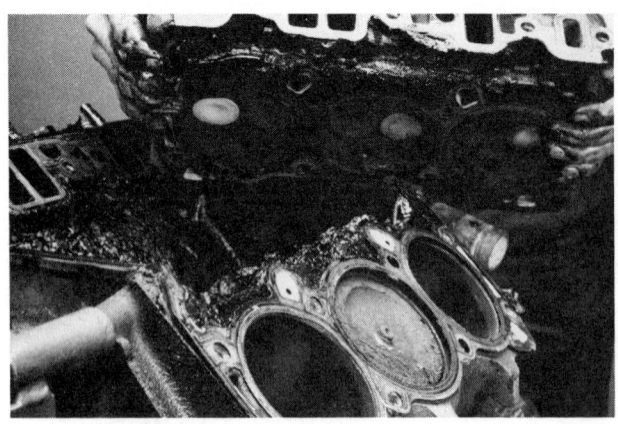

FIGURE 4-24 Removing the manifold *(Courtesy of Perfect Circle/Dana)*

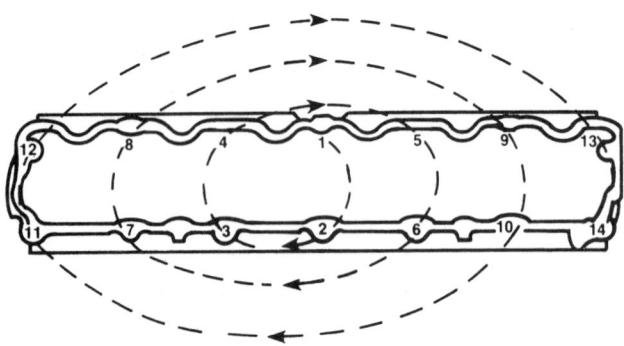

FIGURE 4-25 Cylinder head cover bolts are loosened from the center outward; follow a sequence similar to one shown.

FIGURE 4-26 Removing the rocker arm components is the first step in disassembling the cylinder head. *(Courtesy of Perfect Circle/Dana)*

FIGURE 4-27 Check rocker area for sludge *(Courtesy of Perfect Circle/Dana)*

FIGURE 4-28 Keep pushrods and rocker arms or rocker arm assemblies in exact order if they will be reused. *(Courtesy of Perfect Circle/Dana)*

On standard overhead valve engines, remove the timing cover (Figure 4-29). With certain engines this may require removing the oil pan. With the cover off, the harmonic balancer or vibration damper usually must be removed. This usually requires a special puller designed for the purpose (Figure 4-30).

The camshaft can now be carefully removed (Figure 4-31). Support the camshaft during removal to avoid dragging lobes over bearing surfaces, which would damage bearings and lobes. Do not bump cam lobe edges, which can cause chipping. Some engines might require the removal of the thrust plate before taking out the camshaft.

After the camshaft has been removed, visually examine the camshaft for any obvious defects— rounded lobes, edge wear, galling, and the like. If either the camshaft or lifters are worn at all, both

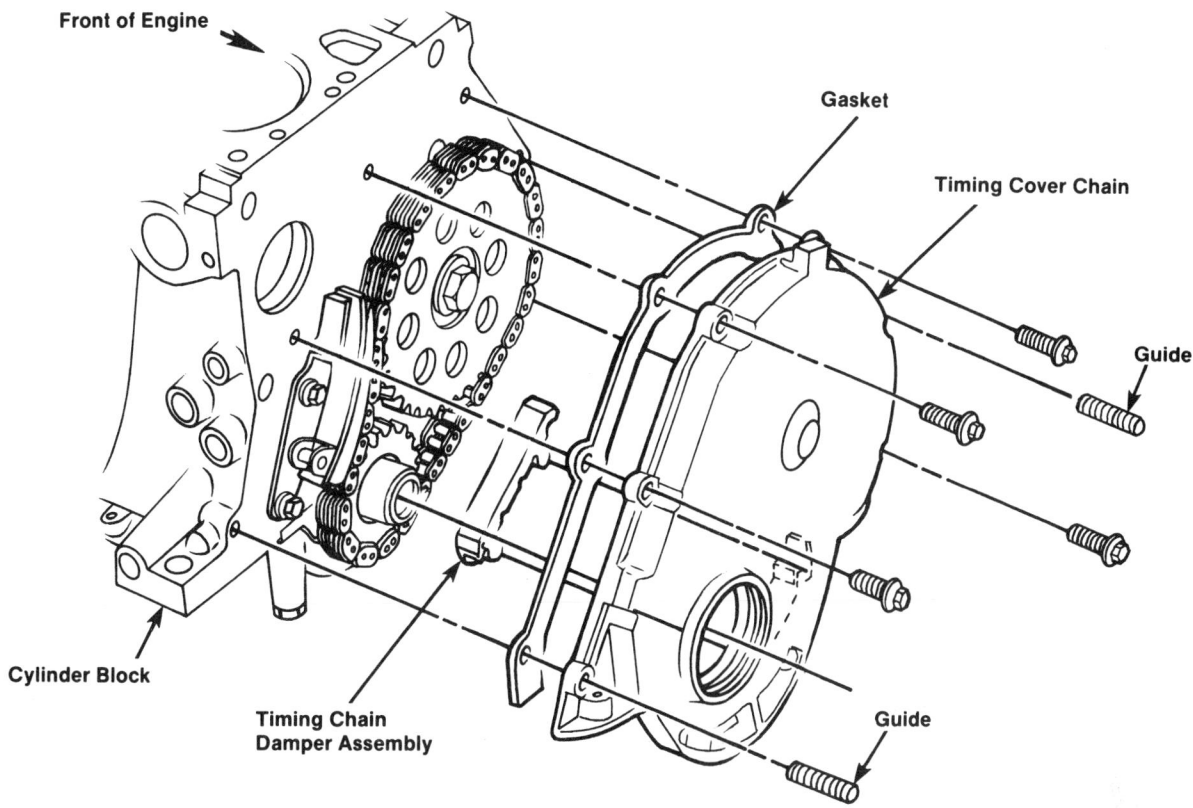

FIGURE 4-29   Timing cover removal

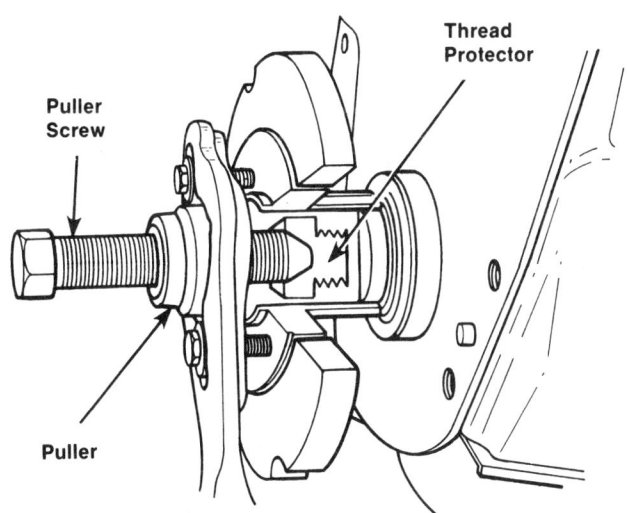

FIGURE 4-30   Use of special puller to remove the vibration damper or harmonic balancer

should be replaced. *Do not* install used valve lifters with a new camshaft. *Do not* install new valve lifters on a used camshaft.

Before tearing down the cam follower assembly, draw a diagram and use a felt-tipped marker to label

FIGURE 4-31   Removing the cam from the engine block *(Courtesy of Perfect Circle/Dana)*

the parts. Since the camshaft and followers are wear-mated parts, this will help insure that each one is returned to the same position.

If the cam has springs beneath it, there is 60 to 80 psi pressure against any spot with the cam lobe against the follower. Random removal of bearing caps can cause the cam to break or spring up, possibly striking the rebuilder. Be sure to follow the automotive service manual for the correct procedure because it is different for the bucket-type lifter design and lash adjuster design.

Next, use a valve spring compressor to begin disassembling the valve train. This compressor will allow the rebuilder to compress the valve springs and remove the keepers (Figure 4-32). With the valves still in the cylinder head, use a valve stem height gauge to measure the stem height for each valve and record it (Figure 4-33). This measurement should be taken from the spring seat to the valve stem tip when the valve is closed. This measurement will be needed during reassembly to determine the installed valve stem height.

Next, remove the valve oil seals and the valves. If a valve seems to be stuck in the guide, the tip might be peened over. If this is the case *do not* drive the valve through the guide; it could score or crack the valve guide and/or head. Raise the stem and file the excess metal until the stem slides through the guide easily (Figure 4-34).

 SHOP TALK _____

Most rebuilders replace valves, valve springs, keepers, and retainers with new ones when reassembling the engine.

While removing valves, look for signs of burning, pitting, cracks, grooves, scores, necking, or other signs of wear. These wear patterns signal other problems in the engine. Valves that cannot be refaced without leaving at least a 1/32-inch valve margin must be discarded. Also discard any valve that is badly burned, cracked, pitted, or shows signs of valve stem wear, bent valve stems, or damaged keeper grooves. Also examine the back side of the intake valves. A black oily buildup in the neck and stem area indicates oil is entering the cylinder through the intake valve (Figure 4-35). Use this method to check for excessive oil consumption.

CYLINDER BLOCK DISASSEMBLY

After the cylinder head has been removed, the cylinder block can be torn down. First remove the oil pan if it was not previously removed. Then remove the oil pump as directed in the service manual.

Continue the disassembly by removing the timing components. There are three timing component

FIGURE 4-32 Valve spring compressor *(Courtesy of Fel-Pro, Inc.)*

FIGURE 4-33 Valve stem height gauge *(Courtesy of Perfect Circle/Dana)*

FIGURE 4-34 Filing excess metal *(Courtesy of Perfect Circle/Dana)*

FIGURE 4-35 Examine the back side of the intake valves.

FIGURE 4-37 Inspect sprockets for wear, cracks, and broken teeth.

assemblies: the chain and sprocket, the gear, and the timing belt and sprocket.

Often the chain and sprocket assembly and the timing belt and sprocket assembly both have tensioners and guides (Figure 4-36). All three types should be replaced as complete assemblies during an engine overhaul. The tensioners and guides wear and should be replaced as well.

Some OHC have complicated driving systems, so if you are unsure of the removal process consult the proper engine repair manual. Inspect the sprockets for wear, cracks, and broken teeth (Figure 4-37). Inspect the timing gears for excessive backlash. Check the chain for slackness and wear.

If the block is to be bored, ridge reaming is unnecessary, but new pistons should be used any-

way. If the engine is to be re-ringed only, carefully scrape the ridge with a ridge removing tool (Figure 4-38). Rotate the tool clockwise with a wrench to remove the ridge. Do not cut too deeply, because it will leave an indentation in the bore. Remove just enough metal to allow the piston assembly to set properly.

If the ridge is too large, the new top rings will hit it and possibly break the ring lands. In this case, the engine should be rebored. The ridge is formed at the top of the cylinder in two cases. Because the top ring stops traveling before it reaches the top of the cylinder, a ridge of unworn metal is left. Carbon also builds up above this ridge, adding to the problem.

After the ridge removing operation, wipe all the metal cuttings out of the cylinder. Use an oily rag to wipe the cylinder and the cuttings will stick to the rag.

Prior to removal, check all connecting rod and main bearing caps for correct position and number-

FIGURE 4-36 Tensioners and guides for timing chain belt and sprocket assembly *(Courtesy of Perfect Circle/Dana)*

FIGURE 4-38 Cylinder ridge removal *(Courtesy of Perfect Circle/Dana)*

ing. If the numbers are not visible, use a center punch or number stamp to number them (Figure 4-39). Caps and rods should be stamped on the external flat surface. Remember that caps and rods must remain as a set.

Remove the piston and rod assemblies as follows:

1. Position the crankshaft throw at the bottom of its stroke.
2. Remove the connecting rod nuts and cap. Tap the cap lightly with a soft hammer or wood block to aid in cap removal.
3. Cover the rod bolts with protectors to avoid damage to the crankshaft journals (Figure 4-40).
4. Carefully push out the piston and rod assembly with the wooden hammer handle or wooden drift and support the piston by

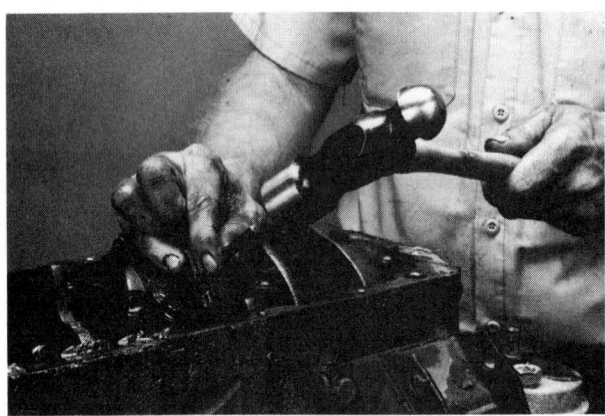

FIGURE 4-39 Stamping bearing caps

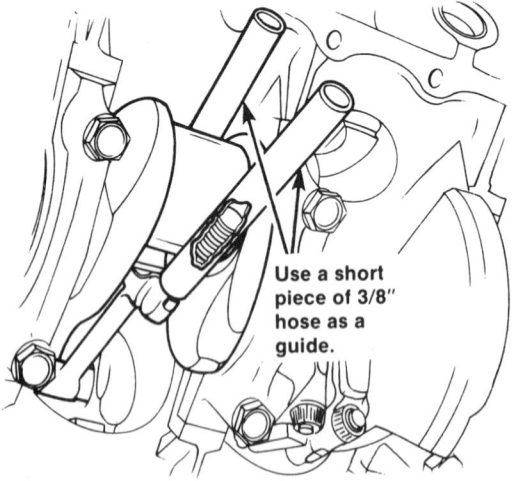

FIGURE 4-40 Install hoses or rod bolt protectors on the rod bolts before removing the piston/rod assembly.

hand as it comes out of the cylinder. Be sure that the connecting rod does not damage the cylinder during removal.

5. With bearing inserts in the rod and cap, replace the cap (numbers on same side) and install the nuts. (Store the piston and rod assembly properly.) Repeat the procedure for all other piston and rod assemblies.

Remove the flywheel or flex plate; a scribe marking the crankshaft and flywheel or plate aids in reassembly. Then remove the main bearing cap bolts and main bearing cap.

After removing the main bearing caps, carefully take out the crankshaft by lifting both ends equally to avoid bending and damage (Figure 4-41). Store the crankshaft in a vertical position to avoid damage (Figure 4-42), or support in a position to avoid sag.

Remove the main bearings from the block and from the main bearing caps. Remove the rear main oil seal from the block and from the main bearing cap. Examine the bearing inserts for signs of abnormal engine conditions such as embedded metal particles, lack of lubrication, antifreeze contamination, oil dilution, uneven wear, and wrong or undersized bearings (Figure 4-43). Also remember that keeping the main bearing caps in order is very important. The location and position of each main bearing cap should be marked. Inspect them for any unusual wear signs, and inspect the main bearing inserts for undersizes (Figure 4-44).

The block cannot be thoroughly cleaned unless all freeze plugs and oil gallery plugs are removed. It is imperative that all plugs be removed to allow for a

FIGURE 4-41 Lifting out the crankshaft *(Courtesy of Perfect Circle/Dana)*

FIGURE 4-42 Crankshaft storage rack *(Courtesy of Atlas Engineering and Manufacturing, Inc.)*

FIGURE 4-43 Check bearing inserts for indications of abnormal engine conditions.

FIGURE 4-44 Inspect each main bearing cap for wear. *(Courtesy of Perfect Circle/Dana)*

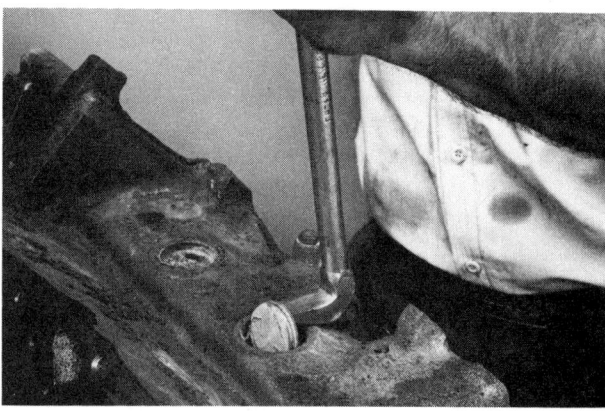

FIGURE 4-45 Removing flat-type core plug *(Courtesy of Perfect Circle/Dana)*

thorough cleaning. To remove cup-type freeze/core plugs, drive them in and then use a pair of channel lock pliers to pull them out. Flat-type plugs can be removed by drilling a hole near the center and then inserting a slide hammer to pull out the plug (Figure 4-45).

Sometimes removing threaded front and rear oil gallery plugs can be difficult. Using a drill and screw extractor can help. In some engines, the cup-type plug can be removed easily by using a slide hammer or by driving the plug out from the back side with a long rod.

 SHOP TALK _____

Using heat to melt paraffin into the threads will make removal much easier. As the part is heated, it will expand and the paraffin will leak down between the threads. Because the paraffin provides a lubrication, you will be able to loosen the two parts. Hot paraffin burns, so wear gloves when handling it.

After the teardown, the cylinder head and block and their parts must be visually checked for cracks or other damage before they are cleaned as described in the next chapter. While inspecting the parts, check to see if the engine has ever been torn down before. To do this, inspect the bearings for undersizes and check to see if the bearings are the original ones. For instance, some bearings like Ford's are date coded. If the engine has been torn down, use caution because a mistake could have been made during the last rebuild. Check for correct sizing before ordering any parts.

ORDERING REPLACEMENT PARTS

In large shops, ordering replacement parts can be done by the parts manager. In a small shop, they can be ordered by the person who is doing the rebuilding.

To get the parts for engine rebuilding from the OEM or a remanufacturer, the following information is very important:

- Make of the vehicle
- Year of the vehicle
- Type of vehicle—car or truck
- Vehicle identification number (VIN). It is located on the front dashboard and is visible through the windshield (Figure 4-46).
- Number of cylinders
- Engine cubic inch displacement
- Block, head, and crankshaft casing number (Figure 4-47).
- Any special factory markings (Figure 4-48).
- Type of transmission (automatic or standard) and the number of speeds.

FIGURE 4-46 Typical VIN plate number

FIGURE 4-47 Camshaft casing numbers give the type and year of cam.

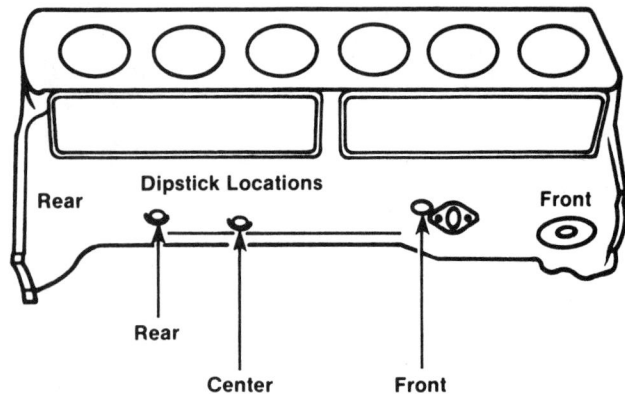

FIGURE 4-48 Special factory markings

When ordering parts it will help the rebuilder to understand the grouping of engines in a so-called family. That is, a family is a group of engines that has evolved from one basic design (Table 4-1). The major difference between the engines within a family is the cubic inch displacement ratings.

REPLACEMENT ENGINES AND KITS

Engine remanufacturers and aftermarket companies offer both partially assembled and complete engines. The partial kits are available in a variety of assembly and part stages.

Lower End Engine Kit

Basically, this kit includes a reground and microfinished crankshaft with main and connecting rod bearings in either 0.010, 0.020, or 0.030-inch

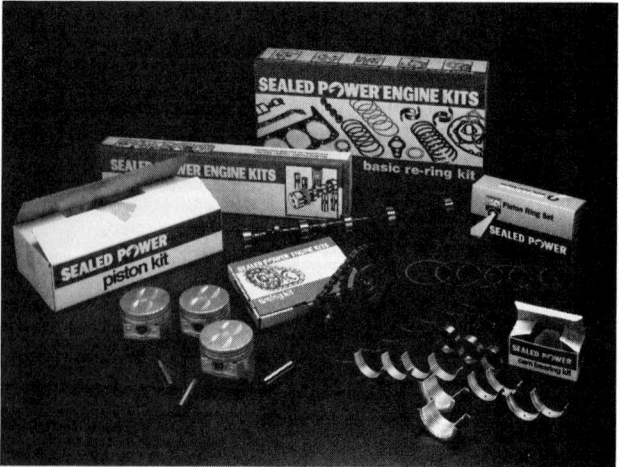

FIGURE 4-49 Engine parts kit *(Courtesy of Sealed Power Corp.)*

TABLE 4-1:   FORD V-8 ENGINE FAMILY CHART

Basic Engine Design	These engines evolved from basic design
MEL	383–430–462 CID
Y	239–272–292–312 CID
90° V	221–260–289–289 HP-302–302 BOSS-351 Windsor
335	351 Cleveland–351 BOSS-400 CID
FE	332–352–360–361 Edsel–390–406–410–427–428 CID
385	429–429 BOSS-460 CID

undersize. This kit is frequently called a *crankshaft kit.*

Engine Parts Kit

An engine parts kit (Figure 4-49) usually includes both the upper and lower engine components plus the following:

- Pistons in either 0.030, 0.040, or 0.060-inch oversize
- Rod bushings (if used)
- Pins
- Rings
- Timing chain and sprockets or cam and crank drive gears

Super or Master Kit

A super or master rebuilder kit usually consists of an engine parts kit (Figure 4-50) plus the following:

- Gasket and seal set
- New reground camshaft
- Cam bearings

FIGURE 4-50   Master rebuilder kit *(Courtesy of Perfect Circle/Dana)*

FIGURE 4–51   Basic short block assembly *(Courtesy of Jasper Engine & Transmission Exchange, Inc.)*

- Valve lifters
- Timing chain gears
- New or reconditioned oil pump (if the pump is part of the front cover, then the gears are furnished)
- Water tube (if used)

Short Block Assembly.   The basic unit (Figure 4–51) usually consists of a remanufactured block, rods, crankshaft, and camshaft plus new:

- Pistons
- Rings
- Bearings
- Timing gear or chain
- Freeze plugs
- Oil galley plugs

Complete Engine Assembly.   In addition to the components found on the short block, the complete engine assembly (Figure 4–52) provided by most remanufacturers includes:

- Remanufactured cylinder head(s)
- Remanufactured rocker arms or assemblies
- Hydraulic lifters
- Valves
- Valve springs
- Oil pan
- Oil pump
- Pushrods

Special Complete Engine Assembly.   This is a new assembly that some remanufacturers have added to their line. It includes everything found in a typical complete engine assembly plus the following:

- Intake manifold
- Valve covers

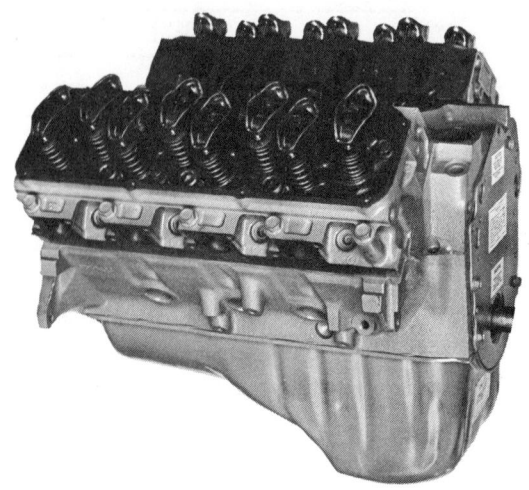

FIGURE 4–52   Complete remanufactured engine assembly *(Courtesy of Jasper Engine & Transmission Exchange, Inc.)*

- Harmonic balancer and pulley bolt
- Fuel pump
- Oil filler base and filter
- Oil pressure sending unit
- Spark plugs
- Set of gaskets

Diesel engines are also available as remanufactured or rebuilt units. A typical long block assembly is shown in Figure 4–53

CORE CREDIT

Engine remanufacturing companies usually offer a full core credit guarantee for nondisassembled engine cores. It guarantees that full credit will be given to the customer regardless of the core's internal condition. Disassembled engines are usually in-

FIGURE 4–53   Diesel engines are also remanufactured. *(Courtesy of Great Lakes Energy Systems, Inc.)*

spected by the company before a salvage value (determined by the amount of usable parts) is issued. Other cores not covered by guarantee include those with broken cranks, visible holes or cracks in the block, or obvious overbores. Individual inspections are performed when individual parts are submitted for core credit. If a rebuildable cylinder head core has three or less combustion chamber cracks, it is usually accepted for full credit. Recognizing a shortage of certain model cores, companies, as a general rule, will only receive in exchange a core that is the same as the one purchased.

Core Suppliers

In some instances, a major component such as a cylinder head, block, or crankshaft will have to be replaced if the engine is to be rebuilt or repaired. These parts are not easy to obtain, especially if the part is a limited production or a high-performance item. The most likely place to find these parts is a specialty company that supplies engine cores and component parts.

Import Vehicle Parts

Some rebuilder shops avoid imported cars because they believe the parts are too difficult to obtain. Although this was true a few years ago, these parts are much easier to find today. When several companies surfaced as national suppliers of a full line of imported parts, a number of domestic aftermarket companies also began manufacturing and distributing imported parts.

REVIEW QUESTIONS

1. Oil cooler lines must be disconnected at the radiator in vehicles with _____ .
   a. fuel injection
   b. turbocharge
   c. automatic transmission
   d. none of the above

2. Technician A removes the air conditioner compressor and the mounting bracket assembly and then disconnects the refrigerant lines. Technician B sets the assembly aside without disconnecting the refrigerant lines. Who is right?
   a. Technician A
   b. Technician B
   c. Both A and B
   d. Neither A nor B

3. Which of the following consists of two gears to provide the necessary mechanical advantage to lift an engine?
   a. cherry picker
   b. mobile crane
   c. canvas hoist
   d. all of the above

4. During engine teardown, Technician A has the engine sitting on blocks. Technician B has the engine securely bolted to an engine stand. Who is right?
   a. Technician A
   b. Technician B
   c. Both and B
   d. Neither A nor B

5. Cylinder head bolts should be _____ .
   a. loosened first, working from the ends toward the center
   b. loosened first, working from the center outward
   c. removed completely one at a time
   d. none of the above

6. If a valve seems to be stuck in the guide, what should be done?
   a. gently drive the valve through the guide
   b. raise the stem and file the excess metal
   c. both a and b
   d. none of the above

7. Which of the following is used to remove gallery plugs?
   a. drill
   b. screw extractor
   c. paraffin
   d. all of the above

8. Which of the following is a timing component assembly?
   a. chain and sprocket
   b. gear
   c. timing belt and sprocket
   d. all of the above

9. If the block is to be bored _____ .
   a. reaming is unnecessary
   b. new pistons should be used
   c. both a and b
   d. neither a nor b

10. To prevent sagging or warpage, store the crankshaft _____ .
    a. on its end
    b. on its side
    c. in suspension
    d. storage method is inconsequential

11. What must be done if the camshaft is worn?
    a. The camshaft must be replaced.
    b. The lifters must be replaced.
    c. Both a and b
    d. Neither a nor b

12. A family is a group of engines _____ .
    a. derived from one basic design
    b. sharing the same cubic inch displacement rating
    c. both a and b
    d. neither a nor b

13. Which kit is also called a crankshaft kit?
    a. engine parts kit
    b. short block assembly
    c. lower end engine kit
    d. master kit

14. Which of the following cores may be accepted for full credit?
    a. Those that have a broken crank.
    b. Those that have obvious overbores.
    c. Cylinder head cores with combustion chamber cracks
    d. None of the above

15. Which of the following is not included in an engine parts kit?
    a. pistons
    b. pins
    c. rings
    d. gasket and seal set

CHAPTER FIVE

CLEANING ENGINE PARTS

Objectives

After reading this chapter, you should be able to
- List the four categories of soil contaminants.
- Identify the three basic processes for cleaning automotive engine parts.
- Explain the six traditional methods of chemical cleaning.
- Describe the three most popular alternatives to traditional chemical cleaning systems.
- Explain the basic thermal cleaning process.
- Explain the various abrasive cleaning methods.
- Describe the types of airless shot and grit materials used in a blaster.
- Identify those situations in which manual cleaning is required.

When the block or cylinder head parts have been removed, they must be thoroughly cleaned (Figure 5-1). The cleaning method depends on the component to be cleaned and the type of equipment available. An incorrect cleaning method or agent can often be more harmful than no cleaning at all. For example, using caustic soda to clean aluminum parts will dissolve the part. So, use aluminum cleaning agents on aluminum parts only.

Only after all components have been thoroughly and properly cleaned can an effective inspection be made or proper machining be done. Therefore, all soils (grease, dirt, oil, scale, gunk, grime, crud, and rust) have to be removed.

REMOVING SOILS

The ability to understand specific soils that might be encountered is the first step toward saving valuable time and effort during the cleaning process. There are four categories of soil contaminants:

1. *Water-Soluble Soils.* The easiest soils to deal with are those that fall under the category of water soluble. All water-soluble soils—including dirt, dust, or mud—are characterized by their tendency to dissolve in the presence of water.
2. *Organic Soils.* Organic soils include those that contain carbon as part of their chemical makeup. In contrast to water-soluble

A

B

FIGURE 5-1 From (A) grime to (B) shine

soils, organic soils cannot be effectively removed with plain water. There are three distinct groupings of organic soils:
- The first group consists of petroleum by-products derived from crude oil. Tar, road oil, engine oil, gasoline, diesel fuel, grease, and engine oil additives are the more common members of this group.
- The second group, by-products of combustion, differs from the first in that combustion must occur before they can form. As combustion takes place,

varying amounts of gasoline, air, and engine oil chemically unite to form other undesirable products such as carbon, varnish, gum, and sludge.

- The third grouping consists of coatings. This category typically covers such items as rustproofing materials, gasket sealers and cements, paints, waxes, and sound-deadener coatings.

3. *Rust.* Rust is the result of a chemical reaction that takes place when iron and steel are exposed to oxygen and moisture. Corrosion, like rust, creates a similar chemical reaction between oxygen and metal containing aluminum. If left unchecked, both rust and corrosion can physically destroy unprotected metal parts quite rapidly. In addition to metal destruction, rust also acts to insulate and prevent proper heat transfer inside the cooling system.

4. *Scale.* When water containing minerals and deposits is heated (as in the cooling system), suspended minerals and impurities tend to dissolve, settle out, and attach to the surrounding hot metal surfaces. This buildup of minerals and deposits inside the cooling system is known as scale. Over a period of time, scale can accumulate to the extent that passages become blocked, cooling efficiency is compromised, and metal parts start to deteriorate.

The process for cleaning automotive engine parts can be divided into three basic categories:

1. *Chemical.* This method of cleaning relies primarily on some type of chemical action to remove dirt, grease, scale, paint, and/or rust (Figure 5–2). A combination of heat, agitation, mechanical scrubbing, and/or washing may also be used in the process to aid in removing surface contaminants. Chemical cleaning equipment includes small parts washers, hot/cold tanks, pressure washers, spray washers, and salt baths.

2. *Thermal.* This method of cleaning relies exclusively on heat to bake off or oxidize surface contaminants (Figure 5–3). Thermal cleaning leaves an ash residue on the surface which must be subsequently removed by an additional cleaning process such as airless shot blasting or spray washing. Thermal cleaning equipment includes conventional ovens and open flame ovens.

3. *Abrasive.* This method of cleaning relies on physical abrasion to clean the surface.

FIGURE 5–2 Chemical hot tank *(Courtesy of Kansas Instruments, Inc.)*

FIGURE 5–3 Thermal ovens can be used to clean a variety of cast-iron and aluminum parts. Here cylinder heads, valve covers, and oil pans are shown following oven cleaning.

This includes everything from a wire brush to glass bead blasting, airless steel shot blasting, abrasive tumbling, and vibratory cleaning (Figure 5–4). Chemical in-tank solution sonic cleaning might also be included here because it relies on the scrubbing action of ultrasonic sound waves to dislodge surface contaminants.

CHEMICAL CLEANING

The first and most traditional line of defense against soils involves the use of cleaning chemicals. However, there is a major concern on the part of rebuilders as to what chemicals to use. Chlorinated hydrocarbons and mineral spirits may have some health risks associated with their use through skin exposure and inhalation of vapors. Hydrocarbon cleaning solvents are also flammable. The use of a

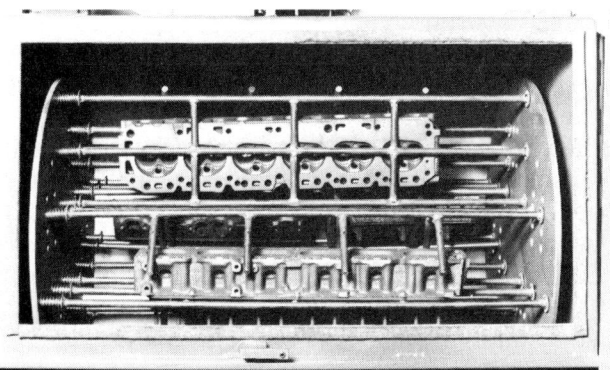

FIGURE 5-4 Heads cleaned in a steel shot blaster *(Courtesy of Kansas Instrument, Inc.)*

water-based nontoxic chemical can eliminate such risks. But those who prefer hydrocarbon solvents say solvents clean better than their water-based counterparts. Others say there is no difference depending on the water-based chemical that is used.

Then there is the question of what to do with the used cleaning solution. Hydrocarbon solvents are labeled hazardous and/or toxic, and require special handling and disposal procedures. The makers of many water-based cleaning solutions claim their products are biodegradable. But once the cleaning solution has become contaminated with grease and grime, it too becomes a hazardous or toxic waste that can be subject to the same disposal rules as a hydrocarbon solvent. Yet because there is so much latitude in local disposal rules, some municipalities allow neutralized water-based biodegradable cleaning solutions to be dumped down the sewer, and others do not.

Another alternative to the liquid waste disposal problem is recycling. Some manufacturers take a "cradle-to-grave" approach with their chemicals, offering waste-handling services. The old solvent is picked up and recycled by a distillation process to separate the sludge and contaminants. The solvent is then returned to service and the contaminants disposed of. (Independent services for maintaining hot tanks and spray washers are also available.)

Oil can also be removed from the solution by skimming it off the top of the tank. Both of these procedures are made easier if the tank has good service access. Note that evaporation is only applicable to water-based solutions.

The choice of any cleaning system must take into consideration the types of chemicals and/or solvents required for the process involved. With more stringent EPA and Occupational Safety and Health Organization (OSHA) regulations, some chemicals are becoming increasingly harder (and more expensive) to dispose of.

There are three types of solutions or solvents currently used in cleaning systems: water-based, mineral spirits (Stoddard solvent), and chlorinated hydrocarbons (carburetor cleaner). Of these, the water-based solutions are the easiest to dispose of because most of the waste volume can be evaporated. The mineral spirit solvents can be pumped out of a tank and recycled through a filter process. They are generally not dumped as a waste product.

Carburetor cleaner solvents, which are used to clean small parts and in some cold immersion tanks, cause the most problems. First, the EPA considers the fumes to be carcinogenic. Second, carburetor cleaner cannot be reprocessed or easily disposed of. Third, the waste sludge must be incinerated at certified facilities. The expense incurred to dispose of the liquid and sludge can be several times that of water-based cleaners.

Common Chemical Cleaning Agents

Since dirt, dust, and mud are the only soils that are naturally soluble (that is, they dissolve in water), all other soils must be made soluble by chemicals before they can be rinsed away. The three types of cleaning chemicals most commonly used in the automotive trade are:

1. Alkaline base
2. Acid base
3. Emulsifiable, solvent type

Alkaline-based chemicals (soaps and detergents) are best suited for removing all forms of organic soil and light rust and tend to be most effective when heated. For example, for every 20-degree Fahrenheit-increase above 140 degrees (up to and generally not exceeding 200 degrees Fahrenheit), cleaning time is reduced by approximately 50 percent. Of course, as the temperature goes up so do energy costs. Since trying to maintain this high chemical temperature can become quite expensive, as a compromise the thermostat can be turned down with effective results if the following conditions are observed:

- Soak time is increased.
- Agitation is increased and/or the concentration of the cleaning product is increased.

CAUTION: There is one caution to mention about all manufactured cleaning materials that cannot be overemphasized: Read the labels carefully before mixing or using (Figure 5-5).

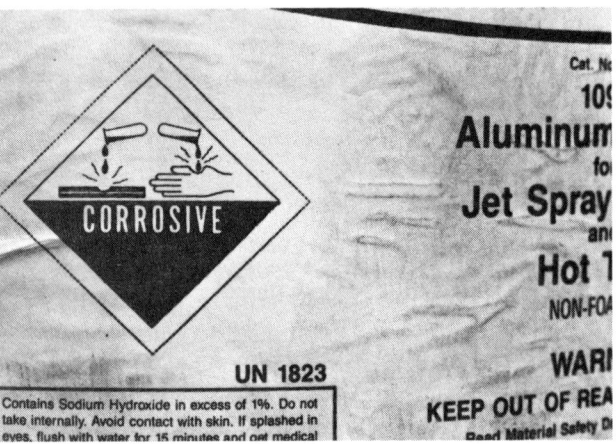

FIGURE 5-5 Read the label carefully before mixing or using. *(Courtesy of Perfect Circle/Dana)*

The three most common types of alkaline-base cleaners are:

1. Resin base
2. Silicate base
3. Phosphate base

Resin base has the longest life. Silicate (also called a chemical inhibitor) added to alkaline cleaners inhibits the chemical reaction on aluminum parts. Phosphate base cleaner, also known as caustic soda, is used in the majority of parts cleaning jobs. It has a short tank life and is dangerous to use and dispose of. Its use is prohibited in some areas.

CAUTION: Remember that when using a phosphate-base cleaner, its waste *must* be disposed of in accordance with EPA regulations. In fact, check with the local EPA office for the proper procedure for disposing all cleaning and hazardous waste that is generated in the shop.

The relative strength of a given alkaline or acid-based cleaning solution is based on a scale that rates chemicals from a strong acid to a strong alkaline. This scale is commonly known as a pH scale. The pH scale runs from 1 to 14 with 1 indicating the strongest acid, 7 indicating the neutral point, and 14 representing the strongest alkaline solution. Anything below 7 is considered acidic and anything above 7 is alkaline. To put things into perspective, most heavy-duty alkaline cleaners range from 10 to 12 on the pH scale, with lye being one of the strongest of the alkaline chemicals. Water is considered neutral and acid used in the automotive field generally falls in the range of 1.5 to 2 on the pH scale.

Acid-based cleaners are primarily used to remove scale or rust. They act to break down scale, making it water soluble. Once in a soluble state, water can be used to flush away the dissolved scale. When acid is used to remove rust, however, it actually dissolves a thin layer of the surface metal that is supporting the rust. Whenever any type of acid-based cleaner is used, keep in mind that acids cannot penetrate organic soils (grease, oil, and the like). If using acid to clean any surface that contains organic material, make sure that it is removed first. As mentioned previously, heat enhances the cleaning ability of most chemicals. Unfortunately, it is not always practical nor possible to heat the chemical solution in every cleaning situation. For that reason emulsifiable solvent chemicals, or degreasers as they are more commonly called, can be used.

Degreasers are designed to clean at room temperature and are most effective in dealing with the by-products of petroleum. The cleaning action of a degreaser starts when wetting agents in the product convert oil binders to soluble soap. Once in a soluble state, the soil is ready to be rinsed away. In many cases where soil is heavily layered, caked on, or difficult to access (as in oil galleries, cooling passages, and so on) they must physically scrub or scrape the surface to ensure adequate chemical penetration.

WARNING: When working with any type of cleaning chemical, be sure to wear protective gloves and goggles and work in a well-ventilated area.

In recent years, the increased use of aluminum and aluminum alloys in parts that were traditionally composed of cast iron has made the use of metal inhibitors mandatory. A metal inhibitor is a chemical additive that helps prevent the acid or alkaline chemical from dissolving the softer aluminum metal. Once a strong acid or alkaline chemical is properly inhibited, the base metal will be protected without compromising the cleaner's effectiveness.

Once the soil's origin is determined and the appropriate chemical cleaner selected, all that is left is to decide on is the most effective method of cleaning the part. But before deciding on the most effective method of cleaning for the situation, keep in mind that cleaning devices serve two purposes. They provide the rebuilder with a method of applying cleaning chemicals and must be able to remove soluble soils after cleaning has taken place. The proper selection of a cleaning machine and cleaning chemical helps reduce labor time considerably and can even make parts cleaning a tolerable event.

SHOP TALK —————

Before using, learn how to operate the equipment. The best way to learn is from someone who already knows the equipment. The next best way is to study the equipment manufacturer's operating instructions. These recommendations also apply to selecting the cleaning compound and preparing the solution.

STEAM CLEANING

Steam cleaning was, at one time, considered the best method of underhood cleaning. This chemical cleaning method is designed specifically to remove both the by-products of petroleum and water-soluble soils (Figure 5-6). Cleaning action is provided when pressurized, superheated water (reaching temperatures of up to 512 degrees Fahrenheit) is allowed to suddenly depressurize as it passes through a calibrated orifice located in the nozzle of the steam gun. The instant the pressure is released, the superheated water explodes as it begins to boil, converting some of the water to steam. This almost instantaneous expansion of the superheated water acts as a propellant to the water droplets that were not converted to steam. As the propelled water droplets hit the part's surface, they blast the deposits away. After the steam cleaning is completed, all surfaces should be flushed with clean water from a high-pressure washer and air dried.

In recent years, the use of steam cleaning in rebuilding shops has declined very rapidly. Although the basic method of steam cleaning is still sound, its decline in popularity among rebuilders is due to the following:

- Environmental considerations require that steam cleaning must be performed in a *closed loop.* That is, the runoff from the cleaning process can no longer be carried off in a public sewer system or on the ground. The runoff must be contained within the steam cleaning system. This is because the runoff contains heavy metal and other hazardous materials that are found in steam cleaning chemicals.
- Steam cleaning usually must be done in an uncongested portion of the shop or in a separate building.
- Care must be taken to protect surrounding painted surfaces and exposed skin that could come into contact with the steam's heat and chemicals. In addition, care must also be taken when working on the slippery floor that this process creates.
- Another hazard associated with steam cleaning is the danger of electrical shock if the machine is not properly grounded.
- Steam cleaning is very labor intensive. Most shops cannot justify the labor cost for using open steam cleaning.

PARTS WASHERS

Parts washers (often called solvent tanks) are one of the most widely used and inexpensive methods of removing grease, oil, and dirt from the metal surfaces of a seemingly infinite variety of automotive components and engine parts. A typical washer setup (Figure 5-7) might consist of a tank to hold a given volume of solvent cleaner and some method of

FIGURE 5-6 The open steam cleaner is no longer as popular a method of engine cleaning as it once was because of EPA regulations that stipulate the system must be operated in a closed loop.

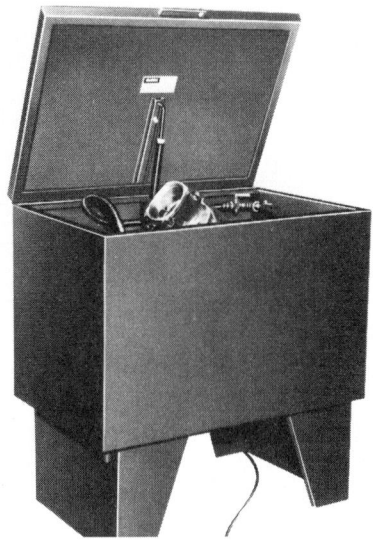

FIGURE 5-7 A parts washer permits hand washing of small engine parts. *(Courtesy of Kansas Instruments, Inc.)*

solvent application (a cleaning tub may or may not be utilized). These methods include soaking, soaking and agitation, solvent streams, and spray gun applicators. While there are minor advantages and disadvantages to each method, all methods generally require a certain amount of brushing, scraping, or agitation to speed up and increase the solvent's cleaning effectiveness. Because the type of solvent cleaners used (liquid petroleum or chlorinated solvent) in most parts washers is highly susceptible to evaporation and fuming, parts washers are restricted to operating at cold or room temperatures.

WARNING: Prolonged immersion of the hands in a solvent can cause a burning sensation. In some cases, a skin rash might develop. When cleaning items in a parts washer, direct the solvent stream away from the hands as much as possible or, better still, wear gloves.

Most small parts washer tanks use some type of fluid circulation system to move the cleaning solvent from the tank's reservoir through at least one filter and onto the part being cleaned. Circulation of the solvent is usually achieved by employing a small motor and pump. The cleaning itself takes place in the tank and, after being used, the solvent is returned to the reservoir to be used again. To prevent solvent circulation problems, heavy deposits of sludge should be removed before putting parts in the washer. This practice also helps to extend the life of the solvent.

Rather than scrubbing engine parts by hand, many rebuilder shops are turning to vibratory cleaners (Figure 5–8). They can do a job on small parts in 5 to 10 minutes.

FIGURE 5–8 Typical vibratory parts cleaner *(Courtesy of L. S. Industries, Inc.)*

FIGURE 5–9 When working in a chemical solvent be sure to wear the safety gloves described in Chapter 1.

SOAK TANKS

There are two types of soak tanks: cold and hot. Cold soak tanks are commonly used to clean carburetors, throttle bodies, and aluminum parts. A typical cold soak unit consists of a tank to hold the cleaner and a basket to hold the parts to be cleaned (Figure 5–9). After soaking with or without gentle agitation is complete, the parts are removed, flushed with water, and blown dry with compressed air. When rinsing parts that have been soaking in a cold tank, never use the parts washer to clean the soaking chemical from the part. The solvent in the parts washer will coat the part with an oily residue and increase the possibility of reintroducing dirt or grit into the clean part.

Cleaning time is short, about 20 to 30 minutes, when the chemical cleaner is new. The time becomes progressively longer as the chemical ages. Agitation by raising and lowering the basket (usually done mechanically) will reduce the soak period to about 10 minutes. Some more elaborate tanks are agitated automatically by a heavy-duty electric motor in conjunction with a cam driven unit attached to the parts platform. This action can be switched on and off as desired.

All cold soak powdered chemicals and most of the liquid solvents must be mixed with water or a petroleum product such as kerosene or diesel fuel to make the cleaning solution. Mixing ratios of both chemicals and solvents vary over a wide range, depending on the manufacturer and the type of cleaning to be done. This information is usually found on the container label.

The best cleaning agents for aluminum parts are those that are designated for use on nonferrous metal. These materials usually contain a chemical inhibitor that is not harmful to aluminum. Products that are made for cleaning aviation parts are generally suitable for cleaning both aluminum and magnesium because many aviation parts are made of aluminum-magnesium alloys. Once the cleaner is mixed as directed by the manufacturer, it is generally kept in a cold soak tank. Parts are soaked in the solution until clean and then removed and rinsed off with water.

 SHOP TALK _____

Some soils on aluminum parts can also be removed by using a steam-cleaning unit and a mild solution (not caustic) that contains a chemical inhibitor. However, this method will not remove carbon and other cooked-on deposits from places such as in piston ring grooves and combustion chambers of aluminum heads.

WARNING: Cold tank descaling solutions are very strong and dangerous to handle. Acid burns will result if descaling solution splashes on the skin or clothes. Cold tank maintenance usually consists only of changing the chemical solutions when needed and doing specified services on any mechanical or electrical accessory attached to it that works in conjunction with it. Check the service manual for complete maintenance instructions.

Cold tank descaling compounds are generally acid and must be kept in a separate cold soaking tank. After the parts are degreased from the tank holding the acid solution, remove the part from the tank and thoroughly rinse away the acid residues with clean water. Then place them in the tank containing the acid descaling solution. After a short period of time (never more than 2 hours), take the parts out of the tank and rinse them again to remove all traces of the acid. Any intermixing of the acid and alkaline solutions results in a chemical reaction that destroys the acid.

Hot soak tanks are actually heated cold tanks. The source of heat is either electricity or gas (either natural gas or propane). The electrical elements are inside the hot tank but are shielded from the solution. The gas heating elements are located outside and underneath the tank. The solution inside the hot tanks usually ranges from 160 to 200 degrees Fahrenheit. Most tanks are generally large enough to hold an entire engine block (and its related parts), and the tanks themselves are filled with a strong alkaline-based cleaner.

WARNING: Hot tank cleaner chemicals should never be handled. Use a scoop and slowly add the chemical to the water or solution to avoid spillover and to give an even distribution.

Hot tanks use a heated cleaning solution (either hydrocarbon or water based) to "boil out" dirty parts. It is a simple immersion process that relies on heated chemical cleaning to lift the grease and grime off the surface. Liquid or parts agitation may also be used to speed up the job. Agitation helps shake the grime loose and also helps the liquid penetrate blind passageways and crevices in the part (Figure 5–10).

Hot tank cleaning is not a labor-intensive process since parts can be left to soak, but soaking takes time, even if the solution is agitated. The actual amount of time it takes to clean an engine block, cylinder head, or other part depends on how dirty the part is, the nature of the grime, and the type of chemical used. Generally speaking, it takes one to several hours to soak most parts clean.

Once the grime comes loose from the part being cleaned, it settles to the bottom of the hot tank and forms a layer of sludge, which must be periodically removed from the tank and disposed of. Oil usually rises to the top, forming a layer of scum on the surface, which can be skimmed off or soaked up.

After removing parts from the hot tank, they should be rinsed immediately with hot clean water or a steam cleaner and then blown dry. Failure to promptly remove any residual cleaning material results in the formation of a hardened film that will be very difficult to remove later.

A broad range of chemicals is available for hot/cold tanks (and spray washers and parts washers). They can be tailored to the specific type of parts being cleaned. Iron, aluminum, or both can be cleaned in the same tank by using the proper chemical.

Hot tank maintenance requirements are basically the same as for cold tanks. Added requirements include servicing and maintaining the heating equipment and guarding more carefully against solution splash that might affect the heating units.

Water cleaning equipment that eliminates or reduces soaps, solvents, and detergents is now in use. Figure 5–11 shows a typical hot water tumbler cleaning unit.

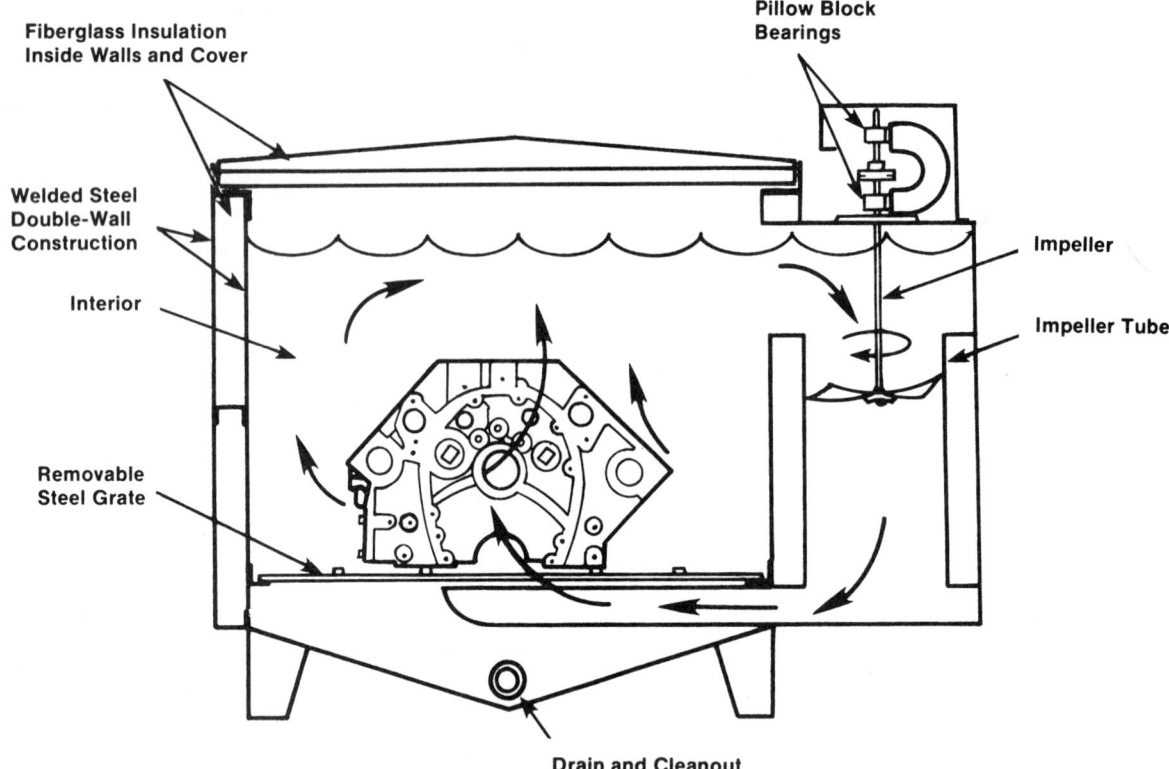

FIGURE 5-10 Typical agitated hot cleaning tank and how the turbulent agitating action is accomplished.

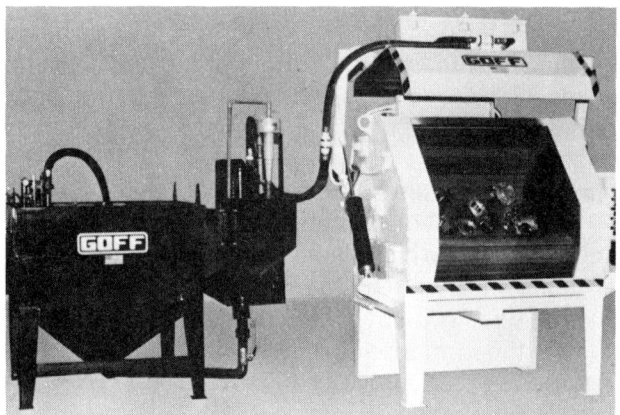

FIGURE 5-11 Typical chemical-free hot-water parts cleaner *(Courtesy of Goff Corp.)*

FIGURE 5-12 Hot spray machines do a fine job of post-machining cleanup. *(Courtesy of Bayco, Inc.)*

HOT SPRAY TANKS

Because of the EPA regulations against open hot or steam cleaning, the spray tank has become popular with rebuilders. The hot spray tank resembles a large automatic dishwasher and is designed to remove organic and rust soils from a variety of automotive parts (Figure 5-12). In addition to parts being bathed and soaked as in the hot soak method, spray washers add the benefit of moderate pressure cleaning. In a typical spray cabinet, linear or circular configurations of spray nozzles apply the hot alkaline-based solution from all directions ensuring more effective cleaning. With some spray washers, the part sits on a turntable that rotates so liquid can be divided at every surface. As the soil is dislodged, it is washed away by the continuous streams of cleaner.

Using a hot jet spray washer can cut cleaning time to less than 10 minutes. Normally a caustic soda or a strong soap solution is used as the cleaning agent. The speed of this system, along with lower operating costs, makes it popular with many machine shop owners.

 SHOP TALK _____

Caustic soda, also known as sodium hydroxide, can be a very dangerous irritant to the eyes, skin, and mucous membranes. These chemicals should be used and handled with care. Because of the accumulation of heavy metals, it is considered a hazardous waste material and must be disposed of in accordance with EPA guidelines.

Spray washers are often used to preclean engine parts prior to disassembly. Though a few shops still use a high-pressure steam spray wand for the same purpose, a spray wand is a one-person, labor-intensive (not to mention messy) job and must be operated under certain restrictions. A pass-through spray washer (Figure 5–13), on the other hand, is fully automatic once the parts have been loaded, and the cabinet prevents the runoff from going down the drain or onto the ground (which is not permitted in many areas because of local waste disposal regulations). Spray washers are also useful for post-machining cleanup to remove machine oils and metal chips.

Spray washers come in two basic varieties: top loading (Figure 5–14) and front loading (Figure 5–15). Most use low-pressure (50 psi) spray nozzles

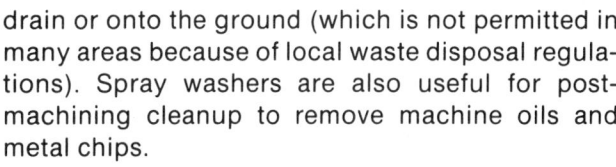

FIGURE 5-14   Top-loading hot spray tank *(Courtesy of Hartridge Equipment Corp.)*

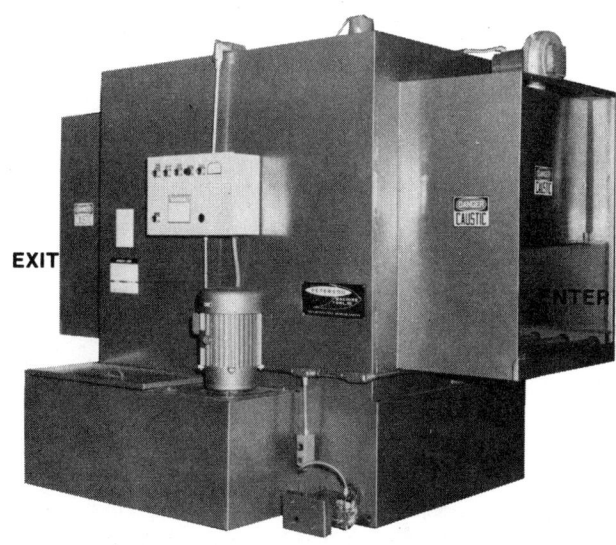

FIGURE 5-13   Typical high-pressure pass-through spray cleaning machine *(Courtesy of Peterson Machine Tool, Inc.)*

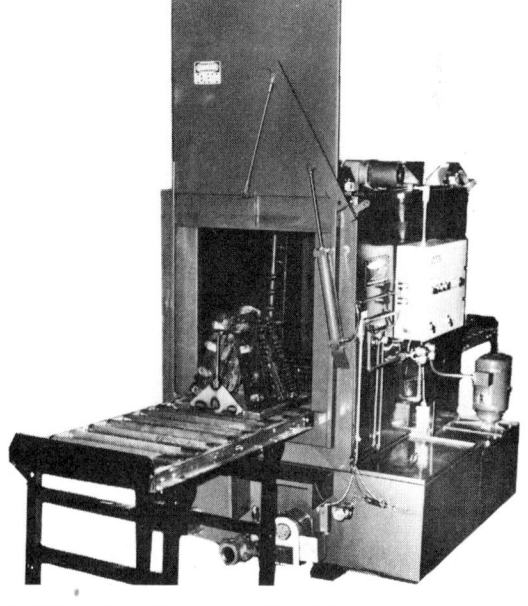

FIGURE 5-15   Front-loading jet spray washer *(Courtesy of Kansas Instruments, Inc.)*

FIGURE 5-16 Typical waste separator *(Courtesy of Kwik-Way Manufacturing Company)*

to wash parts clean, although some spray washers (called power washers) use high pressure (80 to 180 psi or higher) to blast surfaces clean. Like a hot tank, the sludge from the spray washer must be cleaned out periodically and the cleaning solution replenished or replaced. Other maintenance includes cleaning the nozzles and greasing.

Regardless of the chemical cleaning method used, periodic cleaning or replacement of the chemicals and solid material filters is necessary. Failure to "sweeten" or maintain the proper chemical concentration of the cleaner not only reduces cleaning effectiveness but also increases operating costs and labor time as well. Also, the surface of the tank should be skimmed (preferably when cold) daily or weekly to remove any oil, grease, or scum that accumulates at the surface. As far as actual tank cleaning and sweetening processes are concerned, the best source of information will come from the equipment manufacturer or the chemical supplier.

Some spray washers are equipped with a filter system that is capable of processing hundreds of gallons of chemical cleaning solution, removing most sediment, and trapping most oils and miscellaneous suspended waste products. An automatic fluid level sensor disconnects power when it is time for the waste to be emptied. Figure 5-16 illustrates a waste separator that removes hazardous materials from the cleaning solution for disposal.

ALTERNATIVE CLEANING METHODS

The cleaning methods covered so far in this chapter are considered to be traditional chemical

procedures. But in this age of material handling sheets, right-to-know laws, and concerns over hazardous waste generation and disposal, it is impossible to ignore the fact that cleaning equipment manufacturers and chemical producers are currently under much pressure to find alternatives to the environmentally sensitive acid-based, alkaline-based, and solvent-based cleaning chemicals that are currently being used. Chemicals are becoming difficult to dispose of properly and pose a major health risk to those who have to breathe the fumes or handle the cleaning products.

Three of the most popular alternatives to traditional chemical cleaning systems include:

1. *Ultrasonic Cleaning*. This cleaning process has been used for a number of years to clean small parts like jewelry, dentures, and medical instruments. Recently, however, the use of larger ultrasonic units has expanded into small engine parts cleaning (Figure 5-17). Ultrasonic cleaning systems utilize high-frequency sound waves to create microscopic bubbles that burst into energy to loosen soil from parts. Because the tiny bubbles do all the work, the chemical content of the cleaning solution is minimized, making waste disposal less of a problem. At the present time, however, the initial cost and handling capacity of ultrasonic equipment is its major disadvantage.

2. *Citrus Chemicals*. Some chemical manufacturers are starting to develop citrus-base cleaning chemicals as a replacement

FIGURE 5-17 Typical ultrasonic cleaning system *(Courtesy of Ramco Corporation)*

for the more hazardous solvent and alkaline-based chemicals currently used. Because of their citrus origin, these chemicals are safer to handle, easier to dispose of, and even smell good. As to their effectiveness, preliminary results show a great deal of promise, but, like ultrasonics, cost is a factor. At the present time, citrus-based cleaners cost more to produce than conventional chemicals, which is reflected in their high price.

3. *Salt Bath.* Another alternative in chemical cleaning is the salt bath (Figure 5–18). It is a unique process that uses high-temperature molten salt to dissolve organic materials including carbon, grease, oil, dirt, paint, and some gaskets. For cast iron and steel, the salt bath operates at about 700 to 850 degrees Fahrenheit. For aluminum or combinations of aluminum and iron, a different salt solution is used at a lower temperature (about 600 degrees Fahrenheit). The contaminants precipitate out of the solution and sink to the bottom of the tank where they must be removed periodically. The salt bath itself lasts indefinitely as long as the salt is maintained properly. Cycling times with a salt bath are fairly quick, averaging 20 to 30 minutes. Like a hot tank, the temperature of the salt bath is maintained continuously.

Typically, a salt bath is set up like an assembly line. When the parts come out of the salt tank, they are allowed to air cool until they reach the right temperature (about 550 degrees Fahrenheit) to enter a water bath to rinse off the residual salts. The temperature at which the rinse occurs is critical. If the parts are too hot when they go in, there is a danger of cracking due to thermal stress. If they are allowed to cool too much, the salt can harden and bond to the surface. Because of the high initial cost, the salt bath process is limited mainly to high-volume rebuilders. However, the disposal of contaminated salt waste can be as much of a problem as some caustics.

THERMAL CLEANING

Thermal cleaning ovens, especially the pyrolytic type, have become increasingly popular with rebuilders. The main advantage of the thermal cleaning alternative is a total reduction of all oils and grease on and in blocks, heads, and other parts. The high temperature inside the oven (generally 650 to 800 degrees Fahrenheit) oxidizes all the grease and oil leaving behind a dry, powdery ash on the parts. The ash must then be removed by airless shot blasting or washing. Vapors and smoke produced by the baking process are consumed by an afterburner in the oven stack. The parts come out bone dry, which makes subsequent cleanup with shot blast or glass beads easier because the shot will not stick.

Ovens come in two basic varieties:

1. *Convection (Figure 5–19).* Convection ovens are like a kiln—they use indirect heat to bake the parts. That is, by heating the part all the way through, the inside and outside are baked clean.
2. *Open or Direct Flame (Pyrolytic) (Figure 5–20).* These ovens impinge a flame direct-

FIGURE 5–18 Typical salt tank cleaning system
(Courtesy of Kolene Corporation)

FIGURE 5–19 Conventional type of thermal oven
(Courtesy of Bayco, Inc.)

FIGURE 5-20 Direct flame or pyrolytic thermal ovens can handle some large engines. *(Courtesy of Am/Pro Machinery, Inc.)*

ly on the surface of the parts to sear off the contaminants. That is, pyrolytic ovens rotate the parts on a rotisserie-like device to allow heat dissipation throughout the component, regardless of its configuration.

The open-flame or pyrolytic oven action is similar to that of a self-cleaning kitchen oven. The cleaning cycle time is generally 1 to 4 hours, including cooling time. The smoke from the burning process is burnt in an afterburner, leaving carbon dioxide as the exhaust gas that is emitted.

Generally, open-flame cleaning is a three-step process.

Step 1. Parts are put in the oven and heated, which leaves a hard, dry deposit on them.

Step 2. After the parts have cooled somewhat, they are put into an airless blaster where an abrasive, such as steel shot, blasts away the dry residue. This operation takes 10 minutes or less.

Step 3. The parts are shaken in a tumbler to remove the steel shot from the oil/water passageways and the bolt holes. All loose shot can be recycled for reuse. Further details on the airless blaster and the tumbler are given later in this chapter.

 SHOP TALK —————

If a tumbler is not used, the parts must be shaken after blasting to remove excess shot.

Figure 5-21 shows a typical complete thermal cleaning system—pyrolytic oven, airless blaster, and shaker. The metal parts are loaded in specially designed baskets that are easily transferred by hoist from unit to unit. Sometimes large parts are transferred by use of an overhead crane (Figure 5-22).

FIGURE 5-21 Complete thermal cleaning system *(Courtesy of Bayco, Inc.)*

FIGURE 5-22 Large parts are transferred by using an overhead crane. *(Courtesy of Bayco Inc.)*

Some thermal cleaning systems even have a monorail system (Figure 5-23). This keeps individual parts from being loaded and unloaded by hand. All dust from the parts and baking process is collected by a dust collector or vacuum cleaner. In a complete thermal cleaning center such as this, floor-to-floor turnaround time can be as little as 40 to 45 minutes (20 to 30 minutes in the oven, 5 to 7 minutes in the shot blaster, and 5 minutes in the tumbler).

Cycle times for a conventional oven can run from 2 to 6 hours or more depending on the size of the oven and the load (larger loads take longer to reach baking temperature). Cleaning is still required afterward, which can be done by airless shot blasting or a hot spray washer.

 SHOP TALK _____

A slow cooling rate is recommended to prevent distortion that could be caused by unequal cooling rates within complex castings.

Heating aluminum heads is the best and easiest way to install valve guides and seats without galling (Figure 5-24). When the aluminum expands, the guides can be tapped in easily, without forcing. With aluminum parts, the oven temperatures should be kept between 400 and 475 degrees Fahrenheit to eliminate the possibility of warpage that might occur at high temperatures. As described in Chapter 9, ovens can also be used to straighten aluminum head casting.

One of the major attractions of cleaning ovens is that they offer a more environmentally acceptable process than chemical cleaning. But, although there is no solvent or sludge to worry about with an oven, the ash residue that comes off the cleaned parts

FIGURE 5-23 A monorail system featuring a thermal cleaner, blaster, and tumbler/shaker/cooler *(Courtesy of Rogers Machine Company)*

FIGURE 5-24 Heating aluminum heads in a thermal oven is the best and easiest way to install valve guides and seats without galling. *(Courtesy of Bayco Inc.)*

must still be handled according to local disposal regulations. In some areas, steel shot can be thrown in the regular trash. In others, it must be treated as a hazardous waste if lead is present in the residue.

The maintenance procedure given in the owner's manual must be followed if the ovens are to operate properly.

ABRASIVE CLEANERS

Most abrasive cleaner machines are normally used in conjunction with other cleaning processes (such as an oven, hot tank, or spray washer) rather

TABLE 5-1: COMMON MEDIA USED AS ABRASIVE CLEANERS*		
Media	**Available Sizes**	**Use/Result**
Glass beads	8 to 10 sizes; from 30 to 440 mesh; also many special graduations	Blending, light deburring, peening, general cleaning, texturing. Non-contaminating. Peening media for nonmetal removal cleaning (holds tolerances), retains critical part dimensions.
Aluminum oxide grit	10 to 12 sizes; from 16 to 325 mesh	Fast-cutting, matte finish. Do not use if etching is not wanted.
Aluminum shot	10 to 12 sizes; from 8 to 200 mesh	Primarily for cleaning aluminum. Gives bright or satin finish; expensive.
Stainless steel shot	6 to 8 sizes; 16 to 325 mesh	Cleaning or peeling parts where residual iron could be a problem; expensive.
Ceramic shot	6 to 10 sizes; 12 to 320 mesh	Cleaning or peening hard metals.
Salt	10 to 60 mesh	Good cleaning for all soils. No residue, inexpensive, but it can only be used once.
Nut shells crushed	6 sizes; wide-band screening	Cleaning, very light deburring. Good for fragile parts.
Garnet	6 to 8 sizes; wide band screening from 16 to 325 mesh	Noncritical cleaning, cutting, texturing. Noncontaminating for brazing steel and stainless steel.
Crushed glass	5 sizes; wide band screening from 30 to 400 mesh	Fast-cutting, low-cost, short-life, abrasive. Noncontaminating.
Silicon carbide	36 to 220 mesh	Extremely fast cutting.
Steel shot	12 or more sizes; close graduation from 8 to 200 mesh	General-purpose, rough cleaning foundry operations, etc. Peening.
Steel grit	12 or more sizes; close graduation from 10 to 325 mesh	Rough cleaning, coarse textures.
Plastic chips	3 sizes; fine, medium, coarse definite size particles	Cleaning, light deburring
Sand	10 to 50 mesh	General cleaning. Good for all soils.

*Courtesy of Zero Manufacturing Company

than as a primary cleaning process itself. Parts must be dry and grease-free when they go into an abrasive blast machine, or the shot or beads will stick. Table 5-1 lists the media used in abrasive cleaners and the resulting finishing results.

ABRASIVE BLASTER

The airless shot and grit blasters, mentioned earlier, are used best on parts that will be machined after cleaning (Figure 5-25). Rebuilders can choose from the various types of media that are available, such as blasting media, which comes in one of two basic types—shot and grit. Shot is round; grit is angular in shape (Figure 5-26). Steel shot and glass beads are used primarily for cleaning operations where etching or material removal are not desired.

Steel shot and glass beads are also used for peening the surfaces of certain parts to increase fatigue life and resistance to stress corrosion and cracking. Though steel shot is used more often for peening than glass beads because steel is heavier and peens with greater intensity, glass also offers certain peening advantages that are described in Chapter 8.

Steel shot comes in various sizes, with 110, 170, and 230 sizes being the most common for automotive cleaning and peening applications. The 170 size is the most common for the majority of cleaning applications (Figure 5-27). The larger size is used on stampings and brake shoes; the smaller 110 shot size is generally recommended for cleaning aluminum. Glass beads also come in various sizes, with 20-30, 30-40 and 40-60 sieve ranges being the most common. Other types of shot that are available for special cleaning needs include hard steel, stainless steel, ceramic, aluminum, and plastic. Ordinary steel shot has a hardness that ranges from Rockwell C 45

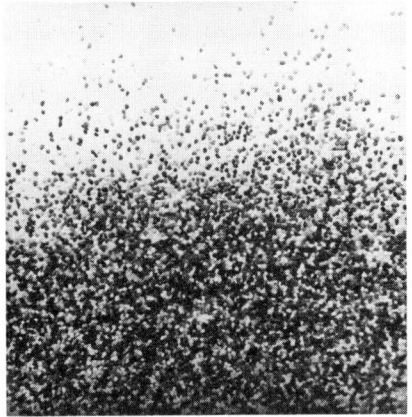

FIGURE 5-26   (A) Blaster shot and (B) grit

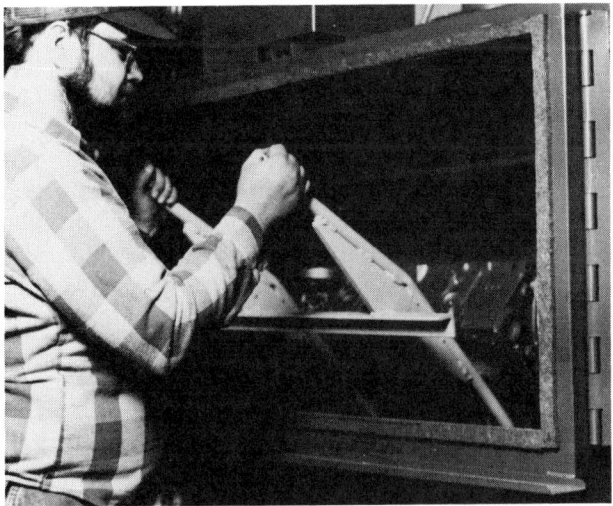

FIGURE 5-27   Loading a V-8 block into a shot blasting machine *(Courtesy of Kansas Instruments, Inc.)*

to 55, the average being about 51 or 52. This is hard enough for cleaning or peening ordinary steels. But for peening heat-treated or hard alloys, the shot must be at least as hard as the steel to get the maximum peening effect. Hard steel shot that ranges

FIGURE 5-25   Airless shot blaster *(Courtesy of Am/Pro Machinery, Inc.)*

A

B

FIGURE 5-28 (A) Valve cover washed in an alkaline cleaner and (B) then given a salt blasting. After blasting, the cover was washed in clean water.

from Rockwell C 55 to 65 with an average of about 61 to 62 should be used on hard metals.

Stainless steel shot can be used for cleaning or peening parts where residual iron left on the surface is not wanted. Ceramic shot, which is very hard (Rc 65), is used primarily for peening heat-treated and hard alloy steels where maximum strength is desired. Ceramic is also "clean" in that it leaves no residue on the surface. Ceramic shot lasts about as long as steel shot, but like stainless it is expensive. Salt is a good and inexpensive cleaner that leaves no residue, but it can only be used once (Figure 5-28).

Aluminum shot, by comparison, is soft and is used primarily to clean aluminum surfaces where a bright or satin finish is desired. Plastic is the softest material of all and is used for cleaning soft metals and plastics and for jobs where shot retention poses a serious problem, such as inside an engine.

The length of time that a given quantity of metal shot or glass bead media lasts in a cleaning or peening operation depends on a number of factors. These factors include the velocity or force with which the media impacts against the surface of the

parts, the angle at which it strikes (90 degrees is the most intense angle), the hardness of the metal being cleaned, and the media being used. The degree of dirt on the parts, the nature of the dirt on the metal, and the volume of parts being processed also help determine how long metal shot or glass bead media lasts.

Grit is used primarily for aggressive cleaning jobs or where the surface of the material needs to be etched to improve paint adhesion. Steel grit and aluminum oxide are the two most common media for this purpose. Because grit cuts into the surface as it cleans, it removes dirt and scale faster than shot blasting or glass beading. But it also removes metal, leading to some change in tolerances. The beneficial effect of grit blasting is that it roughens the surface, leaving a matte finish to which paint or other surface treatments will stick better than a peened or polished surface. On the other hand, grit blasting is an abrasive process that chews out pits in the surface into which pollutants and blast residue can settle. This leads to stress corrosion unless the surface is painted or treated. The tiny crevices also focus surface stresses in the metal, which can lead to cracking in highly loaded parts. Because of that, grit would never be used for peening.

The type of media that is used for a given job depends on the job itself and the type of equipment. Steel shot is normally used with airless wheel blast equipment, which hurls the shot at the part with the centrifugal force of the spinning wheel. Used with air blast equipment, glass beads are blown through a nozzle by compressed air.

The selection of a particular shot size depends on a number of things. First, the media must be small enough to reach into all the corners, crevices, and fillets on the parts. If cleaning gear teeth, for example, the shot must be small enough to penetrate all the way to the root of the gear teeth. With large, flat surfaces like engine blocks, stamped metal covers, and so on, large size shot can be used.

Another thing to consider is the speed at which a particular media cleans. The rate at which deposits are blasted loose depends on the number of impacts per second. The smaller the size of the steel shot or glass beads, the more particles that strike the surface per second. A smaller size, therefore, provides more impacts per pound of media. Smaller beads do the best job of removing light deposits, but larger beads are better for knocking heavier deposits loose.

Depending on the job, large beads might work better than small ones. A combination can also be used—the large beads knock the heavy deposits loose while smaller beads finish scouring the surface. Yet another alternative is to use a combination of beads and aluminum oxide grit if etching of the metal makes no difference.

One of the requirements of shot blast cleaning is that parts must be dry as well as grease and oil free, otherwise shot can stick to the parts. This means that greasy parts must first be washed and dried with shot blasting to remove the hard deposits not touched by washing or removed with a hot tank. Shot blasting is also used for subsequent cleanup operations after thermal cleaning to remove the ash residue and surface oxide left on the parts after baking.

The major problem rebuilders face after shot blast cleaning is the danger of shot retention. If steel shot remains in the parts, it can play havoc inside an engine. Blowing out the shot, shaking it out, or even washing it out are all techniques that can be used to deal with the threat of shot retention. Another approach is to use a soft shot such as aluminum or plastic that does not pose as much of a threat if it gets inside an engine or transmission. Rather than causing damage, soft shot will be crushed if it gets caught between moving parts.

With glass bead blast equipment (Figure 5–29), moisture is a concern because it causes the glass beads to stick together and gum up the equipment. Prevention of this problem requires maintaining the air dryer or desiccant filter on the air compressor so moisture does not enter the system. Moisture can also be a problem with steel shot in wheel blast equipment because the steel shot will rust if left in a highly humid environment for a long period of time. Rusty shot can leave an unattractive residue on parts. If the shot has become rusty, run a load of dummy parts through the equipment.

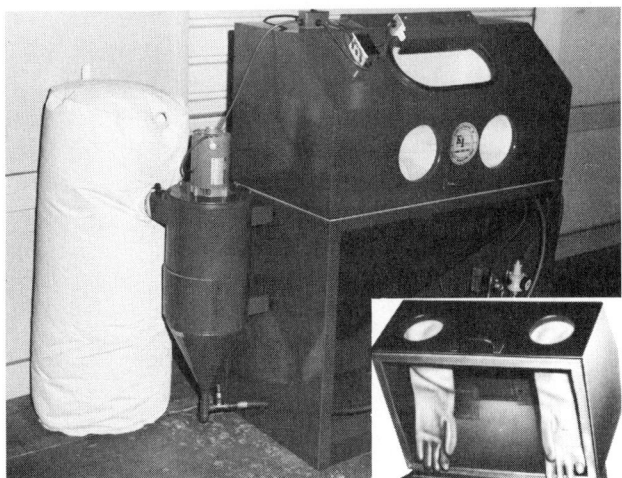

FIGURE 5–29 Glass bead blast equipment in operation. Some machines have the necessary protective gloves built in the machine. *(Courtesy of Kansas Instruments, Inc.)*

GLASS BEAD CLEANING

In glass bead cleaning, microscopic glass beads are blasted by compressed air against the surface to be cleaned. This method is particularly effective in removing carbon and hard, dry deposits from all types of metal. It does not work well on oily deposits. The result of glass bead cleaning is a bright, satin surface finish, making parts look like new.

It will not change the physical dimensions of the parts, so it works well on valve stems and other parts with critical tolerances. The glass beads also micropeen surfaces while blasting. Micropeening provides a better oil-retaining surface. It also eliminates jagged edges, peaks, valleys, or burrs caused by wear and machining. It is important to remove all glass beads from the parts' cavities after cleaning. On aluminum cylinder heads, tape off any oil galley holes and cam bearing surfaces.

Generally, glass beads are recirculated inside the cleaner cabinet by an electrically operated vacuum system. The beads can be used over and over until their dimensions are reduced to ineffectiveness for the type of cleaning being done. The bead sump must be cleaned when needed and new beads added. Electrical units must be serviced and protective equipment kept in repair. Otherwise, very little maintenance is required.

PARTS TUMBLER

A cleaning alternative that can save considerable labor when cleaning small parts such as engine valves is a small tumbler. Various cleaning media can be used in a tumbler to scrub the parts clean. This saves considerable hand labor and eliminates the dust problem created by a grinding wheel. In some tumblers, all parts are rotated and tilled at the same time.

VIBRATORY CLEANING

Shakers, as they are frequently called, use a vibrating tub filled with ceramic steel, porcelain, or aluminum abrasive to scrub parts clean. Most shakers flush the tub with solvent to help loosen and flush away the dirt and grime. The solvent drains out the bottom and is filtered to remove the sludge.

CLEANING BY HAND

Some manual cleaning must be done in every rebuilding job. Regardless of the cleaning process used, it is usually necessary to remove galley plugs and hand clean the oil galleries (Figure 5–30) and cylinders (Figure 5–31). Another often neglected

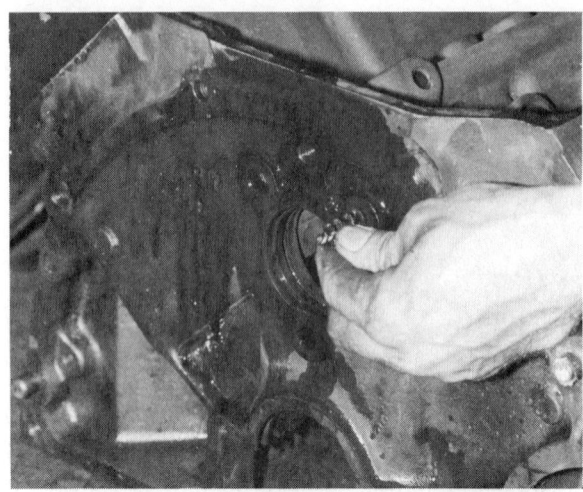

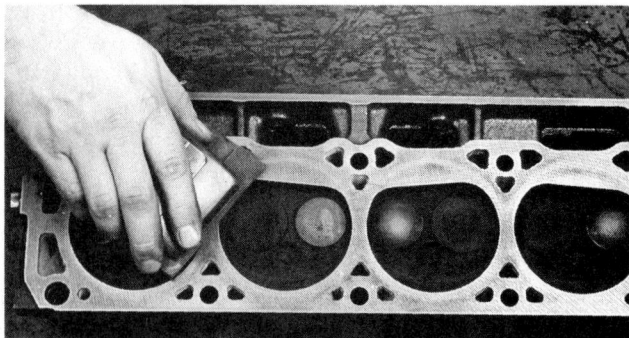

FIGURE 5–32 Cleaning cylinder block with abrasive paper *(Courtesy of Perfect Circle/Dana)*

area is between the heat shield and the bottom of the intake manifold, where carbon and oil deposits collect. The shield should be removed before cleaning the manifold. Residual dirt left in the engine system after cleaning leads to failure, so proper cleaning of all engine components is vital in the rebuilding process. Remove any surface irregularities with a fine, abrasive paper (Figure 5–32), making sure to keep any dirt out of the cylinder bores. The special power scraper pad shown in action in Figure 5–33 is guaranteed not to remove any metal.

FIGURE 5–30 Cleaning the oil galleries

FIGURE 5–31 Cleaning the cylinders with a soft brush

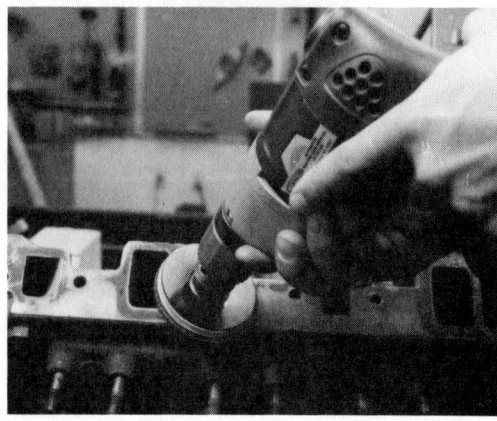

FIGURE 5–33 Electric drill equipped with power scraper pad *(Courtesy of Goodson Shop Supplies)*

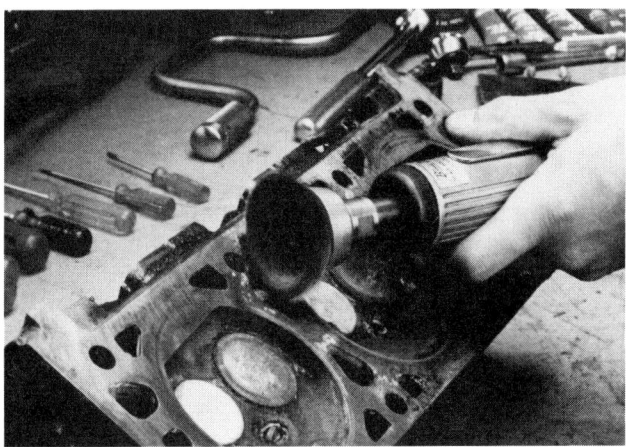

FIGURE 5-34 Using a soft wire to clean foreign material from the engine surfaces *(Courtesy of Fel-Pro Inc.)*

Carbon can be removed from parts using a twist type wire brush driven by an electric or air drill motor (Figure 5-34). Using brushes can often be a time-consuming job, so some shops use the wire brush in addition to another method. Frequently, an air gun with a round-tipped peening tool is used. (Operation of peening repairs is covered in Chapter 8.) Moving the gun in a light circular motion against the carbon helps to crack and dislodge the carbon for easier wire brush cleaning.

WARNING: Always wear a face mask when cleaning parts with a power-driven brush. Wire bristles can be thrown off the brush, so the mask provides necessary skin and eye protection.

Ring groove scrapers can be adjusted to fit the various standard ring groove widths and depths so that carbon can be removed without damage to the base metal. Clean pistons with solvent and a soft-bristle brush (Figure 5-35). Never use a caustic cleaner or wire brush. Another scraper that is effective is the gasket scraper, which is a wide, sharp-edged tool suitable for removing tightly bonded gasket sealer and material (see Chapter 14).

A soft wire wheel, driven by a grinder motor, is sometimes used to clean carbon and other foreign material from various components (Figure 5-36). Combustion chambers and similar areas can be cleaned by using a solid wire-filled brush for heads and a flared type for hard-to-reach places such as ports and cylinder walls. Be sure all surfaces are smooth and shiny.

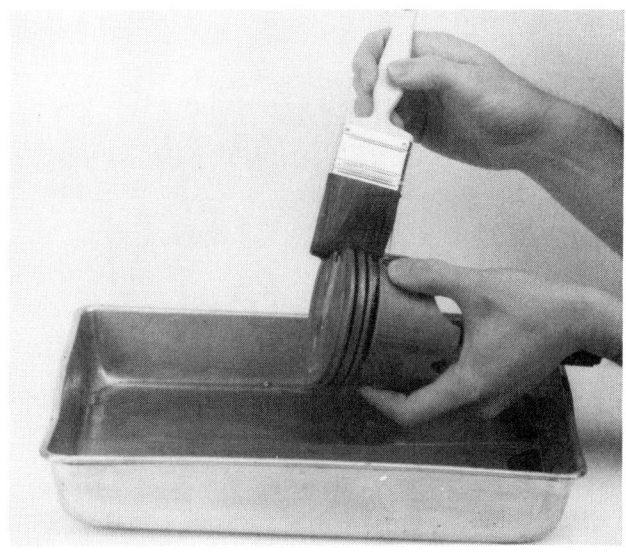

FIGURE 5-35 Careful cleaning of piston grooves with a soft brush, making sure that all abrasive matter and dirt is removed, is most important.

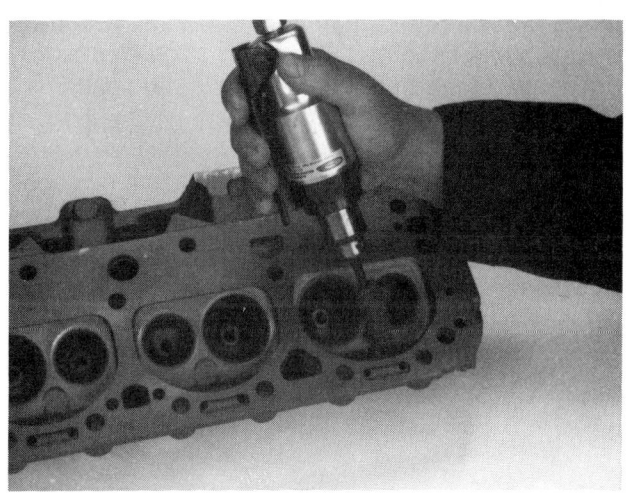

FIGURE 5-36 Electric grinder can be used to clean combustion chambers *(Courtesy of Goodson Shop Supplies)*

To clean valve guides, replace the carbon cleaning brush in the drill with a valve guide cleaner. The flat spring tension blades press against the guides and scrape out carbon without scratching the metal. Next, use a nylon valve cleaning brush to remove loosened carbon and varnish. Remember to thoroughly clean the head, face, and stem of the valve, particularly the area from face to stem. Carbon accumulations insulate valves, preventing them from dissipating heat. Carbon burns and pits. Get rid of all of it in the parts cleaner.

FIGURE 5-37 Marking a cleaned part *(Courtesy of K-Line Industries, Inc.)*

MARKING CLEANED PARTS

After the engine parts have been cleaned and dried, they should be identified by some type of marking system. One of the easiest ways of doing this is to use a self-contained white or yellow roll-on metal tube marker (Figure 5-37). To use, simply press the rubber bulb to prime the rolling ball point and write the desired information on the part or casting. The permanent paint marking will withstand oil, grease, and even wash cycles in caustic or soak cleaning chemicals without losing legibility.

REVIEW QUESTIONS

1. What do all organic soils contain?
   a. oil
   b. tar
   c. carbon
   d. nitrogen

2. Into which of the following categories can the process for cleaning automotive engine parts be divided?
   a. chemical, thermal, scaling
   b. thermal, scaling, abrasive
   c. scaling, abrasive, chemical
   d. abrasive, chemical, thermal

3. Which of the following chemical cleaners is both toxic and flammable?
   a. chlorinated hydrocarbons
   b. mineral spirits
   c. water-based
   d. none of the above

4. What is another name for a degreaser?
   a. alkaline base cleaner
   b. acid base cleaner
   c. emulsifiable cleaner
   d. none of the above

5. What is the thin residue that is formed on the cylinder walls by the combination of combustion heat, wall ring, engine oil, and piston movement?
   a. burnish
   b. glaze
   c. plateau
   d. none of the above

6. At what temperature are parts washers operated?
   a. room temperature
   b. 120 degrees Fahrenheit
   c. 160 degrees Fahrenheit
   d. 200 degrees Fahrenheit

7. What type of solvent cleaners are used in parts washers?
   a. liquid petroleum
   b. chlorinated solvent
   c. all of the above
   d. none of the above

8. Technician A uses cleaning agents designated for use on nonferrous metal when cleaning aluminum parts in a soak tank. Technician B uses a product designated for cleaning aviation parts. Who is right?
   a. Technician A
   b. Technician B
   c. Both A and B
   d. Neither A nor B

9. Failure to "sweeten" the cleaner in a hot spray tank will _____ .
   a. reduce cleaning effectiveness
   b. increase operating costs
   c. increase labor time
   d. all of the above

10. Generally, how many steps are there in an open-flame cleaning process?
    a. two
    b. three
    c. four
    d. five

CHAPTER SIX

INSPECTING THE CYLINDER HEAD AND RELATED PARTS

Objectives

After reading this chapter, you should be able to
- Identify and describe the problems that can be detected by making a valve inspection.
- List the different conditions that can contribute to valve burning.
- List various causes of valve breakage.
- Explain how to perform a valve train inspection.
- Determine the causes of valve guide wear.
- Identify the causes of seat recession.
- Understand the operation and purpose of the valve springs, rotators, and stem.
- Describe the steps of a camshaft drive inspection.

The overhead camshaft (OHC) valve train consists of the following parts:

- Valves
- Valve guides
- Valve rotators, locks, and springs
- Rocker arms
- Pushrods or valve lifters
- Camshafts
- Lifters

Modern automotive engines are highly refined machines. When these engines are properly maintained and operated, they perform efficiently and give dependable service for a remarkably long period of time. However, it would be unrealistic to think that wear does not take place during the normal operation of an engine. As a matter of fact, all mechanisms, simple or complex, will ultimately exhibit wear between moving parts, no matter how well lubricated these parts may be. Engine wear can be classified as either normal or abnormal, depending upon many factors.

Normal wear takes place in any area where there is relative movement, including the following:

- Engine valves and valve guides
- Cylinders, pistons, and piston rings
- Engine bearings
- Crankshaft main journals and crankpins
- Camshaft and valve lifters

Engine bearings are capable of operating under conditions of normal wear for long periods of time, provided that the engine lubricant is maintained according to the engine manufacturer's recommended specifications.

Abnormal engine wear can result from many causes, including the following:

- Improper engine operation
- Failure to maintain engine lubricant
- Inadequate or improper engine maintenance

Operating an engine under these conditions of abnormal wear can lead to premature breakdown of engine parts.

Once the engine parts have been thoroughly cleaned, they should be carefully inspected. All conditions that will lead to abnormal engine wear should be corrected before the existing or new parts are reassembled.

VALVES

The engines of the 1990s demand tougher valves and more of them. It is not unusual for a modern high-performance engine to have as many as thirty-two valves.

Over the past fifty years, a variety of materials has been used in valve manufacturing, including hot and cold rolled steels, high-alloy steel, chrome-plated steel, stainless steel, molybdenum, and stellite. The valve materials of the future are likely to be radically different from those employed today. Currently, automotive engineers are beginning to use pistons made of titanium and ceramics.

Titanium has been used in high-performance engine intake valves since 1969. Intake valves made of titanium-aluminum-vanadium alloy are half the weight of standard valves and have the same head size. They offer engine designers the advantages of light weight combined with superior corrosion resistance. The light weight of titanium valves means less inertia, so less horsepower is required to get the valve train moving.

Currently, most of the titanium valves sold are destined for racing applications. Engine builders for both professional and sports races in the National Association for Stock Car Auto Racing (NASCAR), the International Motor Sports Association (IMSA), and the National Hot Rod Association (NHRA) use titanium intake valves extensively. Titanium alloys are less commonly used for exhaust valves because the explusion of hot gases can affect their durability. Extra care in heat treating, design, and machining is critical in producing titanium exhaust valves. Actually, titanium valves are about six to seven times more expensive than steel valves.

The most important attributes that make ceramics ideal materials for engine valves are their lightness, hardness, and extreme temperature resistance. A one-piece ceramic engine valve weighs only a fraction of a steel valve. This has obvious advantages for reducing the reciprocating mass of the upper valve train, which in turn reduces strain on all the valve train components and increases rpm potential. Ceramics are also being considered for rocker arms, valve guides, valve seats, and lifters. Valve wear with ceramics is almost negligible. They can also be used for both intake and exhaust valves.

One-piece ceramic valves, at present, cost about twice as much as those made of steel.

Another use of ceramics is to apply a thin facing to intake and exhaust valves to help retain heat in the combustion chamber of diesel engines. Retaining heat improves combustion efficiency, so theoretically, ceramic-faced valves as well as pistons and cylinder liners could improve fuel efficiency appreciably.

It might be some time before the rebuilding industry feels the direct impact of the emerging breakthroughs in ceramic valves and other engine parts. Even so, there are many changes taking place in valve technology that are affecting rebuilders today. One change is the increased use of chrome-plated valve stems.

There are two main advantages of chrome-plated valve stems: a harder surface is created as well as a dissimilar surface that resists galling against a cast-iron guide. This is especially important in engines that burn unleaded fuels. Chrome-plated valves are not absolutely necessary from a rebuilder's standpoint, but many original equipment vehicle manufacturers have switched to chrome-plated valves to prolong valve life.

One of the implications of chrome-plated valves for rebuilders is that they cannot be ground down and reused. This means the rebuilder must use new valves or go to rechroming of valve stems after regrinding. The same is true for aluminized valves—the coating will come off when the valve is reground.

Valves can be costly in a rebuilding job. It takes about twenty-eight to thirty different manufacturing steps to produce a typical steel valve (Figure 6–1). Some valves require extra manufacturing steps. A two-piece or wafer tip valve, for instance, must have the head or tip welded to the stem. This weld must be annealed to relieve stress and the head flexed to test the strength of the weld. The valve is then straightened and finished. When a valve is hard faced with stellite, the hard alloy must be welded to it. A new process uses a high-temperature plasma spray to apply molten metal to the valve. A 1/4-inch radius is cut out of the valve face to accept the stellite plasma spray. After spraying, the valve face is recut and finished. Figure 6–2 gives the nomenclature of a valve assembly.

Several valve designs are in use in modern vehicles. The most common are shown in Figure 6–3. Although there are several design shapes to meet various specifications, there are still two basic types: rigid and elastic. The rigid valve is fairly strong, holds its shape, conducts heat very fast, and wears slowly. Unfortunately, it is more likely to leak and burn than the elastic valve. The elastic valve, on the

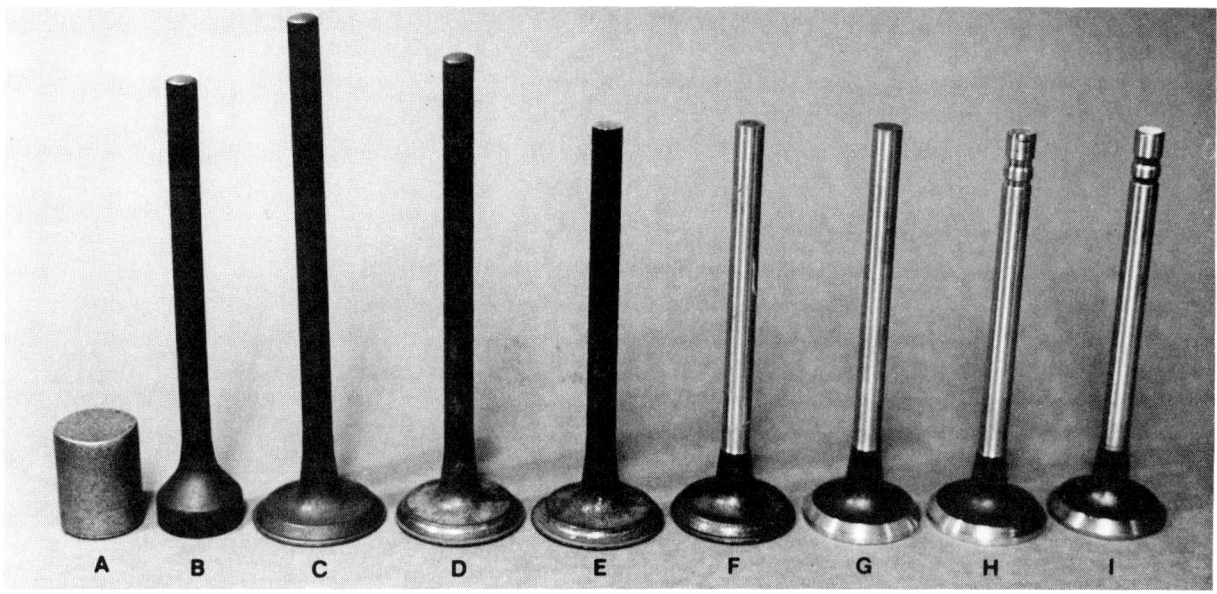

FIGURE 6-1 Steps in making a valve: (A) rough slug; (B) first stamping to draw the stem; (C) second stamping forms the head; (D) stem is straightened; (E) stem cut to length; (F) stem ground to size; (G) seat cut to size; (H) groove cut in the stem; (I) valve is ground and finished.

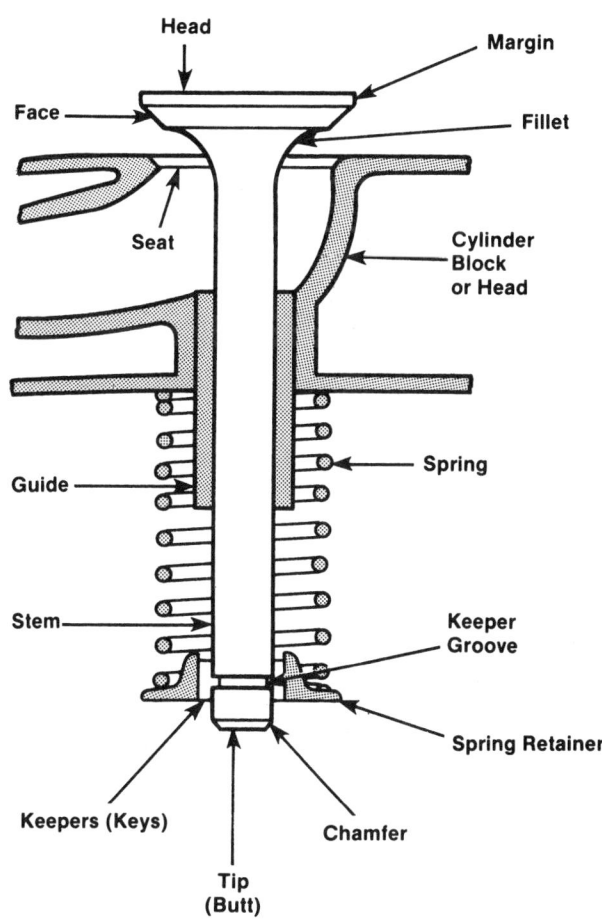

FIGURE 6-2 Valve assembly nomenclature

other hand, is able to flex to fit the shape of the valve seat. This allows it to seal easily, but it runs hot and the flexing can cause it to break. A popular valve head shape is one with a small concave cup or recess in the top of the valve head. It offers a reasonable weight, good strength, and good heat transfer at a slight cost penalty. Elastic heads are generally used for intake valves, and rigid heads are found on exhaust valves.

The sealing force on the valve seat is increased as the valve angle is increased. Forty-five-degree

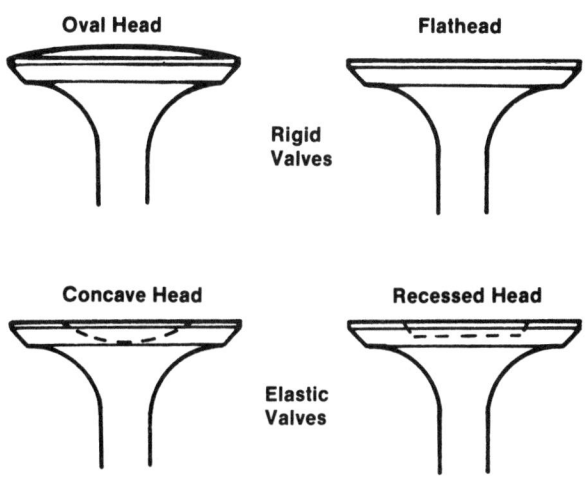

FIGURE 6-3 Valves are designed with different head shapes to meet different purposes.

face angles are used on exhaust valves and on intake valves where high seating pressures are needed. The high pressure will either crush or wipe off deposits to prevent valve leakage. A 30-degree valve face angle is used where high gas flow rates are more important and durability is no problem.

As mentioned in Chapter 2, there are two types of valves: intake and exhaust. Both are subjected to harsh conditions. For example, exhaust valves may be exposed to combustion temperatures of 1300 to 1500 degrees Fahrenheit. This means that they are in fact running red hot. In addition, in 1 hour, running at 55 mph, the valves of a four-cycle vehicle could open and close approximately 90,000 to 100,000 times. Because of the harsh conditions in which valves must operate, they must be inspected carefully.

VALVE INSPECTION

Check each face for burning, pitting, cracks, grooves, or scores (Figure 6–4). Inspect the tip of the valve stem for an even wear pattern, and check the keeper grooves for wear or damage. Discard any valve that is badly burned, cracked, pitted, or shows signs of excessive valve stem wear. Valves with bent stems or damaged keeper grooves should also be discarded.

Two basic types of valve failure are burning and breakage (Figure 6–5).

BURNING

Normally, the valve face is cooled as heat is transferred through the valve seat and on to the

A

B

FIGURE 6–5 Exhaust valves usually fail as a result of (A) burning and (B) breakage; intake valves usually fail because of breakage. *(Courtesy of TRW Inc.)*

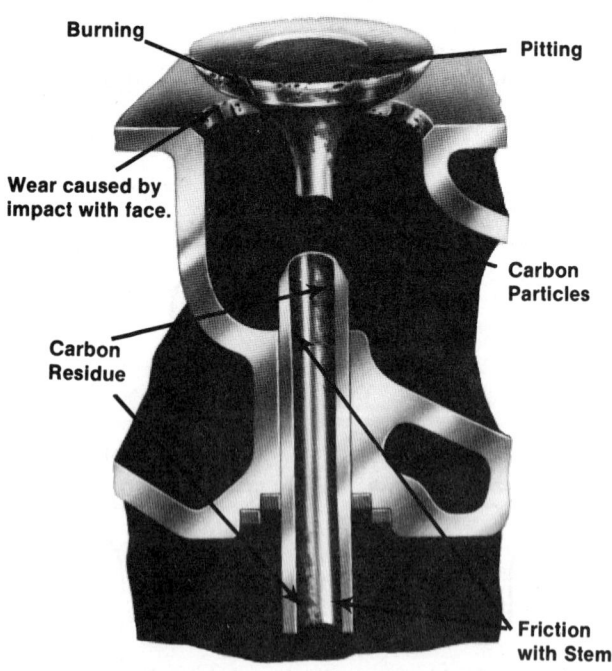

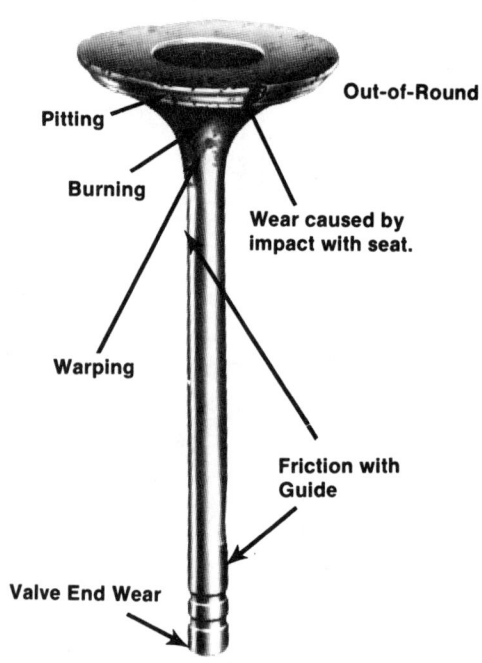

FIGURE 6–4 Common valve problems *(Courtesy of Sioux Tools, Inc.)*

engine coolant in the head. Burning occurs when the valve temperature is raised above the melting point of the metal. Normally, valve burning occurs in a small area on one side of the valve. Burning can result either from the leakage of hot gases or, in some instances, a local hot spot.

The leakage of hot gases past the valve face is due to poor sealing. Poor sealing, in turn, can result from the fact that the valve simply is not seating firmly as far as the valve train is concerned but, due to a local condition on the face or seat, there is still some channeling of hot gases locally at various points around the valve face. Conditions that can contribute to valve burning are channeling, poor seating, local hot spots, and heat.

Channeling

Channeling or local leakage occurs when extreme temperatures are developed at isolated locations on the valve face and seat. Figure 6-6 shows the temperature pattern in a valve head under both normal and abnormal conditions. Under normal

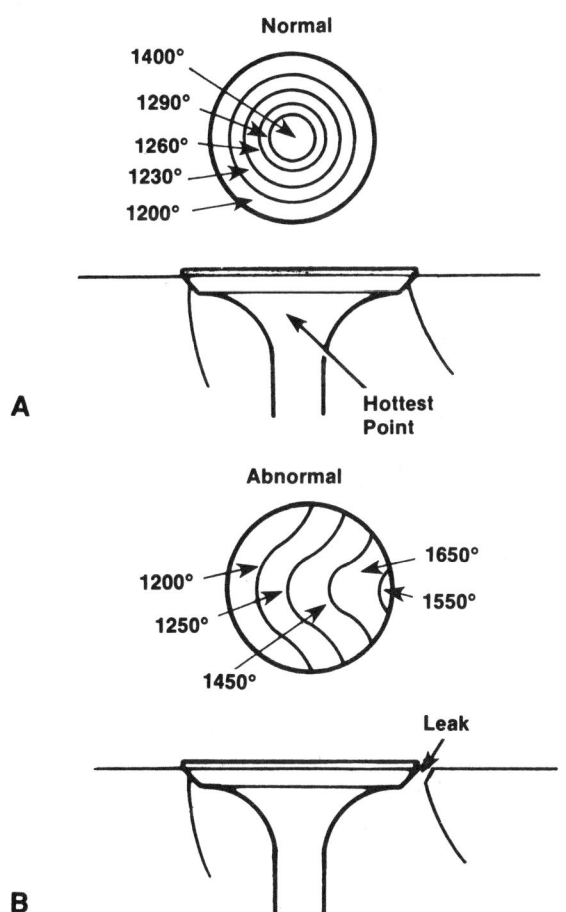

FIGURE 6-6   Valve head conditions under (A) normal and (B) abnormal circumstances

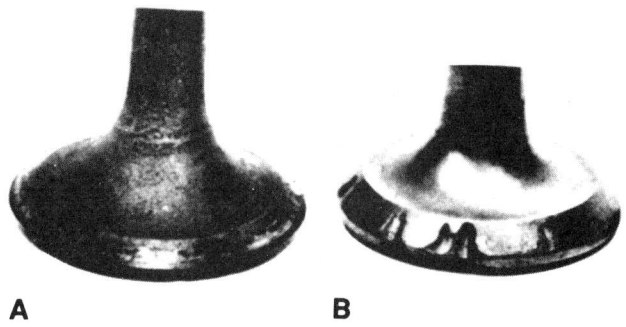

FIGURE 6-7 Channeling can begin due to (A) deposit flake-off and progress to (B) deep face gutters. *(Courtesy of TRW Inc.)*

conditions (Figure 6-6A), the temperature all the way around the valve edge should be about 1200 degrees. Under abnormal conditions (Figure 6-6B), the portion of the valve face that is in the 1650-degree region will progressively burn away causing channeling.

Channeling can begin as a result of the flaking off of combustion deposits from the valve (Figure 6-7A). It can progress into a fairly broad, burned-out area or into deep gutters (Figure 6-7B).

Channeling and leakage past the valve face can also be caused by extensive valve face peening due to the presence of unusually hard combustion deposits or foreign material that can become lodged between the valve face and seat (Figure 6-8). The problem starts in this case when a pattern of these depressions lines up in such a way as to permit a gas escape route. The net effect is then the same as that of a flaked-off deposit. While dealing with the subject of these deposits, in the upper left-hand portion of the chart, it is good to point out that even if deposits do not flake off, the presence of a layer of some thickness on the valve face produces an insulating effect that tends to raise the operating temperature of the valve.

Channeling can also be caused by valve face corrosion in a manner similar to face peening. A corroded valve face is shown in Figure 6-9. This type of corrosion is caused by a chemical reaction between the valve material and necessary gasoline anti-knock additives. It is accelerated by high temperatures.

Hard facing materials ordinarily provide the necessary resistance to face peening, corrosion, and high temperatures to prevent most of the channeling problems referred to. Added to this, rotation will keep face and seat deposits at a minimum, thereby providing a solution to the problems attendant with deposit insulation, flaking, and local overheating.

FIGURE 6-8 Valve face peening due to hard deposits *(Courtesy of TRW Inc.)*

FIGURE 6-9 Valve face corrosion *(Courtesy of TRW Inc.)*

A final and somewhat special form of channeling leakage results from the radial cracking of the valve head due to thermal stresses. Figure 6-10 shows a hard-faced valve in which the hard facing snapped due to hoop stresses. These, no doubt, were generated as a result of heating and cooling

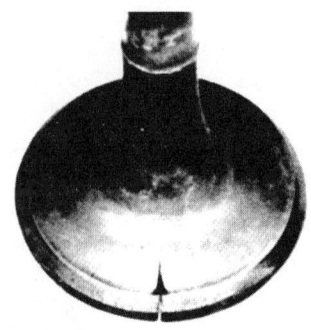

FIGURE 6-10 Snapped hard-faced valve *(Courtesy of TRW Inc.)*

too rapidly. After the facing material snapped, gases were able to wage a blowtorch attack on the base material behind, since it has a considerably lower melting point. These thermal stresses can also cause radial cracking in unfaced valves and can lead either to channeling and progressive burning, which has already been discussed, or to breakage failures, depending upon whether the loss of compression is detected early and the engine taken out of service or whether it is run to ultimate failure. The breakage aspects will be discussed later in this chapter.

Poor Seating

As previously mentioned, poor sealing, leakage, and burning can be caused by the simple fact that the valve is not permitted to properly engage the valve seat. Some of the major reasons for this are indicated in the left-hand portion of Table 6-1.

A valve that has been held slightly off its seat either by insufficient lash or sticking will look like the one in Figure 6-11. The view on the left shows the initial stage. The sooty, radial blowby marks indicate that the valve is fairly square with the seat but is being prevented from making solid contact. At this stage, these marks could be buffed off the valve face and very little damage might be evident. However, with a few more hours of high output operation, the condition shown in the right-hand view will result, at which point the valve is beyond repair.

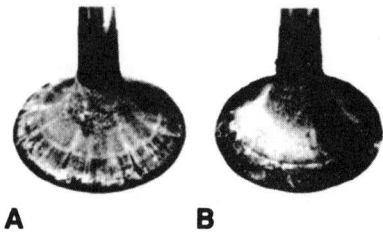

A   B

FIGURE 6-11 Effects of failure to seat: (A) initial stage shows gas leakage; (B) advanced stage of seat burning *(Courtesy of TRW Inc.)*

TABLE 6-1:   FACTORS LEADING TO VALVE FAILURE

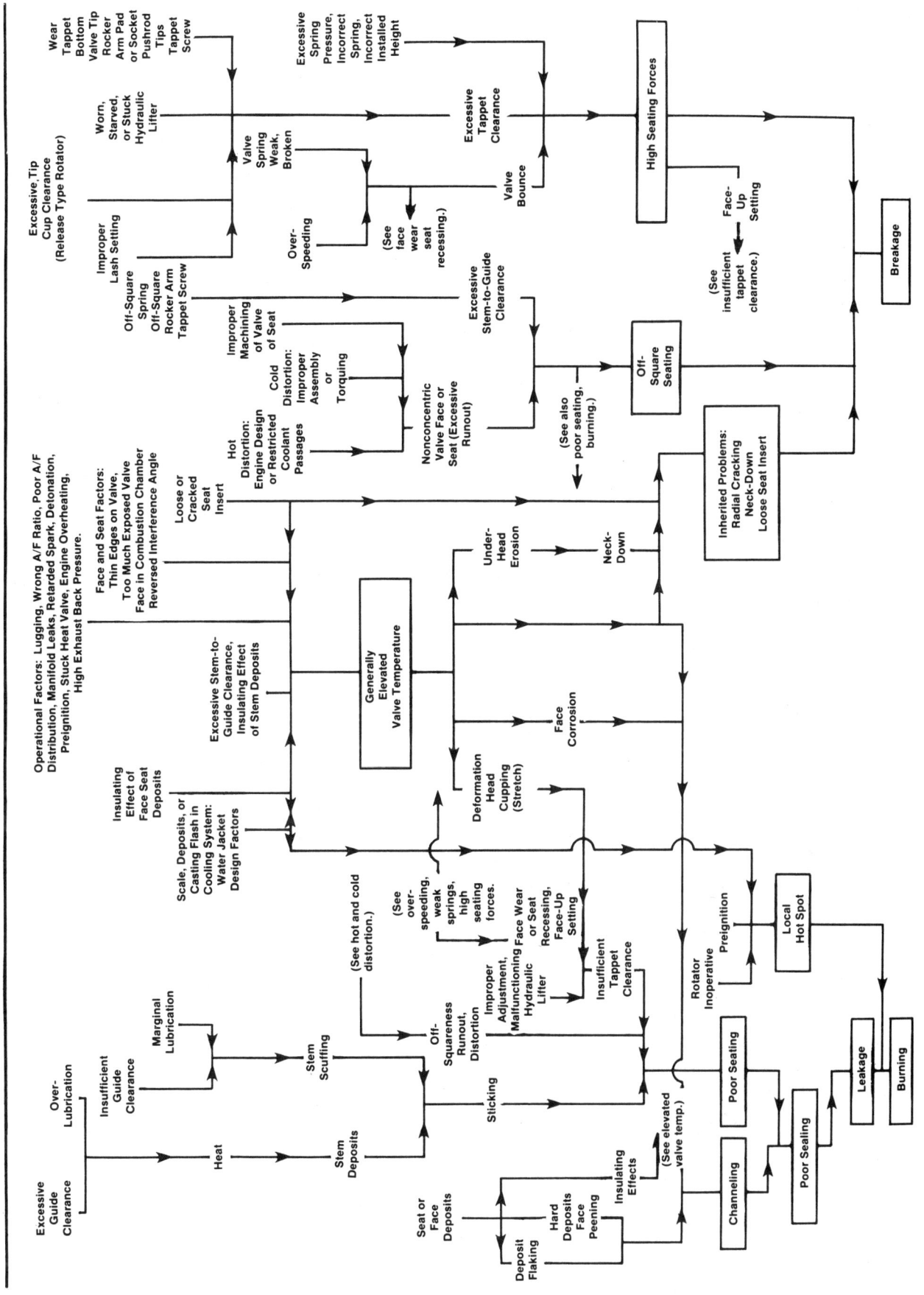

The gradual accumulation of valve stem deposits can render seating less and less positive and can lead to the conditions just illustrated in Figure 6–11. Stem deposits can result from excessive stem-to-guide clearances, which are normally accompanied by an oversupply of oil together with higher-than-normal temperatures, both conducive to deposit formation (Figure 6–12). Also, failure to provide counterbored guides and sharp scraping edges on valve stems or, in some cases, failure to provide rotation, can predispose certain engines to deposit problems.

At the other extreme, insufficient stem-to-guide clearance and the accompanying marginal lubrication can lead to pickup and scuffing, with poor seating and burning. Figure 6–13 shows a valve stem so afflicted. This condition will either hang up the valve or it might ream out the guide, subsequently producing excessive guide clearance.

The effect of insufficient tappet clearance, or lash, in producing poor seating conditions as illustrated in Figure 6–11 is well known. Therefore, perhaps more attention should be given to the causes of insufficient lash. The importance of using proper techniques in the initial adjustment of lash and reapplying them at predetermined periods in the maintenance program cannot be emphasized too much.

In engines equipped with hydraulic lifters, much is said about pump-up and its effect in holding valves open. Investigation has shown that when pump-up occurs, it is usually not the fault of the lifter. For a hydraulic lifter to extend to some length capable of holding a valve open, there must be some float in the valve train to permit this to happen. More

FIGURE 6–13 Valve stem pickup and scuffing *(Courtesy of TRW Inc.)*

often than not, this occurs when engines are operated in excess of their normal speed ranges, and it is further enhanced by weak valve springs. It is also very rare for hydraulic lifters to stick in some extended position that would hold the valves open. The forces prevalent in the valve train will ordinarily prevent a lifter from sticking in the extended position. When lifters cause trouble by sticking, it is usually in the collapsed state. In this condition there are no forces to overcome this sticking except that supplied by the small return spring. So hydraulic lifters, of themselves, are less of an influence in holding valves open than is often indicated.

Further examination of Table 6–1 will show two additional groups of factors that lead to insufficent tappet clearance and hence to valve burning. All of them are progressive effects that can further minimize clearance for each mile driven. One group—valve face wear, seat recessing, and valve face upsetting—derives primarily from mechanical factors dealt with specifically on the right-hand side of the table. The group referred to generally as "deformation" results primarily from valve head operating temperatures and therefore refers to the center of Table 6–1.

Either valve face wear (Figure 6–14) or seat recessing (Figure 6–15) can result from sustained high engine speeds in combination with weak, broken, or improperly installed valve springs. The self-rotation that occurs under such operating conditions causes an excessive amount of grinding action between the valve face and valve seat, and one of the two parts will suffer a high rate of wear, depending upon the materials used.

Valve lash can be reduced as a result of face upsetting from excessively high seating forces (Figure 6–16). The various causes of these high forces are shown on the right-hand side of Table 6–1. Normally, correction of the items shown in that part of the chart will eliminate this condition.

Operation with a valve temperature that is too high will lead to more general deformation of the

FIGURE 6–12 Failure due to stem deposits *(Courtesy of TRW Inc.)*

FIGURE 6-14   Valve face wear *(Courtesy of TRW Inc.)*

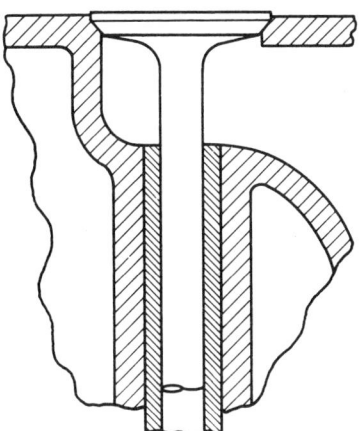

FIGURE 6-15   Valve seat recessing

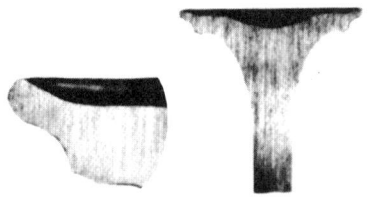

FIGURE 6-16   Valve face upsetting plus edge deformation *(Courtesy of TRW Inc.)*

valve head, involving cupping (Figure 6–17) and perhaps a certain amount of stretch in the underhead region. This, of course, will also act to reduce valve lash. This general deformation can be attributed to the combined effects of excessive operating temperature, combustion pressures on the top of the valve head, and inertia and valve spring forces pulling on the stem. The best cure for this condition is close attention to those maintenance and operating variables in the center of the table that lead to elevated valve temperatures plus the selection of the

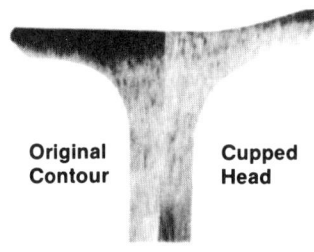

Original Contour   Cupped Head

FIGURE 6-17   Valve head cupping *(Courtesy of TRW Inc.)*

best available valve steel or the application of sodium cooling.

Referring again to Table 6–1, poor seating can also be caused by off-squareness of either the valve face or seat with respect to the guide axis, by runout of the valve face or seat, or by distortion of the cylinder head or block that houses the valve seat. In any of these cases, poor seating and leakage will occur even though the valve is engaging the seat with adequate force and even though there might be no significant deposit problems. The leakage and burning are caused by mismating of the valve face and seat. There is not good, 360-degree physical contact.

A valve burned because of poor mating with the seat (if not rotated) will show a localized burn over a fairly broad area, much like the one illustrated in Figure 6–18. The importance of good valve-grinding and reseating equipment and the checking of finished work with accurate gauges has been emphasized repeatedly and will go a long way toward eliminating burning failures of this kind.

These steps will not, however, eliminate mismating because of subsequent distortion that can occur

FIGURE 6-18   Face burn resulting from seat runout or distortion *(Courtesy of TRW Inc.)*

and also produce such failures. The rebuilder's only approach to the problem of cold distortion is to ensure that cylinder head and block surfaces are flat and burr free and to follow good techniques in cylinder head torquing.

Hot distortion can often be spotted in that the location of the burns on valves and seats might often be in alignment with certain features of the cylinder head or block. For example, if burns are all positioned in the same direction with respect to siamesed ports, regions where there is no water jacketing or the like, thermal distortion is indicated fairly conclusively as a factor in valve failure. When thermal distortion occurs in spite of apparently adequate water jacketing around the valve ports, some obstruction to the proper passage of coolant in that region should be looked for. In any overhaul, removal of grease, scale, and mineral deposits from the entire cooling system should be routine. However, with a local distortion problem, it also pays to investigate individual passages around and between the valve ports with a wire or other suitable probe. The presence of excessive casting flash (Figure 6–19) can introduce severe thermal effects, which will repeatedly ruin valve jobs.

Local Hot Spot

One mode of valve burning that does not appear to require poor sealing or leakage, at least initially, is burning that might be attributed to a local hot spot. Although a valve might be seating well, if one sector of that valve is continuously subjected to excessive temperatures, leakage can develop as a secondary

factor due to melting, warpage, scaling, or localized yielding of the valve steel. Leakage will figure in the failure as it advances but is not an initial factor.

Preignition, for example, can cause rapid burning, as in Figure 6–20, without preliminary leakage. Cause and prevention of preignition is a subject too broad and complex for proper treatment here. It is best approached at the operational level by using suitable fuel and lubricant and avoiding localized glow spots in the combustion chamber. Careful selection of a proper spark plug heat range is important in this regard. Knife-thin margins on valve heads, certain types of combustion chamber deposits, or any feature inside the combustion chamber that can carry over enough heat to ignite the new charge prior to the passage of the normal flame front are potential trouble with respect to preignition.

Table 6–1 shows that some types of local hot spots are controllable as maintenance factors, while others might be inherent in the engine design. In many commercial engines, the manufacturer might provide some form of rotation as a precaution against failures caused by local hot spots. In a case of that kind, a worn-out or malfunctioning rotator could lead to a failure in this category.

Heat

Move to the right on Table 6–1 and note that the entire center portion is devoted to those factors that tend to elevate valve temperature. It is true that modern valve materials have been developed to withstand high temperatures. But sustained high output operation taxes the limits of even these materials. Thus, when some element of design, maintenance, or operation tends to raise temperatures even higher, valve trouble can be expected.

Study of this center portion of the table shows that valves operating at higher-than-normal temperatures can be prone to failure by either the burning or

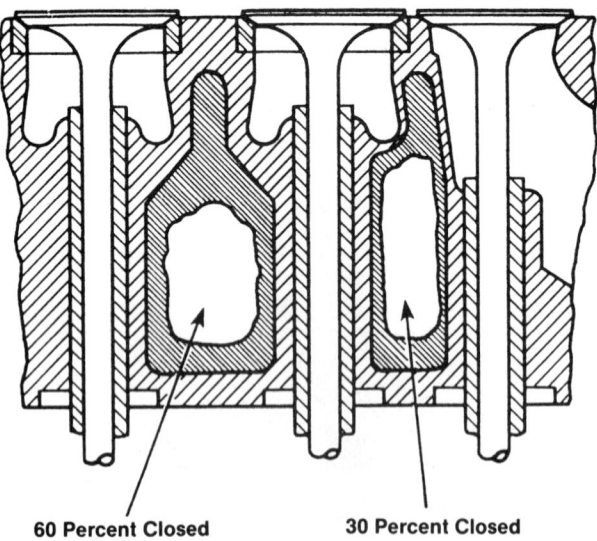

60 Percent Closed   30 Percent Closed

FIGURE 6–19 Cooling system blockage due to casting flash

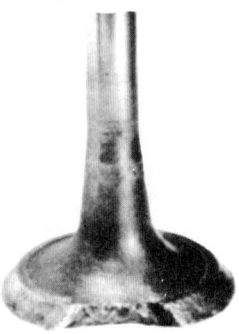

FIGURE 6–20 Effects of preignition *(Courtesy of TRW Inc.)*

the mechanical breakage route through any of a number of possible mechanisms. It is not the intention of this discussion to dwell at length on this part of Table 6–1, except as it supports the items under consideration in the left- and right-hand portions. Suffice it to say that the mechanic should carefully study those variables listed along the top that do lead to extreme valve operating temperatures and have them in mind from the preventive angle.

CAUSES OF VALVE BREAKAGE

Move to the right-hand portion of Table 6–1. You see that factors leading to valve breakage fall into three basic categories. The first of these could be considered a family of inherited problems not directly related to the mechanical action of the valve itself. This includes radial cracking and neck-down inherited from the thermal and chemical environment that might be imposed upon the valve and valve-breaking stresses, which might result from the loss of a seat insert. The other two basic kinds of causes for valve breakage are of a more directly mechanical nature. These are off-square seating and high seating forces. Another common cause of valve breakage is valve bounce or any undisciplined valve motion that causes it to seat "on the fly" without the retarding influence of the cam. Valve bounce is almost always the result of operating an engine beyond its normal speed range, although it can be introduced at lower speeds if valve springs are weak, broken, manufactured, or installed to improper specifications. Alterations to valve train components involving changes in mass or deflection rate can also contribute to bounce.

Loose or Cracked Seat Inserts

If a seat is installed with too much press fit or if it is of such material that it cannot adjust to the thermal variations in the engine, it might crack as illustrated in Figure 6–21. If it stays in position, it will no longer conduct heat as it should, since it will have lost much of its effective contact with the counterbore in the head or block.

In this regard, it can be instrumental in elevating valve temperatures and contributing to the thermal problems already discussed. On the other hand, seat inserts in the condition illustrated will frequently come loose and become dislodged. In this event, the valve will almost certainly be broken after very few engine revolutions.

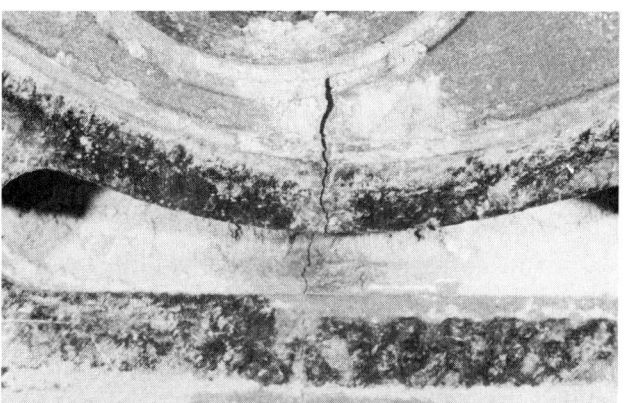

FIGURE 6-21   Seat insert cracking

Radial Head Cracking

As mentioned earlier, the problem of radial head cracks can develop as a result of thermal shock. This can happen if an engine operating at high load is abruptly unloaded or shut down. It will almost certainly happen, in time, with a conventional valve in an engine, which is repeatedly and abruptly subjected to wide fluctuations in output.

A hard-faced valve was previously shown in which this radial cracking had occurred with subsequent burning of the base material. Figure 6–22 shows an unfaced valve likewise subjected to thermal shock. It can be seen that the first stage is the development of radial cracks due to repeated hoop stressing at the outer edge. Further operation will often lead to breakage in the form of pie-shaped chunks as these cracks progress inward to where they intersect or come within close proximity to one another. The best solution to this problem lies in the avoidance of severe fluctuations of engine load. It is recognized, however, that commercial demands do not always permit this. The application of sodium cooling will vastly reduce the temperature gradient across the valve head and generally put an end to this problem. Having applied this expedient, it is still

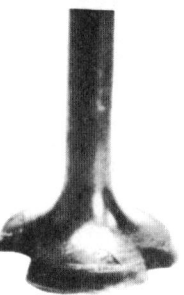

FIGURE 6-22   Failure due to thermal shock *(Courtesy of TRW Inc.)*

good practice to avoid shutting an engine down abruptly from some high operating load.

Neck-Down

The effect of exhaust-valve corrosion from high-temperature exhaust gases has been discussed, insofar as it leads to poor sealing and face burning. This same corrosive action, attacking other parts of the valve, can also lead to physical failures. Underhead erosion as illustrated in Figure 6–23 is a frequent cause of such physical failures. This stem erosion is a special kind of corrosion accelerated by the impingement of the flowing exhaust gases. The mechanical action of the flowing gases, plus their highly corrosive character at the temperatures involved, tend to supplement each other and produce neck-down. The point of maximum neck-down will usually be roughly in line with the angle of the valve face. This is one of the hottest portions of the valve and it is therefore extremely susceptible to this corrosive and erosive action. Also, because the tensile strength of the material is reduced as a result of the high temperatures, it is a region where neck-down is least tolerable from the standpoint of strength. Valve stems frequently fail in this region from the combined effects of reduced cross section caused by neck-down and elevated operating temperatures.

The main approach to preventing recurrence of neck-down failure involves the use of premium valves in installations and precise control of ignition timing, air/fuel ratios, and air/fuel distribution in the manifold. Proper application of the engine with respect to load size and gear ratios should also be checked. All of these factors influence exhaust temperature. For each 25-degree increase in valve operating temperature, the rate of corrosion approximately doubles. From the operator's viewpoint, these variables that influence exhaust gas temperature therefore offer the best means of controlling failures due to any type of valve corrosion.

FIGURE 6-23 Neck-down or underhead erosion *(Courtesy of TRW Inc.)*

Off-Square Seating

Off-square seating has already been discussed in conjunction with the left-hand side of Table 6–1. It appears on the right-hand side of the table as an important possible contributor to valve breakage. Consider the case where the mismatch between the valve and its seat is minor and the valve is flexible enough to provide reasonably good sealing. Under these conditions, low compression, missing, or valve burning might not develop as symptoms of off-squareness. This means that the engine can be run for thousands of miles with the valve flexing to conform to the location of the seat each time it closes. With proper concentricity and squareness, flexing would be kept to a minimum, but in this case, the flexing might be enough to lead to a fatigue failure of the valve stem or head after a definite number of engine cycles (Figure 6–24).

Stem failure due to misalignment will normally occur somewhere between the underhead and the point where the valve stem enters the guide (Figure 6–25). The inset shows the typical appearance of a fatigue failure. The progressive, crescent-shaped lines across the part from the initial point to the final arc of breakage show the creeping nature of such a failure as it transpires in the course of continual working through many, many cycles.

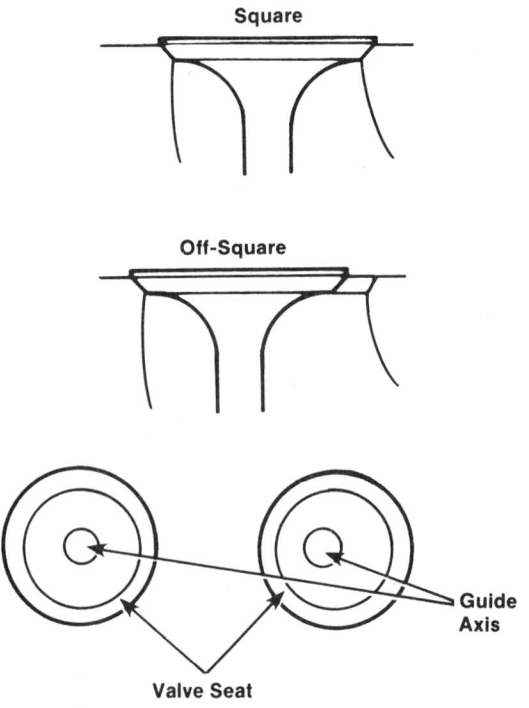

FIGURE 6-24 Poor seating will result when an off-square condition exists.

FIGURE 6-25 Stem failure due to fatigue (with magnified cross section of break)

Table 6-1 indicates numerous possible causes for off-square seating. Both hot and cold distortion have been mentioned before in connection with leakage and burning. They are equally applicable here, especially if the amount of the distortion is small and within the range that can be accommodated by valve flexibility. Lack of precision in the machining of either the valve face or the seat can lead to the same bad effect, namely, valve flexing in service.

Excessive stem-to-guide clearance can also stimulate valve flexing leading to fatigue failures. This is especially true of valve-in-head engines, which require extra vigilance on guide clearance because of the scrubbing action of the rocker arm pad across the valve tip. End squareness of valve springs and proper shape and finish on rocker arm pads are further maintenance factors that help minimize side thrust and cocking forces on valve stems.

To summarize, off-square or nonconcentric valve seating, whatever the cause, can lead to repeated flexing of the valve and ultimately a fatigue type of break either in the upper valve stem or in the head. Concentricity and squareness of valve seats become doubly important if rotation is to be used because a rotated valve can work itself to death trying to accommodate a new situation each time it seats. The one price that must be paid for all the benefits of rotation is extra diligence in generating accurate valve seats, which are square and concentric with the guide axis.

High Seating Forces

Perhaps the most directly understandable cause of valve breakage is that of seating the valve with too much force. This sometimes produces what is called *impact failure*. This type of failure (Figure 6-26) is nearly instantaneous and occurs when the valve is either struck by the piston or when it violently engages the valve seat. Valve-to-piston interference is generally caused by a valve sticking in the guide, broken valve springs, retainer or lock failures, or during valve train surge resulting from weak springs or excessively high engine speeds allowing the valve to remain open or "float" and strike the piston. Impacts due to high seating forces are generally caused by excessive valve lash due to valve train wear, misadjustment, or hydraulic lifter collapse.

Not all valve impact failure involves the valve head. Figure 6-27 illustrates an impact failure due to tip fractures. This results from misalignment problems in the valve train. Sources include worn rocker pads, worn valve guides, excessive valve lash, incorrect rocker arm ratios, incorrect installed valve tip height, and worn valve locks. The break is the result of fatigue and is characterized by contact along the chamfer of the tip (side loading). Due to this side loading, the tip is subjected to alternating tensile

FIGURE 6-26 Valve breakage due to impact failure *(Courtesy of TRW Inc.)*

FIGURE 6-27 Impact failure due to tip fractures *(Courtesy of TRW Inc.)*

and compressive forces causing a break through the lock ring groove.

Failures are not always clearly one type or the other. Varying combinations of high temperature, high seating forces, and off-squareness can produce failures of varying degree and appearance.

Excessive tappet clearances can cause high seating forces and breakage if the valve motion is being controlled by the flank of the cam rather than the closing ramp at the time contact is made with the seat. It is the function of the closing ramp to decelerate the valve just prior to seating, and this keeps seating forces within a reasonable range even at fairly high engine speeds. Extensive operation with too much lash, especially at highway speeds, will almost certainly take its toll in broken valves.

Periodic valve lash checks by proper techniques have already been emphasized for early detection of lash loss in the discussion of valve burning. It is equally valid to suggest periodic lash check to guard against increases in lash, which can very well occur in service. Wear of components such as valve tips, rotator tip cups, rocker arm pads and sockets, pushrod ends, or tappet screws can cause increased lash. Whenever wear of any of these components appears to be progressing at such a rate that increases valve lash, attention should be given to the adequacy of lubrication in that area. For example, blockage of lubrication to a rocker arm can increase wear beyond all normal proportions. Also, the finishes and hardnesses of the parts involved should be checked. Some of these components might be thinly case hardened and once the case has been worn through, replacement is mandatory.

In recent years, there seems to have been an increasing number of cases where worn or spalled tappet bottoms and/or cam lobes have resulted in very sudden increases in tappet clearance. This condition will often involve just one, or perhaps a few, of the valves in an engine. When a radical increase in lash occurs suddenly, as evidenced by loud tappet noise, the engine should be shut down immediately and checked for worn tappet bottoms or cam lobes. Persistence in operating such an engine not only invites valve breakage, but also causes general engine damage as a result of circulating abrasive wear particles throughout the engine. The best preventive measure for worn tappet bottoms or cam lobes is the use of an oil of grade SG (see Chapter 13). Practically all oils of the grades SG, SE, SD, and SC contain the additive zinc dithiophosphate, which suppresses this kind of wear. Such an oil is recommended for use at all times but is particularly essential when a new camshaft or tappets are installed.

FIGURE 6-28 Worn valve stems and tips *(Courtesy of Perfect Circle/Dana)*

WORN VALVE STEMS AND TIPS

A worn valve stem will allow the valve to rock or tip sideways in the cylinder head (Figure 6-28). This results in excessive clearance between the stem and guide, which allows oil to leak into the port. Stem and tip wear is normally the result of poor lubrication, incorrect stem-to-guide clearances, poor valve seating, poor rocker arm contact with the valve tip, or nonsquare valve springs. This wear is measured with a valve stem wear gauge (Figure 6-29).

FIGURE 6-29 A valve stem wear gauge measures wear and taper to 0.001 inch. *(Courtesy of Goodson Shop Supplies)*

SHOP TALK

A rule of thumb for maximum valve stem wear is 50 percent of the maximum guide clearance given in the manufacturer's or service manual's specifications. Never remove more than 0.020 inch of material from the valve stem tip. If more than 0.020 inch must be removed, the valve should be replaced.

Some manufacturers are now using tapered valves because of their high resistance to wear. Valves with tapered stems are normally made with a stellite valve face to be mated with a high chrome seat. They are used primarily in engines with high exhaust port temperatures. Tapered valves also can be made with a chrome-plated stem. This construction allows for greater heat expansion near the valve fillet. It also helps prevent galling and sticking valves.

Actually, the purpose of the tapered stem is to provide additional stem-to-guide clearance near the head of the valve to reduce the possibility of scuffing (Figure 6–30). Typically, the stem diameter at the head is 0.001 inch less than the diameter at the tip end as shown in Figure 6–31. Metals expand as their temperature increases. However, the expansion rate is not the same for all metals. Total expansion de-

pends on the amount of temperature increase. The greater the increase, the more the part expands. In this case, the stainless steel valve stem has a greater expansion rate than the cast-iron head and therefore expands more. An exhaust valve is hotter at the head end, which is exposed to hot combustion gases, than at the tip end. Consequently, it follows that the stem near the head of the valve tends to expand more than the portion near the tip. So the taper helps to provide sufficient clearance at the hot end of the stem. This feature helps to prevent scuffing in situations where high power output demands are placed on a cold engine. Under these conditions, the hot exhaust valve expands much more rapidly than the cylinder head, which reduces stem-to-guide clearance and could cause scuffing. The tapered stem provides additional clearance at the head end to prevent seizure. It should be pointed out, however, that the specified cold clearance must be maintained when installing new valves and/or guides.

Before replacing make sure to check whether or not the original valves are tapered. A tapered valve stem can be easily mistaken for a worn stem.

Tip separation from the valve stem results from misalignment problems in the valve train (Figure 6–32). Sources include worn rocker pads, worn valve guides, excessive valve lash, incorrect rocker arm ratios, incorrect installed valve tip height, and worn valve locks. The break is the result of fatigue and is characterized by contact along the chamfer of the tip (side loading). Due to this side loading, the tip is subjected to alternating tensile and compressive forces causing a break through the lock ring groove.

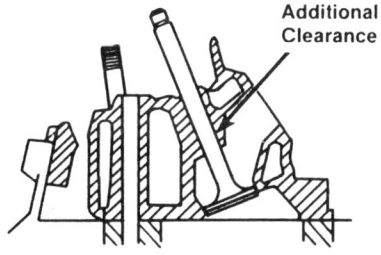

FIGURE 6–30 Tapered stem provides additional clearance

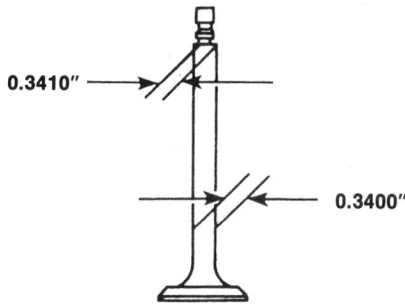

FIGURE 6–31 Stem diameter of head is 0.001 inch, usually less than the tip end diameter

FIGURE 6–32 A "so-called" valve inspector provides a quick check for valve face runout and for bent valve shafts. *(Courtesy of K-Line Industries Inc.)*

VALVE GUIDE WEAR

Valve guides wear from insufficient lubrication, carbon buildup, overheating, or cocked valve springs. Incorrect rocker arm contact across the valve tip can also damage guides because the motion forces the valve stem to move sideways in the guide (Figure 6-33). Excess guide wear causes increased oil consumption, burned valves, or even valve breakage.

Valve guide reconditioning is one of the basics of head rebuilding. There are principally five techniques that can be used to restore worn valve guides:

- Knurling
- Reaming out the old guide and using new valves with oversized stems
- Tapping the guide and threading in a bronze insert
- Boring out the guide and installing a bronze liner
- Replacing the entire valve guide with a new universal one

Instructions for performing these techniques are given in detail in Chapter 10.

The amount of wear between the valve stem and valve guide can be determined in a number of ways. The most common method of measuring valve guide wear is with a split ball or small hole gauge and micrometer. To use the small hole gauge and micrometer method, proceed as follows:

1. Insert the small hole gauge into the valve guide (Figure 6-34).
2. Allow the gauge to expand until it contacts the valve guide bore walls.
3. Rock the gauge from side to side to center it in the valve guide. This motion is similar to the method used to center telescoping gauges and micrometers on the work.
4. When the correct resistance is felt, lock the gauge into that position and carefully withdraw it from the valve guide.
5. Measure the small hole gauge with an outside micrometer (Figure 6-35).
6. Repeat this procedure at the top, middle, and bottom of the guide to determine overall wear. Because valve guides wear more at the ends than in the middle, it is not unusual to obtain a different reading in the middle of the guide.

When determining stem-to-guide wear, measure the valve stem using an outside micrometer and subtract the stem diameter from the three valve guide readings. The resulting figure equals valve stem-to-guide clearance. Use the manufacturer's manual or consult the cylinder head specification manual to determine stem diameter specifications and maximum stem-to-guide clearance. Then add the two specifications to find the maximum guide inside diameter (ID). For example:

$$\begin{array}{l} .3417'' \text{ stem diameter} \\ \underline{+ .0027'' \text{ max. stem-to-guide clearance}} \\ .3444'' \text{ max. guide ID} \end{array}$$

In this example, set the micrometer to the maximum guide ID 0.3444 inch. Use the micrometer setting to adjust the small hole gauge to the same measurement. The split ball is now a go/no go gauge. Insert

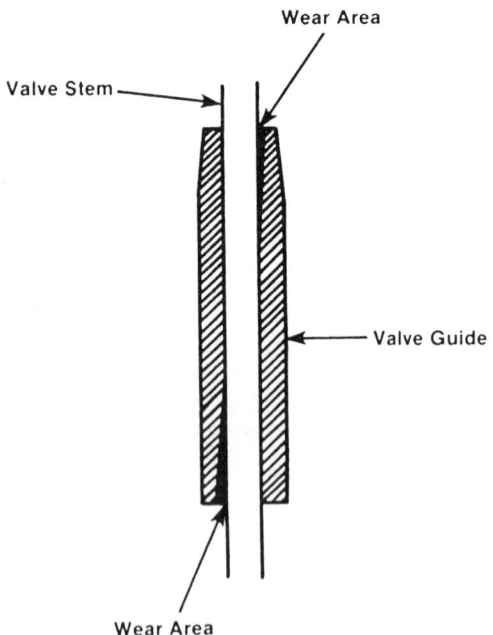

FIGURE 6-33 Valve guides generally wear bell mouthed.

Wear Area

Valve Stem

Valve Guide

Wear Area

FIGURE 6-34 Measuring the valve guide diameter with a small hole gauge *(Courtesy of TRW, Inc.)*

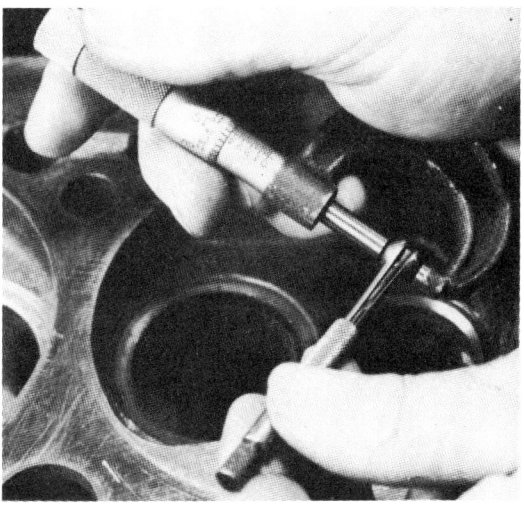

FIGURE 6-35   Measuring the small hole gauge with an outside micrometer *(Courtesy of TRW, Inc.)*

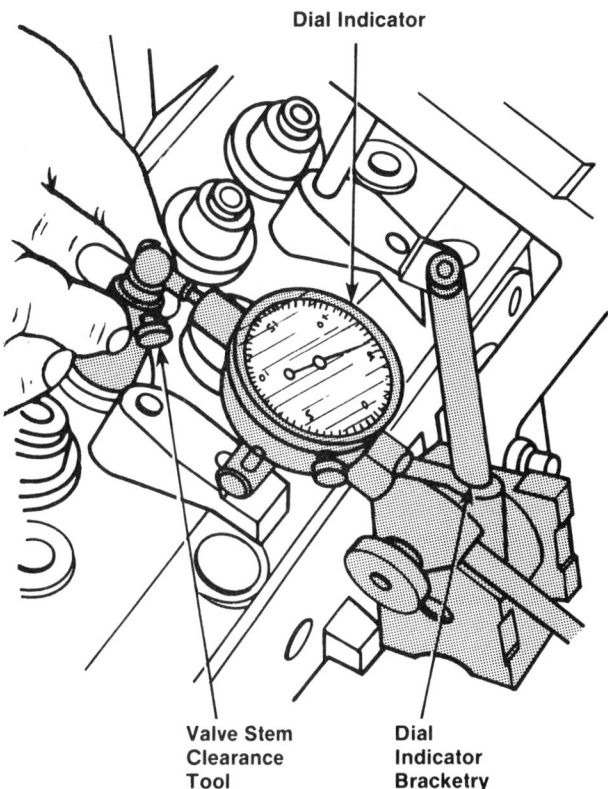

FIGURE 6-37   Checking valve stem clearance—typical

the gauge into the valve guide. If it goes into the guide, the guide is worn beyond maximum limits.

Another method of measuring the stem-to-guide wear is to use a valve guide dial gauge (Figure 6-36). Adjust the gauge so that the dial indicator reads zero at the standard valve guide size. The gauge is then inserted into the guide. As the tool is pushed down the guide, the dial gauge needle registers any oversize in the guide.

The valve stem-to-valve guide clearance of each valve in its respective valve guide is measured with a value stem clearance tool and dial indicator (Figure 6-37). Install the tool on the valve stem until it is fully seated and tighten the knurled set screw firmly.

FIGURE 6-36   Using a valve guide dial indicator *(Courtesy of Hall-Toledo Inc.)*

Permit the valve to drop away from its seat until the tool contacts the upper surface of the valve guide. Then position the dial indicator with its flat tip against the center portion of the tool's spherical section at approximately 90 degrees to the valve stem axis. Move the tool back and forth in line with the indicator stem. Take a reading on the dial indicator without removing the tool from the valve guide upper surface. Divide the reading by two, which is the division factor for the tool.

If manufacturer's specifications are not available, use 0.001 to 0.003 inch for intake valve clearance and 0.002 to 0.004 inch for exhaust valves. For sodium-filled exhaust valves use 0.003 to 0.005 inch clearance. If the indicated clearance exceeds the new specification by 50 percent, replace the guide where possible.

SEAT RECESSION

The results of poor seating have been mentioned several times in this chapter. However, seating itself has not been addressed.

There are two types of valve seats: integral and replaceable insert. Integral seats are part of the casting. Insert seats are pressed into the head and always used in aluminum cylinder heads. Most pre-1978 integral seats are soft cast iron. After 1978, most manufacturers began to produce cylinder heads with induction-hardened cast-iron seats able to withstand the higher heat of exhaust applications.

Insert seats are added to the cylinder head after casting, or as replacements for worn integral seats. Three types of seat inserts are:

1. *Stellite.* Used for extremely high operating temperatures and heavy-duty applications; normally exhaust valves.
2. *Cast Iron.* Used to replace integral cast-iron seats; normally intake.
3. *High Chrome.* Used to replace induction hardened seats, for dry fuel and heavy-duty applications; normally exhaust valves.

Valve seat replacement is needed when a valve seat is cracked, burned and pitted, or recessed (sunken) in the cylinder head. Normally, valve seats can be repaired or reconditioned and returned to service.

The average seat width is 0.060 inch and the seat begins 0.030 inch from the valve margin. A properly ground seat has three angles: top, 30 degrees or 15 degrees; seat, 45 degrees or 30 degrees; and throat, 60 degrees. Typically, the 45-degree angle wedges more tightly than the 30-degree seat, so it is used most often. Using three angles maintains the correct seat width and sealing position on the valve face (Figure 6-38). Correct sealing pressure and heat transfer from the valve through the seat are also affected.

In most vehicles today, the valve face and seat are machined to different angles, usually a variance of 1/2 to 2 degrees (Figure 6-39). This variance forms an interference angle for better sealing at the combustion chamber side of the seat, where combustion pressures try to get between the valve and seat when the valve is closed. With no interference angle provided, the possibility of a negative interference angle exists (open on the combustion chamber side of the seat), allowing combustion pressures to enter and unseat the valve.

Testing Valve Seat Runout

The valve seat runout gauge is similar to the dial indicator. It features a gauge face divided into 0.001-inch (one-thousandth of an inch) increments, an arbor that centers the instrument in the valve guide bore, and an indicator bar that can be adjusted so it bears on the valve seat. The tool is then slowly rotated around the circumference of the valve seat to check its concentricity (runout). To measure for valve seat runout, proceed in the following manner:

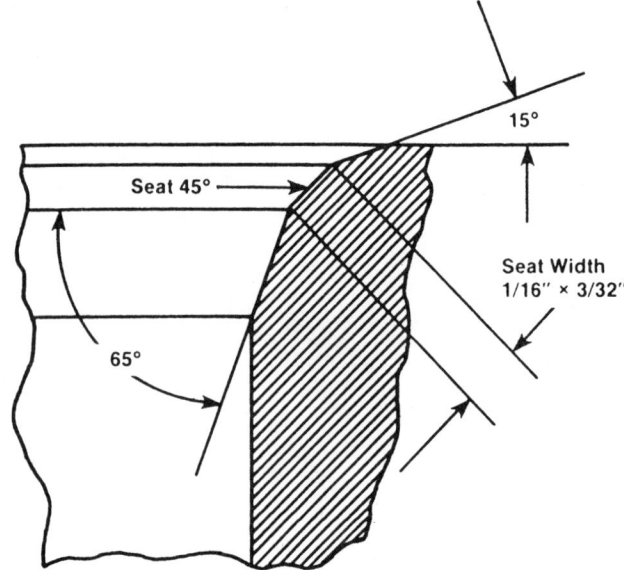

FIGURE 6-38 Typical valve seat dimensions

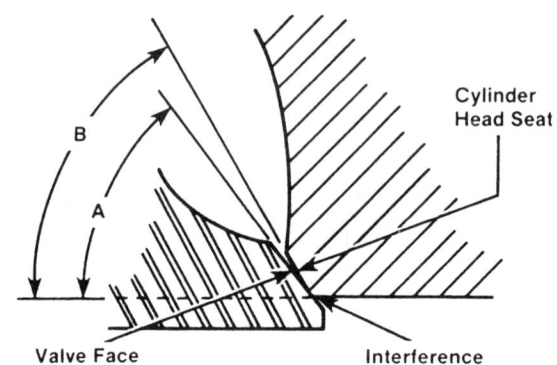

FIGURE 6-39 Valve face angle A is 1/2 degree to 2 degrees less than seat angle B. This interference angle provides a positive seat on the combustion chamber side of the valve and seat.

1. Mount the runout gauge along with the arbor into the valve guide so that the indicator bar contacts the center of the valve seat (Figure 6-40).
2. Set the dial to zero.
3. Hold the dial section and slowly turn the lower section so that the bar travels around the seat.
4. The dial needle will indicate any runout. Typically, the runout should be within 0.002 inch. Refer to specifications in the applicable service publication.
5. If runout exceeds specifications, check the dial indicator setup and repeat the procedure.
6. If runout is still present, the seat must be reground and checked again.

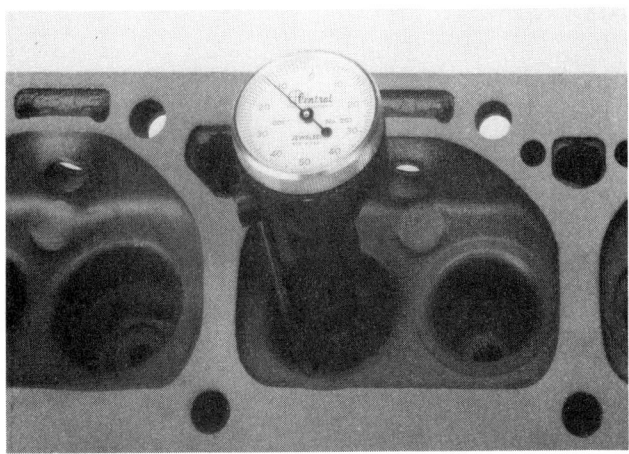

FIGURE 6-40   Measuring valve seat runout

VALVE SPRING

The valve spring performs two functions: it closes the valve and maintains contact in the valve train during the opening and closing of the valve. It has already been seen that insufficient spring pressure can lead to valve bounce and breakage. The fact that too much pressure can also lead to valve breakage points up the general necessity of working between limiting specification values in all phases of engine maintenance. Figure 6-41 shows the components that make up a valve spring assembly.

The common designs of valve springs are illustrated in Figure 6-42. A problem that valve springs might have is spring surge. As the name implies, spring surge is the violent extending motion of the coils resulting in abnormal oscillation. The uniform pitch type is used where surge is usually not encountered. The other designs represent different approaches to surge control. Mechanical spring surge dampers depend on friction to control surge, therefore it is advisable to replace them at the same time the springs are replaced. Always install the closely wound coils of a basket coil-type spring toward the head end of the valve. Mechanical surge dampers (Figure 6-42) should also be installed toward the head end of the valve. To dampen spring vibrations and increase total spring pressure, some engine manufacturers use a reverse wound secondary spring inside the main spring (see Chapter 2).

Spring surge can occur when:

- The springs are weak.
- The installed spring height is improper.
- Engine speeds are excessive.

Whatever the cause, the occurrence of spring surge is visually apparent. The ends of the springs will look

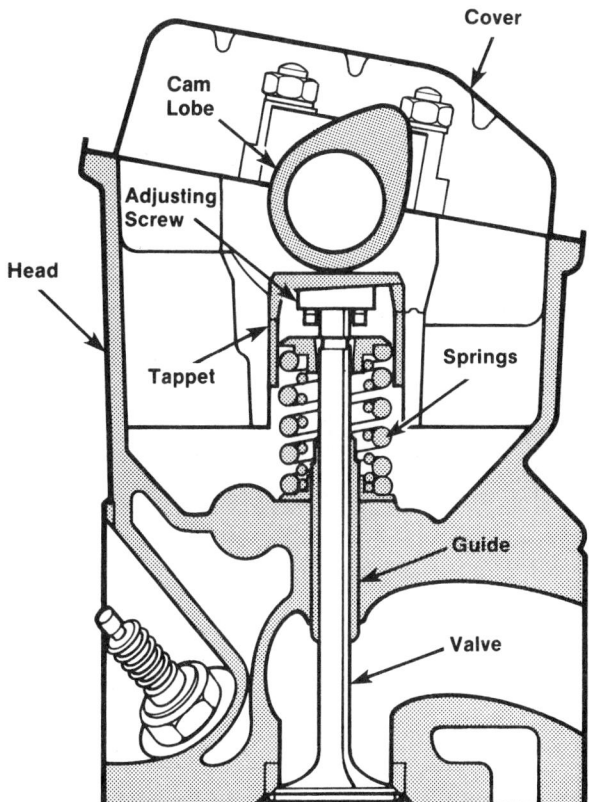

FIGURE 6-41   The valve spring is compressed between the valve seat on the cylinder head and the spring retainer.

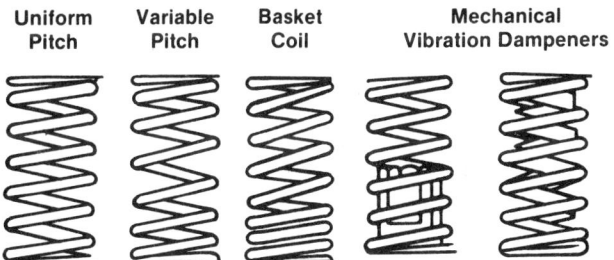

FIGURE 6-42   Valve spring designs

smooth or polished due to their rotation during operation. If left alone and not corrected, spring surge can cause undue damage to the valve train. For example, the self-rotation of the valve springs causes a grinding action between the valve face and seat. As a result, the face will wear down and the seat will recess. Continued operation with spring surge can also cause the springs to break.

The high stresses and temperatures imposed on valve springs during operation cause them to weaken and sometimes break. Rust pits will also cause valve spring breakage. To determine if the

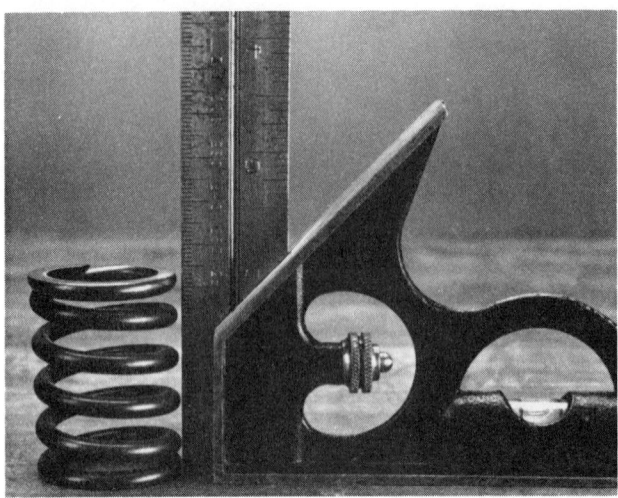

FIGURE 6-43 Freestanding height test *(Courtesy of Perfect Circle/Dana)*

FIGURE 6-44 Spring squareness test *(Courtesy of Perfect Circle/Dana)*

spring can be reused, the following test should be performed:

1. *Freestanding Height Test (Figure 6-43).* Line up all the springs on a flat surface and place a straightedge across the tops. Free length should be within 1/16 inch of OE specifications. Throw away any spring that does not meet this standard. A spring that is too short or too long will not have the correct tension when installed. If shims are used to adjust spring height, care must be taken to make sure that the spring coils do not stack and cause the coil to become solid. If the spring becomes solid, the valve train can be damaged or breakage might occur.

2. *Spring Squareness Test (Figure 6-44).* A spring that is not square will cause side pressure on the valve stem and abnormal wear. To check squareness, set a spring upright against a square. Turn the spring until a gap appears between the spring and the square. Measure the gap with a feeler gauge. If the gap is more than 0.060 inch, the spring should be replaced. This gap is known as side load (Figure 6-45). A side load can cause the valve guide to wear prematurely or it can cause the valve to break.

3. *Open/Close Spring Pressure Test (Figure 6-46).* Use a spring tester to check for open and close spring pressure. Close pressure guarantees a tight seal; the open pressure overcomes valve train inertia and closes the valve when it should close. All spring

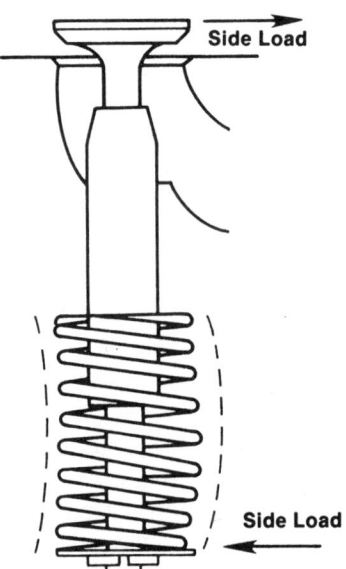

FIGURE 6-45 Example of a side load

tensions must be within ±10 percent of factory specifications. (Specifications are found in manufacturer's manuals, engine repair guides, engine parts catalogs, or cylinder head specification manuals.) For example, if a spring must be installed at 1.700 inches, it will be compressed to that dimension when the valve is closed. Compress the spring to 1.700 inches and note the pressure it takes to force the spring down. Manufacturer's specs state that at 1.700 inches, the pressure should be 80 psi. The spring will pass inspection if it takes anywhere from 72 to 88 psi to compress the spring to a height of 1.700 inches. Similarly, the open pressure test might call

FIGURE 6-46 Open/close spring pressure test *(Courtesy of Perfect Circle/Dana)*

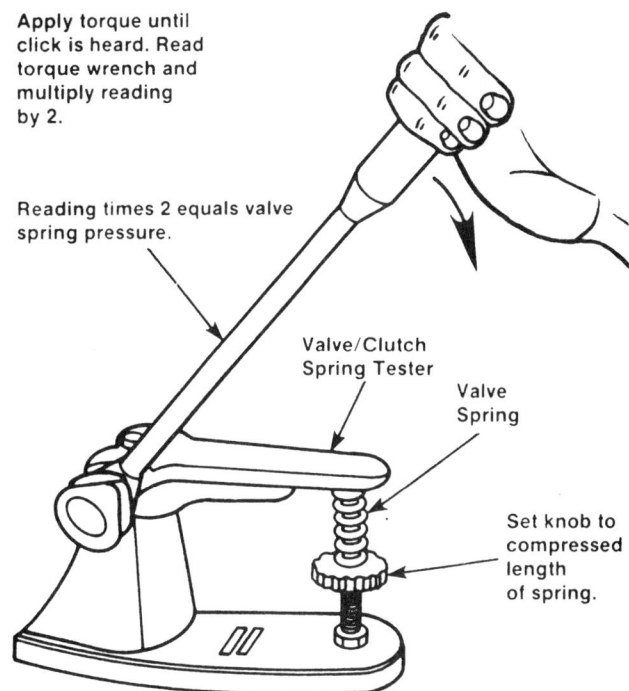

Apply torque until click is heard. Read torque wrench and multiply reading by 2.

Reading times 2 equals valve spring pressure.

Valve/Clutch Spring Tester

Valve Spring

Set knob to compressed length of spring.

FIGURE 6-47 Measuring valve spring pressure with a torque tester

for pushing the spring down to 1.250 inches and require a 200 psi to do so. The ±10 percent rule of thumb allows the spring to pass if it takes between 180 to 220 psi to force the spring to 1.250 inches.

Another method of testing valve spring pressure at the specified spring lengths is to use a torque wrench and valve/clutch spring tester (Figure 6-47). For this method, proceed as follows:

1. Place the valve spring into the checking fixture. Turn the height adjustment knob until the specified compressed spring height is obtained.
2. Attach a torque wrench to the spring tester lever arm.
3. Apply torque to the spring tester until you hear a click.
4. Multiply the torque wrench reading by two to obtain the valve spring pressure.

Other spring testers are shown in Figure 6-48.

VALVE ROTATORS

Valve springs tend to wind and unwind as they are compressed and released. This causes the valve face to turn in relation to the seat, helping even out heat buildup and remove deposits. Some engines have specially made valve rotators in the valve

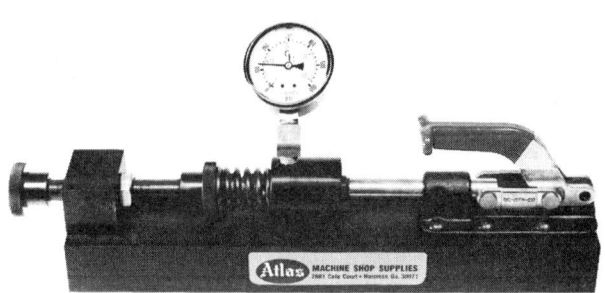

A

B

FIGURE 6-48 Other popular valve spring testers: (A) bench tester *(Courtesy of Atlas Machine Shop Supplies)*; (B) valve spring pressure tester that can be used with head assembly. *(Courtesy of Goodson Shop Supplies)*

spring retainer to rotate the valve in a complete circle. Some engines use rotators only on the exhaust valves; others have them both on the intake and exhaust valves.

Most rotators impart positive rotation to the valve during each valve cycle and improve valve life two to five times, and in some cases, even more. Some fit on the valve tip end of the spring and others at the guide end. This location depends on the engine design. In normal operation, positive rotators will continue to function for more than 100,000 miles and require no maintenance. However, when valves are refaced or replaced at high mileage, the rotators should be replaced because there is no visual means of inspection. Whether or not they rotate when held in the hand is no indication of their functional characteristics in the engine. While rotation can only be checked in a running engine, uneven wear patterns (Figure 6–49) develop at the valve stem tip if the rotators are not functioning properly.

Several types of rotators are available, but the most commonly used are the following:

- *Ball Type.* This employs two small balls and slight ramp. Each ball moves down its ramp to turn the rotator sections in a positive direction as the valve opens (Figure 6–50A).
- *Spring-Loaded Type.* In this design, the spring starts to move down as the valve opens. The spring-loaded rotator can rotate the valve in either direction (Figure 6–50B).
- *Free Type.* This rotator permits a momentary release of the spring tension from the valve during opening so that the valve is free to rotate (Figure 6–51). The action does not cause positive valve rotation. Engine vibration and turbulence of gases contribute to valve rotation during the time the valve is free to rotate.

With any release-type valve rotators, the residual clearance between the valve tip and the inside of

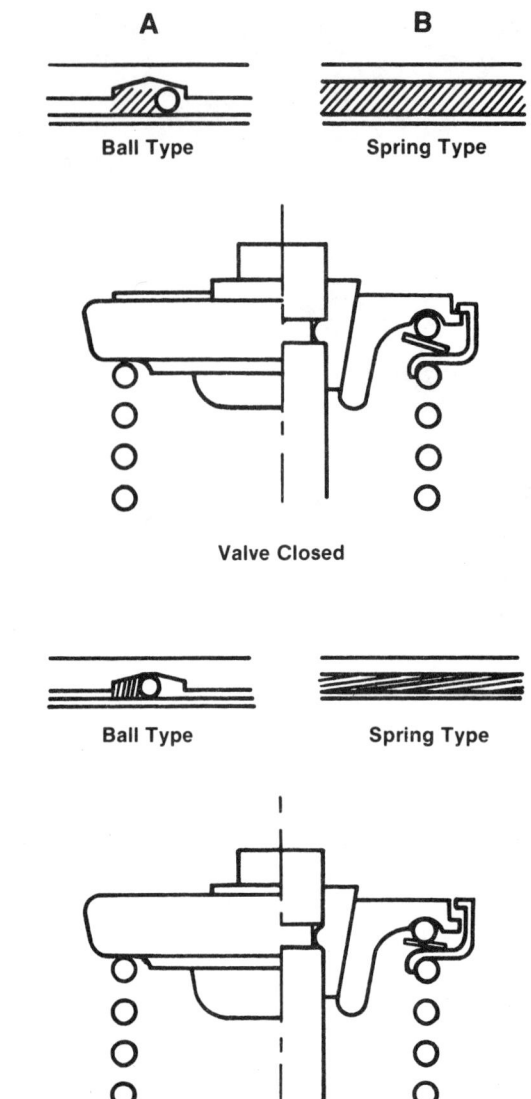

FIGURE 6-50 Valve rotator operation: (A) ball type; (B) spring type

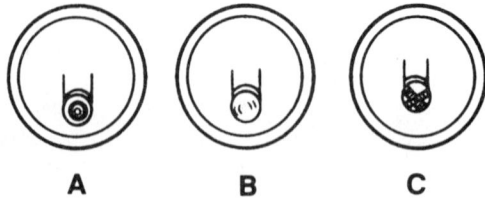

FIGURE 6-49 (A) Proper pattern means rotator functioning properly; (B) no pattern means rotator should be replaced and rotation checked; (C) partial pattern also means replacement and rotation check.

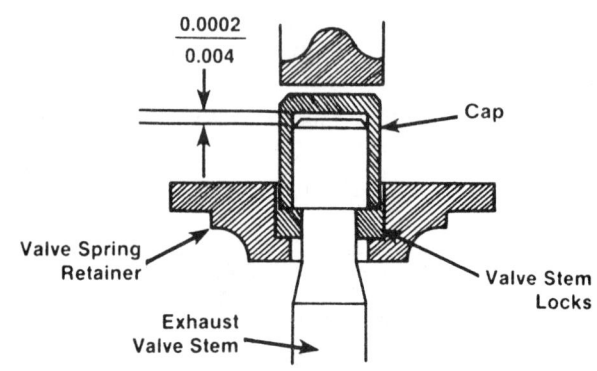

FIGURE 6-51 Free-type valve rotator

the tip cup is often overlooked as a factor in valve lash. In specifying a given tappet clearance, an engine manufacturer must assume that the proper clearance will be maintained in the release-type rotator. One advantage of the positive rotator is that it does not influence tappet clearance.

Some positive rotators are designed to operate between the valve spring and the cylinder head. Others are designed to operate at the valve tip in place of the spring retainer. The free-type rotator is generally installed at the valve tip. Some rotator installations require a shorter valve spring. Make sure the proper spring is installed. Some engines use different length intake and exhaust valve springs and sometimes two different types of spring retainers. Be sure there is no intermixing of parts in these cases. When the rotators are installed at the tip end of the valve, they take the place of the regular spring retainer.

 SHOP TALK ____

In most cases, valve rotation is not recommended with LPG fuels because it aggravates seat recession. However, if seat distortion is bad enough, then rotator caps or coils might make the difference between acceptable and unacceptable life. Uncontrolled rotation and high seat impact loading should be avoided. If an engine requires sodium valves with gasoline, it probably still will with LPG.

VALVE STEM SEALS

Valve stem seals (Figure 6-52) are designed to prevent oil leakage past the valve stems into the combustion chambers. If leakage occurs past the valve guides, oil consumption will rise. Therefore, all the seals should be replaced when a head is rebuilt. Old seals tend to wear, crack, and become brittle due to age and operation at elevated temperatures.

There are dozens of valve stem designs, but they all fall into three general categories:

- *Deflector Seals.* As the name implies, this type of seal deflects oil away from the valve stem. It acts like an umbrella, moving with the valve stem to shield the valve guide from excess oil. This type of valve stem seal is also called an *umbrella seal* (Figure 6-53A).
- *Positive Seals.* This O-ring seal is attached to the valve guide boss (Figure 6-53B). It works like a squeegee, wiping and metering oil on the valve stem. If positive stem seals are in-

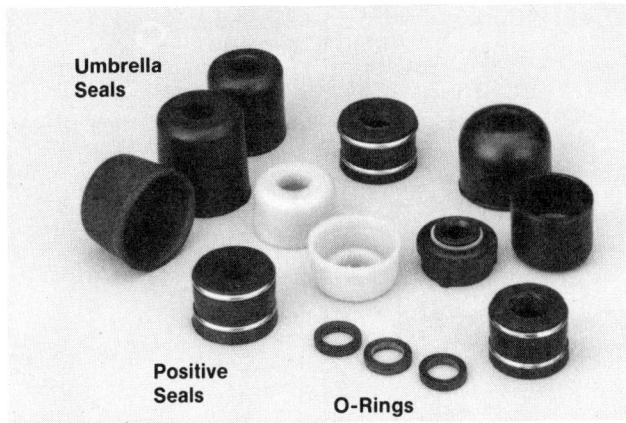
FIGURE 6-52 Various types of stem seals *(Courtesy of Goodson Shop Supplies)*

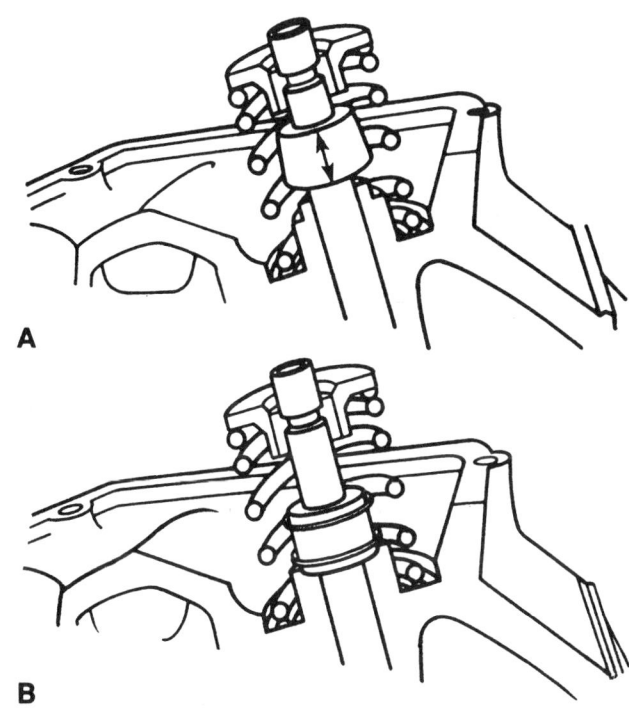
FIGURE 6-53 (A) Deflector or umbrella seal; (B) positive seal

stalled after machining the guide ends, follow the specific instructions covering the use of the original equipment seals and the positive type. The use of both types of seals could reduce oil supply to the valve stem and guide and cause premature wear.

- *O-ring Seals.* Although some experts do not consider O-rings to be valve seals, the rings are sometimes employed to serve the same purpose—they prevent excessive oil consump-

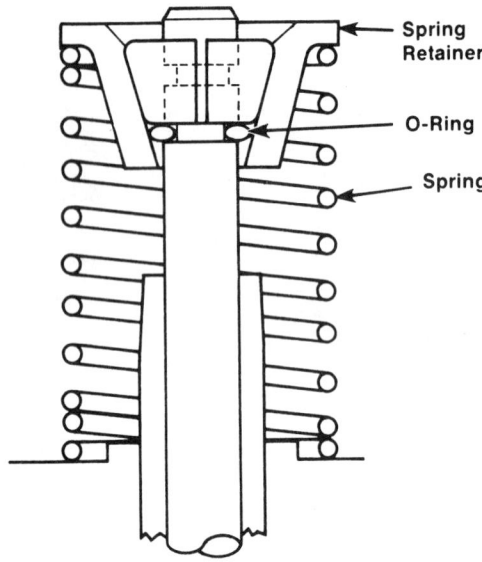

FIGURE 6-54 Installation of an O-ring valve seat

tion (Figure 6-54). They are also frequently used in conjunction with both positive seals and umbrella valve seals.

The most common material used in making valve stems is synthetic rubber because of its reasonable cost, easy installation, and accommodation of oversize valve stems. Three types of synthetic rubber are used (Figure 6-55).

- *Nitrile.* This type works well in older engines, except on certain high-performance applications. It is relatively inexpensive, but when used as a substitute for specified premium grade materials, nitrile can fall apart due to its low heat resistance (up to 250 degrees Fahrenheit).
- *Polyacrylate.* This rubber has a much greater temperature resistance (up to 350 degrees Fahrenheit) than nitrile, and gives a good balance between heat tolerance, oil resistance, and price.

FIGURE 6-55 Synthetic rubber valve stem seals; from left to right: nitrile, polyacrylate, and viton

- *Viton.* This type is currently the best material available for today's hot running (up to 450 degrees Fahrenheit) engine designs, particularly the smaller import and domestic models. While viton is more expensive than the other types, the cost is only a small part of a total valve job. And the extra cost is compensated for with greater reliability, which reduces the chance of a comeback.

 SHOP TALK _____

Always use the type of valve stem seal specified by the original equipment manufacturer. Valve stem seals are significantly different in lubrication properties, and this can cause a failure. Typical valve seal damage is shown in Figure 6-56.

VALVE LOCKS AND KEEPER GROOVES

The valve locks fit in the lock or keeper groove section of the valve (Figure 6-57) and attach the entire spring assembly to the valve stem. A thorough inspection should be made for wear on the lock beads and outside diameters. Worn parts, of course, should be replaced. However, as a precautionary measure, it is advisable to replace all the locks. If an unworn lock is inadvertently mated with a worn lock (Figure 6-58A), the retaining cap may shift, resulting in valve tip breakage (Figure 6-58B).

Excessive keeper groove wear is one of the more common valve failures found. Remember that

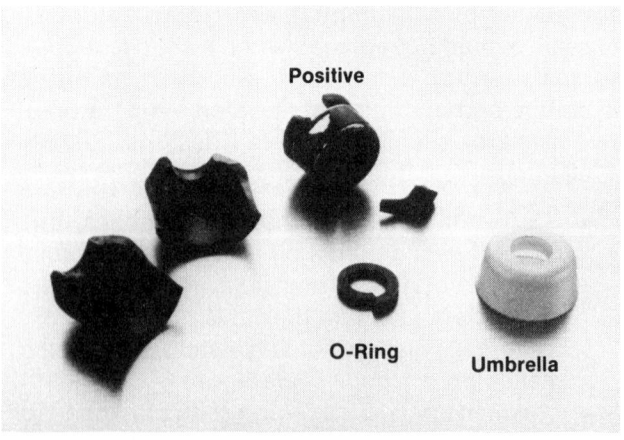

FIGURE 6-56 Typical damaged valve stem seals *(Courtesy of TRW Inc.)*

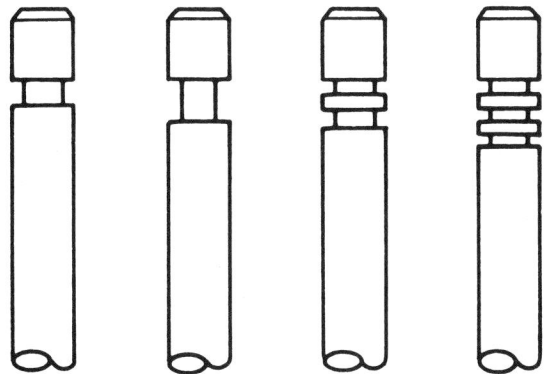

FIGURE 6-57 Single- and multiple-groove valve stems for different keepers or locks

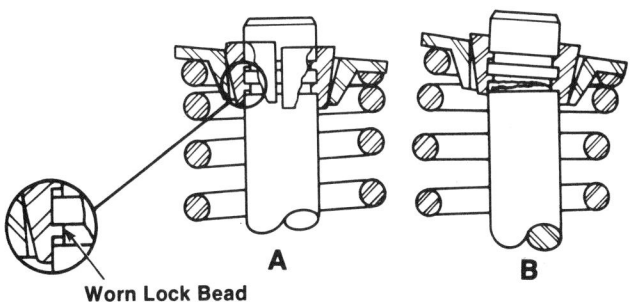

Worn Lock Bead

FIGURE 6-58 Carefully check the valve locks.

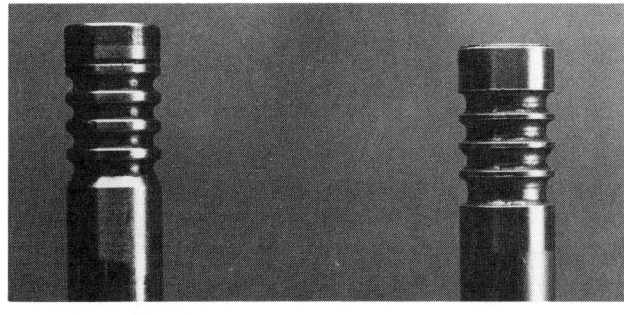

FIGURE 6-59 (Left) Good keeper groove; (right) worn keeper groove *(Courtesy of Perfect Circle/ Dana)*

any wear that occurs on the outside or inside of the keepers also increases the chance of valve stem end breakage. To check keeper groove wear (Figure 6-59), run your finger around the grooves. If they have a sharp or razor-edge feel as you rub your thumb over them, then they should be replaced.

VALVE TRAIN INSPECTION

The valve will operate only as well as the actuating mechanisms (Figure 6-60). This includes the timing gears, chain and sprockets, camshaft, lifters, pushrods, and rocker arm assemblies. For good valve life, each of these must be checked and replaced if worn.

When making an inspection, check for damaged and/or severely worn parts and correct assembly. Ensure the use of the correct parts by proceeding, as follows, with the typical static engine analysis:

1. Individually mounted rocker arm assemblies
   - Check for loose mounting stud and nut or bolt.
   - Check for plugged oil feed in the rocker arm.
2. Pushrods: Check for bent pushrods.
3. Valve spring assembly—with or without damper spring: Check for broken or damaged parts.
4. Retainer and keys—both two-piece and one piece: Check for proper sealing of key's valve stem and in the retainer.
5. Overhead cam follower arm and lash adjuster assemblies
   - Check for broken or severely arts.
   - Check for soft lash adjuster with hand pressure on rocker arm (arm on base circle of camshaft).
6. Camshaft—overhead camshaft applications
   - Check for plugged oil feed.
   - Check for correct cam lift.
   - Check for broken or severely worn parts.
   - Check for soft lash adjuster with hand pressure on rocker arm (arm on base circle of camshaft).
7. Check the timing belt, sprockets and related component.

CAMSHAFT

After the camshaft (Figure 6-61A) has been cleaned and after a visual inspection has been made, check each lobe for scoring, scuffing, fractured surface, pitting, and signs of abnormal wear (Figure 6-61B).

Premature lobe and lifter wear is generally caused by metal-to-metal contact between the cam lobe and lifter bottom due to inadequate lubrication

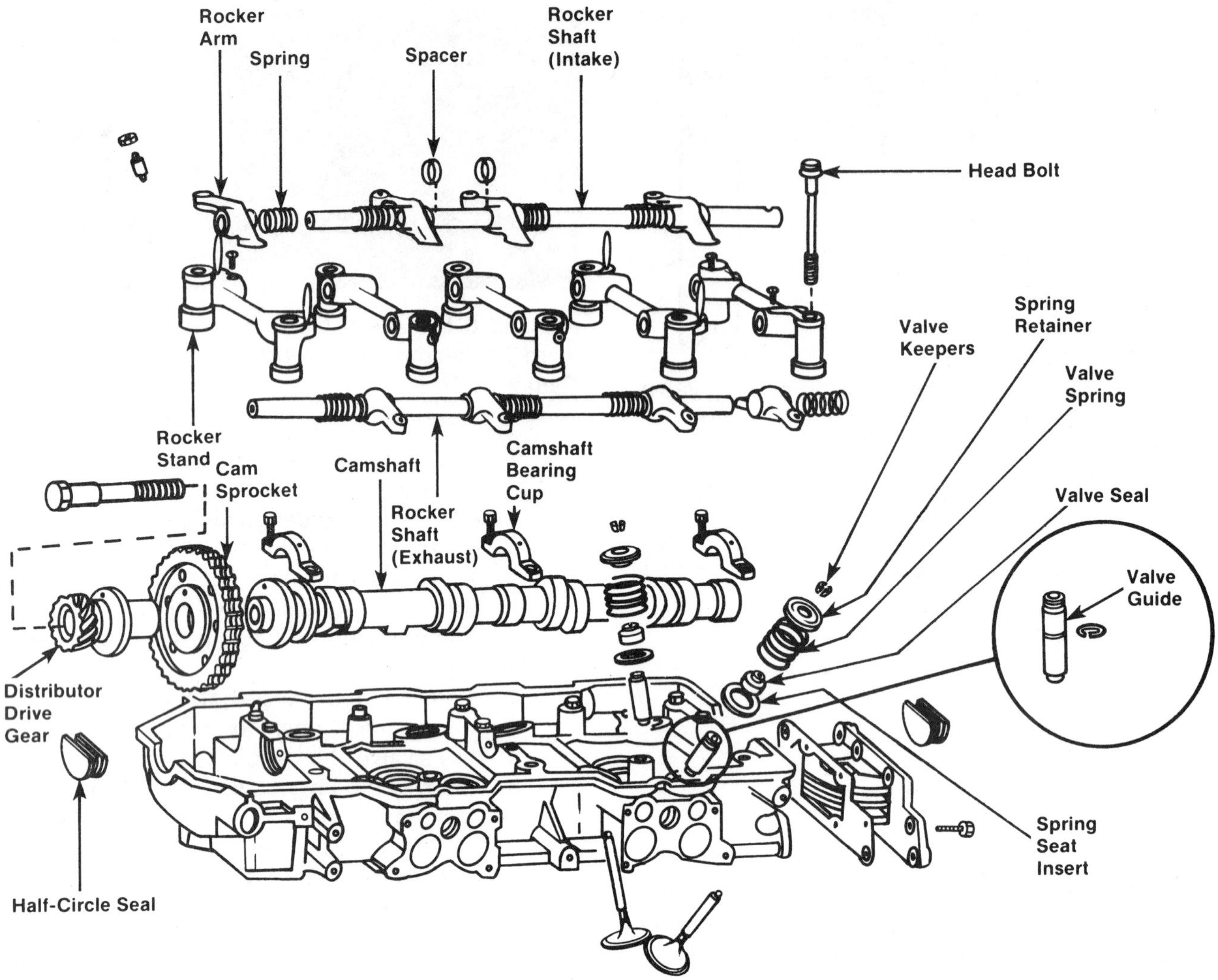

Rocker
Arm

Spring

Spacer

Rocker
Shaft
(Intake)

Head Bolt

Valve
Keepers

Spring
Retainer

Valve
Spring

Valve Seal

Valve
Guide

Rocker
Stand

Cam
Sprocket

Camshaft

Camshaft
Bearing
Cup

Rocker
Shaft
(Exhaust)

Distributor
Drive
Gear

Spring
Seat
Insert

Half-Circle Seal

FIGURE 6-60 Valve train components

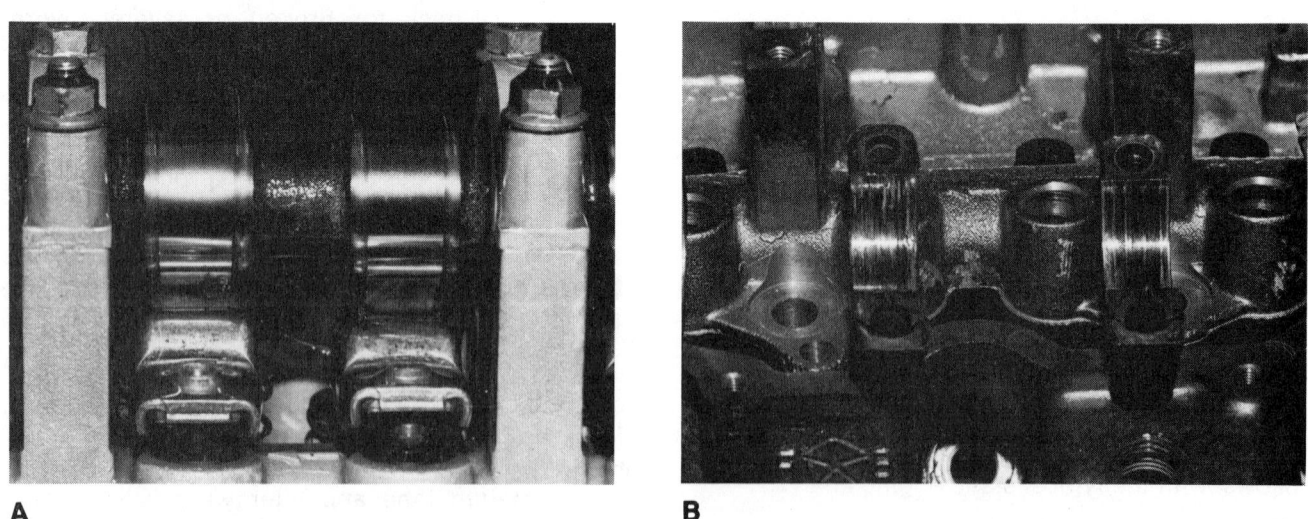

A

B

FIGURE 6-61 (A) Hardened steel camshaft and roller followers; (B) badly burned and worn camshaft bearing bores

FIGURE 6-62 Badly damaged camshaft

(Figure 6-62). The nose will be worn from the cam lobes, and the lifter bottoms will be worn to a concave shape or may be worn completely away. This type of failure usually begins within the first few minutes of operation (Figure 6-63). It is the result of insufficient lubrication or use of an oil that does not meet the engine manufacturer's requirements for viscosity and A.P.I. service grade (see Chapter 13).

There are several methods of measuring cam lobes for wear, but the two most popular with rebuilders are the dial indicator and outside micrometer.

The dial indicator test for worn cam lobes should be conducted with the camshaft in the engine. Check the lift of each cam lobe (Figure 6-64) in

FIGURE 6-63 New lifters should be installed only on new camshafts. Never install new lifters on a worn camshaft or worn lifters on a new camshaft. The mating surfaces between the lifter foot and cam lobe are the most critically stressed areas in an engine. *(Courtesy of Sealed Power Corp.)*

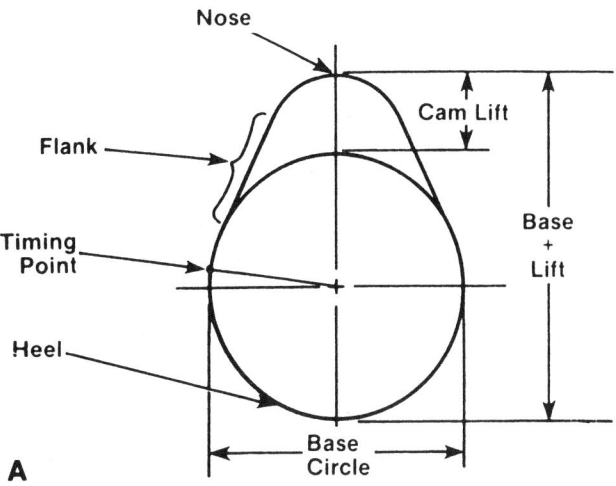

A

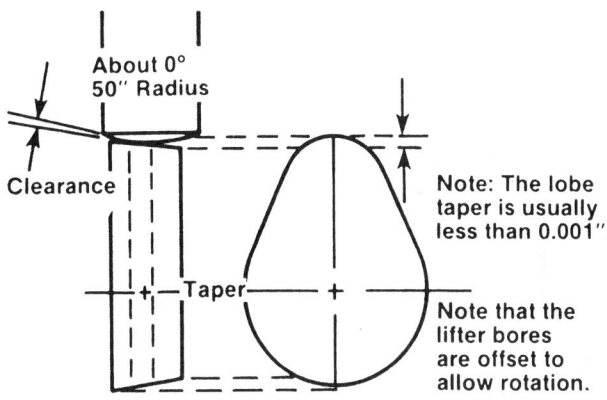

Most late-model automotive cams are tapered to provide lifter rotation. The lifters have a spherical grind so that they do not ride on the edge of the cam lobe. This contact spreads the load of the valve train against more of the lobe face.

B

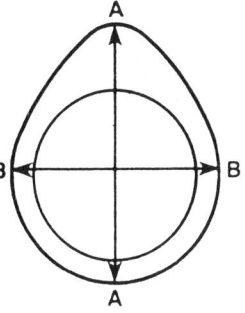

Dimension A minus dimension B equals the cam lobe lift.

C

FIGURE 6-64 (A) Cam lobe nomenclature; (B) correct camshaft pattern; (C) determining cam lobe lift

consecutive order and make a note of the readings. To perform the test, proceed in the following manner:

1. Be sure that the pushrod is in the valve tappet socket. Install the dial indicator so that the cup-shaped adapter fits onto the end of the pushrod and is in the same plane as the pushrod movement (Figure 6–65).

2. Connect an auxiliary starter switch into the starting circuit. Crank the engine with the ignition switch off. "Bump" the crankshaft over until the tappet is on the base circle of the camshaft lobe. At this point the pushrod will be in its lowest position.

3. Put the dial indicator at zero. Continue to rotate the crankshaft slowly until the pushrod is in its fully raised position (highest indicator reading).

4. Compare the total lift recorded on the indicator with specifications.

5. To check the accuracy of the original indicator reading, continue to rotate the crankshaft until the indicator reads zero. If lift on any lobe is below the specified service limits, the camshaft and tappets operating on the worn lobe(s) must be replaced. Any tappet showing pitting or having its contact face worn flat or concave must also be replaced.

To compare the cam lobe height with an outside micrometer, be sure the camshaft is removed from the engine. Place the micrometer in position to measure from the heel to the nose of the lobe and again 90 degrees from the original measurement. Record the measurement for each intake and exhaust lobe. Any variation in heights indicates wear. Also check the measurements taken against the manufacturer's cam lobe heights.

If camshaft wear exceeds the manufacturer's limits, a cam lobe lift loss of no more than 0.005 inch is generally accepted.

Measure each camshaft journal in several places with a micrometer to determine if it is worn excessively. If any journal is 0.001 inch or more below the manufacturer's prescribed specifications, it should be reground to a standard undersize or the camshaft should be replaced.

The camshaft should be checked for straightness with a dial indicator. Place the camshaft on V-blocks. Position the dial indicator on the center bearing journal and slowly rotate the camshaft (Figure 6–66). If the dial indicator shows runout (a 0.002 inch deviation), the camshaft is not straight. If it is not excessively bent, it might be possible to be straightened. Mount the camshaft in a hydraulic press on V-blocks. With the press ram, push down on the high side of the shaft. Remove the camshaft and recheck with the dial indicator. Repeat the procedure as necessary.

If there is any evidence of wear, the lobes must be replaced or reground. Sometimes the lobes are rebuilt by welding on the area of the lobe. The welded area is then reground on the cam grinding

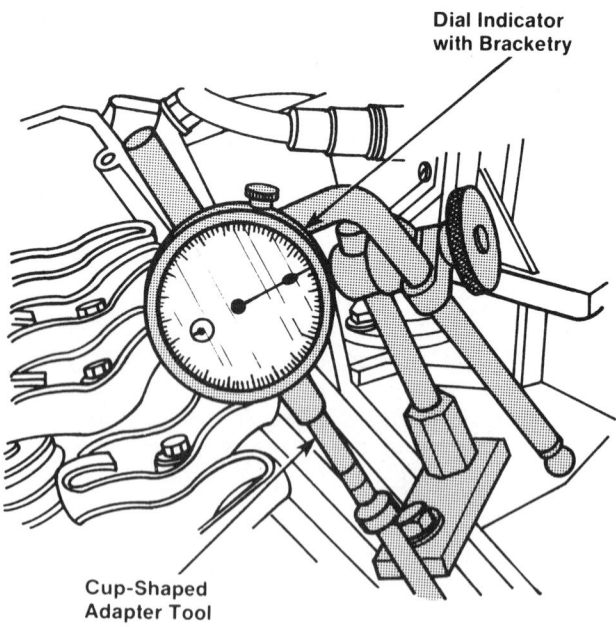

Dial Indicator with Bracketry

Cup-Shaped Adapter Tool

FIGURE 6–65 Checking camshaft lobe using a dial indicator

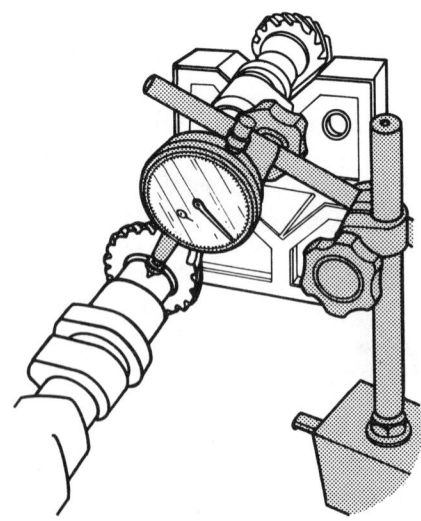

FIGURE 6–66 Checking camshaft for straightness

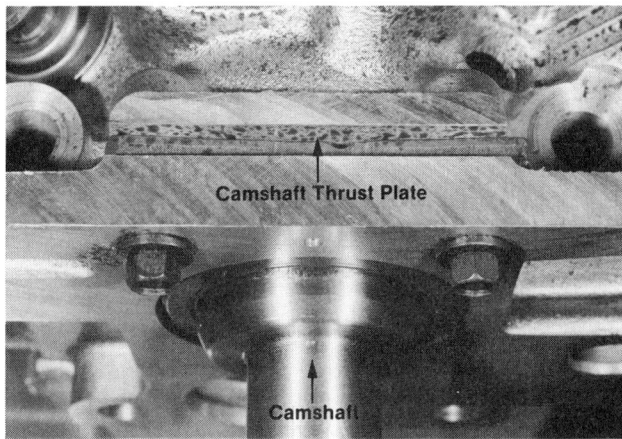

FIGURE 6-67 Excessive camshaft end play can be caused by a worn thrust plate, worn camshaft, or worn cylinder head.

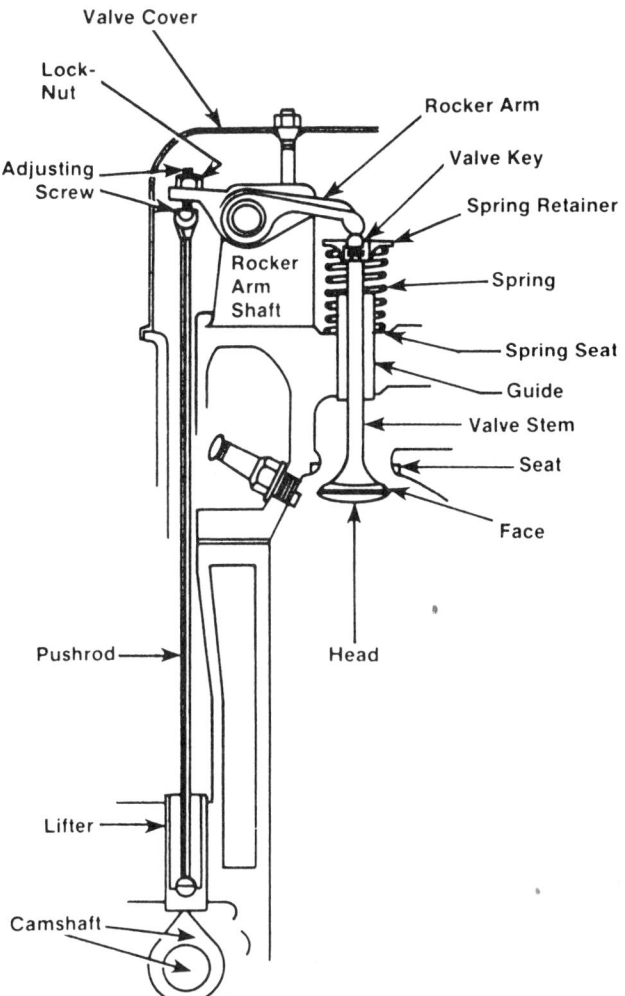

FIGURE 6-68 Parts of a valve train with a solid valve lifter

equipment. After grinding, the camshaft lobe surfaces are covered with a hard surface overlay. The overlay, which is harder and more slippery than the base metal of the camshaft, prevents rapid camshaft lobe wear.

Endwise movement or play of the camshaft is limited by a thrust plate, which is generally bolted to the engine (Figure 6-67). It is usually located between the front bearing journal and the drive sprocket or sprocket. Some engines do not use a thrust plate to hold the camshaft in place. They rely instead on the effects of the spiral teeth of the distributor oil pump drive. The thrust is absorbed between the inner surface of the sprocket and the front of the block.

On all six- and eight-cylinder engines, prying against the camshaft sprocket with the valve train load on the camshaft can break or damage the sprocket. Therefore, the rocker arm adjusting nuts must be backed off, or the rocker arm and shaft assembly must be loosened sufficiently to free the camshaft.

To check the camshaft for end play, push it toward the rear of the engine. Install a dial indicator and bracketry so that the indicator point is on the camshaft sprocket attaching screw. Set the dial indicator at zero. Position a large screwdriver between the camshaft gear and the block. Pull the camshaft forward and release it. Compare the dial indicator reading with the specifications listed in the service manual.

If the end play is excessive, check the spacer for correct installation before it is removed. If the spacer is correctly installed, replace the thrust plate and recheck. Remove the dial indicator.

VALVE LIFTERS

Basically, there are two types of valve lifters: mechanical or solid (Figure 6-68) and hydraulic (Figure 6-69).

All engines that have mechanical lifters use some method of adjustment that is intended to bring valve lash (clearance) back into specification. There are four basic methods for lash adjustment:

1. Rocker arm with adjustable pivots
2. Adjustable pushrods
3. Rocker arms with adjustable screws
4. Adjustable cam follower (using some type of adjustable screw or replaceable shim)

Of these four adjustment types, the first two methods are typically associated with V-type engines. The other two adjustment procedures—rocker arms

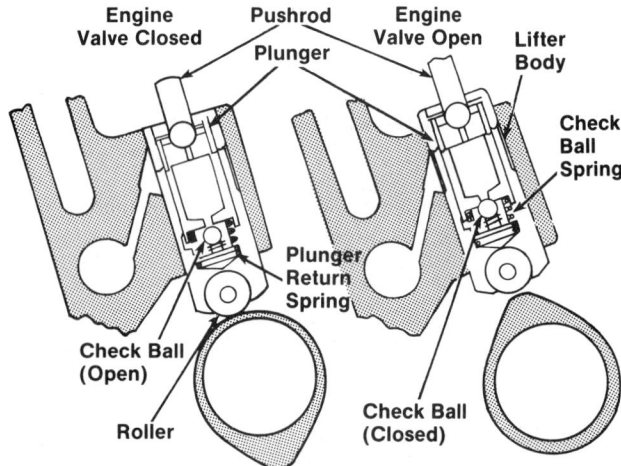

FIGURE 6-69 Exploded view of hydraulic roller-type valve lifter

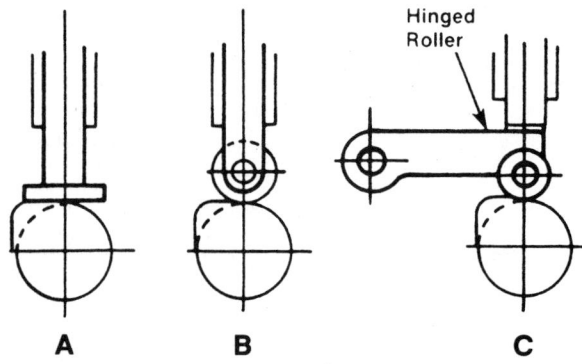

FIGURE 6-70 Typical mechanical valve lifter designs: (A) mushroom; (B) roller; (C) hinged roller

with adjustment screws and adjustable cam followers—are commonly found on four-cylinder OHC designs. Figure 6-70 illustrates the three basic types of mechanical valve lifters.

When inspecting mechanical lifters, keep in mind the bottoms and pushrod sockets. Wear, scoring, or pitting makes their replacement necessary.

The majority of domestically manufactured engines in passenger cars and light trucks use a hydraulic valve lifter. It is one of the engine's most precisely manufactured components, with tolerances to within 35 millionths of an inch. During a lifter's operating lifetime, it will experience loads in excess of 1800 pounds, with temperatures as high as 300 degrees Fahrenheit. Under all conditions, it must maintain a continuous zero clearance with metal-to-metal contact.

A typical hydraulic lifter contains ten parts. There is a plunger, an oil metering valve disc, a pushrod seat, retaining ring, check valve disc or ball, check valve retainer, check valve spring, retainer clip, and plunger return spring all housed in a hardened iron body (Figure 6-71). Some hydraulic lifters (Figure 6-72) use rollers to make them more effective.

By virtue of its name, a hydraulic lifter uses a combination of fluid (oil) and spring pressure to maintain a zero clearance (lash) regardless of wear, tolerance, or temperature changes. The lifter performs this function by using engine oil under pressure to fill a compression chamber located beneath the plunger in the bottom of the lifter. As long as there is no downward force on the plunger, oil (stored in the plunger) flows into the compression chamber through the check valve.

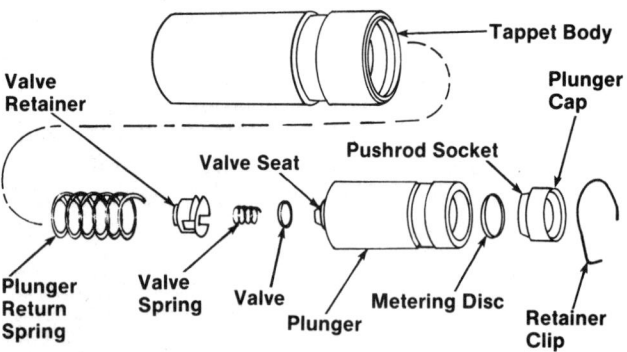

FIGURE 6-71 Exploded view of hydraulic valve lifter

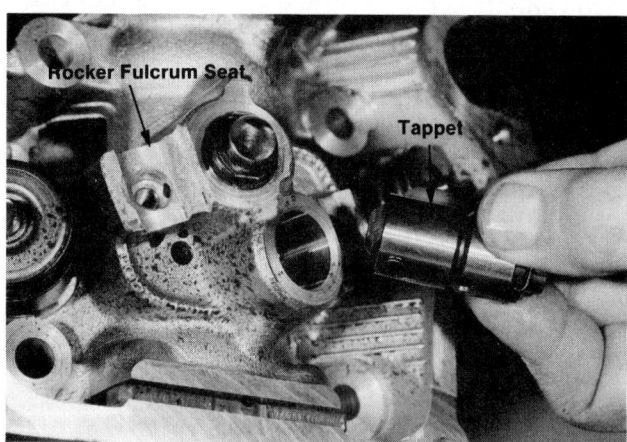

FIGURE 6-72 A stamped metal guide (not shown here) surrounding the roller hydraulic lifters keeps them aligned with the camshaft lobes during operation.

As soon as the upward thrust of the cam lobe pushes the lifter against the pushrod, however, the check valve seats, trapping the oil in the compression chamber. Because oil is a fluid and a fluid cannot be compressed, the lifter begins transmitting the cam's motion (lifter to pushrod, pushrod to rocker, and rocker to valve stem), which eventually forces the engine valve to open against valve spring pressure. Once the valve is opened, the compressed valve spring is capable of exerting enough force to make the lifter's plunger compress slightly. Before this can happen, however, the oil in the compression chamber has to have an escape route (the check valve is still seated). This is where leakdown comes in. As the force of the valve springs tries to push the plunger down, a small amount of oil "leaks" out of the compression chamber past the microscopic space between the plunger and lifter body. Because this is a highly controlled leakage rate, enough oil is left in the compression chamber to support the valve train even after leakdown.

As the cam continues to rotate and the valve closes, the lifter is once again on the cam's base circle, and any oil that has leaked out is immediately replaced from the plunger reservoir to start the process all over. During all of this, the lifter has not let any lash into the valve train, there is a solid column of oil transmitting the cam's motion, and all the parts stay in constant contact regardless of wear or temperature.

Technically, the normal wear of the valve lifters is referred to as *adhesive* or *galling wear*. This is a result of two solid surfaces (camshaft lobe and lifter face) that are in rubbing contact. The two surfaces literally weld together, which results in particles being torn out. This process is considered normal wear between the cam lobe and lifter. Fortunately, proper lubrication retards this process. However, excessive loading will negate the beneficial effects of the lubricant and accelerate the wear process. Some causes of excessive loading would be incorrectly matched valve springs (too much spring pressure), old lifters on a new camshaft, or new lifters on an old camshaft. If a camshaft and lifters are going to be reused, the lifters must remain with their respective lobes. Worn valve lifters and improper camshaft installation are common causes of camshaft/lifter failure. Figure 6-73 shows the ideal contact between the crowned lifter bottom and the tapered cam lobe.

The normal wear path is off center with no edge contact between the lifter and the lobe. The taper on the cam lobe (approximately 0.001 to 0.0002 inch) coupled with the spherical radius (approximately 50 inches) of the lifter bottom is specifically designed to result in an offset contact pattern causing the lifter to rotate. The spinning lifter reduces the sliding friction as well as equalizes the load around the lifter bottom.

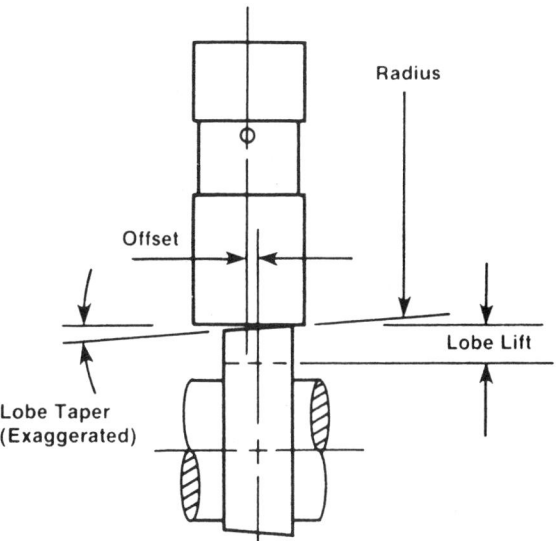

FIGURE 6-73   Ideal contact between crowned lifter and tapered arm lobe

 **SHOP TALK** _____

Using new lifters on worn lobes will result in scuffing of the lifter bottoms and cam lobes. This is because a used camshaft is worn flat during extended service as is the bottom of the old lifter. When installing a new lifter on an old cam, because the lobe is flat, contact takes place at the center of the lifter bottom preventing the lifter from rotating. The opposite is also true. Used lifters cannot be matched with new camshaft lobes. The used lifter bottom is flat or slightly dished. The high edge of the tapered cam lobe is making fine line contact with the worn lifter. The lifter can rotate, but while it is rotating, the edge of the cam lobe is cutting a groove near its edge. As the groove becomes deeper, the edge of the rotating lifter rubs against the edge of the cam lobe. This cutting process continues until the lifter bottom becomes dished and the cam lobe edges are rounded off (Figure 6-74A). Evidence of worn lifters against a new cam lobe is characterized by concentrated wear at the edges of the lobe with a notable absence of the normal wear pattern (Figure 6-74B).

Interference occurs primarily in engines originally equipped with a separate cam sprocket spacer (used to control end clearance). The problem develops when a replacement sprocket with a built-in

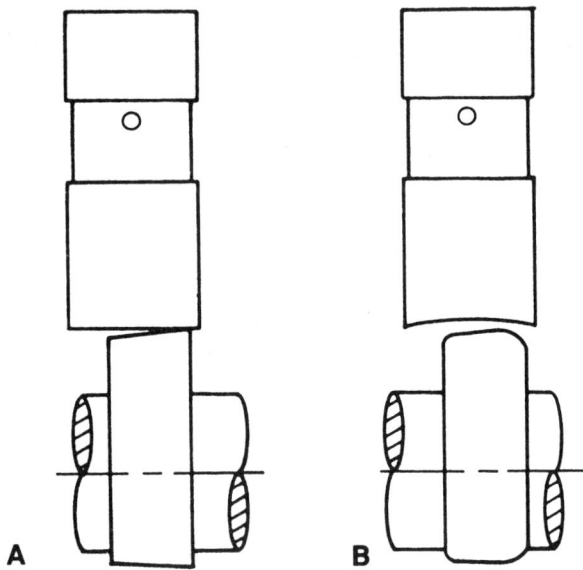

FIGURE 6–74 (A) Cam lobe edges rounded off; (B) worn lifters

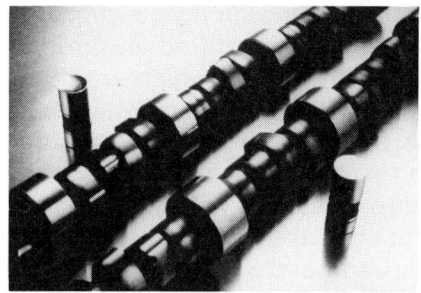

FIGURE 6–76 Example of edge riding *(Courtesy of TRW Inc.)*

spacer is installed without removing the original spacer. This forces the camshaft rearward, allowing the lobes to strike adjacent lifters, chipping the edges of both (Figure 6–75). Interference will also develop if sprocket bolts are not tightened properly or if the cam sprocket/engine block thrust surfaces are worn excessively.

Edge wear on cam lobes occurs when used lifters are installed with a new camshaft (Figure 6–76). The bottoms of used lifters are often flat or slightly concave due to previous wear. As a consequence, the lifters will contact the lobe along a narrow band at the lobe's edge. This tends to create high contact forces and rapid wear results. The cam shown in the right portion of the illustration displays a normal wear pattern. Most new lifters have spherical bottoms (convex), and new cam lobes are tapered slightly. This combination allows the lifters to rotate as they ride on the cam, thereby reducing frictional

and contact forces. As previously mentioned, lifters are designed to rotate in their bores. This enhances the lubrication on the bottoms and sides of the lifters and reduces friction and wear. If mismatched (used) cam and lifters are installed, the lifters might not rotate properly.

Malfunctioning hydraulic lifters can cause valves to burn or break. Whenever the valve train is disturbed, the hydraulic lifters should be removed, disassembled, cleaned, and checked. They should be kept in sequence during removal so that they can be put back in the same place. Lifters should be replaced if the bottoms are worn or pitted or if a new camshaft is installed. It is not good practice to re-grind lifter bottoms because it is difficult to generate a good surface with the proper contour. Lifter bottoms, generally, are spherical (Figure 6–77).

To inspect, disassemble each lifter separately, but make sure not to intermix any parts because they are selectively fitted. Any time hydraulic lifters are removed, the varnish and deposits should be carefully removed from the lifter bores in the engine block, and the galleries should be flushed with pressurized oil to clear any dirt from the holes that feed the lifters. This latter step is of utmost importance

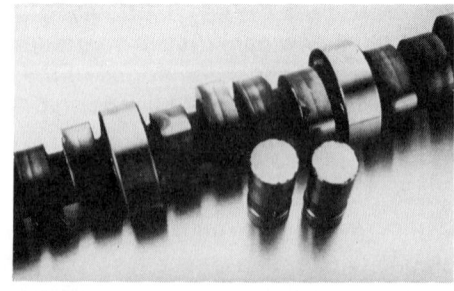

FIGURE 6–75 Example of lobe and lifter interference *(Courtesy of TRW Inc.)*

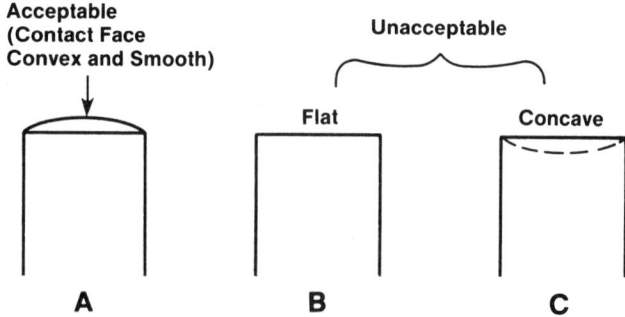

FIGURE 6–77 Inspect valve lifters for damage and wear: (A) acceptable—contact face convex and smooth; (B) unacceptable—flat; (C) unacceptable—concave

because the act of removing the lifters will normally deposit some dirt right where the feed holes break into the bores. If this is not flushed out, the first oil fed to the newly installed lifter will contain this dirt, and trouble might develop immediately.

After cleaning, check the lifter's leakdown with a leakdown tester (Figure 6–78). Lifter (tappet) rate is important; if the tappets leak down too quickly, noisy operation will result. When diagnosis indicates no cause for noisy tappet operation, the condition can sometimes be remedied by checking the tappet leakdown rate and replacing any that are outside specification.

To check the lifter (tappet) leakdown, perform the following procedure:

1. Place the lifter in the tester with the plunger facing upward. Pour hydraulic tappet tester fluid into the cup until the lifter assembly is covered. The fluid can be purchased from the manufacturer of the tester. Using oil, kerosene, or any other fluid will not provide an accurate test.
2. Place the 5/16-inch steel ball provided with the tester into the plunger cap.
3. Adjust the ram length so that the pointer is 1/16 inch below the starting mark when the ram contacts the lifter plunger to permit timing as the pointer passes the start timing mark. Use the center mark on the pointer scale as the stop timing point instead of the original stop timing mark at the top of the scale.
4. Work the lifter plunger up and down until the tappet fills with fluid and all traces of air bubbles have disappeared.
5. Allow the ram and weight to force the lifter plunger downward. Measure the exact time it takes for the pointer to travel from the start timing to the stop timing marks of the tester.
6. A satisfactory lifter must have a leakdown rate (time in seconds) within the minimum and the maximum specified limits. Compare with the reading given in the service manual.
7. If the lifter leakdown rate is not within specifications, replace it with a new tappet. In fact, most shops today usually replace tappets or lifters with new ones when rebuilding an engine because they are time-consuming operations.

 SHOP TALK ————

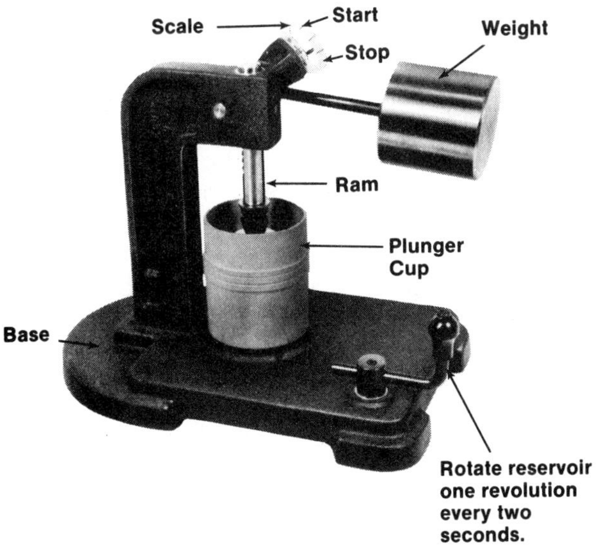

It is not necessary to disassemble and clean new tappets before testing, because the oil contained in new tappets is test fluid.

PUSHRODS

The pushrod fits between the valve lifter and the rocker arm to help transmit the cam action to the valves. Some pushrods might have a groove worn in the area in which they pass through the cylinder head, and some are subject to tip wear. Also check the ends of the pushrods for nicks, grooves, roughness, or signs of excessive wear. Pushrod balls and seats are subject to wear, which results in changes in the valve train adjustment.

Bent pushrods can be the result of valve timing, valve sticking, or improper valve adjustment. Bent or broken pushrods indicate interference in the valve train. Incorrect valve springs or an installed height less than specified can cause coil bind. Also, insufficient valve-to-piston clearance can cause a collision between the valve and piston at high engine speeds.

The pushrods can be visually checked for straightness while they are installed in the engine by rotating them with the valve closed. With the pushrods out of the engine, they can be checked for straightness by rolling them over a precision flat surface such as a surface plate. If a pushrod is not straight it will appear to hop as it is rolled. However, the most accurate way to check for straightness is by

FIGURE 6–78 Hydraulic tappet leakdown tester. Tester uses a special fluid to check lifter leakdown rate. *(Courtesy of Kent-Moore Corp.)*

Scale — Start
Stop
Weight
Ram
Plunger Cup
Base
Rotate reservoir one revolution every two seconds.

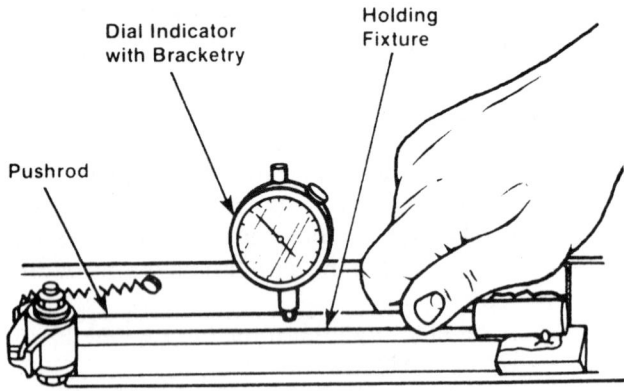

FIGURE 6-79 Checking pushrod runout

using a dial indicator (Figure 6-79). If more than 0.003 TIR is found, the pushrod is not straight and should be replaced.

 SHOP TALK —————

On some overhead cam engines, the lifters are set directly above the camshaft (Figure 6-80). No pushrod is needed.

ROCKER ARMS

The rocker arm translates the upward movement of the valve lifter into downward motion that opens the valve (see Chapter 2). Rocker arms also permit valves to be angled, allowing a hemispherical combustion chamber with one camshaft.

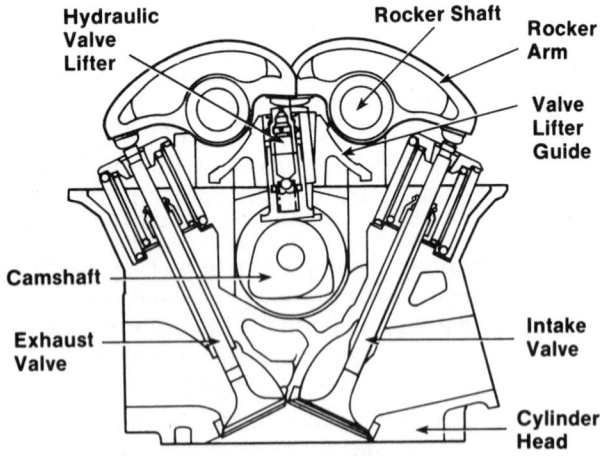

FIGURE 6-80 On an overhead cam engine, the lifters are set directly above the camshaft. No pushrod is needed.

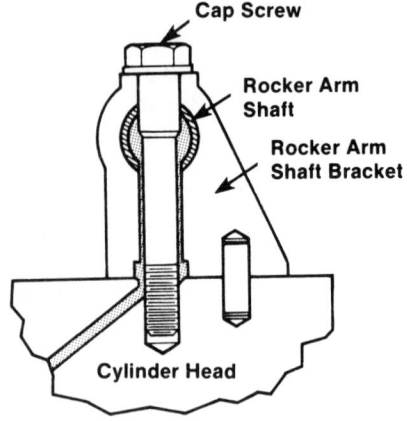

A

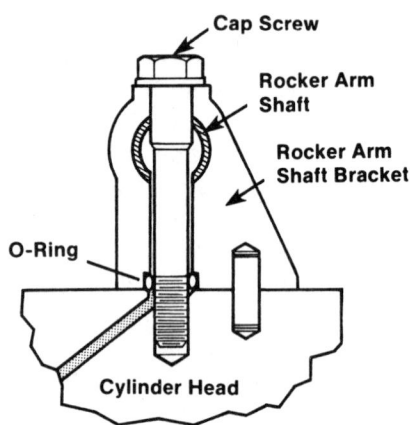

B

FIGURE 6-81 (A) Oil supply open; (B) oil supply shut off

Rocker arms are lubricated in various ways. In some engines oil is forced up around the outside of the cap screw that holds the bracket in place. Oil supply on these engines can be shut off by lifting the bracket around which the oil flows and placing a rubber O-ring over the threaded end of the cap screw. When the cap screw is tightened, a seal is formed between the bracket and the top of the cylinder head (Figure 6-81).

On some overhead valve V-type engines, oil reaches the rocker arm assembly through a drilled passage in the cylinder head and rocker arm shaft bracket (Figure 6-82). In most cases there is only one drilled bracket per cylinder head.

Possibly the most popular method of supplying oil to the rocker arms is through the pushrod valves. Some pushrods are hollow and function as part of the lubricating system for the rocker arm assembly (Figure 6-83). Oil flows under pressure from the block through a hole in the valve lifter and up through the pushrod to the rocker arm. No oil will get to the rocker arms if these pushrods are plugged.

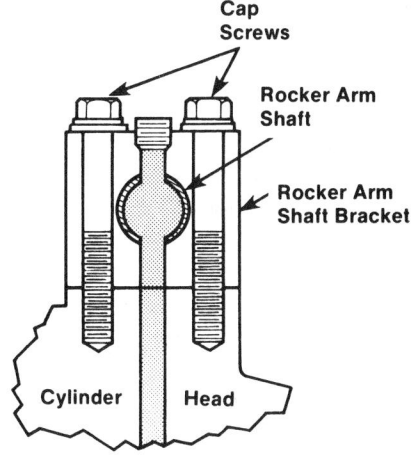

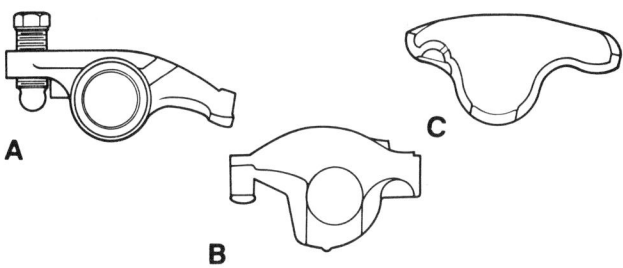

FIGURE 6-84   Three types of rocker arms (A) cast iron; (B) cast aluminum; (C) stamped steel

Look through each pushrod to make sure the oil flow passage is clear.

As mentioned in Chapter 2, rocker arms are made of cast iron, cast aluminum, or stamped steel (Figure 6-84). Cast adjustable rocker arms are attached to a rocker arm shaft (Figure 6-85) that is mounted on the head by rocker arm brackets. Although a cast rocker arm can be resurfaced, a stamped nonadjustable rocker arm that is worn must be replaced. There are several arm designs that are mounted in several different ways (Figure 6-86).

Even though some cast rocker arms are still in use, most domestic engines are equipped with an independent stamped rocker arm assembly for each valve and no rocker arm shaft is used. A stud is either pressed or threaded into the cylinder head and must be replaced if worn, bent, broken, or loose (Figure 6-87). On some engines, the studs are drilled for an oil passage to the rocker arms. Make sure oil can pass through before installing the cylinder head on the block. Replacement press-in studs are available in standard sizes and oversizes. The standard size is used to replace damaged or worn studs and the oversizes are used for loose studs.

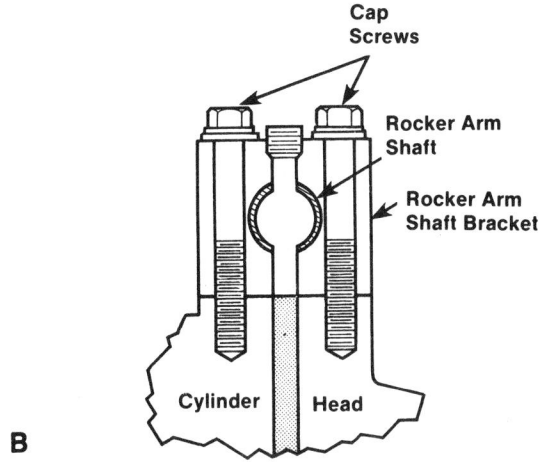

FIGURE 6-82   (A) Oil supply open; (B) oil supply shut off

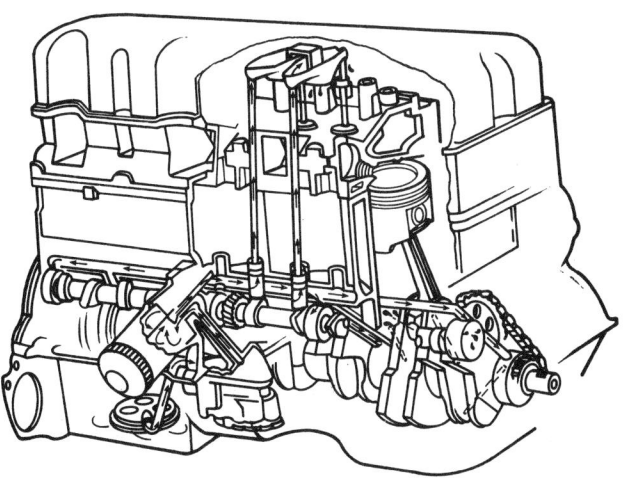

FIGURE 6-83   Pushrods are sometimes hollow and carry lubricating oil to the upper parts of the valve train.

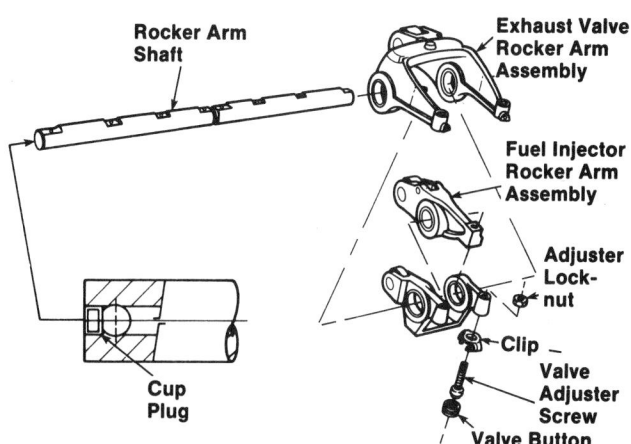

FIGURE 6-85   Typical rocker arm shaft

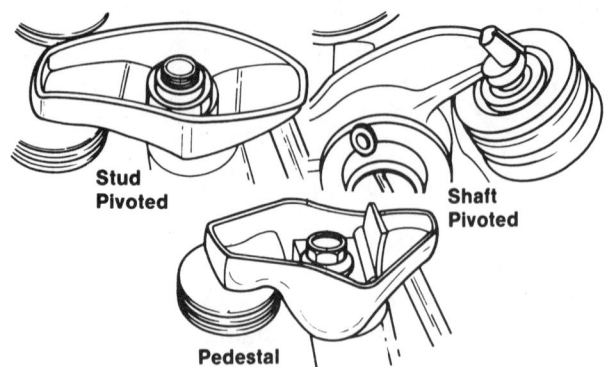

FIGURE 6-86 Rocker arm designs vary and are mounted in several ways.

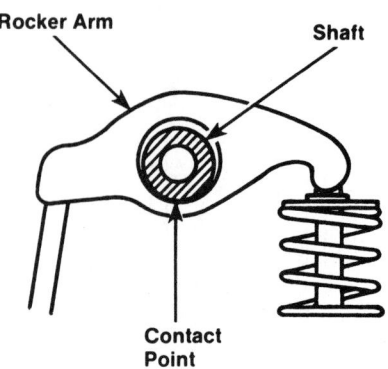

FIGURE 6-88 Wear points: rocker to rocker shaft

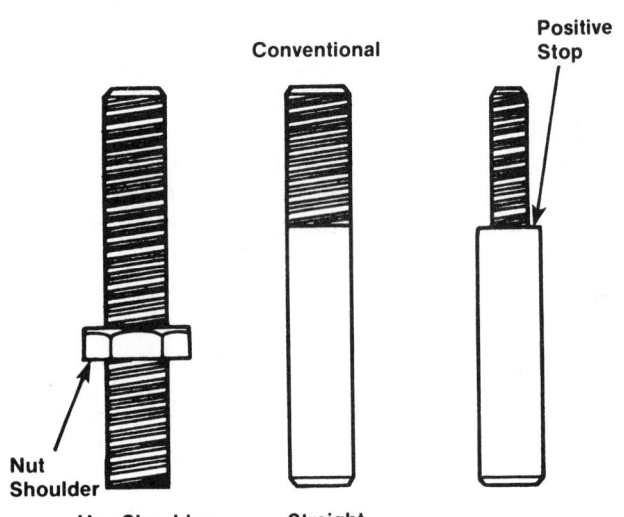

FIGURE 6-87 Three types of rocker arm studs

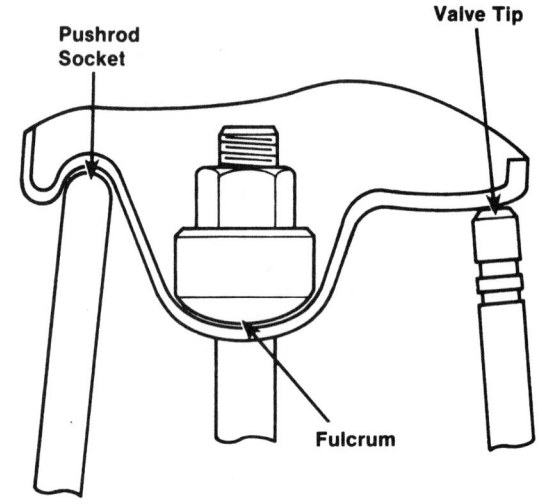

FIGURE 6-89 Rocker arm wear spots

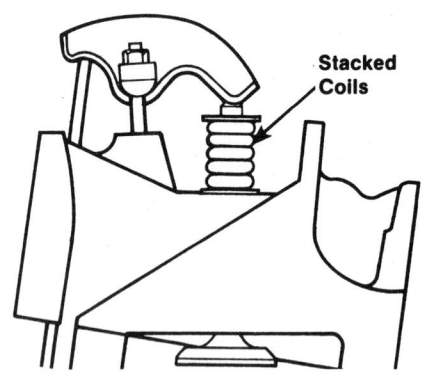

FIGURE 6-90 Spring coil bind

Inspect the rocker shaft assembly for wear, especially at points that contact the valve stem and pushrod. The fit between the cast rocker arm and the rocker shaft is checked by measuring the outside diameter of the shaft and comparing it to the inside diameter of the rocker arm. Excessive clearance requires replacement of the rocker arm or the rocket shaft, or both. Another wear point that should be checked is the pivot area of the rocker to rocker shaft (Figure 6-88). Other rocker arm wear points are shown in Figure 6-89.

Excessive wear of the valve pad occurs when the rocker arm repeatedly strikes the valve tip in a hammer-like fashion. It is able to strike the valve tip in this way when valve train clearance, or lash, is excessive. Excessive valve lash, of course, can occur in several ways, resulting in wear anywhere along the valve train. For example, it will occur when mechanical lifters are not adjusted properly or when

hydraulic lifters are not working properly. In addition, worn rocker arm valve pads can result when there is insufficient lubrication. Proper lubrication transfers heat away from the valve pad and reduces the metal-to-metal contact. This will keep valve pad wear to a minimum.

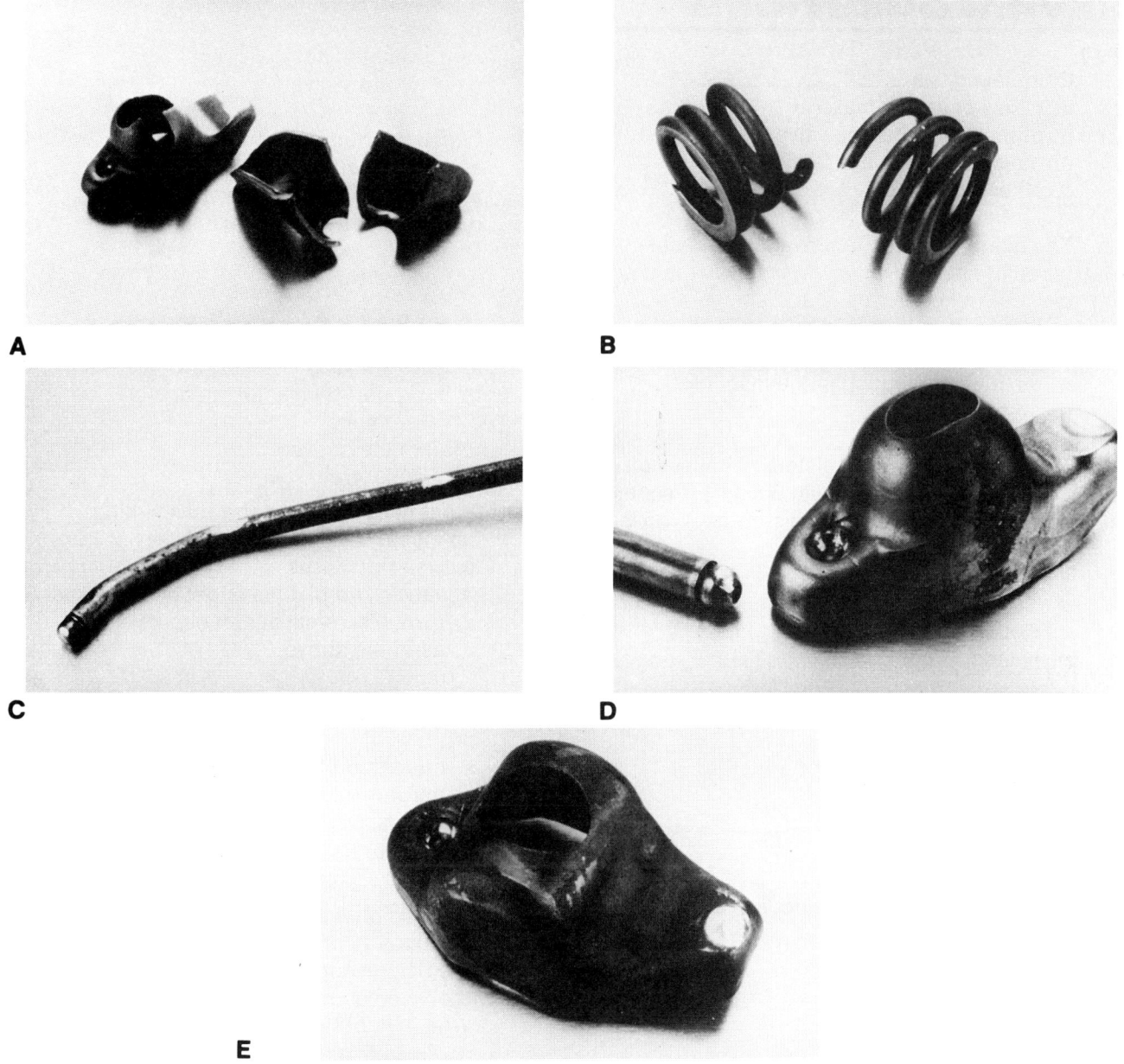

FIGURE 6-91   Common rocker arm and pushrod problems: (A) broken rocker arm; (B) broken valve spring; (C) broken or bent pushrod; (D) worn pushrod end/ball socket; and (E) worn rocker arm pad.

Another problem associated with rocker arms is breakage due to interference in the valve train. There are two ways interference can occur:

1. As shown in Figure 6-90, coil bind is the stacking of the coils so that when the valve spring is fully compressed, the coils contact each other. Coil bind is usually the result of using the incorrect spring or an excessive number of spring shims. Too many shims will cause the spring to be compressed to a lower spring height than

required. Whatever the cause, coil bind makes the spring act like a solid component, thereby leading to valve train damage.
2. The rocker arm can break due to insufficient valve-to-piston clearance. If the clearance is incorrect, an open valve can collide with the piston when it is at top dead center. The force of the collision can cause the rocker arm to break.

Other rocker problems are shown in Figure 6-91.

REVIEW QUESTIONS

1. Channeling is a _____ .
   a. cause of valve breakage
   b. condition leading to valve burning
   c. cause of off-square seating
   d. all of the above

2. The main approach to preventing a recurrence of neck-down failure involves the precise control of _____ .
   a. ignition timing
   b. air/fuel ratios
   c. air/fuel distribution in the manifold
   d. all of the above

3. Technician A says hot distortion causes off-square seating. Technician B says off-square seating is caused by cold distortion. Who is right?
   a. Technician A
   b. Technician B
   c. Both A and B
   d. Neither A nor B

4. Which of the following results from excess guide wear?
   a. increased oil consumption
   b. burned valves
   c. valve breakage
   d. all of the above

5. Which of the following is not a seat insert?
   a. stellite
   b. high chrome
   c. lithite
   d. none of the above

6. Technician A says the function of the valve spring is to close the valve. Technician B says the function of the valve spring is to open the valve. Who is right?
   a. Technician A
   b. Technician B
   c. Both A and B
   d. Neither A nor B

7. Which type of rotator can rotate the valve in either direction?
   a. ball type
   b. spring-loaded type
   c. free type
   d. all of the above

8. Which of the following is a synthetic rubber used in making valve stems?
   a. nitrile
   b. vitrol
   c. polyacetelyne
   d. all of the above

9. Which method of lash adjustment is typically associated with V-type engines?
   a. adjustable cam follower
   b. rocker arms with adjustable screws
   c. adjustable pushrods
   d. none of the above

10. Which of the following causes bent rods?
    a. valve sticking
    b. improper valve adjustment
    c. valve timing
    d. all of the above

11. What happens when the valve timing changes?
    a. wear increases
    b. valves and pistons might be damaged
    c. torque reversal action occurs
    d. all of the above

12. Which reconditioning acceptance factor has priority?
    a. condition of fillets and oil holes
    b. hardness of the rod and main journals
    c. journal size and crankshaft alignment
    d. none of the above

13. Technician A says the Rockwell method for evaluating hardness is more subjective than the Sclerescope method. Technician B says the Sclerescope method is more subjective. Who is right?
    a. Technician A
    b. Technician B
    c. Both A and B
    d. Neither A nor B

14. Engine blocks that are not severely warped can be repaired by an operation called _____ .
    a. plane boring
    b. plane alignment
    c. align boring
    d. none of the above

15. What binds the inertia ring to the hub?
    a. nitrile
    b. viton
    c. elastomer
    d. none of the above

CHAPTER SEVEN

INSPECTING THE CYLINDER BLOCK AND RELATED PARTS

Objectives

After reading this chapter, you should be able to
- Explain what constitutes a crankshaft inspection.
- Make piston measurements and analyze piston condition.
- Check piston ring side clearance.
- Explain how to check piston pins.
- Describe how to check the connecting rods.
- Describe the conditions under which a deck should be resurfaced.
- Inspect the cylinder walls for damage.
- List the factors contributing to bearing distress.

Inspection of lower engine components such as crankshafts, valve timing assemblies, pistons, piston rings and rods, cylinders and cylinder shelves, and the all-important bearings is a crucial part of the engine rebuilder's task.

CRANKSHAFT INSPECTION

The automotive crankshaft assembly includes the crankshaft and bearings, flywheel, harmonic balancer, gear or sprocket (to drive the camshaft), and the front and rear oil seals (Figure 7-1).

After core cleaning, make a cursory visual examination of the crankshaft. Check for the following:

- Are the vibration damper and flywheel mounting surfaces eroded or fretted?
- Are there indications of damage from previous engine failures?

- Do any of the journal diameters show signs of heat checking or discoloration from high operating temperatures?
- Are any of the sealing surfaces deeply worn, sharply ridged, or scored?
- Are there any signs of surface cracks or hardness distress?

If any or all of these conditions are present, document the area(s) of concern on the repair order and mark the areas for further evaluation. Complete information on reconditioning cylinder block and related parts is covered in Chapters 9, 11, and 12.

 SHOP TALK

A rough check for unacceptable ridging or scoring is to run the tip of your fingernail across the journal. Be sure your nail is clean before performing this test. If your fingernail drops into the grooves, they are unacceptably deep.

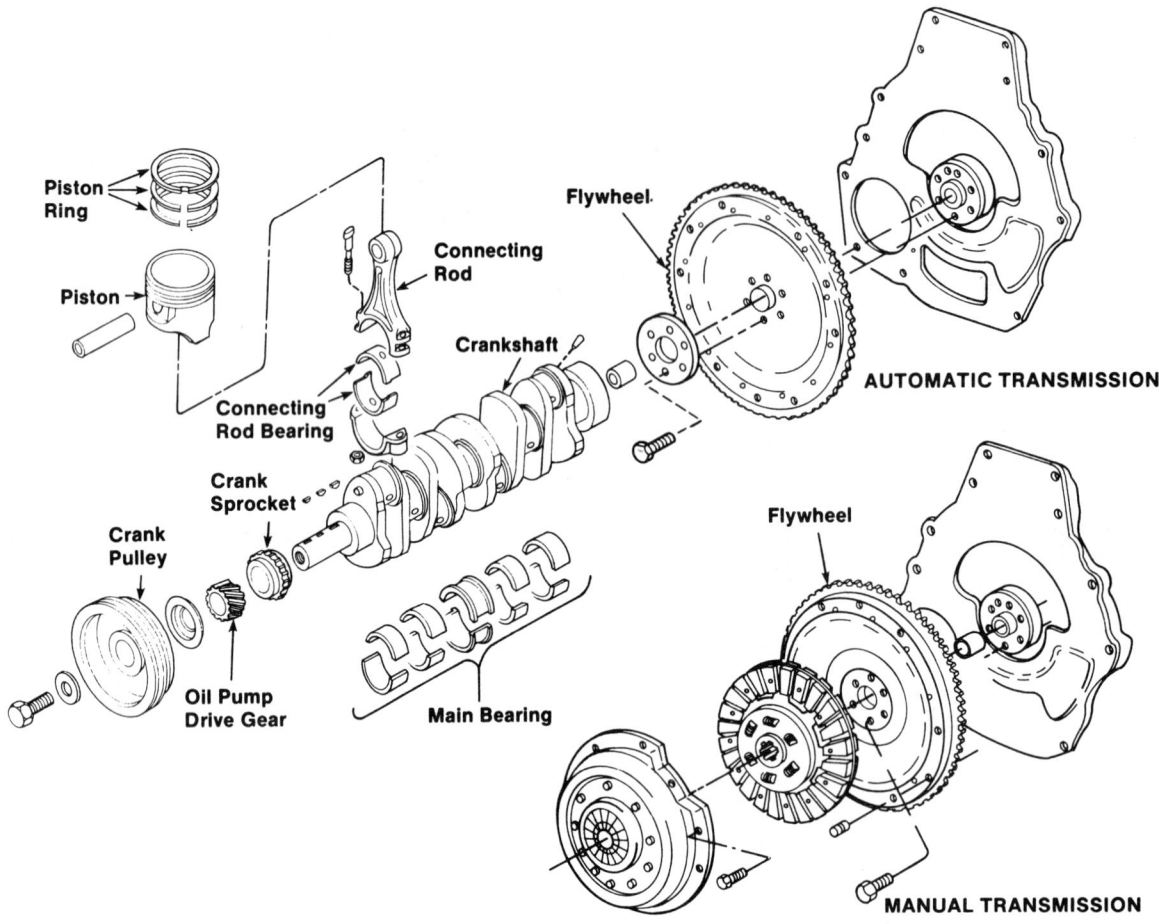

FIGURE 7-1 Typical crankshaft assembly

To measure the rod journal (often called a *throw* [Figure 7-2] or *crankpin*), use a measuring device as shown in Figure 7-3 or an outside micrometer (Figure 7-4). A micrometer can also be used to measure the journals for out-of-roundness and taper (Figure 7-5). Taper is measured from one side of the jour-

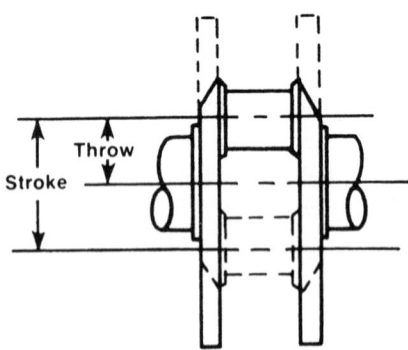

FIGURE 7-2 The stroke of an engine is determined by the crankshaft design and is exactly twice the distance of the crank throw measurement.

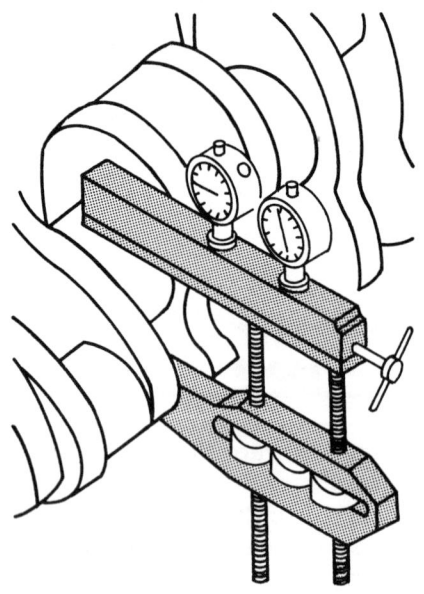

FIGURE 7-3 One method of measuring the rod journal

FIGURE 7-4 Checking a crankshaft with an outside micrometer

nals to the other. The maximum taper is 0.001 inch. The measuring procedure is as follows:

- *Main Bearing Journals.* Measure the diameter of each main bearing journal. Measurements should be taken at each end of the journal, in far enough to clear the fillet radius, to determine if the journal is tapered. Measurements should be taken around the journal in several places to determine if the journal is out-of-round. If any journal measures 0.001 inch less than the manufacturer's specified diameter, has more than a 0.001-inch taper, or is more than 0.001 inch out-of-round, the crankshaft should be reground before reusing in the engine.

- *Crankpins.* Measure each crankpin for taper and out-of-roundness in the same manner as the main bearing journals. The same tolerances apply here and the same corrective action should be taken if necessary.

Compare these measurements with service manual specifications or the same measurement to new standard measurements and establish wear limits to determine the undersize the crankshaft is to be reground. Mark the repair order and the crankshaft to indicate the undersizing requirements and reference source for making these determinations. Generally, rebuilt crankshafts are used for replacement, but be sure that they meet specifications.

CHECKING CRANKSHAFT SADDLE ALIGNMENT

Figure 7-6 is an exaggerated illustration of crankcase housing bores out of alignment. If the engine is operated with a warped housing centerline, the crankshaft will inflict heavy false loads on one side of the main bearings. Depending upon the direction of the bowing or warpage of the main bearing housing bores, any side of the main bearings can become distressed. Engine blocks that are not severely warped can be repaired by an operation called *line boring,* a machining operation in which the main bearing housing bores are rebored to standard size and in perfect alignment (Figure 7-7).

To check the alignment of the crankshaft saddle bore, place an accurately ground arbor into the sad-

A vs B = Vertical Taper
C vs D = Horizontal Taper
A vs C = Out-of-Round
B vs D = Out-of-Round

Check for out-of-round at each end of journal.

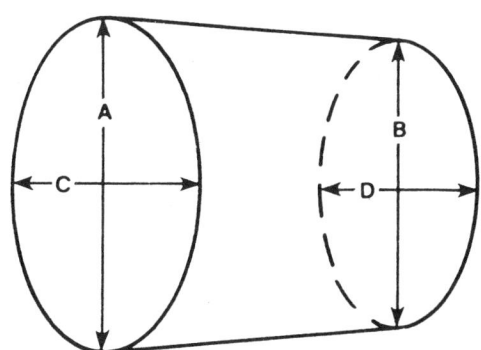

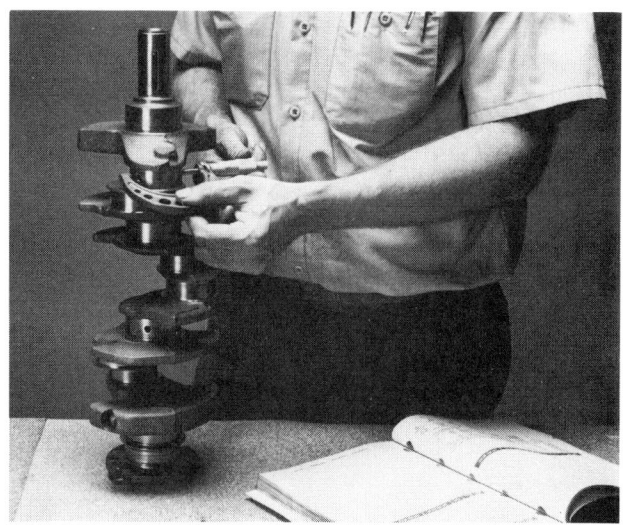

FIGURE 7-5 Crankshaft journal measurement *(Courtesy of Perfect Circle/Dana)*

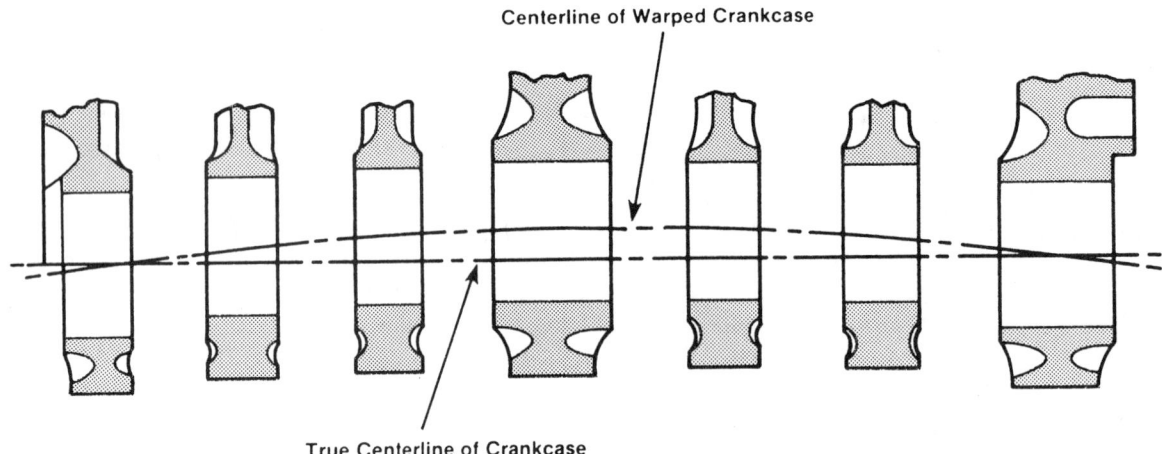

Centerline of Warped Crankcase

True Centerline of Crankcase

FIGURE 7-6 Housing bore misalignment due to a warped crankcase

dles (Figure 7-8). It should be 0.001 inch less in diameter than the low limit of the saddle bore specification and a little longer than the crankshaft. After the arbor is in position, assemble the main bearing caps without their bearings and tighten the cap bolts to the recommended torque specification. Then rotate the arbor using a bar approximately 1 foot long. If it will not turn, one or more of the bores might be out-of-round or the crankshaft could be warped. In either case, the condition must be corrected before the engine is assembled.

If a proper arbor is not available, saddle alignment can be checked with a metal straightedge (Figure 7-9). Place the straightedge in the saddles as shown and, using a feeler gauge that is half the maximum specified oil clearance, try to slide the feeler under the straightedge. If this can be done at any saddle, the saddles are out of alignment and the

block must be rebored. Repeat this procedure at two other parallel positions in the saddles.

Another method of checking the crankshaft saddle bore alignment is through the use of the crankshaft and Prussian blue. When using this method, the crankcase must be checked within the allowable limits indicated in the vehicle's service manual. In addition, the ground surfaces must be free of nicks and scratches. Completely coat all main bearing journals with a thin film of Prussian blue. Install all the main bearings in the block and caps, making sure all locating lugs are properly nested in the machined recess provided. Carefully place the crankshaft in the block. Place the caps with the lower bearings in their respective positions and tighten all bolts alternately to the recommended torque specification. Rotate the crankshaft two revolutions. If the engine is mounted on a stand, turn the engine over

FIGURE 7-7 Checking a diesel block with a line bore machine

FIGURE 7-8 Checking the crankshaft saddle bores with an accurately ground arbor *(Courtesy of J P Industries, Inc.)*

and rotate the crankshaft two more revolutions. By doing this, the weight of the crankshaft will be on the upper and lower halves. Remove the crankshaft carefully from the block in a vertical direction. All areas of contact between the journals and the bearings will be blued. Seventy-five percent of the bearing area should be blued for acceptable alignment; 85 percent is excellent. A small area approximately 3/8 inch on either side of the parting line might not become blued because of bearing eccentricity. Normal blue contact should be in the centerline area covering approximately a 90-degree arc.

 SHOP TALK _____

If the journals measure within specifications but visual pits and gouges exist, polish the worst journal to determine whether or not grinding is necessary (Figure 7-10). If polishing the journal achieves smoothness, then grinding is probably not necessary. If the crankshaft does not have to be reground, check it for straightness.

Out-of-roundness of the saddles can also be checked by bolting on the main bearing caps and checking each bore with a dial bore gauge or an out-of-roundness indicator (Figure 7-11).

CHECKING CRANKSHAFT STRAIGHTNESS

To evaluate the alignment or straightness of the crankshaft, the check should be made with the shaft

FIGURE 7-9 Checking crankshaft saddle alignment with a straightedge and feeler gauge *(Courtesy of TRW, Inc.)*

FIGURE 7-10 Polishing the crankshaft can show possible defects. *(Courtesy of Perfect Circle/Dana)*

FIGURE 7-11 Checking crankshaft housing bores with a dial bore gauge *(Courtesy of Perfect Circle/Dana)*

supported on the main bearing journals, not on the centers of the crankshaft. The recommended location for supporting the crankshaft is on the end main bearing journals.

The journal supports should be a matched pair of well-maintained V-blocks. V-blocks made with roller bearings can introduce erroneous values due to bearing tolerances and clearances. Consequently, the best practice is to avoid the roller type of V-blocks unless they are a calibrated set. Once the crankshaft is located in the V-blocks, select and position the indicator for measuring the alignment/bow of the crankshaft. The selected indicator should have an accuracy of no less than 0.001 inch per full scale deflection. Position the indicator at the 3 o'clock position on the center main bearing journal (Figure 7-12).

Set the indicator at zero and turn the crankshaft through one complete 360-degree rotation. The total deflection of the indicator, the amount greater

FIGURE 7-12 Checking a crankshaft for straightness on a machine *(Courtesy of Storm Vulcan, Inc.)*

than zero plus the amount less than zero, is the total indicator reading (TIR). Bow is 50 percent of the TIR (Figure 7-13). The bow indication at this bearing establishes the bow of the crankshaft. Compare the bow of the crankshaft to the acceptable alignment/bow specifications; accept or reject as appropriate.

Acceptable alignment limits for a fillet hardened crankshaft are usually twice as large as the limits for a nonfillet hardened crankshaft. If bow limits are not available, use the following to evaluate bow acceptability: for fillet hardened crankshafts, allowable bow is 0.001 inch per foot of crankshaft length. For nonfillet hardened crankshafts, allowable bow is 0.0005 inch per foot of crankshaft length.

CHECKING CRANKSHAFT CLEARANCE AND END PLAY

Clearance and end play should be checked before teardown is completed. The clearance check is made with a feeler gauge (Figure 7-14); the end play is checked with a dial gauge (Figure 7-15). Both checks should be within the specified limits given in Table 7-1.

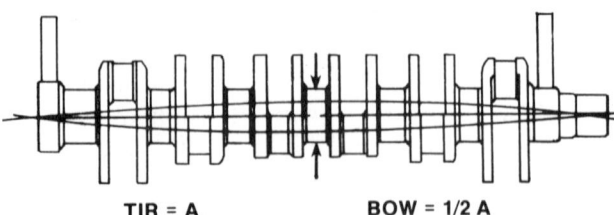

TIR = A   BOW = 1/2 A

FIGURE 7-13 Evaluation of alignment bow should be made with the crankshaft supported on the main bearing journals, not on the centers of the shaft. The recommended location for supporting the shaft is on the end main bearing journals using V-blocks.

To measure crankshaft end play using a dial indicator and bracketry, proceed as follows:

1. Force the crankshaft toward the rear of the engine.
2. Install a dial indicator and bracketry so that the contact point rests against the crankshaft flange and the indicator axis is parallel to the crankshaft axis.
3. Set the dial indicator at zero. Push the crankshaft forward and note the reading on the dial.
4. If the end play exceeds the specified wear limit, replace the thrust bearing. If the end play is less than the minimum limit, inspect the thrust bearing faces for scratches, burrs, nicks, or dirt. If the thrust faces are not damaged or dirty, they probably were not aligned properly.

FLYWHEEL INSPECTION

As mentioned in Chapter 2, the main purpose of the flywheel is to stabilize the speed fluctuations of the crankshaft resulting from power impulses of the engine cylinders. When checking the flywheel in a manual shift vehicle, proceed as follows:

1. Install a dial indicator so that its point bears against the flywheel face (Figure

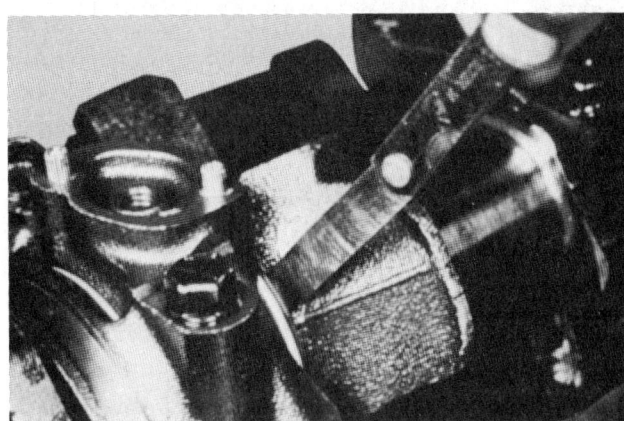

FIGURE 7-14 Using a feeler gauge to check crankshaft end clearance *(Courtesy of Federal Mogul Corp.)*

TABLE 7-1: CRANKSHAFT END PLAY	
Crankshaft Journal Diameter (Inches)	**Crankshaft End Clearance (Inches)**
2 to 2-3/4	0.004/0.006
2-13/16 to 3-1/2	0.006/0.008
over 3-1/2	0.008/0.010

FIGURE 7-15 Measuring crankshaft end play *(Courtesy of Perfect Circle/Dana)*

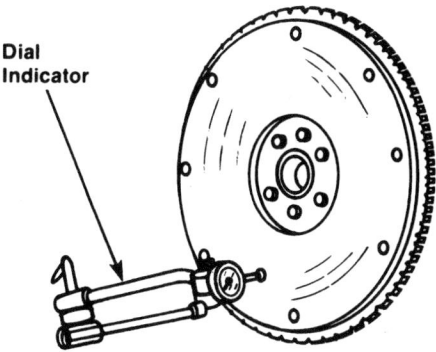

FIGURE 7-16 Checking flywheel face runout in a manual transmission

7-16). Turn the flywheel making sure that it is full forward or backward so that crankshaft end play will not be indicated as flywheel runout.

2. If the flywheel clutch face runout exceeds specifications listed in the specific engine's service manual, remove the flywheel and check for burrs between the flywheel and the face of the crankshaft mounting flange. If none exist, check the runout of the crankshaft mounting flange. Replace the flywheel or machine the crankshaft-flywheel mounting face if the mounting face flange runout is excessive. If the ring gear runout exceeds the specifications listed in the service manual, check installation of the gear to the flywheel flange. If it is not properly seated, reinstall it to the flywheel. If it is properly seated, replace it.

When checking the flywheel in a vehicle with automatic transmission, it must be done with the torque converter bolted to the flex plate. To make the check, proceed as follows:

1. Install a dial indicator so that its point rests on the face of the ring gear adjacent to the gear teeth.
2. Push the flywheel and crankshaft forward or backward as far as possible to prevent crankshaft end play from being indicated as flywheel runout.
3. Set the indicator dial on the zero mark. Turn the flywheel one complete revolution while observing the TIR. If TIR exceeds specifications, the flywheel and ring gear assembly must be replaced or rebuilt.

With both manual shift or automatic transmission, inspect the flywheel for damaged or worn ring

gear. Remember that improper flywheel runout can cause vibrations, poor clutch, and clutch slippage.

VIBRATION DAMPER INSPECTION

A vibration damper or harmonic balancer is necessary to dampen normal torsional vibration of the engine crankshaft.

Basically, the vibration damper consists of two parts: a hub and an inertia ring. A flexible elastomeric (rubber compound) insert bonds the inertia ring to the hub (Figure 7-17). As each cylinder fires, the inertia ring moves slightly in relation to crankshaft rotation. This dampens the torsional vibrations of the crankshaft over a wide range of engine speeds. In some caes, two inertia rings of different sizes are included in the damper design. These different size rings offer more effective control over a wide range of vibrational frequencies.

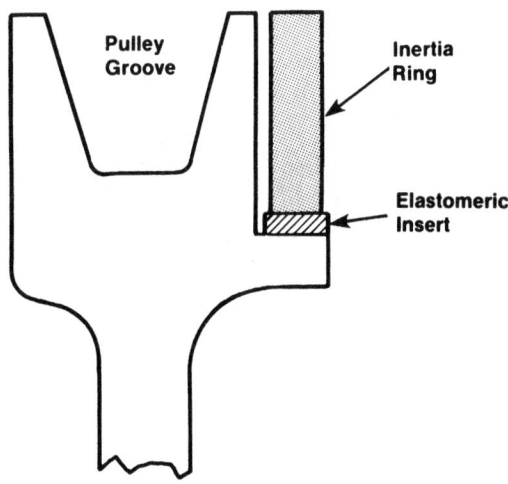

FIGURE 7-17 Cross section of harmonic balancer construction with inertia ring

Over a period of time, the elastomer might deteriorate or the bonding break down. As a result, the damper is ineffective. The damper can even become the source of vibrations itself. Once a damper is damaged, it must be replaced. On damper designs where the hub doubles as a seal journal, the seal can wear a groove in the hub. Oil leakage occurs as a result. In this case, if the damper is in otherwise good condition, a sleeve-type repair can be used to restore it. Some hubs might require machining before this repair can be made.

Check the damper and flywheel mounting threads for condition and size. Verify the thread size with a go/no-go gauge and the condition by comparing the force required to thread a cap screw into each of the threaded holes. Clean or repair threaded holes requiring excessive force to install the test cap screw.

VALVE TIMING DRIVE ASSEMBLY

The valve timing drive assembly (often referred to as the *camshaft drive system*) includes gear drives, chain and sprocket drives, and belt (or chain) and sprocket drives (Figure 7–18). The timing belt (or chain) and crankshaft camshaft sprockets should be inspected and replaced if damaged or worn.

TIMING CHAIN DEFLECTION CHECKS

Machinist's rules are used in procedures such as the timing chain deflection check (Figure 7–19). These checks are performed as follows:

1. Remove the chain tensioner and retaining bolts if the engine is so equipped.
2. Rotate the crankshaft clockwise (as viewed from the front of the engine) to take up slack on the right side of the chain.
3. Mark a reference point on the block at the approximate midpoint of the chain. Measure from this point to the midpoint of the chain.
4. Rotate the crankshaft counterclockwise to take up slack on the left side of the chain.
5. Force the left side of the chain outward with your finger and measure the distance

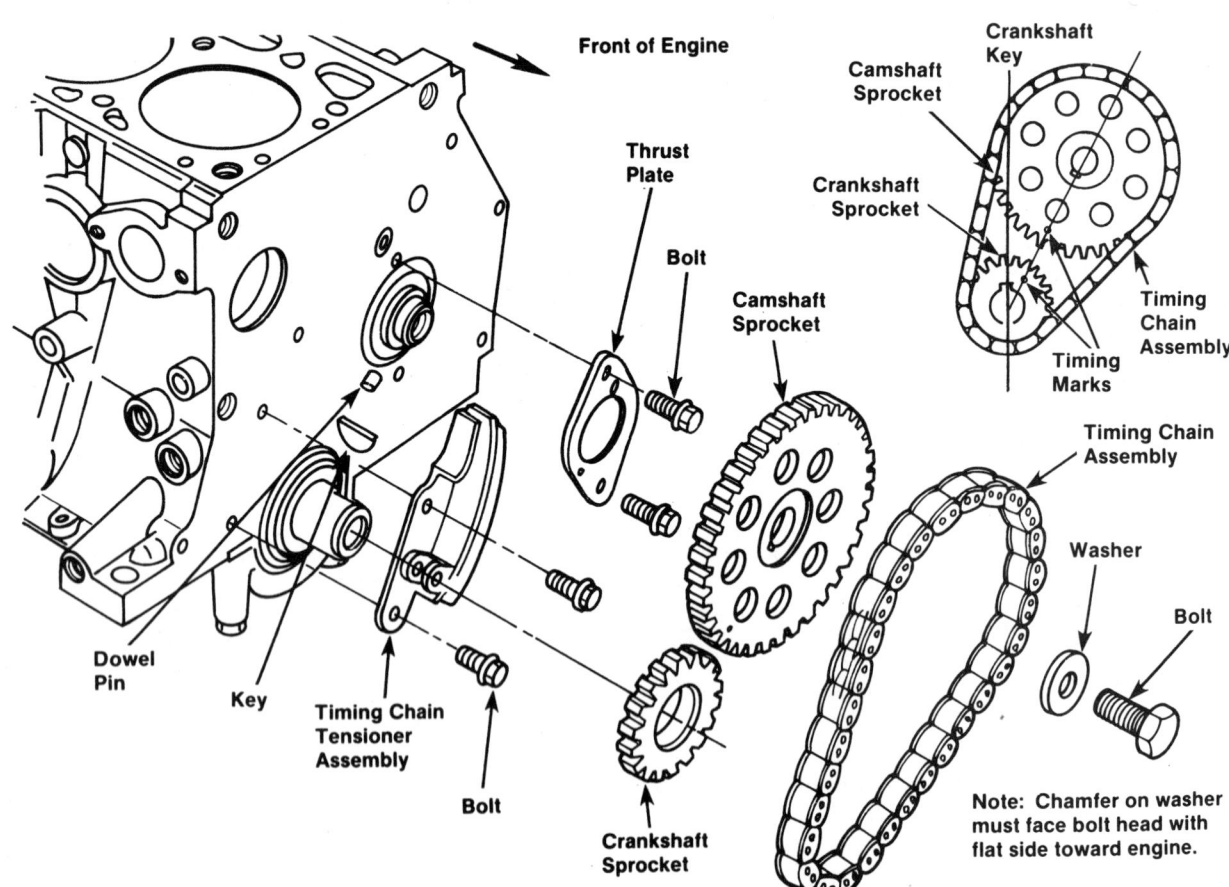

FIGURE 7-18 Typical valve timing drive assembly

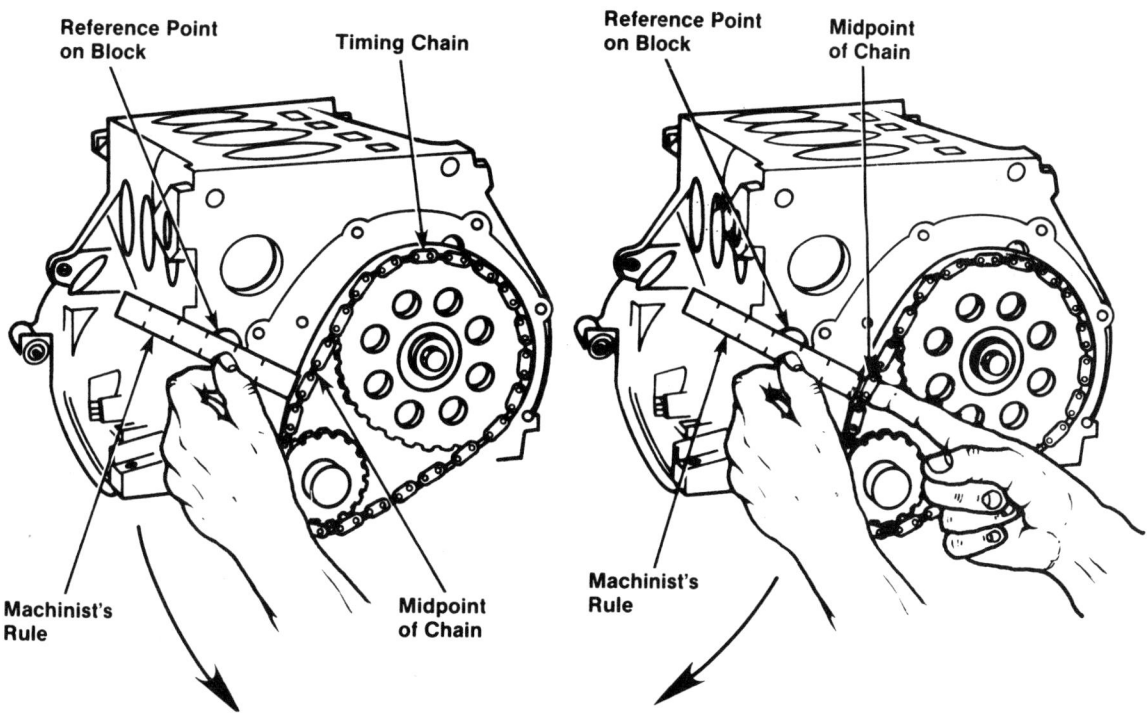

FIRST MEASUREMENT

Reference Point on Block

Timing Chain

Machinist's Rule

Midpoint of Chain

SECOND MEASUREMENT

Reference Point on Block

Midpoint of Chain

Machinist's Rule

First Measurement – Second Measurement = Timing Chain Deflection

FIGURE 7-19   Timing chain deflection check

between the reference point and the midpoint of the chain.

6. Timing chain deflection is the difference between the two measurements. Replace the timing chain and gears if the deflection measurement exceeds specifications.

CHAIN AND SPROCKET BREAKAGE

Causes of chain and sprocket failures generally fall into three categories:

1. *Misalignment.* When the cam and crank sprockets are not properly aligned, the chain must twist to engage the sprocket teeth. This places abnormally high stress on the chain links and also leads to accelerated wear of the sprocket teeth. A failure due to misalignment can usually be identified by a diagonal break across the chain links and heavy wear on the sides of the sprocket teeth. Commonly, alignment problems are the result of failing to tighten

sprocket mounting bolts and excessive wear on the cam sprocket thrust surface.

2. *Overload.* Under normal operating conditions, the loads carried by the chain and sprockets are far below their ultimate strengths. However, foreign objects occasionally find their way into the timing case and lodge between the chain and sprocket.

3. *Improper Installation.* Hammering the sprockets into place can cause cracks to develop through the keyway or possibly break off a tooth.

Although the failure mechanisms differ, the end result is the same—breakage of the chain, sprockets, or both.

Figure 7-20 illustrates a typical overhead camshaft roller chain timing drive. A tensioner provides automatic chain tension adjustment as normal wear takes place. The illustration shows two types: one acts on the chain guide and the other acts independently. The chain guides prevent chain whip. Three common chain types are shown in Figure 7-21. Also check the offset camshaft drive key to be sure that it is correcting the valve timing (Figure 7-22).

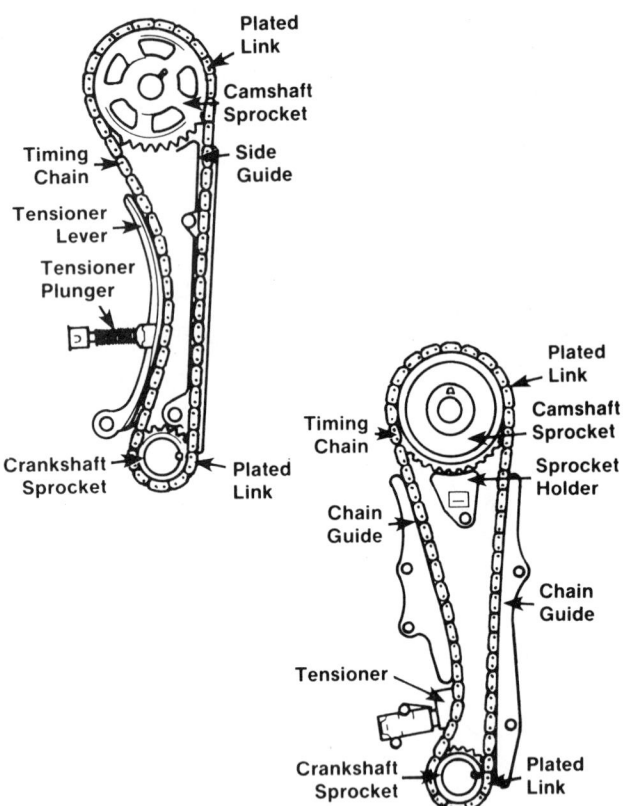

FIGURE 7-20 Typical overhead camshaft roller chain timing drive

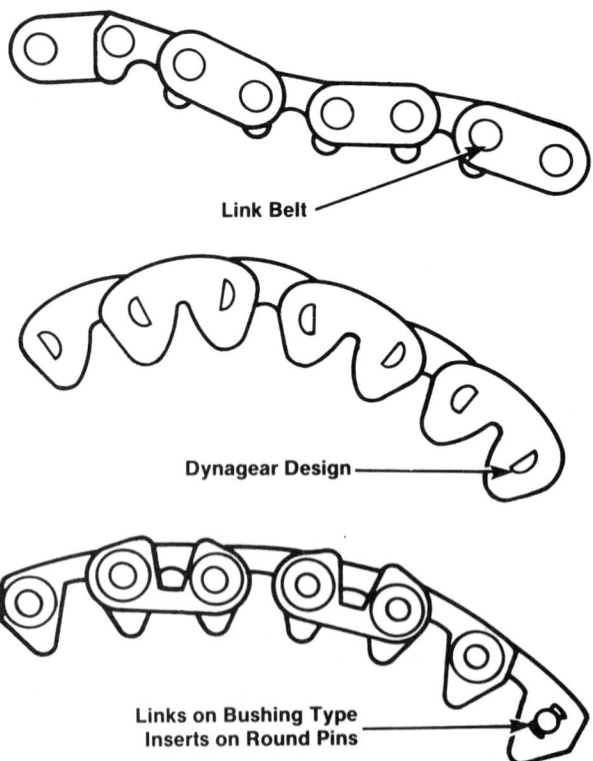

FIGURE 7-21 Common chain types

TIMING GEAR BACKLASH

A gear with cracks, spalling, or excessive wear on the tooth surface is an indication of improper backlash (either insufficient or excessive). With excessive backlash, operation will be noisy because the teeth will make violent impact contact. This overloading when coupled with the normal valve train loads causes accelerated tooth wear and often breakage. Insufficient backlash places a bind on the gears. Also, high contact forces are generated that can rupture the lubrication film between the teeth, causing spalling and wear.

To determine gear backlash, install a dial indicator and dial indicator bracketry on the cylinder block (Figure 7-23). Check the backlash between the camshaft gear and the crankshaft gear with a dial indicator at six equally spaced teeth. Hold the gear firmly against the block while making the check. Refer to specifications in the vehicle's service manual for the backlash limits.

RUBBER TIMING BELT

Stripped/broken rubber belt failure is commonly due to insufficient tensioning, extended service life (belts should be checked for cracking around 50,000 to 60,000 miles), abusive operation, or worn tensioners. Loose timing belts will jump across the teeth of the timing sprockets causing shearing of the belt teeth. Localized tensile overloads from overrevving the engine can lead to belt breakage. Also, those engines equipped with adjustable tensioners should be checked for wear whenever a belt is replaced or retensioned. In addition, check for cord separation and cracks on all surfaces (Figure 7-24). If the belts are damaged, they should be replaced.

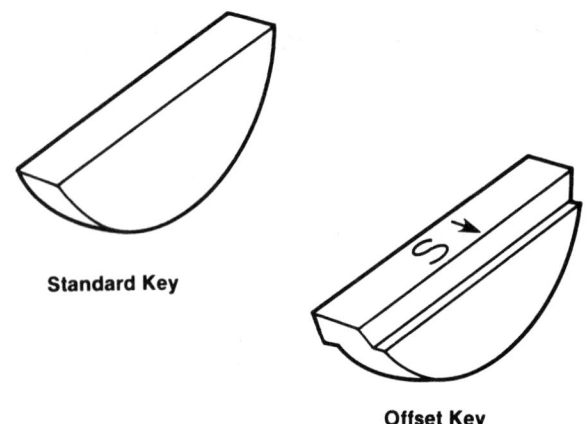

FIGURE 7-22 Offset camshaft drive key that is often used to correct valve timing.

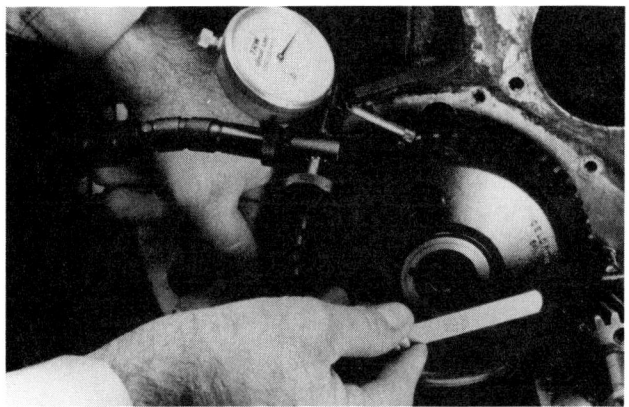

FIGURE 7-23 Check the backlash between the mating gear teeth with a feeler gauge and/or dial indicator *(Courtesy of TRW, Inc.)*

PISTON ASSEMBLY INSPECTION

The piston assembly consists of a piston, piston rings, piston pin, and connecting rod (Figure 7-25). Each of these components must be carefully inspected during the engine rebuilding process.

PISTON INSPECTION

As mentioned in Chapter 2, the piston transmits the force and power of combustion through the wrist pin to the connecting rod and crankshaft. In a four-stroke cycle engine, it also draws air into the combustion chamber, pushes exhaust gases out, and compresses the air prior to the power stroke. Pistons for older engines were generally made from cast iron, but almost all pistons used in today's gasoline engines are made from aluminum alloy. Experiments are being conducted on ceramic pistons.

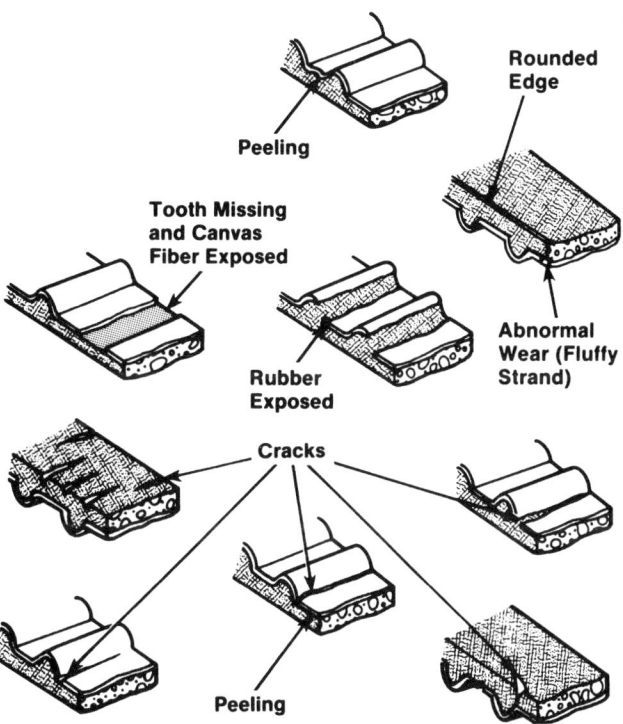

FIGURE 7-24 Timing belt inspection

 SHOP TALK _____

The condition of a rubber timing belt can frequently be checked by the fingernail test. Press your fingernail into the hardened backside of the belt. If no impression is left, the belt is too hard, caused by overheating and aging. Replace the belt.

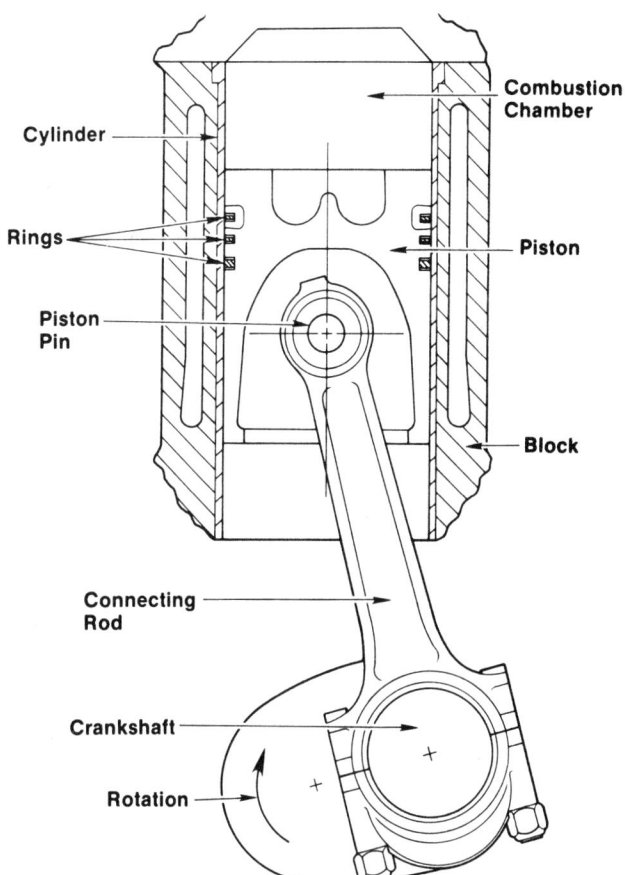

FIGURE 7-25 Typical piston assembly

PISTON DESIGN CHARACTERISTICS

Because the head (top) of the piston forms the bottom of the combustion area, it bears the pressure of compression as it moves up on the compression stroke as well as the even greater pressure as it moves down on the power stroke. It also adds motion, or turbulence, to the air/fuel mixture before and during combustion to make possible the most complete burning in the desired time. To do this, piston heads are designed in many shapes (Figure 7-26).

Although the flathead piston is the simplest and most economical to manufacture, dished or domed pistons are becoming more popular because they give more turbulence to the air/fuel mixture. In a dished design, the top of the piston is scooped out like a bowl. In the domed design, the piston head curves upward into the combustion chamber. These design changes require other changes. If all other dimensions in the combustion chamber area remain constant, changing from a flathead to a dished piston head will reduce the compression ratio. Similarly, changing from a flathead to a domed head will increase the compression ratio.

With any piston head shape, the designer must be careful to avoid interference between the valves and the piston head. Sometimes this requires cutting valve relief areas into the piston head.

 SHOP TALK ⎯⎯⎯⎯

Automobile engines can be classified as either free-running or interference, depending upon what occurs if piston/valve synchronization is lost. As illustrated in Figure 7-27, an interference engine usually sustains damage if synchronization loss occurs. The combustion chamber design of an interference engine (see Chapter 9) is such that an open valve can be struck by a moving piston if timing synchronization is disrupted. This could result in valve and piston damage.

A piston has a major and a minor thrust face (Figure 7-28). The major thrust face is the side of the piston that presses against the cylinder wall on the power stroke. The minor thrust face is the opposite side of the piston. The minor axis runs parallel to the pin axis. An extremely close fit must be maintained between the thrust faces of the piston and the cylinder wall. For this reason, pistons are cam ground to a slight elliptical shape, as shown in Figure 7-29, and constructed so the diameter across the thrust faces

FIGURE 7-26 Various piston head designs

is maintained as the engine warms up. When the piston reaches operating temperature, it expands to a round shape (Figure 7-30).

The skirt of the piston is that outside diameter area that extends from the bottom upward to the lowest ring groove above the pinhole. It is the guide or crosshead whose surface rides along the cylinder wall during engine operation. It distributes the side forces evenly along the cylinder wall.

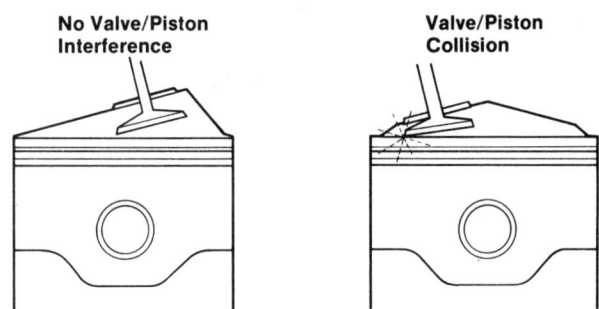

FIGURE 7-27 (A) Free-running and (B) interference types of engine piston/valve synchronization

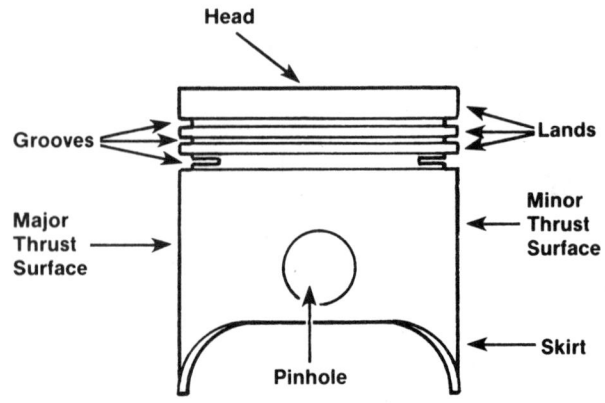

FIGURE 7-28 Names of piston parts

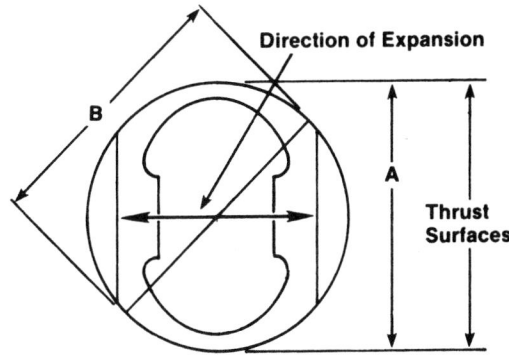

FIGURE 7-29 Elliptical shape of the piston skirt should be 0.010 to 0.012 inch less at diameter B than across the thrust faces at diameter A. Measurement is made 1/8 inch below the lower ring groove.

Sometimes holes or slots are drilled through the body of a piston at the bottom of a ring groove. These holes provide a path for the oil to flow through the piston and drip back to the oil pan after it is wiped off the cylinder wall. Not all piston designs incorporate drain holes. On some designs, the piston and oil ring are made so that all the oil drips down the outside of the piston.

Piston skirts are usually slightly tapered to allow for thermal expansion. The top portion is smaller in diameter (about 0.0015 inch) than the bottom (Figure 7-31). The top of the skirt will expand more than

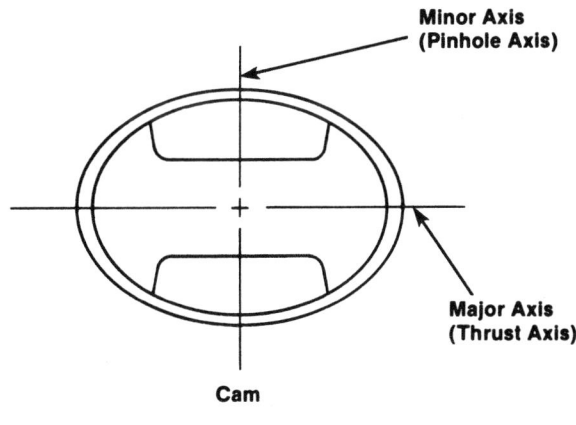

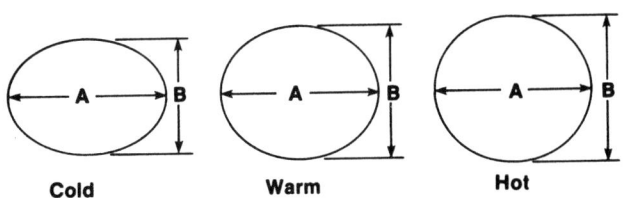

FIGURE 7-30 Cam-ground piston

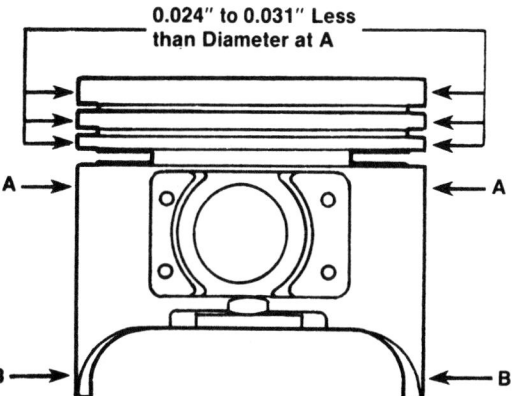

FIGURE 7-31 Piston head and skirt profile: diameters at A and B can be equal or diameter at A can be 0.0015 inch greater than B.

the bottom. To help control piston expansion, belts or struts are inserted inside a steel piston when they are cast (Figure 7-32). A belt is a circular piece inside the piston. A strut is a structural component that resists expansion.

In an operating engine, an oil film protects the skirt from sliding friction. To ensure a consistent oil film on the skirt, a wavy grind finish is sometimes machined in. A wavy grind finish is a thread-like finish that carries oil in its shallow grooves for better skirt lubrication.

Piston Measurements

The piston diameter can be checked by using an outside micrometer (Figure 7-33). The piston diameter should be measured at a point specified by the vehicle manufacturer's shop manual and the measurement compared to specifications. Some manufacturers call for piston measurement across the thrust surfaces at pin height; others specify measurements to be taken just below the oil ring groove or at the bottom of the skirt.

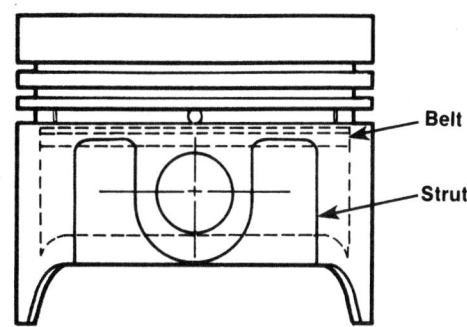

FIGURE 7-32 Piston struts and belts help to control piston expansion.

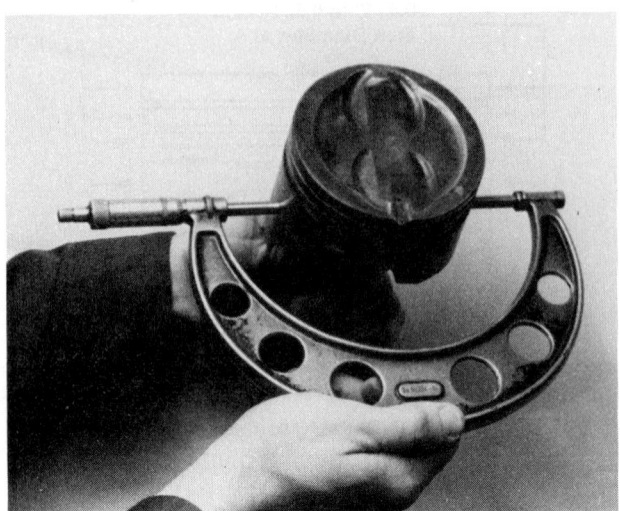

FIGURE 7-33 Piston size should be measured at the height specified in the shop manual and at 90 degrees to the piston pin.

If the piston measurements do not meet the specifications or the skirts are collapsed, the piston should be replaced. To check for a collapsed piston, measure across the widest part of the skirts with an outside micrometer. Most pistons should measure between 0.001 to 0.003 inch smaller than the cylinder bore size. (Measuring the cylinder bore is discussed later in this chapter.)

Carefully inspect the pistons for fractures at the ring lands, skirts, and pin bosses, and for scuffed, rough, or scored skirts. If the lower inner portion of the ring groove(s) has high steps, replace the piston. The step will interfere with ring operation and cause excessive ring side clearance.

Spongy, eroded areas near the edge of the piston top are usually caused by detonation or preignition. A shiny surface on the thrust surface of the piston, offset from the centerline between the piston pinholes, can be caused by a bent connecting rod. Replace pistons that show signs of excessive wear, have wavy ring lands or fractures, or damage from detonation or preignition.

When a piston shows severe damage after an engine failure, it is often accused of causing the trouble. However, the malfunction of a piston alone is rare. Normally, they fail because something else in the engine has created too much heat, too much load, or too little lubrication for the piston. They are designed to take some abuse in these areas, but if pushed too far they will no longer function satisfactorily. Table 7-2 indicates by percentage the principal causes of piston distress. Remember that several causes can combine and be contributing factors toward the total distress of a piston.

Piston Condition Analysis

If faulty pistons are discovered during inspection, it is very important to find out the cause of the damage and correct it during the rebuilding process. Some of the most common causes of failure are the following:

1. *Abnormal Combustion (Detonation and preignition).* To understand why and how detonation and preignition occur requires a basic comprehension of the combustion process. Normal combustion in an internal combustion engine produces a controlled burn of the air/fuel mixture that lasts from 1.5 seconds to 0.004 second from start to finish (Figure 7-34). Immediately after the mixture is ignited by the spark plug, the flame front moves from the point of ignition in increasing circles at the rate of about 50 mph. As the mixture burns at a controlled rate, the gases are heated and the temperature rises to an average of 2000 to 3000 degrees Fahrenheit. Along with this temperature rise is a predictable increase in cylinder pressure. This is normal combustion that converts chemical energy to mechanical energy.

   Anything other than normal combustion is considered abnormal combustion. There are two types:

   • *Detonation.* Often called *pinging*, detonation occurs when the fuel is burned too rapidly (Figure 7-35). The mixture is ignited by the spark plug and combustion proceeds normally until the advancing flame front overheats some of the unburned gases to the point of spontaneous ignition; this is due both to compression as well as radiation. When this occurs, there is no longer a controlled burn, but something much closer to an explosion taking place in the chamber. The secondary flame front advances at supersonic speeds until it collides with the original front

TABLE 7-2: MAJOR CAUSES OF PISTON DISTRESS	
Dirt (abrasive wear)	41.2%
Scuffing, scoring, and overheating	26.7%
Incorrect installation	11.4%
Worn grooves	5.3%
Indeterminate and other causes	15.4%

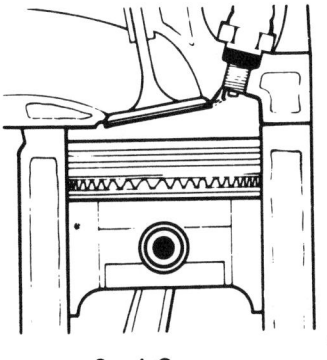 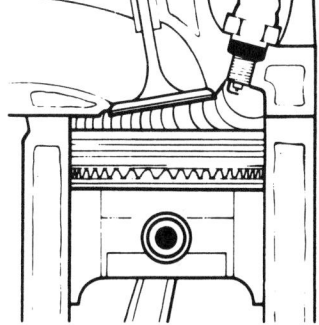

Spark Occurs   **Combustion Begins**   **Combustion Completed**

FIGURE 7-34 Normal combustion

  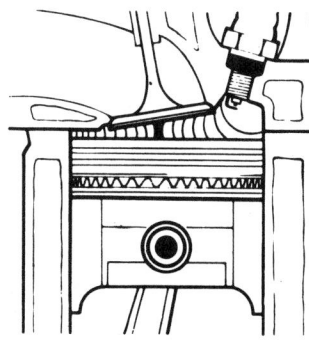

Spark Occurs   **Combustion Begins**   **Detonation**

FIGURE 7-35 Abnormal combustion: detonation or post ignition

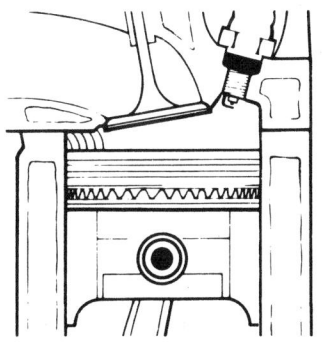 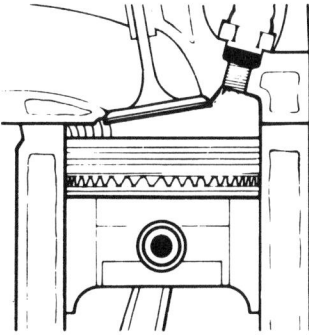 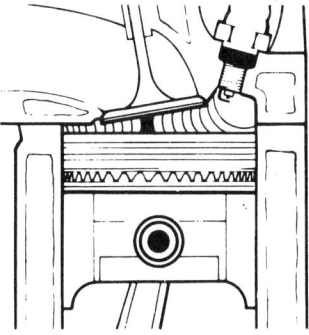

Ignited by Hot Carbon Deposit   **Regular Ignition Spark**   **Flame Fronts Collide**

FIGURE 7-36 Abnormal combustion: preignition

creating the characteristic "ping" that resonates on the walls and chamber surfaces.

Detonation creates an explosion with tremendous pressure and velocity within the chamber. Since the engine cannot effectively use this energy, it is dissipated as heat and high-frequency vibrations that can readily stress the pistons and rings beyond their capacity. Piston tops can be pinholed, crowns eroded, ring lands fractured, and the rings themselves can be broken.

- *Preignition.* Preignition causes the mixture to be burned too soon instead of too fast (Figure 7-36). It occurs when the air/fuel mixture is ignited by an un-controlled source before the spark plug fires.

Preignition can destroy an engine in minutes. It causes a very rapid reaction of the air/fuel mixture with a flame front that moves many times faster than it does with normal combustion. This generates high temperatures, some-times exceeding 4000 degrees Fahrenheit. At the same time, peak pressures are nearly double (about 1200 psi versus 600 psi) that of normal combustion pressures (Figure 7–37).

Detonation and preignition can damage the combustion armor of the head gasket, possibly resulting in the collapse of the armor, which, in turn, can lead to gasket cracks and burn-through (Figure 7–38).

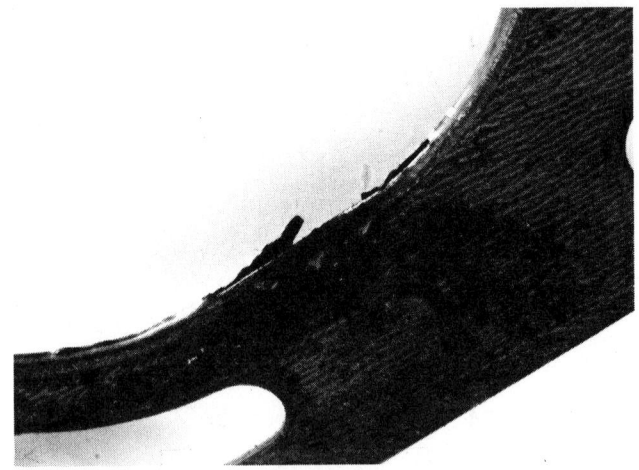

FIGURE 7–38 Typical gasket armor that has been destroyed by detonation or preignition *(Courtesy of Federal Mogul Corp.)*

2. *Broken Second Land and Ring.* This con-dition (Figure 7–39) is normally associated with detonation, but it can also happen, for example, if a 5/64 ring was used in a 3/32 ring groove. The ring will bounce up and down in the groove until it pounds itself apart and also destroys the second land.

3. *Scuffing on the Skirt Near the Pinhole (Signs of distress in the hole).* A pin fit that is too tight in the piston pin bore can cause the piston to seize on the pin if the engine is suddenly cooled down. This prevents the skirt from conforming to the bore and can lead to scuffing on the corners of the skirt and a collapsed skirt (Figure 7–40).

4. *Hole in the Piston Head (Jagged edges and rust on the sides of the rings).* If water from a leaking head gasket fills the cylinder bore, it can crack the piston head and

FIGURE 7–37 Results of preignition *(Courtesy of Federal Mogul Corp.)*

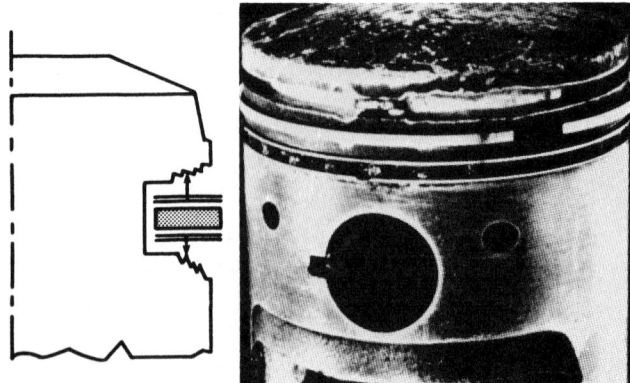

FIGURE 7–39 Broken second land and ring *(Courtesy of Federal Mogul Corp.)*

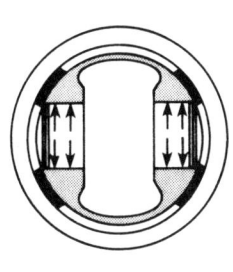

FIGURE 7-40 Scuffing on the corners of the skirt and a collapsed skirt *(Courtesy of Federal Mogul Corp.)*

push a hole completely through it as the piston approaches TDC. Note that the hole is jagged and not smooth (Figure 7-41) as when burned through from preignition. Rust on the side of the rings indicates the bore was filled with water from a leaky head gasket or cracked cylinder block and not from an overflow of fuel.

5. *Top of Piston Badly Pitted and Possibly Cracked (Figure 7-42).* If a valve head should break off, it will be caught between the piston and the cylinder head, causing extensive damage to the top of the piston.

6. *Piston Head Cracked (Piston possibly smashed in many pieces) (Figure 7-43).* If too much material is removed from a cylinder block or head surface during cleaning and the correct cylinder head gasket is not used, the piston reaching the top of its

stroke can hit an overhanging part of the cylinder head. Being of a softer material than the cast iron, the piston will be over-stressed and pounded apart. This same piston head appearance but without the fatigue cracks can result when a connecting rod bolt is improperly tightened, so the rod cap, bolts, nuts, and bearing should be closely examined for clues.

7. *Piston Eroded around the Pinhole (Cylinder bore scored in line with the edge of the pin).* If a lock ring comes loose or the tang breaks off, it can be trapped between the piston and the cylinder bore (Figure 7-44), or it can fall into the oil pan. Bouncing up and down with the piston, it will erode away the soft aluminum piston material. In the meantime, the piston pin can work its way to a position where it can

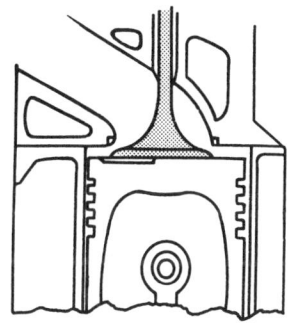
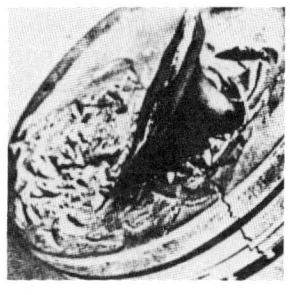

FIGURE 7-42 Top of piston badly pitted and possibly cracked *(Courtesy of Federal Mogul Corp.)*

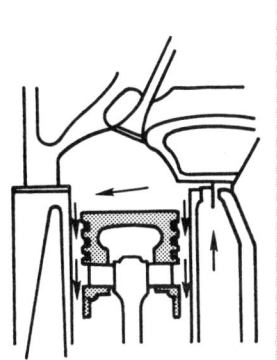

FIGURE 7-41 Jagged-edge hole caused by water from a leaking head gasket *(Courtesy of Federal Mogul Corp.)*

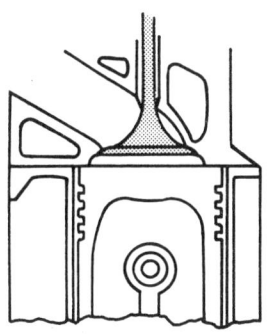

FIGURE 7-43 Piston head cracked *(Courtesy of Federal Mogul Corp.)*

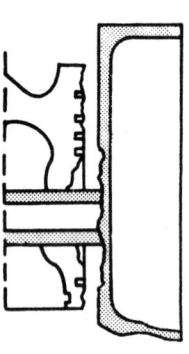

FIGURE 7-44 Piston eroded around pinhole *(Courtesy of Federal Mogul Corp.)*

score the cylinder wall. This problem is caused by improperly seated lock rings, overstressing them during installation, or reusing old ones. The upper and lower end of the connecting rod must be parallel to prevent pounding out of a lock ring.

8. *One Lock Ring Pounded Out (Oblique wear pattern on piston skirt).* If the small and large end bores of the connecting rod are not parallel, a horizontal force will be imparted to the piston pin during the power stroke. This will pound out the lock ring and force the pin up against the cylinder wall where it will machine a vertical groove while traveling up and down. This condition (Figure 7-45) is caused either by a

bent connecting rod or a new connecting rod bushing that was not properly bored parallel to the big end bore.

9. *Clogged and Worn Rings (Poor oil economy).* Blue smoke coming out of the tailpipe and poor oil economy are sure signs of clogged, stuck, and worn piston rings (Figure 7-46). This is usually caused over a period of time by improper oil and filter maintenance.

10. *Heavy Scuffing on Piston Skirt (Figure 7-47).* The damage is caused by overheating of the piston. If only one or two pistons show heavy scuffing, look for too little piston-to-bore clearance.

11. *Heavy Scuffing on Most Pistons (Figure 7-48).* If most or all are scuffed, the cause can be loss of coolant (broken fan belt, radiator leak, cracked heater, or radiator hose) or lack of lubrication (oil leak or

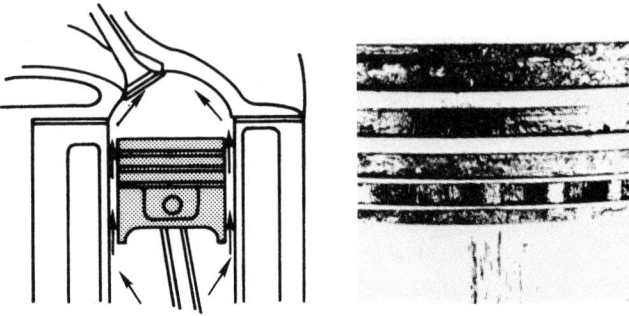

FIGURE 7-46 Clogged, stuck, and worn piston rings *(Courtesy of Federal Mogul Corp.)*

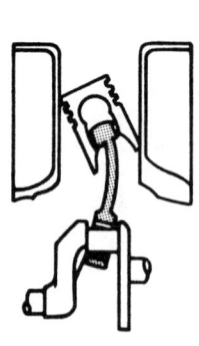

FIGURE 7-45 Pounded out lock ring forces the piston pin against cylinder wall and machines a vertical groove while traveling up and down. *(Courtesy of Federal Mogul Corp.)*

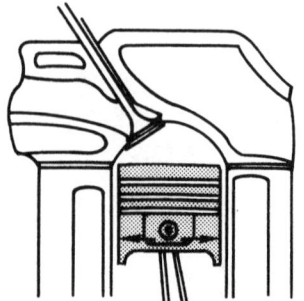

FIGURE 7-47 Heavy scuffing on piston skirt *(Courtesy of Federal Mogul Corp.)*

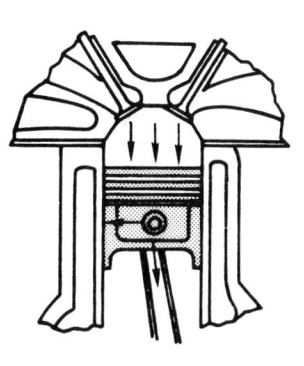

FIGURE 7–48   Heavy scuffing on most of the pistons *(Courtesy of Federal Mogul Corp.)*

0.010 to 0.015 inch than the diameter across the thrust faces. This allows the piston to expand along the pin axis when the piston heats up in the engine. If the piston skirt were round and without a cam profile, the piston would expand outward when heated and, with no space for expansion, push so hard against the cylinder walls that they would wear badly and scuff.

14. *Stuck and Scuffed Piston Rings (Heavy carbon buildup on the piston lands and in the ring grooves and scuffing on the piston skirt).* A too-rich air/fuel mixture will wash the lubricating oil off the cylinder walls, leading to excessive carbon buildup, scuffing, and sticking rings (Figure 7–51).

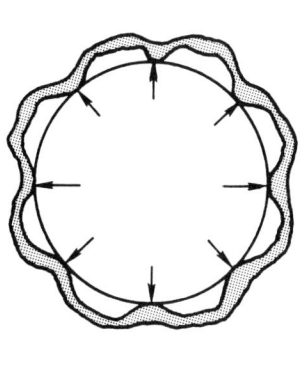

FIGURE 7–49   Vertical scuffing streaks around the piston *(Courtesy of Federal Mogul Corp.)*

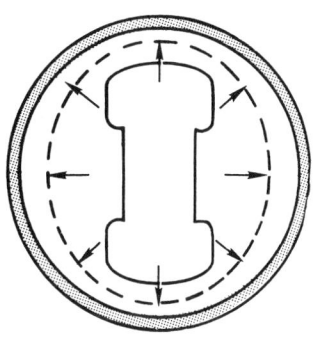

FIGURE 7–50   Wear pattern indicates coolant loss or round instead of cammed piston skirt *(Courtesy of Federal Mogul Corp.)*

failed oil pump). With lack of lubrication, the main and rod bearings will usually be wiped out, especially those farthest from the oil pump.

12. *Vertical Scuffing Streaks around the Piston Skirt (Figure 7–49).* This type of scuffing indicates cylinder bore distortion. It can be caused by excessive torque on cylinder head bolts, improperly assembled sleeves, or even excessive vibration when the cylinder block is being bored out.

13. *Moderate-to-Heavy Wear and Scuffing around the Skirt Including Above and Below the Pinholes.* This wear pattern (Figure 7–50) indicates either a coolant loss or a piston skirt that is round instead of cammed. The skirts of all automotive and truck pistons are finished to a cam shape rather than round, with the diameter across the pin boss axis being smaller by about

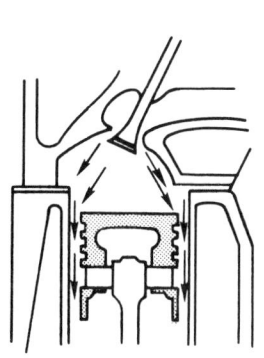

FIGURE 7–51   Excessive carbon buildup, scuffing, and sticking rings caused by a too-rich air/fuel mixture *(Courtesy of Federal Mogul Corp.)*

Scuffed pistons can also be caused by the failure of other automotive systems such as the following:

1. *Lubrication system*
   - Worn oil pump
   - Clogged oil screen or passages
   - Oil pressure relief valve stuck open
   - Incorrect bearing clearances
   - Contaminated oil
   - Low oil level
2. *Carburetion*
   - Too rich or too lean
   - Stuck automatic choke
3. *Cooling system*
   - External or internal leaks
   - Clogged radiator
   - Deposits in water passages
   - Improperly functioning radiator pressure cap
   - Stretched or broken fan belt
   - Corroded water pump impeller

Improperly fitted parts, such as insufficient piston clearance, tight piston pin fit, insufficient ring gap, or incorrect bearing clearances can also cause scuffed pistons.

PISTON RINGS

Piston rings are precision-made seals, with exacting tolerances to seal against the side of the piston grooves and simultaneously fit the cylinder bores into which they are installed. The job of the piston rings is to form a seal between the combustion chamber and the crankcase. This seal prevents the combustion chamber gases from leaking into the crankcase and prevents the lubricating oil from entering the combustion chamber. Typically, modern ring combinations include one, two, or three compression rings, one expander spacer, and one or more oil control rings, depending upon the application (Figure 7-52). Regardless of the number of rings, they always perform the same task.

The compression rings perform the major portion of the work in sealing compression and combustion gases (Figure 7-53). The top ring takes the most punishment and is subjected to the biggest share of the load and heat due to the combustion process. Also, it is exposed to all of the combustion chamber contaminants and operates with the least lubrication of all the rings. The task of the oil rings, regardless of their number, is to meter oil so that the compression rings are lubricated but not overtaxed. Oil flows through the piston holes and runs back in the crankcase. A single slotted or multipiece ring is

used. A typical oil control ring consists of an expandable spring spacer and two rails. When assembled on the piston and in the cylinder, the spring spacer pushes out on the rails. This forces the rails uniformly against the cylinder wall. Several common designs of oil control rings are shown in Figure 7-54.

The compression rings are designed with outward spring tension built into them. During the compression, exhaust, and intake strokes this built-in outward spring tension seals any leakage past the ring faces. However, during the power stroke, the combustion gas pressure is put to work forcing the ring downward and outward. It is this action that helps to create a good side and face seal. Specifically, the second compression ring acts mainly as a backup for both the top compression ring and oil control ring. If the top and oil rings perform their jobs 100 percent, there would be no need for a second ring. Figure 7-55 shows how the positive twist and reverse twist rings perform.

Elimination of the second compression ring in high-performance applications has met with success. However, getting this two-ring combination to perform properly for the mileage put on a normal passenger car has proved to be difficult. It has been estimated that 10 to 20 percent of the blowby and about 30 percent of the oil control is done at the second ring. The purpose of the oil control ring is to meter oil. Actually, the first oil control device in the cylinder is the piston skirt, and a properly designed

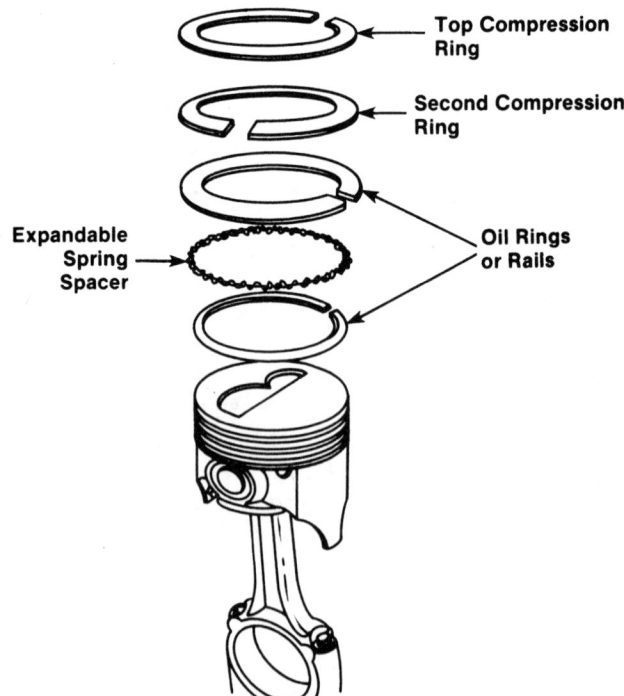

FIGURE 7-52 Typical piston ring assembly

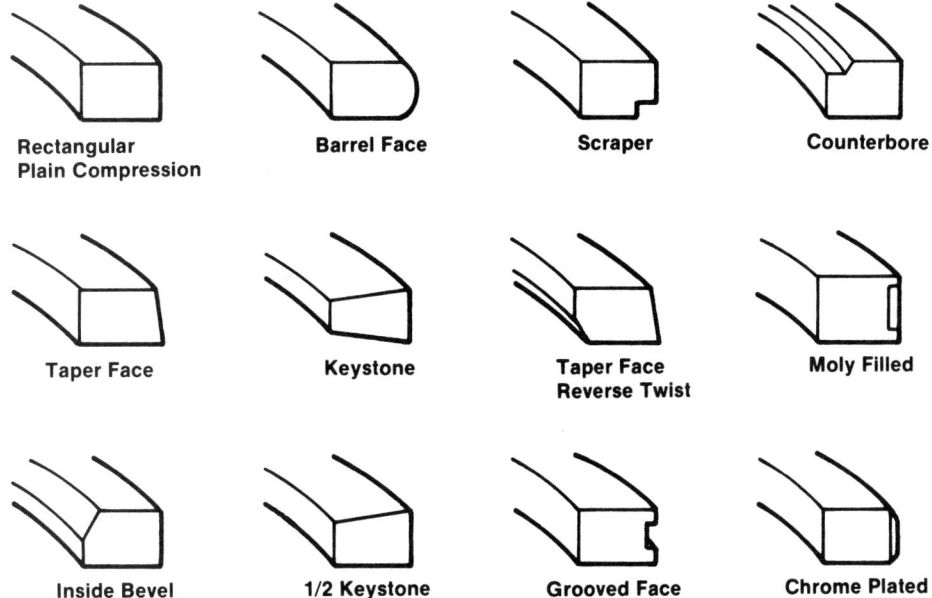

FIGURE 7-53 Typical styles of compression rings

and clearanced piston skirt edge can shear some of the oil film from the cylinder wall, relieving the ring somewhat from a large amount of oil. The oil rings usually consist of two rails or scrapers that remove oil from the cylinder wall and direct it inward through the vented expander where it then returns to the crankcase or, in some cases, oils the piston pin.

Many piston rings have a coating on their surface to reduce friction and prevent wear. Some rings have a chromium-plate coating. Chromium is extremely hard and provides protection against abrasive wear. Another common coating is made from molybdenum. This coating is sprayed onto the ring during manufacturing. These rings, called *moly*

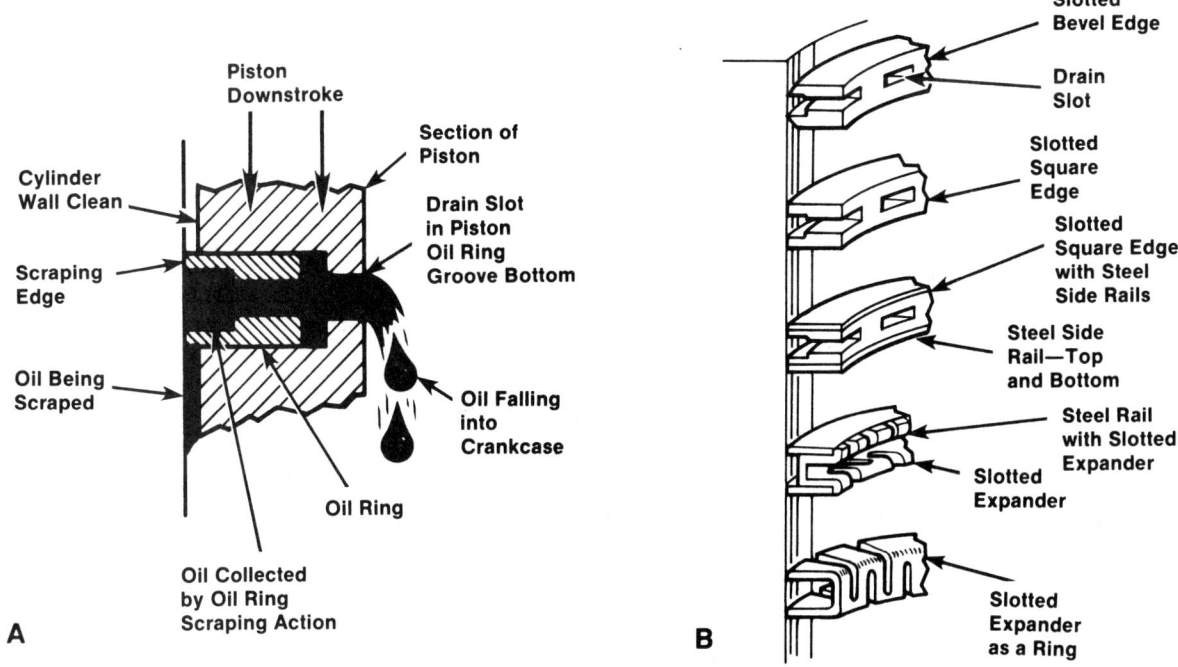

FIGURE 7-54 Oil control rings: (A) operation and (B) variations

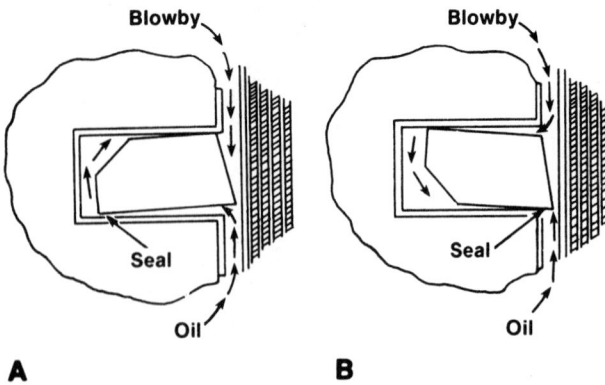

FIGURE 7-55 (A) Positive twist ring used as top combustion ring because it controls blowby effectively; (B) reverse twist ring used as a second ring to control oil flow to the top ring

rings, provide excellent resistance to wear. Aluminum oxide and nonstick coating are used to coat rings because the slippery finish reduces friction and prevents carbon from sticking to the ring.

Low-tension rings are being used with increasing frequency on both domestic and import engines. The term *low tension* refers to the amount of pressure the ring actually exerts against the cylinder wall, not the tangential tension, which is how many pounds of force it takes to bend the ends of a ring together. A ring with low tangential tension can still be a high-pressure ring if it is installed in a small-diameter cylinder bore. Even so, tangential ratings as well as unit pressure are often quoted when talking about low-tension rings.

What exactly is the difference? Conventional oil rings exert somewhere between 180 to 240 psi of unit pressure against the cylinder wall, whereas low-tension rings fall in the 90 to 160 psi range. By comparison, the tangential tension on ordinary three-piece oil rings is normally 18 to 25 pounds, whereas low-tension rings are usually between 6 to 13 pounds.

Low-tension oil rings are fairly easy to identify because of their narrower dimensions. Most currently measure somewhere between 2.75 mm and 4.75 mm in thickness. Oil rings as thin as 1 to 2 millimeters are in the planning stages, but for now the lower limit for mass production appears to be about 2.75 mm. The rails in a conventional 20-pound oil ring can measure 0.024 by 0.150 inch compared to 0.016 to 0.095 inch for a 7-pound low-tension ring.

Low-tension oil rings improve fuel economy and reduce both friction and piston groove pounding. Some late-model engines have been experiencing ring groove pounding problems because of

higher operating speeds and temperatures. Higher temperatures make aluminum soften, allowing rings to pound out their grooves in the pistons. That creates too much side clearance, which leads to ring flutter and breakage.

Checking Piston Ring Side Clearance

Before taking measurements, clean the ring grooves with a piston groove cleaner (Figure 7-56). When the grooves are clean, use a feeler gauge to measure piston ring clearance as follows (Figure 7-57):

1. Install the correct ring into the top groove of the piston.
2. Insert progressively thicker feeler gauges into the space between the top of the ring and the piston groove.
3. When determining the thickest gauge that will fit between the piston ring and ring groove, the thickness of that gauge equals the piston ring side clearance measure-

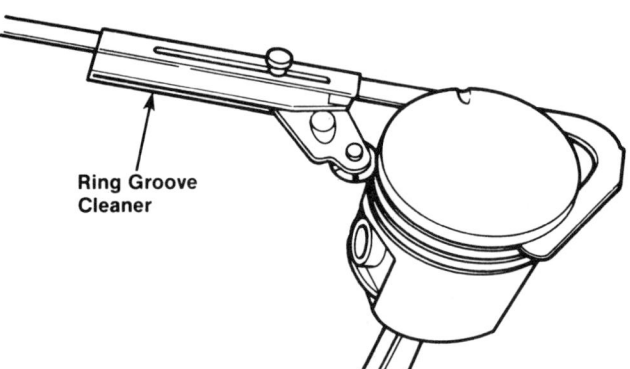

FIGURE 7-56 Cleaning ring grooves

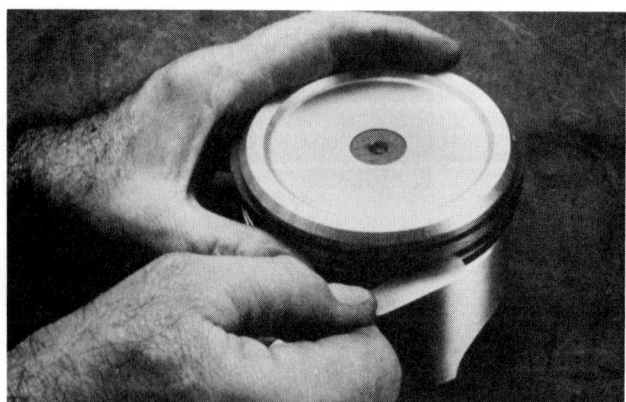

FIGURE 7-57 Using a feeler gauge to measure piston ring clearance.

ment. If it is possible to insert a 0.006-inch feeler gauge between the leaf and land, the piston should be replaced.

It must be remembered that the higher the rpm and compression, the narrower the required gap. For most automotive applications, the recommended end gap for rings up to 4 inches in diameter is 0.010 to 0.020 inch with a mean average of 0.015 inch. For rings above 4 inches in diameter, the end gap is commonly 0.013 to 0.025 inch.

 SHOP TALK _____

Gapless piston rings have been used with varying degrees of success in racing, but OE vehicle manufacturers and rebuilders see little to be gained by using such rings in passenger car applications. The reason for the gap is to provide for the difference in expansion that might occur between the piston ring and cylinder. Otherwise, the ring ends might butt and cause scuffing, scoring, ring breakage, or engine seizure.

Be sure to check to see if the ring is the proper diameter for the bore (Figure 7–58). Also, by measuring the end gap differences between the ring in its top and near bottom positions in the cylinder bore, you can determine if excessive taper exists and block boring is necessary. The ring must be square in the cylinder.

If the inspection indicates any wear beyond that given in the vehicle's service manual or shows any damage, the piston should be replaced.

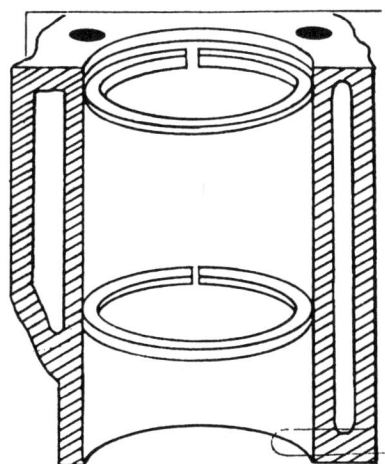

FIGURE 7-58 Checking to determine if ring is the proper diameter

Diesel Pistons and Rings

The pistons and rings in a diesel engine perform like those in the gasoline engine, but with a major difference. The pressures and temperatures in the diesel combustion chamber are much greater, and the rings must be able to handle this difference. This means the pistons and rings are heavier, the metal structure is different, and the tolerances are closer.

To obtain complete combustion in a diesel engine, each droplet of fuel must be surrounded by enough air to burn it completely. This condition is made by creating a violent swirling motion or turbulence inside the combustion chamber. Most of this turbulence is provided by piston crown (or top) designs that have combustion cups. A combustion cup is a machined or cast out volume in the crown of the piston. Two common combustion cup designs include turbulence cup and Mexican hat configurations (Figure 7–59A and B). Most diesel piston crown designs include some configuration that promotes air/fuel mixing. For example, the piston crown can have valve reliefs (or valve pockets) cast or machined into it (Figure 7–59C) to allow it to move closer to the cylinder head, increasing the compression ratio. These reliefs provide clearance for the valve heads when the piston is near or at top dead center and the valves are at various stages of opening and closing.

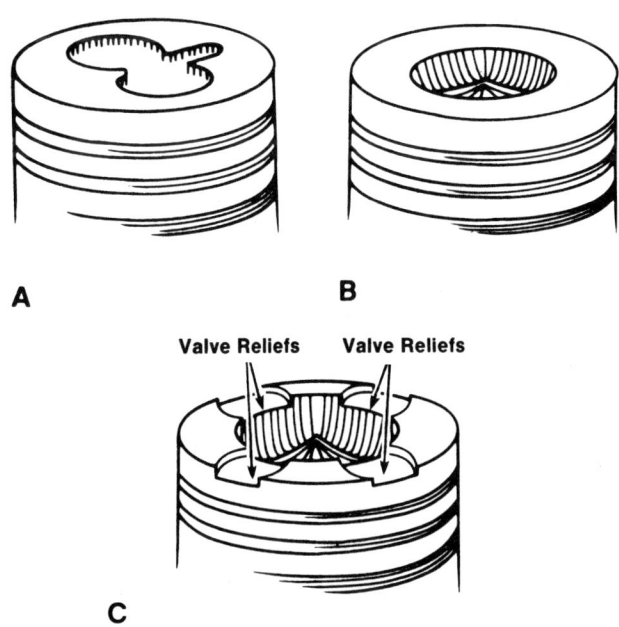

FIGURE 7-59 Three designs of diesel engine piston heads: (A) turbulence cups; (B) Mexican hat; and (C) Mexican hat with valve reliefs

Diesel pistons are also designed with more skirt area than gasoline engine pistons since they are subjected to greater side thrust. Many diesel engine pistons are oil cooled from a spray nozzle in the connecting rod or from a separate tube type spray nozzle aimed at the underside of the piston and connected to the engine oil gallery. Many diesel pistons have the top ring groove equipped with a chrome steel insert or keystone ring to reduce groove wear (Figure 7-60).

Diesel engine pistons are subjected to the same type of loads and stresses as gasoline engine pistons, but the loads and stresses are much greater. Figure 7-61 shows simple compression ring joints formed on some diesel and high-compression gas engines.

PISTON PINS

The wrist pin connects the piston and the connecting rod. In most passenger cars and light truck engines, the pins are press fit in the connecting rod eye (Figure 7-62) and float (or are loose) only in the piston. This means that the piston pin is slightly larger than the hole in the connecting rod. A hydraulic is used to force the pin in the rod. The holes in the piston provide the bearing surface, allowing the rod and pin to oscillate in the piston.

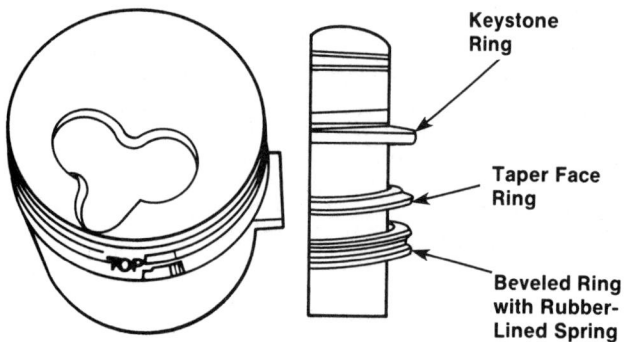

FIGURE 7-60 Typical ring arrangement for a small diesel engine

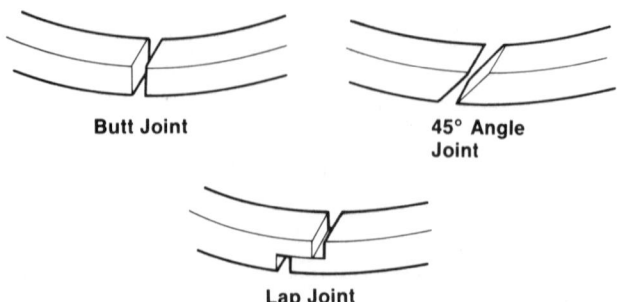

FIGURE 7-61 Diesel and high-compression gas engine compression rings

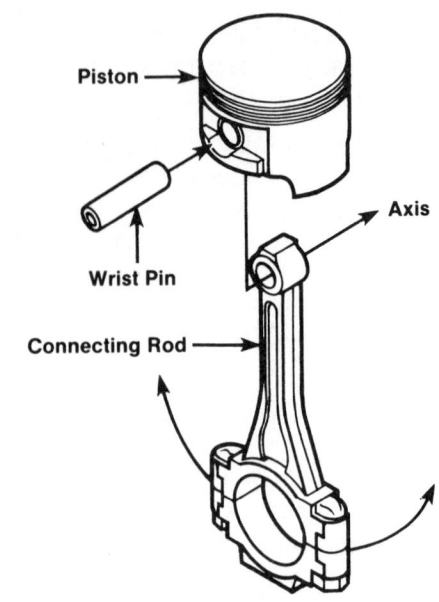

FIGURE 7-62 Connecting the connecting rod to the piston with a press-fit piston or wrist pin

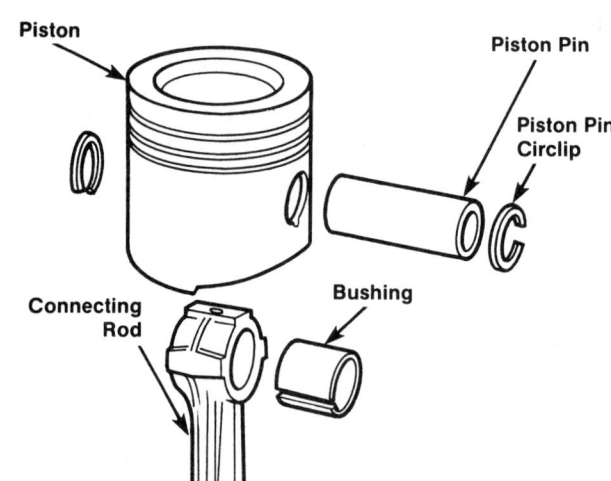

FIGURE 7-63 Free-floating piston pin design

In some engines, the pins float in both the piston and connecting rod eye. In other words, floating pins are free to move rotationally within all bore areas—both those of the rod and of the piston—in a thin lubrication cushion of engine oil (Figure 7-63). The benefit of a floating pin design has to do with engine life and reliability in high-stress situations. A floating pin arrangement does not give any more horsepower than a pressed-pin design in the same motor; it simply adds to the longevity.

But, since the pin is not interference fit to the rod anymore, something has to be added to retain the pin in its proper location. Otherwise, the pin can slide out of the side of the piston, causing damage to the cylinder wall.

Regardless of the type of piston pin system employed, it must be carefully checked. If the press type is to be installed on reused pistons, the pin fit must be checked. When the pin is inserted into the piston, it should move freely in and out of the pinholes, and there should not be any noticeable rocking movement between the pin and piston. If there is, new oversize pins will have to be installed. Piston pins are available in a number of oversizes from 0.001 to 0.005 inch in 0.001-inch increments. If new pistons are used with rebored cylinders, they should be supplied with new, properly fitting pins.

Full floating pins are checked for clearance in both the piston and the connecting rod. The fit in the piston is checked exactly like that for the press-fit design. The fit in the connecting rod is checked by inserting the pin into the connecting rod. Try to rock the pin up and down. Any noticeable movement means excessive wear in the rod bushing—it is most likely that wear will occur in the rod, since there is less bearing surface area there than in the piston. The wear is corrected by fitting the small end of the connecting rod with a new bushing. Maximum allowable pin-to-piston clearance in both systems is usually 0.001 inch.

Keep in mind that piston pinholes cannot be centered in the piston. Offset, they are located toward the major thrust surface approximately 0.062 inch from the piston centerline (Figure 7–64). Pin offset is designed to reduce piston slap and noise that could result as the large end of the connecting rod crosses over upper dead center.

Analyzing Piston Ring Condition

As with pistons, faulty rings are discovered during an inspection. If the rings are bad, they should be replaced. Also be sure to correct the fault during rebuilding. Some of the most common causes of piston ring failure are the following:

1. *Abrasive Wear.* This is the major cause of premature ring failure. The grinding action of hard foreign particles will rapidly remove material from the rings and cylinder walls. Numerous fine vertical scratches, which give the ring faces a dull gray appearance, are characteristic of abrasive wear. The major cause of abrasive wear is the result of honing grit being left in the cylinder bores at the time of rebuild. Other sources of abrasives are infrequent air, oil, and breather filter maintenance; engine vacuum leaks; and the use of a contaminated oil fill container.

2. *Scoring.* Ring faces can scuff or, in extreme cases, score due to metal-to-metal contact with the cylinder wall following breakdown of the oil film. Friction generated by the contact causes instantaneous welding, which is torn loose by the motion of the piston, leaving scratches and voids on the ring faces and cylinder wall. Causes include overheating, cylinder distortion, oil contaminated by coolant or fuel, improper bore finish, lack of lubrication, and insufficient piston-to-bore clearance.

3. *Broken Compression Rings.* Worn grooves, or a combination of worn grooves and worn rings, increase the side clearance between the ring and groove allowing abnormal shock loading during the power stroke. Broken rings have a characteristic area of carbon on both sides of the break.

4. *Insufficient End Gap.* Minimum end gap, as listed on the piston ring instruction sheet, must be maintained. This end gap reduces as the engine warms up and can close completely under some operating conditions. When this happens, the rings

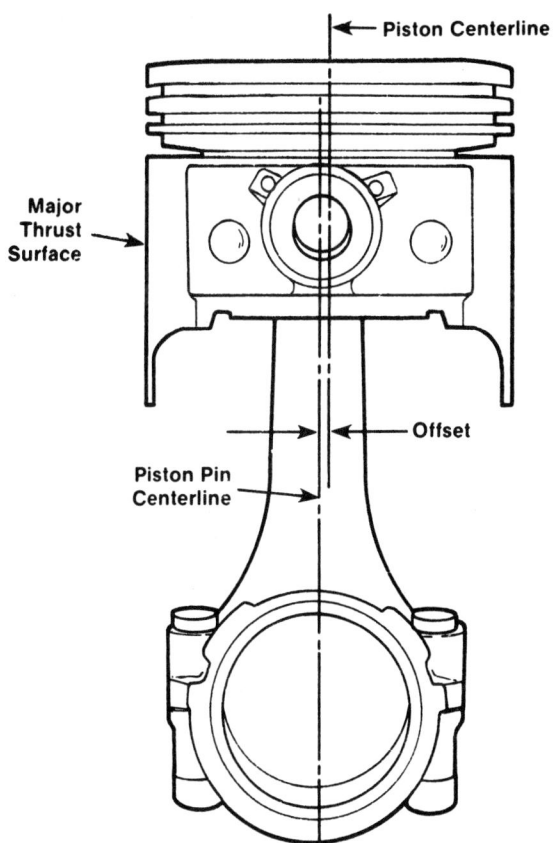

FIGURE 7–64 The piston pin is offset toward the major thrust surface.

lock up in the bore, scuff and score, and fail prematurely. High oil consumption is the first problem noticed. Engine overheating, uncontrolled combustion (preignition or detonation), or high-speed operation can cause end gaps to close and the ends of the rings to contact. Polished ring ends are characteristic of no end gap.

5. *Glazed Piston Rings.* Improper break-in procedures can cause piston rings to glaze in the cylinders and fail to control combustion blowby and oil consumption. Rings seat quickly when subjected to repeated cyclic loading. Glazing occurs when the engine is driven at very low speeds and loads. Follow the manufacturer's recommended break-in procedure. This will produce proper ring seating and prevent glazing (Figure 7–65).

CONNECTING RODS

Connecting rods are the link between the crankshaft journals and the piston. As the crankshaft rotates in the block saddles, the rod (attached at the big end to the crankshaft and at the small end to the piston) transmits the reciprocating force between the piston and the crankshaft journal movement. Because the connecting rod is under a great deal of pressure (3,000 to 4,000 psi of force being applied from the piston end), it must be carefully inspected for signs of fractures and the bearing bores for out-of-round and taper.

Out-of-roundness of the big-end bore can be checked with a dial bore gauge as shown in Figure 7–66. This same basic technique can also be employed using a telescoping gauge or an inside micrometer. Some technicians prefer to make the out-of-roundness with a special bore gauge such as the one shown in Figure 7–67. But, regardless of the

method used, the out-of-roundness is usually the greatest in a horizontal direction. This will also cause the rod bearing insert assembly to be out-of-round when installed. When checking for size and diameter, make the measurements lengthwise. There is no tolerance beyond the manufacturer's specifications (generally no more than 0.001 inch). This means the rebuilder must adhere to the specs given with no minimum or maximum allowable.

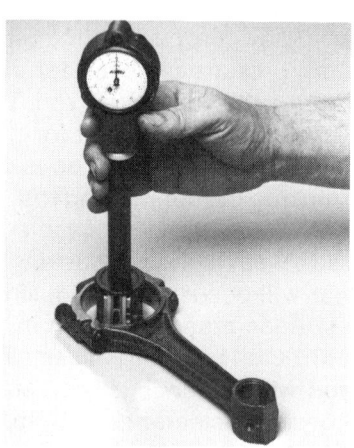

A

B

FIGURE 7–66 (A) Using a dial bore gauge to check for out-of-roundness of the bore; (B) recommended positions for checking out-of-roundness with a bore gauge *(Courtesy of Sunnen Products Co.).*

FIGURE 7–65 Glazed piston rings caused by improper start-up after the engine has been rebuilt.

FIGURE 7–67 Checking the out-of-roundness with a special bore gauge *(Courtesy of Sunnen Products Co.).*

FIGURE 7-68 Checking for bend or twist *(Courtesy of Sunnen Products Co.)*.

Inspect empty bores for bend or twist. One of the special tools on the market designed for this purpose can be valuable (Figure 7-68). If none are available, bend and twist measurements can be made using a dial indicator, V-blocks, and accurately turned mandrels positioned in each end of the rod. Big-end and little-end bores must be parallel within 0.002 inch per 6 inches of rod length. The twist between these bores must not exceed 0.001 inch per 6 inches of rod length (Figure 7-69).

Analyzing Connecting Rod, Piston Pin, and Piston Condition

Connecting rods and piston pins can be sources of problems that must be corrected during engine rebuilding. For instance, rod breakage (Figure 7-70) can occur at any time during the life cycle of the engine, but it generally happens early in the cycle. Some of the more common causes are improper bolt tightening, poor quality bolt, out-of-roundness in big-end bores, bearing failure, and nicking the rod.

Other connecting rod and piston pin problems include:

- *Journal End Break.* A reduction in the clamping forces on the big end of the rod will cause the cap to flex and bend during operation. This constant bending leads to fatigue breaks near the upper and lower bolt reliefs. Major causes of this failure are excessive bearing clearance, spun bearings, improper bolt torquing, and stripped bolts.
- *Overheated Pin End.* Discoloration of the pin end indicates severe overheating during assembly. The use of a torch or excessive heating time in a rod heater can distort the pin end and reduce the hardness. Dimensional instability and deformation reduce the inter-

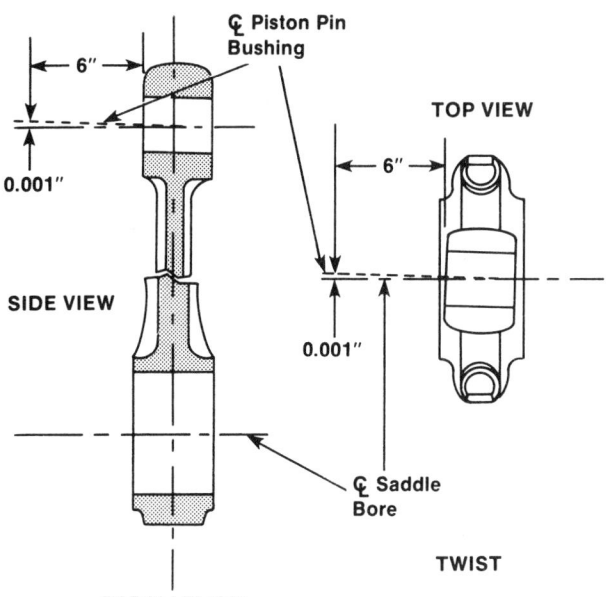

FIGURE 7-69 Important dimension when checking parallelism and twist

FIGURE 7-70 Example of connecting rod breakage *(Courtesy of Perfect Circle/Dana)*

ference fit, and the pin becomes loose and contacts the cylinder wall. Catastrophic failure usually results.
- *Scored Piston Pin.* Scored or seized piston pins normally indicate lubrication failure from overheated pistons. Low coolant level and preignition or detonation are major causes. Distortion of the piston when the pin is pressing into the small end of the rod also contributes to pin failure. The use of a rod heater to properly heat the rod to the correct temperature is recommended (see Chapter 11).
- *Rod Bolt Failure.* This usually occurs due to improper tightening. The bolts are very easi-

ly stretched from overtorquing; not torquing the bolts tightly enough can result in their becoming loose and causing problems. Every time a bolt is loosened and then retorqued, it fatigues slightly. Therefore, it is fairly cheap insurance to replace rod bolts in engines that are expected to be used for heavy service. Always follow the manufacturer's specifications on torque valves. Prior to torquing, pull up both rod bolts evenly so that the cap and bearing are well seated. Use a good torque wrench and check it for calibration often.

- *Bent or Twisted Connecting Rods.* Diagonal skirt wear pattern and connecting rod and bearing side load wear indicate a piston/connecting rod alignment problem. Usually this is caused by a bent or twisted rod. If a misaligned piston and rod assembly is put into service, it will not follow the normal direction of travel. The end result can be a lock ring failure (in pistons so equipped), premature piston skirt or cylinder wall wear, a rod bearing failure, or possibly connecting rod breakage (Figure 7–71). Erratic piston ring action and poor oil control can also occur.

A twist or bend in the con-rod can sometimes be corrected by hand using a connecting rod aligner such as the one shown in Figure 7–72A. The various piston/rod tests and adjustments that can be made on a typical aligner are as follows:

- *Checking the Piston/Rod Assembly.* The setup for checking a piston and con-rod assembly is illustrated in Figure 7–72B. Light will show between the gauge and piston if the rod is bent. The piston can be turned against either side of the rod when checking for twist. A bending bar is employed to straighten the rod.

- *Checking the Rod on the Pin and Correcting the Bend.* Certain types of piston construction make it necessary to check alignment on the pin instead of using the indicator on the piston. Rods are not always assembled with pistons, in which case it is necessary to check with the gauge against the pin. This method (Figure 7–72C) is remarkably accurate because any showing of light between the gauge arms and pin can be readily perceived and as easily corrected.

- *Checking with the Indicator and Correcting the Bend.* The indicator, which the technician slips on the upright of the standard, greatly magnifies the error and can be read with equal facility from either side (Figure 7–72D). The checking gauge arms are turned aside and the technician straightens the assembly while the indicator arm rests on top of the checking vee, which is positioned on the piston.

- *Checking with the Indicator and Correcting the Twist.* When checking for twist, the technician may turn the piston skirt against either side of the rod. With the checking vee positioned on the piston and the indicator arm resting on the vee as shown in Figure 7–72E, the needle of the indicator will point either above or below center, indicating the direction the rod is to be bent to correct its twist. A small hammer may be used if it is necessary to lightly peen and stretch the surface.

- *Checking and Correcting the Bend in the Tractor Rod.* The hand-operated bending bars usually supply sufficient leverage for small and medium rods. Large rods present a more difficult problem. A tractor rod is placed on the sleeve with the bend up. A bending clamp is employed to exert a downward pressure (Figure 7–72F). The rod can be lightly peened at the bend to stretch its surface and overcome any elasticity that might allow it to spring back to its original bend.

- *Checking with a Gauge and Correcting the Twist.* Figure 7–72G shows the rod without the piston mounted on the expanding sleeve. The rod is straightened with the bending bar until the twist is removed. When the rod is straight, the gauge shows uniform contact between both the upper and lower arms and the piston pin.

- *Checking a Rod for Offset.* One arm of the gauge is adjusted to rest on the flange of the

FIGURE 7–71 Example of damage caused by a twisted connecting rod *(Courtesy of Perfect Circle/Dana)*

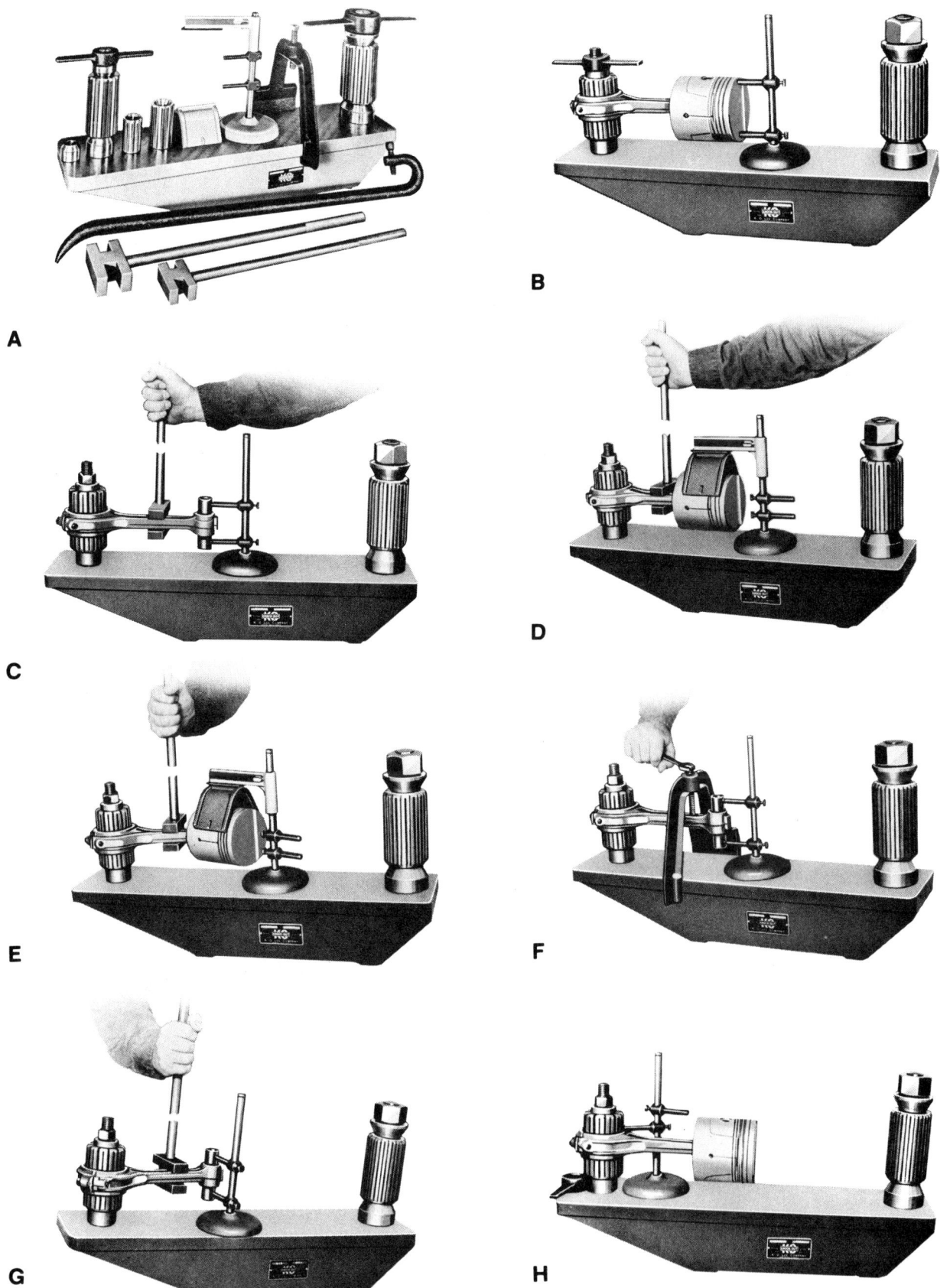

FIGURE 7-72 Various corrective operations that can be performed on a typical connecting rod aligner. *(Courtesy of K. O. Lee Company)*

rod bearing; the other is adjusted to rest on top of the piston skirt (Figure 7-72H). Without disturbing the setting of either gauge arm, the rod is reversed and rechecked upside down from its original position on the sleeve. The first arm is then placed so that it rests on the opposite flange of the rod bearing. The second arm is then placed on top of the piston skirt as before. Whatever it measures above or below the piston skirt represents the offset of the rod. More or less offset can be obtained by bending the rod adjacent to the bearing and then correcting its alignment by counter bending it close to the pin. When checking the offset of a rod without a piston, one gauge arm is adjusted to rest on the flange of the pinhole; the other is on the flange of the rod bearing. When the rod is reversed, as indicated above, the offset will be evident and can then be easily corrected.

Diesel Connecting Rods

Connecting rods for most gasoline engines are cast, whereas those for diesel engines are forged. High-performance gasoline engines also have forged connecting rods. Many diesel engine connecting rods are rifle drilled the length of the rod to provide good piston pin lubrication (Figure 7-73). In some cases, this oil passage is also utilized to provide piston cooling. In this case, the top of the rod has a spray nozzle aimed at the underside of the piston.

The split in the big end of the rod is offset in many cases to allow clearance for rod removal through the cylinder. Some rod caps are attached by cap screws rather than bolts and nuts. The yoke of the rod is threaded in this case. Yokes and caps are often provided with a precisely machined tongue and groove to provide perfect alignment of the upper and lower bearing bore halves.

The diesel engine connecting rod is subjected to the same types of loads and stresses as the gasoline engine rod, but the loads and stresses are much greater (Figure 7-74).

CYLINDER BLOCK INSPECTION

Once the block is cleaned (Figure 7-75) it should be visually inspected for cracks. If cracks were previously discovered by the visual inspection, now is the time to double-check. The block can be checked for cracks using dye or powder, by X-raying, pressure checking, or the magnetic crack check. (For details on crack detection methods, see Chapter 8.)

Check all machined gasket surfaces for burrs, nicks, scratches, and scores. Remove minor imperfections with an oil stone or file.

DECK FLATNESS

The top of the engine block where the cylinder head mounts is called the deck. To check head or deck warpage, use a precision straightedge and feeler gauge (Figure 7-76) to measure the head in

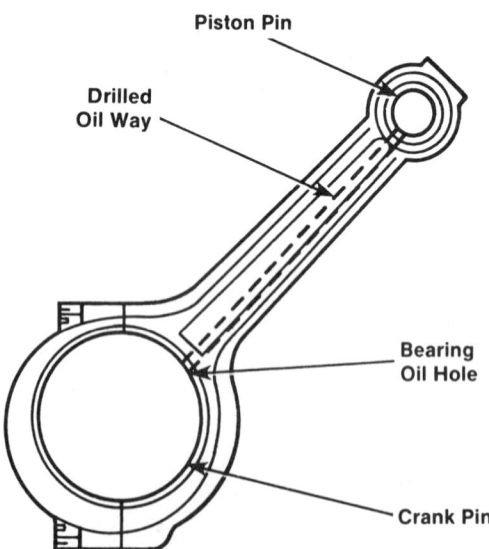

FIGURE 7-73 Typical diesel connecting rod with piston pin lubrication

Piston Pin

Drilled Oil Way

Bearing Oil Hole

Crank Pin

FIGURE 7-74 Assembling a large diesel connecting rod

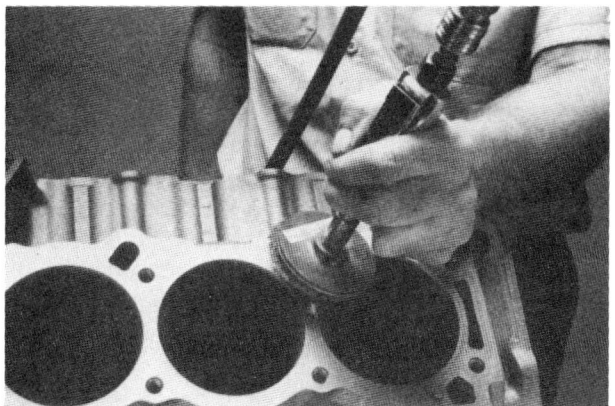

FIGURE 7-75 One method of cleaning the deck surface *(Courtesy of Perfect Circle/Dana)*

FIGURE 7-76 Checking for deck flatness with a precision straightedge and feeler gauge

each direction. Another method of checking a head for straightness is to use a precision arbor and a dial indicator as shown in Figure 7-77.

 SHOP TALK ─────────

Instead of using a straightedge and feeler gauge, many rebuilders prefer to make the warpage test with a measure warp and gauge. Roll this dial indicator across the surface of the block. It can also be used to check the head for warpage (Figure 7-78).

Some engines have special sealing surface flatness requirements; therefore, consult the manufacturer's specifications. If specifications are not available, use 0.003 inch per 6 inches, and no more than 0.006-inch maximum on any length. New or resur-

faced tolerances are usually 0.001 inch. In most cases, if more than one head is used on an engine, all must be resurfaced an equal amount to maintain uniform compression and manifold alignment. If the gasket surface of the block is warped and not corrected, valve seat distortion will occur when the head is tightened. Coolant and combustion leakage can also occur.

Two reasons a rebuilder might elect to resurface are to blend the sleeves into the deck surface after sleeving or to obtain uniform cylinder length for performance or blueprint work. Blocks can be refinished by broaching, milling, or grinding. Any of these methods is acceptable as long as the proper surface finish is obtained. For cast-iron blocks using

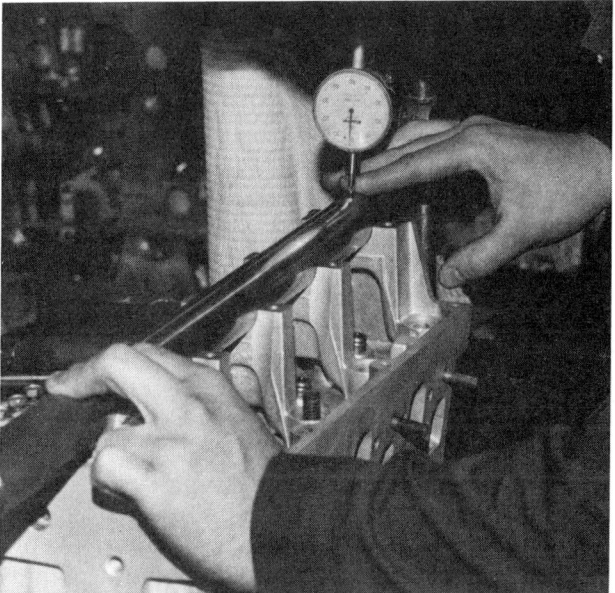

FIGURE 7-77 Checking a head for straightness with a precision arbor and dial indicator

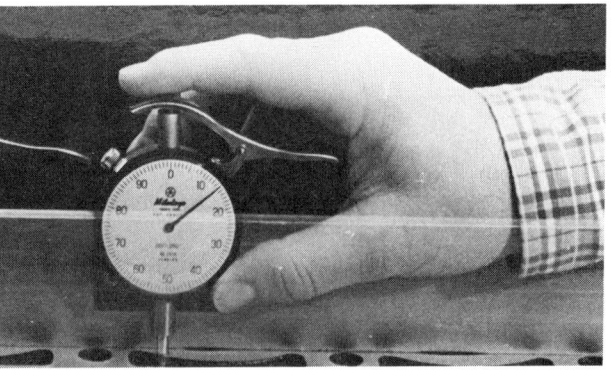

FIGURE 7-78 Another method of checking for deck warpage *(Courtesy of Goodson Shop Supplies)*

composition gaskets, the finish should be 60 to 120 rms microinches (see Chapter 9). The beaded steel head gasket seals by crushing the steel bead against the deck surface and requires a much finer finish to ensure good sealing.

CYLINDER WALLS

Inspect the cylinder walls for scoring, roughness, or other signs of wear. Ring and cylinder wall wear can be accelerated by an abrasive environment. Abrasive particles that get between mated moving parts grind away at the adjoining surfaces and remove material from the parts. Abrasive particles can include metallic debris in the engine, which is a result of wear and nonmetallic dirt particles that entered the engine during operation or maintenance. The source of contamination should be located and corrected to avoid a recurrence of the problem.

Piston ring, piston, and cylinder wall damage can also be caused by scuffing and scoring. Grooves cut in these parts act as passages for oil to bypass the rings and enter the combustion chamber. Scuffing and scoring occur when the oil film on the cylinder wall is ruptured, allowing metal-to-metal contact of the piston and rings on the cylinder wall. The heat generated causes momentary welding of the contacting parts. Reciprocating movement of the piston breaks these welds, cutting grooves in the ring faces, piston skirt, and cylinder wall. Cooling system hot spots, oil contamination, and fuel wash are significant causes of this problem.

A problem affecting the newer thin-wall engines is block distortion. Out-of-round cylinders break the face seal between the rings and cylinder wall permitting the passage of oil into the combustion chamber (Figure 7–79). This was not much of a problem on older engines with more rigidly cast blocks, but the newer, lightened blocks are not as stiff and, conse-

quently, are subject to distortion if proper procedures are not followed during rebuilding.

The cylinder heads and main bearing caps induce stresses in the block. These stresses must be duplicated when the cylinders are being oversized or bored, particularly if the block is of the thin-wall design. To accomplish this, the main bearing caps should be installed and torqued to specifications. Torque or deck plates should be bolted to the block to simulate the effect of the cylinder heads so that the bores will be round when the engine is assembled (see Chapter 9). Omission of these steps on some blocks will almost certainly cause distortion and oil consumption.

CYLINDER BORE INSPECTION

Cylinder bore wear is not uniform (Figure 7–80). Maximum wear occurs at the top of the ring travel area because pressure on the top ring is at a peak and lubrication at a minimum when the piston is at the top of its stroke. Shallow depressions form at the top of the cylinder on the thrust faces, giving the cylinder an oval shape at the top. A ridge of unworn material will remain above the upper limit of ring travel. Below the ring travel area, wear is negligible because only the piston skirt contacts the cylinder wall. Piston slap is one of the major causes of more wear at the top of the cylinder and less at the bottom (Figure 7–81).

There are several requirements for a properly reconditioned cylinder. It must be the correct diameter, have no taper or runout (out-of-round), and the surface finish must be such that the piston rings will

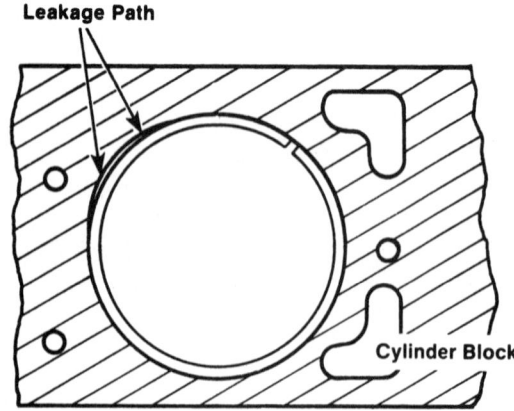

FIGURE 7-79 Checking for oil leakage

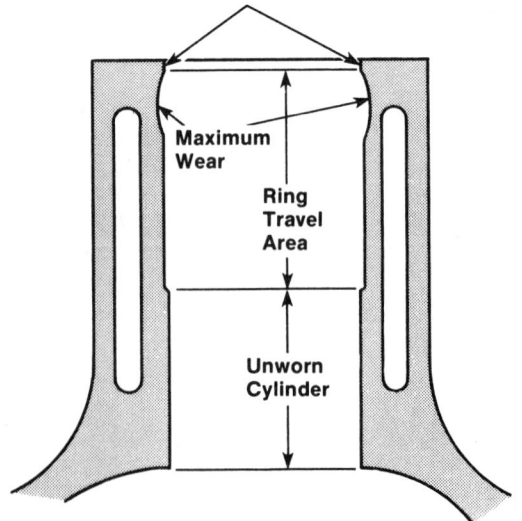

FIGURE 7-80 Worn cylinder cross section

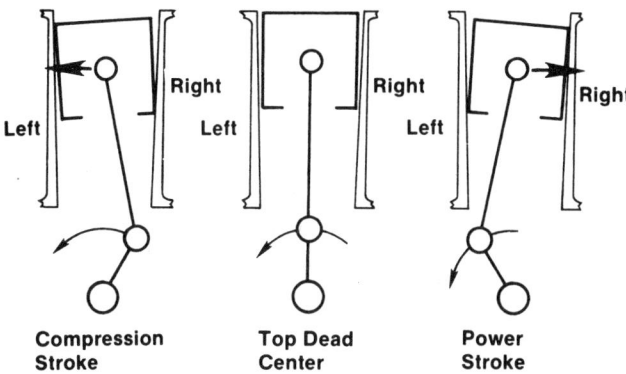

FIGURE 7-81 When piston slap occurs, the bores become tapered since they wear most at the top and least at the bottom.

seat to form a seal that will minimize blowby and control oil.

Taper is the difference in diameter between the bore at the bottom of the hole and the bore at the top of the hole—just below the ridge (Figure 7-82). Subtracting the smaller diameter from the larger one gives the cylinder taper. Some taper is permissible, but not more than 0.006 inch. If taper is less than 0.006 inch, it is possible to get by with just a re-ring job as opposed to reboring.

Cylinder out-of-roundness is the difference between the measurement parallel with the crank and the measurement perpendicular to the crank. Out-of-roundness is measured at the top of the cylinder just below the ridge. The maximum allowable out-of-roundness is 0.0015 inch.

Methods of Checking the Cylinder Bore

Cylinder bores do not always wear in a perfectly round pattern; one side of the cylinder wall usually wears more than the other. For this reason, cylinder roundness must be determined. To measure out-of-round, measure the cylinder bore at a uniform depth in three directions (front/rear, right/left, and diagonally). Then, the smallest reading obtained from the cylinder is subtracted from the largest reading. The resulting figure is the cylinder bore out-of-round.

There are three instruments for measuring a cylinder bore for taper and out-of-roundess:

- Inside micrometer
- Telescoping gauge and outside micrometer
- Cylinder bore dial gauge

The cylinder bore dial gauge is used to determine cylinder bore out-of-round and taper. These two measurements, in addition to main bearing saddle alignment, provide basic information about the condition of the cylinder block. A block outside specification in any of these areas requires machining to restore it to specified tolerances.

Cylinder bore dial gauges are read in the same way that a dial indicator is read (Figure 7-83). As with the inside micrometer and telescoping gauge, the cylinder bore micrometer must be rocked to obtain the correct measurement.

The parts of the cylinder bore dial gauge or cylinder bore micrometer include the handle, guide blocks, lock, indicator contact, and dial indicator. Various indicator contacts can be installed to measure several bore sizes with one gauge.

To measure a cylinder or main bearing bore using a telescoping gauge, proceed as follows:

1. Release the lock screw at the end of the handle.

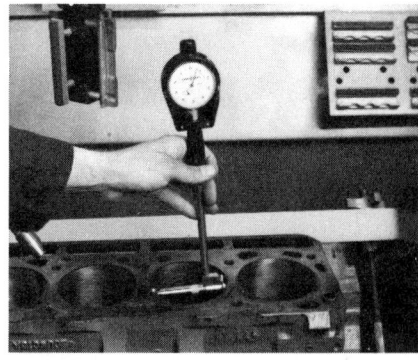

FIGURE 7-82 Checking the cylinder bore taper with a cylinder gauge *(Courtesy of Goodson Shop Supplies)*

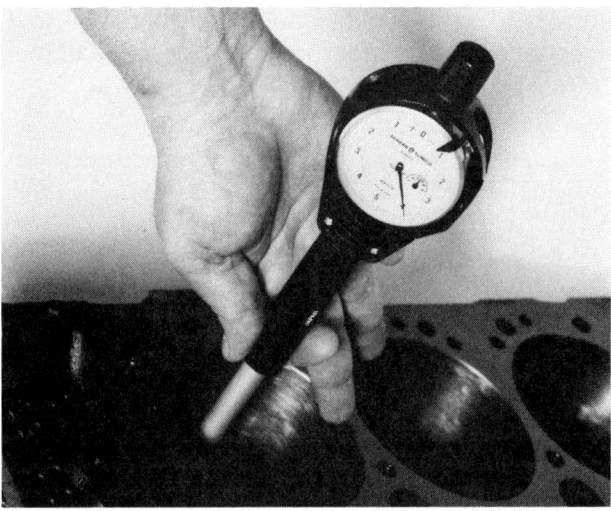

FIGURE 7-83 Checking cylinder bore with a dial gauge *(Courtesy of Atlas Engineering and Manufacturing, Inc.)*

2. Compress the plungers and, using the lock screw, secure them in the retracted position.

3. Place the telescoping gauge into the bore (Figure 7-84). Release the plungers by loosening the lock screw. Allow the plungers to expand until they contact the bore walls.

4. Rock the telescoping gauge back and forth and side to side to check for the correct resistance. These motions are similar to the method used to center a micrometer on the work.

5. When the correct resistance is felt, lock the plungers into position using the lock screw. Then, carefully retract the telescoping gauge from the bore.

6. To complete the measurement, use an outside micrometer to determine the distance between the two plunger faces. Measuring the distance between the extended plungers is the same as measuring the bore.

The procedure for using an inside micrometer to measure bores is shown in Figure 7-85.

Cylinders should also be inspected for any signs of waviness caused by galling or scoring. If taper or out-of-roundness is within maximum allowable specs and pistons are in good condition, maybe only a re-ring is necessary to salvage the engine.

If not in acceptable condition, cylinders that are part of the block casting (as is true of most automobile engines) can be overhauled by boring them out to an oversize. This boring process removes the worn areas on the cylinder walls and allows a new surface to be restored by being finish honed. However, since the cylinder is now larger, a matching

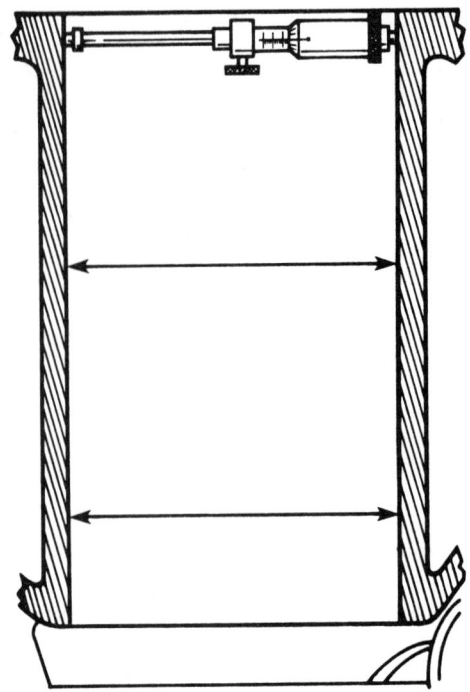

FIGURE 7-85 Cylinder bore taper can be measured with an inside micrometer.

oversized piston and ring set must be installed during the rebuilding process (see Chapter 11).

Another approach to reconditioning cylinders that are formed by the block casting is to bore out the block 3/16 or 1/4 inch larger in diameter, allowing the block to receive a cut-to-length or universal length repair sleeve, which is pressed into the bored out block (Figure 7-86). This sleeve, after being inserted into the block, is then finish bored and honed, generally back to standard size. A piston and ring set of the same size is chosen to match it. Figure 7-87

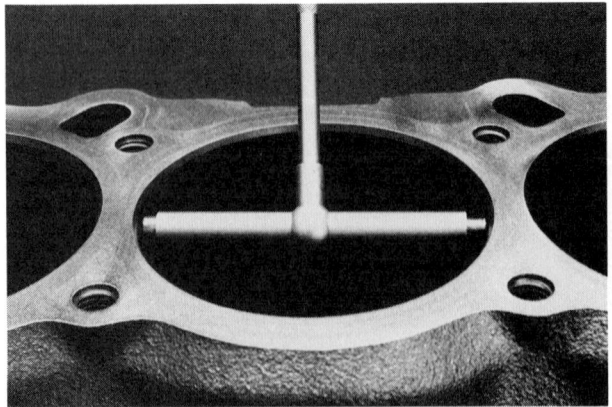

FIGURE 7-84 Checking cylinder bore with a telescoping gauge *(Courtesy of Perfect Circle/Dana)*

FIGURE 7-86 Pressing in a cylinder sleeve in a block

illustrates typical sleeve and sleeve assembly kits that are available. Such kits usually contain the sleeves (wet or dry type), pistons, piston rings, and piston pins and clips (if required). Methods of installing cylinder sleeves are given in Chapter 11.

Cylinder Bore Surface Finish

The surface finish on a properly prepared cylinder wall acts as a reservoir for oil to lubricate the rings and prevent piston and ring scuffing primarily during break-in. Piston ring faces can be damaged if the cylinder wall is too rough, resulting in premature wear. A surface that is too smooth will not allow the rings to seat properly. Figure 7–88 illustrates a good cylinder bore surface finish.

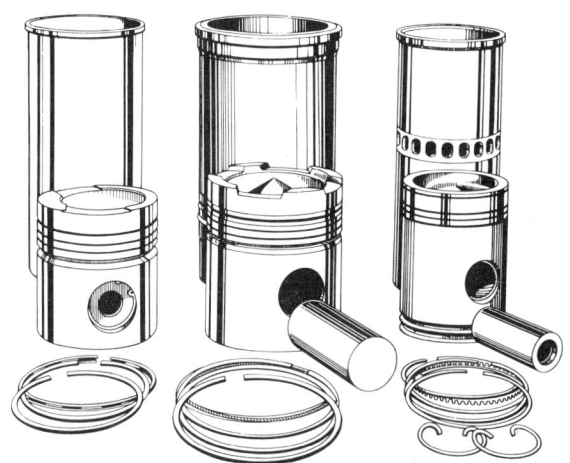

FIGURE 7-87 Typical cylinder sleeve kit

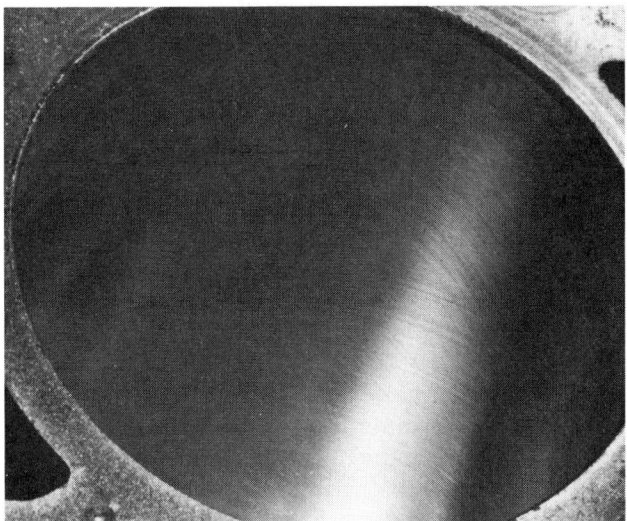

FIGURE 7-88 A good cylinder bore surface finish *(Courtesy of Sealed Power Corp.)*

BEARING INSPECTION

Bearings carry the critical loads in an engine and, like valves and rings, are a major wear item. This is why they should always be replaced when an engine is rebuilt. Yet in spite of the important role they play in engine operation and longevity, bearings do not always receive the attention they deserve.

Two types of bearings are used in automobiles (Figure 7–89):

1. Antifriction ball bearings
2. Sleeve or insert bearings

Antifriction bearings are used in the suspension, transmission, and other parts of the drivetrain, but are rarely found in engines. The sleeve bearing, however, is for use with the components that move within the engine: the crankshaft, connecting rods, and camshaft (Figure 7–90). Since insert or sleeve bearings are the type to be concerned with in engine

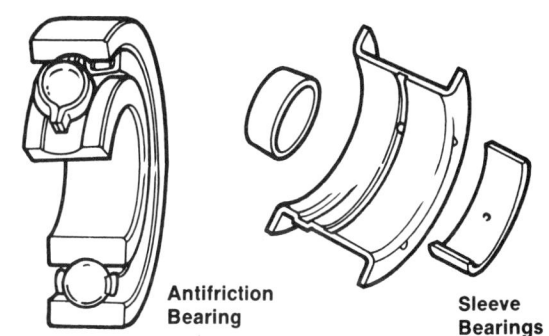

FIGURE 7-89 (A) Antifriction and (B) sleeve bearings

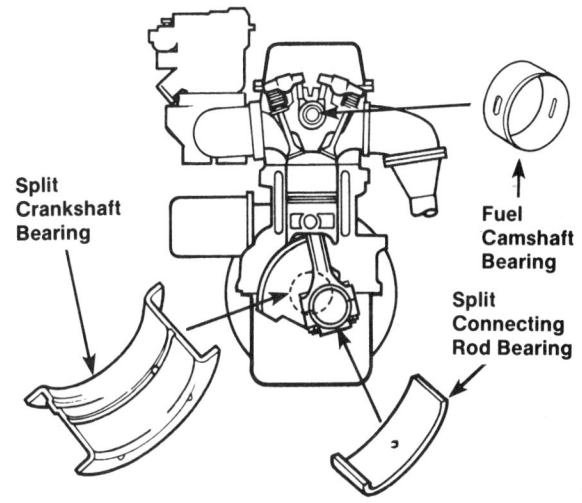

FIGURE 7-90 Engine bearing application

rebuilding, the remainder of this chapter is devoted to their inspection and, if found distressed, how to remedy the cause of the problem.

Bearings can be made from a variety of materials. No one bearing material has all the qualities that can satisfy the four primary requirements of a bearing. These characteristics are:

- *Embedability.* The ability of the bearing to tolerate foreign particles on its surface.
- *Fatigue Resistance.* The ability of the bearing to hold up to stress without failing.
- *Corrosion Resistance.* The ability of the bearing to withstand the corrosive acids inside the crankcase.
- *Surface Action.* The ability of the bearing to withstand metal-to-metal contact without being destroyed.

Since there is no one perfect bearing material, most engine bearings in use today are multilayered. A basic bearing has two layers as shown in Figure 7–91A. One layer is a steel backing and the other is a liner of bearing material. This type is referred to as a bimetal bearing. Some designs incorporate a third layer, which is an overlay on the bearing material liner. The overlay is of a bearing material also but is different than the liner. This type is referred to as a trimetal bearing (Figure 7–91B).

The most commonly used engine bearings can be grouped into four classifications:

1. *Babbitt.* This material can be divided into two categories—conventional and micro or thin babbitt—and can be in either tin or lead base material. Conventional babbitt bearings differ from micro in the amount of babbitt laminated on the steel back: they have a substantially heavier thickness of babbitt deposited on the steel back.
2. *Sintered Copper Lead.* This material is made by sintering metal powders on a steel strip; it is available with or without an overlay.

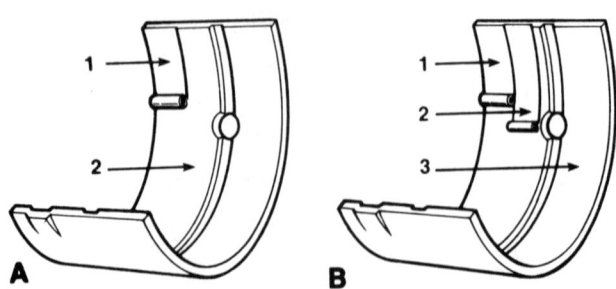

FIGURE 7–91 (A) A bimetal bearing has two layers and (B) a trimetal bearing has three layers.

 SHOP TALK _____

When using sintered copper or one of the harder materials, it is necessary to add a softer layer for direct contact with the journal to provide the necessary embedability.

3. *Cast Copper-Lead.* In this group, the copper-lead alloy is cast on a steel strip; it is available with or without an overlay.
4. *Aluminum.* A widely available, corrosion-resistant material; it is available in solid, bimetal, and trimetal constructions.

There are many different engine bearing designs in use today. These designs vary as far as where in the engine the bearings are used as well as with the dictates of the specific engine. Although the complete spectrum of bearing designs might appear confusing, the picture clears itself when you have a basic understanding of the key bearing design factors that follow.

TYPES OF ENGINE BEARINGS

In the early days of the automotive engine, bearing surfaces were provided by casting a babbitt alloy directly on the supporting surface. This rough casting was then machined to its finished configuration, providing the needed bearing. Bearing replacement, obviously, was a very tedious and time-consuming task. The great majority of modern bearings, however, are the insert type. This name is derived from the fact that the bearing is made as a self-contained part and then inserted into the bearing housing. This type provides many advantages, including relative ease of replacement, greater variety of bearing materials, controlled lining thickness, and improved structure. Bearing thickness can be measured with a micrometer as shown in Figure 7–92.

There are two types of insert bearings:

- *Precision Insert Bearing.* This bearing is manufactured to close tolerances so that no further sizing is required at time of assembly into the engine.
- *Resizable Insert Bearing.* This bearing is manufactured with an extra-thick lining of bearing material on the inside diameter, permitting the bearing to be machined to any desired size up to and including standard size at time of assembly.

There are two basic designs of insert bearings from the standpoint of configuration (Figure 7–92).

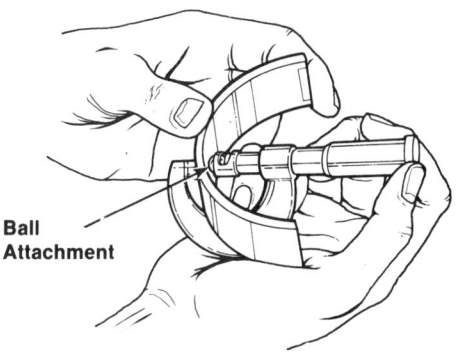

FIGURE 7-92 Measuring bearing thickness

- *Full Round (One Piece) Bearing.* This bearing is used where it is possible to slide the journal into place in the bearing, such as with a camshaft.
- *Split (Two Halves) Bearing.* This bearing is used where the bearing must be assembled around the journal with the bearing housing being of two parts also, including a cap that holds the assembly together. All connecting rod bearings, for example, are of the split type.

The straight shell bearing and the flanged thrust bearing are two other configurations found in an automotive engine. The flanged thrust bearing provides the same support as the straight shell bearing, but also controls any horizontal movement of the shaft. The connecting rod and most main bearings utilize the straight shell design; the flange bearing is used in the thrust position of the block to take the fore and aft movement of the crankshaft. Most thrust main bearings are double flanged; however, there are applications, particularly in heavy-duty engines

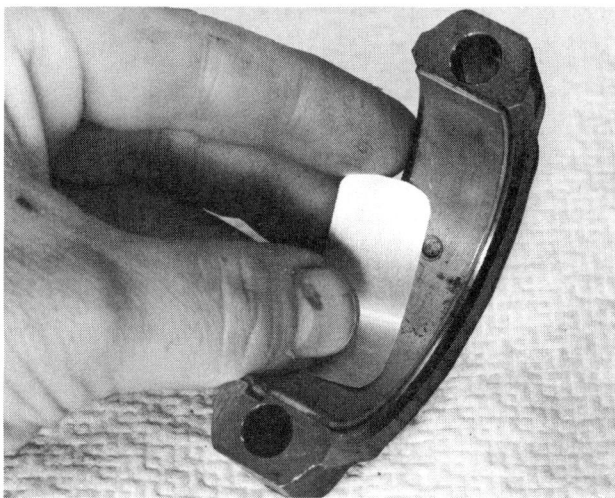

FIGURE 7-93 Use of shims

and in engines built by some import manufacturers, where separate thrust members, called thrust washers, are used. They may be supplied as full round or as halves depending on the vehicle manufacturer's design. Shims may be used to build up thickness (Figure 7-93).

BEARING SPREAD

Most main and connecting rod bearings are manufactured with spread. This means the distance across the outside parting edges of the bearing insert is slightly greater than the diameter of the housing bore (Figure 7-94). To position a bearing half that has spread, it must be snapped into place by a light forcing action. This assures positive positioning against the total bore and assists in subsequent assembly work by keeping the bearing halves in place.

BEARING CRUSH

Each half of a split bearing is made so it is slightly greater than an exact half. This can be seen quite easily when a half is snapped into place in its housing; the parting faces extend a little beyond the seat (Figure 7-95). This extension (as little as 0.001 inch) is called *crush.*

When the two bearing halves are assembled and the housing cap tightened, the crush sets up a radial pressure on the bearing halves so they are forced tightly into the housing bore. This assures that the bearing back is in complete and snug contact with the housing bore surface area—vital to the proper conduction of heat away from the bearing.

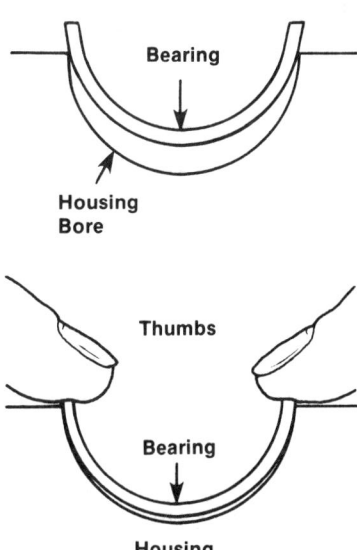

FIGURE 7-94 Spread requires the bearing to be snapped into place.

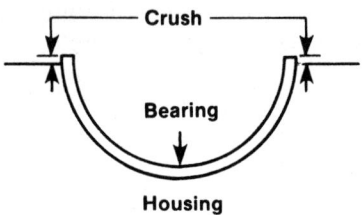

FIGURE 7-95 Crush assures good contact between bearing and housing.

BEARING LOCATING DEVICES

Engine bearings must be provided with some means to keep them from rotating or shifting sideways in their housings. There are a number of approaches to the problem with the specific design utilized being established by the original equipment manufacturer. The most commonly used design is the locating lug. As shown in Figure 7-96, this consists of a protruding lug at the parting face of the bearing that nests into a slot in the housing that has been machined out to receive it.

Another approach is the locating dowel illustrated in Figure 7-97. The dowel can be located on the housing or on the back of the bearing, with the dowel hole being located in the mating part. In some cases, there is a dowel hole in both parts, with the dowel being inserted as a separate piece.

A third design utilizes a flathead screw, which is secured to the housing through a hole in the crown of the bearing.

OIL GROOVES

Providing an adequate oil supply to all parts of the bearing surface, particularly in the load area, is an absolute necessity. In many cases, this is accomplished by the oil flow through the bearing oil clearance. In other cases, however, engine operating conditions are such that this oil distribution method is inadequate. When this occurs, some type of oil groove must be added to the bearing. Some oil grooves are used to assure an adequate supply of oil to adjacent engine parts by means of oil throw-off.

Although oil grooves vary in size and shape, it has been found that a single groove instead of multiple grooves works the best. A few typical oil grooves are shown in Figure 7-98.

OIL HOLES

Oil holes are designed to allow the oil flowing through the engine block galleries to enter the bearing oil clearance space. Connecting rod bearings receive oil from the main bearings by means of rifled oilways in the crankshaft. Oil holes are also used to meter the amount of oil supplied to other parts of the engine. For example, oil spurt holes in connecting rods are often used to direct oil for the purpose of lubricating the cylinder walls (Figure 7-99). Oil spurt

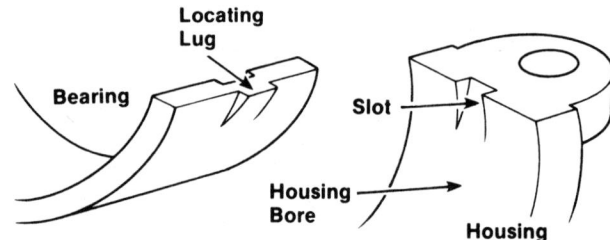

FIGURE 7-96 Locating lug fits into the slot in the housing.

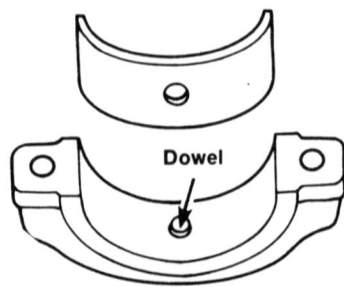

FIGURE 7-97 Locating dowel fits into the dowel hole in the mating part.

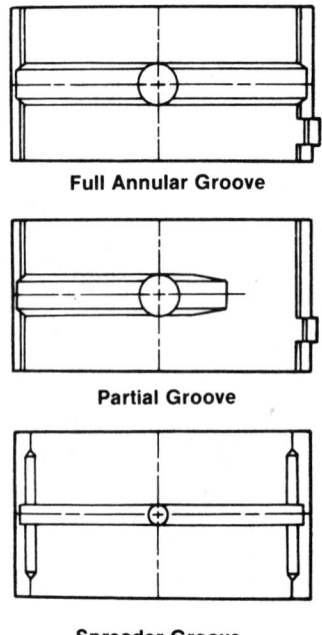

Full Annular Groove

Partial Groove

Spreader Groove

FIGURE 7-98 Typical types of bearing oil grooves

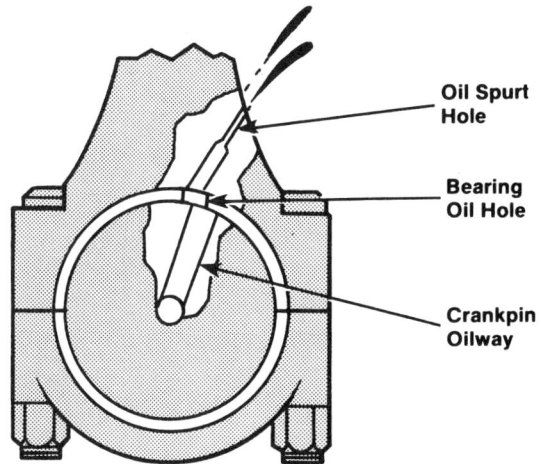

FIGURE 7-99 Oil spurt holes are often used in connecting rods.

holes are also sometimes used in the bearing caps of the crankshaft main bearing (Figure 7-100). When the bearing is equipped with an oil groove, the oil hole normally is in line with the groove.

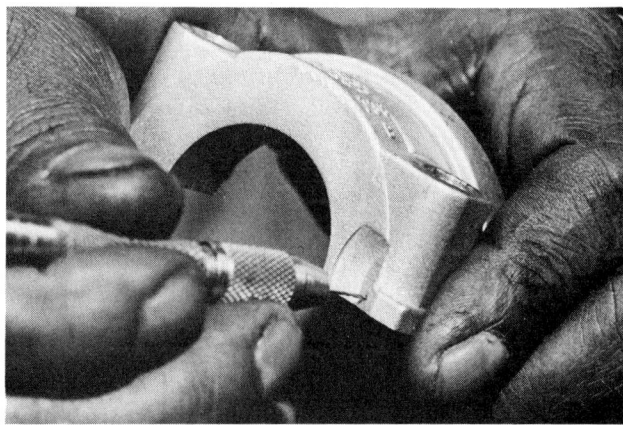

FIGURE 7-100 Oil spurt holes are often used in the main bearing cap to lubricate the crankshaft.

The size and location of oil holes is critical. Thus, when replacement bearings are installed, care must be exercised to be sure that the oil hole matches the original equipment specifications.

In some engine designs, connecting rods are drilled as shown in Figure 7-73 to supply oil from the crankpin to the piston pin bushing. In addition to oil inlet holes, some crankshafts are drilled with all the necessary passages (Figure 7-101) for lubrication. Some camshaft bearings contain oil holes as well (Figure 7-102).

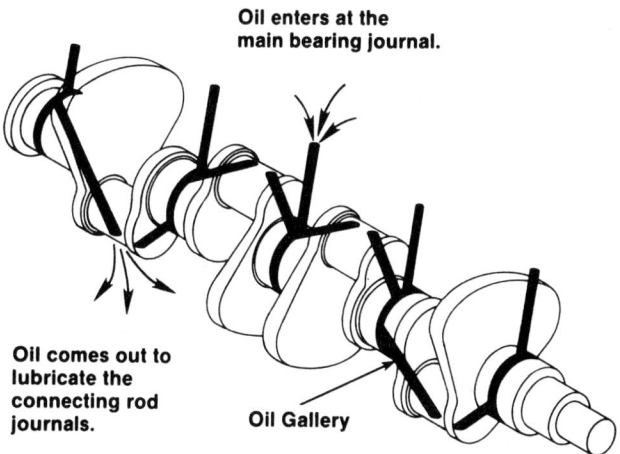

FIGURE 7-101 Crankshaft lubrication passages

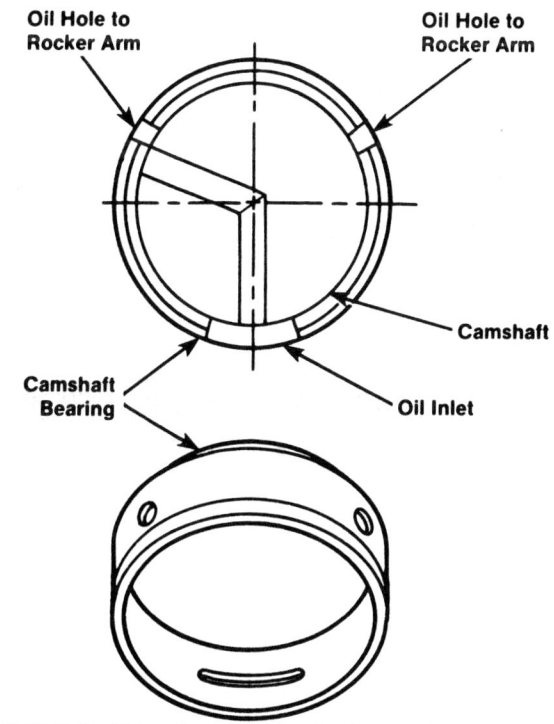

FIGURE 7-102 Some camshaft bearings contain oil holes.

OIL CLEARANCE

Maintaining a specific oil clearance is critical to proper bearing operation. Many times, when wear on the internal engine components such as the crankshaft is negligible, the proper oil clearance specifications can be restored with the installation of standard size replacement bearings. However, if the crankshaft is worn to the point where the installation of standard size bearings will result in excessive oil clearance space, a bearing with a thicker wall must be used to compensate. Although these bearings are thicker, they are known as undersize because the journals and crankpins of the crankshaft are smaller in diameter; in other words, they are under the standard size.

 SHOP TALK —————

Worn crankshaft journals have the same effect on oil consumption as worn bearings. When they are worn, they will not give uniform oil clearance (Figure 7-103).

Undersize bearings are available in 0.001-inch or 0.002-inch sizes for shafts that are uniformly worn by that amount. Undersize bearings are also available in thicker sizes, such as 0.010 inch, 0.020 inch, and 0.030 inch, for use with crankshafts that have been refinished (or reground) to one of these standard undersizes.

Use of undersize bearings with an undersize crankshaft will return the oil clearance to the original factory specifications. In cases where a standard undersize bearing is not desirable, a reborable undersize bearing is a viable alternative. These bearings have a 0.060-inch undersize lining, which allows them to be bored to exact crankshaft undersize requirements.

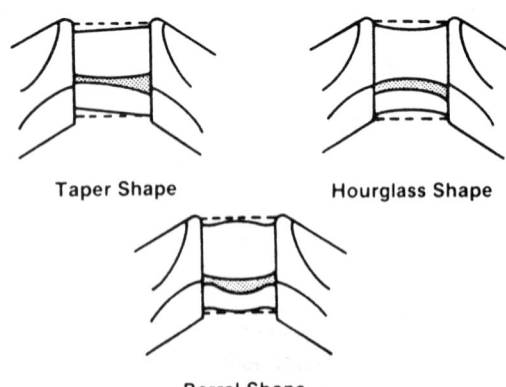

FIGURE 7-103 Out-of-shape crankshaft journals will not provide even oil clearance.

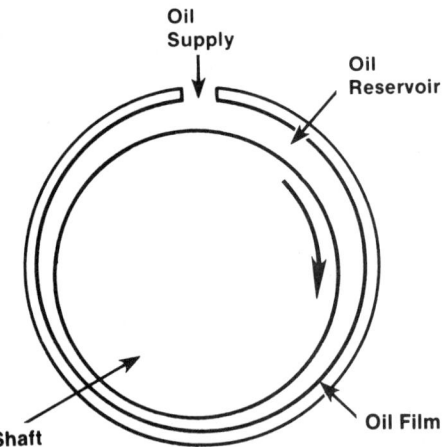

FIGURE 7-104 The eccentric flow of lubricating oil

Most bearings are manufactured with an eccentric wall. This means the wall thickness gradually becomes thinner in a taper from the crown to the parting faces (Figure 7-104). Eccentric wall bearings are produced for two reasons. First, the slight eccentricity built into the bearing wall will help increase the wedge effect to build the oil film under the loaded area. Second, eccentric wall bearings allow a reduction of the vertical oil clearance without reducing the effectiveness of the oil flow through the clearance space for cooling purposes.

BEARING DISTRESS

Table 7-3 indicates by percentage the principal causes of bearing distress. Understand that several causes can combine and be contributing factors toward the total distress of a bearing. Figure 7-105 illustrates the appearance of the more common bearing distresses, and Table 7-4 gives some of the possible causes of distress and corrective action to be taken.

Distress in bearings can be caused by any combination of the possibilities already mentioned in the table. Since dirt plays such a major part in early bearing distress, this can readily be one of the

TABLE 7-3: MAJOR CAUSES OF BEARING DISTRESS	
Dirt (abrasive wear)	43.4%
Insufficient lubrication	17.6%
Misassembly	13.2%
Misalignment	11.7%
Overloading	6.7%
Corrosion	2.0%
Indeterminate and other causes	5.4%

Dirt

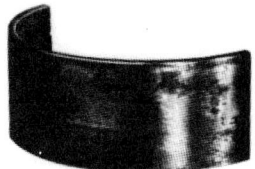

Insufficient Crush

Oil Starvation

Surface Fatigue

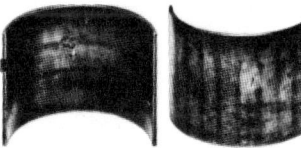

Dirt on
Bearing Back

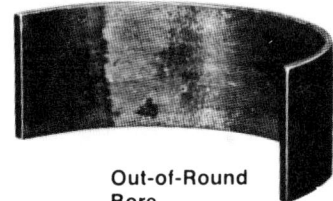

Out-of-Round
Bore

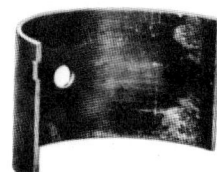

Excessive Crush

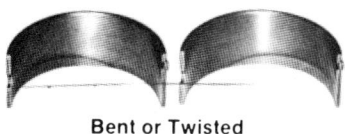

Bent or Twisted
Connecting Rod

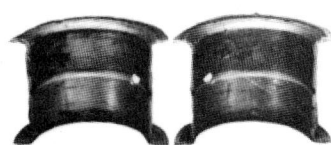

Shifted Bearing Cap

Tapered Journal

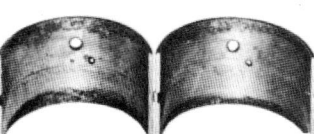

Hourglass-Shaped
Journal

Barrel-Shaped Journal

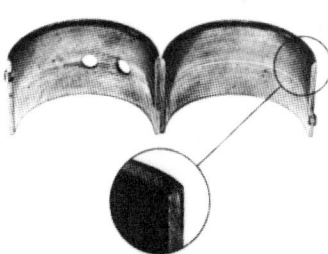

Fillet Ride

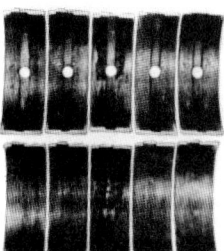

Distorted Crankcase

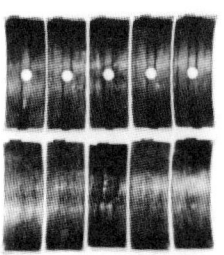

Bent Crankshaft

FIGURE 7-105   Common bearing distresses *(Courtesy of J P Industries, Inc.)*

221

TABLE 7-4: BEARING DISTRESS

Problem	Possible Cause	Remedy
Dirt (Foreign particles in lining)—Localized excessive wear areas are visible near the parting line on both sides of the top and bottom shells.	Improper cleaning of the engine and parts prior to assembly	Install new bearings, being careful to follow proper cleaning procedures.
	Road dirt and sand entering the engine through the air intake manifold	Grind journal surfaces if necessary.
	Wear of other engine parts, resulting in small fragments of these parts entering the engine's oil supply	
Insufficient crush—Highly polished areas are visible on the bearing back and/or on the edge of the parting line.	Bearing parting faces were filed down in a mistaken attempt to achieve a better fit, thus removing the crush.	Install new bearings using correct installation procedures (never file parting faces).
	Bearing caps were held open by dirt or burrs on the contact surface.	Clean mating surfaces of bearing caps and inspect for nicks and burrs prior to assembly.
	Insufficient torquing during installation (make sure the bolt does not bottom in a blind hole).	Check journal surfaces for excessive wear and regrind if necessary.
	The housing bore was oversize or the bearing cap was stretched, thus minimizing the crush.	Check the size and condition of the housing bore and recondition if necessary.
	Too many shims were utilized (if shims are specified).	Correct shim thickness (if applicable).
Oil starvation—When a bearing has failed due to oil starvation, its surface is usually very shiny. In addition, there might be excessive wear of the bearing surface due to the wiping action of the journal.	Insufficient oil clearance—usually the result of utilizing a replacement bearing that has too great a wall thickness. In some cases, the journal might be oversize.	Double-check all measurements taken during the bearing selection procedure to catch any errors in calculation.
	Broken or plugged oil passages, prohibiting proper oil flow	Check to be sure that the replacement bearing about to be installed is the correct one for the application (that it has the correct part number).
	Blocked oil suction screen or oil filter	Check the journals for damage and regrind if necessary.
	Malfunctioning oil pump or pressure relief valve	Check the engine for possible blockage of oil passages, oil suction screen, and oil filter.
	Misassembling main bearings metering off an oil hole	Check the operation of the oil pump and pressure relief valve.
		Be sure that the oil holes are properly indexed when installing the replacement bearings.
		Advise the owner about the results of engine lugging.
Surface fatigue—Small irregular areas of surface material are missing from the bearing lining.	Bearing failure due to surface fatigue is usually the result of the normal life span of the bearing being exceeded.	If the service life for the old bearing was adequate, replace with the same type of bearing to obtain a similar service life.

TABLE 7-4: BEARING DISTRESS (CONTINUED)

Problem	Possible Cause	Remedy
	Excessive loading	If the service life of the old bearing was too short, replace with a heavier duty bearing to obtain a longer life.
	Uneven loading	Replace all other bearings (main connecting rod and camshaft) because their remaining service life may be short.
		Recommend that the owner avoid hot rodding and lugging because these tend to shorten bearing life.
		Check for misassembly. Use proper installation techniques.
Dirt on bearing back—A localized area of wear can be seen on the bearing surface. Also, evidence of foreign particle(s) might be visible on the bearing back or bearing seat directly behind the area of surface wear.	Dirt, dust, abrasives, and/or metallic particles either present in the engine at the time of assembly or created by a burr removal operation can become lodged between the bearing back and bearing seat during engine operation.	Install new bearings following proper cleaning and burr removal procedures for all surfaces. Check journal surfaces and if excessive wear is discovered, regrind.
Out-of-round bore—Localized excessive wear areas are visible near the parting line on both sides of the top and bottom.	Alternating loading and flexing of the connecting rod can cause the bearing seats to become elongated. And because replacement bearing shells, when installed, tend to conform to the shape of the bearing seat, this can result in an out-of-round bearing surface.	Check the roundness of bearing seats before installing the new bearings. If they are found to be out-of-round, recondition the bearing housings (or replace connecting rod). Chck the journal surfaces for excessive wear and regrind if necessary. Install new bearings.
Excessive crush—Extreme wear areas are visible along the bearing surface adjacent to one or both of the parting lines.	Bearing caps were filed down in an attempt to reduce oil clearance.	Rework the bearing housing of the engine block if it has been filed down.
	Bearing caps were assembled too tightly due to excessive torquing.	Replace the connecting rod if its bearing cap has been filed down.
	Not enough shims were utilized (if shims were specified).	Install the new bearing. Check journal surfaces and regrind if necessary. Follow proper installation procedures by never filing down bearing caps and using the recommended torque wrench setting. Correct the shim thickness (if applicable). Check for out-of-roundness of the inside diameter of the assembled bearing by means of an out-of-roundness gauge, inside micrometer, calipers or Prussian blue to assure that any out-of-roundness is within safe limits.

TABLE 7-4: BEARING DISTRESS (CONTINUED)

Problem	Possible Cause	Remedy
Bent or twisted connecting rod—Excessive wear areas can be seen on opposite ends of the upper and lower connecting rod bearing shells. The wear is localized on one portion of the bearing surface with little or no wear on the remainder.	Extreme operating conditions such as hot rodding and lugging.	Inspect connecting rod and recondition or replace if bent or twisted.
	Improper reconditioning.	Check journal surfaces for excessive wear and regrind if necessary.
	Dropping or abusing the connecting rod prior to assembly.	Install bearing.
		Avoid dropping or abusing the connecting rod prior to assembly.
		Use proper installation techniques.
		Check related upper cylinder parts and replace if necessary.
Shifted bearing cap—Excessive wear areas can be seen near the parting lines on opposite sides of the upper and lower bearing shells.	Using too large a socket to tighten the bearing cap. In this case, the socket crowds against the cap causing it to shift.	Check journal surfaces for excessive wear and regrind if necessary.
	Reversing the position of the bearing cap	Install the new bearing being careful to use the correct size socket to tighten the cap and the correct size dowel pins (if required).
	Inadequate dowel pins between bearing cap and housing (if used), allowing the cap to break away and shift	Alternate torquing from side to side to assure proper seating of the cap.
	Improper torquing of cap bolts resulting in a loose cap that can shift positions during engine operation	Check the bearing cap and make sure it is in its proper position.
	Enlarged cap bolt holes or stretched cap bolts, permitting greater than normal play in the bolt holes	Use new bolts to assure against overplay within the bolt holes.
Out-of-shape journal—In general, if a bearing has failed because of an out-of-shape journal, an uneven wear pattern is visible on the bearing surface. Specifically, however, these wear areas can be in any one of three patterns: photo A shows the wear pattern caused by a tapered journal; photo B shows the wear pattern caused by an hourglass-shaped journal; photo C shows the pattern of a barrel-shaped journal.	If the journal is tapered, there are two possible causes: • Uneven wear of the journal during operation (misaligned rod). • Improper machining of the journal at some previous time. If the journal is hourglass or barrel shaped, this is always the result of improper machining or polishing.	Regrinding the crankshaft can best remedy out-of-shape journal problems. Then install new bearings in accordance with proper installation procedures.
Fillet ride—When fillet ride has caused a bearing to fail, areas of excessive wear are visible on the extreme edges of the bearing surface.	Excessive fillets are left at the edges of the journal at the time of crankshaft machining.	Regrind the crankshaft paying particular attention to the allowable fillet radii.
		CAUTION: Be careful not to reduce the fillet radius too much because this can weaken the crankshaft at its most critical point.
		Install new bearings.

TABLE 7-4: BEARING DISTRESS (CONTINUED)

Problem	Possible Cause	Remedy
Distorted crankshaft—A wear pattern is visible on the upper or lower halves of the complete set of main bearings. The degree of wear varies from bearing to bearing depending on the nature of the distortion. The center bearing usually shows the greatest amount of wear.	Alternating periods of engine heating and cooling during operation is a prime cause of crankcase distortion. As the engine heats, the crankcase expands; as it cools, the crankcase contracts. This repetitive expanding and contracting causes the crankcase to distort in time. Distortion can also be caused by: • Extreme operating conditions (for example overheating and lugging) • Improper torquing procedure for cylinder head bolts, particularly with overhead valve V-8 engines	Determine if distortion exists by using Prussian blue or visual methods. Align bore the housing (if applicable). Install new bearings.
Bent crankshaft—A wear pattern is visible on the upper and lower halves of the complete set of main bearings. The degree of wear varies from bearing to bearing depending upon the nature of the distortion. The center bearing usually shows the greatest wear.	A crankshaft is usually distorted due to extreme operating conditions, such as overspeeding and lugging.	Determine if distortion exists by means of Prussian blue or visual methods. Install a new or reconditioned crankshaft. Install new bearings.

causes, perhaps in conjunction with misalignment or misassembly of one bearing or another. The problem then becomes one of determining what is the combination of causes before steps can be taken for their correction.

Corrosion, which at one time was a major distress problem because most engines were operated on a preventive maintenance basis, is almost a thing of the past. Tin-base babbitt bearings are free from any form of corrosion. Lead-base bearings rarely corrode. All of the aluminum alloy bearings also seem to be corrosion free.

REPLACEMENT BEARINGS

The main objective in selecting a replacement bearing is to duplicate the original conditions of the engine when it was manufactured. This dictates that an exact duplicate of the original bearing should be used, with one exception: a smaller inside diameter is usually required to compensate for journal wear. Figure 7-106 illustrates common bearing and bushing locations in a conventional engine.

The ultimate performance of a replacement bearing depends upon using a bearing that provides the proper oil clearance. Thus, using the correct undersize is extremely important. In general, the oil clearance for pressure-lubricated bearings is 0.001 inch for each inch of journal diameter. This figure is modified somewhat by the bearing metal alloy used and the specific engine design. Generally speaking, the exact oil clearance recommended by the manufacturer should be used whenever possible.

Running clearances are particularly critical in high compression passenger car engines. The bearings in these applications are usually made with the wall at the parting line approximately 0.0005 inch thinner than the wall at the centerline. This is to allow additional horizontal clearance for increased oil flow.

Engine bearings will not function properly if they are installed wrong. In many cases, misassembly will result in premature failure of a bearing. Figure 7-107 shows some typical assembly errors most often made in the installation of engine bearings.

DETERMINING THE CORRECT UNDERSIZE

The easiest method for determining the correct undersize of a replacement bearing is as follows:

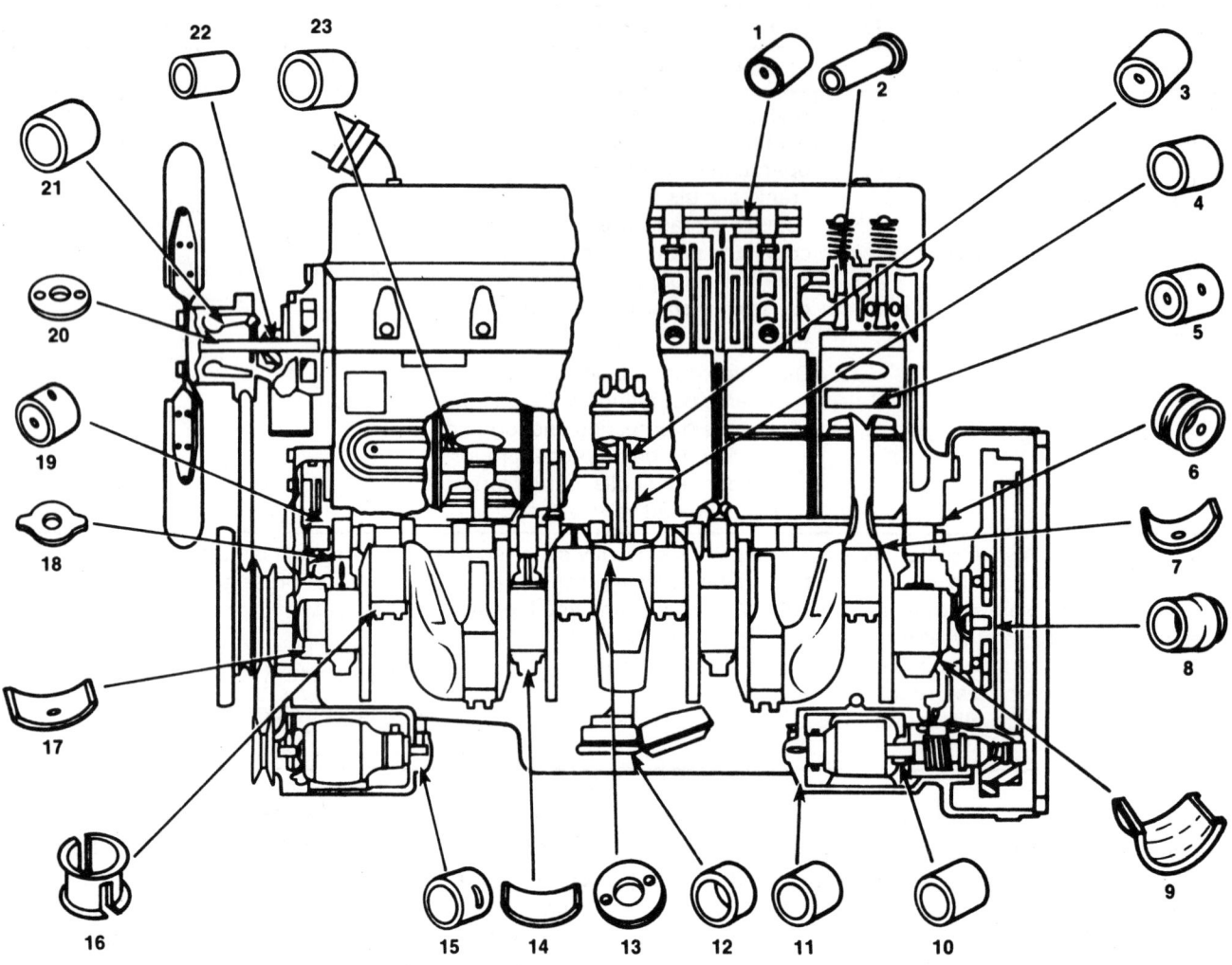

1 Rocker Arm Bushing
2 Valve Guide Bushing
3 Distributor Bushing, Upper
4 Distributor Bushing, Lower
5 Piston Pin Bushing
6 Camshaft Bushing
7 Connecting Rod Bearing
8 Clutch Pilot Bushing
9 Flanged Main Bearing
10 Starting Motor Bushing, Drive End
11 Starting Motor Bushing, Commutator End
12 Oil Pump Bushing
13 Distributor Thrust Plate
14 Intermediate Main Bearing
15 Alternator Bushing
16 Connecting Rod Bearing, Floating Type
17 Front Main Bearing
18 Camshaft Thrust Plate
19 Camshaft Bushing
20 Fan Thrust Plate
21 Water Pump Bushing, Front
22 Water Pump Bushing, Rear
23 Piston Pin Bushing

FIGURE 7-106   Common bearing and bushing locations

1. Refer to a bearing catalog or shop manual to get the standard crankshaft diameter for the engine being repaired.
2. Using a micrometer, measure the actual diameter of the worn journal. If the journal is out-of-round within acceptable limits, use the largest diameter.
3. Subtract the actual diameter (step 2) from the standard diameter (step 1). The answer obtained is the undersize that should be used.

Once the replacement bearing has been selected, it should be compared to the old bearing to make sure it is of the correct design. The oil grooves, oil holes, and means of bearing retention should be the same. If a difference is noted, it might be due to a change in design since the original bearing was manufactured. It is good practice to check this point with the bearing manufacturer before accepting or rejecting the bearing. Also, check the wall thickness of a new bearing and compare it to the value given in the bearing catalog or shop manual.

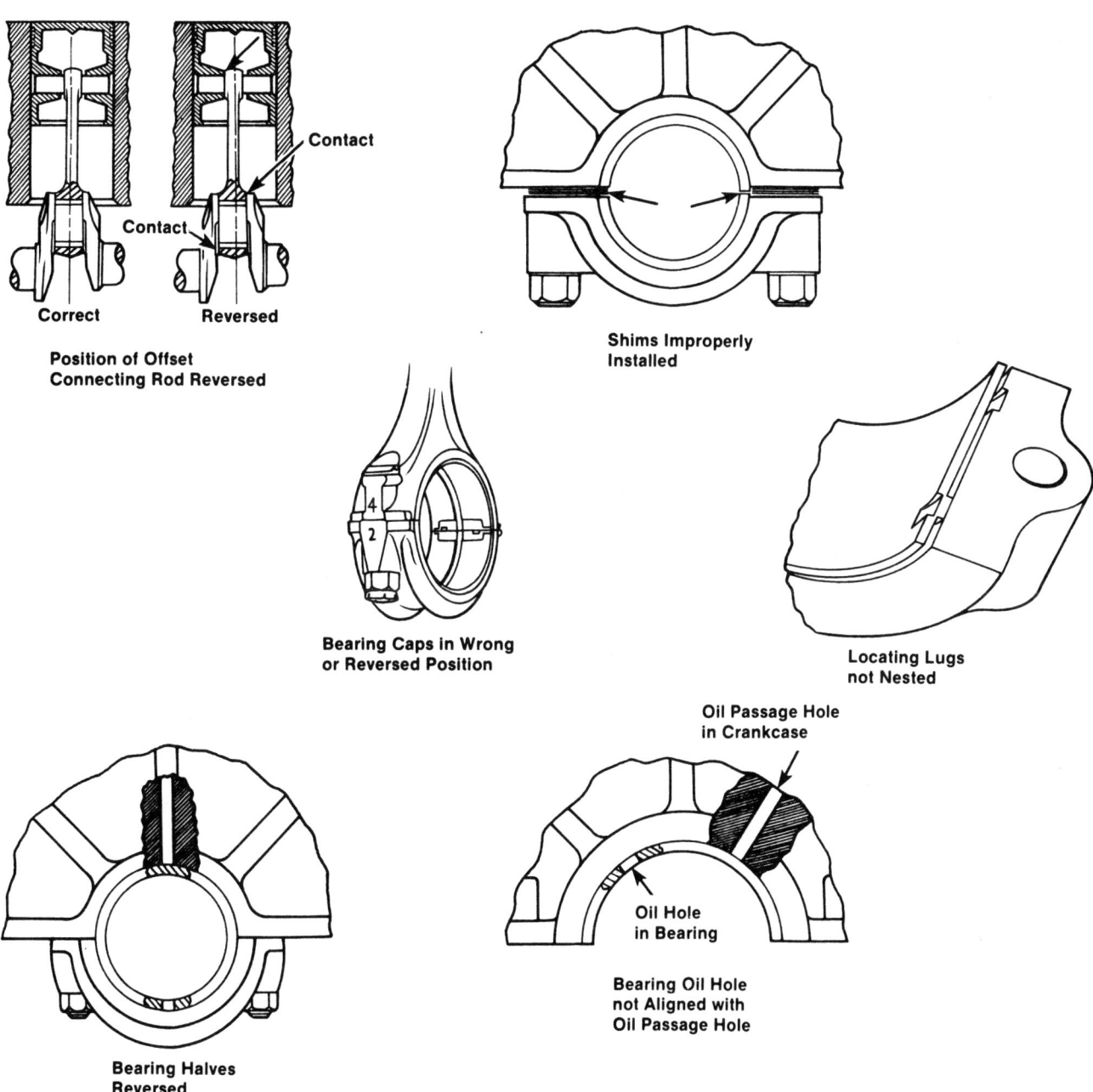

FIGURE 7-107   Typical assembly errors made when installing bearings

SHOP TALK _____

The new replacement bearing should be kept in its carton until needed for installation. Always handle bearings carefully to protect them from burrs, scratches, nicks, or any other distortion. Try not to touch the bearing surfaces with your fingers; even normal hand moisture will cause the surfaces to corrode because of the acid content. At the factory, clean white gloves are worn to guard against this possibility.

REVIEW QUESTIONS

1. Most pistons used today are made from
   _____ .
   a. cast iron
   b. aluminum
   c. ceramic
   d. none of the above

2. Which compression ring is subject to the most punishment?
   a. top
   b. middle
   c. bottom
   d. all rings

3. Low tension oil rings _____ .
   a. must be replaced
   b. improve fuel economy
   c. increase piston groove pounding
   d. both b and c

4. Technician A says a floating pin arrangement gives more horsepower than a pressed-pin design. Technician B says the floating pin adds to the longevity. Who is right?
   a. Technician A
   b. Technician B
   c. Both A and B
   d. Neither A nor B

5. Out-of-roundness of the cylinder bore can be checked with a(n) _____ .
   a. cylinder bore dial
   b. inside micrometer
   c. telescoping gauge
   d. all of the above

6. The block can be checked for cracks using
   _____ .
   a. dye check
   b. pressure checking
   c. magnetic crack check
   d. all of the above

7. The maximum amount of cylinder out-of-roundness allowed is _____ .
   a. 0.0015
   b. 0.006 inch
   c. 0.015 inch
   d. 0.06 inch

8. The maximum amount of cylinder taper allowed is _____ .
   a. 0.0015 inch
   b. 0.006 inch
   c. 0.015 inch
   d. 0.06 inch

9. Technician A says antifriction ball bearings are used in the drivetrain. Technician B says they are used in the components that move within the engine. Who is right?
   a. Technician A
   b. Technician B
   c. Both A and B
   d. Neither A nor B

10. Small irregular areas of surface material are missing from the bearing lining. Technician A says the cause is excessive loading. Technician B says that the bearing has just exceeded its life span. Who is right?
    a. Technician A
    b. Technician B
    c. Both A and B
    d. Neither A nor B

11. What most often causes bearing distress?
    a. inadequate cleaning after machining
    b. insufficient lubrication
    c. insufficient oil clearance
    d. none of the above

12. Which type of bearing is subject to corrosion?
    a. tin
    b. lead
    c. aluminum
    d. none of the above

13. Each half of a split bearing is made so it is slightly greater than an exact half. What is this extension called?
    a. spread
    b. crush
    c. both a and b
    d. neither a nor b

CHAPTER EIGHT

CRACK REPAIR

Objectives

After reading this chapter, you should be able to
- Explain why engine heads and blocks crack.
- List the methods used to detect cracks.
- Describe the pressure testing methods of finding cracks.
- Explain the advantage of the cold process of crack repair over welding.
- Describe the plug pinning process.
- Explain how a ceramic seal is applied.
- Describe the epoxy repair procedure.
- Explain the value of media blast peening.
- Analyze the problems associated with aluminum cylinder heads.
- Describe how to weld aluminum cylinder heads.

Once engine parts have been cleaned and given a thorough visual inspection, actual repair work, if needed, should begin. If cracks in the metal casting were discovered during the inspection, they should be repaired.

Cracks in metal castings are the result of stress or strain in a section of the casting. This stress or strain finds a weak point in that section of the casting and causes it to distort or separate at that point (Figure 8-1). Such stresses or strains in castings can develop from the following:

- Pressure or temperature changes during the casting procedure, which cause internal material structure defects, inclusion, or voids.
- Fatigue resulting from fluctuating or repeated stress cycles, which begin as small cracks and progress to larger ones under the action of the stress.
- Flexing of the metal due to its lack of rigidity.
- Impact damage by a solid, hard object hitting the engine.
- Constant impacting of a valve against a hardened seat, which produces vibrations that could possibly lead to a fracture in a thin-walled casting.

FIGURE 8-1 Example of stress cracks

- Chilling of a hot engine by a sudden rush of cold water or air over the surface.
- Excessive overheating due to improper operation of an engine system.

No matter what caused the crack, the important job is to relieve the stress at the point of distortion or cracking and then add more metal and move it in such a way to close the crack. This can be accomplished by the cold process of pinning or the hot process of welding.

Basically, there are two kinds of cracks:

- *External.* Those that can be seen on the outside of the casting.
- *Internal.* Those that cannot be seen on the surface of the casting.

 SHOP TALK

Installing heat tabs is a good way to protect the rebuilding engine shop against false claims due to engine overheating caused by other engine system failures (Figure 8-2). The tab can be installed with aluminum paste or adhesive. At 255 to 260 degrees Fahrenheit, the center of the tab will melt and fall out—evidence of overheating.

DETECTING CRACKS

With either method of repair, detecting a crack in a metal casting means not only finding that a crack exists but also determining the exact location and the extent of the crack and chalking or otherwise marking it so that it can be repaired.

The three most common methods of crack detection are:

1. Using a magnet and magnetic powder (frequently referred to as the magna flux process)
2. Using penetrant dye (developed specially for nonmagnetic castings such as aluminum heads and blocks)
3. Pressurizing the head or block with a pressure tester

FIGURE 8-2 Installing a heat tab is a good way to be sure that the engine has not overheated.

MAGNETIC CRACK DETECTION

With castings that react to magnetic charges, such as cast-iron engine heads or blocks, cracks often can be located by using a magnet and magnetic powder (Figure 8-3). The procedure is as follows:

1. Thoroughly clean the surface of the casting to be checked. For maximum effectiveness, the casting must be completely free of dirt and grease and must be dry.
2. Plug the magnet input cord into an electric outlet.
3. Position the magnet on the casting in the area to be tested and activate the magnet with the power switch.
4. Using the powder spray bulb filled with magnetic powder, dust powder over the surface of the casting between the two poles or legs of the magnet (Figure 8-4).

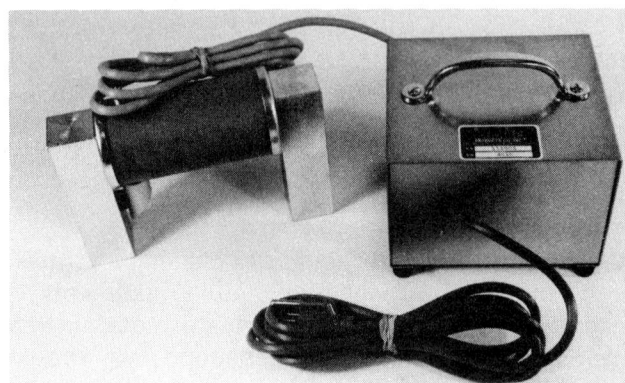

FIGURE 8-3 Permanent magnet crack inspection tester and power source (*Courtesy of Irontite Products Company, Inc.*)

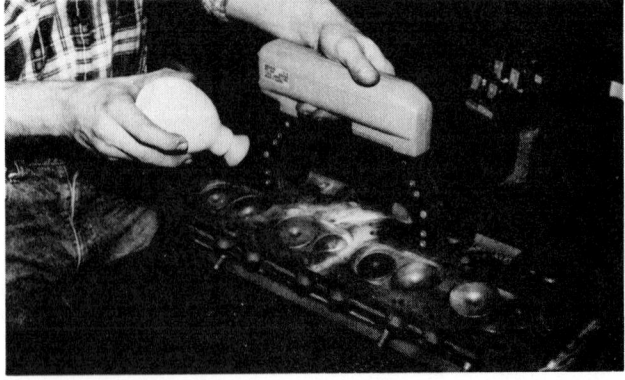

FIGURE 8-4 Applying magnetic power to the surface of a casting (*Courtesy of Magnaflux Corporation*)

The magnet creates a magnetic field in the casting, running from one pole to the other. If there is a crack in the casting between the two poles, it breaks the field and draws an accumulation of magnetic powder, which sharply outlines the crack (Figure 8-5). Since the crack appears only when it runs across the magnetic field between the poles of the magnet, it is necessary to move the magnet to different positions on the surface of the casting to do a thorough job of crack detection.

When the magnetic powder shows the existence and location of the crack, it is very important to carefully outline it with chalk so it can be easily identified after the magnet is removed (Figure 8-6). It is also important to move the magnet along the ends of the crack to find its limits. Frequently the crack that appears initially does not totally relieve the stress in the casting, and it is necessary to go beyond the apparent limits of the crack to relieve the stress.

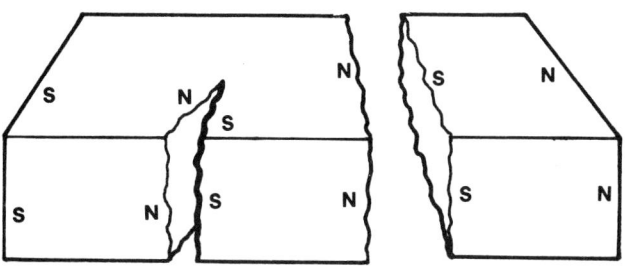

FIGURE 8-5 Since opposite and attracting poles when magnetized will be established across breaks in a casting, the iron powder will be drawn into the crack.

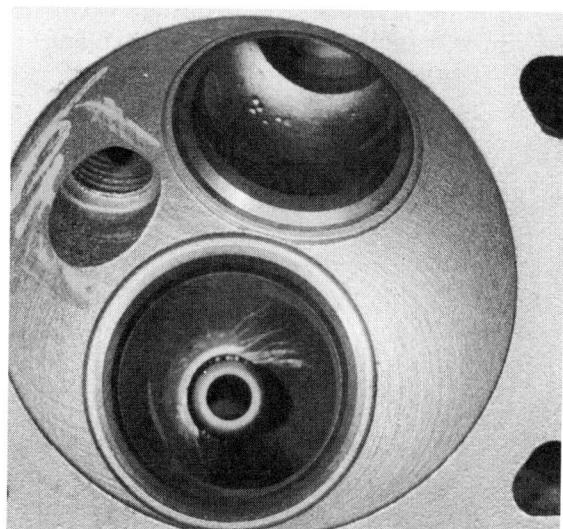

FIGURE 8-6 Mark the cracks with chalk for easy identification.

Magnetic Fluorescent Inspection

This detection method uses a magnetic particle paste rather than dry iron powder. The magnetic/fluorescent paste—mixed with either water or oil as directed on the container—is applied to the part in the suspected area by dipping or spraying. The part is then placed in a magnetic field and viewed under black light in an inspection booth (Figure 8-7). Under black light, any cracks will show up as white, gray, or yellow streaks (Figure 8-8). The magnetic fluorescent method is much more sensitive than dry testing viewed in normal light. More information on the use of black light to detect cracks in cam and crankshaft can be found in Chapter 12.

DYE PENETRANT CRACK DETECTION

Dye penetrant is normally used to find cracks in aluminum heads (Figure 8-9). It can also be used on cast-iron parts, though the cost of test materials makes it more expensive than magnetic particle detection. The dye penetrant process requires four steps:

1. Remove all dirt, grease, moisture, and oil from the surface of the casting to be checked.

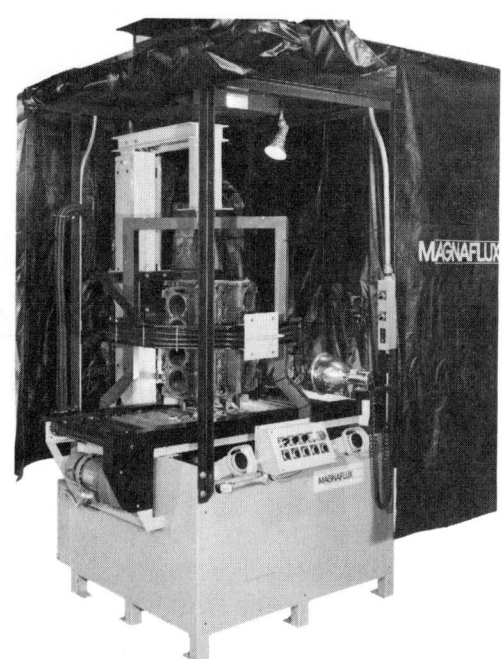

FIGURE 8-7 A magnetic fluorescent inspection booth *(Courtesy of Magnaflux Corporation)*

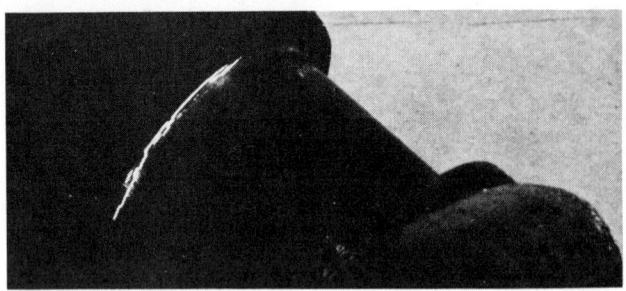

FIGURE 8-8 The white streaks indicate cracking in the fillet area of the crankshaft. *(Courtesy of Magnaflux Corporation)*

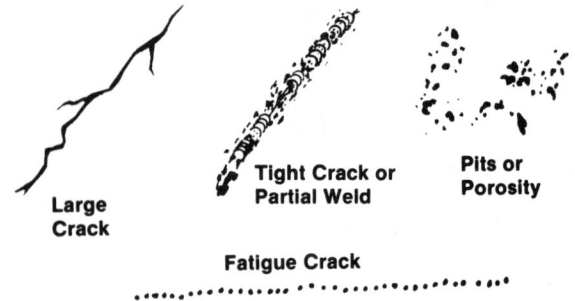

FIGURE 8-11 Patterns of cracks as they appear when penetrating dyes are applied

FIGURE 8-9 Dye penetrant crack detection kit *(Courtesy of Irontite Products Company, Inc.)*

2. Spray dye on the surface and allow it to dry (Figure 8-10A) about 3 minutes (slightly longer in cold weather).
3. Wipe excess dye dust from the casting (Figure 8-10B). Then spray the surface with remover and immediately rinse with clean water and wipe dry.
4. Spray the surface lightly with the developer—just enough to wet the surface of the casting. As the developer dries any dye left in a crack will bleed through the developer, showing up very distinctly as a red line on a white background (Figure 8-10C). Figure 8-11 shows the various patterns of different imperfections when treated with the dye detection system. Another dye system

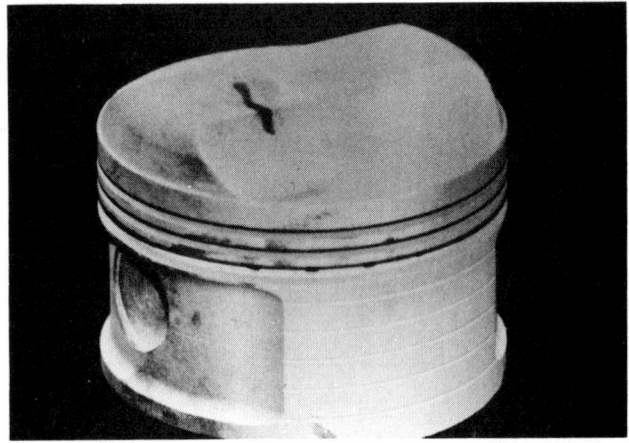

FIGURE 8-12 Liquid penetrant method of testing used on this piston reveals cracks in vivid red. *(Courtesy of Magnaflux Corporation)*

frequently used to detect cracks employs a fluorescent penetrant. It is applied in the same manner as dye penetrant except that the part must be inspected with black light (Figure 8-12).

PRESSURE TESTING

Pressure testing is fast becoming the most popular method used by rebuilding shops to locate cracks, especially for aluminum cylinder heads. Although both magnetic and dye tests do a fine job detecting external cracks, they are not effective for locating internal ones. Pressure testing performs well in detecting both external and internal cracks. This is very important when inspecting thin-walled castings.

There is a variety of pressure testing equipment available, ranging from plates (Figure 8-13) to bench models (Figure 8-14) and combinations thereof. The pressure medium may be air, water, or a combination of both.

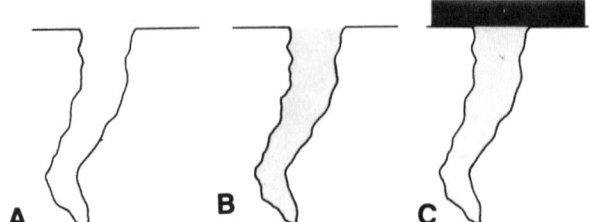

FIGURE 8-10 Method of applying a dye penetrant

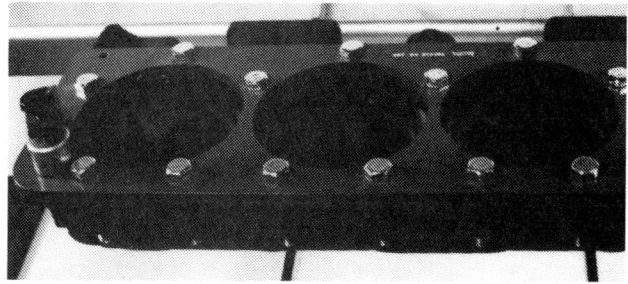

FIGURE 8-13   Pressure testing plates *(Courtesy of Irontite Products Company, Inc.)*

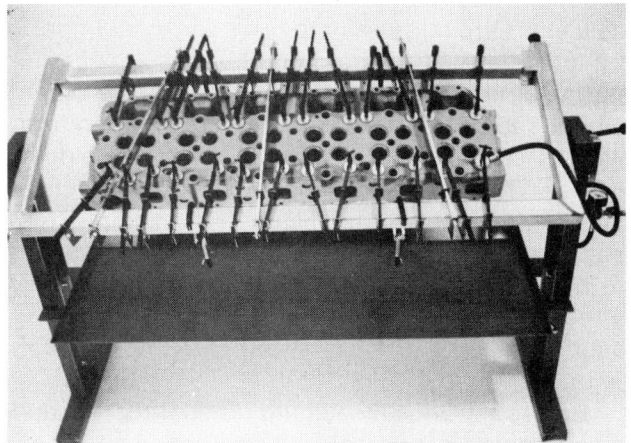

FIGURE 8-14   Pressure testing bench *(Courtesy of Irontite Products Company, Inc.)*

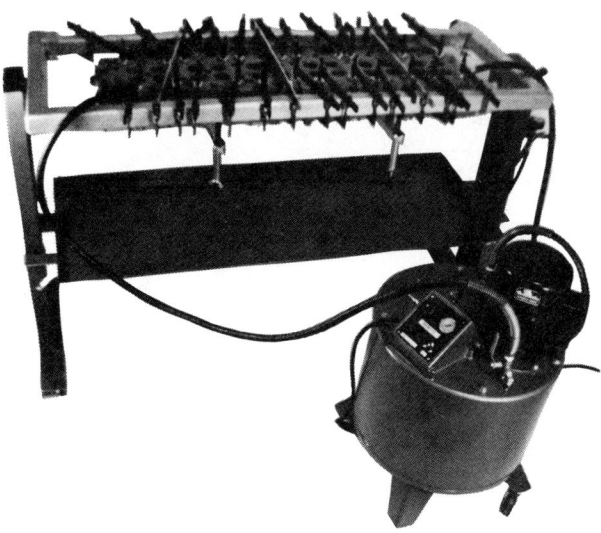

FIGURE 8-15   Typical water pressure tester circulator tank and pump *(Courtesy of Irontite Products Company, Inc.)*

FIGURE 8-16   Test plates are designed especially for the heads. *(Courtesy of T. Hoff Manufacturing, Inc.)*

Air Pressure Testing

When using air, the ports on the head or block are closed off for pressurizing, using a correct head gasket and a bolted-on steel plate. Pressure the casting by applying 40 to 60 psi into the water jackets from an air pressure regulator. Then spray the surface of the casting with a soapy solution. If there are cracks in the casting, the escaping air will cause the soapy solution to bubble up at the crack location. Mark this and any other spots where the bubbles appear.

Water Pressure Testing

Many rebuilders prefer using hot water rather than air to pressurize the casting. The reason for this is that they believe hot water will cause the metal to expand and give a more sensitive test. But, to use the hot water, a circulator tank and pump (Figure 8-15) are needed. Water will appear on the surface of the casting in the area of crack(s).

Universal Pressure Testing

The so-called universal system combines the principles of both air and water testing. That is, once the pressure plate designed for the head being tested (Figure 8-16) is bolted on the head, air is supplied to the assembly (Figure 8-17). The head and plate assembly is submerged in a tank of water. Once the regulated air supply is turned on, watch for a stream of bubbles to point directly to the crack regardless of the location. The universal system is usually a bench model that comes with a tank and all necessary controls (Figure 8-18). Some units are available with a rollover fixture for V-type engines (Figure 8-19).

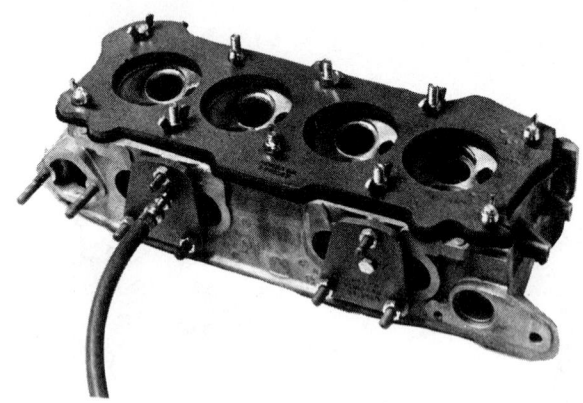

FIGURE 8-17 Test plates installed on the engine and water source connected *(Courtesy of T. Hoff Manufacturing, Inc.)*

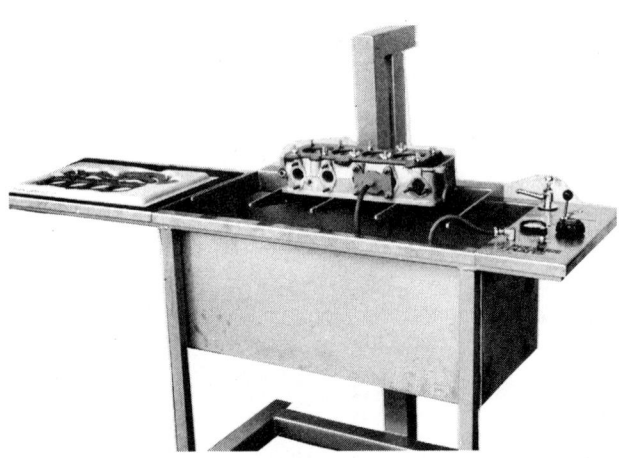

FIGURE 8-18 Universal system bench model *(Courtesy of T. Hoff Manufacturing, Inc.)*

FIGURE 8-19 Bench with a rollover fixture for V-type engines *(Courtesy of T. Hoff Manufacturing, Inc.)*

After the crack(s) have been repaired, pressure test the head or block, following the same procedure as when looking for cracks. This will provide a quick test to make sure that repairs were properly made and that the casting's integrity has been restored. Some repairs can be made with the pressure plates in place.

COLD CRACK REPAIR— PINNING PROCESS

Pinning is a very popular cold crack procedure because it requires no expensive equipment and the basic skills are fairly easy to learn. Another important advantage of the cold crack repair process is that metal is added and moved to close the crack without subjecting the casting to high temperature changes such as those that occur in the welding process. This avoids altering the physical characteristics of the metal in the casting. For example, the heat treating characteristics of the original casting are not altered when a crack in the casting is repaired by the cold process. The hardness of the casting remains exactly as it was before the repair of the crack.

Before starting any pinning procedures, it must be determined if the crack can be stopped and contained within its present confines. Another consideration is whether or not the crack is in a high stress area that could cause it to open up again. This is why combustion chamber roof cracks and cracks subject to valve spring pressures are risky to repair by pinning (Figure 8-20). Consideration must be given to the location of the crack where overheating is a possibility that could cause the crack to open up. Also remember that a pinned repair area must be reached by an air hammer for the final peening necessary in the pinning operation.

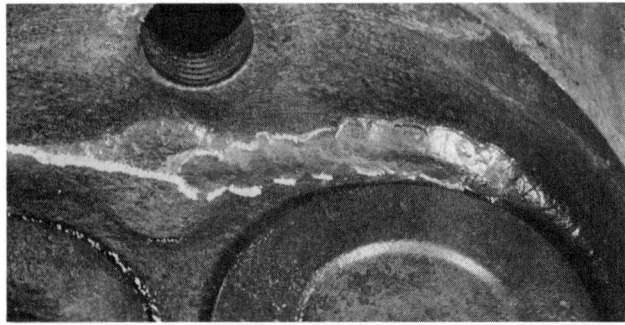

FIGURE 8-20 This head had been previously repaired by pinning. Several thousand miles later, here are the results of high stress. Many rebuilder shops will not attempt to pin this type of crack.

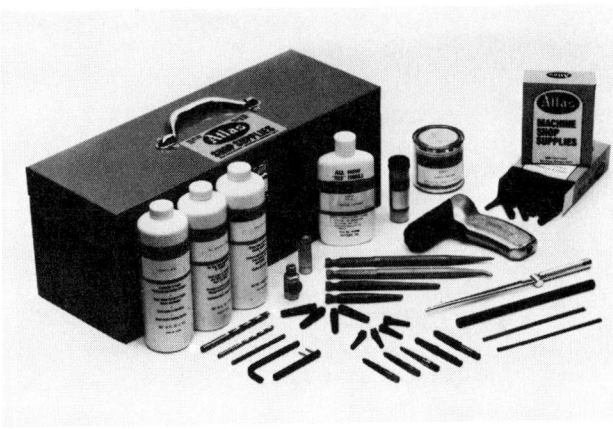

FIGURE 8-21 Typical kit used in the pinning process *(Courtesy of Atlas Engineering and Manufacturing, Inc.)*

FIGURE 8-22 Tapered cast-iron plugs come in various sizes. *(Courtesy of Irontite Products Company, Inc.)*

There are several cold process kits available, all of which contain the same basic items: twist drills, tapered reamers and taps, threaded plugs and rods, peening tools of various sizes, and small air hammers or power driver (Figure 8-21). Tapered cast-iron plugs that are used to repair casting cracks are available in many sizes (Figure 8-22).

Figure 8-23 illustrates the step-by-step procedure for installing a tapered plug in a valve seat crack. Proceed as follows:

1. Prior to drilling for the pinning procedure, a center punch mark should be made in the outside end of the valve seat crack (Figure 8-23A).
2. Start the drill straight into the center punch marking (Figure 8-23B), then drop down at an angle. The plugs should be installed at

an angle to the casting surface, not perpendicular to the casting surface.
3. Observe the angle of the hole (Figure 8-23C). After drilling, check the area for any remaining cracks.
4. Power tap the plug into the hole (Figure 8-23D). Be sure to keep the plug well lubricated.
5. Screw the cast-iron taper plug well into the hole (Figure 8-23E). The plug should be dipped in a ceramic seal before it is torqued into place.
6. The taper plug is installed in the hole (Figure 8-23F).
7. After notching the plug with a "peanut" grinder, use a hammer to break off the plug (Figure 8-23G).
8. Once the taper has been broken off, grind the remaining plug end (Figure 8-23H).
9. Figure 8-24 shows the machined counterbore for a new seat insert. A locking sealant is applied to the bottom of the counterbore just before driving in the new insert. Some final chamber and seat pocket detailing will be required after the insert is installed.

These steps illustrate the repairing of cracks with a single tapered plug. There are times when it is necessary to use several plugs along the line of a larger crack in areas of the casting that are subjected to high pressure and high temperatures. In a situation such as this, the plugs are at an angle in an overlapping fashion (Figure 8-25). Normally, one plug hole is drilled, reamed, and tapped; then the plug is fully torqued in place and the excess cut off before the next hole is started. This permits a better location of the next plug so that it will overlap the preceding plug not only at the surface but below the surface as well.

Where the metal in the casting is relatively thick, after drilling the hole it should be reamed with a tapered reamer before tapping with the tapered tap. With the tapered tap, the full length of the tap is cutting, and the hole must be tapered if the tap is to have a reasonable life.

The next step after installing the plugs and cutting off the excess ends is to peen the crack. The proper use of the air hammer and peening tools to help close the crack is an art that is learned with practice.

When peening the metal in the crack area, always peen in toward the centerline of the crack. Do not peen down the middle of the crack or away from the line of the crack. After the tapered plugs have been installed and cut off, as an added measure,

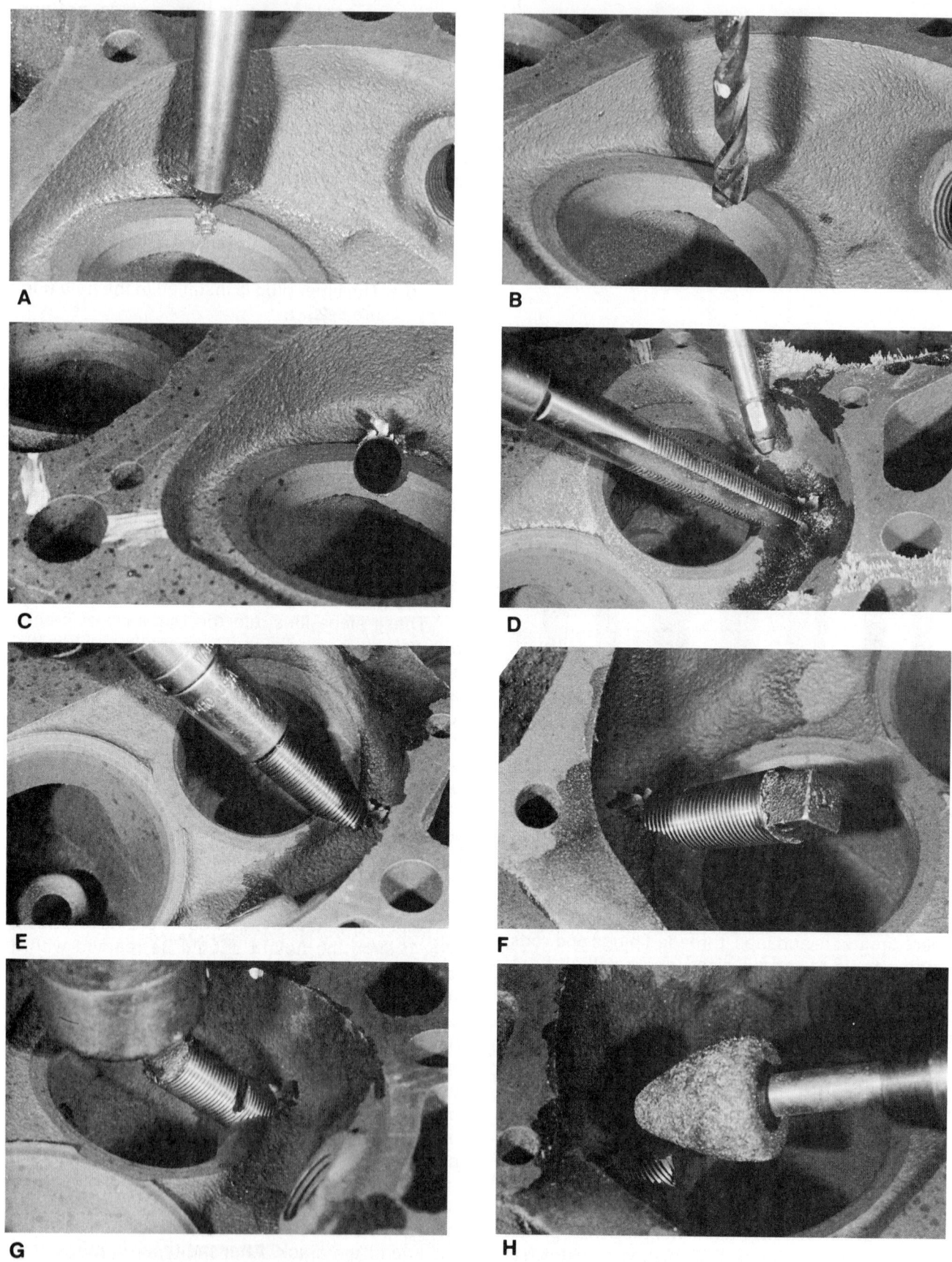

FIGURE 8-23   Steps in pinning a valve seat crack.

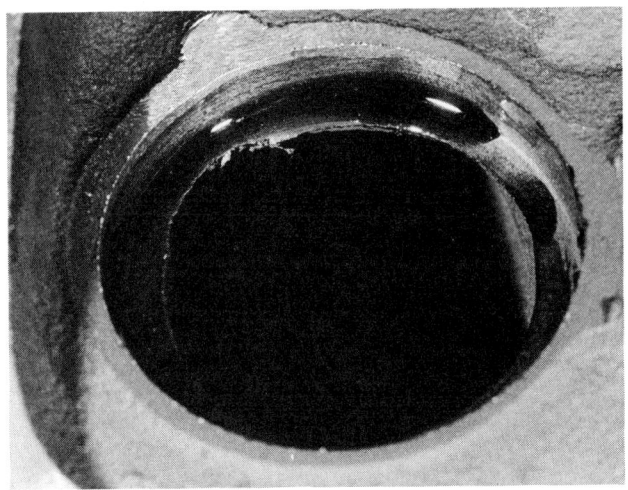

FIGURE 8-24 The machined counterbore for a new valve insert. Proper adhesive is applied to the bottom of the counterbore just before driving in the new insert. Some final combustion chamber and seat pocket detailing will be required after the insert is installed.

FIGURE 8-25 Installing multiple plugs to repair a larger crack

peen the stub end of the plug, always peening outward toward the outside of the plug. It is important to make sure there will be no leak at the thread.

EXTERNAL CRACKS

Tapered plugs can be installed along the line of the crack in readily accessible areas of the casting that are not subject to high pressure and high temperatures. When pinning such cracks, keep in mind that metal castings, like most materials, have a degree of elasticity. The technique installs the plugs along the line of the crack in a manner that takes advantage of the elasticity in the metal. This method is ideal for external cracks.

After "capturing" the crack at both ends and installing plugs at each end, follow along the line of the crack and drill, ream, and tap holes. Space the plugs 1/4 to 1/2 inch apart. Tight interlocking of the plugs that is so necessary for internal crack repairs is usually not needed for exterior ones.

If the metal at the point of repair is thin, normally it is not necessary to use the tapered reamer before tapping. However, if the metal is more than 1/4 inch thick at the location of the drilled holes, the tapered reamer should be used. With the tapered tap, the full length of the tap is cutting, unlike a straight tap where the lead threads do all of the cutting. Accordingly, where the metal is very thick, it is necessary to ream the hole with a tapered reamer if the tapered tap is to last for a reasonable length of time.

Along the line of the crack, torque the tapered plugs concurrently so that none will be loosened as other plugs are torqued. As the plugs are torqued, the crack will be forced open farther. This builds up a counter elastic pressure in the metal to push the crack closed.

After the plugs are torqued into place, cut them off about 1/16 inch above the surface of the casting. Do this by putting a nick in one side of the plug with a small saw and breaking off the upper part of the plug by tapping it lightly with a hammer. Peen the crack thoroughly from both sides, always toward the centerline. Then peen the stub ends of each plug, always peening outward toward the thread of the plug. Here the adding of the metal in the form of the tapered threaded plugs, the moving of the metal from the peening, and the built-up elastic pressure of the metal in the casting to spring back all combine to close the crack.

USING PLUGS ON BOTH SIDES OF A CRACK

It is sometimes possible to lace both sides of cracks in areas that are not subject to high pressure and high temperature. Using a center punch, spot along both sides of the crack where you intend to drill and tap and install the tapered plugs. The holes should be spotted very close to the crack about 1/8 inch apart in a lacing fashion alternately on the two sides of the crack. After spotting the locations for the plugs along the line of the crack and before beginning to drill holes for the plugs, the entire crack should be peened, always peening toward the center.

After the crack has been peened, drill and tap the holes for the plugs. Figure 8-26 shows such a

FIGURE 8-26 Drilling holes on both sides of a crack *(Courtesy of Irontite Products Company, Inc.)*

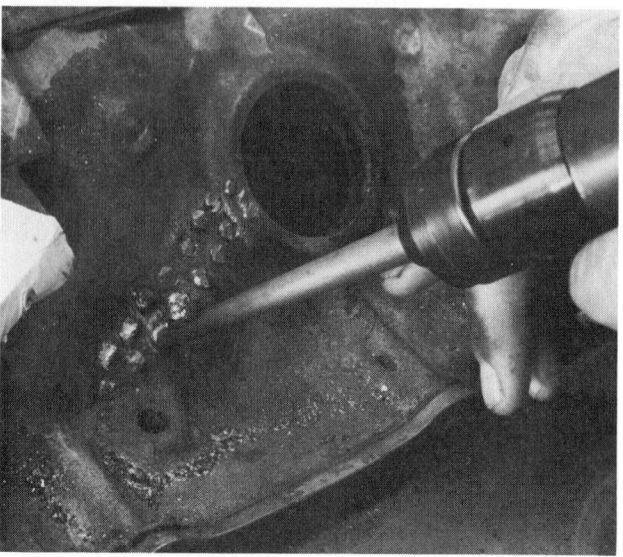

FIGURE 8-27 Peening the stub ends of the plugs *(Courtesy of Irontite Products Company, Inc.)*

repair with the holes drilled and tapped. Here, where the metal in the casting is relatively thin, it is not usually necessary to use a tapered reamer before tapping the holes.

After the plugs are torqued into place, cut off the excess extending above the surface. To do this, notch the plugs with a small saw about 1/16 inch above the surface and break them off by tapping lightly with a hammer. Next, peen the crack again as well as the stub ends of the plugs (Figure 8-27).

In this type of repair, the crack is closed only by moving the metal in the casting caused by torquing the plugs alongside of the crack and by the peening. Hence, the location of the plugs and the peening of the casting are very important.

Some crack repairs use special metal alloy rods available in diameters from 1/8 to 3/8 inch. Once the repair is drilled and tapped, a rod of the correct size for the repair is threaded for a short distance with an adjustable die and screwed tightly into the threaded hole. The unthreaded portion of the rod is then removed and another portion threaded for still another hole beside the first. Especially for repairing small cracks on flat surfaces and down in valve ports, the rods offer a more selective choice as to size and economy than do the tapered threaded plugs.

CERAMIC SEAL

As a part of the overall cold repair process, a ceramic seal should be applied. This can be accomplished by dipping the threaded repair plug or, better still, applying the sealant after the repair process is completed. It gives the added assurance that the crack is closed. To accomplish this task, the head or block is remounted on the pressure tester and the

water ports are sealed so that the casting can be pressurized. After pressurizing the head or block and confirming the high quality of the crack repair, secure the outlet and inlet hoses on the circulator to the opposite ends of the head or block. Activate the circulator and then circulate the ceramic seal through the casting at the temperature that is supplied by the circulator.

After circulating the casting for 15 to 30 minutes, close off the return valve and depress the air pressure button for a few seconds. This pressure forces the ceramic seal in the casting out into the interior of any cracks and into any porosity in the interior of the casting. Next, turn the head over, open the circulator return valve, and press the air pressure button for only a few seconds. This will force the remaining liquid ceramic seal in the casting back into the circulator tank.

Remove the casting and set it to one side to allow the ceramic seal in the interior of the casting to set or fix. After a period of time, the ceramic seal will fix on the inside of the casting and serve as an additional seal to the crack. The time required for the seal to set or fix varies depending upon the temperature and humidity.

EPOXY REPAIR

Another cold process is to use epoxy repair that is suitable for minor external cracks and casting sand holes. Porosity or sand hole(s) causing oil seepage or leakage can occur with modern casting processes. A complete inspection of the engine should be made. If the leak is attributed to the por-

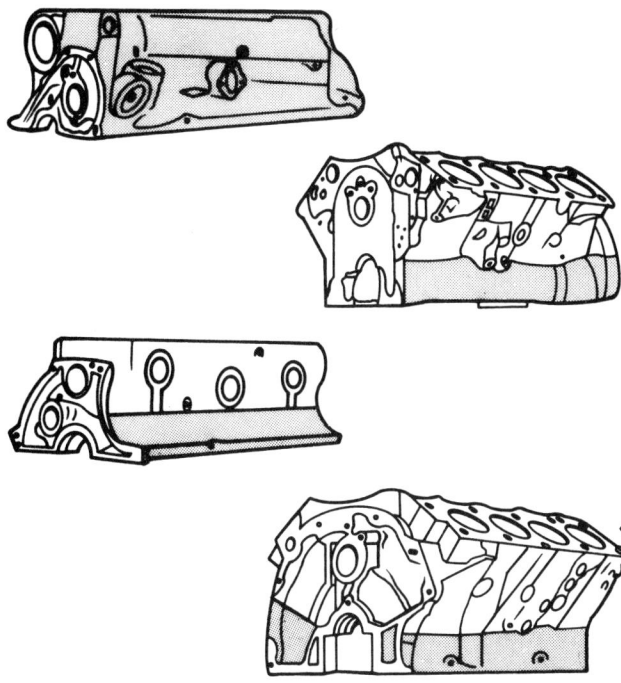

FIGURE 8-28   Shaded areas can often be replaced with epoxy materials.

ous condition of the cylinder block or sand hole(s), repairs can be made with epoxy resin (or metallic plastic). However, do not repair cracks with this material. Repairs with this metallic plastic must be confined to those cast-iron engine component surfaces (Figure 8-28) where the inner wall surface is not exposed to engine coolant pressure or oil pressure. For example:

- Cylinder block surfaces extending along the length of the block, upward from the oil pan rail to the cylinder water jacket but not including machined areas
- Lower rear face of the cylinder block
- Intake manifold casting. Repairs of the intake manifold exhaust crossover section are not recommended because temperature can exceed the recommended limit of metallic plastic (500 degrees Fahrenheit).
- Cylinder front cover on engines using cast-iron material
- Cylinder head, along the rocker arm cover gasket surface

The following procedures should be used to repair porous areas or sand holes in cast iron:

1. Clean the surface to be repaired by grinding or rotary filing to a clean, bright metal surface. Chamfer or undercut the hole or porosity to a greater depth than the rest of the cleaned surface. Solid metal must sur-

round the hole. Openings larger than 1/4 inch should not be repaired using epoxy resin. Openings in excess of 1/4 inch can be drilled, tapped, and plugged using common tools. Clean the repair area thoroughly. Metallic plastic will not stick to a dirty or oily surface.

2. Mix the epoxy resin base and hardener as directed on the container. Stir thoroughly until uniform.

3. Apply the repair mixture with a suitable clean tool (putty knife, wood spoon, or other suitable item), forcing the metallic plastic into the hole or porosity.

4. Allow the repair mixture to harden. This can be accomplished by two methods; heat cure with a 250-watt light placed 10 inches from the repaired surface as directed on the package label, or air dry for 10 to 12 hours at temperatures above 50 degrees Fahrenheit.

5. Sand or grind the repaired area to blend with the general contour of the surrounding surface.

6. Paint the surface to match the rest of the block.

SHOT PEENING

As mentioned in Chapter 5, blasting media can be used to peen metal surfaces as well as clean them. When a metal surface is blasted with round shot or glass beads, the impact of the media dimples or craters the surface, compressing and hardening a thin layer of surface metal. This helps seal microscopic surface cracks while strengthening the metal; greatly improved fatigue life and corrosion resistance result. Figure 8-29 illustrates a machine that is suitable for both blast cleaning and peening.

Though the whole surface of a part can be shot peened, generally only the areas of highest stress are peened. Shot peening the journal fillets on an ordinary cast-iron crankshaft, for example, can make it up to 50 percent stronger, allowing it to be used in place of a more expensive forged steel crank. Peening is also used to improve durability on valve springs, connecting rods, gears, camshafts, rocker arms, rod and head bolts, and other highly stressed parts. Peening can even be used to help seal porosity leaks in aluminum castings, a process that some say is equal to or even more effective than resin or epoxy impregnation. The shallow indentations left in the surface by peening can also serve as lubricant reservoirs on camshafts and gears to reduce galling and friction. Peening can also be used to relieve stress after welding.

FIGURE 8-29 Typical blasting media cleaning and peening machine

Some rebuilders do not consider shot peening as something worth doing except on certain heavy-duty or high-performance engine or drivetrain parts. Yet it can be an effective means of increasing the reliability of used parts, reducing the chance of a failure and solving problem applications where there is a history of breakage. It is important to understand, however, that shot blast or glass bead cleaning is not shot peening. Though there will be some peening effect when cleaning, peening is a controlled process that involves creating a uniformly compressed surface layer in critical areas of a part.

To determine how much force the shot or glass bead is imparting to the surface when it strikes, an Almen test strip is mounted on the part. The test strip is a small piece of metal of known thickness and hardness. As the test strip is peened, the surface layer is compressed, causing the strip to bow upward. The amount of arc in the test strip is then measured with a feeler gauge to determine the peening effect that is being delivered. Once this is known, the speed, size, angle of delivery, and/or type of shot can be varied until the desired results are achieved. The reason for testing is to determine the point at which maximum saturation is achieved. This is when the surface of the metal reaches maximum strength and further peening will not accomplish anything more.

The size and weight of the shot has a lot to do with how quickly the peening effect is achieved (see Chapter 5). That is why steel shot is often used instead of glass. The hardness of the shot is also a factor in influencing the depth of the compressive

layer, with harder shot creating a deeper, stronger layer. Glass beads are used not only for peening jobs where iron residue on the surface would be a problem, but also for peening in a very sharp radius such as a thread root.

Some general rules on peening:

- Consider peening anything that moves, is heavily loaded, and has a history of failure.
- Always peen crankshafts, if the crank is being chrome plated, prior to plating. It is also a good idea to re-peen the journal fillets on reground cranks.
- Consider peening welded repair areas on aluminum heads, blocks, and housings to relieve stress and reduce the chance of re-cracking. Peening can also be used to relieve surface stress after grinding.
- Do not remove more than 10 percent of a peened surface by grinding or polishing.
- Do not heat peened parts above their stress-relieving temperatures or the peening effect will be relaxed and lost (around 250 degrees Fahrenheit for aluminum and 500 degrees Fahrenheit for carbon steel).

REPAIRING CRACKS BY WELDING

There are several welding processes that can be found in the automotive machine shop. These processes include:

- *Gas-Oxyacetylene.* This process is one of the more common methods for joining metals (especially for cast-iron and steel parts) in the rebuilder shop (Figure 8–30). Basically, the process uses oxyacetylene gas and oxygen with a torch to heat the base material and melt the filler rod that is to be added to the weld joint. This process is commonly used to braze weld cast iron or steel with bronze filler rod, which contains about 60 percent copper and 40 percent zinc and has a melting point of approximately 1600 degrees Fahrenheit. This process is more of a joining process than a welding process, since the base material is not melted. The gas-oxyacetylene process is also used in the fuse welding of gray cast iron, using gray cast-iron filler rod, and is commonly used to repair such items as heads, blocks, crankshafts, and many other components. The part to be welded is preheated 900 to 1200 degrees Fahrenheit, which is indicated by the metal turning a dull red in the 900- to

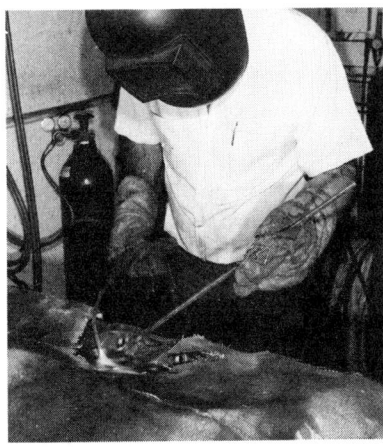

FIGURE 8-30 When oxy-welding, keep the welding rod hot and down in the heated area, ready to flow into the head casting.

FIGURE 8-31 Welding an aluminum head by the TIG welding process

1050-degree Fahrenheit range. This process gives a weld deposit similar to that of the base metal, since cast-iron filler rod is used. It is important to slowly heat and cool cast iron to prevent cracking and hardening.

- *Shielded Metal Arc.* A second welding process generally found in some shops is the shielded metal arc or stick electrode. An electric arc is struck between the rod and the metal to be welded. The metal core of the rod provides the filler material and the coating produces a gas atmosphere and slag coverage to protect the molten weld deposit. Alloying elements may be added to the weld deposit through the coating. This process is commonly used on carbon and alloy steels and also may be used for nickel rod welding on cast iron. It is important to keep the electrodes dry and prevent the coating on the rod from absorbing moisture. This can be done by keeping the electrodes in a drying oven. A simple drying oven can be made by installing a light bulb in an old refrigerator that will stay warm enough to keep the electrodes dry.

- *Tungsten Inert Gas.* Another welding process is the tungsten inert gas process, also known as the TIG process, used in many shops to do aluminum welding such as on heads (Figure 8-31). An electric arc is struck between a tungsten electrode in the torch and the base material when the filler material in the form of wire or rod is fed into the arc (Figure 8-32). The molten weld deposit is protected by an inert gas such as helium, which flows through the torch, around the tungsten electrode, and over the weld deposit. The TIG process can be used on a variety

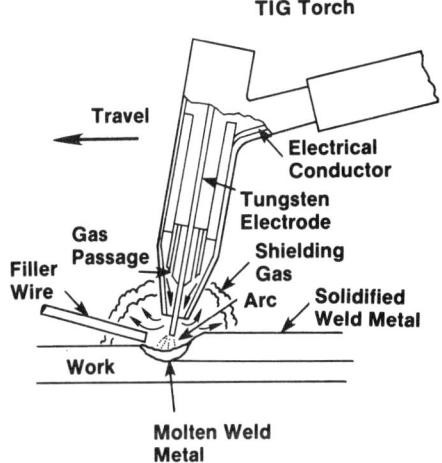

FIGURE 8-32 Principles of the TIG process. If a filler metal is required, it is fed into the pool from a separate filler rod.

of metals such as steel and aluminum, as well as other metals. It also has the advantage of producing a good quality weld that is clean and free of slag.

- *Gas Metal Arc.* The gas metal arc welding process, or MIG, is used in many shops to build up crankshafts. It is similar to the TIG process except that the electrode is the filler wire that is to be added to the weld deposit and the arc is struck between the filler wire and the base material. The weld deposit is also protected by a gas coverage such as argon or some other protective gas. MIG can be used on steel, aluminum, and several other metals. The process produces a fairly clean weld deposit and is relatively fast with a high deposition rate. It is possible to add alloying elements through the filler wire, which might have a center core. The MIG process is very popular in auto body repair, especially for work on unibody cars.

- *Submerged Arc.* The submerged arc welding process can also be found in some shops for joining and for the buildup of shafts. It is similar to the MIG process in that the filler wire acts as the electrode and an arc is struck between the base metal and the wire. However, in this case, protection of the weld deposit is provided by a flux, which flows around the wire to cover the weld areas. The process has a very high rate of deposition and is used on steels, especially where thick sections are to be welded and a large amount of weld is required. It produces a very high-quality weld deposit, and alloying elements can also be added through the flux. Further details on submerged arc welding in crankshaft repairs can be found in Chapter 12.
- *Flame Spraying.* Flame spraying surface buildup is used in a variety of applications such as surfacing the decks on blocks and repairing bearing bores. It is not a joining process, but rather a surface buildup method using a powder that is melted in a flame and sprayed onto the surface of the metal to be built up. It gives a partially fused and mechanical bonding to the surface of the metal that it is applied to. Plasma arc spraying is another process similar to flame spraying but gives a much higher quality deposit.

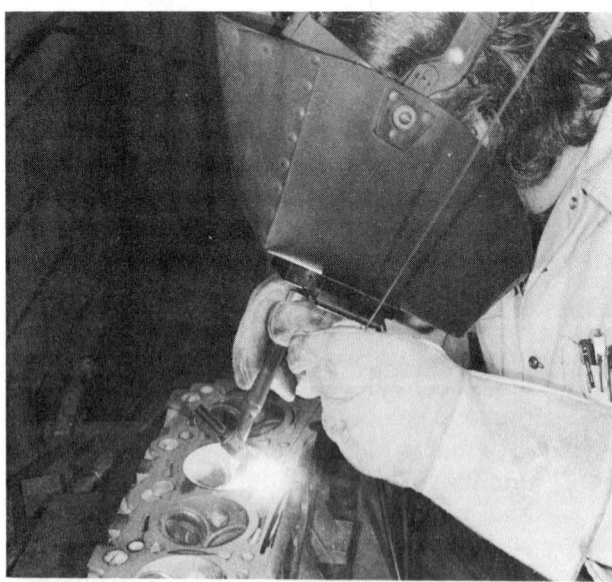

FIGURE 8–33 When welding be sure to wear eye protection.

WELDING SAFETY

When doing welding work, the following safety precautions must be kept in mind:

- Under no circumstance should a technician ever perform any kind of welding without proper eye and skin protection (Figure 8–33). This means welding masks and heavy, fire-resistant gloves and clothing must be worn.
- Exposure to arcs in arc welding, even in MIG and TIG welding, can give a technician what is known as "welder's flash" or "arc eye," a painful condition in which the eyes are burned by ultraviolet radiation. Repeated exposure can damage eyes permanently.
- Severe burns can occur from flying sparks and welding sparks, and from trying to pick up hot workpieces without gloves. An operator can also burn the skin through exposure to ultraviolet rays.

Besides burning the eyes and skin, there is the ever-present danger of accidents causing electrocution and explosion because of the high voltages and gases. To prevent accidents, follow these precautions:

- Make regular equipment inspections and replace old or damaged cables, connections, and gas hoses immediately.
- Never weld or cut near explosive liquids or vapors, gas tanks, oil barrels, or dirty rags.
- Always weld behind flame-resistant screens or in booths. This will protect other workers from flying sparks and flash burns.
- Never weld without adequate ventilation and make sure the welder is positioned so that any moving air takes fumes away from, not at, the operator.
- Always have a fire extinguisher nearby. In fact, it might be a good idea to have an extinguisher mounted on all welding carts.
- Mark all hot workpieces clearly so that no one else will accidentally pick them up and get burned.

When welding with gas, observe the following precautions:

- Do not store gas tanks near any heat sources.
- If you have oxygen tanks, store them at least 20 feet apart and keep them at least 35 feet away from oil and greases.
- Shut off all tank valves tightly when you are finished, and keep tank valve caps in place when tanks are not in use.

- Know the color codes for tanks. It is very important to know what kind of gas is inside each.
- Secure tanks with a chain or strap so they cannot fall. A tank with a broken valve is a lethal missile.

When arc MIG and TIG welding, observe the following:

- Spread out electrical cables while doing a job.
- Allow no splices within 10 feet of the electrode holder.
- Always check connections, ground the weldment, never weld in a wet area, and never use a wet welding machine until it has been dried out and tested.

HEAT OR FURNACE WELDING CRACK REPAIRS

Heat or furnace welding is considered by many people to be the best way to repair cracks in a cast-iron head. By preheating the entire casting the problem of stress cracks forming during the cooling-off period is eliminated. Heat welding, however, requires a good heat source and proficient welding skills. As a rule, it is difficult for the novice.

A description of the key elements of the heat welding crack repair process is as follows:

1. Slowly heat the casting to about 1400 degrees Fahrenheit using a firebrick and "log lighter" setup such as shown in Figure

8–34. When heating the casting, cover it with a welding booth curtain to help hold the heat in. Then make an access hole in the curtain over the area to be welded. When welding, try to stay in the 1200- to 1300-degree-Fahrenheit range. The casting will be a dull, cherry red color when it is ready to be welded.

2. Weld using an acetylene torch adjusted to a neutral flame and a pure cast-iron rod. Use a high-quality flux. Puddle the base metal and then melt the rod into the puddle.

3. After welding is completed, bring the temperature back up to 1400 degrees Fahrenheit for 30 minutes. Then shut down the heat and allow the part to cool at not more than 100 degrees Fahrenheit per hour. This procedure allows for proper stress relieving and normalizing.

WELDING ALUMINUM HEADS

Aluminum heads, as stated in Chapter 2, have become popular with vehicle manufacturers in recent years, primarily because of the weight saving they offer. However, there are many problems associated with an aluminum cylinder head. The typical rebuilder shop is most likely to encounter the following problems:

- Cracks in the aluminum between the valve seat rings (Figure 8–35). These cracks, usual-

FIGURE 8–34 Setup for heat welding cast-iron heads. The heads are set over the natural gas pipes and surrounded with firebricks.

FIGURE 8–35 Small casting cracks between the center two valve seats (see arrows) are common with aluminum heads. Small cracks coming from the coolant holes are also common.

ly quite small, require close inspection to find. They very seldom leak and can be closed by a light peening. Some shops make no repairs to them.

- Bottomside cracks coming from the coolant passages. These cracks can be repaired by veeing out the damaged area and welding with an aluminum filler rod.

- Topside cracks across the main oil artery. These cracks, although not too common, are usually very visible (Figure 8–36). Most authorities recommend replacing the head completely if such a crack is found. The length of time required to make the repair is not reasonable. The labor cost is not worth the risk of possible failure, and it is more expensive than the cost of purchasing an uncracked core.

- Detonation damage can occur on any cylinder (Figure 8–37). Repairs can be made by

welding and freehand machining with a rotary burr in a die grinder.

- Meltdown damage is a somewhat common occurrence on the high-swirl combustion chamber heads (Figure 8–38). Again, repairs can be made by welding and freehand machining.

- Coolant related metal erosion. If damage around coolant passages is excessive, if the side of any valve seat has been exposed, or if the combustion chamber shows erosion, the head must be repaired or replaced (Figure 8–39). Coolant erosion such as shown in Figure 8–40 can be easily fixed by welding and resurfacing.

While cast-iron head surface cracks can be successfully repaired by drilling and pinning, TIG welding is the preferred repair technique for aluminum heads (Figure 8–41).

FIGURE 8-36 A topside oil artery crack. The crack appeared when an oxyacetylene flame was passed over the casting. Carbon in the flame was trapped in the crack, highlighting it.

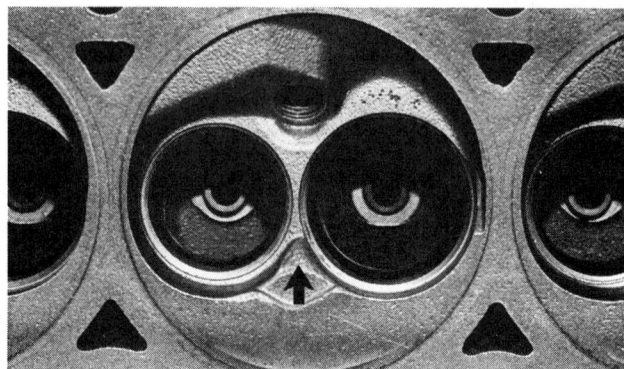

FIGURE 8-38 Typical example of meltdown damage that can be seen in the combustion chamber

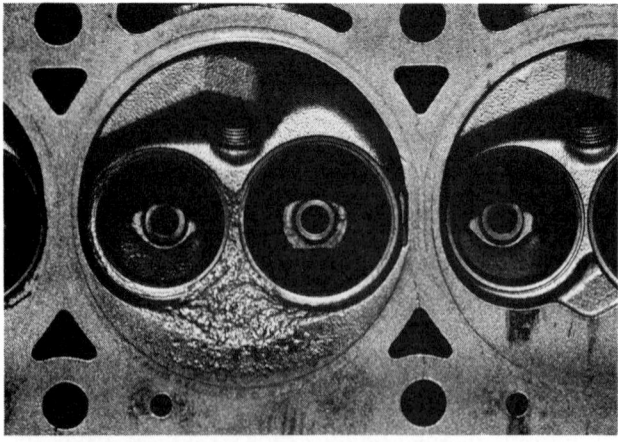

FIGURE 8-37 The effect of detonation on a combustion chamber. This can occur in any cylinder.

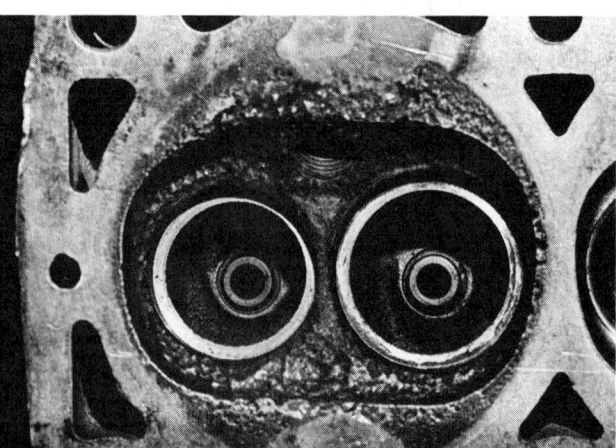

FIGURE 8-39 An example of severe coolant-caused damage to the combustion chamber. Such damage can be repaired, but it is very time consuming.

FIGURE 8-40   Coolant erosion damage (indicated by the arrows) that can be easily fixed by welding and resurfacing

FIGURE 8-41   TIG welding aluminum repair

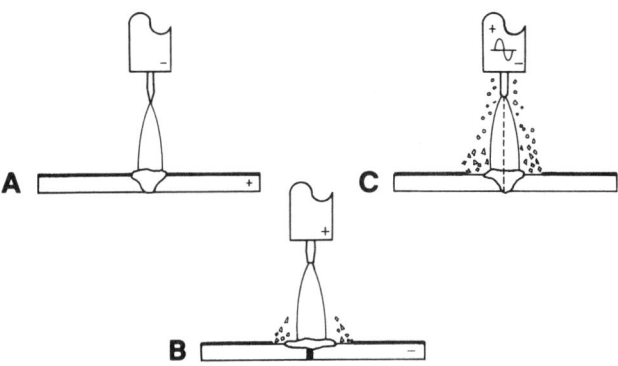

FIGURE 8-42   Types of TIG welding: (A) TIG welding with DCSP produces deep penetration because it concentrates the heat in the joint area. No cleaning action occurs with this polarity. (B) TIG with DCRP produces a good cleaning. Argon ions strike the surface, breaking up oxides. Electrons cause a heating effect at the electrode, making weld penetration shallow. (C) TIG with ACHF combines good weld penetration on negative and desired cleaning on positive half cycles. High frequency reestablishes arc, which breaks each half cycle.

Welding aluminum is often considered difficult because it welds differently than iron or steel. When exposed to air, aluminum forms an oxide coating on the surface that helps protect the metal against further corrosion. The oxide layer makes welding difficult because it interferes with fusing and weakens the weld. Cleaning the surface can remove the oxide, but as soon as the metal is heated it reforms—unless the weld is bathed in a constant supply of inert gas.

When an alternating current from the TIG welder is applied to the surface of the aluminum (Figure 8-42), it heats the metal during half of the voltage cycle (electrode negative) and cooks off the oxide during the reverse portion of the cycle (electrode positive). This back and forth heating and cooking action keeps the weld free from contamination and makes for a strong weld.

Aluminum welds are extremely vulnerable to any kind of contamination, so a constant supply of inert argon or helium shielding gas is needed to keep air away from the metal (Figure 8-43). It is important to preclean the surface area thoroughly with a stainless steel wire brush. This brush, by the way, should not be used for anything else. Use it only to clean aluminum.

A variety of different types of welding electrodes can be used with a TIG welder. Tungsten thorium (color-coded green) electrodes work best with aluminum. Zirconium tungsten works even better with aluminum but costs five times as much.

Depending on the technique, different rebuilders claim different filler rods should be used (Figure 8-44). However, one of the most popular filler rods used with aluminum-silicon alloys is ER4043.

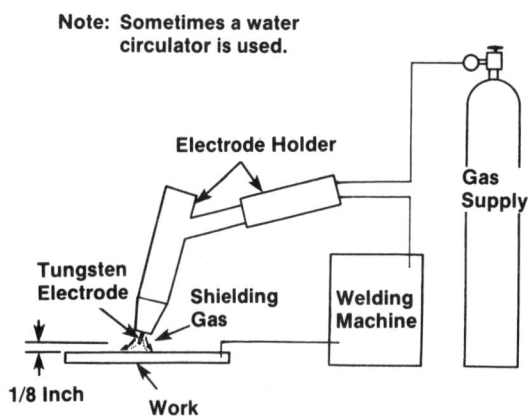

FIGURE 8-43   Schematic of TIG welding

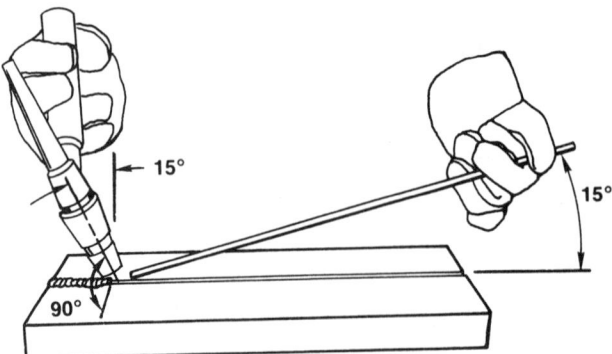

FIGURE 8-44 Proper positions of torch and filler rod for TIG welding

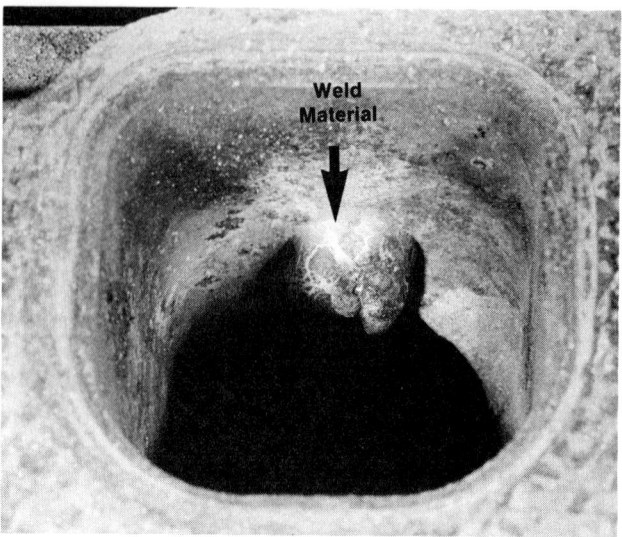

FIGURE 8-45 The welding material extends from the bottomside of the head (in a water jacket area) and out the exhaust port. When welding in a water jacket area, try to blow the crud out through the bottom of the area that has been ground out.

When repairing a crack in an aluminum head, it is very important that the full extent of the damage is identified so the crack can be completely ground out. Most shops use a carbide cutter with a lubricant for this procedure. Extend the cutting or grinding a short distance beyond the visible ends of the crack. As to depth, many shops will leave as much as 1/32 to 1/16 inch so the weld will sag and achieve good penetration. There are those experts, however, who say that the entire crack should be eliminated.

CAUTION: Aluminum shot is often used as the cleaning media in blasting machines for aluminum heads. If a bead remains in an engine, it will crush if it contacts a moving part.

After the crack has been cut out, any remaining oil or glass bead must be removed. An acetone or lacquer thinner can be applied to clean the head before preheating for welding. The important point is to eliminate as much contamination as possible. This can be a difficult problem, especially when making cuts between valve seats that can reach to the water jackets (Figure 8-45). If it becomes obvious that the customer has used a coolant sealant, for example, it might be better to reject the head rather than attempt to weld through heavy contaminant concentrations.

Prior to welding, the head must be preheated. Contrary to what some people think, the heat from the torch itself is not sufficient to raise the temperature of the head enough to make a quality weld. Preheating can be accomplished by placing the head in an oven at 350 to 400 degrees Fahrenheit for about 1-1/2 hours. (Use temperature-sensitive heat crayons to keep an eye on the temperature.) Pre-

heating accomplishes several things. It helps reduce the thermal stress during the welding process itself, and it also helps to cook out any oil or grease that might contaminate the weld.

Following preheating, welding should be done in an area designed to inhibit drafts from cooling the head too quickly, which could cause additional cracking and warpage. While welding, it is important not to touch the metal with the tungsten electrode; this can contaminate the electrode (Figure 8-46). Using the high-frequency AC setting, the electrode should be held about 1/8 inch above the working surface while welding. Any contaminants that bubble to the surface can usually be brushed away with a stainless steel wire brush used solely for this purpose.

 SHOP TALK

Repairing extensive crack conditions can be difficult. For this reason, many shops that find cracks extending from the valve guide out and down into the valley under the cam will not attempt to make these welds. They believe these welds take too long and new cracks can develop while trying to make the repair. If the head begins to cool before welds are completed, it should be reheated or new cracks might appear.

FIGURE 8-46   When welding with TIG equipment, do not let the electrode touch the metal.

After welding is completed, the head should be reheated in the baking oven to 350 degrees Fahrenheit for 15 to 20 minutes. Then shut off the oven and allow the heads to cool until they can be touched by hand. To increase productivity, some shops preheat three heads before welding. The oven is then shut off while the first head is removed for welding. The two remaining heads are kept in the oven until ready for welding.

 SHOP TALK _____

Many machine shops perform crack repair welding before attempting to straighten a warped cylinder head because welding can increase warpage (see Chapter 9). On the other hand, some rebuilders say it is better to straighten the head first to see if it is salvageable before spending a lot of time trying to weld cracks. Follow the procedure recommended in the shop where the work is being done.

Before milling the head to finish the repair (Figure 8-47), a light coating of wax or silicone spray can help prevent the aluminum filings from sticking to the cutter bits. Material removal should be limited to 0.001 inch per pass.

After the machine work on the repaired area is done (Figure 8-48), the head should once again be pressure checked to make sure the original crack has been sealed and no new cracks have opened up. This might seem like an unnecessary step, but if it can prevent comebacks, it is worth it.

FIGURE 8-47   After welding, combustion and seat recesses are machined back to the original contours.

FIGURE 8-48   This photo shows that the intake valve seat counterbore has been machined. Use lubrication on the cutter to prevent burring and enlarging the bore.

REVIEW QUESTIONS

1. Detecting a crack in a metal casting involves _____ .
   a. finding that a crack exists
   b. determining the exact location and extent of the crack
   c. chalking it
   d. all of the above

2. Which of the following is not a common method of crack detection?
   a. using an ion detector
   b. using the magna flux process
   c. pressurizing the head or block with a pressure tester
   d. both a and c

3. Cracks can be located by using a magnet and magnetic powder. Technician A says that a crack only appears when it runs in line with the magnetic field. Technician B says that a crack only appears when it runs across the magnetic field. Who is right?
   a. Technician A
   b. Technician B
   c. Both A and B
   d. Neither A nor B

4. Technician A says the magnetic fluorescent inspection method is more sensitive than dry magnetic testing. Technician B says the dry magnetic testing method is more sensitive than the magnetic fluorescent inspection. Who is right?
   a. Technician A
   b. Technician B
   c. Both A and B
   d. Neither A nor B

5. Why would a dye penetrant probably not be used on cast-iron parts?
   a. It is less effective on cast-iron parts.
   b. It stains cast iron.
   c. It is too expensive.
   d. None of the above.

6. Which of the following is the most effective method for locating cracks in iron and aluminum cylinder heads and engine blocks?
   a. magna flux process
   b. dye penetrant process
   c. pressurizing the head or block with a pressure tester
   d. none of the above

7. Besides air, which of the following is used to pressurize a head or block?
   a. helium
   b. hydrogen
   c. hot water
   d. none of the above

8. Stresses in castings can develop from which of the following?
   a. fatigue that results from fluctuating stress cycles
   b. temperatures that change too rapidly
   c. flexing of the metal due to the lack of rigidity
   d. all of the above

9. Technician A peens down the middle of a crack. Technician B peens in toward the centerline of a crack. Who is right?
   a. Technician A
   b. Technician B
   c. Both A and B
   d. Neither A nor B

10. How are plugs installed along the line of a crack in areas of the casting that are subjected to high pressure and temperatures?
    a. at angles to the casting surface
    b. in an overlapping fashion
    c. both a and b
    d. none of the above

11. Tight interlocking of the plugs is necessary for _____ .
    a. internal crack repairs
    b. exterior crack repairs
    c. both a and b
    d. neither a nor b

12. Applying a ceramic seal is part of the _____ .
    a. cold process repair
    b. exterior process repair
    c. both a and b
    d. neither a nor b

13. Which welding process is not a joining process, but a surface buildup method?
    a. shielded metal arc
    b. submerged arc
    c. flame spraying
    d. none of the above

CHAPTER NINE

RECONDITIONING CYLINDER HEADS AND BLOCKS

Objectives

After reading this chapter, you should be able to
- Explain how to straighten an aluminum head.
- Give three reasons for resurfacing the deck surface of a cylinder head.
- Name the four machines used for resurfacing.
- Explain the advantages and disadvantages of the four resurfacing machines.
- List the problems that result from excessive surfacing.
- Explain how to measure the engine's clearance volume.
- Explain how to machine V-6 and V-8 engine cylinder heads.
- Describe deglazing and cylinder honing.

The reconditioning of cylinder heads and engine blocks (Figure 9-1) requires the skills of a rebuilder or machinist. Before the engine can be reassembled, a rebuilder or machinist might be required to straighten cylinder heads, resurface the deck of the engine block, hone or bore the cylinders, obtain the correct cylinder wall surface, and repair and install the cylinder sleeve. The procedures for these operations are covered in this chapter; additional work for the rebuilder or machinist is described in Chapters 10 to 12.

Of course, before any reconditioning or rebuilding work is started, threaded holes should be cleaned with the correct size tap to get rid of any and all burrs or dirt, and to allow for proper torquing (Figure 9-2). Use a bottoming tap in any blind holes. Chamfering or counterboring will eliminate thread pulls and jagged edges. Also all burrs and casting slag should be removed from inside the block with a high-speed grinder.

STRAIGHTENING ALUMINUM CYLINDER HEADS

As the use of bimetal engines (aluminum cylinder and cast-iron block) increases, so do the problems for the rebuilder. As described in Chapters 5 and 8, aluminum head work is different from cast iron in several important ways. For instance, aluminum is an entirely different metal. It is softer than cast iron, it cannot take abuse like an iron head, and any abuse can lead to distortion, cracking, and warpage.

Warpage is usually the result of overheating (low coolant, uneven coolant circulation within the head, a too lean fuel mixture, incorrect ignition timing, and so on). The hotter the head runs, the more it wants to pull away from the block. Since it is bolted

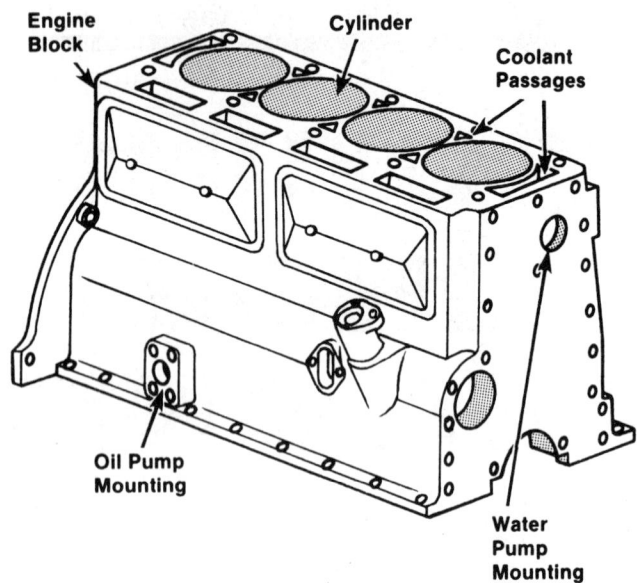

FIGURE 9-1 In-line and V-engine blocks

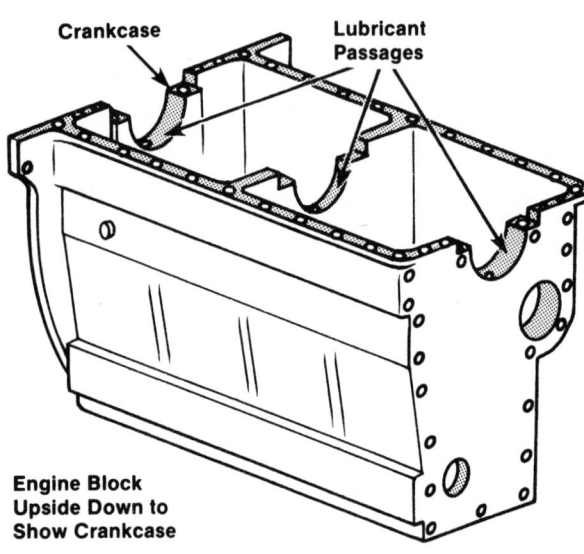

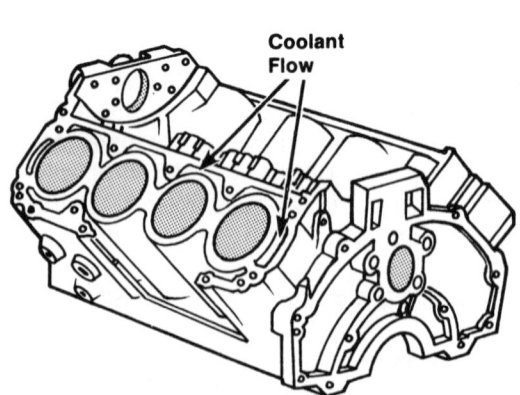

FIGURE 9-2 Cleaning threaded holes *(Courtesy of Perfect Circle/Dana)*

FIGURE 9-3 Arrangement used for straightening a typical cylinder head. Straightening minimizes surfacing and can correct camshaft misalignment and eliminate recession caused by grinding misaligned valve seats.

in place, the only direction it can move is up, so the head bows in the middle. This is why scuffed cam bores are found in many aluminum heads with overhead cams. Sometimes the warpage is so bad the cam will not turn in the head once the head is unbolted from the block (Figure 9-3).

Alignment of the cam bores in an overhead cam head must be checked with either a straightedge and feeler gauge (Figure 9-4) or with a dial indicator. If off by more than 0.002 to 0.003 inch, corrective action is required.

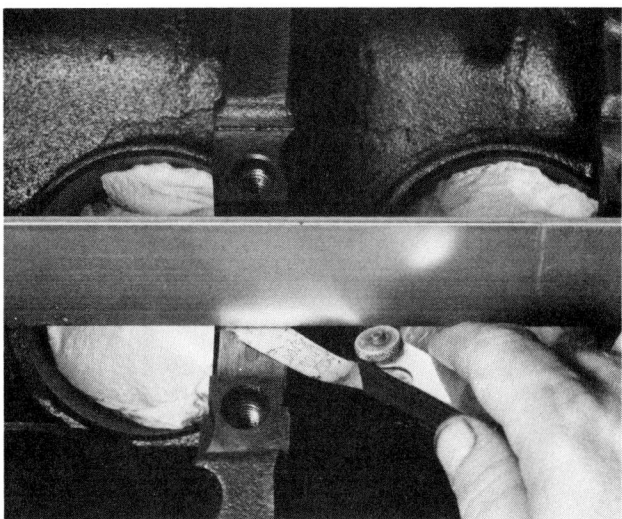

FIGURE 9-4 Checking cam bore alignment with a straightedge and feeler gauge. Plug cylinder bore with clean, lint-free cloths to prevent dirt from entering.

 SHOP TALK _____

Although specifications vary according to the application, traditionally, the maximum acceptable limit for cast-iron heads is 0.005 inch. Aluminum is not as forgiving, so 0.002 inch to 0.003 inch is a more realistic upper limit.

Although align boring and the installation of oversized cam bearings in the head or a cam with oversized journals might restore alignment in the cam bores, a badly warped head might be too far gone to allow the face to be resurfaced. Removing metal from the face of the head also alters the valve train geometry, which limits the amount of aluminum that can be removed.

A number of precautions should be taken, however, before the stress-relieving process is begun:

1. Remove bearings, core plugs, adjusting screws, and other hardware before heating the head (Figure 9-5).
2. Check the valve guides. If they are worn, they must be replaced before heating past 250 degrees Fahrenheit (Figure 9-6). Guides will be difficult to remove after heating at 400 to 450 degrees.
3. Most experts agree that it is wiser to straighten a warped head before machining takes place to minimize the amount of metal that has to be removed. In addition,

FIGURE 9-5 Remove adjusting screws and other hardware before heating the head.

FIGURE 9-6 Worn valve guides should be replaced before heating the head past 250 degrees Fahrenheit.

all straightening should be done before any welding repair is undertaken.

A straightening fixture can be made by using 1-1/2-inch-thick iron or mild steel plate surfaced on both sides and drilled and tapped to accept head bolts. These are all that is required for clamping the head to the bare plate (Figure 9-7). Reinforcing the plate from underneath can add to its rigidity. Using colloidal copper antiseize compound will prevent binding of the threads at high temperatures. Ready-made straightening fixtures are also available (Figure 9-8).

All straightening is done in reference to the cam side of overhead cam heads. Warpage is determined by using a straightedge and feeler gauges at each

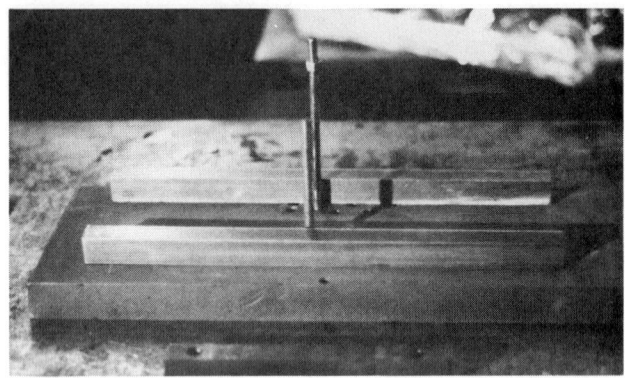

FIGURE 9-7 Typical straightening fixture

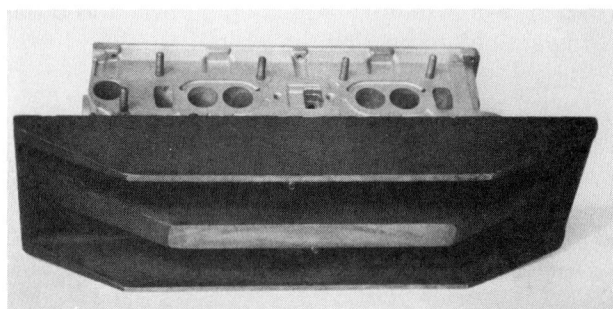

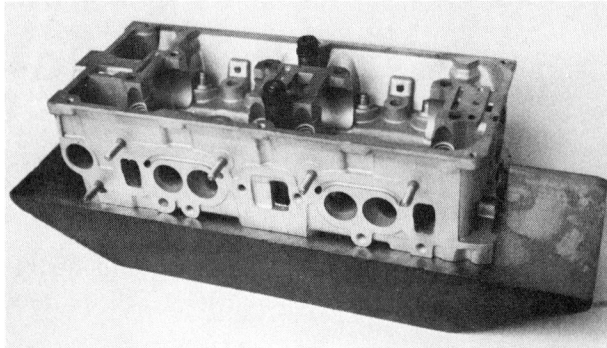

FIGURE 9-8 Construction of ready-made straightening fixture and how the cylinder appears on it *(Courtesy of T. Hoff Manufacturing)*

FIGURE 9-9 Placing brass shims at each end of the cylinder head

FIGURE 9-10 Tightening center studs

end of the cylinder head. Brass shims measuring half the total warpage at one end are then placed at each end (Figure 9-9). On pushrod cylinder heads, it might be desirable to straighten in reference to the deck side to further minimize stock removal.

By comparison, warpage on the deck side is usually greater than on the cam side. The deck side will also reveal more localized distortion because of hot spots around paired exhaust valves.

The center studs should be tightened to no more than 25 foot-pounds (Figure 9-10). Multiple runs through the oven are preferable to overtighten-

ing. Studs are only positioned in the center of the head to allow for expansion over the full length of the cylinder head. Place the head and fixture in a thermal oven (Figure 9-11). After keeping the heads at 400 to 450 degrees for 4 or 5 hours, the oven should be shut off and the heads allowed to cool slowly. Slow cooling prevents stresses induced from non-uniform cooling of the castings. It is usually recommended not to remove the head until it has cooled below 250 degrees Fahrenheit. Straightening can remove as much as 0.030 to 0.040 inch of warpage in a single treatment and as much as 0.900 inch in multiple treatments.

FIGURE 9-11 Placing the head and fixture in a thermal oven

In a case where the head has been overcorrected, simply reverse the head in the setup and tighten in place over 1-inch-square cold rolled steel parallels. The parallels allow reversing the setup without having to strip the head of dowel pins or other hardware. Little or no shimming is usually required. In most cases, overcorrections are no more than a few thousandths of an inch.

RESURFACING CYLINDER HEADS

There are three reasons for resurfacing the deck surface of a cylinder head:

1. To make the surface flat so that the gasket seals properly
2. To raise the compression ratio
3. To square the deck to the main bores

This can be done by grinding, milling, belt surfacing, and broaching.

As engines undergo heating and cooling cycles over their life span, certain components tend to warp. This is especially true of cylinder heads. Too much warpage can prevent a good seal. By using a precision straightedge or flatness bar and feeler gauge, as mentioned in earlier chapters, the amount of warpage can be easily measured. While the manufacturers' specifications should always be checked, maximum head warpage for cast-iron limits are generally recognized as follows:

- In-line 6-cylinder—0.006 inch
- In-line 4-cylinder—0.004 inch
- V-6—0.003 inch
- V-8—0.004 inch
- Aluminum heads—0.002 to 0.003 inch

 SHOP TALK _____

Some manufacturers allow aluminum heads to warp more than cast iron before recommending replacement.

The component surface should be checked both across the head as well as lengthwise. Be sure to check flatness of the intake and exhaust manifold mounting surface on the head. In general, maximum deformation allowed here is 0.004 inch.

In addition to warpage inspection, check for dents, scratches, and corrosion around water passages. This is especially important on aluminum heads. Heads that are deformed beyond specifications must be surfaced. As mentioned, this can be accomplished by cutting, grinding, broaching, or sanding. The finished surface, however, should not be too smooth. It must be rough enough to provide "bite," but not enough to cause a poor seal and leakage.

SURFACE FINISH

No surface, no matter how it appears, is perfectly smooth. When viewed in cross section (Figure 9-12), a surface consists of a series of peaks and valleys. A special instrument, called a profilometer, which incorporates a stylus that moves across the area to be checked, is used to check surface roughness and to measure the distance between the peaks and valleys (Figure 9-13). It calculates the average value, which is about one-third of the peak-to-valley depth.

The unit of measurement used for checking surface finish is the microinch. One microinch is equal to one-millionth (0.000001) of an inch. Average surface finish can be expressed using either the RMS

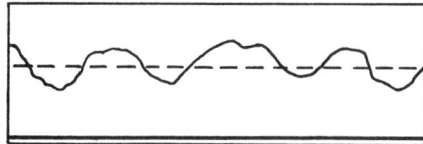

FIGURE 9-12 Cross section of a surface. Dotted line indicates about one-third of the maximum peak-to-valley depth. In this case, if the distance from the lowest valley to the highest peak was 75 millionths, the surface analyzing equipment would register about a 25-microinch surface.

FIGURE 9-13 Measuring surface roughness with a profilometer

(root mean square) or AA (arithmetic average) methods. RMS readings are approximately 11 percent higher than AA readings, but for all practical purposes this difference is negligible when checking machine finish.

 SHOP TALK _____

A profilometer is the ideal instrument to measure surface finish; however, it is expensive. Another way to measure surface finish is to use a specimen block kit. This kit contains a set of blocks that have various microinch roughnesses. By using a fingernail to compare the finish on the head with the many blocks in the kit, it is possible to estimate the head surface smoothness.

For proper head gasket seating, a cylinder head surface finish range of 60 to 120 microinches is generally recommended. This finish consists of shallow scratches and small projections that allow for gasket support and sealing of voids. A surface finish greater than 120 microinches is too rough. Such a finish has scratches and projections that allow only the gasket to be supported at a few points. Even with the head bolts torqued to proper specifications, there is improper loading on the gasket bore flange leading to leaking combustion gases.

Some new and late-model Chevrolet heads have an extremely smooth finish of approximately 32 microinches. This smooth surface is the tolerance to which Chevrolet manufactures its cylinder heads. Chevrolet claims this finish allows for mass production and the use of steel-embossed head gaskets.

Each of the four machines for resurfacing—belt surfacer, milling, broaching, and wet grinder—has advantages and disadvantages.

WARNING: Before attempting to test or operate any surfacing machine, be sure to become familiar and follow all the cautions and warnings given in the machine's operation manual. Also, when operating these machines, you must wear safety glasses, goggles, or a face shield.

Belt Surfacers

Introduced to the rebuilding market in 1980, the belt or sanding surfacer (Figure 9-14A) is quickly becoming the most popular method of resurfacing. Many in the industry believe the two key elements for effectively operating a belt surfacer are adequate horsepower and downward pressure. Without these essential ingredients, belts can bog down and/or lose their sharpness. Both conditions lead to various surfacing inconsistencies. Others believe that too much horsepower can tear a belt. Those with this perspective recommend lower horsepower surfacers as well as dressing the belts frequently to keep them from glazing. The main point is that a sharp belt surface is essential for consistent surfacing and belt life.

 SHOP TALK _____

On V-heads, insert the end of the coolant hose into a valve guide on the right-hand combustion chamber. Open the right-hand valve to allow coolant to be dispersed into the head without spraying the surrounding area and the operator. On flat heads, place the hose into a spark plug hole.

One of the greatest advantages of the belt surfacer is its quick setup time. An operator merely places the part to be surfaced on the belt against the restraint rail (Figure 9-14B).

CAUTION: Never start the belt until the workpiece is firmly against the restraint rail. Also, do not place the piece on a moving belt. Belt motors are activated by foot switches. Be careful with belt edge setups as shown in Figure 9-15. "Catching" the belt edge can cause belt fragmentation and operator injury.

A

B
FIGURE 9-14 (A) Belt surfacer and (B) how it is used

FIGURE 9-15 Catching the edge of a belt can cause fragmentation as well as injury to the operator. Be careful with belt edge setups.

Cylinder heads should be moved in a back-and-forth (front-to-back) motion across the belt width in short, 3- or 5-second intervals (Figure 9–16). Hold the work with both hands when operating with or without an air hold-down fixture (Figure 9–17). Do

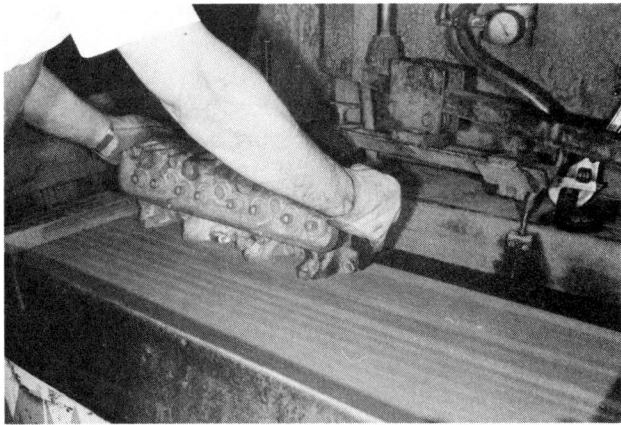

FIGURE 9-16 Surface sanding the manifold side of a cast-iron cylinder head

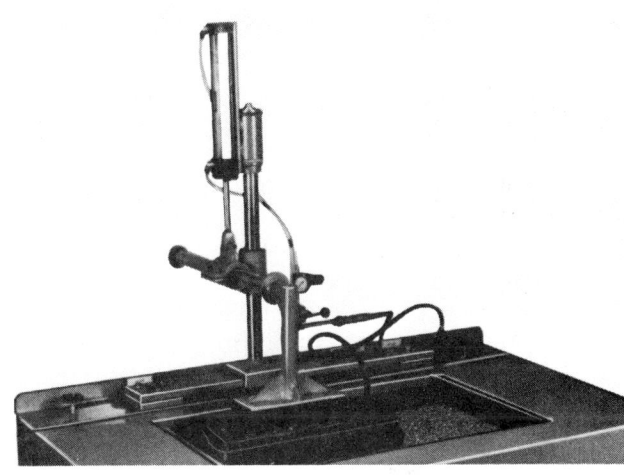

FIGURE 9-17 Resurfacer with an air hold-down system *(Courtesy of Kansas Instruments, Inc.)*

not allow the work to move off the belt; even a small amount will cause an uneven grind.

Keeping the head moving in short bursts is critical for proper surfacing. By leaving the head against the rail, rounded corners can result. This is a well-known problem on Omni heads where the belt will push up the edge of the head, rounding off the corners. Rounded corners can also result from using worn or improperly tightened belts.

Keeping the head on the belt for extended periods of time also causes heat distortion, leading to a wavy surface finish. If steam rises from the component when coolant is applied, the head has been left on the belt too long. Keeping the head in the same position also causes the belt abrasive to wear unevenly, knocking the grit lower in one spot. The abrasive must wear across the belt at an even rate. If it does not, an irregular surface finish results.

When surfacing long or very large items that have a tendency to warp when heated, use the maximum coolant flow permissible. Grind for short periods. With the belt stopped and the work tilted up between grinds, use the coolant hose on the surface being ground to reduce the temperature. Grind these larger surfaces with a new belt; it allows quicker cutting and less heat buildup. If extreme precision is required, let the item cool off completely after initial surfacing, then resurface for only a few seconds. This will remove any imperfections before heat builds up again.

On some manifolds and top-heavy blocks, leave the object in one place. Do not push back and forth and grind for short periods of time. If much of this type of grinding is to be done, vary the location on the belt to even out wear.

When using counterbalance fixtures, lift the blocks by a central main bearing cap. Insert the end of the counterbalance rod into a central cylinder hole and set the horizontal bar welded to the rod across two main caps as shown in Figure 9–18. This will balance the block lengthwise. Slide the counterweight along the rod until the surface to be ground is balanced; tighten the locking bolt. Set the block on the belt and grind no longer than 4 to 5 seconds; then wash off and inspect. With a new sharp belt, be alert for the rapid stock removal.

Although operation of belt surfacers does not require skilled machinists, the quality of the final surface finish is extremely operator-sensitive. Those operators who do not pay attention to proper surfacing procedures, that is, the length of time the component is on the belt, proper movement of the component on the belt, condition of the belt surface, and appropriate downward pressure, will not generate quality surface finishes. The aluminum head shown in Figure 9–19A was belt sanded and found to be flat with a straightedge. "Bluing" was then applied to the sanded surface to make a quality control check. The "blued" head pictured in Figure 9–19B shows the surface finish results obtained with a different operator. The blue dycum in the center has not even been touched.

Types of Belts. There are generally three types of abrasives used on belts.

- *Silicon Carbide Belts.* This is the most commonly used type of belt. It has good belt life and is reasonably priced. It works well on aluminum.
- *Aluminum Oxide Belts.* They provide neither optimum performance nor belt life. Their main purpose is to exist as a competitive lower-priced alternative.

A

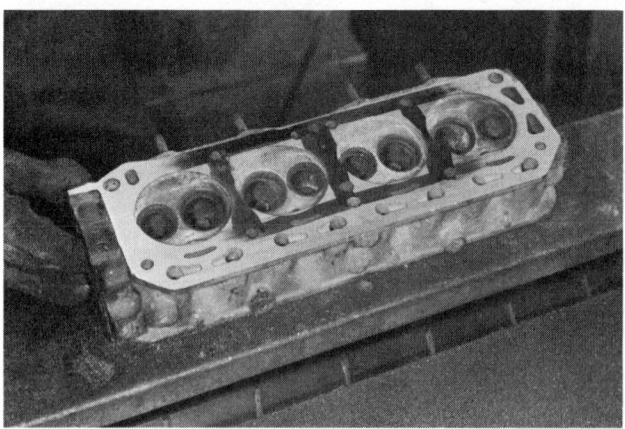

B

FIGURE 9-18 The counterweight balance maintains even pressure on the V-type block.

FIGURE 9-19 (A) Aluminum head, after being sanded flat, with bluing applied to the sanded surface; (B) results of the bluing quality control check.

• *Zirconia Alumina Belts.* When properly used, they have excellent belt life but are more expensive than the other two. They do not perform as well on aluminum as belts of silicon carbide.

Most belts offered today use 40 grit abrasive. This seems to provide the best overall surface finish.

Construction of belt backing is also important. Polyester backings on belts that are fully waterproof will resist stretching, clean easily, and will not wear out before the abrasive is fully used. Cotton-backed or poly-cotton-backed belts can stretch and wear prematurely. Remember, a belt that cannot maintain its tightness will result in surface inconsistencies.

Regardless of the type of surfacer used, belts should be dressed. Dressing the belt after it has been in use unclogs grit and metal accumulated between the abrasive particles (Figure 9–20).

Contrary to the belief of many people, dressing the belt will not reduce its life. How often a belt is dressed depends on the types of components being surfaced. Some shops dress after every few units; others dress less frequently. The most important point to remember is that a glazed belt clogged with grit is not necessarily a worn-out belt; it will not, however, be able to surface components correctly with a belt in this condition.

Sharp belts are capable of removing about 0.001 inch of material per second on cast iron.

Belt surfacers can be used effectively to surface aluminum heads. The greatest danger is removing too much material too quickly. Although a back-and-forth movement for 5 seconds is still recommended, very light or no downward pressure should be applied to the component. Weighted bags con-

taining minimum amounts of lead shot are the maximum amount of pressure that should be used with aluminum (Figure 9–21). Some belt surfacers are equipped with an air hold-down to accomplish the same objective. Figure 9–22 shows a special shop-made fixture used to bridge the head and distribute downward pressure for truing up the unit in specific areas. The handle is returned to the center position for even distribution of pressure for final surfacing. Many rebuilders use no extra weights at all.

Although many rebuilders use the same belt for surfacing both aluminum and cast iron, others use a belt with a finer grit for surfacing aluminum to achieve a smoother, specified RMS finish (Figure 9–23). Some rebuilders, while using the same grit for both metals, use only worn belts (although still containing usable abrasive) on aluminum to avoid removing material too quickly. Other rebuilders run a

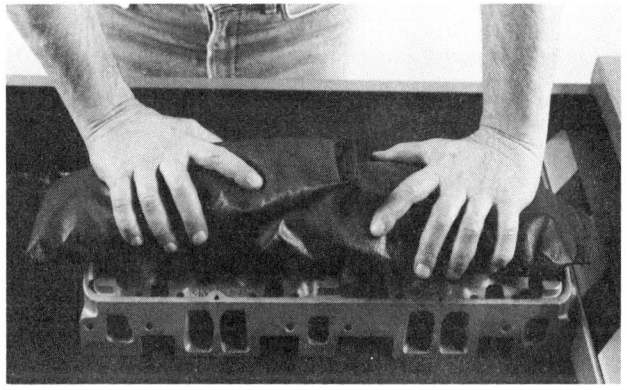

FIGURE 9-21   Weighted bags containing minimum amounts of lead shot are the maximum amount of pressure that should be used with aluminum. *(Courtesy of Kansas Instruments, Inc.)*

FIGURE 9-20   Dressing the belt to unclog accumulated grit and metal

FIGURE 9-22   Special fixture bridges the head and distributes pressure evenly.

FIGURE 9-23 Different belt grits can be purchased for aluminum and cast-iron materials

wax stick across the belt surface to prevent the belt from removing material too aggressively. The latter procedure is similar to one that has been used for many years by those employing stone grinders to surface aluminum components.

Although a finer grit or semiworn belt is often used for surfacing aluminum, when it comes to surfacing diesel heads, a sharp belt is a must. Many rebuilders and even some suppliers of belt surfacers feel that belt surfacers should not be used for surfacing diesel heads, especially long in-line six cylinders. Some rebuilders and manufacturers say that on long diesel heads, material abraided on the leading edge of the head tends to become a cushion on the opposite end, preventing precise truing. Turning the head from end-to-end seems to solve some of the problem.

Inability to control the exact amount of material removed and achieve specific RMS finishes, combined with cosmetic problems resulting from the long straight lines left on the heads by belt abrasives are other reasons some diesel rebuilders refuse to use belts. There are, however, diesel engine rebuilders who use belt surfacers and claim they get good results.

Belt sanding surfacers are versatile. Although primarily used for cylinder heads, some rebuilders also surface passenger car blocks, along with a variety of odd jobs such as manifolds, oil pans, and so on. Keep in mind, however, that belt surfacers were never designed for heavy stock removal. Those rebuilders consistently removing excessive amounts of material end up having surfacing problems along with shortened belt life. Quality control of surface finishes can also be a problem. These machines, depending on the operator's technique and atten-

tion to the machine and belt variables, may not consistently produce a flat surface. Therefore, the finished surface should always be checked for flatness.

On the negative side, belt surfacers are noisy. Unlike traditional surfacing equipment, they also cannot be set up to remove precise amounts of material. Another caution to keep in mind is belt surfacers require ample voltage to operate effectively. Where low voltage exists, belt surfacers will not have enough horsepower available to surface components correctly.

Milling Machines

Milling machines are widely accepted in the rebuilding industry. Setup is usually made by leveling the head until the cutters touch all four corners. Then the cutters are moved to the center of the head to find the low spot; this enables the operator to easily determine how deeply a cut must be made. It is common practice to make a rough (deep) cut first to clean up the head and then follow with a finish (shallow) cut to improve flatness and the microinch finish.

 SHOP TALK _____

The operation and setup procedures given in this chapter are considered typical. For more detailed operational procedures, consult the machine's service manual.

Table traverse speeds of the typical vertical head and block milling machine (Figure 9-24) are 4 and 9 inches per minute (ipm). The table traverse reversing dogs are adjustable and located on the lower front of the table edge. The dogs contact limit switches located on the top of the electrical control box to reverse the table direction. The cutterhead is raised and lowered using the handwheel on the right of the column. The cutterhead vertical lock is located just below the raise/lower handwheel. Be sure the table is level (Figure 9-25). The following are the various methods used to hold the head and block configurations encountered (Figure 9-26):

 1. Generally in-line cylinder blocks (for example, four and six cylinder) are positioned on the machine with the pan rails on the table surface. The small edge clamps are then used to clamp the block to the table. In cases where it is desired to leave the main bearing caps attached to the block, it might be necessary to use two

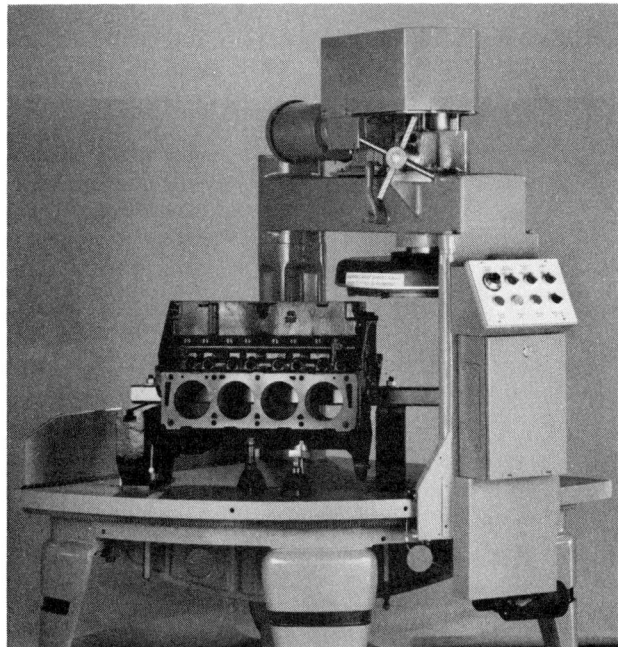

FIGURE 9-24 Cylinder head and block milling machine *(Courtesy of Storm Vulcan, Inc.)*

FIGURE 9-25 The table must be level. *(Courtesy of Storm Vulcan, Inc.)*

T-slotted parallels on which to rest the pan rails between the main caps. The precision level provided should be used to check the alignment of the workpiece. Unlevel readings are usually caused by dirt between the contacting surfaces if the machine is level. Be sure pan rails are clean and free of gasket material.

2. V-type blocks are mounted on the machine using the optional rollover fixture. A cylinder block hold-down bar is provided and should be installed whenever machining

V-type blocks and features a turnbuckle that can be attached to the rollover bar after the cylinder block is in place.

CAUTION: Never machine V-8 blocks without a hold-down bar.

---

3. Alignment is again achieved by using the precision level. Adjustment is normally only made on the front to rear alignment as the rollover bar provides accurate alignment to the main bearing bores. Use the fine feed adjustment knob for accuracy. At least one support jack should be used under each of the front and rear sides of the block to assure stability. Do not remove the level until all support jacks have been adjusted.
4. Parallels must be mounted on table T-slots. In this instance the parallels are used to bring the surface into the working height range of the cutterhead.
5. V-type and other angled cylinder heads are machined after mounting on the optional rollover fixture. The aluminum mounting plate is attached to the rollover bar.
6. Cylinder heads can then be attached to the mounting plate by fastening them to either the intake or exhaust manifold surface. After attachment of the head to the mounting plate, the precision level is used to align the head. Front to rear adjustment is accomplished by tightening the adjustment arm clamp, loosening the two mounting block clamps, and moving the adjustment by means of the screw provided. A support jack should be positioned under the free edge of the cylinder head to prevent movement during machining. Left to right alignment of the cylinder head is accomplished by rotating the eccentric adjuster located at one end of the aluminum mounting plate. Recheck front to rear alignment and contact of the support jack after adjusting the eccentric. Be sure all nuts, bolts, and clamp knobs are tight before beginning to cut.
7. After securing the workpiece to the machine table, the table reversing dogs should be adjusted to provide travel adequate to allow the cutterwheel to clear the

FIGURE 9–26 Methods for holding head and block configurations depend on the type of milling machine used. Check the manufacturer's instructions for the proper setup. Some of the more common holding methods are shown above.

work completely on the left end and to complete its cut on the right end.

8. The surface finish can be varied by changing the table traverse rate. Increasing the rate creates a more coarse finish; decreasing the rate results in a smoother finish. The cutterhead should only be fed downward when the workpiece is on the right side of the cutter.

Milling machines offer several advantages: they can remove more stock per pass (up to 0.050 inch); they have low maintenance requirements; and they can provide a variable finish range. With a milling machine, hard spots in castings are not, however, easily machined. The hard spots will dull the cutters and require that they be resharpened (Figure 9-27A).

Aluminum can also be a problem to mill (Figure 9-27B). It sometimes tends to "chunk out." To help prevent this and achieve a good finish, penetrating oil is often sprayed on the aluminum surface to be machined.

A

B

FIGURE 9-27 (A) Milling an aluminum cylinder head; (B) surfacing an aluminum head on a milling machine with a single-point flycutter

FIGURE 9-28 Flywheel surfacing can be done on some milling machines *(Courtesy of Rogers Machine Company)*

FIGURE 9-29 Electronically controlled milling machine

With some milling machines it is possible to mill or surface flywheels (Figure 9-28). Most modern milling machines are electronically controlled by an operator (Figure 9-29).

Broaching Machines

Broaching machines using an underside rotary cutter (broach) are also popular (Figure 9-30). A block, cylinder head, or intake manifold is held in an inverted position as the broach passes underneath. Electronic controls are now available to provide different cutter speeds and traverse rates simply by turning dials instead of changing driving belts. Machines of this style have a drawback, though, because a mirror is required to see if the surface has cleaned up. Also, overclamping of the mounting fixture can distort a head.

To machine a cylinder head with a rotary broach (Figure 9-31), proceed as follows:

A

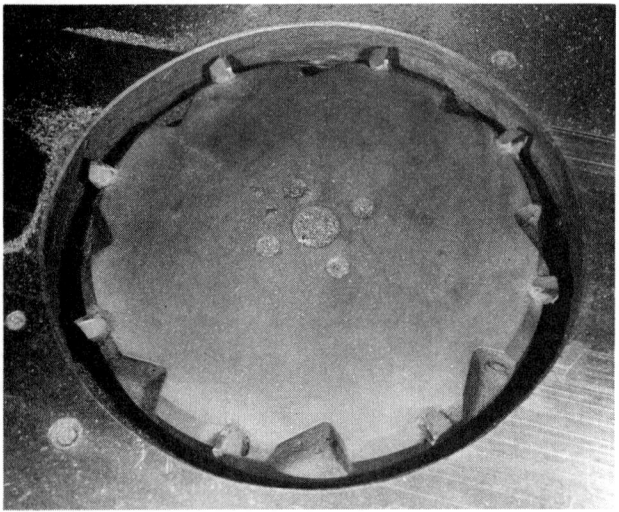

B

FIGURE 9-30 (A) Typical broaching machine and (B) closeup of an underside rotary cutter *(Courtesy of Winona Van Norman Machine Company)*

FIGURE 9-31 Cylinder surfacing on a rotary broach *(Courtesy of Winona Van Norman Machine Company)*

1. Before laying the cylinder head on the precision locating table, remove all nicks, burrs, and gasket material from the cylinder head. This can be done with a file.
2. Lay the clean cylinder head on top of a locating table and place four shims approximately 0.015-inch thick under each corner. Slide the head holding clamps in position, select the correct fingers, and place in arms.

CAUTION: Clamp fingers must be directly opposite each other to prevent the head from tilting when it is clamped.

---

3. Lightly tighten screws "A" (Figure 9-32) for aligning purposes. Lock the bridge clamps to the rails with eight nuts "B." Back off screw "A" one-half turn and check to see if the head is lying flat on the four shims. Retighten screw "A" and make a final check to be sure all shims are tight. This checking operation is very important— it assures the operator that the head is lying flat. It also assures the operator that the uniformity of the combustion chamber will be retained. Also, when the table is moved past the cylinder head, there will not be any actual contact between the cylinder head and the top of the precision locating table. This will eliminate scoring and wear, which could affect the accuracy of the ma-

FIGURE 9-32 Angle adapter, attached to a cylinder head *(Courtesy of Winona Van Norman Machine Company)*

chine. Always use shims for the reasons mentioned.

4. With the cylinder head clamped in the machine, the head is now ready for machining. Start the motor so that the cutter is rotating, raise the cutter until it contacts the head lightly, then manually traverse the cutter to the right from under the cylinder head. Feed the desired amount into the cutter with the micrometer feed wheel, which is graduated in 0.001 inch. Start the feed by tightening the feed knob "C" and take a cut.

5. If the motor slows down or acts overloaded, it indicates that either the cutters are dull or the cutting edge is too wide and should be resharpened or replaced.

6. After the first pass, check the finish with a mirror. The mirror can be passed under the head at any convenient position.

7. Normally one cut is sufficient, but if the mirror shows that more material must be removed, disengage the feed, manually traverse the cutter to the right, raise the cutter the desired amount, and take a second cut.

8. After the head is finished, lower the cutter below the tabletop, stop the motor, manually traverse the table under the head, place shims under the head, loosen the head, and remove it from the machine.

9. Always put shims under the head before loosening so that it will not drop and damage the table.

Broaching machines readily accommodate in-line as well as V-configuration blocks for surfacing. Preset feeds and speeds of the cutter head and table traverse can provide a consistent 80 to 90 microinch finish. Also, several hundred heads can generally be surfaced before cutter resharpening is required.

Surface Grinders

Typically, this type of resurfacing machine sets up like a mill. First, an operator has to level the head and then make a fast pass over the head to be sure the grinding wheel is spotting all four corners. If it is not, adjustments must be made.

The grinding wheel must be sharp, since a dull wheel will just polish the head surface and leave burned spots. A segmented wheel seems to work best—it does not load up, works well with iron or aluminum, and provides maximum control of surface finish. Surface grinders can be used on small or large engines (Figure 9-33).

A

B

FIGURE 9-33   Surface grinding a (A) small and (B) large engine

As with the mill, there are several methods used to hold the various workpieces in a typical surface grinder (Figure 9-34). These methods depend on the head configurations as follows:

1. Generally, in-line cylinder blocks (for example, four and six cylinder) are positioned on the table surface. The small edge clamps are used to clamp the block to the table. In cases where it is desired to leave the main bearing caps attached to the block, it might be necessary to use two T-slotted parallels on which to rest the pan rails. The precision level provided should

A

B

C

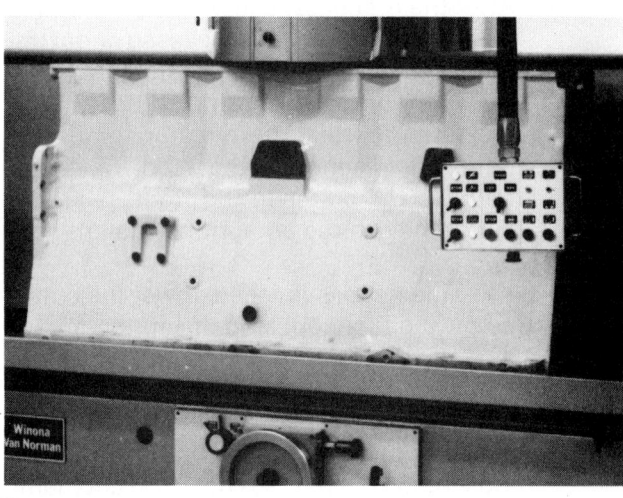

D

FIGURE 9–34 Workpieces that can be held in a surface grinder: (A) V-type cylinder heads; (B) V-type cylinder block; (C) diesel cylinder head; and (D) Caterpillar block *(Courtesy of Winona Van Norman Machine Company)*

be used to check the alignment of the workpiece. Unlevel readings are usually caused by dirt between the contacting surfaces if the machine is level. Be sure the pan rails are clean and free of gasket material.

2. V-type blocks are mounted on the machine using the rollover fixture. Alignment is again achieved by using the precision level. Adjustment is normally only made on the front to rear alignment as the rollover bar provides accurate alignment to the main bearing bores. At least one support jack should be used under the front and rear side of the block to assure stability. This machine features a turnbuckle, which may be attached to the rollover bar after the cylinder block is in place. This feature

allows rapid changeover from blocks to heads. The pivot bolts need only be finger tight.

3. Cylinder heads having two sides parallel may be mounted. In this instance the parallels are used to bring the surface into the working height range of the wheel head. The cylinder head is clamped using toe clamps. At least one jack should be used to support long cylinder heads by placing it in the approximate center of the head and adjusting it upward until it contacts the head but does not distort it upward.

4. V-type and other angled cylinder heads are ground after mounting on the rollover fixture. The aluminum mounting plate is attached to the rollover bar. Cylinder heads may then be attached to the mount-

ing plate by fastening to either the intake or exhaust manifold surface. After attachment of the head to the mounting plate the precision level is used to align the head. Front to rear adjustment is accomplished by tightening the adjustment arm clamp, loosening the two mounting block clamps, and moving the adjustment by means of the screw provided. A support jack should be positioned under the free edge of the cylinder head to prevent movement during machining. Left to right alignment of the cylinder head is accomplished by rotating the eccentric adjuster located at one end of the mounting plate.

5. Recheck front to rear alignment and the contact of the support jack after adjusting the eccentric. Be sure all nuts, bolts, and clamp knobs are tight before beginning to grind.

6. After securing the workpiece to the machine table, the table reversing dogs should be adjusted to provide travel adequate to allow the grinding wheel to clear the work completely on the left end and to complete its grind on the right end. Position the coolant deflection curtains at the most convenient height above the workpiece to prevent splashing outside the machine.

7. All grinding on a typical machine is done with the wheel head switch in the 1600 rpm position. The surface finish may be varied by changing the table traverse rate. Increasing the rate creates a coarser finish; decreasing the rate results in a smoother finish. The wheel head should only be fed downward when the workpiece is on the right side of the grinding wheel. Maximum wheel head downfeed is 0.002 inch per cycle. The wheel head elevation wheel is graduated in 0.0025-inch increments.

Many rebuilder shops have a so-called "machining center," which has the capability to accurately true an engine block surface and bore to the crankshaft centerline (Figure 9–35). Such machines will accurately machine anything from a small cylinder gas engine to a large diesel (Figure 9–36). Many of these machines have a digital readout (Figure 9–37).

STOCK REMOVAL GUIDELINES

The amount of stock removed from the head gasket surface has to be limited. Excessive surfacing can lead to the following problems:

FIGURE 9–35   Typical machining center *(Courtesy of Winona Van Norman Machine Company)*

- *Spark Knock.* This is due to a higher compression ratio.
- *Piston-to-Valve Interference.* When the block and/or head is surfaced, the piston-to-valve clearance during the overlap period becomes less. To prevent the valves from making contact with the piston, a minimum of 0.070 inch piston-to-valve clearance is recommended.
- *Valve Train Problems.* Surfacing will cause valve tips, rocker arms, and pushrods to be dimensioned closer to the camshaft. This will cause a change in rocker arm geometry and can also cause hydraulic lifters to bottom out.
- *Misalignment.* When metal is removed from the block or heads on a V-type OHV engine, the heads will be positioned closer to the crankshaft. This downward movement will cause the intake manifold to fit differently between the heads. As a result, ports might be mismatched and manifold bolts might not line up. In order to return the intake manifold to its original alignment, corrective machining is required.

Another important consideration when surfacing cylinder heads is the size of the clearance volume, which is the area above the piston that is formed by the combustion chamber. As material is removed from the cylinder head surface, the clearance volume is reduced. The size of the clearance

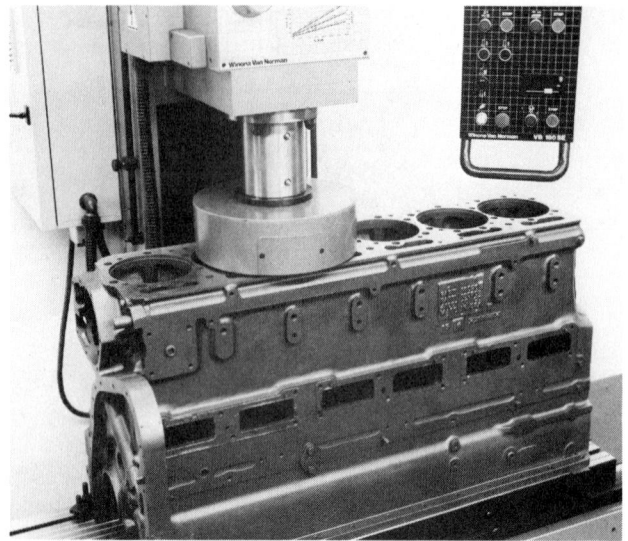

FIGURE 9-36 Operation of machining center *(Courtesy of Winona Van Norman Machine Company)*

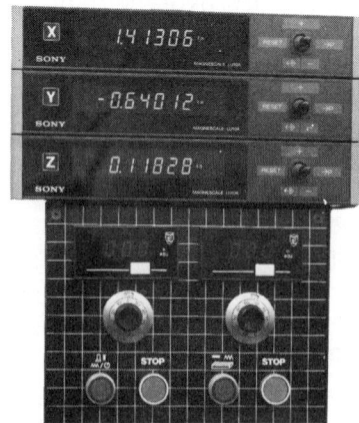

FIGURE 9-37 Typical machining center digital readout device *(Courtesy of Winona Van Norman Machine Company)*

volume determines engine compression. Removing an excessive amount of material during surfacing will increase the compression ratio beyond the manufacturer's specifications. Removing more material from one side of a V-engine than another will result in unequal clearance volumes.

Engine's Clearance Volumes

It might be necessary to measure and adjust the size of an engine's clearance volumes. Clearance volume is the volume of the combustion chamber plus the volume of the cylinder when the piston is at TDC.

Measuring the clearance volume is called cc-ing the cylinder head. Cc-ing is done with the valves and

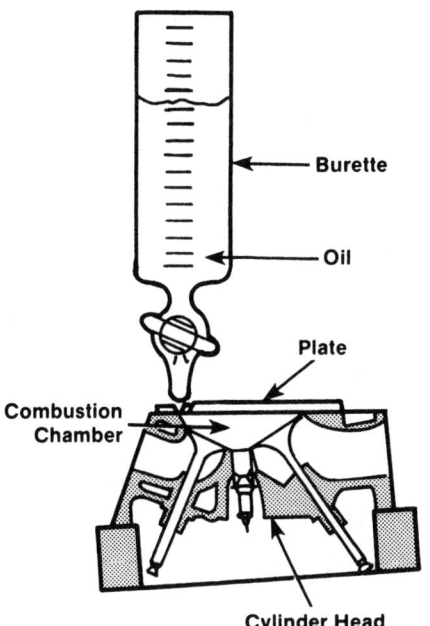

FIGURE 9-38 Cc-ing a cylinder head to find the combustion volume

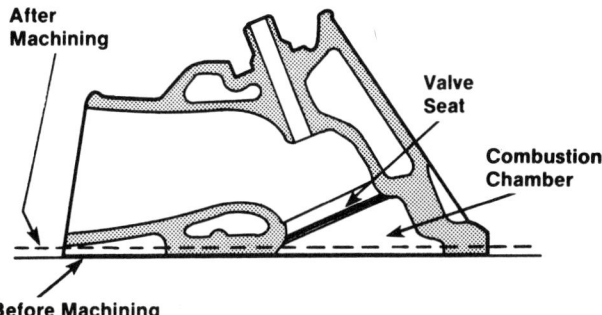

FIGURE 9-39 Rebuilders must compensate for machining where necessary to prevent preignition and detonation problems.

spark plugs installed. The cylinder head is mounted upside down, and a glass or plastic plate is installed over the combustion chamber. A graduated container called a burette is used to fill the combustion chamber with thin oil. The oil is poured through a hole in the plate, as shown in Figure 9-38. The amount of oil that enters the combustion chamber is noted by observing the cubic centimeter (cc) markings on the burette. The number of cc's used is equal to the clearance volume (in cubic centimeters) for this cylinder.

The clearance volumes for an engine can be adjusted in several ways. They may be reduced (as described) by surfacing the cylinder head. This, of course, reduces all the clearance volumes for that cylinder head. Individual clearance volumes can be increased by grinding the valve seats to sink the valves and by grinding and polishing metal from the combustion chamber surface. Either method can be employed to equalize all the clearance volumes and adjust them to the manufacturer's specifications.

Compression Ratio

Both the rebuilder/installer and the automotive machinist can prevent most of the causes of abnormal combustion (see Chapter 3). Each must be concerned with specific areas. First consider what the rebuilder does to an engine that can directly affect the combustion process. Since both detonation and preignition can be related to compression temperatures, the compression ratio and chamber configu-

ration become critical. This requires careful consideration, because there are many things that can be done to an engine that will have a direct effect on the compression ratio (Figure 9-39). The following are prime examples:

- Boring the block oversize changes the swept volume and the compression ratio. Generally speaking, the compression ratio increases at the same percentage that the displacement increases. Boring an engine 0.060 inch oversize will increase the displacement 9 cubic inches or slightly more than 3 percent. Assuming the replacement piston has the same compression height as the original, the initial 9.0:1 compression ratio will be increased by 3.0 percent to 9.27:1. As a rule of thumb, there is a 2 percent increase at 0.030 inch oversize, a 4 percent increase at 0.060 inch and a 6 percent increase at 0.125 inch. With today's gas, anything more than 9.0:1 is suspect without a detonation sensor.
- Decking the block changes the compression ratio (Figure 9-40). Removing 0.010 inch from the deck surface of an engine with a 4-inch bore, 76-cc head, 0.060-inch head gasket, 0.080-inch deck height would raise the compression ratio by 0.14:1.
- Resurfacing the head also increases the compression ratio. Though the effect varies for every type and size of chamber, a good rule of thumb is that when a head is surfaced 0.010 inch, the chamber is reduced by 1.50 cc for a 60-cc head and by 2.50 cc for a 90-cc head. This will increase the compression ratio by about 0.141:1 to 0.20:1, depending on the specific head configuration and actual swept volume.
- Variations in head gasket thicknesses affect the compression ratio (Figure 9-41). There can be as much as 0.040 inch difference be-

FIGURE 9-40 Decking cylinder heads can increase compression ratios

FIGURE 9-42 Some rebuilders use copper shims between block and gasket to return engines to proper compression ratios.

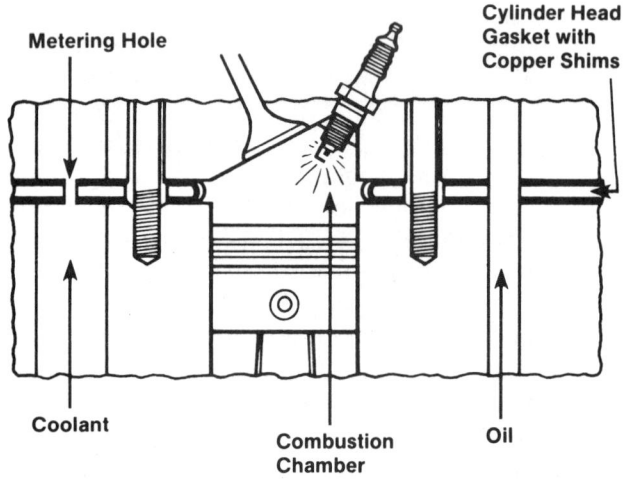

FIGURE 9-41 Cylinder head gaskets form the most critical seal on an engine between the cylinder head and the engine block deck.

tween various types and brands of gaskets. For instance, changing from a soft-faced to a steel or copper shim gasket can increase the compression ratio by as much as 0.50:1 (Figure 9-42).

• The piston itself can have an effect on the compression ratio. Replacing a dished piston with a flat top will increase the compression ratio significantly. Using the wrong piston with a particular chamber will raise (or lower) the compression ratio. The difference of 10 cc in chamber size can have a dramatic

effect on the compression ratio and might require the use of a different piston.

Fortunately, most aftermarket suppliers either deck or destroke their oversize pistons to enable the rebuilder to reduce or maintain the compression ratio instead of increasing it. These pistons should be used whenever available.

If nothing was done to reduce the compression ratio and the engine was rebuilt by boring to 0.060 inch oversize, the block was decked 0.010 inch, the heads were surfaced 0.010 inch and a thinner (0.020 inch) head gasket was used, how much would the compression ratio increase? Add 0.27 for boring, 0.14 for decking, 0.14 for surfacing, and 0.35 for the difference in head gaskets for a total increase of 0.9:1. Even if the engine were only at 8.0:1 originally, this would still be too close to the danger level, given today's gas.

Although not all of these steps are done every time an engine is rebuilt, the lesson is clear: the rebuilder must pay attention to the cumulative effects on compression when rebuilding engines today.

Although they do not directly affect the compression ratio, there are two other areas of concern regarding heads and head gaskets. Both involve the end gases. In normal combustion, self-ignition is prevented by removing heat from the end gases through quenching, squish, and turbulence (Figure 9-43). As mentioned in Chapter 2, squeezing the end gases into a small area is called *quenching*. This thins out the end gases, exposing them to proportionately more cooling surfaces and assuring better

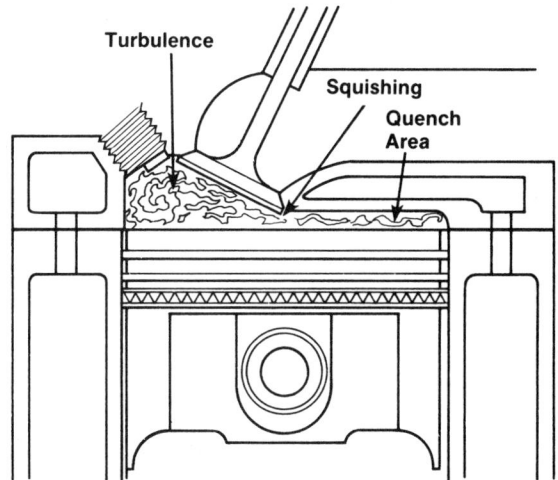

FIGURE 9-43   Removing heat from the end gases by quenching, squish, and turbulence prevents self-ignition.

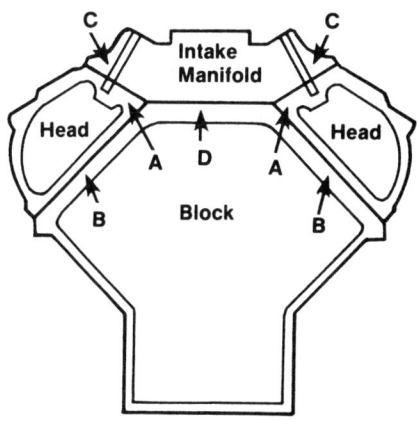

A Angle between combustion surface and intake manifold surface
B Total amount removed from combustion chamber surface of the head and cylinder block deck
C Amount to be removed from the intake manifold side of the head
D Amount to be removed from the base of the intake manifold

FIGURE 9-44   Machining V-6 and V-8 engine cylinder heads

heat dissipation. At the same time, some of the end gases are squished or squirted back toward the open combustion chamber, creating an additional turbulence in the chamber that helps eliminate stagnant air and pockets of heat. By moving the compressed mixture around, more of it contacts the outer surfaces and is cooled. This turbulence also promotes a more homogeneous mixture, creating a faster and more uniform burn.

When the head is surfaced, any existing quench areas will be altered. This can affect quench, squish, and turbulence. At the same time, head gasket shape and diameter are often altered by aftermarket suppliers to provide the clearance needed for oversize bores and the standardization required for broader application coverages.

These changes also affect the end gases. Coupled with boring, decking, and piston changes, it is possible to alter the normal combustion process for better or worse. With lean burn, high swirl chambers, and pistons that protrude above the deck on some of today's engines, any changes a rebuilder makes can create more significant effects in the future. The rebuilder should be cautious when surfacing heads and replacing gaskets.

The rebuilder should also take care when switching cylinder heads from one year to another because of possible differences in the chamber design in general and the quench area in particular. Emissions engineers have discovered that though quenching is a good way to help manage normal combustion, it also increases the skin effect, which adds to emissions. Whenever the end gases are cooled to the point that they do not burn, they are exhausted as unburned hydrocarbons. By changing, minimizing,

or even eliminating quench areas, emissions are better controlled. Usually these changes are made together with other modifications to the engine so there is a risk involved when switching heads.

Machining V-6 and V-8 Engine Cylinder Heads

Each time the head gasket surface of the head or the block of a V-type engine is machined, the heads are lowered and brought toward the center of the engine. This can cause a misalignment of the manifold bolt holes and ports. This problem can be corrected by machining the intake surfaces of the head or manifold (Figure 9-44). This will allow the manifold to reach a lower position. However, it will also reduce the manifold end gaps. The ends of the manifold or block also need to be machined to maintain the original gap size, or crushing of the end strip seals will result.

Table 9-1 will help determine the amount to be removed from either the intake manifold side of the head or the amount to be removed from the base of the intake manifold.

On most OHV engines, it is recommended that the pushrods be shortened an amount equal to the amount removed from the cylinder head. For example, suppose 0.015 was removed from surface B on a cylinder head and the angle to surface A is 10 degrees. Using the formula given in Table 9-1, proceed as follows: Multiply the 0.015 removed by the 10-degree multiplier, which is 1.233, and the proper

TABLE 9-1: HEAD ANGLE IN DEGREES

A	Amount to be Removed at C	Amount to be Removed at D
0° or 90°	B × 1.000	B × 1.4
5°	B × 1.100	B × 1.4
10°	B × 1.233	B × 1.4
15°	B × 1.414	B × 1.4
20°	B × 1.673	B × 1.4
25°	B × 2.067	B × 1.4
30°	B × 2.733	B × 1.4
35°	B × 4.072	B × 1.4
40°	B × 8.113	B × 1.4

amount to remove from the intake manifold side of the cylinder head is achieved.

Example:
$$
\begin{array}{r}
.015 \\
\times 1.233 \\
\hline
45 \\
45 \\
30 \\
15 \\
\hline
.018495
\end{array}
$$
Amount to be removed on intake side "C"

The same amount must be removed from both cylinder heads or the holes in the intake manifold will not line up properly.

REVIEW QUESTIONS

1. Which of the following does not cause overheating?
   a. uneven coolant circulation
   b. too lean fuel mixture
   c. incorrect ignition timing
   d. none of the above

2. Before the stress-relieving process begins, Technician A removes bearings, core plugs, and adjusting screws. Technician B checks the valve guides. Who is right?
   a. Technician A
   b. Technician B
   c. Both a and b
   d. Neither a nor b

3. Which of the following is not a reason for resurfacing the deck surface of a cylinder head?
   a. makes the surface flat so that the gasket seals properly.
   b. lower the compression ratio
   c. square the deck to the main bores
   d. all of the above
   e. none of the above

4. What is becoming the most popular method of resurfacing?
   a. belt surfacers
   b. milling machines
   c. broaching machines
   d. wet grinders

5. Which of the following belts is the most expensive?
   a. silicon carbide
   b. aluminum oxide
   c. zirconia aluminum
   d. none of the above

6. Which type of belt is the most commonly used?
   a. silicon carbide
   b. aluminum oxide
   c. zirconia aluminum
   d. zirconia carbide

7. Which of the following is an advantage gained by using a milling machine?
   a. Hard spots in castings are easily machined.
   b. Aluminum is easily machined.
   c. Variable finish range is provided.
   d. All of the above
   e. None of the above

8. Which of the following easily machines hard spots in castings?
   a. belt surfacers
   b. milling machines
   c. broaching machines
   d. wet grinders

9. Which of the following is caused by excessive surfacing?
   a. spark knock
   b. misalignment
   c. valve train problems
   d. all of the above
   e. none of the above

10. For which of the following is a mirror required to see if the surface has cleaned up?
    a. belt surfacer
    b. milling machine
    c. broaching machine
    d. wet grinder

CHAPTER TEN

RECONDITIONING VALVE TRAIN COMPONENTS

Objectives

After reading this chapter, you should be able to
- Explain how to grind valves.
- Describe how valve guides can be restored.
- Explain how to install valve seats.
- Describe how to install valve stem seals.
- Explain how to replace valve train components.
- Determine how to assemble the valve train components.

When reconditioning the valve train, there are many important details that must be given due consideration, such as the valves, valve springs, stem clearance, hydraulic lifters, rocker arms, and all the functional operations that make up the valve train. In other words, the entire valve train must be reconditioned so that when the valve is properly seated and closed, the valve stem will be centered through the least worn portion on the valve guide. This can only be successfully accomplished with the use of a valve tool reconditioning system that will restore the valve operation to its original dimensions.

In Chapter 6, the details of how to look for valve train problems and their probable causes were fully explained. This chapter covers the reconditioning of valve train components to insure top-quality engine performance (Figure 10-1).

GRINDING VALVES

Valve grinding or refacing is done by machining a fresh, smooth surface on the valve faces and stem tips. As described in Chapter 6, valve faces suffer from burning, pitting, and wear caused by opening and closing millions of times during the service of an engine. Valve stem tips wear because of friction from the rocker arm.

Before grinding, inspect each valve face for burning and each stem tip for wear. Replace any valves that are badly burned or worn. Reusable valves are cleaned by soaking them in solvent to soften the carbon deposits. The deposits are then removed with a wire buffing wheel. Extra care must be taken to insure that the valves do not become

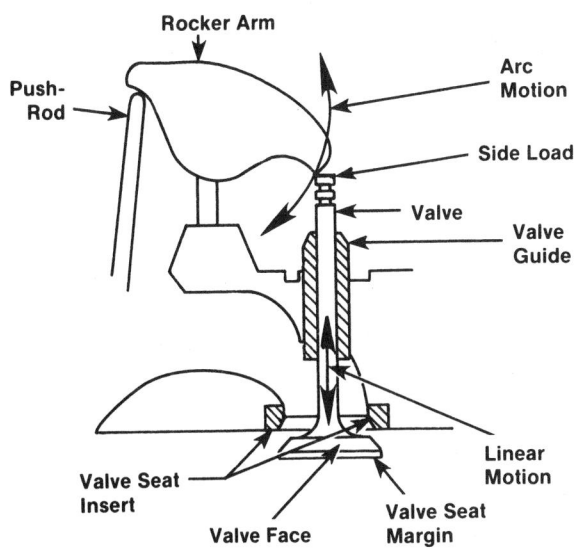

FIGURE 10-1 Many rebuilders say most warranty claims are due to valve train and guide problems.

FIGURE 10-2 Grinder type valve resurfacer *(Courtesy of Sunnen Products Co.)*

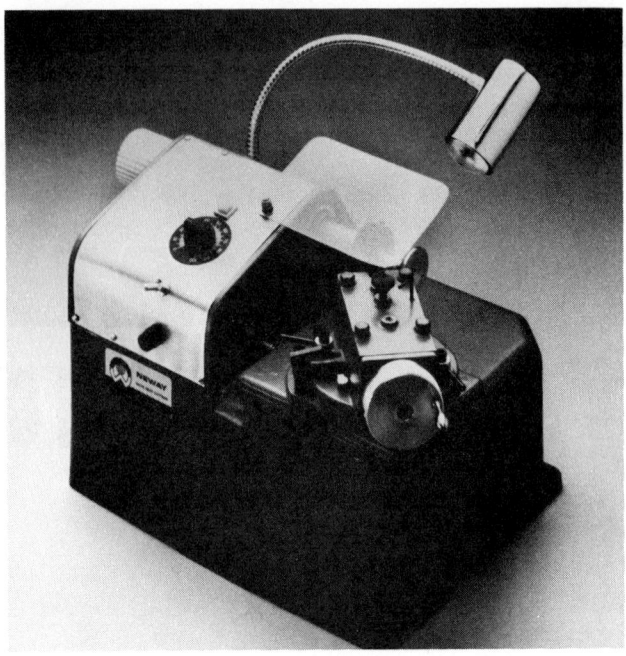

FIGURE 10-3 Cutter-type valve resurfacer *(Courtesy of Neway Manufacturing, Inc.)*

nicked when using the wire wheel; otherwise, valve breakage might occur later in operation.

Once the deposits are removed, the valve can be resurfaced. Two surfaces on the valve that are reconditioned by grinding are the valve face and valve tip. The face is reconditioned prior to grinding the valve tip.

Valves can be refaced on either grinding (Figure 10-2) or cutting (Figure 10-3) machines. Although both can reface valves and smooth and chamfer valve stem ends, the traditional grinder method of refacing is still the most popular. Larger valve grinder machines (Figure 10-4) are capable of doing valve train reconditioning tasks such as refacing certain types of rocker arms and other repair jobs described in this chapter. There are also computerized valve refacing grinder machines.

When using a grinder valve refacing machine, always be sure that the grinding wheel is properly dressed as directed by the manufacturer in the instruction manual. To dress a typical left grinding wheel, proceed as follows (Figure 10-5):

1. Position the chuck carriage to the extreme left. Adjust the diamond holder in the post so that the diamond has about 3/8 inch overhang in front of the post.

2. Place the attachment with the diamond slightly clearing the wheel periphery, point-

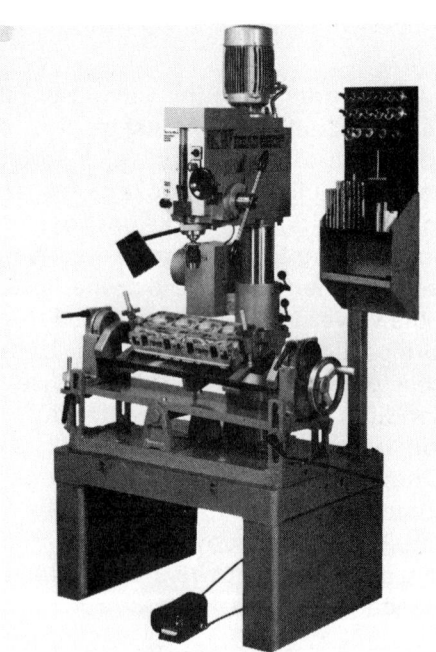

FIGURE 10-4 Typical complete grinder "head" shop *(Courtesy of Kwik-Way Manufacturing Company)*

ing at a slight angle to the wheel face and about 1/8 inch to the left of the wheel.

3. Firmly tighten the attachment to the chuck carriage plate. The amount of diamond

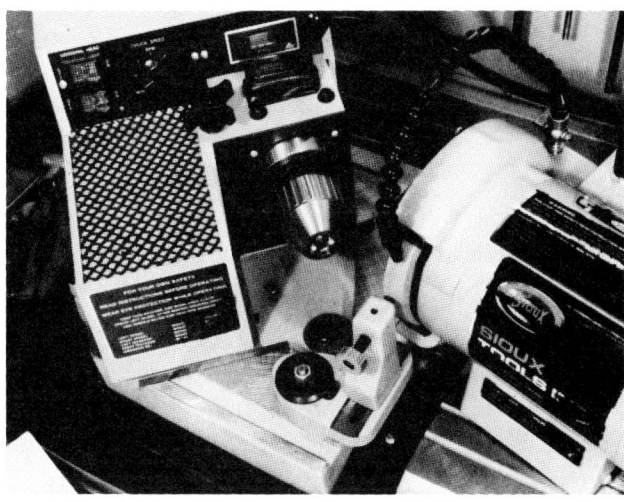

FIGURE 10-5   Dressing a grinding wheel on a computerized machine *(Courtesy of Sioux Tools, Inc.)*

overhang should be kept to a minimum in order to maintain as rigid a support as possible. The rubber chuck shields should be used to protect the chuck from wheel grit while dressing.

4. Start the machine and advance the grinding wheel carefully to prevent gouging.

5. After adjusting the diamond for dressing, apply coolant.

6. Pass the diamond over the wheel while feeding cuts of 0.0005 inch or less per pass (Figure 10-6). The feed screw micrometer thimble is graduated in increments of 0.001 inch. The diamond should occasionally be rotated slightly to present a new cutting edge. A rapid traverse of the diamond will result in a rough condition, which is excellent for fast stock removal but poor for

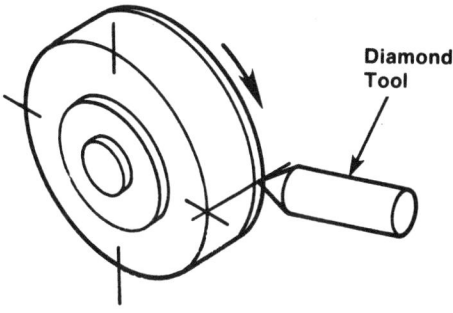

FIGURE 10-6   Dressing the stone means using a diamond tool to clean and restore the stone's sharpness.

finish yet is sometimes used to make a hard wheel cut more freely. However, if this is continually necessary, a softer grade wheel should be used. For best results, turn off the chuck head motor while dressing the wheel.

7. Dress the wheel each time the grinding head is repositioned or when a new wheel is installed. Be sure the grinding head clamp is securely tightened before dressing or grinding.

 SHOP TALK _____

Be certain the chuck is clean before inserting each valve. The valve refacing equipment must be in good working condition if a quality face is to be generated. More information on grinder wheel dressing can be found in Chapter 12.

To start the valve grinding operation, chuck the valve as close as possible to the valve head to eliminate stem flexing from wheel pressure (Figure 10-7). Take light cuts using the full wheel width (Figure 10-8). Make sure coolant is striking the contact point between the valve face and grinding wheel. Remove only enough metal to clean up the valve face. A knifelike edge will burn easily or might cause preignition (Figure 10-9). As a general rule, it is not advisable to grind a valve face to a point where the margin is reduced by more than 25 percent or to where it is less than 0.045 inch on the exhaust and 0.030 inch on the intake valves. After grinding, check the runout; it should not exceed 0.002 inch total indicated reading (TIR). The face should not show any chatter marks or unground areas. After grinding, reexamine the valve face for cracks. Sometimes fine cracks are visible only after grinding; sometimes they occur during grinding due to inadequate coolant flow or excessive wheel pressure.

_____

WARNING: Always wear eye protection when operating any type of grinding equipment.

_____

If an interference angle is to be used, the grinder is set 1/2 to 1 degree less than the standard 30- or 45-degree face angle. Always consult the manufacturer's specifications to determine whether an interference angle is to be used. Grinding an interference

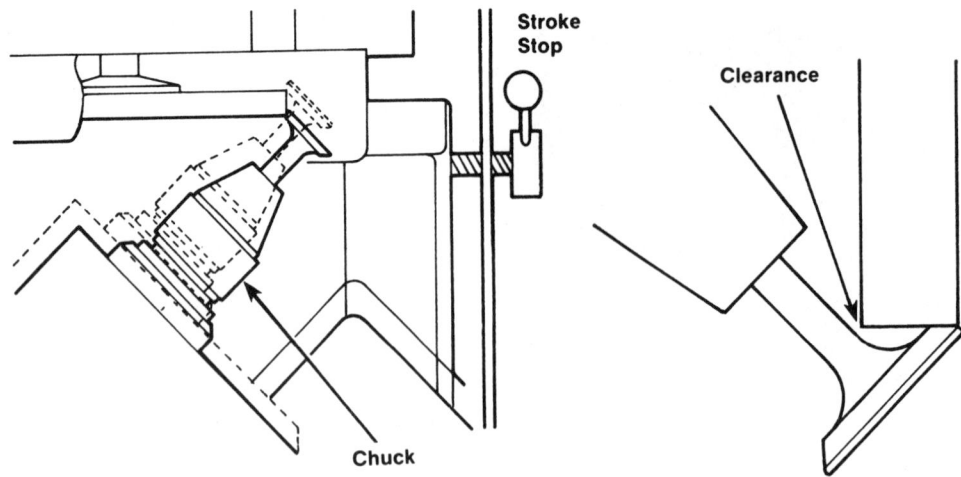

FIGURE 10-7 Once the grinder is set up, the valve is secured in a chuck and the proper stroke is set, making sure there is clearance between the stone's edge and the valve.

FIGURE 10-8 Taking light cuts on the valve *(Courtesy of Sioux Tools, Inc.)*

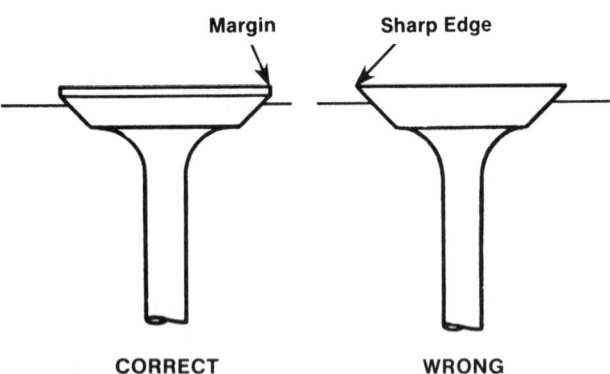

FIGURE 10-9 Correct versus wrong edge of valve face

angle produces a face angle close to 29 or 44 degrees. For the valve to seat properly, the face angle cannot be less than 29 or 44 degrees (Figure 10-10). The valve seals tightly only at the outer edge of the seat because of this interference angle. The narrow contact between the valve and the seat means the unit pressure is high between the narrow contacting surfaces. Because of this high pressure, any particles lodging on the face or the seat are likely to be broken up and blown away.

On large-diameter and hard-faced valves, it might be necessary to make a finish dress of the grinding wheel for a finish grind. Do not remove the valve from the chuck. Position the dressing tool between the valve and wheel so that a complete traverse of the grinding wheel can be made without contact of the valve to the grinding wheel.

CAUTION: Use the proper attachment equipment recommended by the manufacturer when dressing the stone or grinding. Never attempt to do this freehand.

Once the face is ground, the valve tip might also have to be ground. This is best determined by placing the valve in the cylinder head to check the stem height. Using the manufacturer's specifications, determine whether and to what extent the tip must be ground. As shown in Figure 10-11, the tip is ground using a stemming stone that has been dressed. Just like the facing stone, the stemming stone is dressed with a diamond tool. The valve is secured in a V-holder and clamp, and a coolant is used to cool the valve tip and remove grit during grinding.

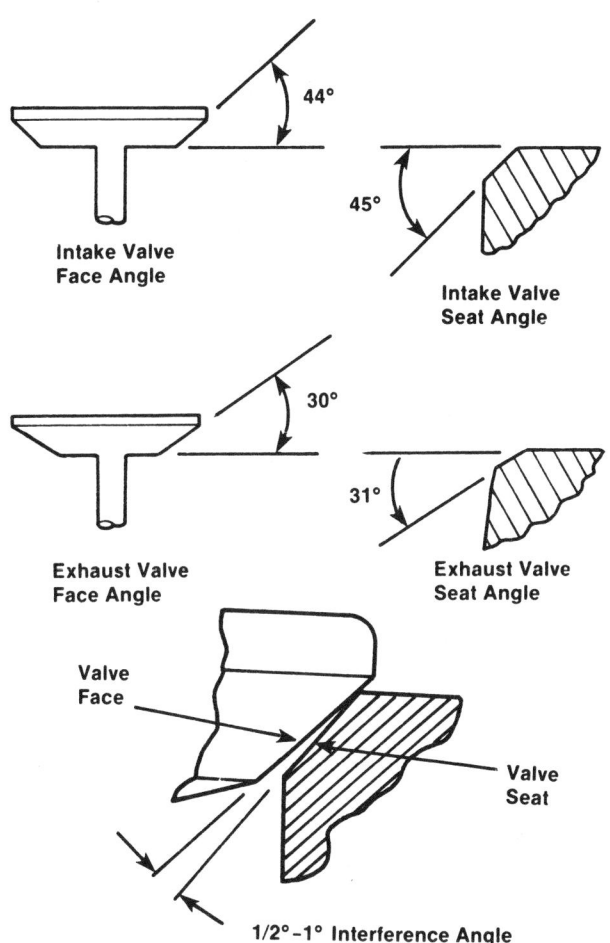

FIGURE 10-10   The difference between the seal angle and the valve face angle is known as the interference angle. This results in a line contact.

The valve tip is ground so that it is exactly square with the stem. Because valve tips have hardened surfaces that are up to 0.030 inch in depth, only 0.010 inch can be removed during grinding. If more than 0.010 inch is removed from the tip, the valve must be replaced. Also, if too much material is removed from the valve tip, there might be interference between the rocker arm and spring retainer or valve rotator. Follow the manufacturer's specifications for the allowable limits.

There is little basic difference in the refacing procedures for a grinder and cutter machine. In fact, most of the guidelines given for grinding a piston facing are the same for cutting. Actually, the only difference is in the cutting tool—a grind wheel or a cutting tool such as shown in Figure 10-12.

SHOP TALK _____

Some valve refacing equipment requires a procedure that is the reverse of the one just given in the text. That is, the valve tip is dressed and chamfered before it is refaced. If this is not done, the valve face with this equipment will be ground off center. The equipment manufacturer's instructions for procedures and sequence should be followed.

The valve stem length must be measured on engines that do not have a valve lash adjustment. Material removed from the valve face and valve seat increases the amount of valve stem length on the spring side of the cylinder head when the valve is seated. If excessive, this could cause the hydraulic lifter plunger to bottom out and prevent the valve

FIGURE 10-11   Grinding the tip of a valve *(Courtesy of Sioux Tools, Inc.)*

FIGURE 10-12   The cutting method of refacing a valve *(Courtesy of Neway Manufacturing, Inc.)*

from being fully seated when it is closed. A special tool is used to measure the stem length on the spring side of the head when the valve is held in its seated position.

Grinding material from the stem tip will generally correct any stem length problems and will center the hydraulic lifter plunger. Do not remove more material from the valve stem than the maximum allowed in the vehicle's shop manual. If the valve stem height specifications are not given by the manufacturer, there are several different types of gauges available for checking valve stem height. An instruction manual for use and height specification is generally supplied with the gauge. In any case, measurement must be done after the valve and seat have been reconditioned.

Do not grind badly worn tips on release-type rotated valves. If adjustment is required with this type of valve, remove metal from the open end of the tip cup to reduce the clearance. A few strokes over a piece of fine emery cloth placed on a smooth, flat surface, such as a piece of plate glass or surface plate, is usually all that is needed. The cup must be held flat while doing this. Grind the valve tip to increase the clearance. Flat and square grinding is a must for proper adjustment and good results. If specifications are not available, a tip cup clearance of 0.002 to 0.004 inch can be used.

When grinding valves, be sure to follow these guidelines:

1. Always check the setups with the chuck rotation and grinding wheels off. This will avoid damaging the chuck, wheels, or valves.
2. Valves with minimum overhang should be ground to make them consistent with the proper setup. On valves that are undercut behind the head, make sure the chuck is clamping on the stem, not undercut.
3. Do not draw the wheel off the valve.
4. Never attempt to remove the valve chuck while the motor is still turning. The piston must not be removed from the chuck until grinding is completed. To inspect the valve face, back the stone away from the valve, then move the carriage and the valve away from the stone.
5. Adjust the coolant flow onto the face of the valve or when dressing the diamond tool. Never dress the wheel without first covering the chuck. Never dry grind.
6. Generally it is wise to grind valve face angles and valve stem butts equally. For example, if 0.003 inch is ground off the face angle, grind 0.003 off the butt.

7. Do not cut into, or "neck," the valve fillet with the grinding wheel.
8. Grind the valve face until it is smooth. Always back the wheel away from the valve to prevent damage to the face. Never run the valve off the edge of the wheel. Both the valve and the stone become rounded if this is allowed to happen.
9. Replace the valve if the margin is less than 1/32 inch after grinding.
10. Refinish the stone face as required to assure the angles remain constant. Replace worn or damaged grinding wheels. Wheels that are grossly undersize will no longer grind properly.
11. On hard materials it might be necessary to put a coarser dress on the grinding wheel, slow down the chuck rotational speed, or change to the optional stellite wheel. The general rule in grinding is to use a soft, coarse wheel on hard materials and a harder, more smoothly dressed wheel on soft valve material.
12. Good housekeeping is essential to keep any precision tool in condition. Use the rubber shields or protector boots when grinding or dressing to keep grit and coolant out of the chuck. The chuck on your machine has been factory adjusted to grind valves within 0.001 inch TIR concentrically.

VALVE GUIDES

Why do valve guides fail? That is a question many rebuilders have asked themselves when trying to figure out where they went wrong. Valve guide problems can be lumped into one of three basic categories: inadequate lubrication, valve geometry problems, and wrong valve stem-to-guide clearance (too much or too little).

Inadequate lubrication can be caused by oil starvation in the upper valve train due to low oil pressure, obstructed oil passages, improper operation of pushrods, using the wrong type of valve seat, and so on. Insufficient lubrication results in stem scuffing, rapid stem and guide wear, possible valve sticking, and ultimately valve failure due to poor seating and overheating.

Geometry problems include the wrong installed valve height and off-square springs, rocker arm tappet screws or rocker arms that push the valve sideways every time it opens. This causes uneven guide wear, leaving an egg-shaped hole. The wear leads to increased stem-to-guide clearance, poor valve seating, and premature valve failure.

As for valve stem-to-guide clearance, a certain minimum amount is needed for lubrication and thermal expansion of the valve stem. Exhaust valves require more clearance than intakes because they run hotter. Clearance should also be close enough to prevent a buildup of varnish and carbon deposits on the stems, which could cause sticking. Insufficient clearance, however, can lead to rapid stem and guide wear, scuffing, and sticking, which prevents the valve from seating fully. This, in turn, causes the valve to run hot and burn.

If the clearance is too great, on the other hand, oil control will be a serious problem. Contrary to what some technicians might think, oil can be drawn past both the intake and exhaust guides. Though oil consumption is more of a problem with sloppy or worn intake guides because the guides are constantly exposed to vacuum, oil can also be pulled down the exhaust guides by suction created in the exhaust port. The outflow of hot exhaust creates a venturi effect as it exits the exhaust port, creating enough vacuum to draw oil down a worn guide (Figure 10-13).

Oil in the exhaust system of late-model vehicles with catalytic converters can cause the converter to overheat and suffer damage. On the intake side, oil drawn into the engine past worn intake guides can foul the spark plugs, cause the engine to emit higher than normal unburned hydrocarbon (HC) emissions, and contribute to a rapid buildup of carbon deposits on the backs of the intake valves and in the combustion chamber.

Carbon deposits in the combustion chamber can raise compression to the point where detonation occurs under load. Deposits on the backs of the intake valves in engines equipped with multipoint fuel injection can cause hesitation and idle problems because the deposits soak up fuel and interfere with proper fuel vaporization.

Inadequate valve cooling is another problem that results from too much valve stem-to-guide clearance. The valve loses about 15 to 30 percent of its heat through the stem, so if the stem fits poorly in the guide, heat transfer will be reduced and the valve will run hot. This can contribute to valve burning. Remember that one of the functions of a valve guide, besides positioning and supporting the valve stem, is to help cool the valve (Figure 10-14). Most of a valve's heat is conducted away through the valve seat, but between 15 to 30 percent of the heat is conducted up through the valve stem to the guide.

Another problem created by excessive clearance is air leakage. Worn intake guides or ones with too much clearance allow unmetered air to be drawn into the intake ports. The effect is similar to that of

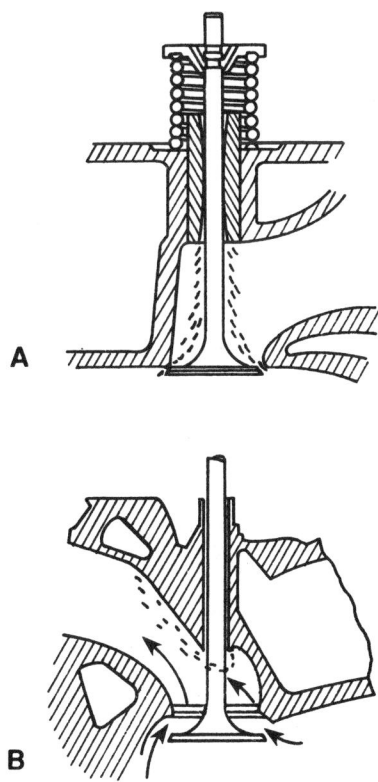

FIGURE 10-13   (A) Worn intake guides allow intake vacuum to suck oil down the guide. (B) Worn exhaust guide.

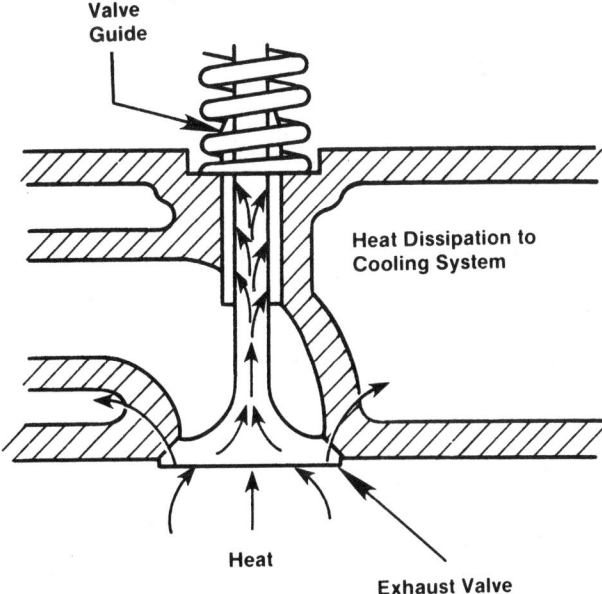

FIGURE 10-14   The valve guide helps to dissipate heat from the valve.

277

FIGURE 10-15 Several designs of valve guides *(Courtesy of Goodson Shop Supplies)*

worn throttle shafts on a carburetor. The extra air reduces intake vacuum and upsets the air/fuel calibration of the engine at idle, creating a lean misfire problem and a rough idle. Both intake and exhaust guide clearance must also be within certain tolerances to support the valves as they open and close. Too much clearance allows the valves to wobble, which leads to accelerated stem and guide wear as well as poor valve seating. Valve wobble also flexes the head of the valve, which may eventually lead to breakage.

One other cause of premature valve guide failure is lack of adequate lubrication after rebuilding and during initial start-up. If the assembly oil has drained off (or was forgotten), dry scuffing will doom the guides from the first crank. Ordinary motor oil can be used to prelube the valve stems and guides, but oil will not stay on the stems forever. An engine assembly lubricant such as a lubriplate is a much-used alternative since it has greater staying power. Heavy gear oil should be avoided because some contain additives that are incompatible with the valve seals or bronze valve guides. The best approach, particularly if the head sits around for a while before being used, is to apply a light coating of engine assembly lube.

The best results are usually obtained by carefully following the factory specifications. Generally speaking, factory specs call for passenger car intake valve stem-to-guide clearance in the 0.001 to 0.003-inch range. Exhaust valves as a rule tend to be given more clearance—about 0.0005 inch over the specs for the intake side. Diesel engines generally have looser fits on both intake and exhaust. On heads with sodium-filled exhaust, an extra 0.001 inch of clearance is needed to handle the additional heat conducted up through the valve stems.

There is no uniform amount of valve stem-to-guide clearance because of the following variables:

- *Thermal characteristics of the engine.* Some engines run hotter than others (diesels, for example, and those subjected to heavy-duty or high-performance use), which requires additional clearance for heat expansion. The location of the guides in the head with respect to the cooling jackets is also an important factor.

- *Thermal characteristics of the valves.* Sodium-filled valves conduct more heat up through their stems and consequently swell more than ordinary exhaust valves.

- *Thermal characteristics of the guides.* Bronze has a higher rate of thermal conductivity than cast iron so bronze guides do a better job of carrying away heat.

- *Type of valve guide material used (Figure 10-15).* Bronze guides, as a rule, can handle about half the normal minimum factory clearance specified for cast-iron guides or integral guides because of the antiseize characteristics of the material and its superior oil retention qualities.

- *Whether or not the guide has oil retention grooves.* A knurled guide, one with oil retention grooves, or a bronze threaded liner all provide better lubrication than a smooth guide. Consequently, the amount of stem-to-guide clearance need not be as great. Half the factory minimum specified clearance is usually possible. At the same time, oil retained in the grooves forms a wet seal that effectively controls oil usage. This means such guides can also handle greater than normal clearances.

- *Type of valve seal used.* Positive valve seals reduce the amount of oil that reaches the valve stem compared to deflector or umbrella-type valve seals. A guide with a deflector valve seal might need somewhat tighter clearances than one with a positive valve seal to control oil consumption.

Considering all the variables involved, it is impossible to come up with a standard clearance that would apply to all guides in all engines. Therefore, most rebuilders use the factory specs as a guide, going with the lower end of the specified range for light-duty applications and the mid to upper limit for heavy-duty or high-performance applications—or they develop their own specs based on experience.

When it comes to valve guides themselves, as mentioned in Chapter 2, there are two basic original equipment types:

- Cast-iron integral
- Replaceable—either cast iron or bronze.

Some guides might have internal rifling or spiral grooves to improve oil retention. These are usually used with exhaust valves and on diesel engines or dry fuel engines, such as those built to run on propane.

Most people believe bronze guides are better material. The antiseize and antiwear characteristics of bronze generally enable a bronze guide to last two to five times longer than a cast-iron guide. Unfortunately, bronze guides are more expensive so their use in original equipment applications is limited. Bronze is also a more difficult material to machine so tooling does not last as long with bronze as it does with cast iron. Because of these drawbacks, bronze is frequently used as an aftermarket replacement item.

Bronze can make a valve guide better than new because it provides better thermal conductivity than cast iron, helping the valves to run cooler and last longer. Bronze also has a lower coefficient of friction and retains oil better than cast iron. This allows closer stem-to-guide clearances for improved valve stem cooling and reduced oil consumption.

KNURLING

Knurling is one of the fastest techniques for restoring the ID dimensions of a worn valve guide. The process raises the surface of the guide ID by plowing tiny furrows through the surface of the metal (Figure 10-16). As the knurling tool cuts into the

FIGURE 10-16 A typical knurl tool tip (top) and the cut it makes (bottom) *(Courtesy of Goodson Shop Supplies)*

guide, metal is raised or pushed up on either side of the indentation. This effectively decreases the ID of the guide hole. A burnisher is used to press the ridges flat and is then used to shave off the peaks of these ridges to produce the proper-size hole and restore the correct guide-to-stem clearance.

Knurling can be done with either a tap-type knurling tool (Figure 10-17) or a wheel-type or roller knurler (Figure 10-18). To use the latter, proceed as follows:

1. Insert the roller knurling tool into the valve guide. Use a drift and hammer to start it into the guide (Figure 10-19A).
2. Use a hex driver and a drill motor to "roll" the knurling tool through the valve guide

FIGURE 10-17 Tap-type knurling tool *(Courtesy of TRW Inc.)*

FIGURE 10-18 Roller knurling tool *(Courtesy of Sunnen Products Co.)*

A

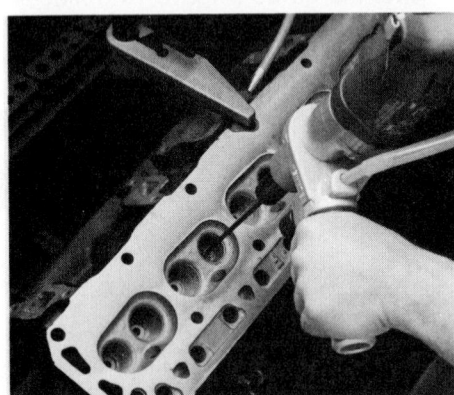

B

C

FIGURE 10-19 Steps in using a roller knurling tool *(Courtesy of Sunnen Products Co.)*

(Figure 10-19B). Several passes through the guide are often required to achieve the desired guide diameter. Use plenty of oil while cutting.

3. Gauge the guide diameter to check if the raised guide wall is sufficient for reaming or honing (Figure 10-19C).

Either knurling tool can be hand or power driven. When power knurling, a drill speed reducer will usually be necessary (Figure 10-20). When clearances do not permit the resizer arbor of the speed

reducer to be removed through the valve guides, remove the arbor from the knurler and back it through the resized guide. (Never back the knurling tool itself through the resized guide.) Then select the reamer and continue the operation as shown un Figure 10-21.

 SHOP TALK _____

On L-head engines, place a cloth over the tappets during the operation to prevent chips from getting into the engine.

One of the main advantages of this technique is that it does not change the centerline of the valve stem appreciably, so it reduces the amount of work necessary to reseat the valve. Knurling also allows a rebuilder to reuse the old valve if wear is within acceptable limits, helping to reduce rebuilding costs. But, in spite of its speed and simplicity, knurling is not a cure for restoring badly worn guides to their original condition. Considerable guide wear from hole to hole also can remain a problem.

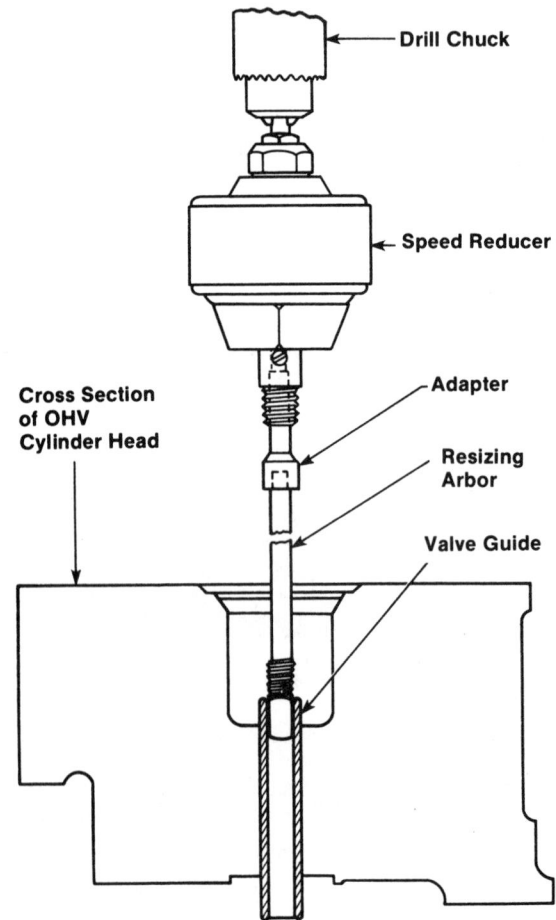

FIGURE 10-20 Drill/speed-reducing operation

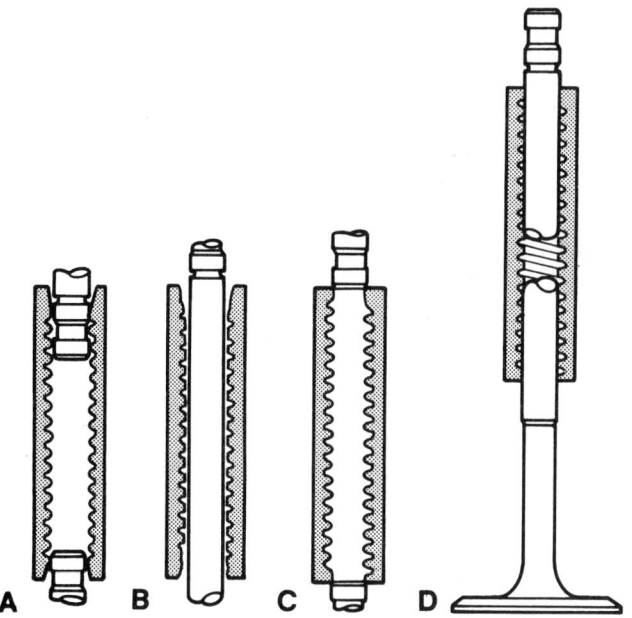

FIGURE 10-21 Valve guide knurling procedures; (A) first knurl; (B) first ream; (C) final knurl; (D) finished guide

The first step in the knurling process is to determine how badly the guides are worn. This might be fine in a small shop with a rebuilder who can take the time to do it, but for a rebuilder in a large, busy shop, it can be very time-consuming and therefore not cost-effective. And unless each guide is checked, knurling all the same amount may or may not restore them to the desired size.

As described in Chapter 6, to determine the amount of guide wear using a telescoping or split ball gauge, subtract the smallest valve stem diameter measured at the end of the guide rub area (most worn area) from the bell-mouth reading to determine the extent of the wear. Do not use a valve seat grinding pilot to check the guide. A valve guide pilot will fit snugly in the unworn center section of the guide and not give a true indication of the amount of bell-mouth wear at each end of the guide. There are several other methods of checking clearance between valve guides and valve stems. One of the most accurate tools is the valve guide gauge set shown in Figure 10-22. Another test is the so-called wobble test using a dial indicator (Figure 10-23).

Opinions vary as to how much wear is acceptable for knurling. Most automotive experts seem to agree that the maximum amount of acceptable wear ranges from 0.004 to 0.007 inch. Oversized knurling tools are available to handle guides with greater wear, but most rebuilders agree that once a guide is worn more than 0.004 to 0.007 inch, some other

A

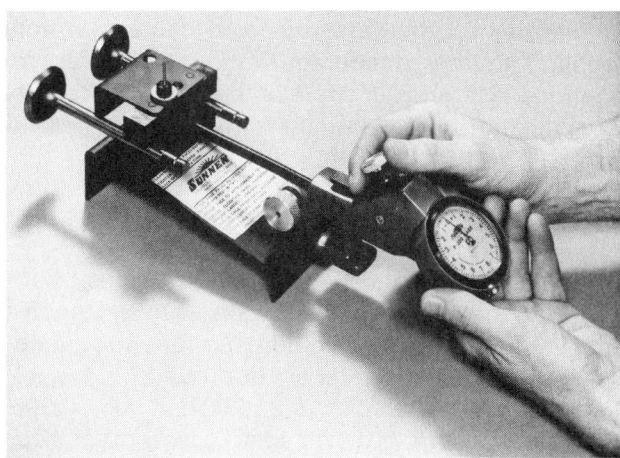

B

FIGURE 10-22 (A) Checking valve guide with gauge; (B) setting gauge with setting fixture *(Courtesy of Sunnen Products Co.)*

FIGURE 10-23 Dial indicator for wobble test

technique besides knurling should be used to restore it.

Some rebuilders do not consider knurling to be a long-term repair because it restores only a portion of the original surface. As the raised ridges are worn, wear accelerates and clearances become excessive. This occurs long before it would have if the entire ID of the guide had been restored.

Another consideration with knurling is that guides do not wear evenly from top to bottom. They wear the least in the middle and the most toward either end. The tapered wear pattern means a knurling tool will do the least where it is needed the most—toward the ends. Knurling a bell-mouthed guide, therefore, only succeeds in restoring the middle section of the guide. Excessive clearance might remain at either end. This has the same effect as shortening the valve guide: it decreases stem cooling and can encourage wobbling. The recommended clearance is one-half the minimum factory specification when knurling worn guides.

REAMING AND OVERSIZE VALVES

Reaming is used to repair worn guides by increasing the guide hole size to take an oversize valve stem or by restoring the guide to its original diameter after installing inserts or knurling.

Standard oversizes are 0.003, 0.005, 0.006, 0.010, 0.013, and 0.015. In some cases, other sizes also may be available. Be sure to check their availability before doing any machining.

When using a decimal-size reamer, 0.343 inch for example, determine the reamer size needed by adding the valve stem diameter to the stem to the guide clearance desired. For example:

$$0.3447'' \text{ stem diameter}$$
$$+0.0013'' \text{ clearance desired}$$
$$\overline{0.3460'' \text{ reamer size needed}}$$

Some reamers are sized by fractions. An 11/32-inch reamer, for example, is equivalent to a 0.343-inch decimal-size reamer. However, it actually measures 0.3437 inch. Therefore, an 11/32-inch plus 0.003-inch clearance measures 0.3467 inch. If the reamers are sized by fractions, be sure to take the size difference into account when making the calculations.

While reaming (Figure 10–24), limit the amount of metal removed per pass to 0.005 inch. If more metal has to be removed, make the cuts in progressive steps, but never exceed 0.005 inch per pass. Always reface the valve seat after the valve guide has

FIGURE 10-24   Limit the amount of metal removal per pass while reaming. When power reaming is used, a speed reducer is also employed *(Courtesy of Goodson Shop Supplies)*

been reamed and use a suitable scraper to break the sharp corner (ID) at the top and bottom of the valve guide.

The advantage of reaming for an oversized valve is that the finished product is totally new: the guide is straight, the valve is new, and the clearance is accurate. The use of oversized valve stems is generally considered to be superior to knurling; yet, like knurling, it is relatively quick and easy. The only tool that is required is a reamer. The valve centerline is maintained so the work required to finish the seat is reduced. However, since reaming requires the use of new valves, it can be more expensive on an engine with many worn guides. Its use is also limited to heads where the guides are not worn beyond the limits of the maximum oversize valve that is available. Because of these limitations, many rebuilders prefer more cost-effective alternatives such as guide liners, inserts, or replacement guides.

To install an oversize stem valve using the reaming method, clean out the valve guide hole with a solid valve guide reamer. Then proceed as follows:

1. Align the guide assembly with the original valve guide bore using a self-expanding pilot and 3/8 inch aligning bushing (Figure 10–25A).
2. Insert the bushing and reamer in the guide assembly after aligning the guide bore (Figure 10–25B).
3. The reamer is guided true with the reaming bushing. A T-wrench is used for hand operation (Figure 10–25C).
4. The power drive is adaptable as a power unit for the reaming operation (Figure 10–25D).

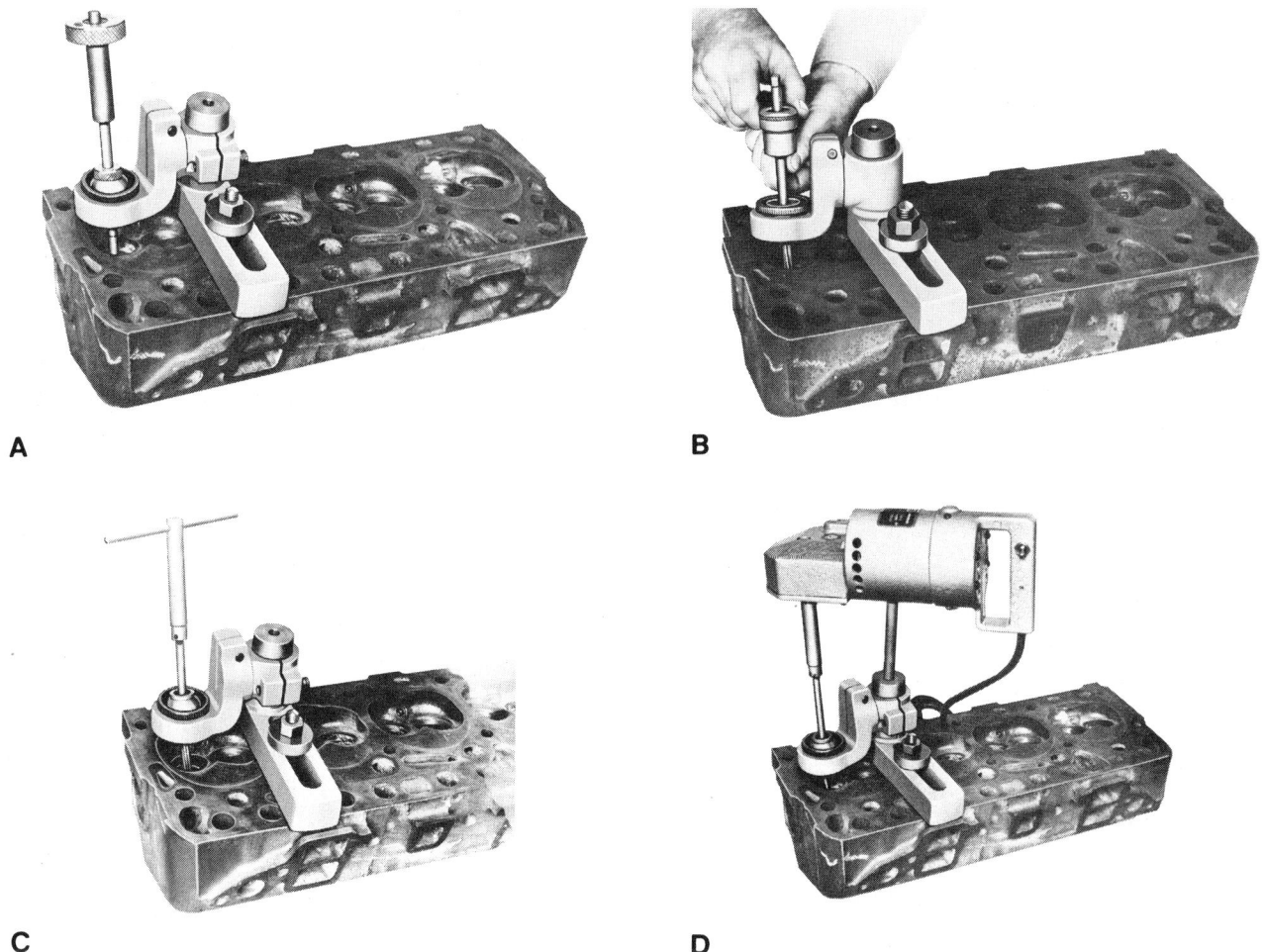

A

B

C

D

FIGURE 10-25 Steps in reaming to install an oversize stem valve *(Courtesy of K.O. Lee, Co.)*

The same procedure can be used when installing replaceable valve guides. The OD of the valve guide and hole must be lubricated before attempting to install the guide.

THIN-WALL GUIDE LINERS

The thin-wall guide liners repair technique offers a number of important advantages and is also popular with many production engine rebuilders (PER's) as well as smaller shops. It provides the benefits of a bronze guide surface (better lubricity, wear, and tighter clearances), it can be used with either integral or replaceable guides (cast iron or bronze), it allows the use of standard or undersized valve stems (reground valve stems for added savings), it is faster, easier, and cheaper (by more than half) than installing new guides in heads with replaceable or integral guides, and it maintains guide centering with respect to the seats.

Thin-wall guide liners are manufactured from a phosphor-bronze or silicon-aluminum-bronze material. The popular liner size is the so-called "0.502-inch universal guide." This liner has 0.502-inch OD, and its ID varies from 0.312 to 0.375 inch. These liners can be cut to almost any length. They are designed for a 0.002- to 0.0025-inch press fit. A tight fit is essential for proper heat transfer to the head and to prevent the liner from working loose.

The liners are installed by first boring out (Figure 10-26) the original guides to 0.030-inch oversize with a special piloted boring tool (Figure 10-27A) pressed into the guide using a driver and air hammer (Figure 10-27B). On guides not precut to length, the excess must be milled off before finishing (Figure 10-27C). The liner is then wedged in place and sized in a single operation by passing a ball broach down through it (Figure 10-28A). This eliminates the need to ream it to size and assures a tight fit between the liner and guide. If a ream finish is desired (Figure 10-28B), it can be obtained by lubricating the ream-

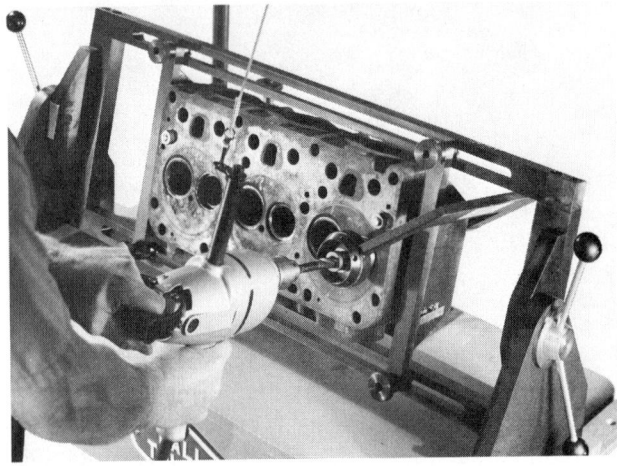

FIGURE 10–26 Two setups for boring thin-walled liners

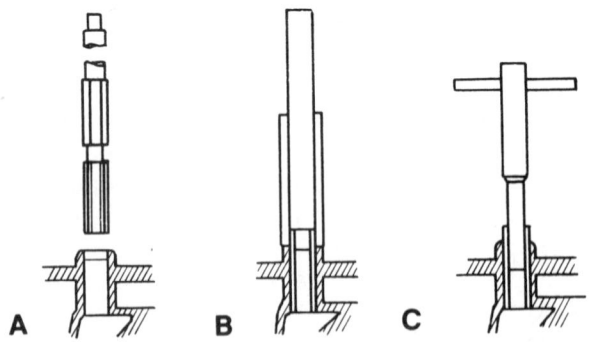

FIGURE 10–27 Bronze guide liner installation: (A) bore guide oversize; (B) install bronze liner in guide; and (C) trim the liner

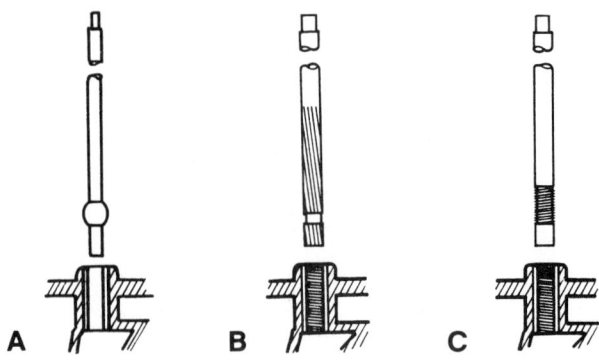

FIGURE 10–28 ID finishes used for guides: (A) ball broach; (B) finish ream, and (C) spiral

er with a bronze-lube and then running it through the guide. For closer than normal stem-to-guide clearance, spiraling or knurling is suggested for added lubrication (Figure 10–28C).

The only trick to using liners is to make sure the hole is round and the correct size. If the hole is distorted excessively or if it is too large, the liner will not fit properly and will cause problems. Running a ball broach or a roller burnishing tool down the liner helps to compensate for hole distortion and presses the liner for a tight fit.

 SHOP TALK

It must be remembered that valve guide honing is ideal for sizing both integral and insert guides. Very close tolerances, ± 0.0002-inch, can be obtained with this process. With reaming, for example, it is difficult to have any clearance closer than 0.0005 inch. These close tolerances give better oil control and heat dissipation.

THREADED BRONZE INSERTS

Another approach that can be used to restore worn guides is to tap the guides and thread in a spiral wound bronze insert. This process uses inserts that are similar to a helicoil insert (see Chapter 14).

The step-by-step procedure for installing threaded bronze inserts is as follows:

1. Thread the valve guide with a self-piloting tap (Figure 10–29A). This tap assures true alignment and a cylindrical bore regardless of a bell-mouthed or egg-shaped condition of the guide.

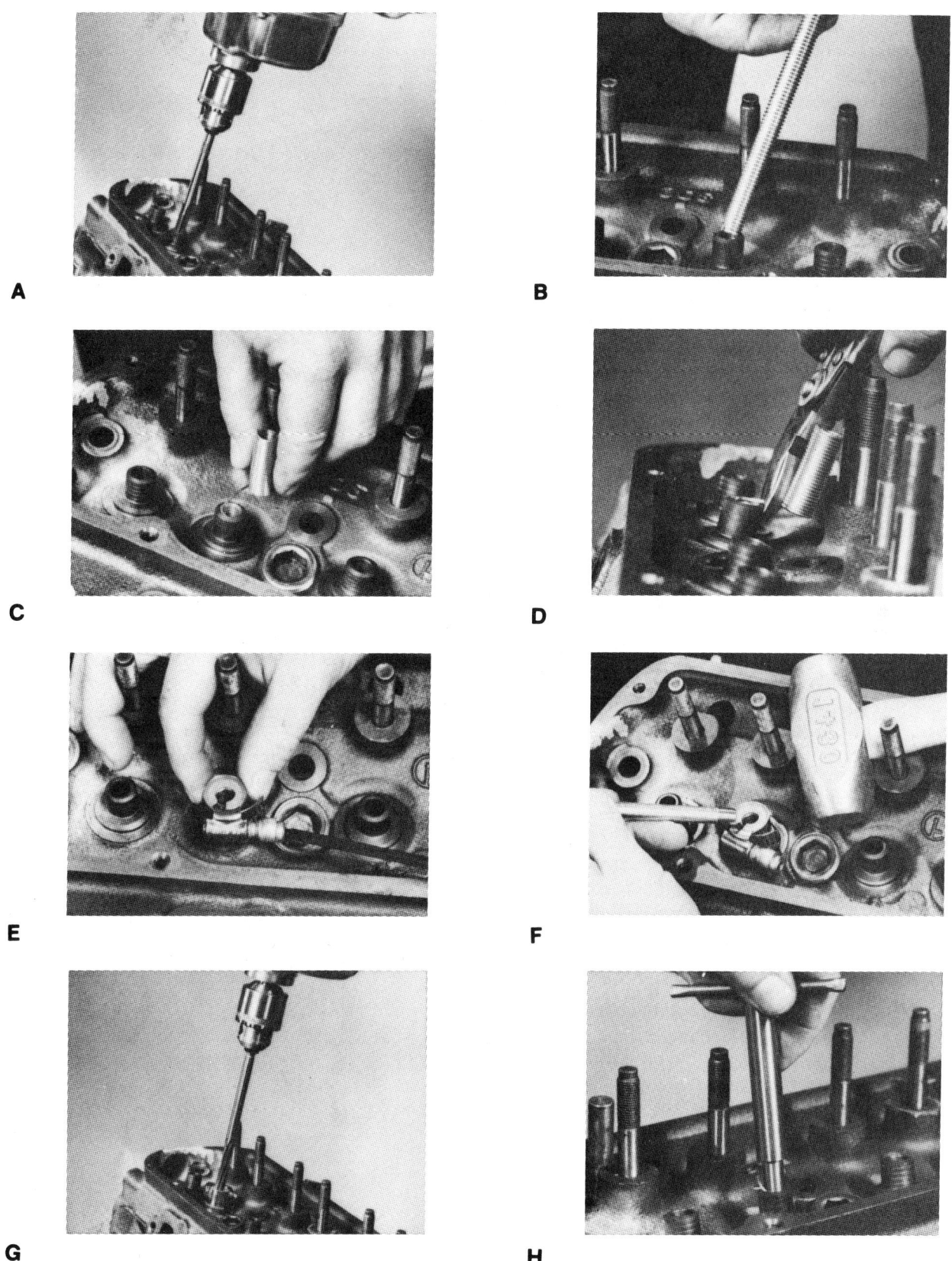

A

B

C

D

E

F

G

H

FIGURE 10-29   Procedure for restoring worn guides with threaded bronze inserts *(Courtesy of Winona Van Norman Machine Company)*

2. Screw the bronze insert bushing into the guide using a special inserting tool (Figure 10-29B). One length of bushing accommodates all lengths of guides.

3. Unwind the excessive portion of the bushing away from the guide (Figure 10-29C).

4. With side cutter pliers, cut the bushing so that it is flush with the OD of the valve guide (Figure 10-29D).

5. Place a plastic serrated bushing retainer and worm gear clamp over the guide tower and tighten with a wrench (Figure 10-29E). This will prevent the bushing from turning while reaming.

6. Drive the broach through the guide to firmly set the bushing (Figure 10-29F). This assures the proper seating of the bushing in the threads and presizes the guide for reaming.

7. Ream the guide with a self-piloting reamer (Figure 10-29G). This will assure a round concentric bore with no taper. Remove the plastic bushing retainer.

8. Insert the cutoff pilot in the guide bore and turn to cut the bushing flush with the top of the guide (Figure 10-29H).

9. The valve guide is now rebuilt. However, if it ever becomes necessary to replace a threaded bronze insert, simply remove the old one and repeat steps 2 to 8.

Regardless of the type of repair technique used to restore valve guides to their original specs, there are several pitfalls that can snare the unwary rebuilder. Occasionally a head will be encountered with different sized valve stems. The factory will usually attempt to mark the head by placing marks on the intake side casting indicating the oversize used (Figure 10-30). But sometimes there will be no markings, especially if someone else has already done a valve job on the head and resized the guides. Always check the valve head for oversize markings.

If the rebuilder makes the mistake of assuming all the valves are the same size and redoes the guides accordingly, he or she will be in for a nasty surprise when attemping to put the head back together. The valves with the oversize stems will not fit the restored guides, and the standard size valves will slip right in. Make sure to check the specified clearances at the right positions on the stem and restore the guides accordingly. It is also important to check the guide ID correctly. Remember that the guides wear the least in the middle and the most toward the end.

Another thing to watch out for is tapered valve stems. Some newer model vehicles have valves

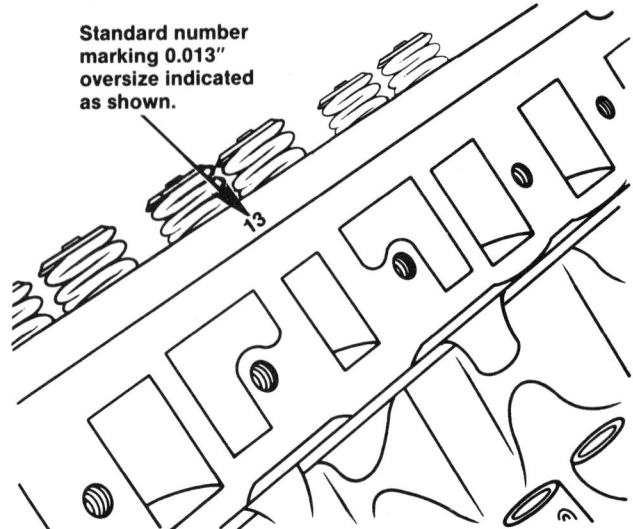

FIGURE 10-30 Factory heads usually have one or more oversize valves.

which are 0.001 inch thicker at the top of the stem than at the bottom to compensate for increased thermal expansion closer to the valve face (Figure 10-31). With tapered guides, care must be taken to install them from the correct side.

VALVE GUIDE REPLACEMENT

Replacing the entire valve guide on heads with replaceable guides is yet another method of repair.

FIGURE 10-31 With a taper stem, dimension B is always smaller than dimension A. The taper can be measured by an outside micrometer.

Pressing out the old guides and installing new ones can be difficult with some aluminum heads where the interference fit is considerable. Cracking the head or galling the guide hole is always a risk. One recommendation is to preheat the heads in an oven prior to guide removal and to lubricate dry liners before driving them out.

 SHOP TALK

An oven should be used to bring the head to 250 degrees Fahrenheit. Guides can be difficult to remove after heating to 500 degrees (the temperature at which most straightening of aluminum heads takes place). A torch should never be used because it can burn through the head. Localized heating can also induce considerable distortion.

As mentioned in both Chapters 2 and 6, there are two basic types of valve guides: integral and insert (Figure 10-32). Since aluminum is much softer than cast iron, cylinder heads made of aluminum require a valve seat insert of a harder material.

Integral Guides. To replace integral guides, bore (Figure 10-33) or drive (Figure 10-34)

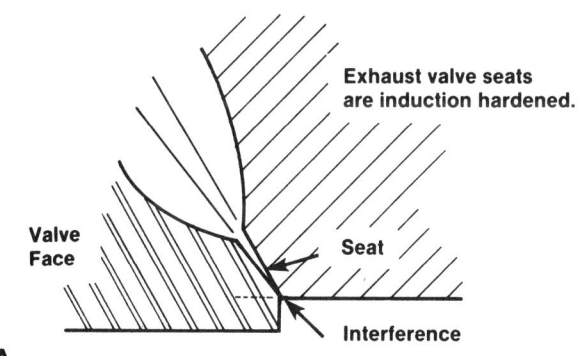

A

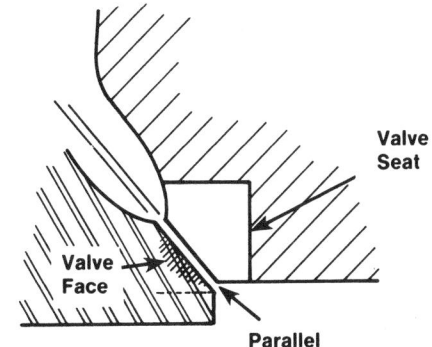

B

FIGURE 10-32   Two types of valve guides: (A) integral seat and (B) replaceable insert seat

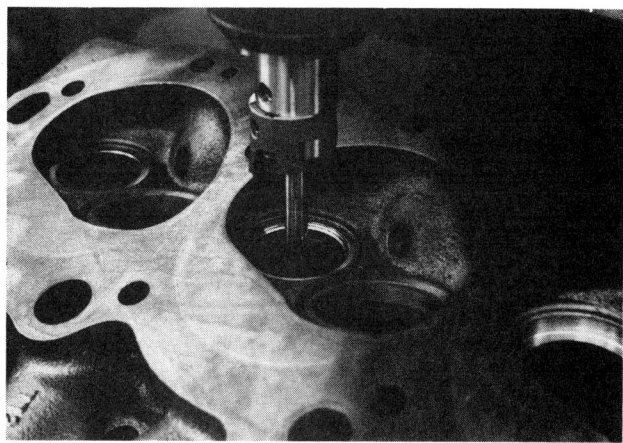

FIGURE 10-33   Boring out an old integral guide

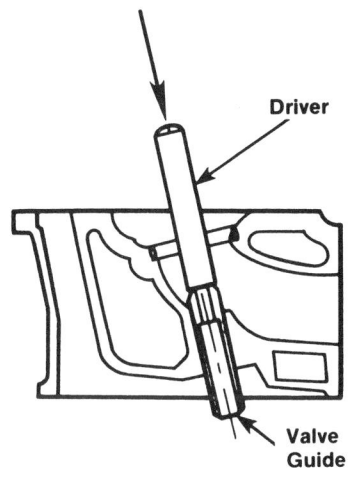

FIGURE 10-34   A driver is used to drive out the old guide and drive in a replacement.

the old guide out and drive a thin-wall replacement guide into the hole. Many shops use a seat and guide machine (see Figure 10-4) for this process, although it can be done with portable equipment. Drive the replacement guides in cold with approximately 0.002 inch press fit. Use an assembly lube to prevent galling. It is necessary to keep the centerline of the guide concentric with the valve seat so that the rocker arm to valve stem contact area is not disturbed (Figure 10-35).

Occasionally, a new guide will not be concentric with the valve seat. Install a new seat to correct the problem and check the rocker arm to stem contact area (Figure 10-36).

Insert Guides. To press out an old valve guide, place a proper-sized driver so that its end fits snugly into the guide. The shoulder on the driver must also be slightly smaller than the OD of the

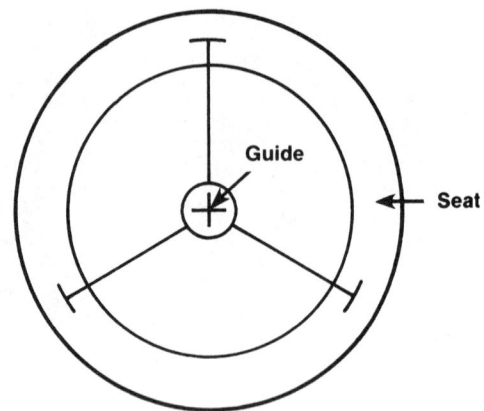

FIGURE 10–35 The seat should be concentric with the guide.

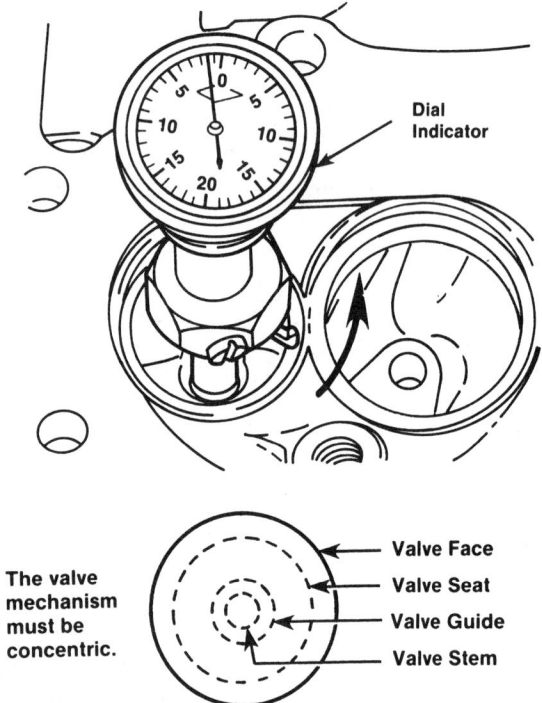

The valve mechanism must be concentric.

— Valve Face
— Valve Seat
— Valve Guide
— Valve Stem

FIGURE 10–36 Checking valve seat concentricity with a special gauge. The gauge is turned in the direction of the arrow. As indicated, the valve face, seat, guide, and stem must all be concentric.

guide, so that it will go through the cylinder head. Use a heavy ball peen hammer to drive the pressed guide out of the cylinder head.

Pressing or driving out and installing new guides is not difficult, but there is always the danger of breaking the guide or tearing up the guide hole in the head. Cast-iron guides in particular have a tendency to gall aluminum heads and once the hole is damaged, it must be bored out and an oversized

guide installed—assuming one is available to fit the application. New guides should be chilled prior to installation because of the interference fit between the guide and head. Chilling them in a freezer or with dry ice works well. Lubricant also helps prevent galling.

Use caution when driving or pressing replacement insert guides—the wrong procedure can easily damage the head. For example, some engines use a stepped exhaust guide. It has to be driven out from the combustion chamber side or the head could be cracked and ruined. Also, many import heads have flanged guides that must be driven out from a particular side. Be sure to check the manufacturer's manual for any head that is unfamiliar. As a rule of thumb, if the guide is straight (no taper or flange) on an aluminum cylinder head, drive out the guide toward the combustion chamber.

Bronze guides, or valve guides on aluminum heads, can be difficult to remove. If a problem is encountered, the head can be heated for a short time (approximately 15 minutes at 400 degrees Fahrenheit). The guides will release quickly and easily because of head expansion. When the head cools, it will return to its previous size.

🔧 SHOP TALK _____

When replacing valve guides in aluminum head engines, many experienced rebuilder/ machinists tap a screw into the bottom of the guide and install a screw so that the driver is against it (Figure 10-37). Failure to do this might cause the top of the guide to mushroom which would ruin the guide bore resulting in a loose new guide.

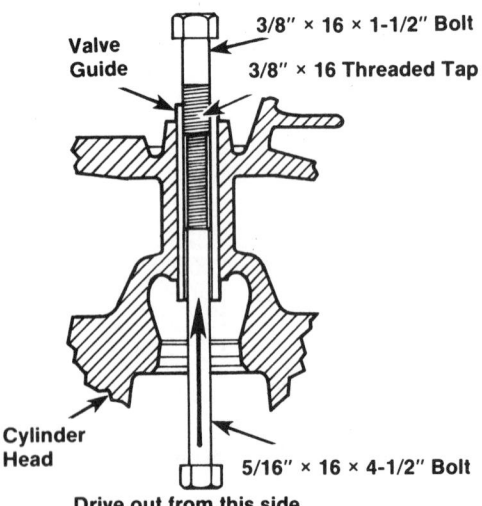

Valve Guide

3/8" × 16 × 1-1/2" Bolt
3/8" × 16 Threaded Tap

Cylinder Head

5/16" × 16 × 4-1/2" Bolt
Drive out from this side.

FIGURE 10–37 Removing a guide from an aluminum head

When installing new guides, be careful not to damage them. Use a press and the same driver that was used to remove the old guide. Align the new guide and press straight down (Figure 10–38A), not at an angle. An air hammer (Figure 10–38B) and special driver can also be used to install new guides.

At least one engine manufacturer uses guides that are cut off at an angle on the combustion chamber end. Others may also be found to have guides cut at an angle at one end or the other. No attempt should be made to press or drive against the angled end. Ascertain the correct amount of guide protrusion. Guide height is important to avoid interference with the valve spring retainer on engines with high valve lift. The guide must also fit the hole tightly or it can work loose. The manufacturer's specifications give the correct valve guide installed height, but it is good practice to measure how far the old guides stick out of the cylinder head and to use this measurement as a reference. As each guide is installed, the technician should measure the installed height with a scale (Figure 10–39). After the new guides have been installed, insert a valve and check for any stem interference.

FIGURE 10–39 Determining the correct amount of guide protrusion *(Courtesy of TRW Inc.)*

Valve guide honing is ideal for sizing both integral and insert guides. Very close tolerances, ± 0.002 inch, can be obtained with this process. With reaming, for example, it is difficult to have any clearance closer than 0.005 inch. These close tolerances give better oil control and heat dissipation.

Pressure checking after assembly of the valves is also strongly recommended (Figure 10–40) to check the quality of the seal. One of the main drawbacks of replacing guides is that it changes the valve centerline, particularly when drilling out heads with integral guides. This can sometimes lead to seat refinishing problems if the centerline has moved too far off center.

A

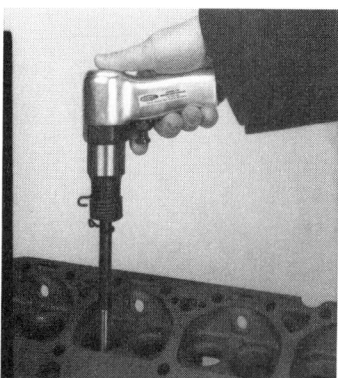

B

FIGURE 10–38 Installing new guides (A) with press and driver *(Courtesy of TRW Inc.)* and (B) with air hammer *(Courtesy of Goodson Shop Supplies)*

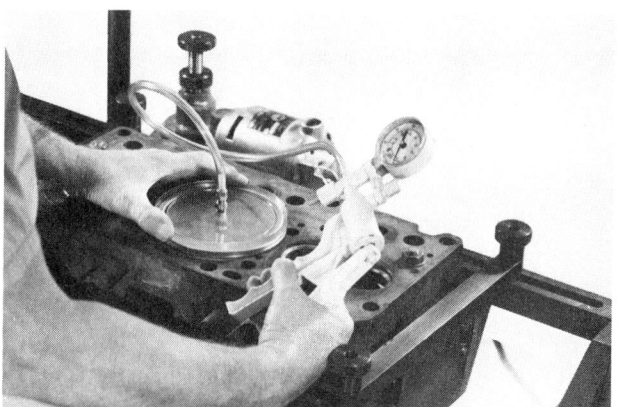

FIGURE 10–40 Pressure checking a valve guide. To use it properly, be sure to follow the manufacturer's instructions. *(Courtesy of Hall-Toledo Inc.)*

VALVE SEATS

The most critical sealing surface in the valve train assembly is between the face of the valve and its seat in the cylinder head when the valve is closed. Leakage between these surfaces reduces the engine's compression and power and can lead to valve burning. To ensure proper seating of the valve, the valve seat must be:

- Correct width (Figure 10–41)
- Correct location on the valve face
- Concentric with the guide (less than 0.002 inch runout)

The ideal seat width for automotive engines is 1/16 inch for intake valves and 3/32 inch for exhaust valves. Maintaining this width is important to insure proper sealing and heat transfer. However, when an existing seat is refinished to make it smooth and concentric, it also becomes wider. Wide seats cause the following problems:

- Seating pressure drops as seat width increases.
- Less force is available to crush carbon particles that stick to the seats.
- Seats run cooler, allowing deposits to build up on them.

The seat should contact the valve face 1/32 inch from the margin of the valve. When the engine reaches operating temperature, the valve expands slightly more than the seat. This moves the contact area down the valve face. Seats that contact the valve face too low might loose partial contact at normal operating temperatures.

Like valve guides, there are two types of valve seats—integral and insert. Integral seats are part of the casting. Insert seats are pressed into the head and are always used in aluminum cylinder heads. Most pre-1978 integral seats are soft cast iron. After

1978, most manufacturers began to produce cylinder heads with induction hardened cast-iron seats able to withstand the higher heat of exhaust applications. Insert seats are added to the cylinder head after casting, or as replacements for worn integral seats.

RECONDITIONING VALVE SEATS

Valve seats can be reconditioned or repaired by one of two methods, depending on the seat type:

- Machining a counterbore to install an insert seat
- Grinding, cutting, or machining an integral seat

But before starting the seat work, carefully check the seats for cracks (Figure 10–42). Cracked integral seats sometimes can be repaired by installing inserts, if the crack is not too deep. Cracked insert seats must be replaced. Check insert seats for looseness with a small pry bar (Figure 10–43). Replace if movement is noted.

FIGURE 10-42 Cracked seats

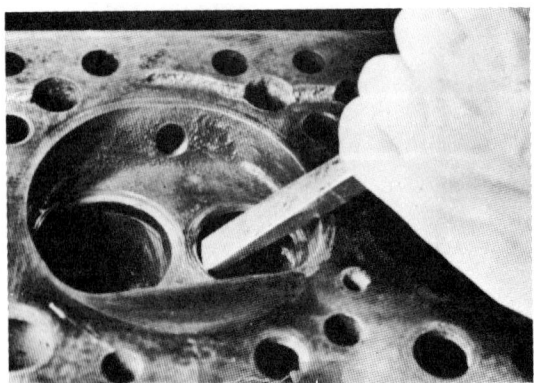

FIGURE 10-43 Insert seats being checked for looseness with a small pry bar *(Courtesy of TRW, Inc.)*

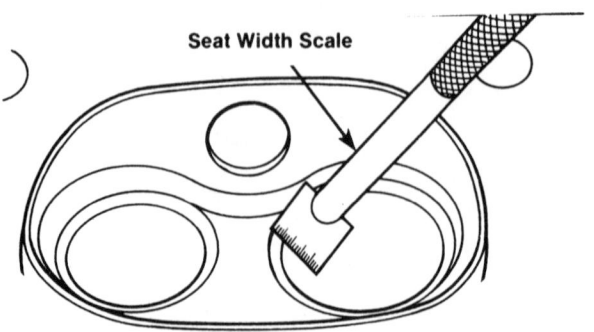

FIGURE 10-41 Checking valve seat width

Seat Width Scale

Installing Seat Inserts

Valve seats can be installed in cylinder heads in a variety of ways. One of the most common and successful procedures is as follows:

🛠 1. To remove a damaged insert, use a puller (Figure 10-44) or a pry bar (Figure 10-45). Although the puller bar can be used to remove the inserts from most any type of head, when using a pry bar extra care must be taken not to damage aluminum heads. In fact, many experienced technicians, rather than prying on the seat, cut it out with the next smallest seat cutter available.

FIGURE 10-46 Counterbores are cleaned up or recut to accommodate oversize inserts *(Courtesy of TRW, Inc.)*

This will leave a thin piece that can be easily removed. The seat can also be removed by welding. Weld an old, smaller valve to the seat and knock the seat out by driving it against the stem or weld a bead on the seat. When the weld cools, it will shrink. This will loosen the seat for easy removal.

2. After removal, clean up the counterbores or recut to accommodate oversize inserts (Figure 10-46). The bores must be clean and round with 0.001 inch TIR, and the bottoms must be clean and flat within 0.001 inch TIR with no tool chatter marks for good heat conductivity. Use either the engine manufacturer's recommendations or Table 10-1 for interference fit.

3. Insert the counterboring pilot into the valve guide. Then mount the base and ball shaft assembly to the gasket face angle of the cylinder head (Figure 10-47).

4. Use an outside micrometer to accurately expand the cutterhead to a predetermined

FIGURE 10-44 Puller removes inserts *(Courtesy of TRW, Inc.)*

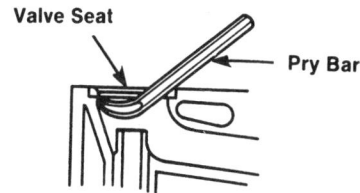

FIGURE 10-45 A damaged valve seat can be removed with a pry bar.

TABLE 10-1: RECOMMENDED INTERFERENCE FITS

Outside Diameter (inches)	Insert Depth (inches)	Interference Fit (inches)
0–1	0–1/4	0.001–0.003
1–2	1/4–3/8	0.002–0.004
2–3	3/8–9/16	0.003–0.005
3–4	9/16–16/16	0.004–0.006

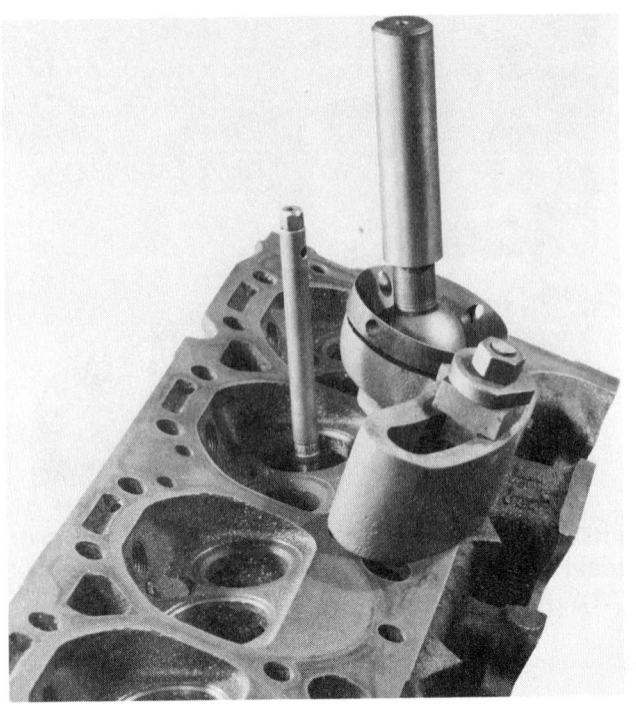

FIGURE 10-47 Mounting the base and ball shaft assembly to the gasket face angle of the cylinder head *(Courtesy of Hall-Toledo, Inc.)*

FIGURE 10-48 Using an outside micrometer to expand a cutterhead accurately *(Courtesy of Hall-Toledo, Inc.)*

size of the counterbore (Figure 10-48). Remember that the counterbore should have a slightly smaller ID than the OD of the insert.

5. Place the valve insert counterboring tool over the pilot and ball-shaft assembly. Preset the depth of the valve seat insert at the feed screw (Figure 10-49). A magnetic ring is often used to trap the dust created when grinding valve seats (Figure 10-50).
6. Cut the insert by turning the stop-collar until it reaches the preset depth (Figure 10-51). Use a lubricant on the cutters for a smoother finish.
7. To install the insert, heat in a parts cleaning oven to approximately 350 to 400 degrees Fahrenheit. Chill the seat insert in a freezer or with dry ice before installation.

WARNING: Wear the proper gloves when handling dry ice.

---

8. Press the seat with the proper interference fit using a driver (Figure 10-52).
9. When the installation is complete, the edge around the outside of the insert is staked

FIGURE 10-49 Presetting valve seat insert depth at the feed screw *(Courtesy of Hall-Toledo, Inc.)*

as shown in Figure 10-53. By doing so, the inserts will be secured more effectively in the counterbore.

FIGURE 10-50   A magnetic ring collects grinding dust *(Courtesy of Goodson Shop Supplies)*

FIGURE 10-51   Cutting the insert *(Courtesy of Hall-Toledo, Inc.)*

FIGURE 10-52   Inserts being installed and seated firmly *(Courtesy of Sunnen Products Co.)*

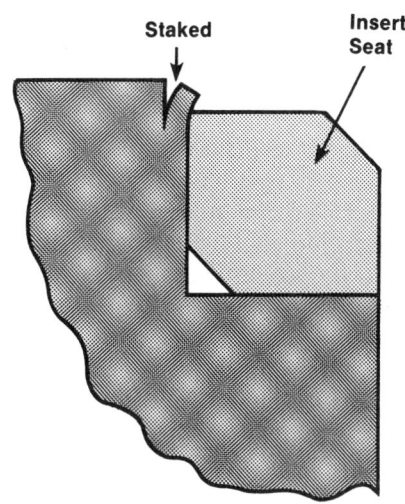

FIGURE 10-53   Staking the outside of the insert

SEAT GRINDING

As stated earlier, it is possible to recondition an integral seat by grinding, cutting, or machining. To grind an integral valve seat, it is very important to select and install the correct size pilot and stone.

When selecting a stone, choose the appropriate grade as indicated in Figure 10-54. Keep in mind that

FIGURE 10-54   Grades of seat grinding stone: *Nickel Chrome* (NC) will not groove or load up; *General Purpose* (GP) removes material from cast-iron seats or inserts; finishes (use light pressure); *Finishing* (F) gives an ideal surface on most seats and inserts and removes very little material for a fine finish; and *Stellite* (S) is a special composition grind wheel for stellite and other hardened valve seat inserts *(Courtesy of Goodson Shop Supplies)*

for hard seats use a soft stone and for soft seats (cast iron) use a hard stone. The use of cutting oil will aid in grinding some materials. The stone should be carefully dressed before grinding each seat to insure a clean smooth cut (Figure 10-55). In some cases, it might be necessary to dress the stone several times for each seat. Use a dial indicating runout gauge to check the seat (Figure 10-56). Try to obtain a seat with no more than 0.001 inch TIR, although 0.002 inch is permissible. If the full 0.002-inch tolerance is also permitted on the valves, a mismatch of 0.004 inch between the valve and seat is possible; this, of course, is not desirable. The stone should be approximately 1/8 inch larger than the seat.

 SHOP TALK —————

Before grinding, many rebuilders clean the seats by placing a piece of fine emery cloth between the stone and the seat and giving the surface a hard rub. This will help to prevent contamination of the seat grinding stone with any oil or carbon residue that might be present on the valve seat. This contamination could cause glazing and is ineffective for seat grinding.

Insert a pilot shaft of the correct diameter into the valve guide of the seat to be ground (Figure 10-57). The pilot should fit snugly in the valve guide and should not wiggle. The pilot is used to guide and

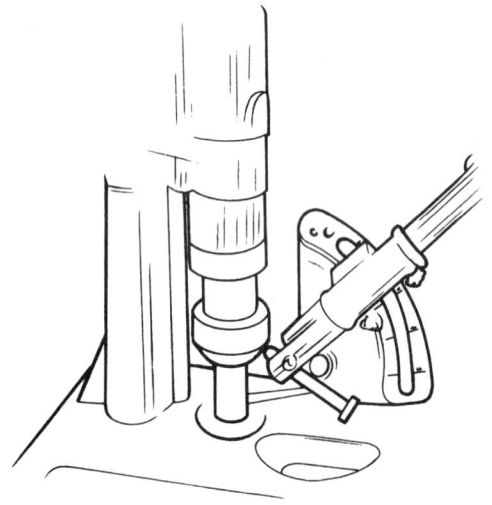

FIGURE 10-55 Stone should be carefully dressed before grinding each seat to insure a clean, smooth cut. *(Courtesy of TRW Inc.)*

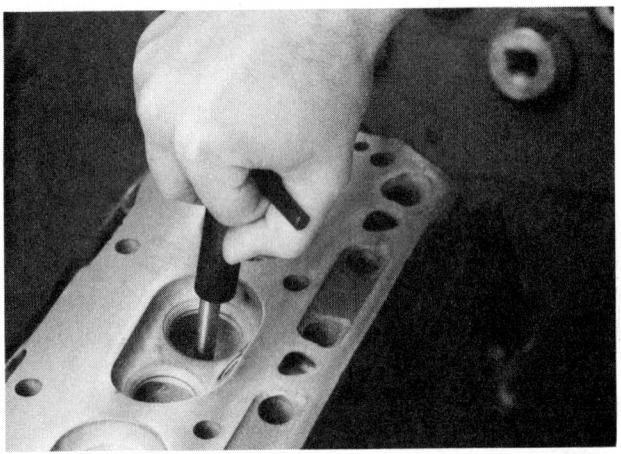

FIGURE 10-57 Installing a pilot guide in preparation for grinding *(Courtesy of Sunnen Products Co.)*

FIGURE 10-56 Check seat with dial indicating runout gauge *(Courtesy of TRW Inc.)*

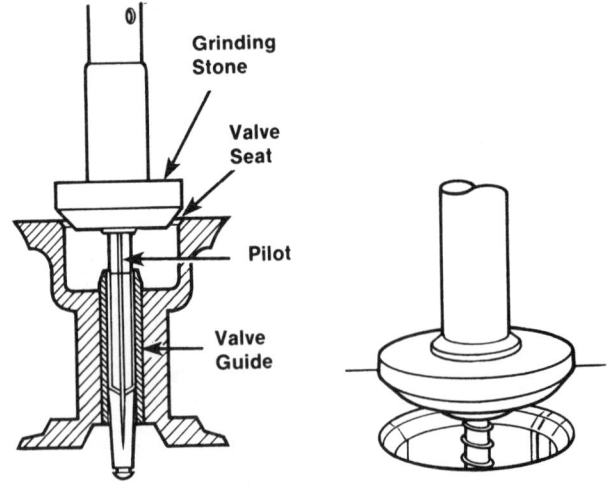

FIGURE 10-58 When grinding the valve seat do not remove any more material than is necessary to produce a good valve seat surface *(Courtesy of Sioux Tools Inc.)*

center the grinding wheel. Since the valve guide is used for centering, all valve guide service or replacement must be performed before seat grinding (Figure 10-58).

Grind the seat by continually raising and lowering the grinder unit on and off the seat at a rate of approximately 120 times per minute (Figure 10-59). This way, the wheel will maintain its maximum speed and also be given a chance to free the cuttings and

FIGURE 10-59   Grinding the seat *(Courtesy of Hall-Toledo Inc.)*

FIGURE 10-60   Grinder with console controls *(Courtesy of Hall-Toledo Inc.)*

wheel particles from the seat and the wheel. Some grinders are equipped with console controls that permit the operator to set the grinding speed (Figure 10-60).

Repeat the raising and lowering process until the seat is clean and free of defects. Remove only enough material to clean the seat and provide a good finish of sufficient width all the way around the seat. Avoid any side pressure during the grinding process. Also make sure the stone pressure against the seat is carefully controlled to prevent any chatter. Excessive pressure or chatter can damage a valve seat very quickly. A drop or two of oil on the pilot and hex or star drive can reduce chatter. But avoid getting any oil on either the seat or stone.

Usually the manufacturer sets the specifications as to the amount of face-to-seat contact. When they are not available, a general rule is that the contact area should be 1/16 inch for intake valves and 3/32 inch for exhaust valves.

As can be seen in Figure 10-61, a proper overhang is necessary. Briefly, the overhang is the area of the face between the contact area and the margin. It should be 1/64 inch for both intake and exhaust valves.

When the contact area is too wide and there is not enough overhang, the seating pattern can be corrected by overcutting or topping. Overcutting involves grinding the seat top at a slight angle, usually 15 degrees. As can be noted in Figure 10-62, this decreases the contact area and increases the overhang.

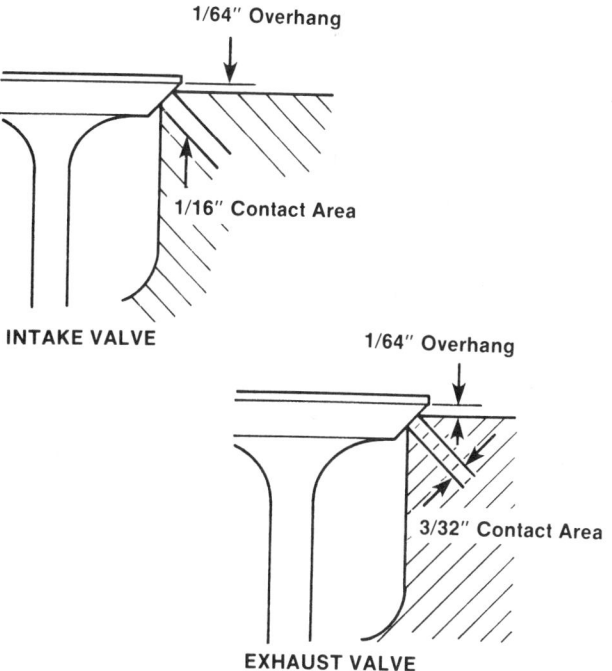

FIGURE 10-61   Proper valve overhang

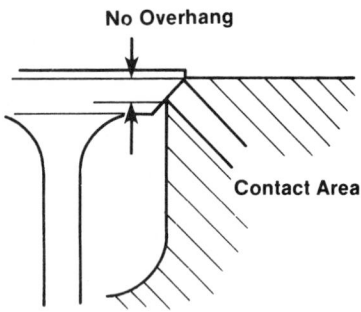

No Overhang

Contact Area

BEFORE OVERCUTTING

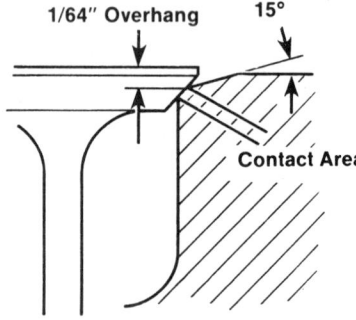

1/64" Overhang   15°

Contact Area

AFTER OVERCUTTING

FIGURE 10-62   Results of overcutting

If the contact area is still too wide after overcutting, the seat bottom can be lowered by undercutting or throating. Undercutting is done in the same manner as seat grinding, except that a 60- to 70-degree stone is used to narrow the seat from the port side.

When the valve and seat are ground with an interference angle, the result will be a line contact

FIGURE 10-63   Measuring seat width if an interference angle is used. *(Courtesy of TRW Inc.)*

and it is necessary to measure the seat width as shown in Figure 10-63. If no interference angle is used, the result will be a seat width, usually about 1/16 inch. The width can be measured at the same time the seat location is checked (Figure 10-64).

To determine the seating location, coat the valve face lightly with Prussian blue, as shown in Figure 10-65. Insert the valve in place and lightly bounce it on the seat (Figure 10-66). Remove the valve and wipe it clean. Again, insert the valve in place and lightly bounce it on the seat. Remove it

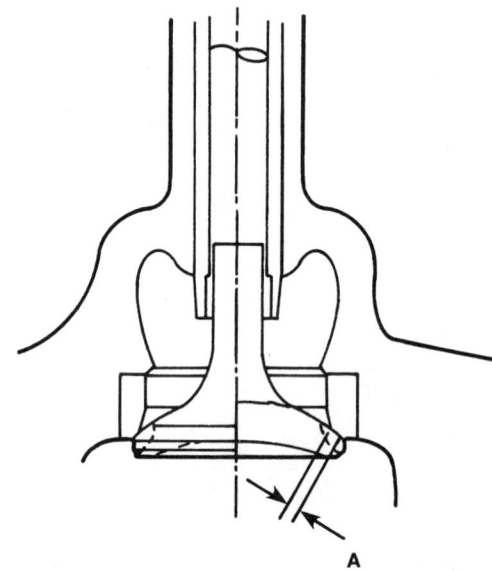

A

FIGURE 10-64   When the valve face angle is the same as the seat angle, a specified seat width results.

FIGURE 10-65   Valve face is coated lightly with Prussian blue. *(Courtesy of TRW Inc.)*

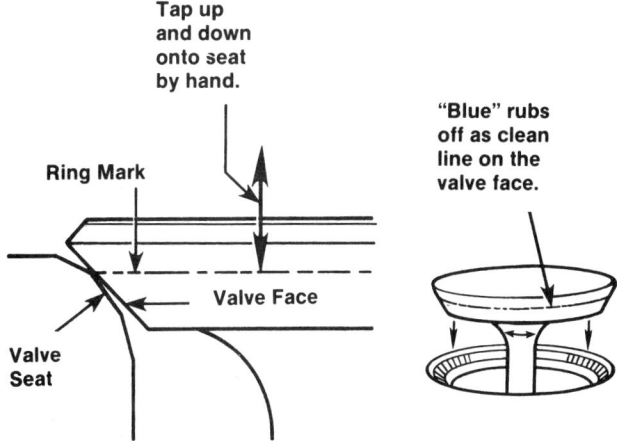

FIGURE 10-66   Valve is inserted in place and lightly bounced on seat. *(Courtesy of TRW Inc.)*

FIGURE 10-67   A correctly formed seat will display no more than 0.002 inch total indicator deflection.

and note the seating pattern on the face. Check to be sure that seats are well seated (Figure 10–67).

 SHOP TALK

A felt-tipped marking pen or a lead pencil works well when Prussian blue is not available. When the valve is pressed against the seal, the imprint of the contact area will be visible on the face. The width of the contact area on the face can then be measured.

Some problem in measuring might be encountered when using aluminized valves because the faces might not be smooth enough to get a good imprint with Prussian blue. Do not grind off the aluminum coating because it is important for corrosion resistance. Use an old refaced valve or select another new valve with the same dimensions for checking the seating pattern.

SHOP TALK

Never lap an aluminized valve. Most rebuilders do not recommend lapping any valves because it can cover up excessive runout and irregularities or result in rapid wear of the seats, valve faces, valve stems, and guides if the lapping compound is not completely removed.

CUTTING VALVE SEATS

Cutting valve seats differs from grinding only in the equipment used (Figure 10–68). Hardened valve seat cutters replace grinding wheels for seat finishing. The basic seat cutting procedures are the same as those for grinding, except that the steps are in a slightly different order. That is, it is usually recommended that the bottom is narrow cut first, followed by the top narrow cut, then the final seat cut of either 45 or 30 degrees. This procedure seems to give a better seat finish. The cutter is used over a pilot, as with a grinding stone. Check the width, location, and concentricity. Cutting machine manufacturers claim this method has advantages such as speed, simplicity, and no grinding dust.

FIGURE 10-68   Cutting a seat valve *(Courtesy of Neway Manufacturing, Inc.)*

MACHINING VALVE SEATS

As stated earlier in this chapter, a valve system rebuilding machine such as the one shown in Figure 10-69 can be used to install valve guides (Figure 10-70) and to machine valve seats (Figure 10-71). Some have optional seat cutters that make three-angle cuts (Figure 10-72). The cutters are set to either an intake or exhaust valve face. Once set, they machine the seat as well as the top and throat angles. Two primary advantages of these cutters over other methods are high speed and precision. A set of

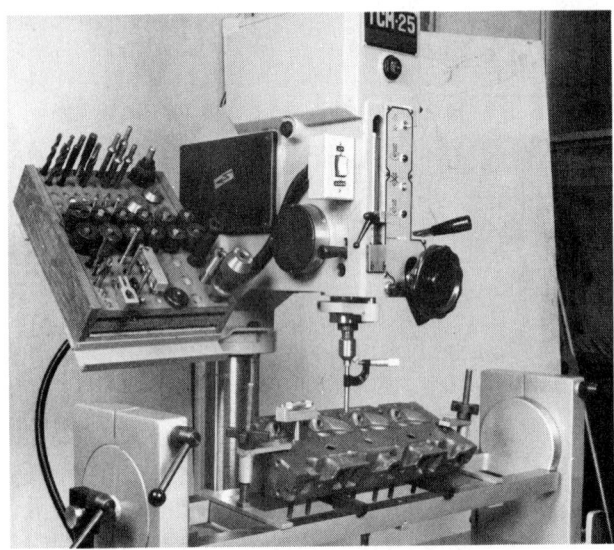

FIGURE 10-69 Typical work area of valve guide and seat machine *(Courtesy of Peterson Machine Tool Inc.)*

FIGURE 10-70 Self-centering core drill provides machining on worn guides *(Courtesy of K. O. Lee Company)*

FIGURE 10-71 Machining a valve seat *(Courtesy of Kwik-Way Manufacturing Company)*

FIGURE 10-72 Machining all three angles at once. *(Courtesy of Peterson Machine Tool, Inc.)*

heads for a V-8, for example, can be finished in 15 to 20 minutes with every seat precisely located. Concentricity should still be checked to insure excellent performance.

The average seat width is 0.060 inch and the average seat begins 0.030 inch from the valve margin. A properly ground seat has three angles: top, 30, or 15 degrees; seat, 45, or 30 degrees; and throat, 60 degrees. Typically, the 45-degree angle wedges tighter than the 30-degree seat, so it is used most often. Using three angles maintains the correct seat width and sealing position on the valve face (Figure 10-73). Correct sealing pressure and heat transfer from the valve through the seat are also affected.

The heads can be held in various ways (Figure 10-74) in valve guide and seat machines. Most of

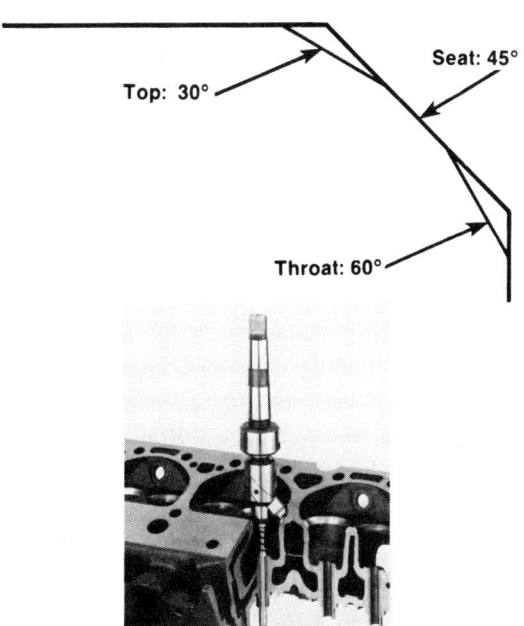

FIGURE 10-73 Making a three-angle cut

A

B

FIGURE 10-74 Two ways of holding a head in a valve guide and seat machine: (A) four-way tilt table and (B) revolving fixture *(Courtesy of Peterson Machine Tool, Inc.)*

them have a four-way tilt table and a fixture that revolves 360 degrees to permit the technician to work on both sides. They also feature some type of easy leveling device that aligns the valve guide in a vertical plane to match the spindle centerline (Figure 10-75). Be sure to check the machine's service manual for proper setup and operation.

In addition to simple valve seat and guide installation procedures, a typical head shop machine can do such operations as cutting the three angles of a guide seat, installing bronze inserts, knurling valve guides, establishing alignment on studs (Figure 10-76), facing stud boss, and tapping for screw-in studs.

FIGURE 10-75 Leveling device helps to align the valve guide in a vertical plane *(Courtesy of K. O. Lee Company)*

FIGURE 10-76 Drilling and rapping for manifold studs *(Courtesy of Peterson Machine Tool, Inc.)*

VALVE RECESSION

Tetraethyl lead compounds, formerly found in gasoline, provide vital lubrication to the valve seat and valve stem in addition to raising the octane rating. Essentially, the modifications in engine design to reduce emissions and the use of unleaded fuel have greatly increased valve and cylinder head temperatures and virtually eliminated all lubrication from the valve seat and valve stem. This has caused a considerable increase in valve stem and guide wear and exhaust valve recession into the seat. The extent of this wear depends greatly upon the severity of engine operation and the length of time it has been in service. Older four-cylinder import engines are especially prone to valve recession problems because these engines tend to operate at higher rpms than a V-6 or V-8, and many do not have hydraulic valve lifters. Such engines require periodic valve lash checks and adjustments (which are often neglected), so as the exhaust seats wear away and the valves sink into the head, clearances tighten up and the valves burn.

The only sure way, at present, to prevent valve recession is to install hardened exhaust valve seats in the head. Use a premium-grade insert for the exhaust valves on engines built prior to 1976 because ordinary cast-iron seats are not hard enough, though cast iron can still be used for the intakes. As shown in Figure 10-77, the difference in valve recession on low lead fuel is obvious between the valve with the hardened seat and the one with the soft seat.

LAPPING VALVES

Lapping valves is a valve finishing operation in which a fine wet grinding compound is placed between the valve face and seat and the valve turned back and forth. The procedure can be accomplished by hand using lapping sticks or by a powered lapping tool (Figure 10-78). Although both methods attach themselves to the valve head by using a suction cup, the latter, driven by a power drill, is much faster than the stub method.

Although a few rebuilders believe that a quality valve job is not complete until all valves have been properly lapped, most shops consider lapping an obsolete practice. True, it does produce a smooth idle when the engine is first started, but once it is warmed up, the valves expand so that the lapped surface of the valve does not line up with the lapped area of the seat. Some engine manufacturers still recommend valve lapping.

If lapping is to be done, it is best that the seats are lapped with an old valve that has been refaced. In doing this, the seat finish can be improved without undercutting on the face of the valve to be installed in the rebuilt engine.

VALVE STEM SEALS

Valve stem seals (Figure 10-79) are used on many engines to control the amount of oil allowed between the valve stem and guide. The stems and

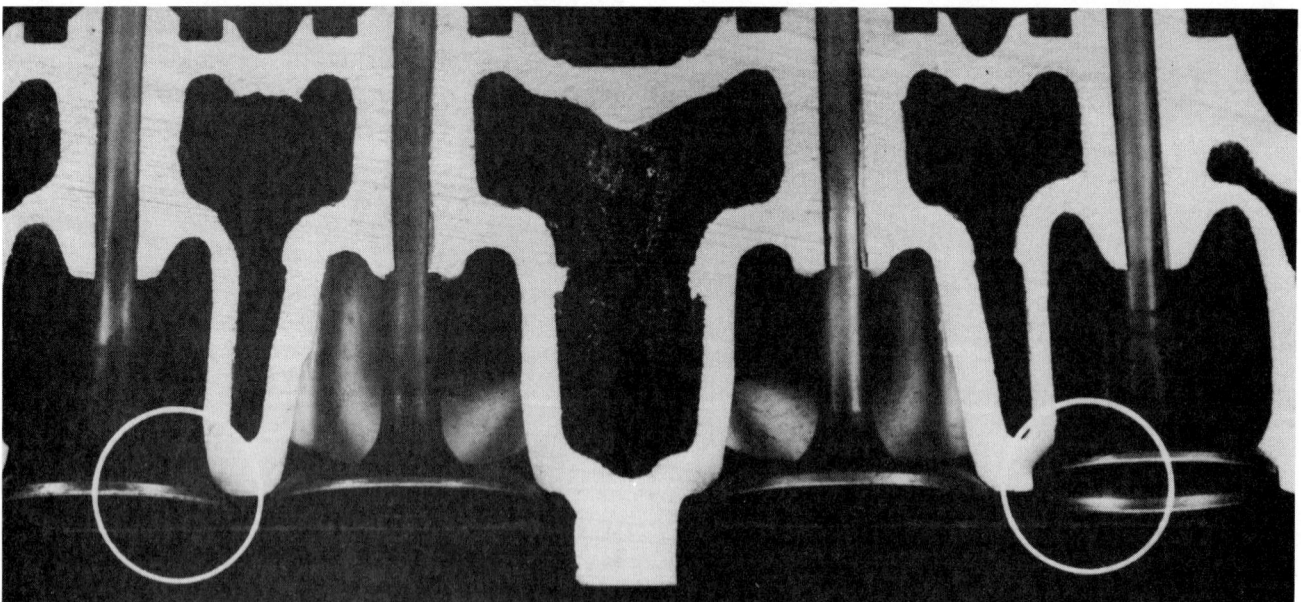

FIGURE 10-77 Photograph shows how far the valve seat has receded into the cylinder head (circled areas). As the valves recede, this closes up the valve lash, causing the valve to hang open and burn. *(Courtesy of Unocal)*

FIGURE 10-78 The lapping tool changes the rotary action into an effective oscillating motion. *(Courtesy of Goodson Shop Supplies)*

guide will scuff and wear excessively if they do not have enough lubrication. Too much oil produces heavy deposits that build up on the intake valve and hard deposits at the head end of the exhaust valve stem. Worn valve stem seals can increase the oil consumption by as much as 70 percent.

There are basically two types of seals: positive seals that fit tightly around the top of the stem and

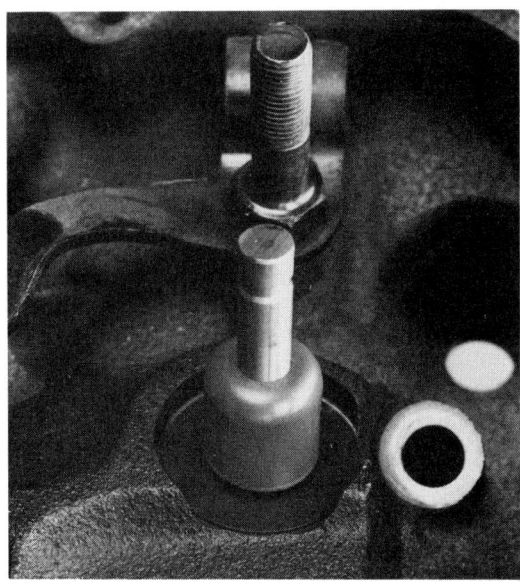

FIGURE 10-79 The valve stem oil seal on the guide chimney

scrape oil off the valve as it moves up and down, and deflector, splash, or umbrella-type seals that ride up and down on the valve stems to deflect oil away from the guides.

The type of seal that is used in a particular engine, like the stem-to-guide clearance, depends upon a number of variables. These include all the things that determine guide clearance as well as the oiling characteristics of the head and upper valve train, and the type of PCV system used. If the design of the head is such that the guides tend to be flooded with oil and oil consumption is a problem, positive-type seals might be specified.

Some rebuilders say positive valve seals are the only way to go in spite of the added cost and machining required to install them because they provide a tighter oil seal and greatly reduce the amount of oil that reaches the guides. Positive seals also allow some engines to handle a little extra stem-to-guide clearance in high output or heavy-duty applications without fear of increasing oil consumption. Another benefit is that they prevent air leaks past the guides. But positive oil seals can cause problems when used in certain applications. One seal manufacturer has issued a bulletin warning against the substitution of positive seals for deflector seals on the basis that positive seals might starve the guides for adequate lubrication, which accelerates stem and guide wear. This would be more of a concern on engines with integral cast-iron guides than those with bronze guides. It is usually wise to use a replacement seal of the same material and design as the original. Engine manufacturers have done extensive research and testing to come up with the particular seal they recommend for a given application.

 SHOP TALK _____

It is recommended that seals be installed on both intake and exhaust valves. The intake valve is the most critical since oil can be drawn into the intake manifold by manifold vacuum and then passed to the combustion chamber where it is burned. However, the high-velocity exhaust gases in the exhaust port can also create a vacuum and draw oil from the overhead past the valve stem and be burned.

The ultimate effectiveness of the valve stem seal depends entirely on the way it is secured to the guide. Many guides require machining, and this must be done using the proper tools. A special valve guide machining tool is available for valve seat cutting (Figure 10-80). Such a tool is made up of a

FIGURE 10-80 Machining tool for seals *(Courtesy of Goodson Shop Supplies)*

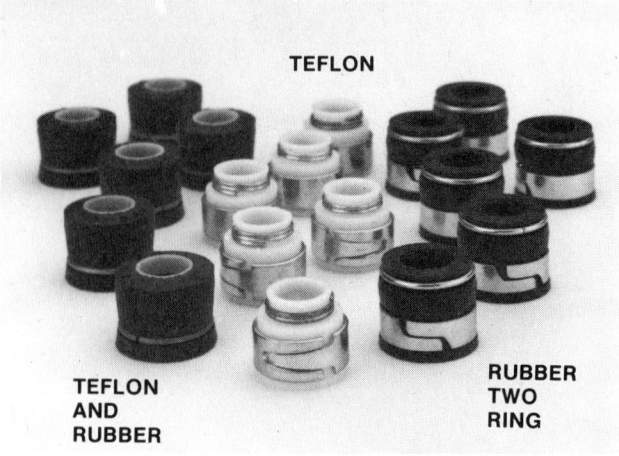

TEFLON

TEFLON AND RUBBER

RUBBER TWO RING

FIGURE 10-81 Designs of positive valve seals *(Courtesy of Goodson Shop Supplies)*

cutter and pilot, with sizes that vary according to the valve stem diameter and desired guide OD. The pilot is inserted into the guide, and the cutting tool machines the top of the guide.

Apply heavy pressure to the cutting tool to machine the guides to specification. Do not bounce the tool on the guide. Be sure to wipe and lubricate the tool pilot after machining each guide to prevent buildup of metal particles. Many original equipment cylinder heads are already machined to receive some type of valve stem seal; however, the guide OD is not always concentric with the valve stem hole. If the ST-type seals were installed with this condition, it would result in the seals being distorted and pulled to one side, opening up a path for oil to get by. It is imperative, therefore, that guides are machined where specified. If a guide is machined with the tool indicated but does not clean up, the next smaller OD tool must be used.

INSTALLING POSITIVE VALVE SEALS

To install a positive valve seal (Figure 10-81), proceed as follows:

1. Place the plastic sleeve in the kit over the end of the valve stem (Figure 10-82). This will protect the seal as it slides over the keeper grooves. Lightly lubricate the sleeve. If it extends more than 1/16 inch below the lower keeper groove, you might want to remove the sleeve and cut off the excess length for easier removal.

2. Carefully place the seal on the cap over the valve stem and push the seal down until it touches the top of the valve guide (Figure

FIGURE 10-82 Plastic sleeve goes over the end of the valve stem. *(Courtesy of Fel-Pro Inc.)*

10-83). At this point the installation cap can be removed and placed on the next valve. A special installation tool (Figure 10-84) can be used to finish pushing the seal over the guide until the seal is flush with the top of the guide.

If an installation tool is not available, two small screwdrivers placed on either side of the seal can be used to pull the seal downward until it is in place on the valve stem (Figure 10-85). Often, a twist with the fingers will be enough to pop the seal down onto the guide. Do not push the seal down any further than flush with the top of the guide or the seal will open up. After all valves have seals installed, replace the valve springs, retainers, and keepers.

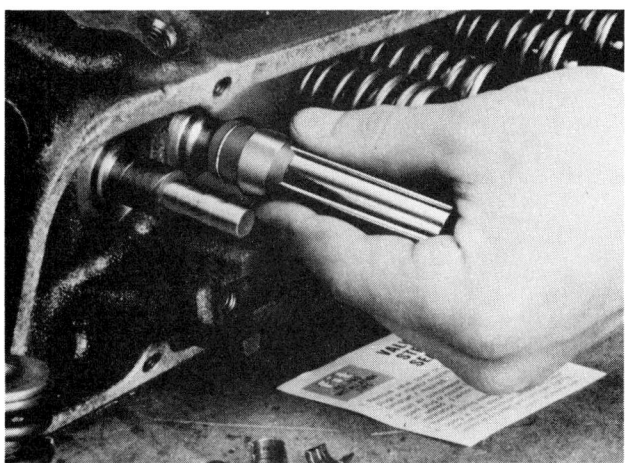

FIGURE 10-83   Push the seal down until it touches the top of the valve guide. *(Courtesy of Fel-Pro Inc.)*

FIGURE 10-84   Special installation tool to finish pushing the seal over the guide *(Courtesy of Perfect Circle/Dana)*

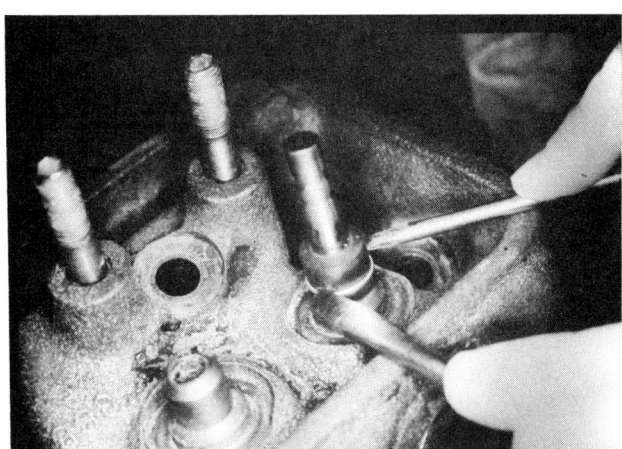

FIGURE 10-85   Installing a positive valve seal with two screwdrivers *(Courtesy of Sealed Power Corp.)*

INSTALLING UMBRELLA-TYPE VALVE SEALS

An umbrella-type seal is installed on the valve stem before the spring is installed. It is pushed down on the valve stem until it touches the valve guide boss (Figure 10-86). It will be positioned correctly when the valve first opens.

INSTALLING O-RINGS

When installing O-rings, use engine oil to lightly lubricate the O-ring. Then install it in the lower groove of the lock section of the valve stem (Figure 10-87). Make sure that the O-ring is not twisted.

 SHOP TALK _____

Regardless of the type of valve stems used, they should never be employed to reduce excessive oil consumption due to worn guides or valve stems. Replace the worn parts.

VALVE TRAIN COMPONENTS

Valve train components include all those already described in this chapter plus the followers,

FIGURE 10-86   Installing an umbrella valve seal *(Courtesy of Fel-Pro Inc.)*

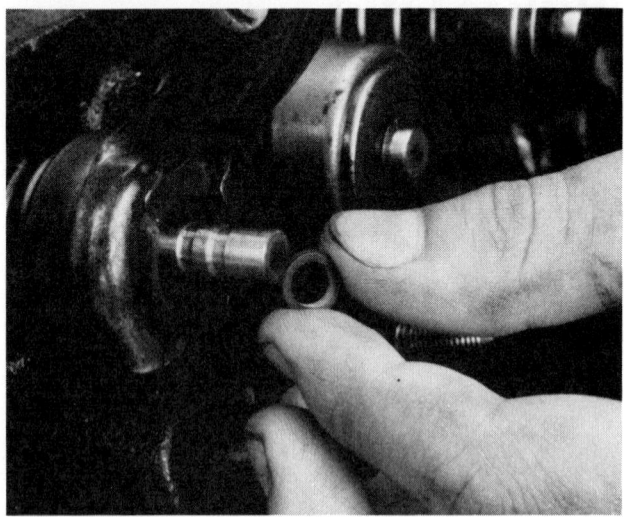

FIGURE 10-87 Installing an O-ring *(Courtesy of Fel-Pro Inc.)*

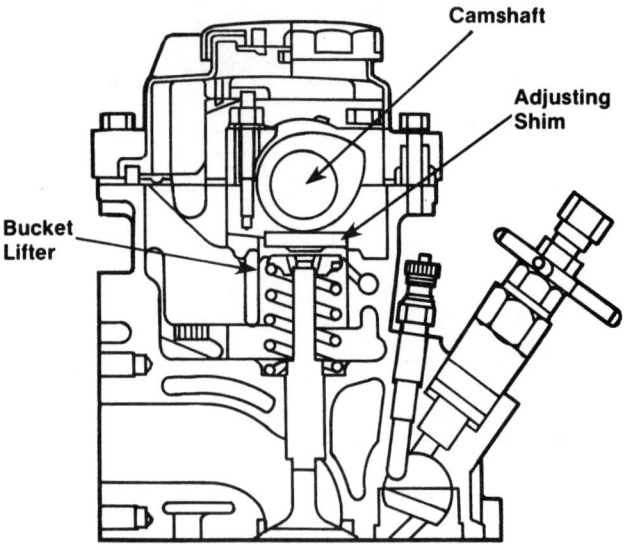

FIGURE 10-88 The camshaft pushes on a bucket lifter to move a valve assembly.

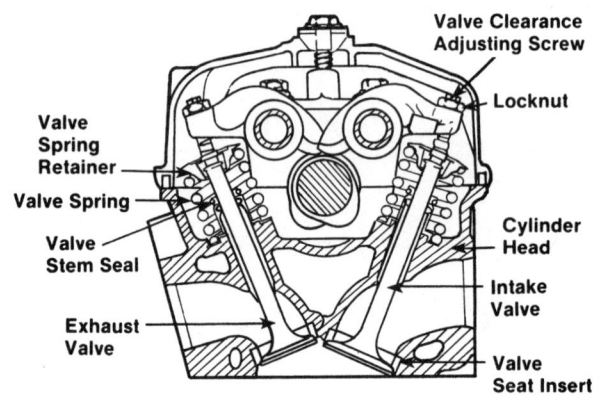

FIGURE 10-89 Parts of a valve assembly that use rocker or finger followers

pushrods, and rocker arms. Most valve train components are generally replaced rather than repaired. The exception is the machining of some types of rocker arms.

FOLLOWERS

The followers, as mentioned in Chapter 4, can be a lifter (tappet), bucket follower, or finger follower. On an overhead camshaft (OHV) engine, the movement of the lifter-type follower will move a rocker arm directly. The movement of the lifter-type follower must go through a pushrod to move the rocker arm when the camshaft is in the block. Inverted bucket followers (Figure 10-88) and rocker or finger cam followers (Figure 10-89) are only used on overhead camshaft (OHC) engines.

Lifters

As mentioned previously, there are two types of lifters: mechanical (solid) and hydraulic. All engines that use mechanical lifters employ some method of adjustment that is intended to bring valve lash back into specification (see the vehicle's service manual). There are four basic methods of lash adjustment: rocker arms with adjustable pivots, adjustable pushrods, rocker arms with adjustment screws, and adjustable cam followers (using some type of adjustment screw or replaceable shim). Mechanical lifters have two points of major concern: the bottoms and pushrod sockets. Wear, scoring, or pitting makes their replacement necessary.

OHC Cam Followers

In an OHC engine, the camshaft and followers (Figure 10-90) are wear-related parts that must be reassembled in the same positions. Replace both the cam and followers as a set if they are too worn to be reused.

When the camshaft in an OHC engine is installed, some clearance might be needed between the cam and cam followers. The installed valve stem height is critical to avoid incorrect valve train geometry and to correctly set the valve lash.

Many OHC engines use shims placed above the cam follower to adjust the lash. For example, shims might be needed between the bucket style follower and cam lobes to set the valve lash. These shims

FIGURE 10-90   Typical cam followers

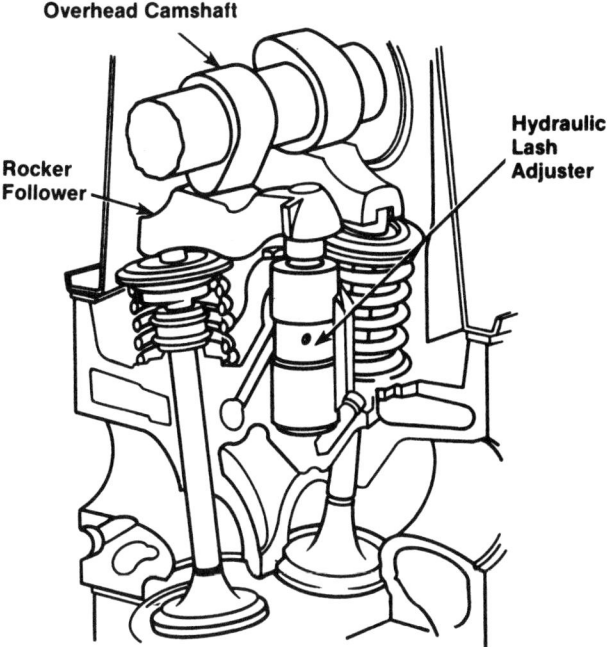

FIGURE 10-91   Hydraulic lash adjuster is often used to maintain valve assembly clearances.

usually come in a set of twenty-six thicknesses and require special tools for adjustment. Specs given in the service manual are usually very exact for OHC engines, so be precise in checking clearances.

Some OCH engines have an allen adjusting screw in the cam followers. Turning the screw changes the valve clearance. Other OHC engines, primarily import aluminum heads, have an adjustable rocker arm. In these engines, a jam nut and bolt allow for mechanical tappet adjustment. Measuring the distance between the bottom of the bolt and the valve stem tip helps determine the correct adjustment.

New OHC engines use hydraulic lash adjusters (Figure 10-91) in conjunction with the cam lifters. Lash is adjusted automatically with every engine revolution; however, the correct valve stem height is critical. Regular inspection of the valve stem height is recommended. Check the manufacturer's service manual for suggested inspection intervals. Keep in mind that on some OHC engines, one or two specialty tools might be needed to assist in adjusting valve lash (Figure 10-92).

PUSHRODS

As previously mentioned, some pushrods are hollow and act as part of the oil supply line to the rocker arms and valves. No oil will get to the rocker arms if these pushrods are plugged. Some pushrods have a groove worn in the area in which they pass through the cylinder head, and all are subject to tip wear. Badly worn or plugged pushrods should be replaced. Check all pushrods for bend, and if more than 0.030 inch TIR is found, replace them. Bent pushrods deflect in operation, hindering proper valve action.

FIGURE 10-92   Adjusting valve lash with a specialty tool

Figure 10-93 illustrates typical pushrod configurations. If a pushrod is found to be faulty, replace it. Some pushrods are adjustable.

ROCKER ARMS

Worn rocker arms, rocker arm shafts, and adjusting screws must be replaced. It is good practice to disassemble rocker arm assemblies and thoroughly remove all sludge and varnish. Replace all worn parts and lubricate before reassembling. In fact, many rebuilder shops, because of the time involved in cleaning, replace the old rocker arm assembly with a new one.

Some engines are equipped with an independent rocker arm assembly for each valve and no

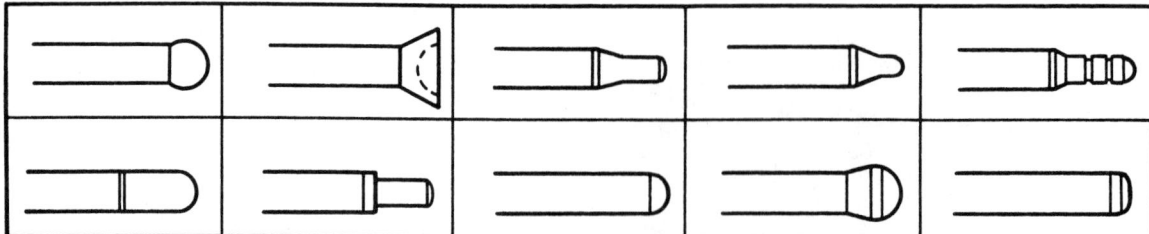

FIGURE 10-93 Typical pushrod end configurations

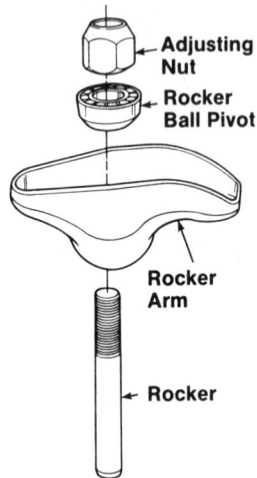

FIGURE 10-94 Independent rocker arm assembly for valve *(Courtesy of TRW Inc.)*

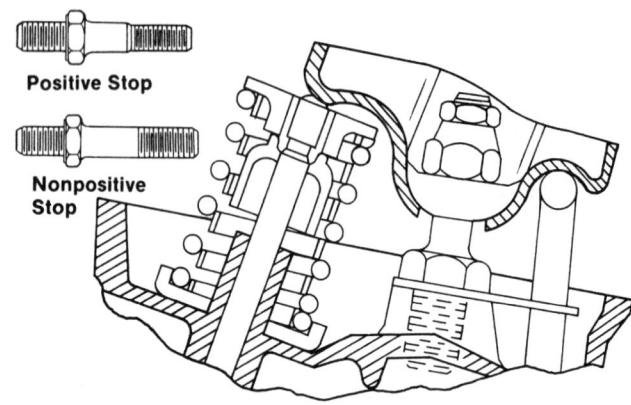

FIGURE 10-95 Positive and nonpositive stop screw-in studs

rocker arm shaft is used (Figure 10-94). A stud is either pressed or threaded into the cylinder head and must be replaced if worn, bent, broken, or loose. On some engines, the studs are drilled for an oil passage to the rocker arms. Make sure oil can pass through before installing the cylinder head on the block. Replacement press-in studs are available in standard and oversizes. The standard size is used to replace damaged or worn studs and the oversizes are for loose studs. Rocker arm studs that are broken or have damaged threads can be replaced with standard studs. Loose studs in the head can be replaced with 0.006-, 0.010-, or 0.015-inch oversize studs that are available for service.

Threaded rocker arm studs are frequently substituted for damaged pressed-in studs. Threaded studs may be either a positive or nonpositive stop type (Figure 10-95) and are made with a jam nut or have a jamming shoulder. For an accurate change from factory pressed-in studs to screw-in studs, a tapping machine should be employed (Figure 10-96).

When a jam nut is used, the metal surface of the stud boss must be reduced by milling to the thickness of the nut (Figure 10-97). Then each stud hole is drilled and tapped. Care must be exercised to maintain correct alignment when drilling and tap-

ping stud holes. Any stud angle other than the original can destroy the rocker arm geometry. Studs with a jamming shoulder do not require boss milling.

Standard and oversize studs can be identified by measuring the stud diameter within 1-1/8 inch from the pilot end of the stud. The stud diameters usually are 0.006-inch oversize, 0.010-inch oversize, or 0.015-inch oversize.

When going from a standard size rocker arm stud to a 0.010- or 0.015-inch oversize stud, always

FIGURE 10-96 For an accurate change from factory pressed-in studs to screw-in studs, a tapping device should be used. *(Courtesy of Kwik-Way Manufacturing Company)*

FIGURE 10-97 Milling tool for machining the top of the stud boss *(Courtesy of Kwik-Way Manufacturing Company)*

A

B

FIGURE 10-98 Removing the rocker arm stud with a stud remover *(Courtesy of Atlas Engineering and Manufacturing, Inc.)*

use the 0.006-inch oversize reamer before finish reaming with the 0.010- or 0.015-inch oversize reamer.

 1. Position the sleeve of the rocker arm stud remover over the stud with the bearing end down (Figure 10-98A). Thread the puller into the sleeve and over the stud until it is fully bottomed. Hold the sleeve with a wrench, then rotate the puller clockwise to remove the stud (Figure 10-98B). If the rocker arm stud was broken off flush with the stud boss, use a screw extractor (commonly called an *easy-out*) to remove the broken stud. A hole of the correct size is drilled in the center of the broken stud. The extractor is then installed in the drilled hole (following the instruction of the tool manufacturer) and turned to remove the remaining portion of the stud.

⚙ SHOP TALK ─────

Another popular type of stud puller is shown in Figure 10-99. The collet on this tool slips in the inner bore of the lifter. The slide hammer shaft now creates a solid gripping pressure by expanding the collet outward. Thus, the collet lip is locked into the inner groove of the lifter, eliminating the possibility of slippage. After one or two taps of the slide hammer, the lifter can be removed.

2. If a loose rocker arm stud is being replaced, ream the stud bore using the proper reamer (or reamers in sequence) for the

FIGURE 10-99 A lifter puller that is popular on smaller engines *(Courtesy of Goodson Shop Supplies)*

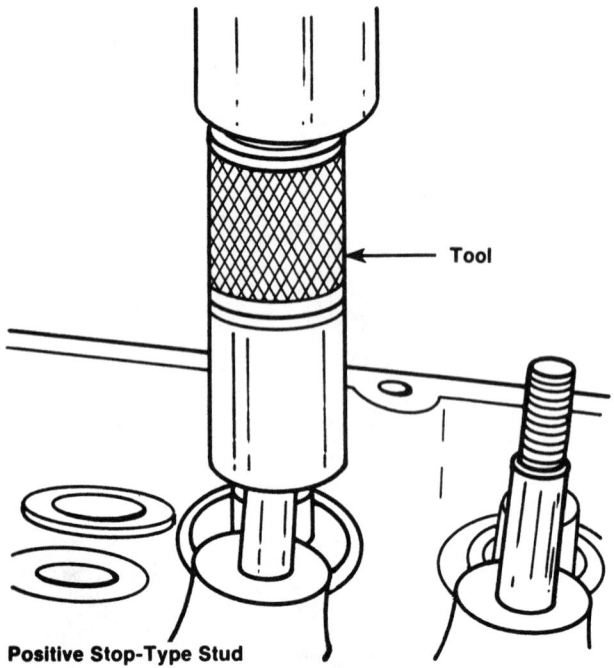

Positive Stop-Type Stud

FIGURE 10-100 Installing the rocker arm stud

FIGURE 10-101 To keep valve train parts organized for reassembly, it is wise to use a tray like the one shown. *(Courtesy of Goodson Shop Supplies)*

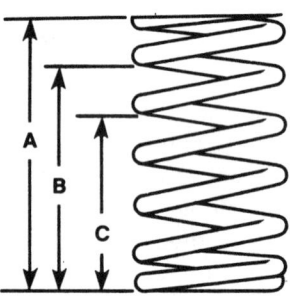

FIGURE 10-102 Valve spring height terms: (A) free height; (B) valve closed spring height; (C) valve open spring height

selected oversize stud and make sure the metal particles do not enter the valve area.

3. Coat the end of the stud with a polyethylene grease. Align the stud with the stud bore, then tap the sliding driver until it bottoms (Figure 10-100). When the driver contacts the stud boss, the stud is installed to its correct height.

ASSEMBLING THE CYLINDER HEAD

When all of the valve train parts have been cleaned, inspected, reconditioned (or replaced), a portion of the cylinder head is ready to be reassembled (Figure 10-101). It makes sense that all valve train parts are reinstalled in the reverse order that they were removed (see Chapter 4). But before starting to assemble the valve train there are two critical measurements that must be carefully checked: the installed stem height and the installed spring height.

Installed Stem Height

Installed stem height is determined by measuring the distance between the spring seat and stem tip. Since this measurement directly influences rocker arm geometry and installed spring height, accuracy and precision are important (Figure 10-102).

This is especially true when the valve or valve seat has been ground. There are a number of tools that can be used to obtain an accurate stem height reading. These include a depth micrometer, vernier caliper, and telescoping gauge (Figure 10-103). There are several specially designed stem height gauges available (Figure 10-104). Use these as directed by their manufacturers. Similar measuring devices are shown in Figure 10-105.

Although it is easy to measure stem height, it is often difficult, as stated earlier in this chapter, to obtain a list of published installed height specs. With the exception of a few aftermarket companies that have assembled their own stem height data, it is sometimes difficult to find OE-endorsed installed stem heights.

This is the reason it is suggested in Chapter 4 to record stem heights when disassembling the engine. These measurements will act as the original

A

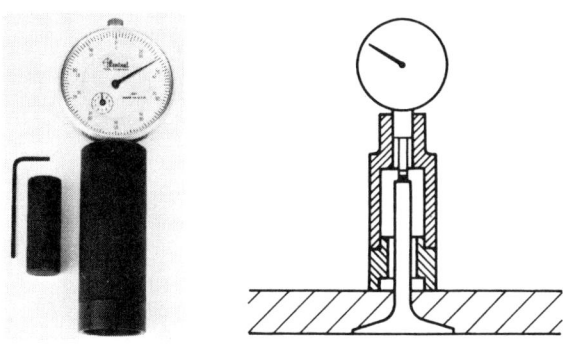

FIGURE 10-104   Specially designed stem height gauges *(Courtesy of Goodson Shop Supplies)*

B

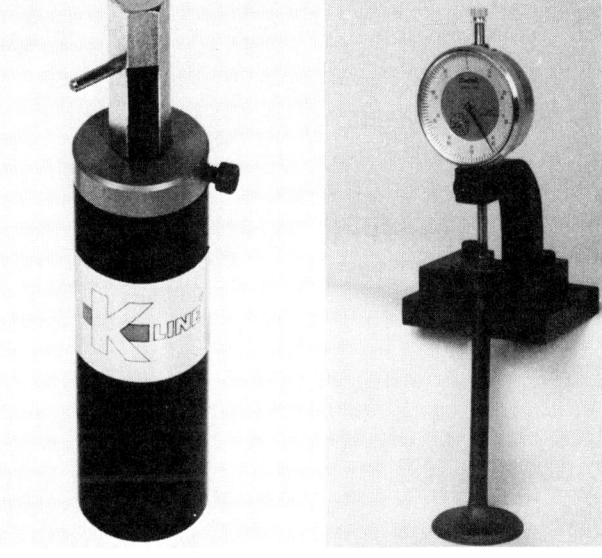

FIGURE 10-105   Other spring height measuring gauges

stem height (Figure 10-106). Keep the reconditioned stem height near the measurement ± 0.030 inch by grinding the valve tip or installing a new seat. However, this method works only if the head measured is in good condition. If it has had several valve jobs or has seats that show substantial wear, then this method may not be very reliable.

Installed Spring Height

Another specification can be used that corresponds directly to installed stem height—installed spring height. Installed spring height is measured from the spring seat to the underside of the retainer when it is assembled with keepers and held in place (Figure 10-107). This measurement, which can be

C

FIGURE 10-103   Three of the most common methods of checking stem height: (A) depth micrometer, (B) vernier caliper, and (C) telescoping gauge *(Courtesy of TRW Inc.)*

Cylinder	318-Moper			
	Before Intake	After	Before Exhaust	After
1	2.050	2.061	2.054	2.062
2	2.037	2.043	2.046	2.049
3	2.032	2.046	2.036	2.046
4	2.052	2.072	2.043	2.052
5	2.041	2.042	2.030	2.037
6	2.037	2.050	2.031	2.039
7	2.046	2.052	2.048	2.061
8	2.039	2.048	2.051	2.059

FIGURE 10-106 Keep a record of stem heights

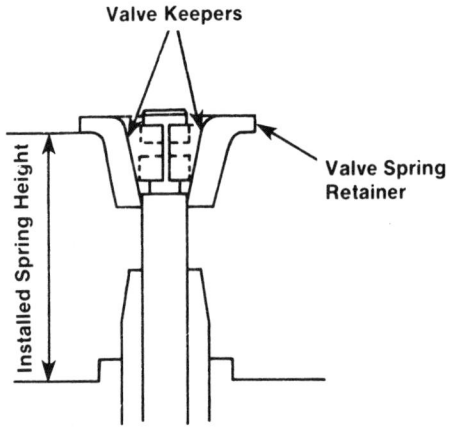

FIGURE 10-107 Installed spring height

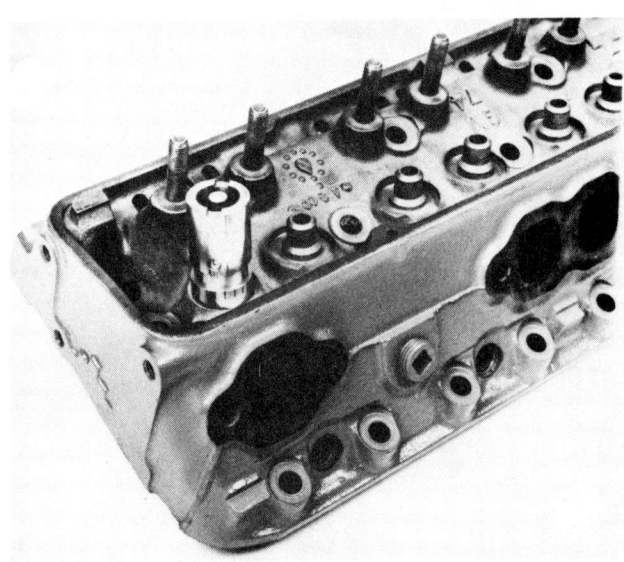

FIGURE 10-109 Two types of commercial spring testing gauges *(Courtesy of Goodson Shop Supplies)*

made by using a set of dividers or scales (Figure 10-108), telescoping gauge, or spring height gauge, should be taken only after valve and seat work is completed, valves are installed in their guides, and

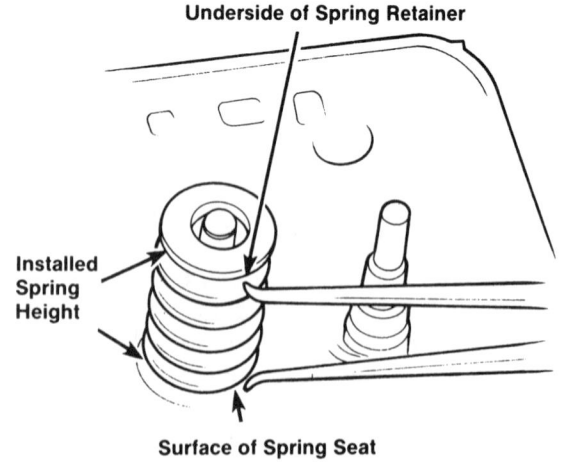

FIGURE 10-108 Measuring installed spring height with a set of dividers

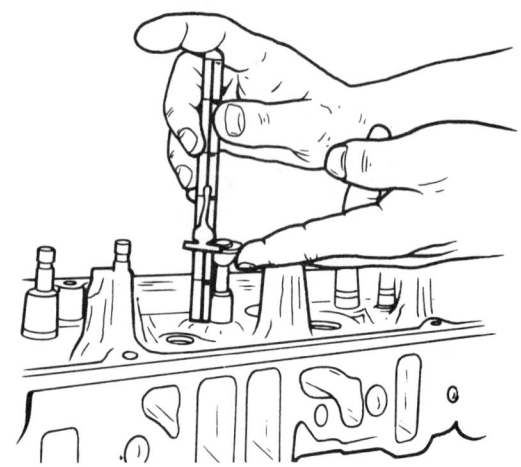

retainers and keepers are assembled. Special shop gauges are also available (Figure 10–109).

By comparing the measurement to factory specs, it is possible to determine the increase needed in installed stem height. For example, if the installed spring height for an exhaust valve is 1.600 inches, and the measurement is 1.677 inches the increase in spring height is 0.077 inch. This means installed stem height also has been increased by 0.077 inch.

Adjustments to valve spring height can be made with the aid of valve spring inserts, otherwise known as spring shims (Figure 10–110). Even though valve shims come in only three standard thicknesses— 0.060, 0.030, and 0.015 inch—using combinations of different shims gives the correct amount of compensation (within 0.005 or 0.010 inch). For reference, a 0.060-inch shim will provide a 12-pound increase in spring pressure, whereas the 0.030-inch and 0.015-inch varieties provide a respective 6- and 3-pound pressure increase.

By comparing the reconditioned spring height to published specifications, the desired amount of tension correction can be easily determined (Figure 10–111). For example, if reconditioned spring height is 0.180 inch and the specs call for 0.149 inch, a 0.030 shim (0.149 + 0.030 = 0.179) would be needed. If more than one shim is required, place the thickest one next to the spring—not the head. If one side of the shim is serrated or dimpled, place that side

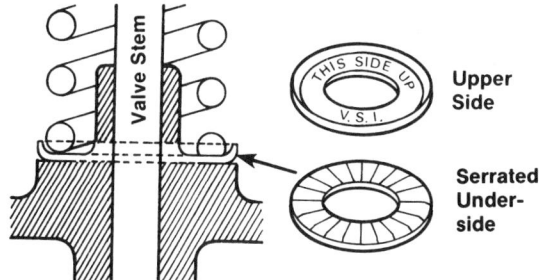

FIGURE 10–110 Valve spring shims are usually available in three sizes: Type A (0.060 inch thick) for springs in service; type B (0.030 inch thick) for new springs; type C (0.015 inch thick) for balancing. *(Courtesy of Perfect Circle/Dana)*

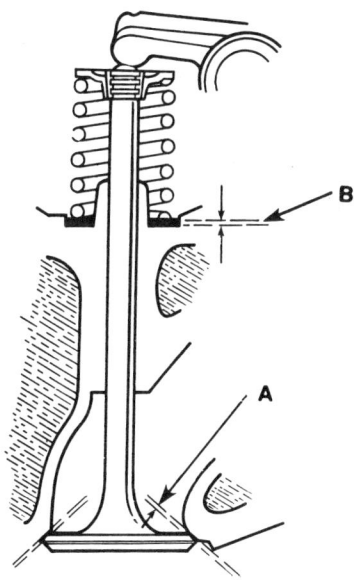

FIGURE 10–111 Valve spring shims are used to correct installed spring height. Loss of metal at A is compensated for at B.

against the head. Doing this will minimize heat transfer to the spring.

It is important to keep in mind that increases in installed stem height should be kept to ± 0.030 inch of the manufacturer's specifications or the technician's original figures to prevent altering the alignment of the rocker arm and stem contact point. Such alterations in rocker arm geometry can lead to early guide and stem wear, breakage, and other problems. For this same reason, if the cylinder head is surfaced more than 0.15 inch, subtract the additional amount surfaced from the 0.030-inch safety margin.

If installed valve stem height is excessive, it might be necessary to grind part of the tip. However, do not remove more than 0.020 inch to prevent cutting through the hardened portion of the stem. Excessive stem grinding can cause rocker arms to contact the retainer and release a valve during engine operation. If correct installed stem height cannot be obtained by grinding the tip, a new seat must be installed.

SHOP TALK

Install valve rotators above and below the spring as directed by the manufacturer. Variable-rate springs (see Chapter 6) are usually installed with the closest spaced coils toward the cylinder head.

After the seals are installed as described earlier in this chapter, insert the shims and place the

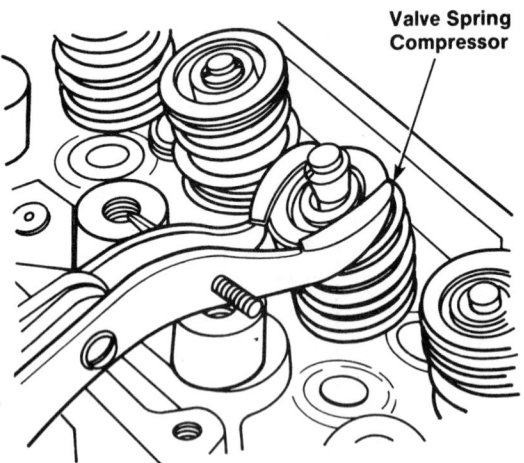

FIGURE 10-112 Compress the spring just enough to install the keeper.

spring(s) over the valve stem and onto the spring seat. There are two words of caution:

1. If the springs have close-spaced coils, the close end should be placed next to the head.
2. If the spring has an OD that is slightly tapered, the end with the greater diameter should be facedown.

Compress the spring only enough to install the keepers (Figure 10-112). Excess compression can cause the spring retainer to damage the valve seal.

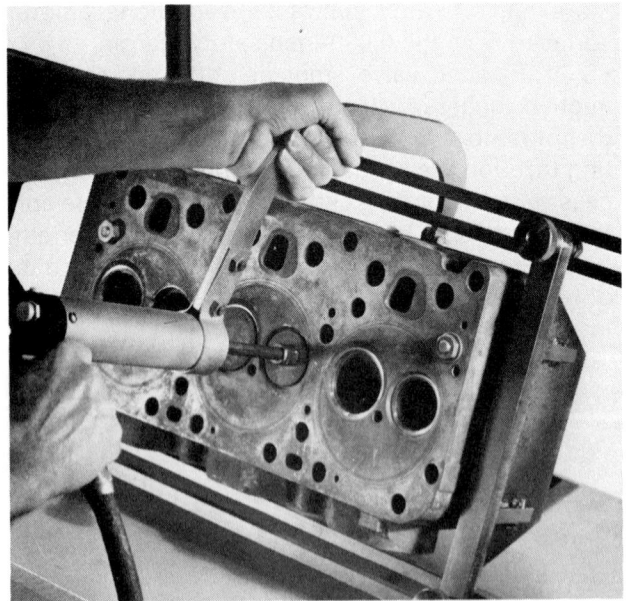

FIGURE 10-113 Air-operated valve spring compressor *(Courtesy of Hall-Toledo, Inc.)*

This is especially important when using an air-operated compressor (Figure 10-113). Also be sure that the inside of the spring does not contact the valve seal. Tap the stem end with a rawhide or rubber mallet to seat the keepers.

A universal valve spring compressor, such as the one shown in Figure 10-114, is employed with many import vehicle models, especially those with water-cooled engines. The base of this tool bolts securely to the head while the lever mechanism pivots from the base, thus compressing the spring. Removal and installation of the spring retaining keepers is then easily accomplished through the large

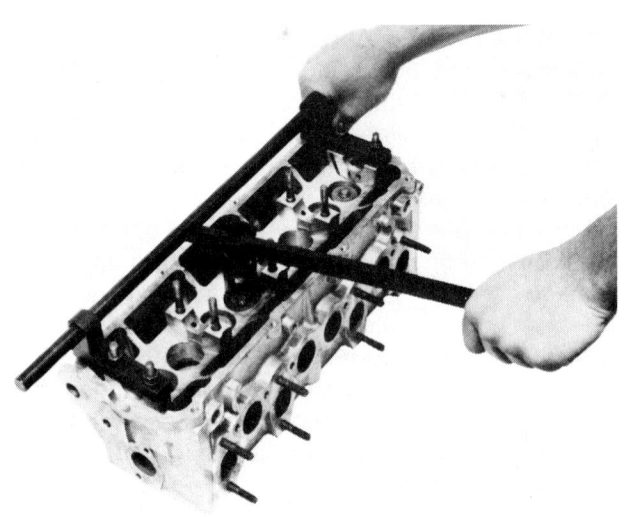

FIGURE 10-114 A spring compressor is used on many import engines *(Courtesy of Atlas Engineering and Manufacturing, Inc.)*

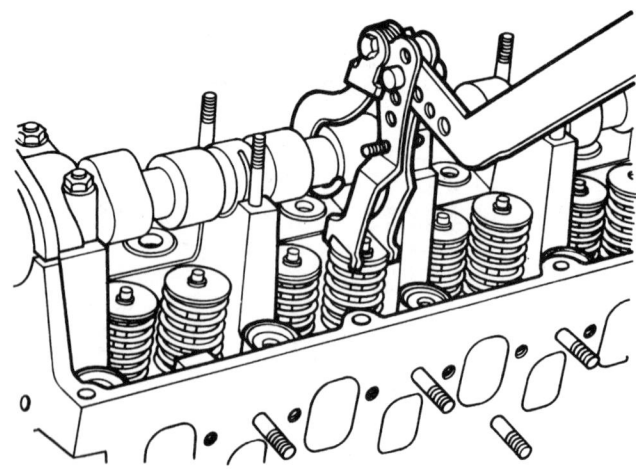

FIGURE 10-115 OHC engine's valve spring compressor

openings on both sides of the thick-walled compression tube. Although it is not mandatory, most rebuilders make it a policy to automatically replace the keepers on each job. This rationale is based on the notion that replacing half the source of keeper groove wear is 50 percent better than doing nothing—and a lot less expensive than repairing the damage caused by a dropped valve. A valve spring compressor for an OHC engine (Figure 10–115).

 SHOP TALK _____

Remember that new keepers must be installed in pairs. If a new keeper is mated with a used keeper, the spring retainer may cock and break off the valve tip.

Details and instructions for the completion of the cylinder head assembly are given in Chapter 15.

REVIEW QUESTIONS

1. Which of the following can be reconditioned by grinding?
   a. valve face
   b. valve tip
   c. both a and b
   d. neither a nor b

2. Which of the following occurs due to excessive valve stem-to-guide clearance?
   a. inadequate valve cooling
   b. air leakage
   c. accelerated stem and guide wear
   d. all of the above

3. After grinding the valve, runout should not exceed _____ .
   a. 0.002 inch (TIR)
   b. 0.020 inch (TIR)
   c. 0.030 inch (TIR)
   d. 0.045 inch (TIR)

4. Technician A says that bronze valve guides retain oil better than cast-iron ones. Technician B says that cast-iron valve guides are easier to machine. Who is right?
   a. Technician A
   b. Technician B
   c. Both A and B
   d. Neither A nor B

5. Which of the following is not true of knurling?
   a. It is one of the fastest techniques for restoring the ID dimensions of a worn valve guide.
   b. It reduces the amount of work necessary to reseat the valve.
   c. It is useful for restoring badly worn guides to their original condition.
   d. None of the above

6. Which of the following is the maximum amount of acceptable wear for knurling?
   a. 0.005 to 0.009 inch
   b. 0.006 to 0.008 inch
   c. 0.004 to 0.007 inch
   d. 0.003 to 0.006 inch

7. To insure proper seating of the valve, the valve seat must be _____ .
   a. correct width
   b. correct location on the valve face
   c. concentric with the guide
   d. all of the above

8. When grinding valve seats, _____ .
   a. the stone should be approximately 1/8 inch larger than the seat
   b. a hard stone should be used on a hard seat
   c. a soft stone should be used on a soft seat
   d. all of the above

9. The removal of tetraethyl lead from gasoline _____ .
   a. increases valve and cylinder head temperatures
   b. eliminates lubrication problems to the valve stem and valve seat
   c. decreases exhaust valve recession into the seat
   d. both b and c

10. Technician A says that positive valve stem seals fit tightly around the top of the stem and scrape oil off the valve as it moves up and down. Technician B says that positive seals are more expensive and require more machining than umbrella-type seals. Who is right?
    a. Technician A
    b. Technician B
    c. Both A and B
    d. Neither A nor B

11. Which of the following is usually replaced rather than repaired?
    a. keepers
    b. pushrods
    c. rocker arms
    d. all of the above
    e. none of the above

12. When adjusting spring height, Technician A places the thickest insert next to the spring. Technician B places the thickest insert furthest away from the spring. Who is right?
    a. Technician A
    b. Technician B
    c. Both A and B
    d. Neither A nor B

13. Which of the following is a method of lash adjustment?

    a. mechanical lifters
    b. rocker arms with adjustable pivots
    c. adjustable cam followers
    d. all of the above
    e. both b and c

14. Which of the following is true concerning lapping?
    a. It is a valve finishing operation.
    b. It is done with lapping sticks.
    c. It is done by a powered lapping tool.
    d. All of the above

15. A pushrod should be replaced when bent more than _____ .
    a. 0.002 inch (TIR)
    b. 0.020 inch (TIR)
    c. 0.030 inch (TIR)
    d. 0.045 inch (TIR)

CHAPTER ELEVEN

RECONDITIONING RELATED CYLINDER BLOCK PARTS

Objectives

After reading this chapter, you should be able to
- Determine the condition of an engine cylinder.
- Inspect and measure engine cylinders.
- Understand the methods for boring and honing cylinders.
- Install cylinder sleeves.
- Adjust the piston clearance on a reconditioned engine.
- Identify the four ways in which piston ring oil, vacuum, compression, and combustion leakage can occur.
- Install piston rings.
- List the conditions that must be met for the crankshaft to operate properly.
- Recondition a connecting rod bore.
- Reassemble a piston and connecting rods using the press method and the heat method.

In the previous chapter, the reconditioning of the valve train network and related upper engine parts was fully explained. In this chapter, the reconditioning of various components of the lower engine—including the cylinders, pistons, piston rings, connecting rods, and piston pins—will be covered.

CYLINDER RECONDITIONING

Proper cylinder reconditioning can make the difference between a smoothly performing engine and one that burns oil. Attention to detail and use of correct techniques make the difference.

Obtaining the optimum cylinder surface finish is one of the most important operations of an engine overhaul because it is a primary factor in good oil control. There are probably almost as many theories

on cylinder finishing as there are those doing the work. Methods used range from the latest in automatic equipment to sandpaper. But, unfortunately, many of these theories are not supported by facts.

There are several requirements for a properly reconditioned cylinder. It must have the correct diameter, it must be free from taper and runout (out-of-round), and the surface finish must be such that the piston rings will seat to form a seal that will minimize blowby and control oil. The surface finish on a properly prepared cylinder wall acts as a reservoir for oil to lubricate the rings and prevent piston and ring scuffing primarily during the break-in period. Piston ring faces can be damaged if the cylinder wall is too rough, resulting in premature wear. A surface that is too smooth will not allow the rings to seat properly.

To determine whether or not a block must be resized, check the bores for size with a bore gauge (see Chapter 7). If the reading is more or less than specs, then boring or honing will be necessary. Regardless of the method used, make sure to leave sufficient material (0.003 inch is recommended) to finish the hone. But before reconditioning, remove any sharp edges and recut the chamfer around the bores with an abrasive wheel (Figure 11-1).

CYLINDER DEGLAZING

If the inspection and measurements of the cylinder wall prove that surface conditions, the taper, and out-of-round are within acceptable limits, all that is necessary is to deglaze the cylinder walls. Combustion heat, wall ring, engine oil, and piston movement combine to form a thin residue on the cylinder walls that is commonly called *glaze.*

The two most common types of glaze breakers are shown in Figure 11-2. With either type use an abrasive with about 220 or 280 grit. The driver end of the glaze breaker is installed in a slow-moving electric drill that operates at 300 to 500 rpm or in a honing machine. While both types of glaze breakers can be used, most rebuilders prefer the resilient-based design (Figure 11-3). Resilient-based hone glass breakers are available in several sizes (Figure 11-4).

WARNING: Always wear eye protection when operating deglazing, honing, or boring equipment.

The glaze-breaker, as it rotates, is moved up and down. Proper deglazing of the cylinder provides the satin finish and diamond crosshatch pattern (Figure

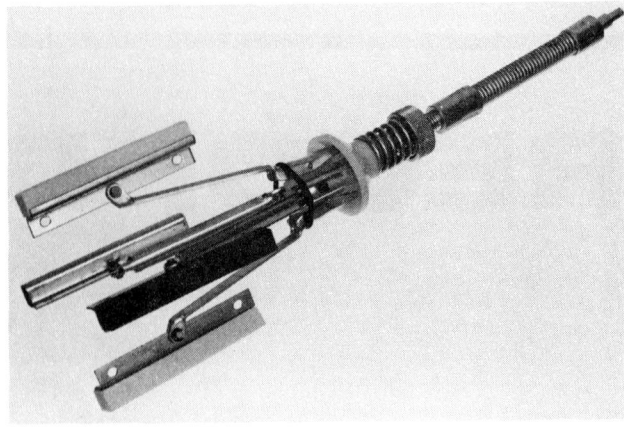

A

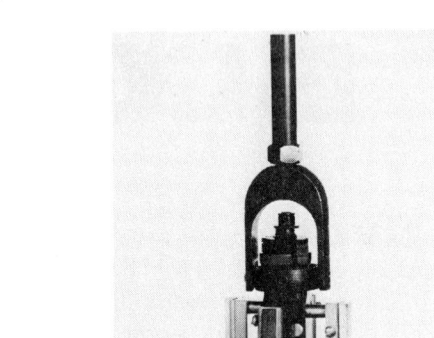

B

FIGURE 11-2 Glaze breakers: (A) resilient-based hone; (B) rigid hone type *(Courtesy of K-Line Industries, Inc.)*

FIGURE 11-1 An abrasive wheel removes sharp edges and recuts the chamfer around the bores. *(Courtesy of Goodson Shop Supplies)*

FIGURE 11-3 Deglazing a cylinder with a resilient-based hone *(Courtesy of Brush Research Manufacturing Company, Inc.)*

FIGURE 11-4 Various sizes of resilient-based hone-type brushes *(Courtesy of Brush Research Manufacturing Company, Inc.)*

11-5) necessary to prevent early break-in scuffing and to seat the rings correctly. Wet deglazing should be used whenever possible to reduce working heat, eliminate loading of stones, give cleaner cutting, and a finer finish. Rough hone or deglaze the cylinder to within 0.0005 to 0.001 inch of finish size. Use a fast and steady up-and-down movement of the hone, so as to produce stone marks with a well-defined diamond-shaped pattern. Expand stones gradually; do not use excessive pressure because it can rupture the granular structure of the surface metal. When stones "squeal" excessively they are either glazed or too hard. Glaze can be broken by sliding stones up and down in the bore without rotating them if sufficient pressure is applied. Score

FIGURE 11-5 Cylinder walls should be carefully honed or deglazed to produce a proper crosshatch pattern.

marks in the finish indicate loaded stones and loaded areas should be rubbed off with another stone.

Roughing Stones Suggested for Wet Honing: Unhardened sleeves and blocks: C150JV or KV; C180JB or KV

Hardened sleeves: C100JR or KV; C120JV or KV

Finishing Stones Suggested for Wet Honing: Unhardened sleeves and blocks: C320IV or JV; C400IV or JV

Hardened sleeves: C180HV or IV; C220HV or IV

Explanation of the stone designation: The letter C means silicon carbide; the numbers 80, 100, and so on, indicate the size of the grit; letters J, K, etc., indicate the hardness; V means vitrified bond.

Deglazing residue and metal fragments adhering to the cylinder walls are abrasive and will quickly damage the rings, pistons, and cylinders if not removed. Gasoline, kerosene, or solvents will not work. One of the best methods is to use plenty of hot, soapy water, a stiff bristle brush, and a soft, lint-free cloth (Figure 11-6). Then rinse the block with clear water and dry thoroughly. Coat with clean, light engine oil to prevent rust. Cylinder cleanliness can be checked by wiping them with a clean white cloth after oiling. The cylinder walls can be considered free of abrasive residue when the cloth remains clean.

 SHOP TALK _____

If the cylinders have varnish deposits, they can be swabbed with lacquer thinner to remove the varnish. Should the cylinders be lightly scuffed or scratched, light honing is recommended (Figure 11-7). Also remember that rebuilders should adhere to the engine manufacturer's instructions.

CYLINDER BORING

When cylinder surfaces are badly worn or excessively scored or when a perfectly straight (no taper) cylinder is desired, a boring bar is needed to bore the cylinders for oversize pistons or sleeves. A boring bar leaves a pattern on the cylinder wall similar to uneven screw threads (Figure 11-8). The tool causes surface fractures in the metal. These fractures generally extend to a depth of 0.0005 to 0.001 inch below the surface. This makes it necessary to finish hone after rough sizing the cylinder to remove

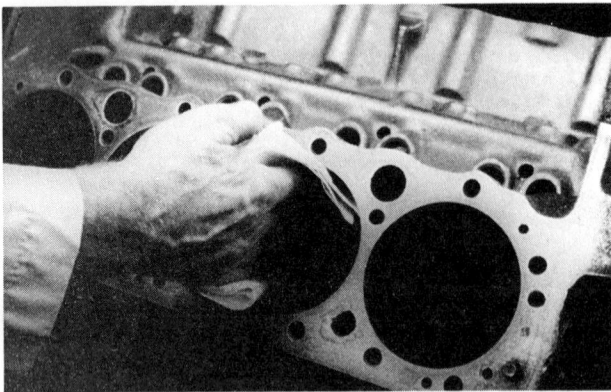

FIGURE 11-6 Cleaning must be thorough to be sure that all abrasive material is removed from cylinders. Use soap and hot water or a special cleaning compound designated for this purpose. *(Courtesy of TRW, Inc.)*

FIGURE 11-7 Lightly hone lifter bores to remove varnish. *(Courtesy of Perfect Circle/Dana)*

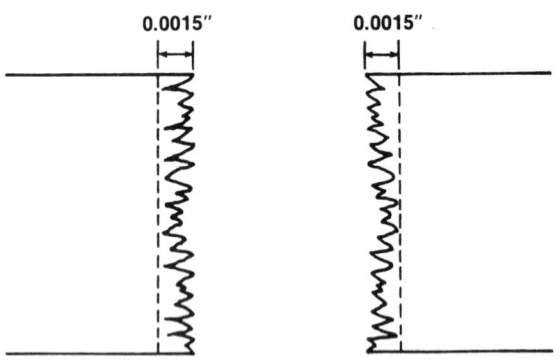

FIGURE 11-8 Although the bores might look smooth inside, there are probably screw-like threads, ridges, craters, gouges, or possibly tears on the surface that cannot be seen by the naked eye. The depth of this fractured metal averages from 0.001 to 0.0015 inch per side. By leaving 0.003 to finish hone out, the unwanted rough surfaces of the boring operation are removed.

fractured metal fragments so that the proper finish can be applied to the base metal.

There are two types of boring machines.

1. Productive engine rebuilding shops usually have large stationary boring equipment (Figure 11-9). The cylinder block is mounted on the machine's table and a rotating cutting bar is centered over the cylinder. Then the bar driven by an electric motor enters the cylinder to cut an oversized bore.

2. The portable unit is mounted on the engine block (Figure 11-10). The block is solidly placed on the floor or in a stand (Figure 11-11). Because most portable units are quite heavy, a chain crane or hoist is necessary to lift the unit to the top of the block.

Since there are many types of cylinder borers and each has slightly different operation instructions, it is important to follow the manufacturer's directions. The following procedures for operation are general in nature.

1. After the block surface has been cleaned and is free of nicks or burrs, center the bar over the cylinder to be bored. For normal boring, centering is done in the unworn lower part of the cylinder. This area provides the greatest accuracy in centering.

FIGURE 11-9 Typical stationary boring machine *(Courtesy of Sunnen Products Company)*

FIGURE 11-10   Typical portable boring bar that mounts on the cylinder block *(Courtesy of Sunnen Products Company)*

FIGURE 11-11   V-8 cylinder block mounted on cradles for honing *(Courtesy of Goodson Shop Supplies)*

FIGURE 11-12   Boring bar is centered with three centering fingers or plungers. *(Courtesy of Kwik-Way Manufacturing Company)*

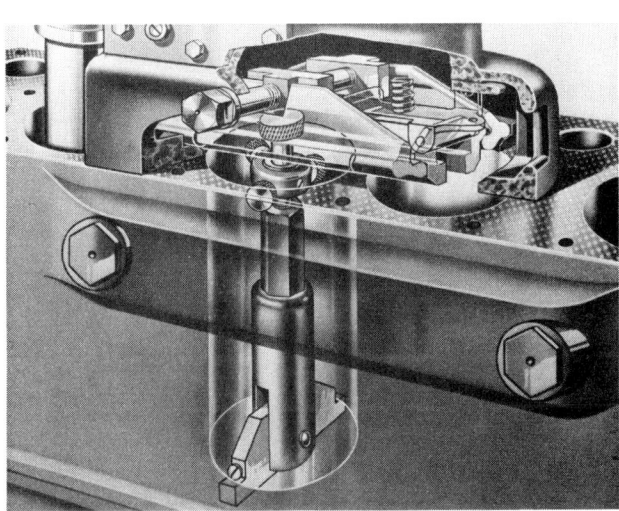

FIGURE 11-13   After centering, an anchor is used to clamp the boring bar in position *(Courtesy of Kwik-Way Manufacturing Company)*

2. Most boring bars have three or four centering fingers or plungers on the end (Figure 11-12). Turning a control knob pushes these three or four fingers out of the bar. The fingers contact the cylinder wall at three or four equidistant spots. When the fingers are pushed tightly against the cylinder, the boring bar is positioned in the exact center of the cylinder.

3. Once the bar is centered, the boring bar assembly is clamped into place on the block. This involves inserting an anchor assembly into the cylinder next to the one being bored. The anchor assembly is then used to secure the boring bar to the block (Figure 11-13). After centering and anchoring, the boring bar can be raised out of the cylinder.

4. Prepare the boring tool bit. Many tool bits have a tungsten carbide tip on the cutting end. Each time it is used, the cutting tip must be sharpened on an iron disc whose surface is charged with diamond dust. The disc is usually mounted on and driven by the boring-bar motor. The angles on the tip of the tool bit must be carefully maintained for proper cutting. The cutter is installed in a special fixture that is positioned on a pilot shaft over the disc (Figure 11-14).

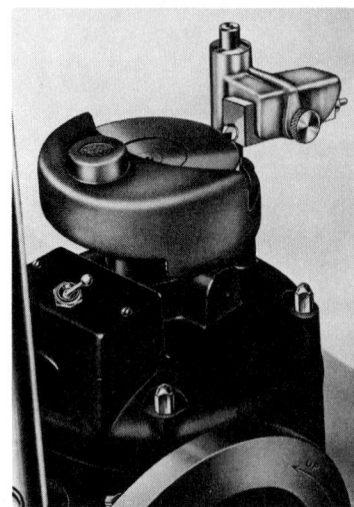

FIGURE 11-14 Tool bit is sharpened with a disc on top of the boring bar. *(Courtesy of Kwik-Way Manufacturing Company)*

The fixture controls the position of the tool bit and sets the angle to which it is sharpened.

5. Set the tool bit to machine the cylinder to the desired size. The size the cylinder will be bored depends on what sizes of oversize pistons are available. Late-model engines have very thin cylinder walls, and some engines cannot be bored at all. Others have oversize pistons available in 0.010-inch steps. An oversize piston set should be chosen that allows the cylinder to be completely remachined. For example, if cylinder wear is in excess of 0.010 inch, a 0.020 inch oversize piston should be used. On the other hand, boring the cylinder 0.020 inch oversize when 0.010 inch would be enough is not a good idea because it would eliminate the possibility of another rebore when the engine is worn again.

🔧 SHOP TALK _____

In most rebuilder shops, it is a general practice to purchase and have on hand the pistons to be used. The rebuilder or machinist then measures the diameter of the new pistons. The boring bar is set to machine the cylinders to 0.003 of this measurement. The proper clearance between the cylinder and piston is achieved by polishing the rebored cylinder with a hone.

6. The dimension to which the cylinder is machined is determined by the position of the tool bit in the boring bar. The position of the tool bit is accurately set with a special micrometer that is part of the boring bar. In many types of boring bars, the tool bit fits in a holder, which in turn fits into the boring bar. The tool bit is installed in the holder using a micrometer that has been adjusted to the desired dimension of the cylinder. The tool bit and holder are then adjusted to this dimension, and the holder is mounted into the boring bar. On other types of boring bars the micrometer is designed to adjust the tool bit when it is in the boring bar (Figure 11-15).

7. With the tool bit properly set, the machining operation can begin. The motor is turned on, causing the bar and tool bit to rotate. A feed lever is engaged to move the bar slowly down the cylinder. The rotating and advancing tool bit machines the cylinder. Motor speed and feed speed controls can be used to vary the quality of the finish.

8. When the tool bit has advanced through the bottom of the cylinder, the bar is switched off. The tool bit is removed from the bottom to avoid scratching the new cylinder as the bar is raised out of the cylinder. A special, long tool bit is then used to put a slight chamfer in the top of the cylinder. This helps in placing the piston

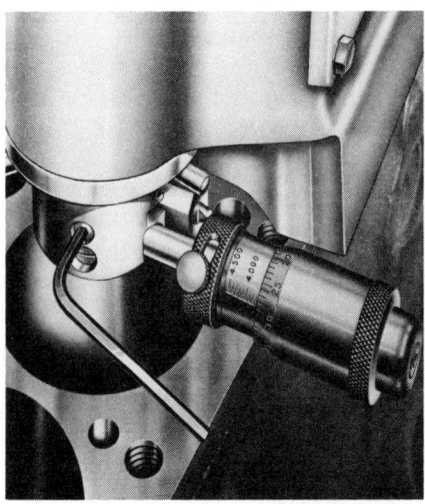

FIGURE 11-15 The tool bit is adjusted to the correct size using the boring bar micrometer. *(Courtesy of Kwik-Way Manufacturing Company)*

and piston rings in the cylinder. This complete procedure is repeated to bore each of the other cylinders.

Boring bars and boring-bar operators vary in their ability to make a good cylinder wall finish. It is possible to make a very highly finished cylinder by boring. However, no matter how fine a boring job looks to the naked eye, there are sharp peaks and valleys. Under a microscope, the bored surface appears torn and fragmented with many relatively "deep" holes (0.001 inch). No natural bearing surface exists, thus the rings would attempt to prepare their own surface. Metal-to-metal contact would result and cause high frictional temperatures with subsequent surface welding and scuffing.

CYLINDER HONING

The honing operation after boring cylinders is vitally important in obtaining proper ring and cylinder seating. Failure to remove adequate stock (0.003-inch minimum) from the cylinders by honing can result in leaving traces of the boring bar cutter machine markings. These peaks and valleys or torn and fragmented cylinder surface can retain sufficient lubrication to retard seating (wear-in) of the cylinder surface. Such a surface can only result in poor oil control, blowby, and a dissatisfied customer. That is, honing must leave the cylinder with the proper surface so that it will distribute oil, serve as an oil reservoir, and provide a place for worn metal and abrasive particles. At the same time, it must have sufficient flat areas or plateaus to act as a bearing surface on which an oil film can form.

A conventional rigid hone usually consists of two stones mounted opposite each other on a holder with guides located at right angles between them (Figure 11-16). The hone rotates at a selected speed. It also reciprocates in the cylinder either manually or automatically, depending on the type of equipment. Outward pressure is applied to the stones so that they will remove stock from the cylinder while turning. Coolant, which is a special honing oil, flows over the stones and cylinder to regulate the temperature and flush out metallic and abrasive residue. It is necessary to select the proper stones to obtain the correct cylinder surface finish. Honing stones are classified by grit size, although honing equipment manufacturers can also have their own designation. The lower the grit number, the coarser the stone. The hone-stop shown in Figure 11-17 takes full strokes without damaged stones from running them out the cylinder bottom. The stop-disc rotates when touched by the hone; it prevents the stones from going too far and allows full, even strokes for improved honing quality.

A honing stone has thousands of tiny cutting edges that leave a multitude of crisscross grooves in the cylinder wall (Figure 11-18). These cutting edges continually break away from a properly operated hone to expose new, sharp cutting edges. Millions of tiny diamond-shaped areas are generated during the honing process, which serve as lubricant reservoirs to maintain the oil film on which the piston rings ride.

Cylinder hone machines are available in manual and automatic models (Figure 11-19A). The major advantage of the automatic type is that it allows the machine operator to dial in the exact crosshatch angle required in both the upper and lower stops of the bore (Figure 11-19B).

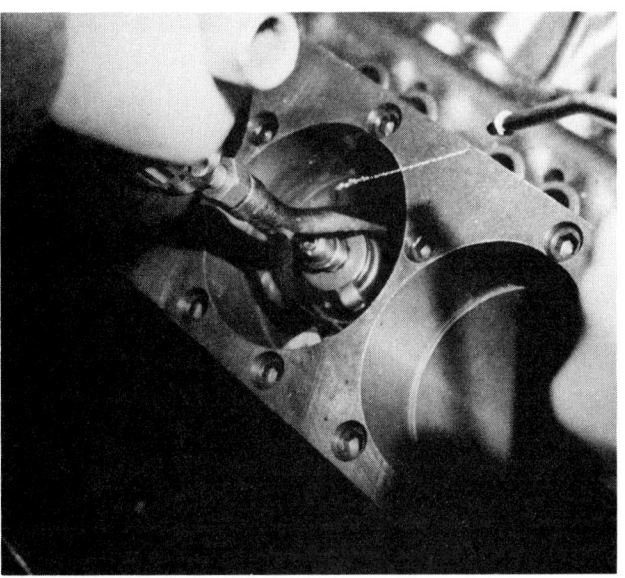

FIGURE 11-16 Typical rigid hone installed in a machine *(Courtesy of Sealed Power Corp.)*

FIGURE 11-17 Typical hone-stop *(Courtesy of Goodson Shop Supplies)*

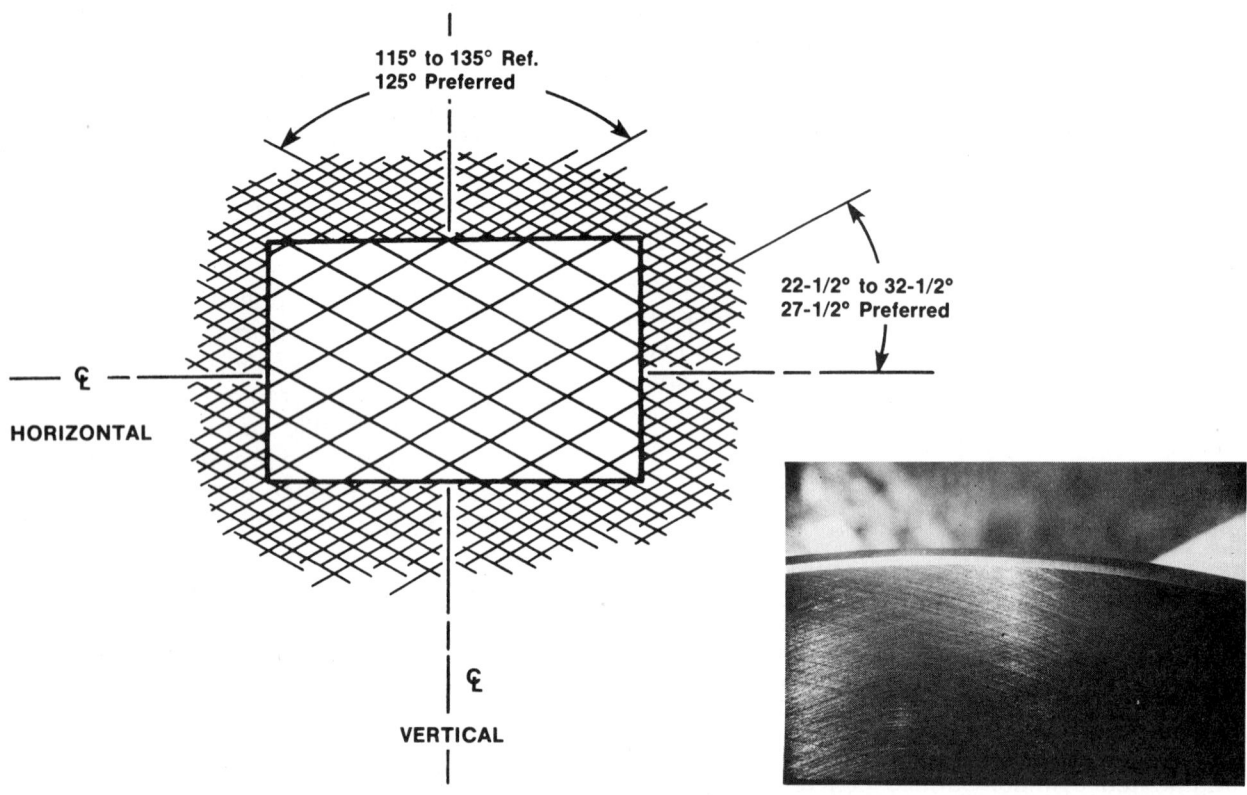

FIGURE 11-18   Ideal cylinder honing pattern finish *(Courtesy of Sealed Power Corp.)*

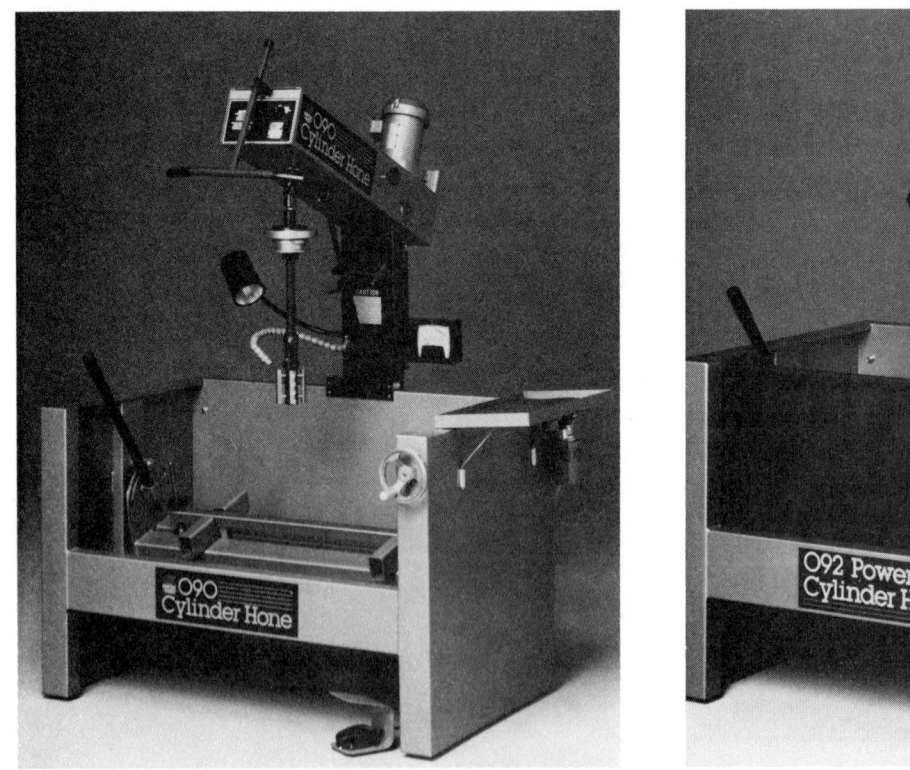

A

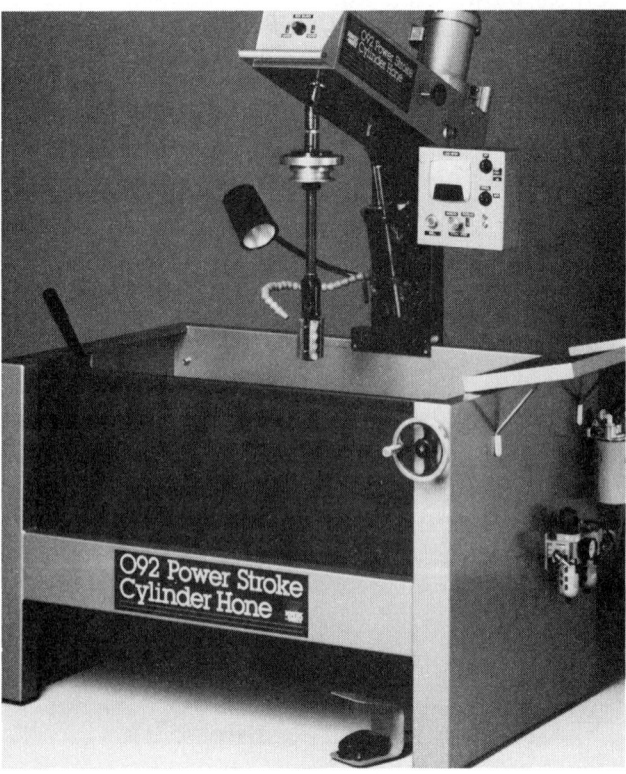

B

FIGURE 11-19   (A) Automatic cylinder hone and (B) manual cylinder hone *(Courtesy of Kwik-Way Manufacturing Company)*

The procedure recommended by most OEM and aftermarket manufacturers is as follows:

1. Install and torque the main bearing caps. The block being complex in design and having a variety of thick and thin sections will twist and distort as pressures from bolt torquing begin to set up internal stresses.

2. Use a torquing plate. (Sometimes referred to as a deck, stress, or honing plate.) The top of the cylinder block is affected by the torquing of the cylinder heads (Figure 11-20). The cylinders can be finished to the finest specifications and all bores checked to be round and straight. But after the cylinder heads and main bearing caps are torqued in place, the picture has changed. Instead of perfectly round bores, they can now be distorted and irregular in shape. To counteract these additional mechanical stresses, it is necessary to prestress the top of the cylinder block, similar to that which was done at the lower end by torquing the mains. This can be accomplished by using a torquing or deck plate as shown in Figure 11-21. The plates can be made from steel that is 1-1/2 inches thick or cast iron having a minimum thickness of 2 inches or can be purchased from specialty high-performance shops. Place a new head gasket under the plate before boring or honing. Be sure the gasket is the same as that used in the final assembly. When selecting head bolts for the deck plate, make sure that they extend the same distance beneath the plate as the regular head bolts do beneath the cylinder head. If

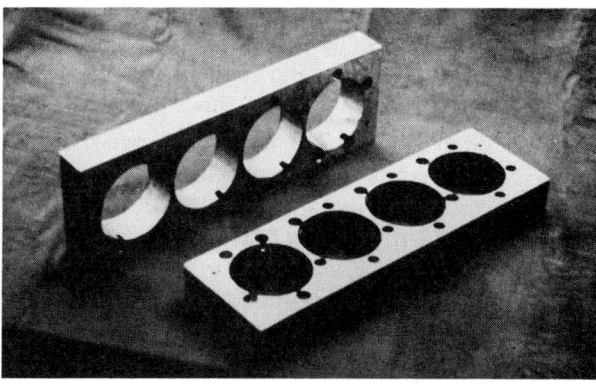

A

B

FIGURE 11-21   (A) Torquing plates made from 2-inch steel and (B) attached to both deck surfaces *(Courtesy of Sealed Power Corp.)*

deck bolts are too long, they will go deeper into the unused portion of the tapped threads in the cylinder block and result in permanent cylinder distortion at the bolt location. The purpose of the deck plates could be defeated entirely by improper lengths of deck bolts. Since all cylinder boring is done with the plate installed, the clamping of the boring unit might have to be modified slightly to hold the boring bar in place on top of the torquing plate during the boring operation.

SHOP TALK _____

Piston rings should be fitted to the desired end gap with the torquing plate attached. Cylinders that are round and free from distortion allow rings to seat quickly. Near perfect roundness at the top of the cylinder bore is extremely critical for all types of compression rings and greatly increases the effectiveness of the popular L-shaped head land ring.

FIGURE 11-20   Installing a torquing plate *(Courtesy of Atlas Engineering and Manufacturing, Inc.)*

3. Leave at least 0.003-inch stock for honing. Rough hone to within 0.0005 inch of finished size using 150 to 220 grit stones. The final 0.0005 inch should be removed with 280 grit stones for moly, chrome, or plain-faced rings (14- to 23-microinch finish). Some special cases might require a finer, 5- to 13-microinch finish obtained wth 400 grit stones. Currently, most of the original passenger car engine manufacturers are finishing their bores in the 10- to 20-microinch range. Table 11-1 gives approximate surface finishes on cast-iron blocks using both automatic and hand-operated equipment.

 SHOP TALK _____

As stated in Chapter 9, the profilometer is the best method of determining surface roughness. However, because of their high cost, profilometers are not commonly found in a rebuilding shop. But there is a correlation between hone grit designation and microinch finish on base metal, which simplifies the selection of stones for the desired final finish. A base metal finish is that of the selected stone after it has cut away all other scratches and grooves formed by previous operations. At this point, removing more metal will not produce a finer finish. For example, if a 20 finish is desired, use a 280 grit stone, which produces a 14 to 23 base metal finish. Some technicians have the idea that a given manufacturer's Number 550 stone is also 500 grit. Although it is true that "the higher the number, the finer the grit," Number 500 is actually 280 grit, 300 is 220 grit, 200 is 150 grit, and 100 is 70 grit.

4. Use a good flow of honing oil to keep the work cool and to flush away abrasive particles. Saturate each cylinder wall and the honing stones with a good grade of honing oil. Kerosene, mineral spirits, or a light-bodied mineral oil can be used. A good flow of clean coolant should be directed into the cylinder bore even before stone rotation starts. A continuous flow of lubricant should be supplied to the honing stones and cylinder walls while honing (Figure 11-22). Filters should be part of every automatic hone to keep the coolant clean. This avoids bore scratching, protects coolant pump parts, and increases the useful life of the coolant. The use of clean coolant on hand honing equipment is equally important. It is desirable that the stone breaks down during the honing process to expose a continuous supply of sharp, clean cutting edges. The use of honing oil will flush the loose abrasive and metal particles from the stones and cylinder wall. Equally important, it will cool the work and keep the stones clean and cutting freely. Surface temperatures are greatly reduced by the use of coolant. It eliminates the possibility of stones becoming loaded, which causes many of the small cutting edges to push rather than cut, and produces a cylinder wall finish having deep scratches, smeared grooves, torn and fragmented material, and glazed and burnished areas.

5. Expand the stones gradually to avoid excessive pressure, which can cause excessive stone wear and loaded stones. Loaded stones will decrease the speed of stock removal and produce a nonuniform and

TABLE 11-1: APPROXIMATE SURFACE FINISHES		
	Microinches Finish	
Grid Size	Automatic Equipment	Hand-Operated Equipment
70	85 to 105	135 to 170
150	—	25 to 40
220	25 to 35	20 to 25
280	14 to 23	15 to 20
400	8 to 13	5 to 10
600	4 to 8	3 to 5

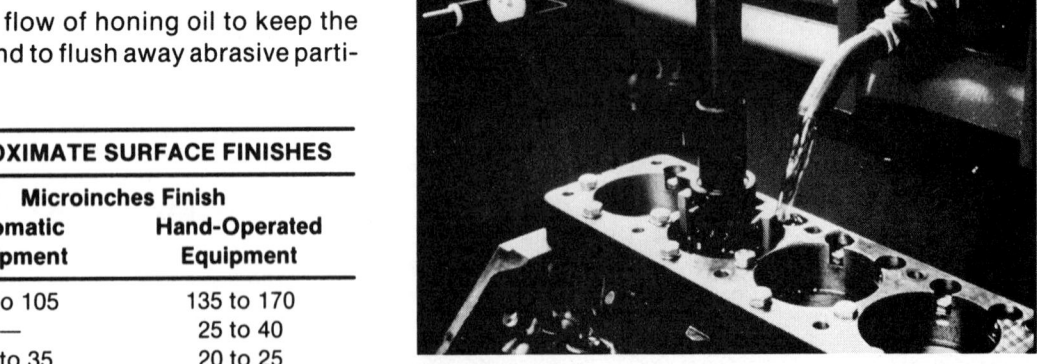

FIGURE 11-22 Use a good flow of honing oil to keep the work cool and flush away abrasive particles *(Courtesy of TRW, Inc.)*

burnished surface. Start the stones with a firm cutting pressure. Excessive pressure will only increase stone wear and will not make them cut any faster. Use a fast and steady, up-and-down movement to produce a well-defined diamond-shaped pattern. A good pattern is to have each side of the diamond shape to be 20 to 30 degrees from an imaginary horizontal line.

6. Allow the stones to run free of heavy drag after they have cut the bore to the desired size. On manual equipment this polishing operation, which is reciprocating the hone for several strokes at the finished diameter, produces very desirable flat areas or plateaus to act as bearing surfaces on which a film of oil can form. When honing with an automatic honing machine, the final honing pressure can be controlled by adjusting the feed rate to a low setting to produce the pleateau surface.

7. Clean the cylinders thoroughly after honing is completed. Cleaning is most essential after the honing operation to remove abrasives and loose metal particles. As when cleaning up after a deglaze operation, use hot, soapy water and scrub vigorously with a stiff, nonmetallic bristle brush. Scrub until the soapsuds remain white, then swab each cylinder wall with the hot, soapy solution to float out all remaining foreign matter. Next, wipe out the bores with paper toweling until clean towels show no dirt. A generous coat of engine oil should be applied to all cylinder surfaces to prevent rust. During assembly, piston rings and pistons must also be coated with oil because dry starts raise surface temperatures and cause scuffing.

Some piston ring and oil consumption problems can be traced to improper cylinder finish. Cylinder finish problems can be grouped into three general categories.

1. Deeply grooved coarse finish with torn and folded edges
2. Fine finish with closely spaced grooves and insufficient plateau
3. Burnished bore

A coarse finish with torn and folded metal (Figure 11-23A) can be recognized by its nonuniform crosshatch pattern and variable width grooves. The grooves are the result of the peaks in the cylinder surface being folded over by the hone, rather than cut away. This creates an excess amount of friable (easily pulverized) metal, which must be worn away to provide an operating surface for the piston rings. The wide, deep grooves carry oil in excess of the amount needed for ring lubrication. Torn and folded metal causes excessive ring face wear as it is broken away. Consequently, oil consumption will be excessive. One cause is using honing stones that are too coarse. Excessive cutting pressure can also be a cause.

At the opposite end of the cylinder finish spectrum is a fine finish with closely spaced grooves and very little plateau (Figure 11-23B). The surface has a sawtooth appearance when viewed in cross section. As with the coarsely torn and folded finish, an excessive amount of friable material is created. Insufficient plateau prevents the formation of an adequate oil film. Consequently, the rings can easily rupture the marginal oil film during operation.

If the honing stones have been cut as they were intended, they will leave the high and low areas. When honing off the high peaks for more ring contact, leave the low spots for oil retention. This process helps seat the rings faster and will also extend the ring life since they will not have to wear in to seat. Be sure to maintain the 45-degree crosshatch pattern for oil control. Piston ring engineers say that if the crosshatch is too steep the oil will migrate downward, causing the oil film to be too thin. This results in piston scuffing and/or piston scoring. The engineers also say that if the crosshatch is too flat, the oil film will be too thick. This causes rings to hydroplane or skate on the oil film, resulting in poor oil economy. Often a resilient-based brush is used to break down a plateau surface situation (Figure 11-24).

Although line boring is a more versatile procedure, line honing is also an accepted method in

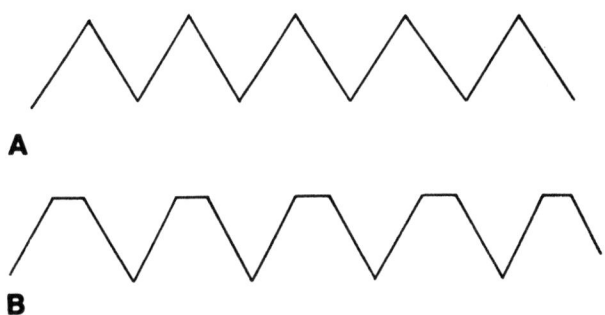

FIGURE 11-23 Comparison of (A) rigid honed and (B) plateau honed

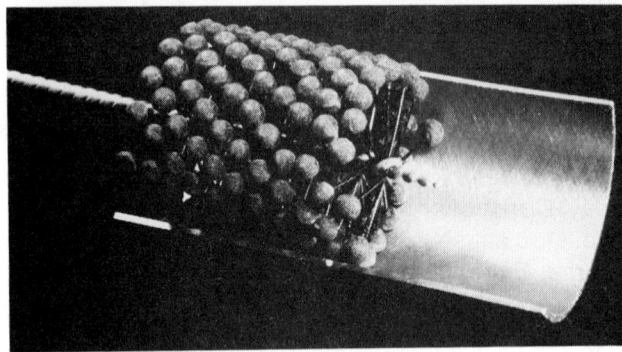

FIGURE 11-24 Using a resilient-based stone to cut down plateau surface *(Courtesy of Brush Research Manufacturing Company, Inc.)*

FIGURE 11-25 Line boring procedure for main bores *(Courtesy of Sunnen Products Co.)*

correcting main bearing bore misalignment (Figure 11-25). In align honing, less material has to be ground from the cap parting lines than with the boring method. Align honing allows you to precisely remove high spots. Size should be checked frequently throughout the honing process.

 SHOP TALK _____

When honing, keep the following points in mind:

- *Crosshatch pattern is directly affected by the rate at which the hone head is traversed up and down. A fast traverse rate results in a steep, angled crosshatch pattern; a slow rate results in a crosshatch pattern with lines more closely parallel to the block deck surface.*
- *Always use a very steady stream of honing oil. The honing oil not only maintains the block at a cool temperature (thus assuring precision dimensions), it also flushes away broken stone particles and metal chips. If fluid flow decreases, check the level in the tank and/or the filter element for clogging.*
- *Hone stones are available in several grit sizes. Always select a grit size appropriate for the amount of material to be removed and the finish desired. Where large amounts of material are going to be removed, select a coarser grit (80 to 150), then finish the bore with a finer stone, for example, 280 grit. Under normal circumstances, no more than approximately 0.001 inch on the bore diameter would have to be removed with the finer stones.*
- *Check the bore frequently while honing to be sure the desired size is not exceeded.*

CYLINDER SLEEVE INSTALLATION

If a cylinder cannot be cleaned sufficiently to fit the largest overbore for which pistons are available, or if it is damaged, a dry or wet cylinder sleeve or liner can often be installed (see Chapter 2). Remember, driving or pressing the sleeve in distorts the cylinder bores on either side. So, install the sleeve before finishing the other cylinders. Some engines, especially diesel types, are equipped with removable cylinder sleeves of either the dry or wet design.

 SHOP TALK _____

At the present time, there is only one reliable method of accurately measuring the cylinder wall casting thickness. This ultrasonic device measures the casting thickness by sound waves. This nondestructible test ensures the presence of an adequate thickness of the wall to allow an overbore. Often called the machinist's sonar, the device looks like a small electric test meter with a probe. The probe is set into position on the cylinder wall with cup grease. Accuracy of the metal thickness reading is within ± 0.002 inches.

Dry Sleeves

Cylinder sleeves, both the dry and wet types, are usually made of cast iron. The simplest approach in sleeving a block is to use an integrally cast block along with a dry sleeve or dry cylinder liner, which is designed to fit into the bore of each cylinder. The piston then moves inside the dry sleeve, wearing against it instead of the cylinder block casting. Since the dry sleeve rests against the cylinder block and has the block to support it, it can be very thin. At the time of reconditioning, the liner is removed and a new one inserted. The dry sleeve usually is retained in the block by a small lip or flange at the top, which drops into a mating counterbore in the top of the deck of the block. This flange prevents the sleeve from dropping downward into the crankcase. The cylinder head with the head gasket, when bolted in place over the block with the sleeve inserted, holds the sleeve firmly in position. As shown in Figure 11-26, the water jacket where the coolant circulates is completely enclosed within the block casting. And since the sleeve makes no contact with the engine coolant, it is referred to as a dry sleeve. It is imperative that the dry sleeve fits snugly (0.001 inch tight to 0.001 inch loose) in order to obtain effective heat transfer through the sleeve into the block and then into the water jacket. A poor fit can lead to poor heat transfer and engine failure.

Dry sleeves are generally available in 3/32-inch or 1/8-inch wall thickness and various lengths. Boring for the 3/32-inch wall sleeve requires the hole to be approximately 3/16 inch (0.1875) over the standard size for the cylinder that is to be sleeved. Boring for the 1/8-inch wall sleeve requires going approximately 1/4 inch (0.250) over cylinder standard size.

The outside diameter (OD) of the sleeve is measured with an outside micrometer (Figure 11-27).

The cylinder diameter should be slightly smaller than the outside diameter of the sleeve. Measure the sleeve in several spots to determine its diameter. An interference fit of 0.0015 to 0.0025 inch is desirable. Reach the desired size with at least two passes of the boring bar. That means taking about 0.030 inch or less out on the second pass to ensure a bore that is smooth and dimensionally correct.

After boring out the cylinder to the desired size replace the boring bar with a special cutoff tool (Figure 11-28). This action will leave a slight lip at the bottom of the cylinder for the sleeve to seat against. The top of the cylinder is chamfered to help start the sleeve into the cylinder.

Immediately prior to the installation, heat the block and pack the sleeve with dry ice. Also coat the sleeve with a lubricant. Then press or drive the sleeve into the cylinder until it touches the lip at the bottom of the bore. To drive the sleeve down into the cylinder use a hammer and sleeve driver (Figure 11-29) or a hydraulic driver (Figure 11-30). When installed, the sleeve will extend slightly above the deck of the block. The boring bar with the face tool installed in the cutterhead is used to machine the top of the sleeve flush with the deck. The sleeve is then bored to the desired size, just like any other cylinder. The bored and sleeved cylinder is then chamfered and washed as described for deglazing.

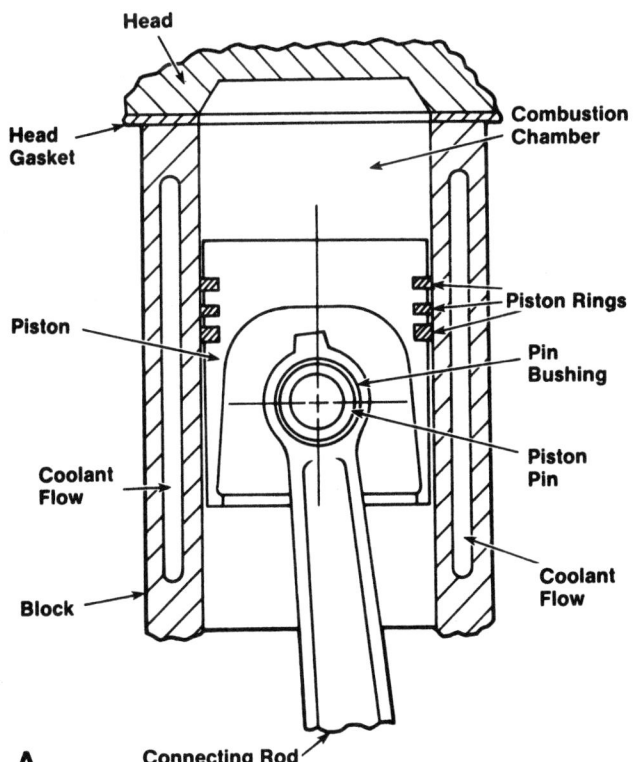

A

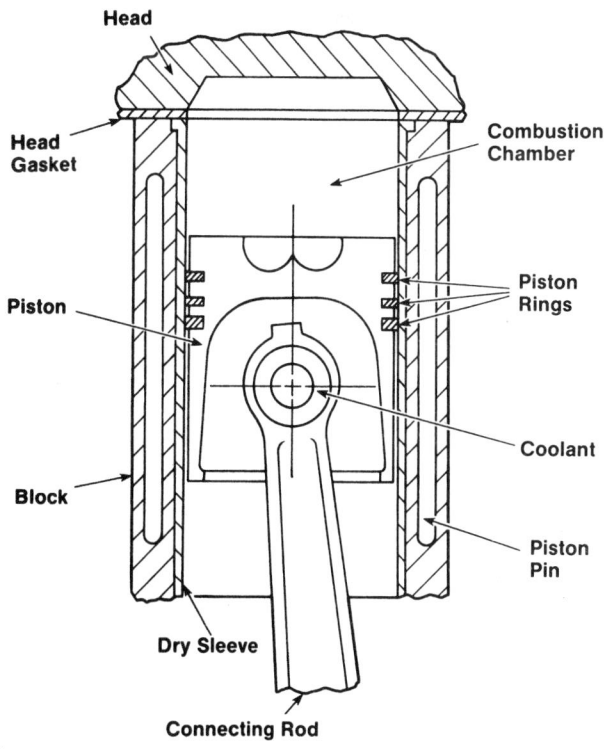

B

FIGURE 11-26 (A) Typical non-sleeve cylinder and (B) a typical installed dry sleeve engine

FIGURE 11-27 Measuring the outside diameter of a sleeve with a micrometer *(Courtesy of Perfect Circle/Dana)*

FIGURE 11-28 Special cutoff tool replaces the boring bar *(Courtesy of Perfect Circle/Dana)*

 SHOP TALK —————————

When block repairs are required in the upper counterbore or lower block sealing area to provide good sealing surfaces, parts manufacturers offer block repair bushings to restore those sealing surfaces. Block repair bushings represent a simple, relatively inexpensive method of salvaging worn or cracked cylinder blocks. Where the bottom ledge or shelf of the counterbore has been damaged but can be restored by machining off a small amount of metal, a flange shim is available. The flange shim is installed under the flange of the sleeve and restores the sleeve to its proper protrusion above the deck of the block. Flange shims are now used by some engine manufacturers new from the factory.

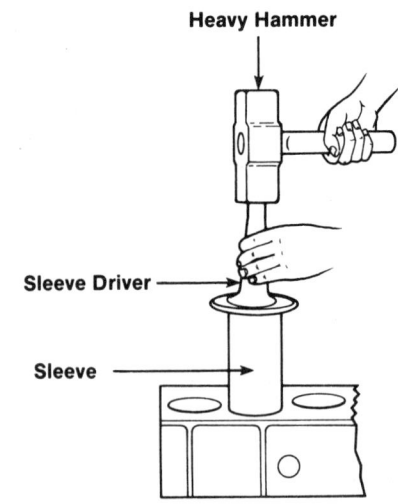

FIGURE 11-29 Installing a sleeve with a sleeve driver and hammer.

Wet Sleeves

Although the water jacket in an integrally cast block is cast in one piece with the cylinder itself, a wet sleeve block is cast to require that a sleeve be inserted into it to close or complete the forming of the water jacket (Figure 11-31). These open-style block castings are designed to take a replaceable wet sleeve, which is called wet because the coolant makes direct contact with the outside diameter of the sleeve. Wet sleeves have two critical sealing areas that enclose the water jacket: the top flange area (which fits into the counterbore of the block) and the lower sealing ring area, where O-ring seals are used to prevent coolant from leaking into the crankcase oil. Since the water jacket is open in the block and is completed by the wet sleeve, the sleeve

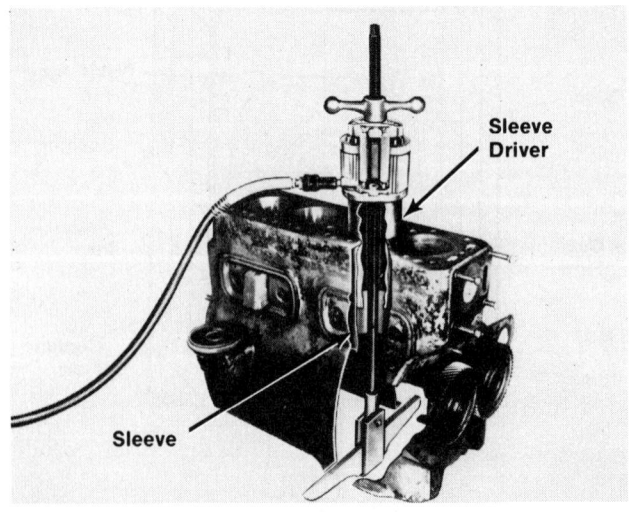

FIGURE 11-30 Installing a sleeve with a hydraulic driver *(Courtesy of Tool Division—SPX Corp.)*

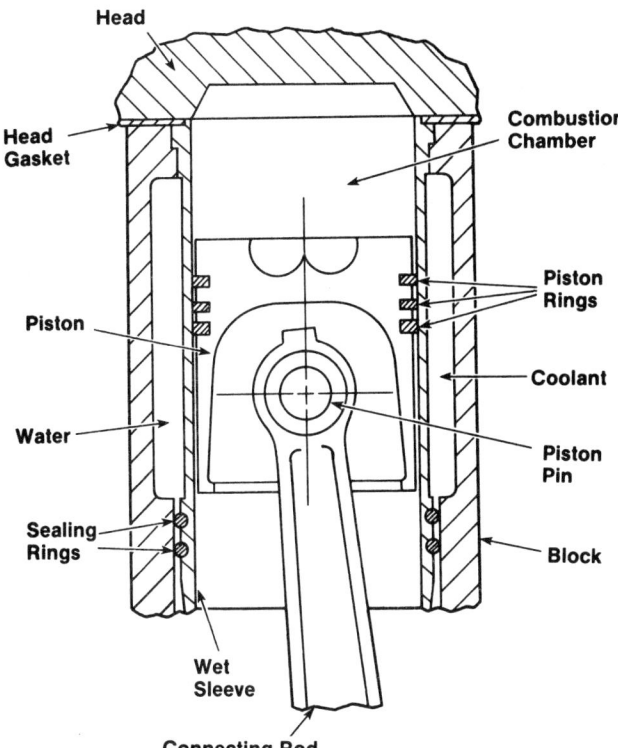

FIGURE 11-31 Typical installed wet sleeve in place in a cylinder

gets no central support from the block. For this reason, wet sleeves must be made thicker than dry sleeves.

To rebuild, the wet-sleeved block requires resleeving and normally takes a sleeve assembly or liner kit with a new wet sleeve, piston, and ring set. While the procedure for installing a dry sleeve and a wet sleeve is basically the same, be sure to check the installation instructions furnished by the manufacturer very carefully.

Some liners or sleeves contain fire walls to protect head gaskets from the direct impact of combustion. The fire wall is a small, annular ring protruding above the flange, next to the sleeve bore. The fire wall greatly reduces the likelihood that the head gasket flange will be burned. Also, crimping beads are often machined into liner flanges to promote sealing against the head gasket. A crimping bead is a narrow, short annular ring located on the top side of the flange. When the head gasket and cylinder head are bolted down, the crimping bead bites into the gasket, affecting a positive seal in that area.

PISTONS

Although the piston may be cast from one of several metals (see Chapter 2), it is precision ma-

 SHOP TALK _____

It is important to keep in mind that wet cylinder sleeves can be a source of what may appear to be piston ring trouble. Errors in installation that lead to buckling of the liner are to be guarded against. One point that needs more attention is wet liner rubber sealing O-rings. If these rings are rolled, twisted, or sheared, the sleeve may buckle, as illustrated. Of course, this means scoring, eventual fracture, and coolant leakage. Before putting the seal in place, check the seal area—on the lower deck especially—for scale and deposits that would prevent proper seal seating. Round off the top edge of the lower deck. Lubricate the lower deck and sealing rings so the sleeves enter the lower deck with just a slight push. When in place, check the inside diameter of the liner with a plug gauge 0.001 to 0.0015 inch smaller than the sleeve's inside diameter. If it binds, note the spot, pull the sleeve, and find out why. Also check for coolant leakage by the rubber sealing ring.

chined, made to withstand considerable stress, but there are limits that the piston cannot exceed. If subjected to mishandling, excessive heat or load, or inadequate lubrication, a piston can be damaged, resulting in an engine failure.

The piston-to-cylinder-wall clearance is one of the most important measurements in an overhaul. This measurement is the clearance between the piston skirt thrust faces and the cylinder wall. Precisely the right amount of piston clearance is necessary in an engine (Figure 11-32). If the clearance is too

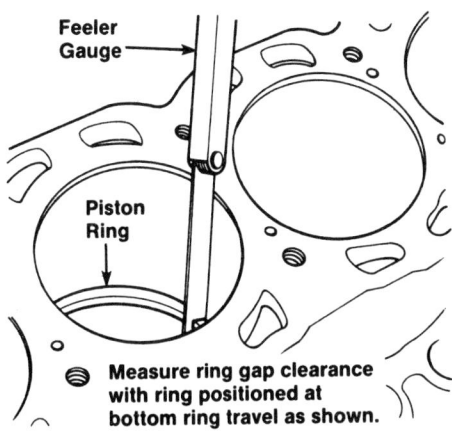

FIGURE 11-32 The amount of clearance between the piston skirt thrust faces and the cylinder wall must be exact.

small, there will be no space for a film of oil between the piston skirt and cylinder wall. Without an oil film the piston will heat up, begin to scuff and score, and possibly expand so much that it sticks or seizes in the cylinder. Too much piston clearance can allow the piston to rock, or slap, in the cylinder (Figure 11–33). The skirts of the piston can break if they slam against the cylinder walls.

Adjusting the piston clearance on a reconditioned engine is called fitting the piston, and there are several ways of accomplishing this. The method used is determined by the manufacturer's recommendations and the method used to recondition the cylinder walls. The rebuilder should look up the manufacturer's specifications for piston clearance as well as the recommended procedure for fitting pistons.

When replacing the original pistons in deglazed cylinder walls, the clearance should be the same as when the engine was new. All that is necessary is to check the clearance. Use the piston size measurements and compare them against the cylinder measurements. For example, assume the piston measured for cylinder number 1 was 3.500 inches across the skirt thrust faces and the cylinder measurement for cylinder number 1 was 3.503 inches. To find the clearance, subtract the piston measurement from the cylinder measurement. The clearance for this cylinder is 3.503 inches – 3.500 inches = 0.003 inch. Then compare this measurement to the manufacturer's specifications.

When the cylinders are honed to remove taper or honed to polish after boring, a different procedure for fitting pistons is generally used. To remove taper, the cylinders are honed to oversize, so the pistons would have an excessive amount of clearance in the honed cylinder. Expand or resize the pistons to correct this.

RECONDITIONING PISTONS

Pistons are expanded or resized on a piston knurlizer (Figure 11–34). When the knurlizing tool contacts the piston skirt, controlled air pressure is applied. The piston is supported on the underside by the anvil and roller so that the shape and diameter across the pin sides of the piston will not be altered or distorted. The teeth of the tool displace the metal of the piston skirt by controlled pressure on the knurling tool. The displaced metal is forced upward between the tool's teeth, which produces an interrupted surface and increases the effective diameter of the piston skirt.

The piston can be indexed to position the knurlizing tool at any point along the skirt, which makes it possible to increase the skirt diameter at any point desired between the top and bottom. The diameter can be controlled to within plus or minus one-half a thousandth of an inch. Pistons for many makes of engines can be knurlized with connecting rods attached.

CAUTION: It is not recommended that any cast steel or semi-steel thin-walled piston be knurlized. The walls may possibly collapse during the knurlizing process.

A feeler strip and a kilograms- or pounds-pull specification are provided by the manufacturer to check the fit between knurlized pistons or new pistons and polished cylinders. A feeler strip is a long piece of feeler gauge material. To use it, the piston is installed in its cylinder upside down. The recommended thickness feeler strip is then installed

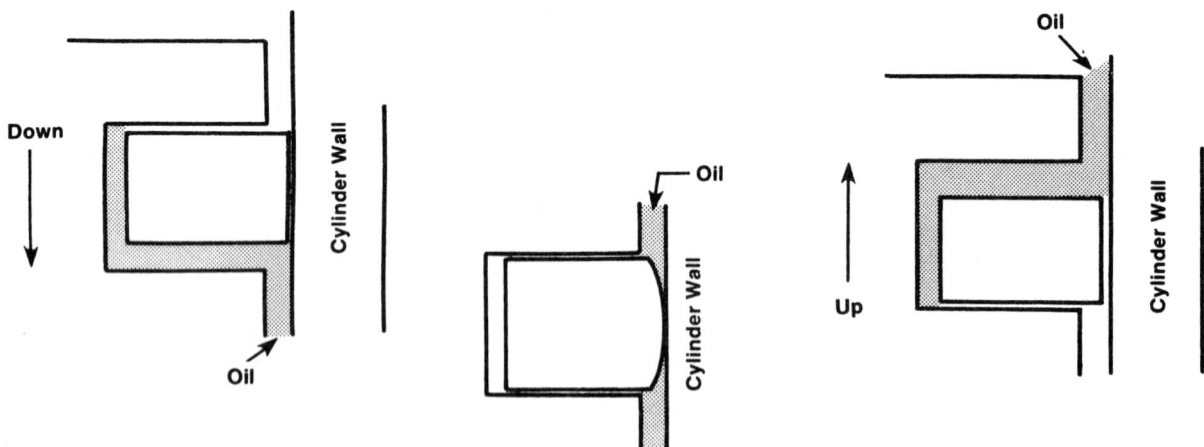

FIGURE 11-33 Excessive clearance causes the pistons to rock or slap in the cylinder

FIGURE 11-34 Piston knurling machine and a knurled piston *(Courtesy of Atlas Engineering and Manufacturing, Inc.)*

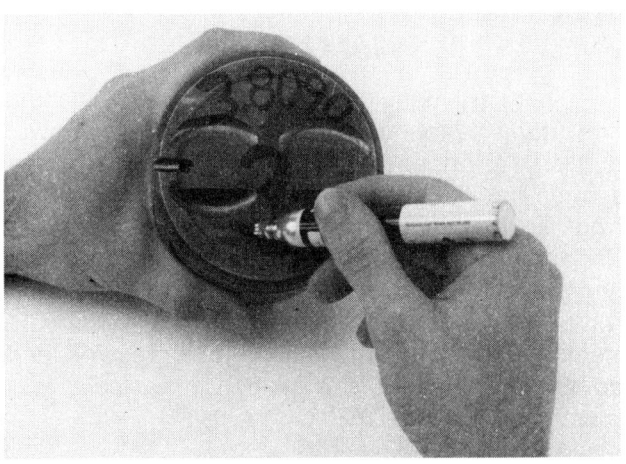

FIGURE 11-36 Mark a piston for use in each bore honed. *(Courtesy of Botts Auto Parts Company)*

in the cylinder at the same time as the piston, along the piston skirt.

A scale called a fish scale is attached to the end of the feeler strip by a hole or clamp. The technician pulls the feeler strip out from between the piston and cylinder wall and observes the scale (Figure 11-35). If clearance is correct, the feeler strip will pull out at the specified pull on the scale. If too much pull on the scale is required, clearance is too small. If little or no pull on the scale is necessary, clearance is too large.

The feeler strip and scale are used to check the clearance during honing or polishing. The technician begins honing or polishing and frequently stops to check the clearance. Each piston is matched to each cylinder (Figure 11-36). After one piston is correctly matched, it is marked as to the cylinder it fits, and the next piston is matched and marked the same way.

Another way to make a rough check on piston clearance is to check the piston in the cylinder. Place the piston in the cylinder upside down. Allow the piston to slide down the cylinder, being sure to catch it at the bottom or it could be damaged. The piston should be able to slide down an oiled cylinder under its own weight. A knurlized piston requires light finger pressure to push it down the cylinder.

At one time, it was common practice to machine (turn or grind) oversize pistons back to their standard size. But, most of today's pistons should not be machined to a size smaller because the walls would be too thin for safe operation. The exceptions are pistons that have a predrilled head. Generally, they can be remachined to a small size on a piston machine and turning tool. When machining a piston, (Figure 11-37), the machine's manufacturer's instructions must be carefully followed.

FIGURE 11-35 Checking clearance with a minimum clearance feeler gauge. The gauge should come out when pulled gently.

FIGURE 11-37 A vise such as shown above is valuable for precision flycutting of reliefs. *(Courtesy of Atlas Engineering and Manufacturing, Inc.)*

PISTON RINGS

There are many considerations when installing and fitting piston rings for maximum horsepower output, performance, and service. From the passenger car to the high-performance engine, the rings' sealing ability depends on the fit between the ring face and the cylinder wall (back clearance and end gap). The fit between the sides of the ring and the piston ring groove (side clearance) is also extremely important. Oil, vacuum, compression, and combustion leakage can happen in basically four ways:

1. Past the ring face
2. Around the back and sides of the rings
3. Through the ring end gap (if too excessive)
4. Failure to follow proper break-in procedure

Leakage past the top ring can result from out-of-round cylinder walls due to wear, heat expansion, or pure mechanical distortion. This can result from block deformation caused by high horsepower output or from just bolting on cylinder heads. The incidence of bore distortion problems greatly increased in recent years since manufacturers introduced the thin-wall blocks.

Piston rings do their best to conform to slight cylinder wall deformation and machining imperfections; however, they are by design fairly rigid and in most cases unable to compensate for bore irregularities.

To insure that the proper ring groove side and back clearance is maintained, the right groove sides must be smooth, flat, and the same angle as the ring. Excessive clearances can result in gas and oil leakage around the ring sides and back. Too much back clearance can affect the face seal of the rings by creating a large nonfunctional space (volume) behind the ring that would decrease the force exerted by the compression gases on the backside of the ring to force it out against the cylinder wall during the power stroke. Excessive side clearance can and usually does result from worn, wavy, or irregular ring grooves. New piston rings in a worn groove will twist, flex, and pound against the worn surface and make for a very poor combustion seal. If ring side lapping is necessary for custom fitting piston rings, it should be confined to the top side of the ring. Otherwise, the bottom side seal is apt to be impaired due to the altering of the ring flatness.

Clearance is also critical at the ring end gap (Figure 11-38). The ring expands when subjected to heat, so clearance must be built in to insure that the ends do not contact each other during operation.

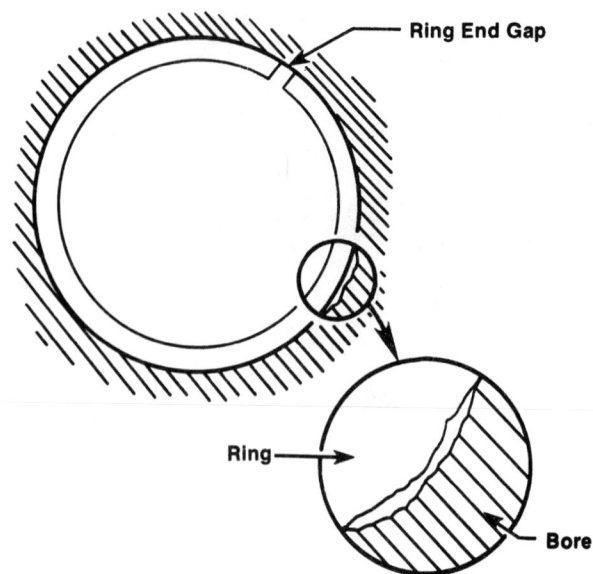

FIGURE 11-38 When new rings are placed in an engine, they are not seated. The shape of the rings is not exactly the same shape as the cylinder bore.

Insufficient clearance can cause the ring to butt and break contact with the wall, which creates a leakage path. Too much end clearance will cause excessive leakage through the end gap. Follow the end gap recommended by the ring manufacturer.

The break-in procedure for a new set of piston rings is crucial to any installation. Once the engine is fired, it must be placed under a load and not allowed to idle for any appreciable length of time. Under power, the expanding combustion gases will force the rings against the cylinder wall so that the two surfaces can lap together. The cylinder walls and rings are lubricated by the splash of the connecting rod and crankshaft in the sump. Failure to force the rings out against the cylinder wall will allow the face of the piston rings and the cylinder wall to polish or glaze. When this occurs, blowby and high oil consumption result and will continue until the rings are replaced. Full details on engine break-in are given in Chapter 15.

INSTALLING PISTON RINGS

Before placing the rings on the pistons, thoroughly clean all ring grooves and drain holes. To do this, use a ring groove cleaner tool as described in Chapter 7. Be sure to select the correct cutter blade. Also be careful not to make the groove wider or deeper with this tool.

Piston rings should be installed according to the ring manufacturer's instructions (Figure 11–39). Specifications such as placing the correct ring in its proper ring groove right side up and recommended ring gap positioning around the piston must be followed to ensure proper ring operation. As just mentioned, end gap is important; it is the distance between the right ends when the ring is confined in a cylinder of the specified diameter. Table 11–2 gives the minimum end gap clearance.

All rings should have at least the minimum specified end gap to allow for the expansion between piston ring and cylinder. If no gap exists, scuffing, scoring, ring breakage, or engine seizure can occur. The tolerances for maximum end gap are not as critical as those for minimum. However, a conservative maximum gap is 0.040 inch greater than the minimum gap.

When the end gap is more than the specified limit, a larger oversized ring set should be used. Check for worn top grooves by installing a new ring in the piston top groove while holding the face of the ring flush with the land. Use a feeler gauge to determine the side clearance.

FIGURE 11–39 Install the segmented oil rings by placing the expander-spacer in the piston groove. Follow the instructions that come in the box. *(Courtesy of Perfect Circle/Dana)*

TABLE 11-2: MINIMUM END GAP CLEARANCE	
Cylinder Ring Diameter (Inches)	**Minimum End Gap (Inches)**
Less than 3	0.007
3–3-31/32	0.010
4–4-31/32	0.013
5–6-31/32	0.017
7–8	0.023

FIGURE 11–40 Check for proper ring gap with a feeler gauge by pushing the ring into the bore. *(Courtesy of Perfect Circle/Dana)*

The end gaps of each ring must be arranged so that they are not aligned vertically (Figure 11–40). Aligning all ring end gaps vertically might cause a compression leak through the end gap spacings. Follow the manufacturer's directions or space the gaps 120 degrees apart on the piston. Typical piston ring spacing is shown in Figure 11–41.

If the ring gap is satisfactory, install the rings on the piston. The plain compression ring is nondirectional, which means that it can be installed with either side up. Today, however, most compression rings are directional, which means they have a distinctive top side and bottom side. Directional compression rings can be identified by a taper on the ring face or a bevel on the inner (IB) or outer (70) face. The ring with the inner bevel, unless specifical-

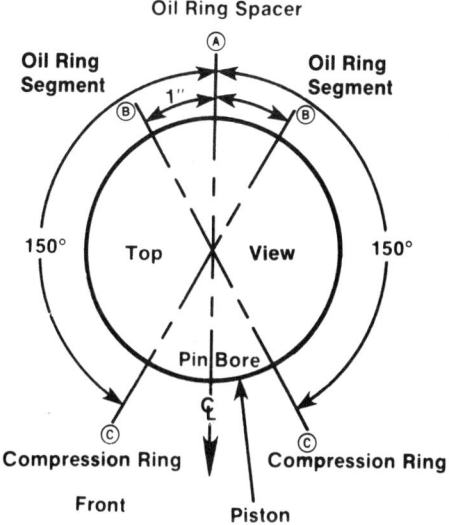

FIGURE 11–41 Piston ring spacing

ly marked to the contrary, must be installed with the bevel up. There is also a ring used that has the inside bevel on the lower inner corner. This ring is known as a reverse twist and has a very pronounced top marking, as does the taper-faced ring. All rings that have a channel on the outer face are installed with the channel down. Many rings have standarized markings, such as those shown in Figure 11–42, that show which direction is up or toward the piston head.

With the number and variety of pistons changing, double-check the application for correct ring sizes. Some engines will require different size rings, depending on the year the engine was manufactured and the make of piston. Check for correct piston and rod assemblies also.

When installing rings, use a ring-expander tool (Figure 11–43) to avoid overspreading them. Select the correct ring for the correct ring groove. Expand the piston ring on the ring-expanding tool, then place the ring and tool over the piston, set the ring into the groove, and release the tool (Figure 11–44). Double-check the installation instructions to make sure the ring has gone into the right groove and is not upside down.

FIGURE 11–42 Typical markings to show the top of the ring when installed.

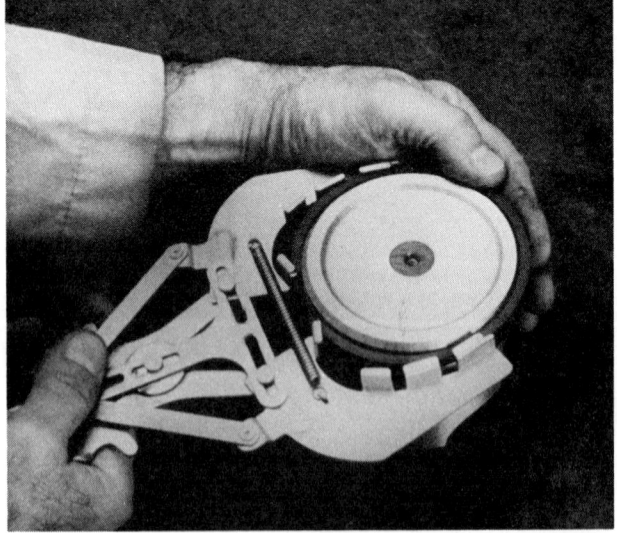

FIGURE 11–43 A ring expander must be used to install the rings on the piston.

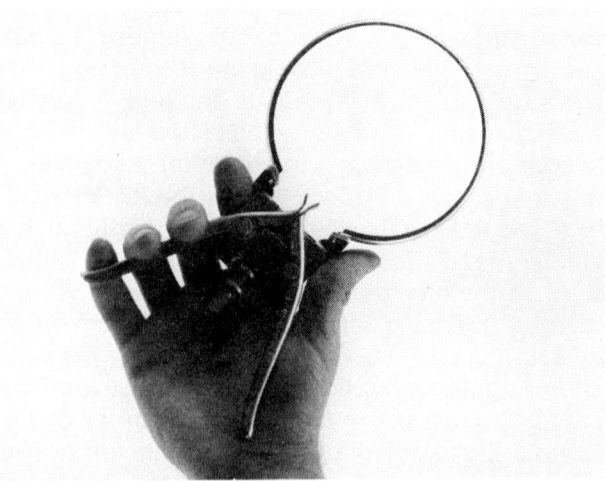

FIGURE 11–44 Carefully install new rings per the manufacturer's instructions using a proper installation tool. *(Courtesy of Botts Auto Parts Company)*

 SHOP TALK _____

Do not attempt to install rings by hand because they can become distorted and begin small metal fractures opposite the gap that can lead to ring breakage eventually.

The top ring and top groove of most aluminum pistons that have been used for a length of time become worn due to abrasives and high temperature in the top ring land area. This increases the top ring side clearance and causes increased blowby and oil consumption. The maximum allowable side clearance between a top ring and top groove is 0.006 inch.

If a new top compression ring is installed in a worn groove, it is evident that a proper seal cannot be formed with either the ring face or ring sides and poor performance results. The worn groove forces the upper corner of the ring face to contact the cylinder wall, causing the oil to be wiped into the combustion chamber instead of down into the crankcase. In addition, continued deflection of the new top ring in the worn groove results in ring fatigue and breakage. Therefore, it is important that the top grooves be checked and remachined when worn to prevent excessive oil consumption, ring breakage, and high blowby. The easiest way to check for wear is to clean the grooves with a ring groove cleaner (see Chapter 7) or an abrasive cord made for this purpose (Figure 11–45).

The top groove wear gauge is one of the easiest ways to check this possible problem. Each gauge flange is machined 0.006 inch oversize so that if the

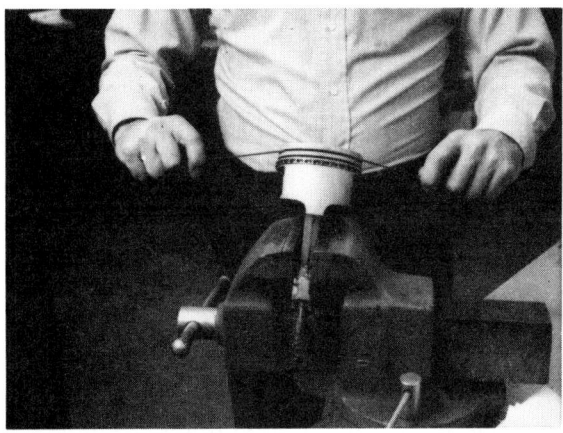

FIGURE 11-45 Cleaning a top groove with an abrasive cord *(Courtesy of Goodson Shop Supplies)*

gauge enters the groove as shown, the piston needs regrooving. Because of the variation in temperatures on the piston head, the top ring groove does not wear evenly around its circumference. The unevenness reduces the effectiveness of the seal between the new ring and the bottom of the ring groove. For this reason it is necessary to check the grooves for possible variation in wear at several points around the piston. The maximum amount of wear is usually found on the side of the piston closest to the exhaust valve. If the gauge enters the groove at any point, the piston must be replaced or regrooved to insure good ring performance.

Excessive top groove clearance can be corrected by purchasing new pistons or by regrooving the old ones with a ring groove reconditioner (Figure 11-46) to accommodate a new standard width ring and a 0.024-inch-thick steel top groove spacer above the ring. Accuracy in regrooving the top

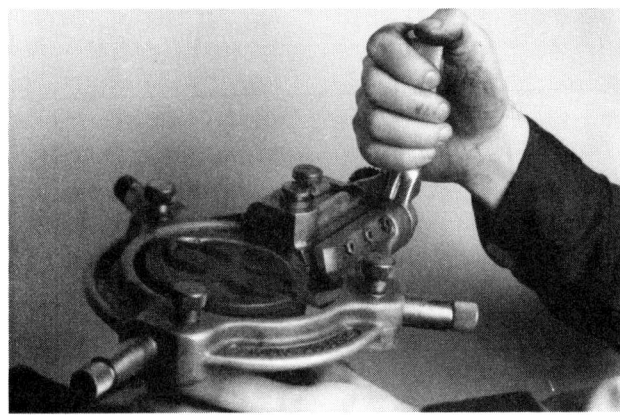

FIGURE 11-46 One type of piston ring groove reconditioner.

grooves with this machine is assured because the cutting tool is guided by pilots that ride in the second ring groove, ensuring that the new top groove will be parallel to the other grooves. The constant feed of the cutting tool insures proper finish on the sides of the groove. This operation can be performed quickly without removing the rods from the pistons, regardless of whether the pistons have centers or not.

After the regrooving operation, the steel spacer (usually 0.0300 inch) must always be installed above the new top ring (Figure 11-47). The combination of the new top ring and a thin flat spacer will retard the rate of ring and groove side wear. Failure by the technician to correct groove wear always results in unsatisfactory engine performance. Almost all aluminum pistons removed for reringing have excessive top groove clearances that must be corrected for satisfactory ring performance.

CONNECTING RODS

Standard connecting rods (sometimes called con-rods) are usually made of cast or forged steel in an I-beam construction. A few high-performance rods are available in forged aluminum or reinforced plastic, some of which are round.

The great stress and strain put on standard connecting rods by the modern engine unfortunately is not usually visible in most cases. For this reason it is very important that connecting rods be carefully inspected for out-of-roundness, bending, or twisting as described in Chapter 7, or tested on a precision

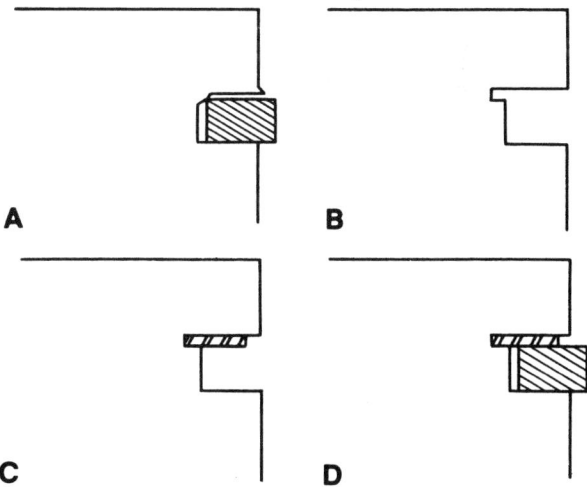

FIGURE 11-47 Making ring groove correction: (A) new ring in worn groove; (B) groove widened; (C) insert installed and locked in place; (D) new ring installed in reconditioned groove

electronic con-rod gauge (Figure 11–48). The advantages of using any con-rod gauge are:

- It checks center to center distance, bend, and twist at the same time with one setup (Figure 11–49). The center-to-center distance is the length of the con-rod from the center of the small end to the big end of the rod.
- It is calibrated to show total piston (wrist) pin displacement from parallel, assuming a constant 8-inch piston pin length.
- It shows out-of-parallel to the nearest 0.001 inch for bend and twist on two separate indicators.

The rods should be checked for cracks using the tests described in Chapter 8. If any cracked rods are found, they should be replaced.

 SHOP TALK _____

When installing piston rings, keep the following in mind:

- *Always thoroughly lubricate rings and cylinder walls before installing rings.*
- *Piston ring grooves vary in width, even on the same piston. Groove width specifications are given as ring widths in any engine parts manufacturer's catalog. They are listed along with pistons and/or ring sets.*
- *Also remember that ring grooves must be completely clean before installing rings in them. Even a slight film of carbon or grease in the groove will cause the rings to stick in the groove or press abnormally tightly against the cylinder wall.*
- *If there are two oil rings on each piston and only one of the rings has an expander in the package, that ring should be installed in the top oil ring groove. The bottom oil ring, on a two oil ring setup, serves mainly as a flexible piston stabilizing force and as a backup to the top oil ring.*
- *When installing rings on pistons, always install the lower ring first. Never try to expand a ring sufficiently to pass over an already installed ring.*
- *Groove widths can also be checked with a machinist's 6-inch steel rule that is graduated in 1/4, 1/8, 1/32, and 1/64 inch. The measurement must be care-*

FIGURE 11-48 Electronic out-of-round bore gauge system *(Courtesy of Sunnen Products Co.)*

fully made. It is easy, for instance, to mistake 13/64 for 7/32 inch.
- *When installing spacers, expanders, or rails in slotted oil ring grooves, always take care to place the two free ends of the item on a solid part of the groove. If that is not done, the item can work through the slot and down inside the piston.*
- *Expanders are never used behind a top compression ring. Because of less lubrication at the top, an expander might cause the top ring to run dry and quickly scuff or score the cylinder. An expander is sometimes used behind the second compression ring, especially if the ring set is designed to be installed in worn cylinders.*

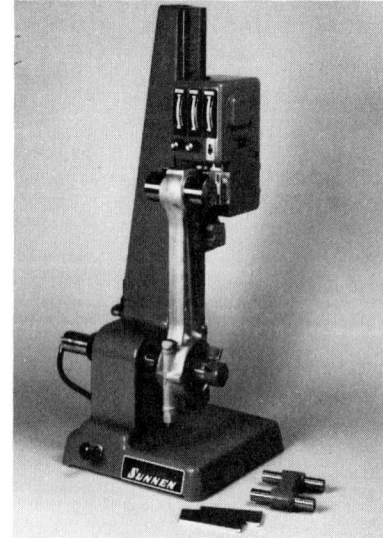

Figure 11-49 With one setup, this electronic gauge checks center-to-center, bend, and twist at the same time. *(Courtesy of Sunnen Products Co.)*

For the crankshaft to operate properly, the following conditions must be met (Figure 11–50):

1. The crankshaft journal must be smooth, round, straight, of specified size, and correctly aligned with the other journals of the shaft (see Chapter 12).
2. The connecting rod and its bearing must be free from bend and twist, must be square with the cylinder bore, and must have the specified oil clearance.
3. The rod big-eye end bore must be perfectly round and straight, have the proper rms finish (usually 60 to 90 microinches, depending on the manufacturer), and be of a specified size (Figure 11–51). The bearing inserts must have the proper crush in the rod bore when torqued to specifications.
4. The small end bore must be finished according to the type of piston pins being used. (This is covered later in the chapter.)
5. The proper amount of lubrication, at the correct pressure, must be supplied to the con-rod bearings at all engine speeds.

Reconditioning Connecting Rod Bores

If an inspection of the connecting rods shows the saddle or big-eye bore is out-of-round, it must be

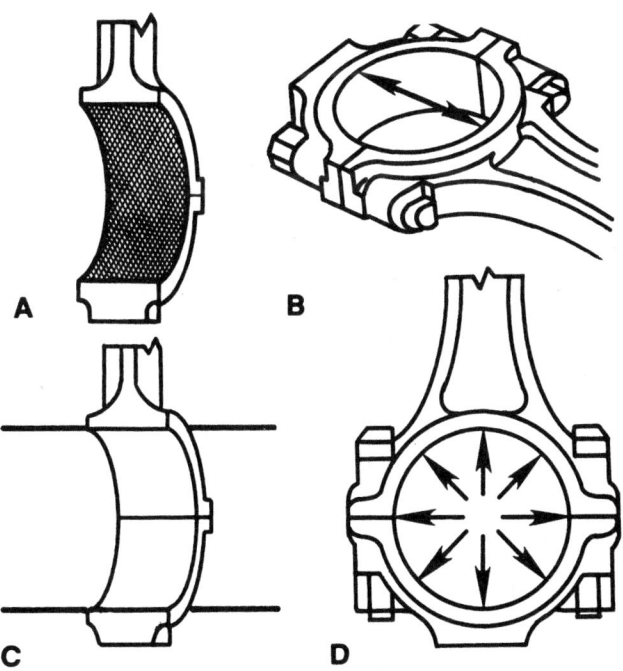

FIGURE 11–50 For proper operation, the crankshaft journal must be (A) round; (B) straight; (C) smooth; (D) correct size.

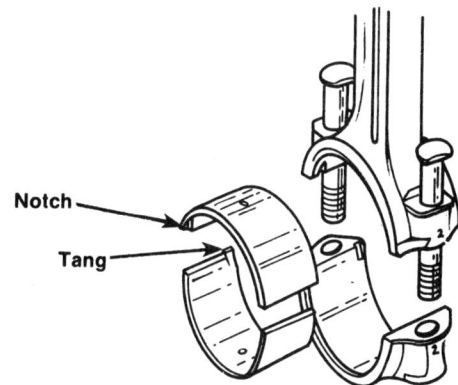

FIGURE 11–51 Locating lugs or bearing tangs fit into notches in the bearing bore to prevent movement of the bearing inserts.

 SHOP TALK _____

The amount of crush is the amount the ends of the inserts extend above the edges of the rod and cap when installed. A normal amount of crush is around 0.001 inch. Too much crush causes the insert to buckle into the rod bore, causing bearing and journal failure and hindering or destroying oil distribution to other vital parts. Too little crush will allow the inserts (bearing assembly) to be loose in the bore, leading to early bearing failure.

resized. To do this, the following three pieces of equipment must be available:

- A precision cap and rod grinder (Figure 11–52) to grind the rod and cap parting surfaces square

FIGURE 11–52 Cap and rod grinder *(Courtesy of Sunnen Products Co.)*

- A precision dial gauge such as the one that is described in Chapter 7 to accurately read the size of the bore
- Either a boring machine (Figure 11–53A) or honing machine (Figure 11–53B) to finish the inner surfaces of the bore

To start the saddle bore reconditioning procedure, separate the connecting rod from the mating surfaces where the con-rod and cap connect. If studs are used to retain the cap, remove them from the rod with a stud driver. The parting faces and side of the rod and cap should be deburred, if necessary, before grinding the parting faces. Then the cap can be placed in a connecting rod vise.

To operate a typical saddle-bore grinder, clamp the rod to a plate on the grinder as instructed by the manufacturer. Turn on the grinder and, using the feed handle, advance the rod toward the grinding wheel. The clamp assembly is attached so that it can be rotated back and forth across the grinding wheel (Figure 11–54). As the feed handle is advanced, the rod is moved across the grinding wheel. The mating surface is precision ground by an amount read on the feed handle. The same procedure is followed to remove material from the mating surface on the bearing cap. Just enough material should be removed from the rod and cap parting or mating to make the saddle bore about 0.002 inch smaller.

The connecting rod and cap are reassembled using a connecting-rod vise (Figure 11–55). The new cap nuts are torqued the same amount as required for final assembly. The reassembled rod will have a smaller vertical dimension, but it is still out-of-round.

To bring the connecting rod back to the original specified diameter, it can be bored or honed. Be sure to follow the manufacturer's instructions when operating the boring or honing machine (Figure 11–56). The finished surface should be 60 to 90 microinches for correct bearing contact and heat transfer.

Because the center-to-center distance is shortened, the compression height of the piston in the bore will change and the compression ratio will be reduced slightly. The rod center-to-center distance can vary over a 0.010-inch range in most gasoline engines without a noticeable effect in performance. With diesel engines, a 0.010-inch change in deck clearance will change the compression ratios between one and two full points. The rod center-to-

A

B

FIGURE 11–53 (A) Boring machine; (B) honing machine *(Courtesy of Sunnen Products Co.)*

FIGURE 11–54 The mating surface of the cap and rod are ground to make the saddle bore smaller. *(Courtesy of Sunnen Products Co.)*

center distance can be checked on a machine such as shown in Figure 11-57.

To complete the rod reconditioning, remove any burrs on the sides of the con-rods by *lightly* sanding the surfaces. This deburring can help to prevent scuffing between rods on the crankshaft. Another minor machining operation that helps engine performance is to rechamfer the rod's housing bore. This can be accomplished by using a chamfering tool in the honing machine (Figure 11-58). This will restore the original oil throw-off properties as designed by the manufacturer.

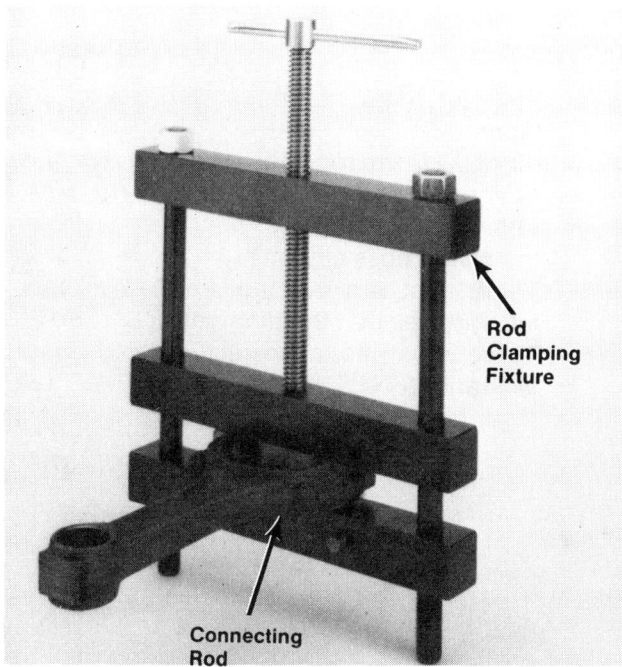

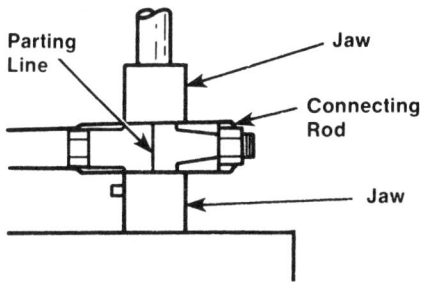

FIGURE 11-55   Rod clamping fixture or vise *(Courtesy of Goodson Shop Supplies)*

FIGURE 11-57   The three-point expanding mandrel shown here provides exact alignment for precise con-rod remanufacturing. The preset centering rpm is activated at the control panel without changing the boring speed. *(Courtesy of Rogers Machine Company)*

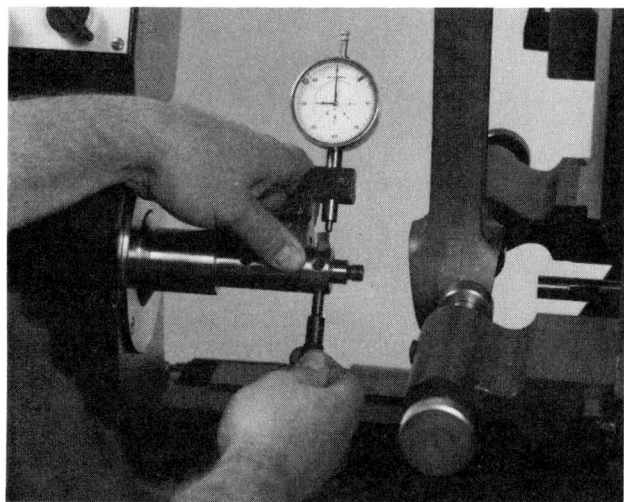

FIGURE 11-56   After setting the bore diameter with a micrometer, the measurement is transferred to the boring spindle and the boring tool is secured. *(Courtesy of Rogers Machine Company)*

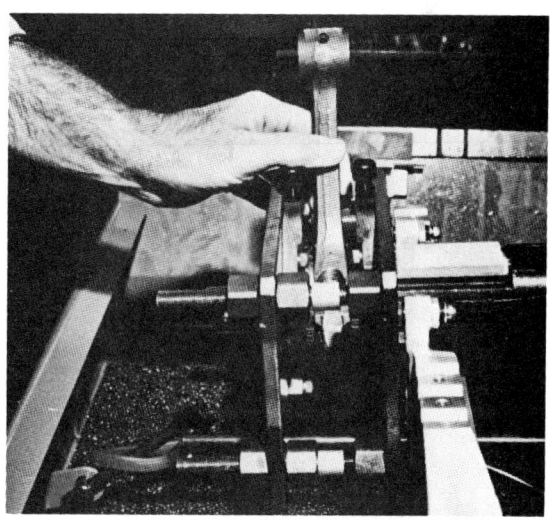

FIGURE 11-58   Chamfering rod housing bores with a chamfering tool *(Courtesy of Sunnen Products Co.)*

After all machine reconditioning work has been completed, the connecting rod should be reinspected as described in Chapter 7. After the connecting rod has passed inspection, it should be stored on the wall or on a rod caddy (Figure 11–59).

PISTON PINS

Piston pins connect the upper (small) end of the con-rod to the piston. This must be a flexible connection that allows the piston to work freely back and forth on the pin. The pins are made from high-quality steel and are case hardened during manufacturing to increase their wear resistance. To reduce pin weight, they are generally hollow.

To assure a good fit, the following must be achieved:

- Pinholes must be straight and free of any taper.
- Pinholes in the rod ends and/or bosses must be round.
- There must be perfect alignment between the piston boss pinholes.
- There must be a proper oil finish on all bearing surfaces that the pin rides on.
- There must be a correct amount of oil clearance.

A

B

FIGURE 11-59 After connecting rods have been machined, they can be stored (A) on pegs located on a wall, or (B) on pegs of a movable caddy.

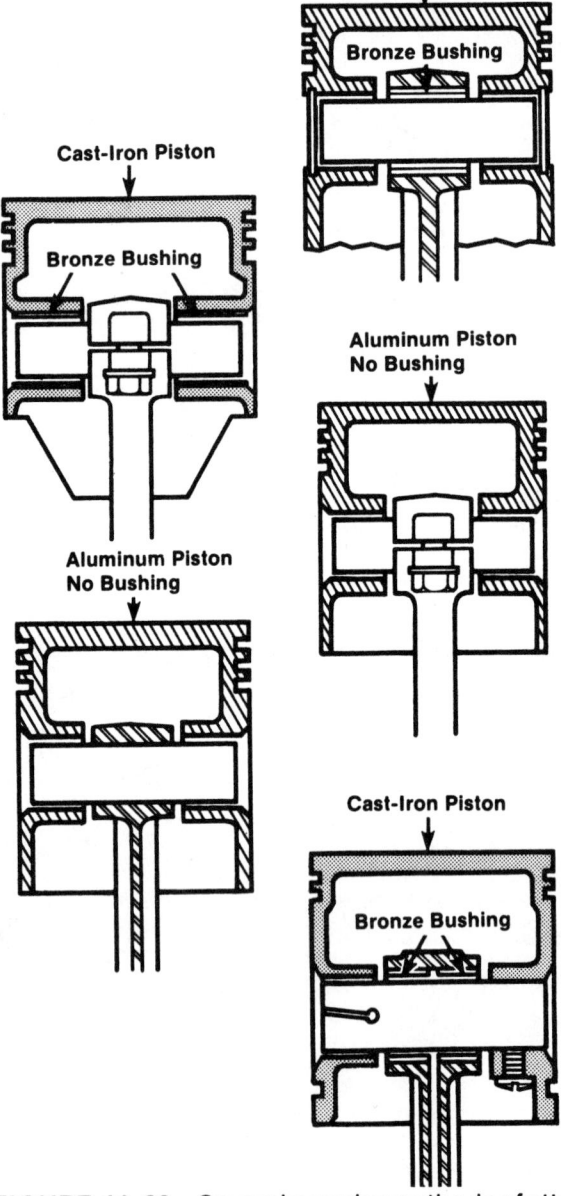

FIGURE 11-60 Several popular methods of attaching the connecting rod to the piston using piston pins

As shown in Figure 11-60 there are several methods of using piston pins to attach the connecting rod to the piston. As mentioned in Chapter 7, the two most common methods of connection are:

- *Full-Floating Pins.* In this method of attachment, the piston pin is inserted in the holes in the piston and the bushing in the connecting rod. The pin can pivot freely in both the connecting rod and piston. The pin is held by retaining rings in the ends of the piston pin bore.
- *Oscillating Pins.* In this method of attachment (also known as semi-full-floating design), the pin is fixed to either the connecting rod or to the piston. It must be clamped to the connecting rod with or without a bushing, press fit in the rod (the most common way), or locked to the piston with a set screw. (The latter is usually found only in older engines.)

Oscillating pin attachments do not require spring retainers to hold the pin in place.

Table 11-3 gives the recommended pin clearance for both types of attachment methods. Because of the difficulty of determining the piston pin clearance in assembly, it is necessary to use special, highly accurate measuring devices such as the one mentioned in Chapter 7. The old method of checking new pin fits by feeling them is not reliable enough for today's modern engines. The machine used for gauging the piston for correct piston pin clearance can also measure the amount of interference in a connecting rod (Figure 11-61).

WARNING: When using the heat method, handle the rods very carefully. They get extremely hot and might burn the flesh.

TABLE 11-3: RECOMMENDED PIN CLEARANCES

Precision Pin Fits on Engines with 3/4"- to 1-1/4"-Diameter Pins

Description	Aluminum Piston (inches)	Cast-Iron Piston (inches)	Connecting Rods (inches)
Full-floating	0.0001 to 0.0003 clearance	0.0003 to 0.0005 clearance	0.0003 to 0.0005 clearance (all pressure feed 0.0005 to 0.0007 clearance)
Oscillating in bushed position		0.0003 to 0.0005 clearance	Clamped in rod
Oscillating in piston (no bushing)	0.0003 to 0.0005 clearance	0.0006 to 0.0008 clearance	Clamped in rod
Oscillating in piston, press fit in rod	0.0003 to 0.0005 clearance		0.0008 to 0.0012 press fit
Set screw type piston	Screw side 0.0002 to 0.0003 press fit / Free side 0 to 0.0001 clearance	Screw side 0.0001 to 0.0002 press fit / Free side 0 to 0.0001 clearance	When locked in piston and all pressure feed: 0.0007 to 0.0009 clearance

Precision Pin Fits on Engines with 1-1/4"- and 1-1/2"-Diameter Pins*

Description	Aluminum Piston (inches)	Cast-Iron Piston	Connecting Rods (inches)
Full-floating (1/4"-dia. pinholes)	0.0003 to 0.0005 clearance		0.0007 to 0.0009 clearance (all pressure feed 0.0009 to 0.0011 clearance.)
Full-floating (1-1/2"-dia. pinholes)	0.0005 to 0.0007 clearance		0.0010 to 0.0012 clearance (all pressure feed 0.0013 to 0.0015 clearance)

*On large-diameter pins, check the engine manufacturer's manual for recommended clearances.

FIGURE 11-61 Gauging the rod for correct interference *(Courtesy of Sunnen Products Co.)*

Full-Floating Pin

In many instances, all that is required when using full-floating pins is to replace the old con-rod bushing. That is, press out the old bushing and press in a new one. The bushing driver is installed in the ram press. Then the new bushing is pressed into the rod. If the bushing has an oil hole, it must be properly aligned with the oil hole in the connecting rod. If the bushing has a parting line or split in it, it must be positioned to the side—not in the direction in which the rod is stressed when the piston applies pressure during the power stroke.

The new bushing must be sized to fit the piston pin with a clearance of 0.0003 to 0.0005 inch. The bushing may be machined by using either of two types of equipment. The connecting rod can be mounted on a piston pinhole boring machine. The connecting rod is positioned and a boring bar, set to

the correct dimension, is used to machine the hole (Figure 11-63). The pin fit is correct if an oiled pin can slide through the connecting rod bushing by itself at room temperature.

The bushing can also be machined on the same honing equipment used to hone the saddle bores. The operator moves the rod back and forth over expandable honing stones (Figure 11-64). A precision dial gauge on the unit is used to measure the hole. Both the boring bar and hone may also be used to machine the pinholes in the piston if an oversize pin is needed.

Whether the connecting rod bushing is bored (Figure 11-65) or honed, a precision gauge is necessary. This can be done only by carefully following the directions given in the manufacturer's instruction manual. Frequently when the small end of a rod is repaired on full floating types, a bronze bushing is pressed into place and then expanded (Figure 11-66). The bushing is faced off flush with the rod, and the bore is then honed to the proper size.

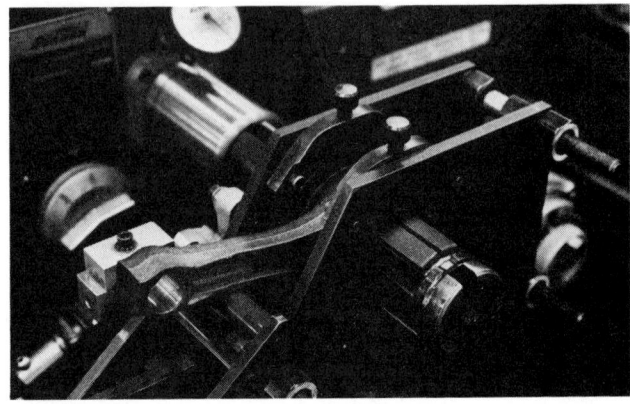

FIGURE 11-63 Machining with a boring bar *(Courtesy of Perfect Circle/Dana)*

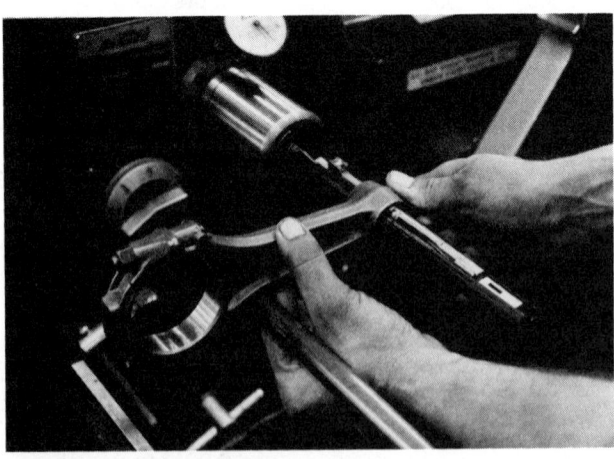

FIGURE 11-62 Pressing in a new piston-pin bushing *(Courtesy of Perfect Circle/Dana)*

FIGURE 11-64 Honing a piston to the correct clearance for an oversized piston pin. *(Courtesy of Perfect Circle/Dana)*

FIGURE 11-65 Boring piston pin bushing to size *(Courtesy of Sunnen Products Co.)*

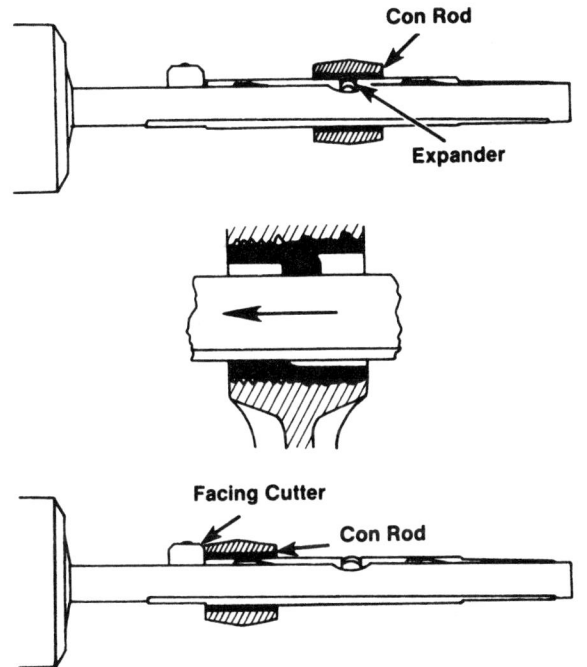

FIGURE 11-66 The rod bushing is expanded to seat it into the rough bore.

WARNING: Always wear eye protection when working on any of these machines.

Once the bushing is in place in the rod, insert the piston pin into the piston and rod holes and press it in with thumb pressure or by *lightly* tapping it with a plastic-faced hammer. The pin and holes should be well lubricated with light oil before the insertion is made. If the pinholes are chamfered only on one side, make sure to press the piston into the chamfered side first.

With the piston pin in place, install a new pin retainer snap or lock rings. Make sure that the snap rings are fully seated, but squeeze them only enough to get them into the pinhole (Figure 11-67). Squeezing more can overstress them and lead to early failure.

 SHOP TALK

With the rod evenly spaced between the bosses, the pin should stick out equally on both sides. Regardless of the method used to attach the piston and connecting rod together, check the piston for free movement around the pin.

Before assembling the piston and connecting rod, inspect the rod caps again and clean each bore surface. Make sure the back of each rod bearing is clean, then install the bearing halves in the rod caps. Make sure that bearing locking lips are nested in place. Liberally spread oil on the bearing surfaces. Install the rod bearing halves in the connecting rod blades. Again, make sure the locking lips are seated, then coat the bearing surfaces with oil.

Oscillating Pins

When assembling the piston and connecting rods, piston pins can be installed with a hydraulic press or by the heat method.

Press Method

The press method requires a special pin inserter tool and adapters. The tool helps to prevent piston distortion and rod cocking during assembly, and it centers the pin in the boss. Chamfering the small end of the rod and lubricating the pin will make

FIGURE 11-67 When using lock rings, lubricate the pinhole in the rod and piston with a light oil.

installation much easier. Honing might be necessary to acquire the correct interference. To use a typical bench-type ram press, proceed as follows:

1. Clean the piston, rods, and pins.
2. Bevel the edges of the pinholes in the rods using a proper size countersink.

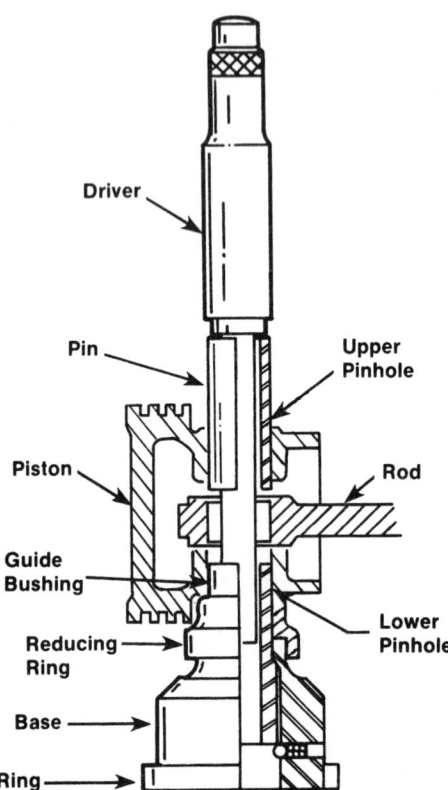

3. Inspect the pins and remove any sharp edges or burrs.
4. Lubricate the pins and the pin bore with a good lubricant.
5. Select the proper pin inserter tool included with the press.
6. Push the guide locating plate against the side of the press frame to position the side without the hole under the hydraulic cylinder.
7. Insert the guide bushing in the base from the bottom. Push the bushing all the way into the base until it clears the detent ball. If required, place a reducing ring on top of the base.
8. Slide the pin into the upper pinhole in the piston until it comes in contact with the rod (Figure 11–68A).
9. Place the piston on the base (reducing ring) and insert the driver into the pin.
10. Grasp the driver and push downward to align the rod and pin. Tap the driver with a soft hammer to start the pin into the piston.
11. Place the entire assembly (piston and pin inserter tool) in the press with the base on the solid portion of the locating plate.
12. Grasp the driver and push downward to keep the assembly properly positioned until the pressure is applied to the driver (Figure 11–68B).
13. If not in place, install the saddle by snapping into the lower end of the ram.

FIGURE 11–68 Guide bushing and pin driving arrangement *(Courtesy of Sunnen Products Co.)*

FIGURE 11–69 When using a typical oven, put the small end of the rod in the correct position on the heat unit.

WARNING: Remember that the average bench press develops 10 tons of force. Keep your hands clear of the punch and workpiece. Wear proper safety items such as safety glasses, full face mask or shield, and other safety equipment as necessary or required. Improper use of the press could result in personal injury.

14. Move the control lever to the full *down* position and hold it there until the pin is pushed into the rod and the guide bushing bottoms on the solid locating plate. Then return the lever to the *up* position. This automatically centers the piston pin in the rod.
15. Repeat the procedure for the remaining pistons and rods.

CAUTION: Do not use excess pressure. It can crack the pin bosses. Also, the piston pin must be pressed from one side only so that the rod is centered on the pin. Attempting to correct the position of a piston pin that has been pressed through too far by pressing the opposite side could result in piston distortion.

Heat Method

Because of the variety of pistons and pins used today, many rebuilder shops consider heat the better method. To use a typical heat or heater oven (Figure 11-69), place a small end of the rod in the proper position on the unit. As the rod eye heats up, it will expand. Depending upon the size, the rod is heated 400 to 500 degrees Fahrenheit. If the rod starts to turn blue, it is too hot. Use of the heater makes the oscillating piston pin press fit to a slip fit. The piston pin at room temperature is quickly inserted with the help of a special fixture through the heated rod (Figure 11-70).

WARNING: When using the heat method, handle the rods very carefully. They get extremely hot and might burn the flesh.

When reassembling the piston to the connecting rods the first step is to determine the position of each. Pistons must be installed in the block in the correct position. Most of them are marked with an "F," an arrow, a notch, or the word *front* (Figure

FIGURE 11-70 Pin is in position, ready for heat treatment. *(Courtesy of Sunnen Products Co.)*

11-71). This means that this indicator should go toward the front of the engine. A few engines use top or cam side markings rather than front markings. Some truck pistons have a notch on the side of the head that faces the intake manifold. If the piston is installed incorrectly, there might be a problem with piston-to-valve clearance or noise can result.

To avoid engine failure, the connecting rods must be oriented correctly. The direction will vary from manufacturer to manufacturer, so if you are not certain, consult the specific manufacturer's manual.

FIGURE 11-71 The word "front" on a piston means that this indicator should go toward the front of the engine.

Complete Automotive Engine Rebuilding and Parts Machining

REVIEW QUESTIONS

1. Which of the following is a consequence of too much piston clearance?
   a. The piston will expand so much that it sticks or seizes in the cylinder.
   b. The piston rocks or slaps in the cylinder.
   c. There is no space for a film of oil between the piston skirt and the cylinder.
   d. All of the above

2. Pistons are expanded or resized on a

   _____ .
   a. piston skirt
   b. fish scale
   c. heater oven
   d. piston knurlizer

3. Technician A knurlizes a semi-steel thin-walled piston. Technician B knurlizes a cast steel thin-walled piston. Who is right?
   a. Technician A
   b. Technician B
   c. Both A and B
   d. Neither A nor B

4. When measuring piston clearance, the feeler strip is pulled out from between the piston and cylinder wall with too much pull on the fish scale. Technician A says this means the clearance is too large. Technician B says it means the clearance is too small. Who is right?
   a. Technician A
   b. Technician B
   c. Both A and B
   d. Neither A nor B

5. Which of the following statements concerning diesel pistons is incorrect?
   a. Diesel pistons use heavier piston pins.
   b. Many diesel pistons have their top ring groove equipped with a chrome steel insert to reduce groove wear.
   c. Diesel pistons are designed with less skirt area than gasoline engine pistons.
   d. Diesel pistons are subjected to greater loads and stresses than gasoline engine pistons.

6. Leakage past the top piston ring can result from out-of-round cylinder walls due to

   _____ .
   a. wear
   b. mechanical distortion
   c. heat expansion
   d. all of the above

7. Installing new piston rings in a worn ring groove _____ .
   a. makes for a very poor combustion seal
   b. causes the rings to twist, flex, and pound against the worn surface
   c. both a and b
   d. neither a nor b

8. Technician A uses a ring-expander tool to install piston rings. Technician B installs piston rings by hand. Who is right?
   a. Technician A
   b. Technician B
   c. Both A and B
   d. Neither A nor B

9. Which of the following statements concerning piston ring installation is incorrect?
   a. Always install the upper ring first.
   b. Never use expanders behind a top compression ring.
   c. Thoroughly lubricate the rings and cylinder walls before beginning the installation.
   d. Piston ring grooves can vary in width, even on the same piston.

10. What is the proper crush for the bearing inserts in the connecting rod bore?
    a. 0.001 inch
    b. 0.005 inch
    c. 0.01 inch
    d. 0.05 inch

11. When reconditioning a connecting rod bore, the reassembled rod will have a

    _____ .
    a. larger vertical dimension and still be out-of-round
    b. larger vertical dimension and no longer be out-of-round
    c. smaller vertical dimension and no longer be out-of-round
    d. smaller vertical dimension and still be out-of-round

12. The two most common methods of attaching the connecting rod to the piston are with

    _____ .
    a. full-floating and oscillating pins
    b. press and full-floating press pins
    c. oscillating and chamfered pins
    d. boring and honing pins

CHAPTER TWELVE

RECONDITIONING CRANKSHAFTS, CAMSHAFTS, AND ENGINE BALANCING

Objectives

After reading this chapter, you should be able to
- Describe how crankshafts are classified.
- Describe how to verify crankshaft capability.
- Explain what might be involved in reconditioning the crankshaft.
- Explain how to recondition a camshaft.
- Explain what is meant by engine balancing and describe how it is accomplished.

The crankshaft and camshaft are rotating parts of the engine. Because of their motion, the operation of the valve train, pistons and rods, and other engine parts is possible. Their motion is used to turn the drive shaft and move the conventional rear-wheel-drive vehicle. In front-wheel-drive vehicles, it is the action of these parts that provides the motion and power necessary for the combination transmission-differential components to operate the short drive shafts that extend to the front-wheel gearbox.

CRANKSHAFT

As mentioned in Chapter 2, crankshafts are either forged from steel alloys or cast from nodular or malleable iron. Although forged crankshafts are considered to be stronger than cast, they are more expensive to manufacture. At one time, cast shafts were brittle and easily broken. However, with today's materials and casting techniques, cast crankshafts can almost be machined and repaired as easily as forged types. Diesel engine crankshafts are still forged rather than cast. They are generally larger in diameter and wider than those for a similar size gasoline engine. The major difference between forged and cast crankshafts is the method of repair and machining, both of which are described later in this chapter.

 SHOP TALK _____

To identify the difference between the two crankshaft types, lightly tap one of the shaft's counterweights with a ball peen hammer. The forged shaft will have a sharp, ringing sound; the cast type will produce only a dull thud.

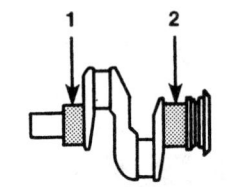

Two Main Bearing Crankshaft
(Two Cylinders)

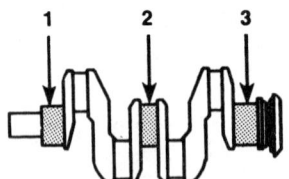

Three Main Bearing Crankshaft
(Four Cylinders)

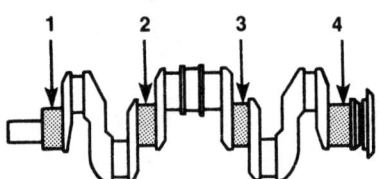

Four Main Bearing Crankshaft
(Six Cylinders)

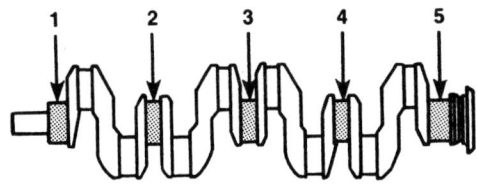

Five Main Bearing Crankshaft
(Eight Cylinders)

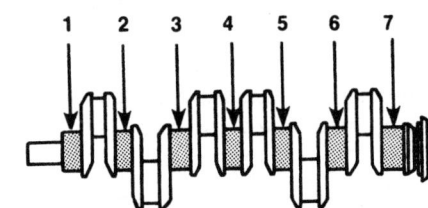

Seven Main Bearing Crankshaft
(Six Cylinders)

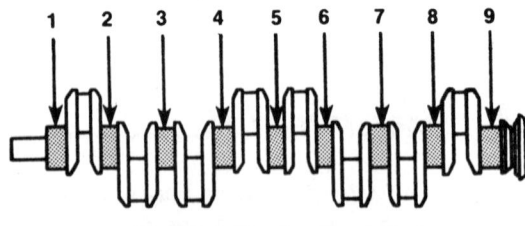

Nine Main Bearing Crankshaft
(Eight Cylinders)

FIGURE 12-1 Crankshaft classifications. Shaded areas are main bearing journals.

CRANKSHAFT CLASSIFICATIONS

As mentioned in Chapter 2, crankshafts are generally classified by the number of main bearings (Figure 12-1). Of course, crankshaft configurations vary from engine to engine. Six-cylinder in-line crankshafts, for example, usually have six connecting or seven main bearing journals. The four-main-bearing crankshaft must be of heavier construction than the seven-main-bearing crankshaft.

The V-6 crankshaft has four main journals and six crank throws spaced for even firing. The V-8 crankshaft has five main journals and four connecting rod journals (Figure 12-2). Each connecting rod journal has two connecting rods attached to it (Figure 12-3).

Cylinder Numbering

The way the cylinders are numbered and the firing order affect how the crankshaft is shaped (see Chapter 2). The number used for each cylinder is extended to the crankshaft to identify the corresponding connecting rod journals.

From the driver's seat of a rear-wheel-drive car with the engine in front, the rear of the engine is nearest the driver. This is where the flywheel is. Farthest away is the front of the engine. From this same position, the left and right sides and banks of a V-type engine are identified.

In transverse engines, the standard identification is also used. The flywheel is at the rear of the engine. Standing near the flywheel, facing the engine, the left side is to the left and the right side to the right. Most transverse engines are installed with the

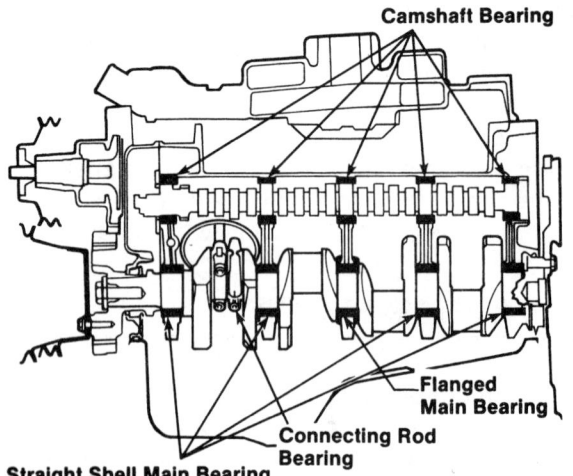

FIGURE 12-2 Bearing locations (five-main-bearing V-8 engine)

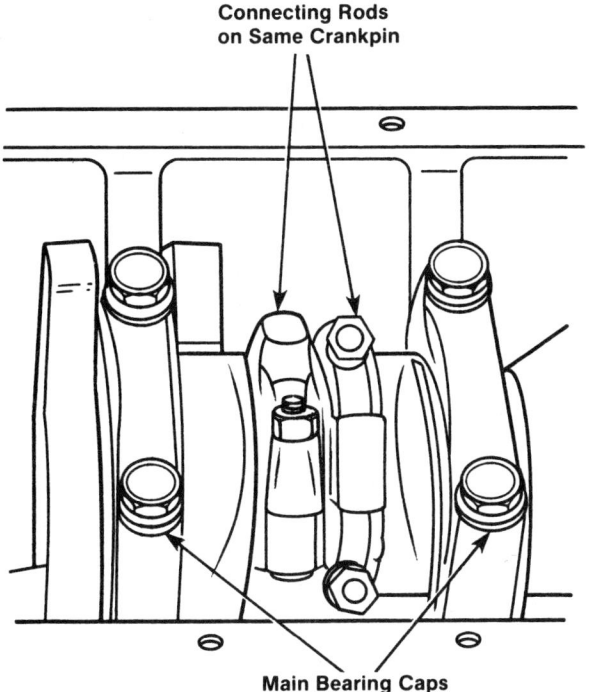

FIGURE 12-3   Two connecting rods can share a single throw or offset portion of a crankshaft.

flywheel and the transaxle to the left of the center of the car. Most service diagrams for transverse engines identify the front of the vehicle (Figure 12-4).

In-Line Cylinder Numbering.
The cylinder numbering system for the in-line engine is simple and almost universal. The number one cylinder is at the front, farthest from the flywheel. The number two cylinder is next counting back toward the flywheel (Figure 12-5). The exception to

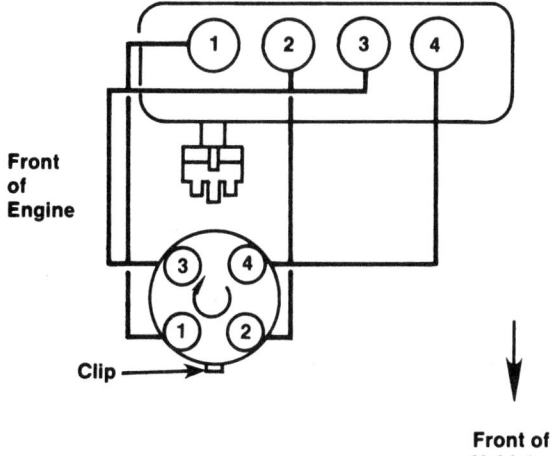

Firing Order 1-3-4-2

FIGURE 12-4   Engine layout for a transverse-engine front-wheel-drive car

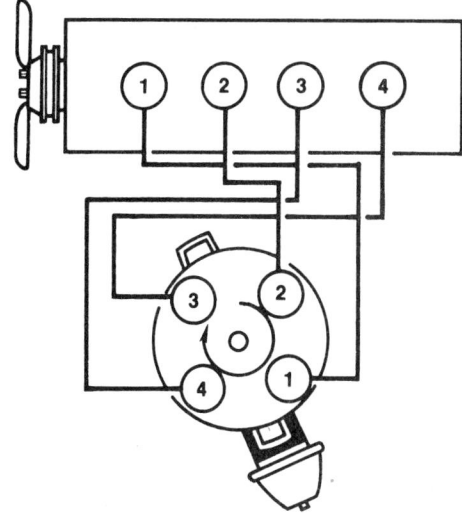

Firing Order 1-3-4-2

FIGURE 12-5   Most engines are numbered with the No. 1 cylinder furthest from the flywheel.

this is some imports whose engines are numbered with number one being the nearest to the flywheel.

Trying to remember that the number one cylinder is closest to the fan might have worked in the past, but it does not help at all with transverse engines and others where the fan is in a different location now.

V-Engine Cylinder Numbering.
With V-engines, there is no universal sequence. The most common system used is by American Motors, Chrysler, General Motors, and some imports. With this system, the cylinder farthest from the flywheel is the number one cylinder. This is usually, but not always, in the left bank. The next cylinder closest to the front is the number two cylinder. This will be the most forward cylinder in the opposite bank. Then, moving from bank to bank, the numbering continues taking the cylinders in order front to back but alternating from one bank to the other (Figure 12-6).

The slight offset of the cylinders in one bank from those in the opposite bank allows the connecting rods to be placed one after the other on the crankshaft. The pattern in numbering cylinders in this order is just like in the in-line engine, in consecutive order from front to flywheel as the connecting rods appear on the crankshaft. Most, but not all, V-engines have the left bank of cylinders slightly forward.

Ford designates the most forward cylinder number one. But then, the Ford system numbers all cylinders on the right bank in consecutive order from number one through number three on V-6 engines,

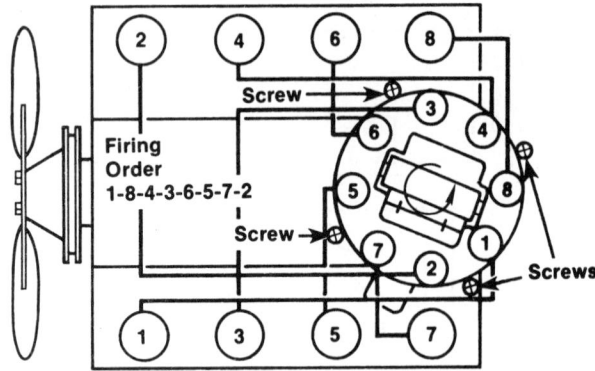

FIGURE 12-6 Most V-8 engines are numbered with the odd-number cylinders on the left and the even numbers on the right side.

and through number four on V-8 engines. Moving to the left bank, the front cylinder is number four on V-6 engines and number five on V-8 engines, and the sequence continues. These cylinders can also be numbered in their banks, 1L, 2L, 3L, 4L, and 1R, 2R, 3R, 4R. Some European imports use the Ford system, but with number one being the closest to the flywheel.

CONNECTING ROD JOURNAL NUMBERING

Journal numbering corresponds to the connecting rod and cylinder numbering. In the in-line engines, the rod journals will be numbered in sequence from front to rear. But in the V-engines, it depends on the manufacturer's system.

In the American Motors, Chrysler, and General Motors system, the rod journal nearest the front of engine will be number 1-2. The next will be number 3-4, followed by number 5-6 and 7-8. In the Ford system the sequence will be number 1-5, 2-6, 3-7, and 4-8. This is for a standard V-8 engine crankshaft, one where two rods are connected to one journal. This is found in all V-8 and some V-6 engines with a 90-degree V-angle. But in some V-6 engines, an offset rod journal is required to improve the spacing of the firing order. Here, each journal assumes the number of its cylinder.

Main journal numbering is always in consecutive order from front to rear. Number one main journal is closest to the front of the engine, the end opposite the flywheel. The next main journal is number two and so on to the rear main journal, closest to the flywheel. For a four-cylinder in-line engine, the main journals will be numbered one through five. A V-8 is also numbered one through

five, with a main journal on either side of each pair of connecting rod journals. In a V-6 engine, there will be four main journals with number one at the front.

Firing Order

The question of standard or offset connecting rod journals in a V-engine is directly related to firing sequence and V-angle. If the firing is evenly spaced so the amount of time between successive firing is equal, the result will usually be smoother operation than if the time intervals are unequal. The geometry of the rod journal position on the crankshaft is determined by the time the various pistons fire to maintain a smooth firing sequence.

One complete revolution of the crankshaft is 360 degrees. Twice that, or 720 degrees of revolution, represents two complete revolutions of the crankshaft. For one piston to come down once on a power stroke, the crankshaft must travel through 720 degrees of revolution.

If the engine has two cylinders, in each 720 degrees there will be two power strokes. Rather than having the pistons operating so they fire at the same time or within a short time span and then wait for two more revolutions before they fire again, the operation is smoother if the firings are evenly spaced so there is equal time between each power stroke. With two pistons, one cylinder will fire each 360 degrees. If another cylinder is added, for a total of three, the firing impulses, to be evenly spaced will be every 240 degrees. (720 degrees/3 = 240 degrees). In a similar way, the cylinders for a four-cylinder engine should fire at 180-degree intervals; a six-cylinder at 120-degree intervals (Figure 12-7), and an eight-cylinder at 90-degree intervals (Figure 12-8). An eight-cylinder engine is inherently smoother than a six-cylin-

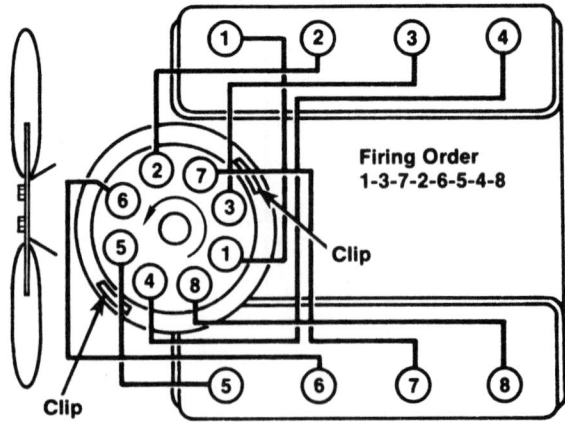

FIGURE 12-7 Ford V-8 engines are numbered sequentially from right to left.

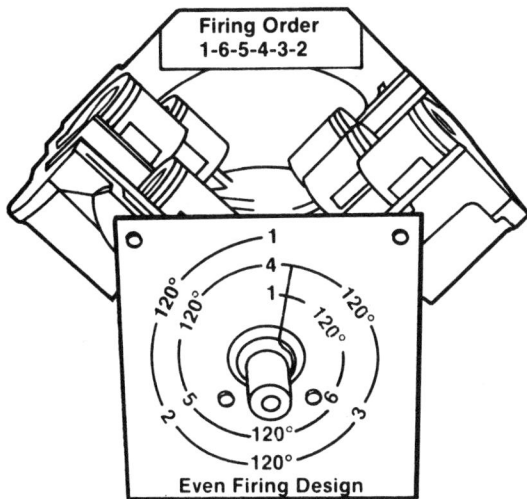

FIGURE 12-8   Firing order and even firing design of 90-degree V-6 GM engine with offset rod journals

der or a four-cylinder because there are more power strokes in each 720 degrees of crankshaft rotation.

There is a definite relationship in the angles of the V-block, the firing sequence, the rod journals, and the firing interval. The crankshaft of a V-8 engine has a common rod journal for two opposite connecting rods. If the firing sequence is to be evenly spaced, the V-angle must be 90 degrees. Any other angle will require either offset rod journals or uneven intervals between firing impulses.

In-Line Engine Interval.   In an inline engine, when the crankpin is straight up, the piston is at top dead center (TDC). When it is straight down, the piston is at bottom dead center (BDC). If the engine is a four-cylinder engine, there will be only two crankpin positions. When the two crankpins are straight up, two will be straight down, allowing a firing pulse every 180 degrees, or every half turn of the crankshaft.

V-Engine Interval.   In a V-engine, the piston is at TDC when the crankpin is at some angle off straight up and at BDC when the crankpin is at some angle off straight down. If the engine has eight cylinders and the two opposite connecting rods are connected to a common rod journal, the angle of the V is 90 degrees.

When the crankpin is at 45 degrees after straight up, one piston will be at the top of its stroke. Exactly 90 degrees later, the opposite piston will be at the bottom of its stroke. In another 90 degrees, the first piston will be at the bottom and in the next 90 degrees, the second piston will be at the top.

This allows a V-8 engine to operate with one cylinder firing every 90 degrees, with opposite pistons connected to a common rod journal. If the cylinder on the right fires when the first crankpin is at 45 degrees, this will be followed by other cylinders firing when the first crankpin is at 135 degrees, 225 degrees, 315 degrees, 45 degrees, 135 degrees, 225 degrees, and 315 degrees, for a total of eight evenly spaced explosions in 720 degrees. The sequence gives eight firings in every two revolutions of the crankshaft, one every 90 degrees.

The ideal V-angle for a V-6 engine is 60 degrees. With that angle, one cylinder will fire each 120 degrees (720 degrees/6 = 120 degrees). For this, the engine needs a six-throw crankshaft, one every 60 degrees. There are, however, many 90-degree V-6 engines, derived from the 90-degree V-8. This can affect firing intervals. It can also affect ordering parts.

Offset rod journals can smooth out firing impulses in a 90-degree V-6 engine. With a V-8 built on a 90-degree V-angle, the firing interval was a regular 90 degrees. Changing the engine to six cylinders and using a common rod journal for each pair of opposite connecting rods meant an irregular sequence with alternate intervals of 150 degrees and 90 degrees, as shown in Figure 12-9. The result was a rough operating engine. It took a massive flywheel to smooth out the roughness, and even then the result was unsatisfactory.

For 1978, the automotive engineers made further changes, the major ones in the crankshaft, as shown in Figure 12-10. Instead of continuing with a crankshaft with common rod journals for opposite connecting rods, the shaft was redesigned to provide an individual journal for each rod. This replaced the three-journal crankshaft with a six-journal shaft. The four main bearings remained unchanged. The rod journals of this engine are now made of three pairs of two journals each, but the two journals in each pair are offset 30 degrees from each other. This change has resulted in a crankshaft that provides an even firing interval of 120 degrees between each of the six cylinders (Figure 12-11).

Other changes were then necessary in the camshaft and distributor. Flywheel weight reduction was possible because the smoother firing interval results in smoother engine operation.

 SHOP TALK _____

Since there are no universal standards for cylinder numbering, connecting rod journal numbering, and engine firing order, check the service manual for the engine being worked on for such details.

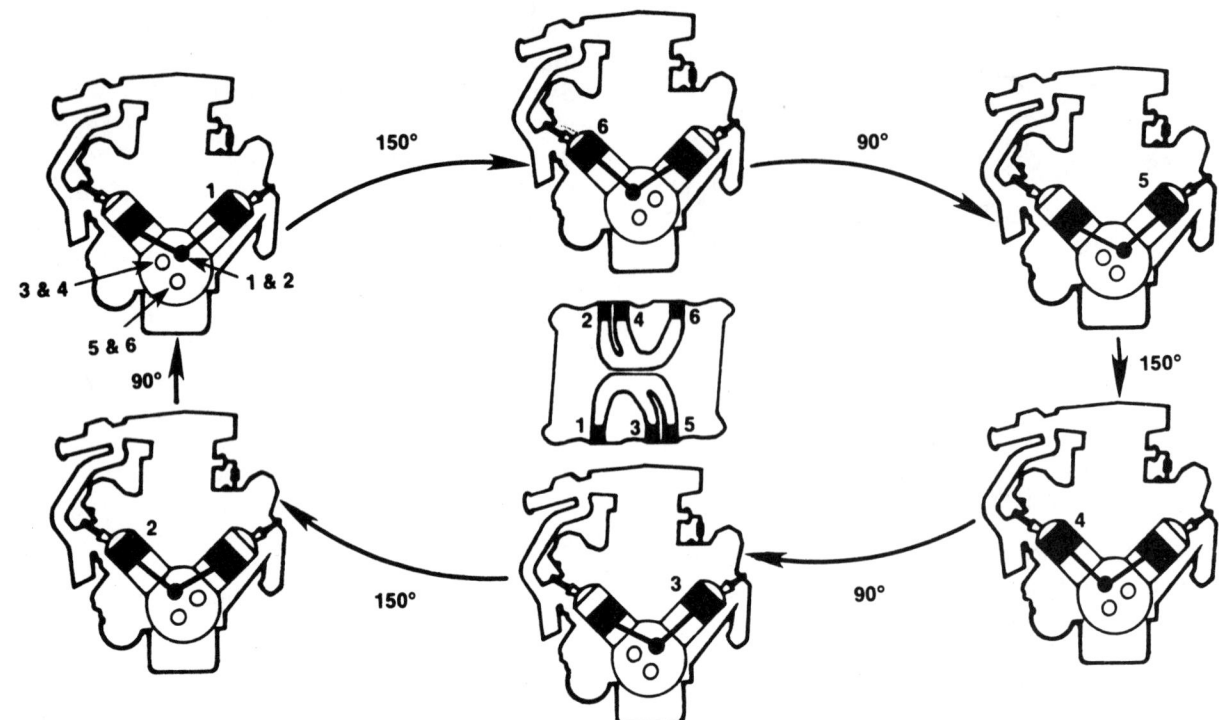

FIGURE 12-9   A 90-degree V-6 engine with common connecting rod journals results in an uneven firing sequence.

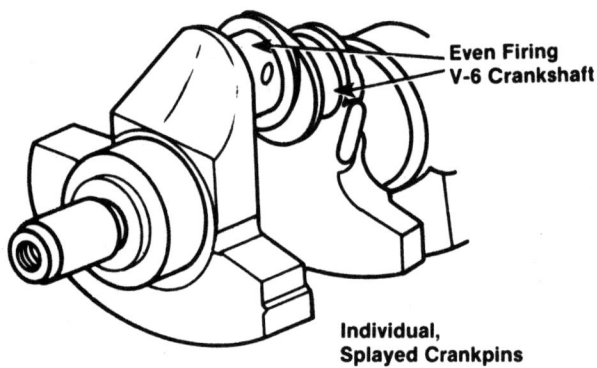

Even Firing
V-6 Crankshaft

Individual,
Splayed Crankpins

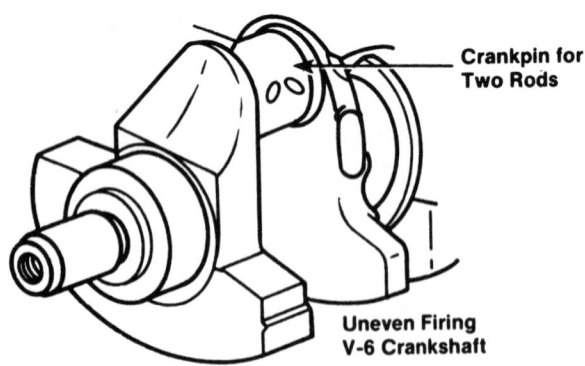

Crankpin for
Two Rods

Uneven Firing
V-6 Crankshaft

FIGURE 12-10   Crankpin offset

The nomenclature of a crankshaft is shown in Figure 12–12.

RECONDITIONING CRANKSHAFTS

Before starting any crankshft reconditioning work, evaluate the shaft in question as detailed in Chapter 7. In other words, measure the diameter of the bearing journals, the seal diameters, and the gear fit areas. Compare these measurements to new standard measurements and establish wear limits to determine the undersize to which the crankshaft is to be reground. Mark the repair order and the crankshaft to indicate the undersizing requirements and reference source for making these determinations.

Evaluate the alignment/bow of the crankshaft for acceptability. This evaluation should be made with the crankshaft supported on the main bearing journals, not on the centers of the crankshaft. The recommended location for supporting the crankshaft is on the end main bearing journals.

The journal supports should be a matched pair of well-maintained V-blocks (Figure 12–13). V-blocks made with roller bearings can introduce er-

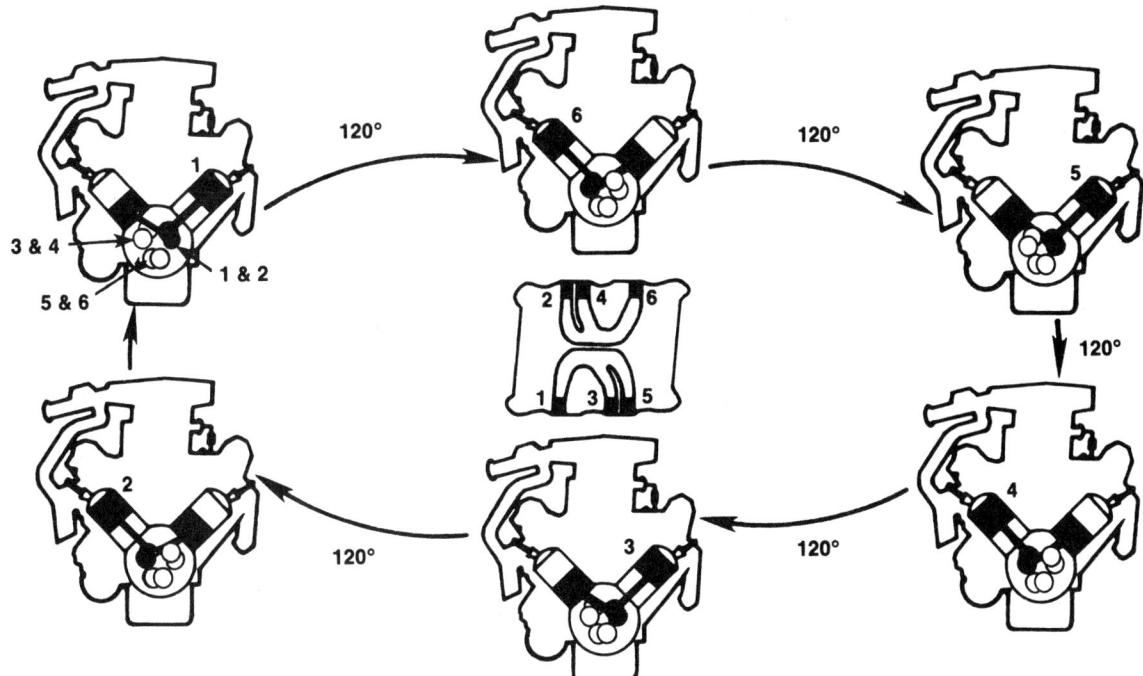

FIGURE 12-11   Offsetting the crankshaft journals 30 degrees results in a 90-degree V-6 with an even firing sequence.

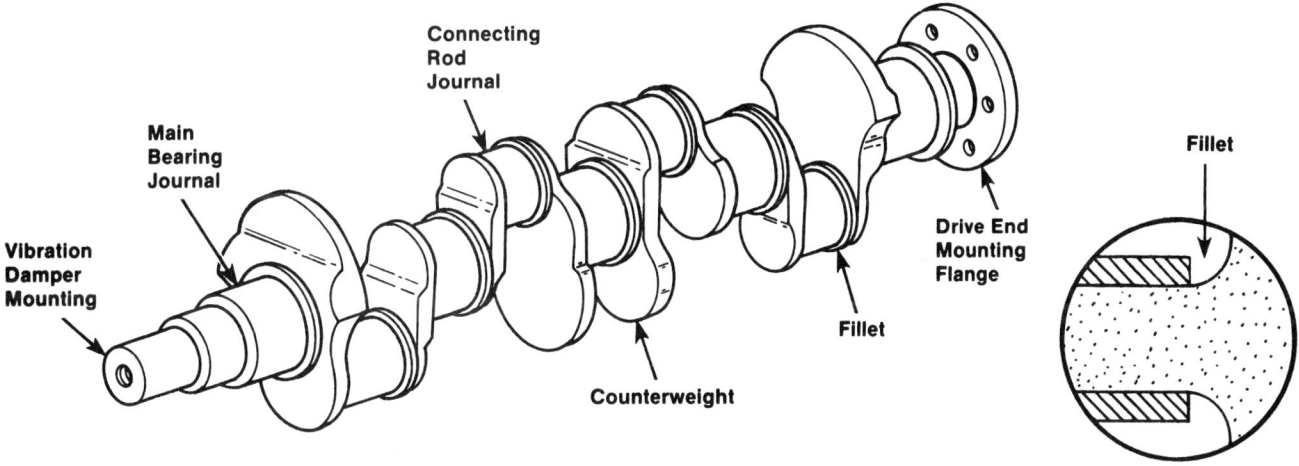

FIGURE 12-12   Crankshaft nomenclature

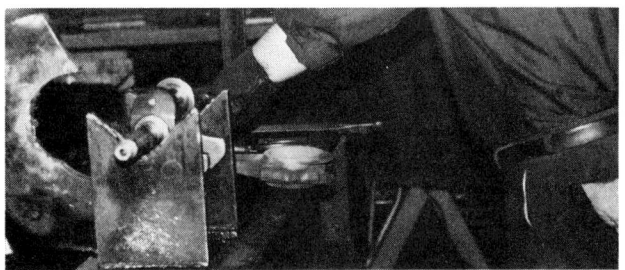

FIGURE 12-13   Supporting the crankcase journal in V-blocks

roneous values due to bearing tolerances and clearances. Consequently, the best practice is to avoid the roller type of V-blocks unless they are a calibrated set. Once the crankshaft is located in the V-blocks, select and position the indicator for measuring the alignment/bow of the crankshaft. The selected indicator should have an accuracy of no less than 0.001 inch per full scale deflection. Position the indicator at the 3 o'clock position on the center main bearing journal.

Set the indicator at zero and turn the crankshaft through one complete 360-degree rotation. The total deflection of the indicator, the amount greater than zero, plus the amount less than zero, is the TIR. Bow is 50 percent of the TIR. The bow indication at this bearing establishes the bow of the crankshaft. Compare the bow of the crankshaft to the acceptable alignment/bow specifications; accept or reject as appropriate.

Acceptable alignment limits for a fillet-hardened crankshaft are usually twice as large as the limits for a non-fillet-hardened crankshaft. If bow limits are not available, use the following to evaluate bow acceptability: for fillet-hardened crankshafts, allowable bow is 0.001 inch per foot of crankshaft length. For non-fillet-hardened crankshafts, allowable bow is 0.0005 inch per foot of crankshaft length.

Check the damper and flywheel mounting threads for condition and size. Verify the thread size with a go/no-go gauge and the condition by comparing the force required to thread a cap screw into each of the threaded holes. Clean or repair threaded holes requiring excessive force to install the test cap screw.

To recondition the crankshaft it is usually necessary to correct any conditions that the inspection or evaluation proved were not up to specifications. Most of these problems will generally require machining. For example, crankshafts and flywheels can be reground to provide a smooth surface. Severely damaged journals can be rebuilt by an automatic welding process and then machined to the proper size. Crankshafts can also be straightened if required (Figure 12-14).

Visual inspection of the crankshaft usually is not as reliable as the magnetic particle test when looking for structural faults. As with the fluorescent inspection technique described in Chapter 8, no hood is required with the crankshaft technique. It can be performed with a machine (Figure 12-15) or by hand as shown in Figure 12-16. In either case, the basic procedure is the same.

1. To magnetize, apply power to the coil field indicator for about 2 seconds. This will create a magnetic field (Figure 12-16A).

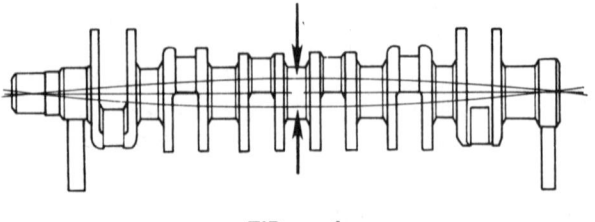

TIR = A
BOW = 1/2 A

FIGURE 12-14 Evaluating alignment bow

FIGURE 12-15 Checking for crankshaft structural faults with magnetic detector *(Courtesy of Magnaflux Corp.)*

2. Spray an aerosol can of glow spray over the complete work area. Check the reading on a magnetic field indicator (Figure 12-16B).

3. Use a 100-watt black light to inspect the areas. Depending on the spray glow material, the cracks will show up in bright colored lines (Figure 12-16C).

4. To demagnetize the crankshaft, place the coil over the workpiece, apply power, and move the coil over the shaft until it is approximately 2 feet away. Then turn off the power (Figure 12-16D).

It is important to determine the hardness of the connecting rod and main journals and evaluate the indicated values for acceptability. It is important that the rebuilder specialist knows what range of values to expect. (Hardness values are expressed using the Rockwell 'C' [Rc] scale.) Most current crankshafts are manufactured using some type of fillet hardening. However, until the "age of electronics," this had to be done by nitral acid etching. With modern hardening gauges such as those shown in Figure 12-17, fillet hardness can be accomplished without an acid.

If hardness guidelines are not available, apply the following for evaluating journal hardness acceptability: for crankshafts with fillet hardened journals, accept only hardness values of 36 Rc and above; for crankshafts without fillet hardened journals, accept only hardness values of 30 Rc and above. Do not recondition any crankshaft with journals that do not meet the hardness minimums.

CRANKSHAFT GRINDING

When grinding a crankshaft, no more material is removed than is required to produce a good journal

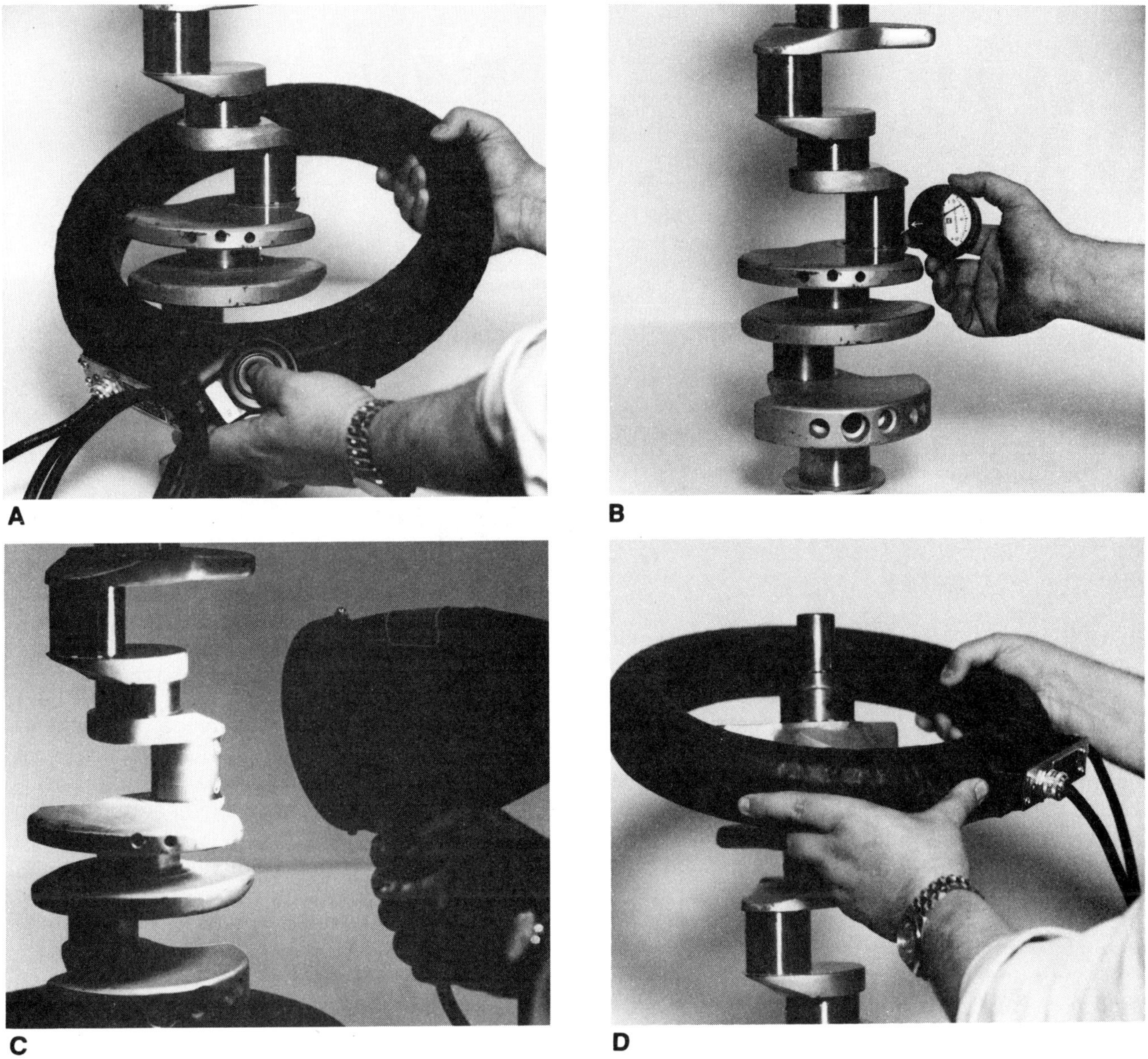

A

B

C

D

FIGURE 12-16   Performing the magnetic particle test by hand *(Courtesy of Atlas Engineering and Manufacturing Co.)*

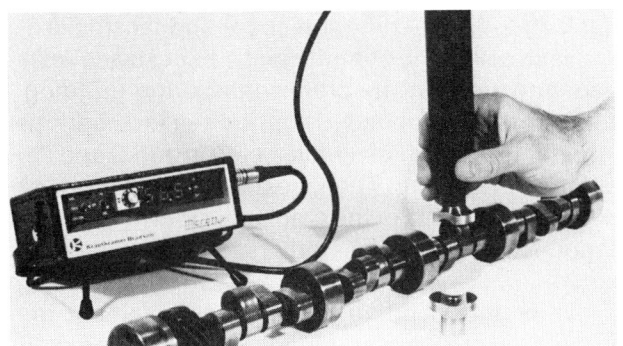

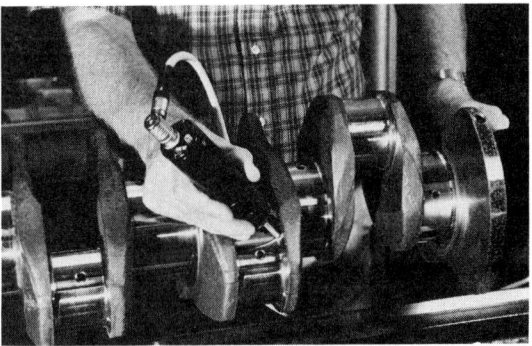

FIGURE 12-17   Crankshaft and camshaft electronic hardness gauges *(Courtesy of Krautkramer Branson, Inc.)*

FIGURE 12-18   Typical crankshaft grinder machine *(Courtesy of Rogers Machine Company)*

surface. Journals are generally to normal undersizes of 0.010, 0.020, or 0.030 inch. It is possible to leave the main bearings at their standard size and just regrind the connecting rod journals. It is also possible to regrind just one journal on a crankshaft and leave all the others at the standard size. Journals may be ground to undersizes of 0.030 inch or more. The more accepted practice, however, is to grind both main and connecting rod journals to 0.010 or 0.020 inch undersize. Crankshafts are ground on large crankshaft grinding machines like the one shown in Figure 12-18.

Begin the grinding process by selecting and installing the grinding wheel that is to be used during the operation. The selection process should address two key elements:

* Effective grinding width of the wheel
* Hardness of the bonding material and the size grit used in the manufacturing of the wheel

The recommended choice by most rebuilders is a wheel that has sufficient width to completely grind the entire connecting rod journal and fillet areas in a single cut. If a full width wheel is not available or is impractical due to the variety of cranks being done, be sure the wheel has sufficient width to *minimally* cut the fillet area on one side of the journal as well as at least 70 percent of the width of the journal. Wheels with less widths should not be used.

Verify that the grinding machine coolant temperature control system is operational and coolant concentration percent is within manufacturer specification. Recommended coolant temperature is 60 to 70 degrees Fahrenheit.

Confirm that the centerline of the grinding machine headstock and tailstock are in alignment and that this centerline is also parallel to the axis of the grinding wheel.

Set the throw offset of the grinding machine headstock and tailstock to complement the throw of the crankshaft (Figure 12-19). The throw of a crankshaft is half (50 percent) the stroke of the crankshaft. (The scale on a few crankshaft grinders reads stroke not throw length.) Follow the grinding machine manufacturer's procedure to install the crankshaft on the chuck and for offsetting the head- and tailstocks.

Next, select the quantity and size of steady rests needed to support the crankshaft during grinding. Steady rests are needed to support the rod bearings of the same orientation (while being ground) and the center main bearing. Then, post the size and surface finish requirements of the crankshaft journals, fillets, sealing surfaces, and gear fit areas in an easily seen location.

Dress the grinding wheel to complement the journal fillet radii requirements. The fillet is the radius at each side of the journal where the journal surface ends (Figure 12-20). The radius must be right. If it is too large, there will be rapid bearing

FIGURE 12-19 Crankshaft set up in chucks in a crankshaft grinder *(Courtesy of Storm Vulcan, Inc.)*

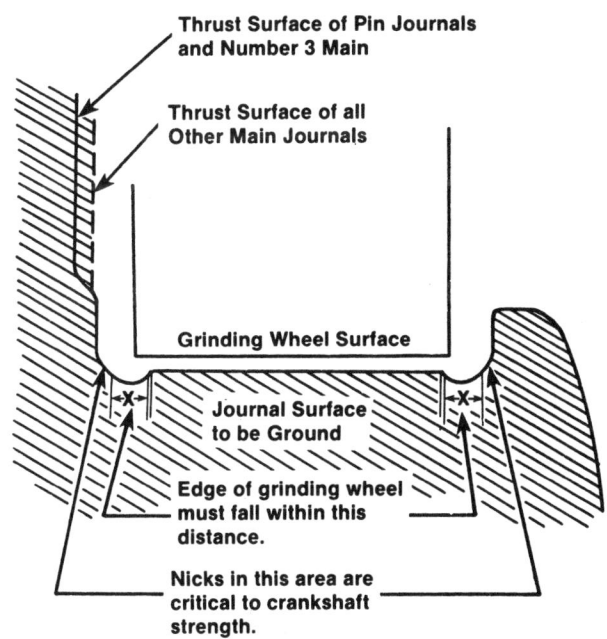

Thrust Surface of Pin Journals and Number 3 Main

Thrust Surface of all Other Main Journals

Grinding Wheel Surface

Journal Surface to be Ground

Edge of grinding wheel must fall within this distance.

Nicks in this area are critical to crankshaft strength.

FIGURE 12-20 Crankshaft journal must be machined with a radius, known as a fillet, to prevent stress cracks.

wear. If it is too small, stress will develop and could lead to a broken crankshaft there. The radius should never be less than 1/64 inch below the original radius.

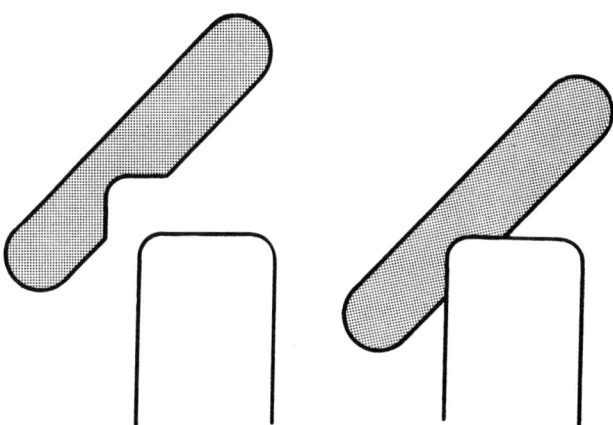

FIGURE 12-21 Wooden tongue depressor placed at a 45-degree angle to evaluate dressing

Follow the grinding machine manufacturers' recommendations, but be sure to check the results of the dressing operation before actual use. A wooden tongue depressor is an excellent aid in evaluating the size of the actual radii on the grinding wheel. To make this check, hold the tongue depressor at a 45-degree angle to the grinding wheel and allow the wheel to cut the dressed radius into the depressor (Figure 12-21). Use radius blocks to measure the cut radius. Compare the size of the dress radius to the radius size specifications to verify if it is acceptable.

Adjust the rotational speed of the headstock to approximately 50 revolutions per minute; be sure the direction of rotation is opposite that of the grinding wheel. In most instances the rotation of the crankshaft should be away from the machine operator.

Some grinding machines are equipped with an automatic wheel dresser (Figure 12-22). Mounted on a heavy-duty wheel slide, the massive dresser body provides greater accuracy when dressing the grinding wheel and permits dressing the grinding wheel without removing the work. Fingertip controls provide a variable rate of diamond travel during dressing. Positive regulation of the diamond feed assures longer life of the diamond and grinding wheel. A side crankmount wheel dresser (Figure 12-23) is often used for truing the sides of the grinding wheel with the face and sizing the width. Being magnetic, it mounts on the side of the counterweight.

 **SHOP TALK** _____

The best results from any dressing operation with a diamond tool are obtained by keeping the following in mind:

FIGURE 12-22 Automatic wheel dresser *(Courtesy of Rogers Machine Company)*

FIGURE 12-23 Side crankmount wheel dresser *(Courtesy of Goodson Shop Supplies)*

1. *Set the Tool at a Proper Angle. Position the tool at 10 to 15 degrees from the surface. Smaller angles can cause the diamond to be blunted. Larger angles can cause the wheel to remove metal holding the diamond in.*

2. *Prevent Vibration in the Holder. Any looseness in the tool holder will allow vibration of the tool, resulting in a poor finish on the wheel.*

3. *Turn Coolant On before Dressing. Wheels used wet should be dressed wet. To keep the diamond cool, turn on the coolant before beginning to dress. (If a wheel is dressed dry, allow the diamond to air cool between cuts but do not quench it with water.)*

4. *Begin to Dress at Highest Point. Wheels often wear unevenly and might have high and low spots. Set the tool to dress high spots first and feed it in about 0.001 inch per pass until finished. This will prevent overheating and damage to the diamond.*

5. *Use Steady Travel across the Face. Uneven travel across the face will cause an uneven finish. The best rate is found by experiment, but whatever the rate, it must be even.*

6. *Turn the Tool in the Holder. For longer life and faster dressing, turn the tool often to expose new diamond surface to the wheel.*

7. *Do Not Bump the Diamond. Diamonds are hard but brittle and can be harmed if dropped or bumped.*

The sequence of grinding the different areas of the crankshaft is as follows:

1. Rod bearings
2. Main bearings
3. Thrust bearings
4. Sealing surfaces (if required)
5. Flywheel and damper mounting surfaces

A common approach can be applied to grinding all of these areas except the flywheel and damper mounting surfaces.

When using a full width wheel, advance the grinding wheel until extremely light contact with the journal is made. Move the wheel laterally until the sidewall of the wheel lightly contacts the thrust wall of the journal. Be sure the grinding wheel does not initially abruptly contact the fillet areas of the crankshaft during the grinding operation. Turn on the machine coolant, set on automatic feed, and grind to the desired size. Maintain tension on the steady rest as the journal changes in size (Figure 12-24).

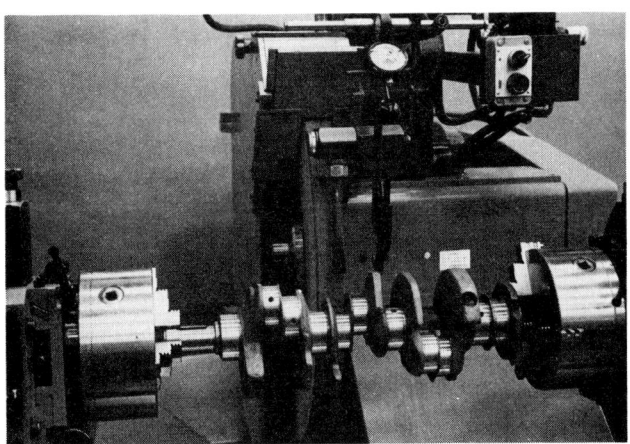

FIGURE 12-24 Crankshaft grinder in action

 **SHOP TALK** _____

Good grinding is almost impossible without a good coolant and its proper use. Be sure to have the proper coolant and keep it clean by filtering or changing as it becomes dirty. Use plenty of coolant when grinding and while dressing the grinding wheel. One of the most common causes of poor grinding is due to the improper use of or the wrong type of coolant. The coolant must flow between the work and the grinding wheel, not on top of the work or on the face of the wheel (Figure 12-25).

If the grinding wheel is not a full width wheel, the journal must be double cut. To double cut the journal, back the wheel away from the surface of the journal, traverse to the other side of the journal, and repeat the sizing process. Reduce the size of this half of the journal until the entire journal is uniformly sized.

When grinding the connecting rod journals, use the gear step diameter and the flywheel mounting

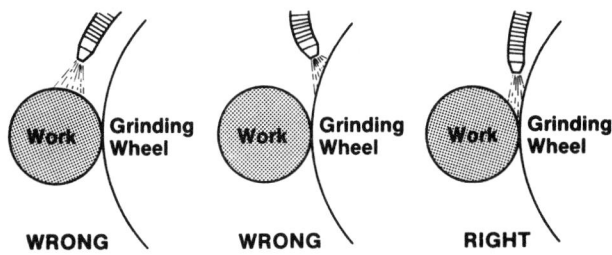

FIGURE 12-25 Coolant application

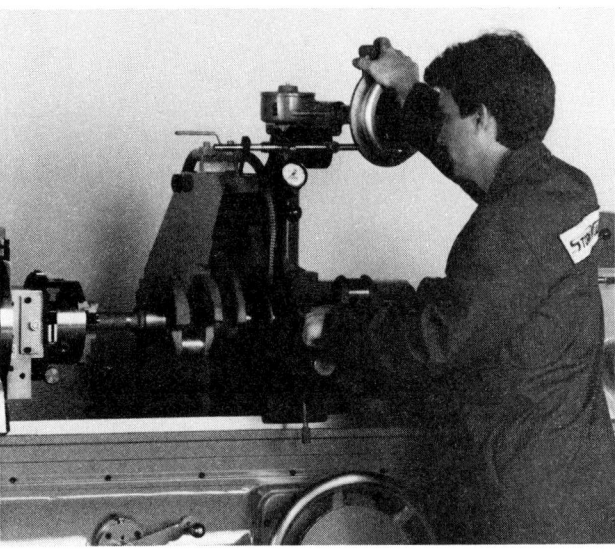

FIGURE 12-26 A dial indicator is mounted integrally with the steady rest. *(Courtesy of Storm Vulcan, Inc.)*

flange to hold and locate the crankshaft. Adjust the position of the crankshaft until the TIR of these diameters does not exceed 0.002 inch. The order of processing is staggered from the end journals toward the center. For example, when an in-line six-cylinder crankshaft is to be reconditioned, the grinding order would be 1, 6, 2, 5, 3, 4 or 6, 1, 5, 2, 4, 3. Staggering the grinding sequence improves the journal to journal alignment during the grinding operations. Install steady rests on the journals that have the same orientation, i.e. journal numbers 1 and 6 of our sample crankshaft. Move the steady rests to the next set of journals to be ground and continue processing (Figure 12-26).

When grinding the main bearing journals, use the gear step diameter and flywheel mounting flange to hold and locate the crankshaft (Figure 12-27). Adjust the position of the crankshaft until the TIR of these diameters does not exceed 0.002 inch. The order of processing is staggered from the center journal toward the end journals. Again, staggering the grinding sequence improves the journal to journal alignment during the grinding operations. Determine the runout of the center main bearing, which is half the bearing TIR. Be sure to measure the TIR of the bearing at the 3 o'clock position on the bearing. Install a steady rest on the center journal; that is, journal number 4 of our sample crankshaft. This runout measurement defines the undersize to which the crankshaft will have to be sized. The runout of the bearing *must* be less than half of the chosen

FIGURE 12-27 Setup for grinding mains on dead centers *(Courtesy of Winona Van Norman Machine Company)*

undersize. Do not remove this steady rest when the resizing is complete—it should remain in position until all grinding on the main bearing journals is complete.

The headstock and tailstock on some grinding machines are equipped with cross slide heads, which permit faster setup and quicker changeover from mains to rods (Figure 12-28). Main journals can be aligned from flywheel flange and timing gear location. This method of truing assures that main bearings are ground in line. Even crankshafts with damaged or untrue centers can be trued by simple adjustment of the cross and vertical slides.

There are two commonly used methods of regrinding crankshaft journals.

1. *Sweep Grinding.* As its name implies, the grinding wheel is moved or swept slowly

FIGURE 12-28 Four-way cross slide heads *(Courtesy of Storm Vulcan, Inc.)*

up and down the length of the journal until it is ground to the desired size. After the first journal has been ground to size (as determined with a hand-held outside micrometer), the Arnold journal sizing gauge (Figure 12-29) should be placed on the journal and set at zero. However, if it is known that 0.010 inch must be ground from the first journal, the Arnold gauge can be placed in position on the journal immediately and then set at zero after the 0.010 inch has been removed. This zero setting is used for grinding all of the remaining journals to size. Be sure to touch up the fillets of the journal so that they will be square and true.

 SHOP TALK _____

Many crankshaft grinder operators use a combination of plunging and sweeping on journals that are wider than the grinder stone. Plunge the first side to plus 0.001 inch over the finish size, then plunge the second side to size and sweep across the journal to the first side.

2. *Plunge Grinding.* In this method, the grinding wheel is dressed to the exact width of the journal to be ground and has the proper fillet radius. The wheel is then aligned perfectly with the journal and fed straight in until the journal is ground to the proper

FIGURE 12-29 The Arnold journal sizing gauge gives accurate readings of journal size while grinding. *(Courtesy of Winona Van Norman Machine Company)*

undersize. When plunge grinding is used, it is unnecessary to move the grinding wheel from side to side, which means this method is faster. But if a wheel of the proper width is not available, the sweep method is easier.

With either method, the grinding wheel must be allowed to spark out when finishing to obtain the highest quality finish. Otherwise, the journal would be marked and scratched. The journal should also be checked to make sure that it is tapered. A tapered journal can cause uneven bearing wear or failure. Check the vehicle's service manual for allowable tolerances. As a general rule, no allowable taper is recommended for a reground journal. It is good practice to check for taper on the first rod journal and again on the first main journal.

When thrust bearing surfaces are ground, they are usually sized as a separate operation. A 30-degree angle grinder is recommended. If an angle grinder is available, position, hold, and grind the crankshaft thrust areas using the same process and cautions described in grinding main bearings. Reconditioning convention suggests the thrust surfaces should be undersized to the same undersize as the main bearings.

If an angle grinder is not available, use the gear step diameter and the flywheel mounting flange to hold and locate the crankshaft. Adjust the position of the crankshaft until the TIR of these diameters does not exceed 0.002 inch. When correctly positioned, the thrust bearing surfaces can be ground by lightly and intermittently forcing (bumping) the side of the grinding wheel to contact the surfaces. Care must be exercised when bump grinding to avoid grinder burn and grinding wheel breakage. Continue the bumping operation until the thrust surface is correctly sized. The thrust surfaces should be undersized the same as the main bearings.

Grinding the sealing surfaces and the flywheel or damper mounting surfaces requires holding and locating the crankshaft differently. Use the centers of the crankshaft for location and a grinding dog as a driver for rotating the crankshaft. Adjust the shape and contour of the centers in the crankshaft until the TIR of the gear step and the flywheel mounting flange does not exceed 0.002 inch. Adjusting the centers of the crankshaft is best accomplished with a small rotary file with a conical shaped burring bit.

When correctly positioned, the sealing surfaces can be ground using the same technique as described for rod and main bearing journals. The flywheel and damper mounting surfaces are reconditioned by bump grinding, which is accomplished by

lightly and intermittently forcing (bumping) the side of the grinding wheel to contact the mounting surfaces. Care must be exercised when bump grinding to avoid breaking the grinding wheel. Heavy or excessive contact between the wheel and the end of the crankshaft must be avoided.

Crankshaft Polishing

Journal diameters and sealing surfaces of the crankshaft are polished to improve their surface finish and appearance after being ground. The polishing process should not be construed as a final sizing operation; it is not. In fact, it is generally accepted that the amount of material removed by polishing should be limited to 0.0002 inch.

Crankshaft polishing can be done on the crankshaft grinder with a portable polisher (Figure 12–30), separately on a special polishing machine (Figure 12–31), or by hand lapping (Figure 12–32). To perform the latter, cut a piece of emery paper to fit the journal, wrap a rawhide bootlace around it, then pull alternately on the ends. Start with a medium grit, then finish up with #320 grit. A micro finish of approximately 15 microinches can be achieved with this grit. Best results from any of the polishing methods are obtained by polishing the shaft in the direction of the crank rotation. Recommended surface finish of bearing journals is 32 microinches; sealing diameter is 15 microinches.

Next, polish the blend radius of the oil hole cross drillings of the crankshaft. The polishing process should improve the surface finish and blend the radius of the oil holes. Jewelers rouge can be an effective aid in optimizing the polishing and blending efforts. Surface finish should be 10 microinch.

FIGURE 12–30 Portable crankshaft polisher

FIGURE 12-31 Crankshaft polishing machine *(Courtesy of Storm Vulcan, Inc.)*

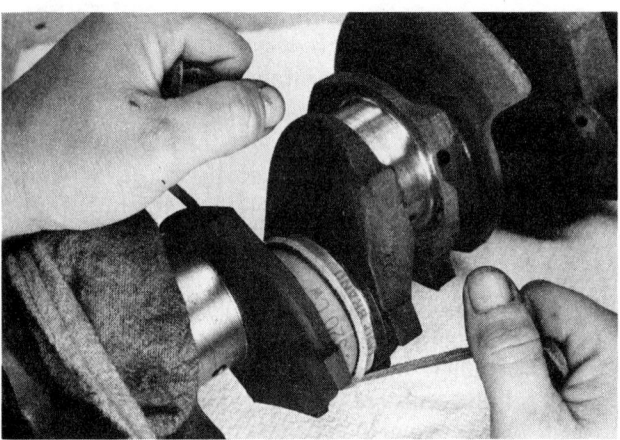

FIGURE 12-32 Hand lapping a crank journal

Crankshaft Journal Hardening

Crankshaft journal wear can be reduced by hardening. Either of the following two methods is usually employed to surface harden journals:

- *Heat Treating.* With this method, the basic structure of the metal is changed by the heat treating. Under controlled conditions, the metal is hardened to a specific depth. On journal surfaces this is 0.005 inch or more. By hardening to this depth, there is enough hardened material available for grinding to a smaller size after the initial surface has worn.
- *Salt-Bath Nitriding.* In this process, carbon and nitrogen are released in a liquid bath into which metal parts are dipped. The nitrogen combines with the carbon and iron to form a tough, wear-resistant surface of carbon-bearing iron nitride. This process has been known as *Tufftriding*, but a variation called the *Melonite* process is now gaining favor. Melonite involves fewer problems during manufacture in complying with EPA waste disposal requirements (see Chapter 1).

Neither Tufftriding nor Melonite-treated journals lend themselves to grinding. To check to see if a crankshaft has received a nitriding treatment, file a small portion of the shaft other than the journal (such as a counterweight). If metal can be removed under light pressure with a sharp, medium-fine mill file, the shaft has not been treated in a salt-bath nitriding process, so the journals were heat treated and therefore can be ground. If the file does not remove metal from the surface, the entire shaft has been treated and the hardening on the journals will not be deep enough to accept grinding.

Another check for the same purpose is to place a drop of 10 percent watery solution of copper ammonium chloride on the shaft. If it takes longer than 10 seconds for the solution to change from its normal light blue color to a reddish brown copper color, the shaft has probably been Tufftrided or Melonized. If the color change occurs in less than 10 seconds, it has not received the salt-bath nitriding treatment and can be ground with standard grinding equipment.

Shot Peening. Shot peening—fully described in Chapter 8—is another method of journal hardening. The crankshaft is placed on a turntable within the cabinet of a blast cleaner. Fine steel shot is then blasted at the crankshaft. Some rebuilders mask off surfaces that do not require strengthening.

Flywheel Grinding

If the flywheel is not up to manual specifications or if the face runout is more than 0.0005 inch TIR per inch of diameter, the flywheel can be resurfaced on a flywheel grinder (Figure 12-33). Best results will be obtained from this grinder when the grinding wheel is kept sharp (clean). When the wheel is clean and free cutting, a large shower of sparks will be evident from the grinding wheel and very little heat will build up in the workpiece. Overall grinding time can in many cases be reduced by taking a little extra time to dress the wheel. The cutting rate of the wheel will be so much better when the wheel is kept sharp that the time spent in dressing will be more than made up by the decrease in the grinding time.

WARNING: As with any crankshaft grinding (including the flywheel) always wear the proper eye protection.

FIGURE 12-33 Flywheel grinder *(Courtesy of Rogers Machine Company)*

A

B

C

FIGURE 12-34 Setting up a flywheel: (A) place crankshaft flange adapter ring on table; (B) lay flywheel on adapter ring; (C) tighten hold-down bolt. *(Courtesy of Rogers Machine Company)*

To set up a flywheel for a surface finish, proceed as follows:

1. Select the correct size crankshaft flange adapter ring and place it on the machine's table (Figure 12-34A).
2. Lay the flywheel on the adapter ring referencing the crankshaft flange surface (Figure 12-34B).
3. Select the centering cone and bolt (Figure 12-34C). Tighten the bolt and the machine is ready for grinding.

When surface finishing, keep the grinding wheel cutting at all times; never allow the wheel to coast for a long time without feeding it downward. Long periods of coasting will cause the grinding wheel to glaze and load up. The only exception to this is when a very smooth finish is desired. A variety of surface finishes is obtainable using the machine without varying the grit size of the grinding wheel. This range of surface finish is obtained by varying the pressure exerted on the grinding wheel. A coarse finish is obtained by feeding heavily then backing off the workpiece quickly. Smooth finishes are obtained by backing off the workpiece then coming down with a very light feed pressure.

SHOP TALK _____

Most flywheels are medium-hard ductile iron, but some are softer cast iron and some are harder steel. Flywheel stones work better and last longer if the proper stone is used for the material being ground. Therefore, the technician has two options to improve stone life:

- *Keep different types of stones on hand and change to match the flywheel*
- *Experiment with different types of stones to find one well suited to the range of flywheel that is ground most*

If a stone loads up and requires frequent dressing, it is probably too hard for the flywheels. If the stone wears rapidly, it is probably too soft. The grinding pressure also affects stone life. Thus, by experimenting with different stones and pressures, it is possible to find a combination that extends stone life significantly. Also remember that a flared-cup stone resurfaces flat or recessed flywheels. A straight-side stone is for flat flywheels only, but it lasts longer.

A

B

There are, of course, various designs of flywheels (Figure 12-35). Most can be ground on flywheel grinders. When grinding recessed flywheels, the depth dimension from the pressure plate mounting surface to the clutch friction surface should be restored after grinding the friction surface. This is done by grinding the pressure plate mounting surface an amount equal to that ground off the friction surface. The procedure for grinding recessed flywheels is as follows:

1. Mount the flywheel on the machine.
2. If original equipment specs are unavailable, use a depth micrometer to measure the depth from the pressure plate mounting surface to an unworn area of the clutch friction surface.
3. Grind the friction surface to clean up.
4. Use a depth micrometer again to measure the new depth. Subtract the original depth from this measurement to obtain the amount to be removed from the pressure plate mounting surface.
5. Put a dial indicator on the pressure plate mounting surface; set the dial at zero.
6. Grind the surface until the dial indicator reading changes by the proper amount. The same basic procedure can be employed when checking to make sure that

C

D

FIGURE 12-35 Flywheel designs: (A) Volkswagen, (B) flat; (C) recessed; (D) center plate *(Courtesy of Rogers Machine Company)*

FIGURE 12-36   Checking a flywheel

the original flywheel step for proper clutch performance is retained (Figure 12-36).

Special cutting tools are available to cut radii and undercuts or recesses on flywheels when needed (Figure 12-37). Flywheel grinding can be accomplished on certain milling machines using the proper accessories.

Vibration Damper Reconditioning

A faulty vibration damper or harmonic balancer can be replaced with OEM or a remanufactured unit. The balancer can also be reconditioned as mentioned in Chapter 7. That is, instead of replacing the grooved part (Figure 12-38A) use a repair sleeve to restore the seal. Before doing this, check the crankshaft for nicks and burrs. If necessary, use a fine emery cloth to remove the burrs. If the shaft has deep scores, they should be filled with metal epoxy filler. Then install the sleeve according to the manu-

FIGURE 12-37   Cutting a recess on a flywheel to reduce weight for balancing.

A

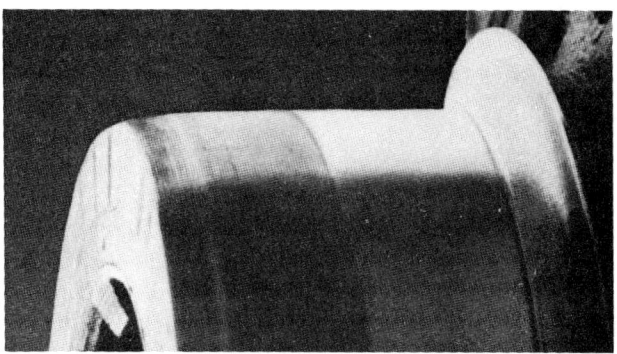

B

FIGURE 12-38   (A) Groove in harmonic balancer comes from the old seal and contaminated oil creating abrasion against the shaft. (B) Economical repair sleeve restores sealing surface and takes only a few minutes to install. *(Courtesy of CR Industries, Inc.)*

facturer's instructions. The repair sleeve, in most cases, offers a sealing that is as good as or better than the original crankshaft finish (Figure 12-38B).

Building Up the Crankshaft

Crankshafts can be reconditioned by adding metal to their bearing surfaces. Although there are several ways this can be accomplished, the two most common methods are:

1. *Chromium Plating.* In this process, pure hard chromium is plated electrolytically onto the main journal or connecting rod surfaces. This nonporous, well-bonded coating is then ground and polished to complete the reconditioning job.

   Metal spraying is another method of adding metal to the crankshaft in which molten steel is sprayed on the bearing surfaces while the crank is rotating. Chrome and metal spraying are seldom used today.

2. *Electro-Welding.* A process in which a bead of alloy or high-carbon steel wire material is deposited by an electric arc welding process onto the journal surface (Figure 12–39A). This continuous weld allows the buildup of any crankshaft journal with a minimum. Some crankshaft welders, such as the one shown in Figure 12–39B, can be used in two welding technologies— submerged arc and inert gas shield (see Chapter 8). For welding nodular cast-iron crankshafts and/or diameters less than 1-1/2 inches, the greatest success rate is

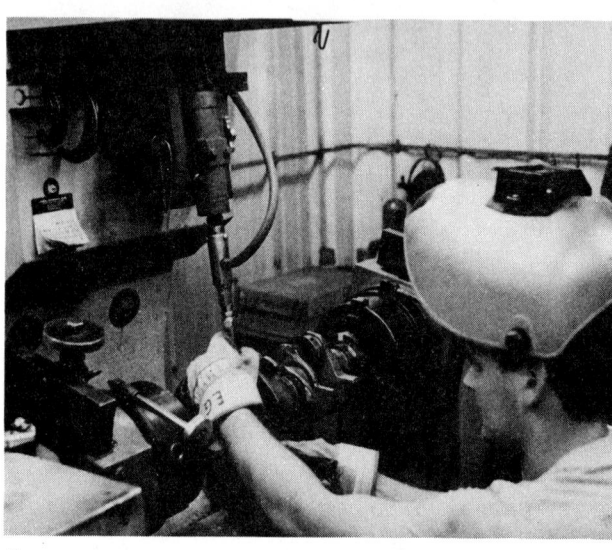

A

B

FIGURE 12–39 (A) Operating a crankshaft welder and (B) crankshaft welder *(Courtesy of Storm Vulcan, Inc.)*

achieved by using the inert gas shield welding procedure with a mixture of argon, helium, and carbon dioxide.

When operating chrome plating or electro-welding equipment to recondition crankshafts, make sure to follow the manufacturer's instructions to the letter.

After the crankshaft has been built up, it is ground to the original standard size main bearing journal diameter and standard size connecting rod bearing journal diameter. The disadvantage of this process is that it is much more expensive than grinding. In addition, the equipment required to build up a crankshaft is expensive. It is usually found only in large rebuilding and remanufacturing shops.

Some manufacturers and remanufacturer shops use chemical procedures to increase wear resistance and make fatigue endurance better. Some of these crankshafts can be reconditioned; others cannot. A few shafts are still shot peened. That is, the shaft to be shot peened is placed on a turntable in a cabinet and fine shot is blasted on the areas that are to be peened. Crankshafts that have been treated in this manner are said to have 100 percent better fatigue properties than crankshafts that were not. The operation is similar to the airless shot blasting cleaning technique (see Chapter 5).

Straightening the Crankshaft

Straightening a damaged crankshaft is a very critical step in the reconditioning process. Since straightening requires deformation of the crankshaft metal, the risk of breakage is always present. Improper straightening techniques can lead to disastrous results. Proper methods will allow reconditioning of shafts that otherwise would be scrapped and also minimize the amount of material that must be removed to clean up other shafts. The following procedure is a general guideline for using a crankshaft straightening press such as the one shown in Figure 12–40. Extreme care must be exercised when working with nodular iron crankshafts and other shafts the operator is not familiar with. In some instances original manufacturers have specific recommendations for straightening their shafts. Whenever possible these recommendations should be sought out and followed.

The general crankshaft straightening procedure is as follows:

1. Place the crankshaft on the press with the V-blocks under the front main journal and the rear main seal surface. If the main seal surface is unable to be used or is damaged, utilize the rearmost available main journal.

FIGURE 12–40 Crankshaft straightening press *(Courtesy of Winona Van Norman Machine Company)*

2. Swing the end stop into position against the front nose of the crankshaft to prevent end to end movement.
3. Position the dial indicator over the center main bearing and rotate the shaft to determine the point of maximum runout. Mark this point and record the TIR on a notepad.
4. Repeat step 3 for all intermediate main bearing journals, marking the point of maximum runout for each and recording the TIR reading in its proper sequence.
5. With all maximum runout points marked on the shaft, the general plane of the bend should be evident. In most cases all the maximum readings will be up at the same time, with the highest of the maximum readings indicating the point of the actual bend (Figure 12–41). In some cases more than one bend might be present. In this instance the shaft should be treated as if it were multiple short shafts and each bend removed separately.

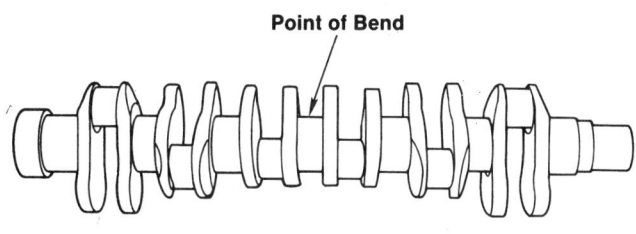

Point of Bend

| TIR | 0.005 | 0.008 | 0.012 | 0.007 | 0.003 |

FIGURE 12–41 Highest maximum reading indicates the point of the actual bend

6. Rotate the shaft so that the maximum runout point (previously marked) is down. Slide the clamp trolleys into position over the main journals adjacent to the actual bend location. Tip the clamp forward and engage the threaded rod on the front. Tighten the four hand knobs until the clamps contact the main journals. Slide the hydraulic cylinder under the bend location (rod journal between the two clamps) using the proper length ram block to contact the journal.
7. Pump the handle on the hydraulic power unit to apply force to the bend. The dial indicator should be in position while applying force to the shaft. Apply force until the indicator passes the zero runout position. (This is overbend and will vary depending on the type of shaft and amount of bend present.) Once the indicated overbend has been noted and written down, swing the indicator backward off the journal. Place the bronze-tipped chisel into the fillet of the bent journal and strike with a hammer. Repeat this blow on the opposite fillet (Figure 12–42). Release the hand knobs and tip the clamps to the rear.
8. Swing the dial indicator forward and rotate the crankshaft to check its straightness. Indicate all journals once again. Repeat the straightening process (steps 3 to 7) if all journals *do not* indicate within pregrind tolerance (usually 0.001 to 0.002 inch).
9. Position the dial indicator over the front nose of the crankshaft and measure the

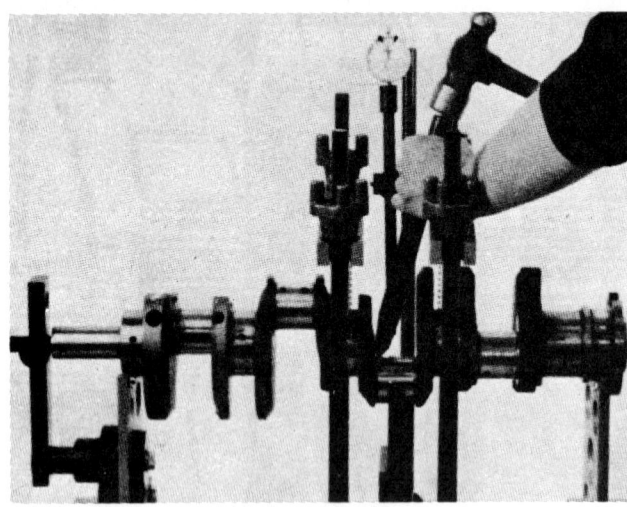

FIGURE 12-42 Striking the fillet of the bent journal with a hammer *(Courtesy of Winona Van Norman Machine Company)*

runout (Figure 12-43). Nose runout is normally corrected by peening in the fillet area (Figure 12-44).

WARNING: Use extreme care when straightening nodular iron shaft noses because they are easily broken.

---

10. Reposition the dial indicator to measure flywheel flange runout. If the shaft has been straightened 0.001 to 0.002 TIR on all mains, flywheel flange runout usually will not exceed 0.002 or the manufacturer's specification, whichever is larger. Remember that the end stop must contact a smooth machined surface on the end of the crankshaft nose.

 SHOP TALK _____

Keep a log of crankshaft model or forging numbers, the amount of bend before and after straightening, and the amount of pressure and overbend applied to the shaft during straightening. This type of log will improve efficiency on repeat jobs.

ALIGN BORING AND HONING

If the main bearing bores are not aligned, the condition can be corrected by align boring the saddles. After placing the main bearing caps in position, torque them to the required specs. Each bore should then be measured for size, stretch, and misalignment with a dial gauge (Figure 12-45). Compare these specs to the manufacturer's specs, and this

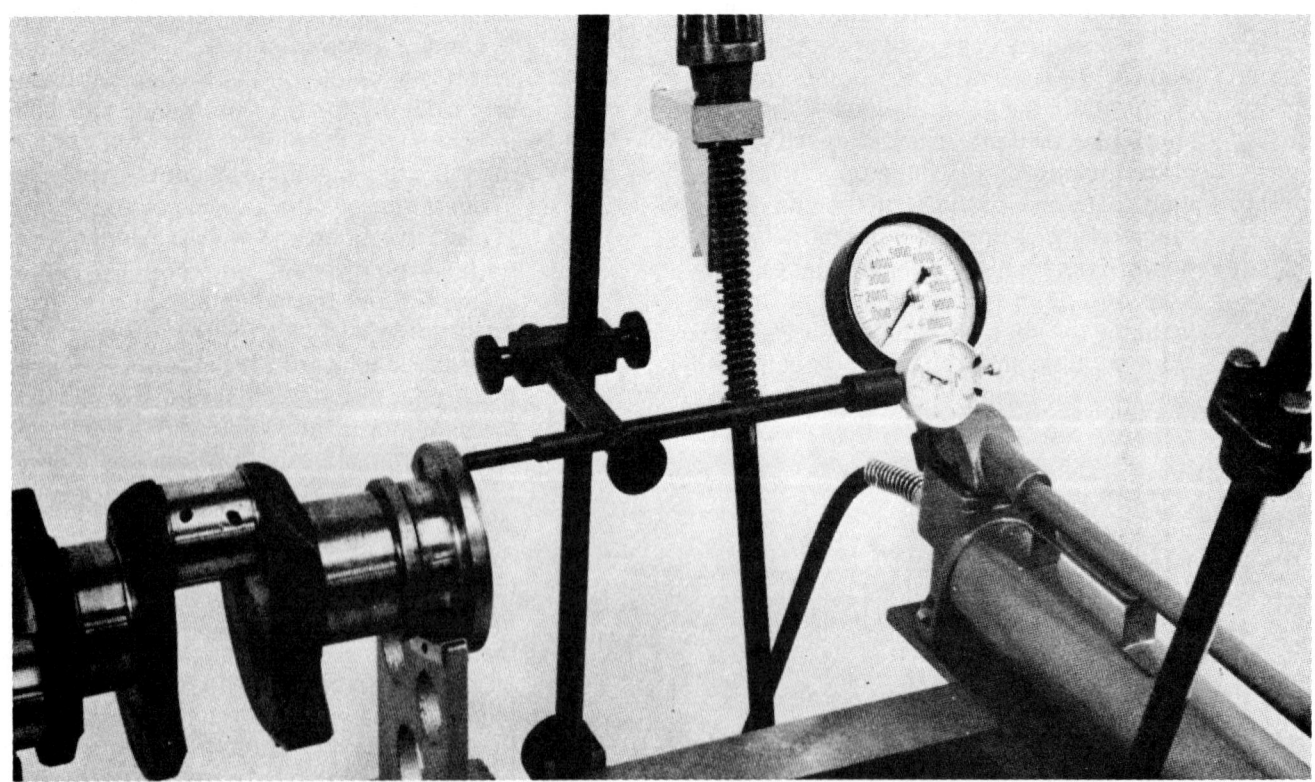

FIGURE 12-43 Measuring runout with a dial indicator *(Courtesy of Winona Van Norman Machine Company)*

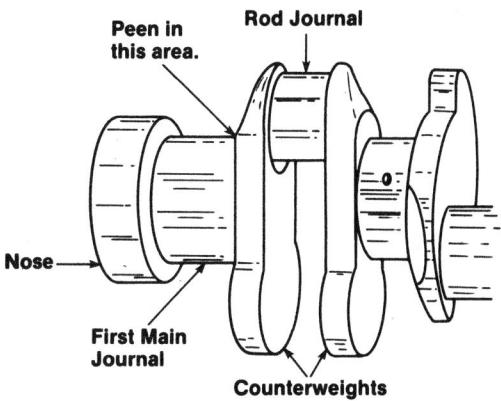

FIGURE 12-44 Peen in the fillet area to correct nose runout

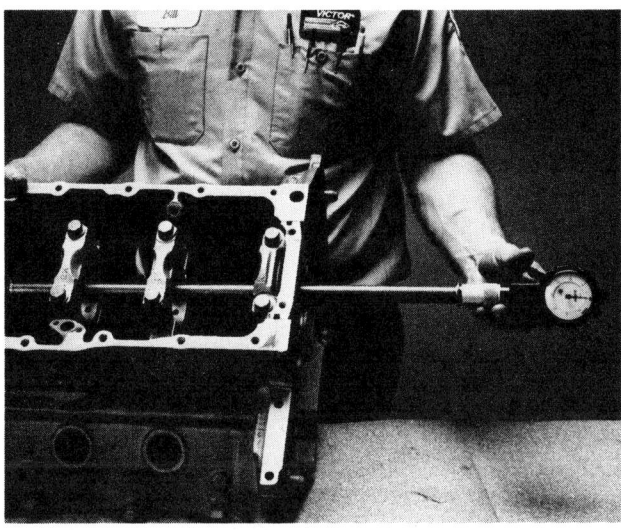

FIGURE 12-45 Checking the alignment of the bearing *(Courtesy of Perfect Circle/Dana)*

FIGURE 12-46 Small line boring machine *(Courtesy of Sunnen Products Co.)*

will determine the amount of the surface that needs to be ground off. After boring, check each bore with a dial gauge to determine where they are in relationship to the specs. Boring machines can be rather small (Figure 12-46) to rather large (Figure 12-47).

Although align boring is more versatile, align honing (sometimes called *line honing*) is also an accepted method for correcting main bearing bore misalignment (Figure 12-48). In align honing less material has to be ground from the cap parting lines than with the boring method. Align honing allows the technician to precisely remove high spots. Size should be checked frequently throughout the honing process.

The three basic causes of main bearing bore misalignment are the following:

- *Warpage.* When an engine runs thousands of miles, stresses created by continual heating and cooling cause block warpage and distortion. The most serious result is misalignment of the main bearing bores (Figure 12-49A). As warpage takes place slowly over a period of time, the original main bearing inserts and crankshaft compensate for the warpage through gradual wear. However, if a reground crankshaft and new inserts are installed in the warped block, this misalignment (even though slight) is all that is needed to make the crankshaft bind. The inserts and crankshaft try to compensate immediately for the existing warpage and they wear out quickly.

- *Stretch.* High loads usually cause the main bearing caps to become larger, to stretch in a vertical direction, and usually pinch in at the parting line (Figure 12-49B). Again, the original main bearing inserts compensate for this distortion through wear. Installing new inserts in a block with stretched caps makes the crankshaft bind and results in failure of the new inserts or crankshaft.

- *Spin.* If an engine block is subjected to excessive heat and loads, the bearings might seize to the crankshaft and spin with it. The result is a scored main bearing bore housing with burned bearing and crankshaft.

FIGURE 12–47 Large line boring machine *(Courtesy of Rogers Machine Company)*

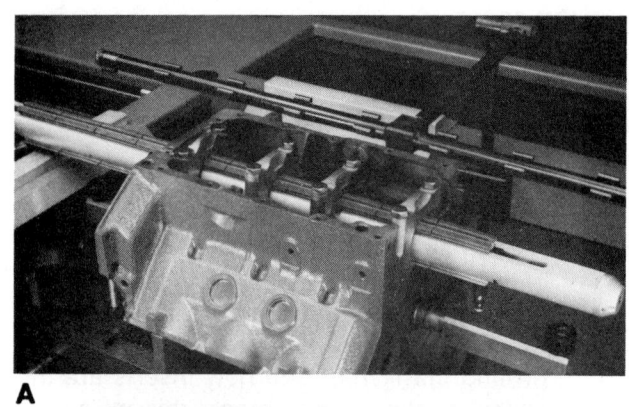

A

B

FIGURE 12–48 Line honing (A) small engine and (B) a larger one *(Courtesy of Sunnen Products Co.)*

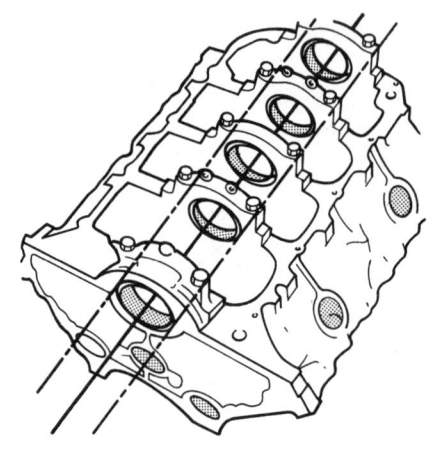

A

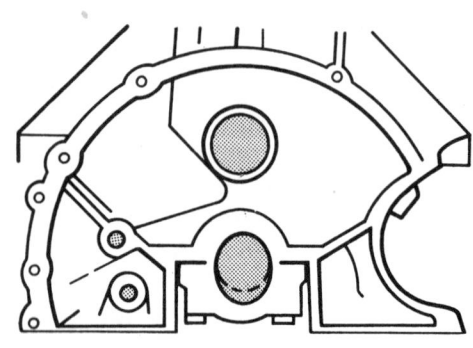

B

FIGURE 12–49 Block showing: (A) misalignment caused by warpage; (B) cap stretch

The preliminary step to any align boring or honing operation is to determine the condition of the existing block bores. Initial inspection consists of measuring the bore sizes using inside micrometers, telescoping gauges, or dial bore gauges (see Chapters 6 and 9). Bore measurement is followed by an alignment check, which can be done with test bars as follows:

 1. Remove all main bearing caps.

2. Carefully deburr the pan rails because most bearing devices use them as the base reference point.

3. Pan rail straightness should be verified with a straightedge. Most blocks are machined from these surfaces with the cam tunnel, block deck, and mainline to be parallel with them. The cylinder bores are machined at 90 degrees to the centerline.

4. Place the bearing alignment device over an end bearing bore with the indicator stem resting at the lowest point (Figure 12–50).

5. Adjust the indicator to give a convenient reading, slide the device across the bore to find the lowest point, and set the indicator at zero.

6. Repeat steps 4 and 5 in each bore and compare the readings. These readings will be used to determine how deeply into the block you will have to cut to clean up the bores. An example of readings found in an eight-cylinder diesel block could be:
Bore 1–0 inch
    2–0.002 low
    3–0.001 low
    4–0
    5–0.003 low
    6–0.001
    7–0.001

7. With bore number one as the starting point, all other bore depths are in relation to it. Bore number 5 is the lowest, so the bar will have to be set down to clean up this bore. Therefore, when positioning the bar with the alignment fixture, the bar will have to be set to cut into the block approximately 0.004 inch. In the above example, this will clean up all of the bores (Figure 12–51).

8. Out of alignment situations can also exist horizontally or side to side. It is suggested to get the block and boring bar completely set up with the centering fixture dial indicator reading zero when swept from block parting line to parting line (Figure 12–52). Repeating this at each bore will reveal any side-to-side misalignment of the bores. If a

FIGURE 12–50 Line boring of camshaft bores for overhead cam cylinder heads *(Courtesy of Rogers Machine Company)*

FIGURE 12–51 Precise setting of the boring tool in the collar installed on the bar *(Courtesy of Rogers Machine Company)*

shift in the bore exists, the bar will have to be moved sideways to average out the side shift condition. Failure to do so might result in not cleaning up the sides of all bores.

9. The machinist will have to determine the point at which the mainline must have ma-

FIGURE 12-52 Centering device for an accurate positioning of the boring bar *(Courtesy of Rogers Machine Company)*

jor repairs such as sleeving or buildup. Scrapping the block is sometimes necessary, especially if an align bore has been performed previously. Many engines, especially diesels, cannot tolerate moving the main bearing centerline into the block more than a few thousandths of an inch. The effect on deck height, gear train backlash, flywheel housing alignment, and so on must always be considered.

10. After the bar is positioned and the amount of stock to be removed has been determined, the main caps can be machined or ground in preparation for boring. This is necessary on main caps because they are often very rough on the sides and not machined square to the parting line. The caps must always be cut or ground on the same plane as the original parting line and bearing bore. This is very important on the thrust cap because just a slight tipping of the cap caused by improper grinding can create binding and misalignment of the thrust bearings. On automotive blocks, 0.005 to 0.010 inch is adequate stock removal for boring. On line bores like the larger diesels, 0.010 to 0.020 inch might have to be removed from the caps so that proper bore cleanup occurs. Minimum stock removal is always preferred. A part of cap preparation is checking for proper

press fit of the cap into the block. Caps should never be loose so they can slide in the block. Loose caps must be repaired or replaced before any line boring assembly preparation. Many caps with up to 0.007 inch press fit in the block. Any bad bolts or threads must also be taken care of to insure proper torque settings. After these steps, the caps can be installed and the block rechecked for proper positioning. Only at this point can boring proceed. Where new caps have been installed or where heavy cuts must be taken, two passes are preferable: a first rough cut followed by the final finish pass of 0.004 to 0.005 inch. On blocks where the bores are large or the unsupported span of the bar is great, the steady rest should always be used to insure consistent bore size and finish.

11. Tool bit sharpening has a great effect on consistent bore size and finish. The preferred method is to sharpen the bit to the proper angles on the grinder and then grind a small, smooth radius to the tip area on the bit. A sharp tip will cut one or two bores to the calibrated size and then break down, cutting undersize. A bit with a radius is more stable on size and finish. Also make sure the bar oilers are on any time the bar is in operation. The steady rest should also be lubricated manually while the machine is running to prevent seizure or scoring of the bar.

Mount the engine block on the machine using the appropriate parallels, V-block adapters, and hold-downs. These devices can be used in a variety of ways to hold the engine on the machine (Figure 12-53). Certain odd-shaped blocks might even require the fabrication of simple mounts and/or hold-down devices.

When mounting the engine block on any alignment machine, the main points to consider are:

- Rigidity of the block
- Even stressing to prevent distortion when clamped
- The block should be positioned close to the support columns for maximum bar support. However, be certain to leave sufficient room between the block and support bushings to insert the centering devices.
- With the spindle retracted, position the block left to right to take full advantage of the spindle stroke. The tool bit mounting hole should

FIGURE 12-53 A V-block mounting fixture accommodates many types of engine blocks *(Courtesy of Rogers Machine Company)*

FIGURE 12-54 Measuring sag with a variable sag control *(Courtesy of Rogers Machine Company)*

be near the first bore area. This will minimize the number of times the tool must be set up.

- The engine block should be centered and parallel to the spindle, and the vertical height should be such that the drive head and support columns can be adjusted to the same centerline as the bearing bores. Very low profile workpieces might require the stacking of parallels to accomplish this.
- Ideally, the block should be positioned so that when the boring bar is installed, the drive coupling U-joints will have little or no angularity.
- When using long bars with a short workpiece, it is best to position the support columns so that the center area of the bar is being used for the boring operation. The overhang of unused bar at either end will then be equalized resulting in a minimum bar sag or deflection (Figure 12-54).

When align boring or honing, always follow the manufacturer's instructions and safety precautions. Boring speeds vary with the material, bore diameter, and finish desired. The following can be used as a general guide in choosing speeds:

Cast iron: 200 to 400 rpm
Large diesel blocks: 250 rpm min.
Babbitt/aluminum: 300 rpm max.

Feed rates from 2 to 6 are suitable for most finish bores. Faster feed rates are used for roughing and cutting of soft materials at high boring speeds.

Crankshaft seals and bearing inspection and installation are covered in Chapters 6, 7, 14, and 15.

CAMSHAFT

As previously mentioned in Chapter 2, the camshaft controls the operation of the valve system, the oil pump, and, on some engines, the fuel pump and ignition distributor. A typical overhead valve (OHV) engine's camshaft is located in the center of the block directly above the crankshaft. The overhead cam (OHC) type of camshaft is in the cylinder head. Overhead valve engines used to be the prevailing type, but the current trend is toward OHC engines.

PREPARING TO RECONDITION THE CAMSHAFT

Before undertaking any camshaft reconditioning job, it is necessary to know how to distinguish and classify the camshaft material used. In fact, it is almost impossible to go through all the stages of reconditioning a shaft if you do not know what material it is made of.

By and large, as mentioned in Chapter 2, the camshafts can be divided into two basic groups: those made of steel and those made of iron. And these in turn can be subdivided into two types:

- *Steel Shafts.* Low carbon and high carbon.
- *Iron Shafts.* Chilled and malleable or spheroidal.

When the type of material is not known, the following spark test can be made.

Spark Test

This method does not guarantee that the exact composition of the material will be established, but it gives a good guide as to the rough quality. It consists of examining with the naked eye the band of sparks that comes from a piece of steel or iron contacting a grinding wheel at high speed (Figure 12–55). For the sparks to be regular and sufficiently intense, the wheel must be of medium grain and have a peripheral speed of about 20 m/sec.

Once the quality of the material is known, it is possible, with a fair degree of certainty, to establish the reconditioning procedure to follow. Although there are certain guidelines, they are by no means definite.

Low-Carbon Steel Shafts

These types of shafts are gradually disappearing. Built in low-carbon steel, their surfaces are hardened by case hardening. The term *carburizing* or, more simply, *case hardening* means the treatment by which carbon is introduced into the steel by diffusion. The aim of case hardening is to create, on the steel surface, a high-carbon layer. After hardening, this layer develops high degrees of hardness and good resistance to wear and stress. The hardness of the surfaces is normally very high, but it is diminishing and, except for special cases, the thickness does not exceed a few tenths of an inch. The result is that when some material is removed from the surface of the cams, with the aim of eliminating their wear and shape imperfections, the hardened surface degrades its hardness or disappears completely, and this makes further use of the shaft impossible, if not dangerous. In such cases, the case-hardening operations would have to be repeated, but in practice this is not feasible because the high temperatures required (1500 to 1600 degrees Fahrenheit) would only damage the main journals (Figure 12–56).

According to recent experiments, the cheapest and most feasible system is to coat the surface of the cams with certain materials, by a microdynamic diffusion process between the weld material and the base material. This is accomplished by using, for example, the Eutalloy-Microflo process. With this

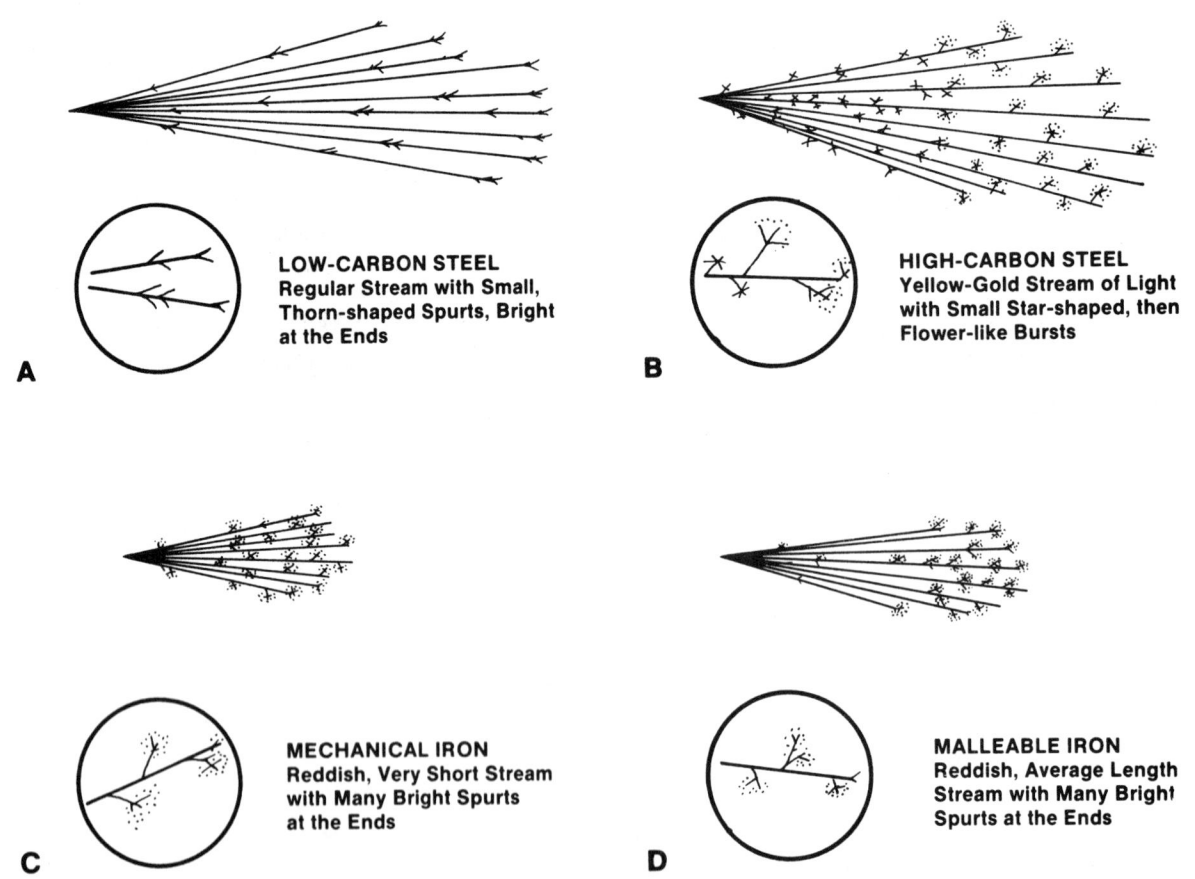

A

LOW-CARBON STEEL
**Regular Stream with Small,
Thorn-shaped Spurts, Bright
at the Ends**

B

HIGH-CARBON STEEL
**Yellow-Gold Stream of Light
with Small Star-shaped, then
Flower-like Bursts**

C

MECHANICAL IRON
**Reddish, Very Short Stream
with Many Bright Spurts
at the Ends**

D

MALLEABLE IRON
**Reddish, Average Length
Stream with Many Bright
Spurts at the Ends**

FIGURE 12-55 Various results of a spark test: (A) low-carbon steel; (B) high-carbon steel; (C) mechanical iron; (D) malleable iron

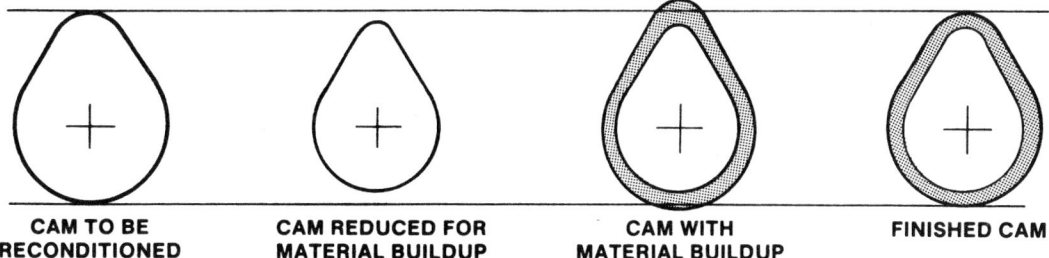

FIGURE 12-56   Repair of a low-carbon steel shaft

system, it is possible, with great precision, to deposit measured quantities of special powdered alloys starting from a thickness of 0.0004 inch with hardness up to 65 Rc and minimum distortion of the workpiece, as long as the surface being coated is not brought to the melting point but remains solid. The whole process is based on a special type of oxyacetylene torch and cartridges containing the powders. Sucked in by the oxyacetylene gas mixture, the powders cross the flame where they are heated and then blasted to the surface of the base metal. The heat of the flame melts them and joins them to the workpiece by microdynamic diffusion.

High-Carbon Steel Shafts

These steel shafts have a carbon content of 0.40 to 0.60 percent. Their main feature is that the active surfaces of the cams can be hardened directly without any preliminary carburizing treatment.

Hardening is achieved by heating the area involved over its critical point, between 1500 to 1600 degrees Fahrenheit. Then, after keeping this temperature long enough for the heat to be diffused on the inside as well, cool the area in a medium that rapidly removes the heat, such as quenching water or oil. Using water as a quenching medium gives greater surface hardness but also makes the material more brittle. Using oil causes brittleness, but consequently, less hardness.

The choice of medium should be based on two factors:

1. The carbon content of the steel and the hardness of the surface, which should not generally be less than 55 points of the Rc scale
2. The heating temperature of the surface

While there are various types of pyrometers and thermocouples that can measure the heating temperature, most rebuilder shops, because of the cost of these devices, use temperature-sensitive marking pencils. The special feature of these pencils is that the dry, chalky line made before or during the heating operation on the surface of the workpiece becomes liquid when the preset temperature is reached. In these types of shafts, the hardening depth is usually very high, never restricted to a single lobe, and extends below the 45-degree line (Figure 12-57A). This means that the shafts, if there is no chipping or abnormal wear on one or more cams, can be reconditioned directly with one single grinding operation. But if all the cams are worn and wear is due to poor surface hardness, the cams should be rough ground until their sizes and profiles are equal and then rehardened and finish ground after the desired hardness has been reached. When only one or some of the cams are chipped to such a point that it is not worthwhile reducing the others, material buildup is the only procedure to adopt if you want to salvage the shaft.

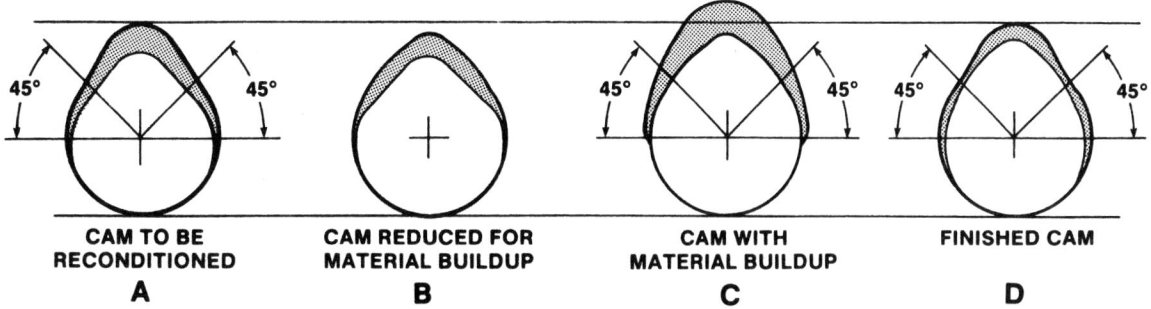

FIGURE 12-57   Repair of a high-carbon steel shaft

The first thing to do is to remove some material in the points to plate weld. For this purpose, the camshaft should be set up in the grinder off center a few degrees, first in one direction and then in the opposite direction, so that the wheel works over only the worn surface of the cam (Figure 12–57B). Then, weld plating will have to be done with a type of electrode that ensures a good bond to the base material (with either TIG or MIG procedures) and sufficient hardness (Figure 12–57C). Finally, the shaft should be straightened carefully and reground so that the cams are the same shape and size (Figure 12–57D).

Chilled Cast-Iron Shafts

In light vehicles, iron shafts have almost completely taken the place of steel ones because they are much cheaper and accurately cast, while steel shafts require considerable machining. The surface hardening itself is obtained during casting, thus another necessary process can be avoided.

Partial or total surface hardening of the cams is obtained by locating metal chills at desirable positions in the mold. When in action, these chills act like quenchers because they remove heat and rapidly cool the contact material, which becomes much harder than the material not directly associated with the metal chills. The hardening depth is generally quite considerable in chilled iron shafts, which means that unless one or more of the cams are chipped or abnormally worn, they can be reconditioned merely by regrinding. If several or all of the cams are worn, however, repair work will have to include metal buildup with a suitable quality powdered material or electrodes, following the operating sequence and suggestions given concerning steel shafts.

The only difference is that the straightening of the shaft and the preparation and cleaning of the surfaces to be weld plated have to be performed with extreme care. These surfaces must be degreased with a solvent like trichloroethylene and then heated so that all traces of oil are completely eliminated from the pores in the material.

Malleable or Spheroidal Cast-Iron Shafts

These shafts are between high-carbon steel shafts and chilled cast-iron shafts. They are tougher and more ductile, and their surfaces can be rehardened by using an oxyacetylene torch and quenching in oil or water. The hardening depth of the cams is generally very considerable and they can therefore be reconditioned by regrinding. Salvage, in cases of abnormal wear or chipping of one or more cams, is always possible by microdynamic diffuson as described under "Low-Carbon Steel Shafts" and by buildup with electrodes as described under "High-Carbon Steel Shafts" and "Chilled Cast-Iron Shafts.

CAMSHAFT STRAIGHTENING

If bend in the shaft is found during the camshaft inspection, it can usually be straightened without much difficulty. As described in Chapter 6, set the camshaft on a V-block fixture and determine the height on the high side. Remove the camshaft from the block and check it with a dial indicator. Repeat the procedure until the shaft is straight. Never install a camshaft that is bent.

All of the necessary operations required to prepare a camshaft for grinding—de-burring, straightening and re-centering cams, cutting centers and checking lift and runout—can be performed on a camshop bench like the one in Figure 12–58. The camshop can also be used to remove lubrite from the mains after Parkerizing treatment.

CAMSHAFT GRINDING

If the camshaft lobes or journals do not meet the vehicle manual specifications, the shaft must be reground. At one time grinding lobes was considered difficult. But, with today's modern camshaft grinders, this task is much easier. Some machines are computer controlled to guide the grinding head around the lobe in the correct profile.

Once it has been determined by careful inspection that the shaft is salvageable, the next decision is

FIGURE 12–58 Typical camshop bench *(Courtesy of Storm Vulcan, Inc.)*

FIGURE 12-59   This gauge arrangement is one of the simplest ways to find uneven cam lobes. *(Courtesy of Goodson Shop Supplies)*

whether the lobes should be reduced or built up (Figure 12-59). That is, the camshaft, like a crankshaft, can be built up by welding on the area of the lobe that is under specs. The welded area is then reground on the cam grinding machine to the original equipment specifications for lift and duration. When reducing the lobe to its original equipment specs, the amount of material removed is very limited (0.001 to 0.002 inch). This is especially true in the case of nonadjustable valve trains.

The first step in regrinding camshaft lobes on the typical cam grinders like the one shown in Figure 12-60 is to make a master (profile) cam in the machine's master making attachment. Once it is ground to the original equipment specifications, it can be installed on the keyed spindle. The shaft can be installed on the machine (Figure 12-61) and indexed as described in the owner's manual. After touching the shaft on each side of the lobe with the grinding wheel, retract it one turn of the micro-feed wheel. Start the grinding wheel motor, headstock rotation, and the coolant flow and bring the grinding wheel into contact with the cam lobe. Grind until all the lobes are cleaned up and are at OE specs.

Camshaft journals can be reground in the same way as crankshaft journals. Once the journals are reground, the proper undersize bearings must be installed. Frequently it is necessary to align bore for these cam bearings. This is done in the same manner as for the main bearings, except that it is necessary to slant the support bars, brackets, and trunnions to center the tool bar in the cam bearings. Once the

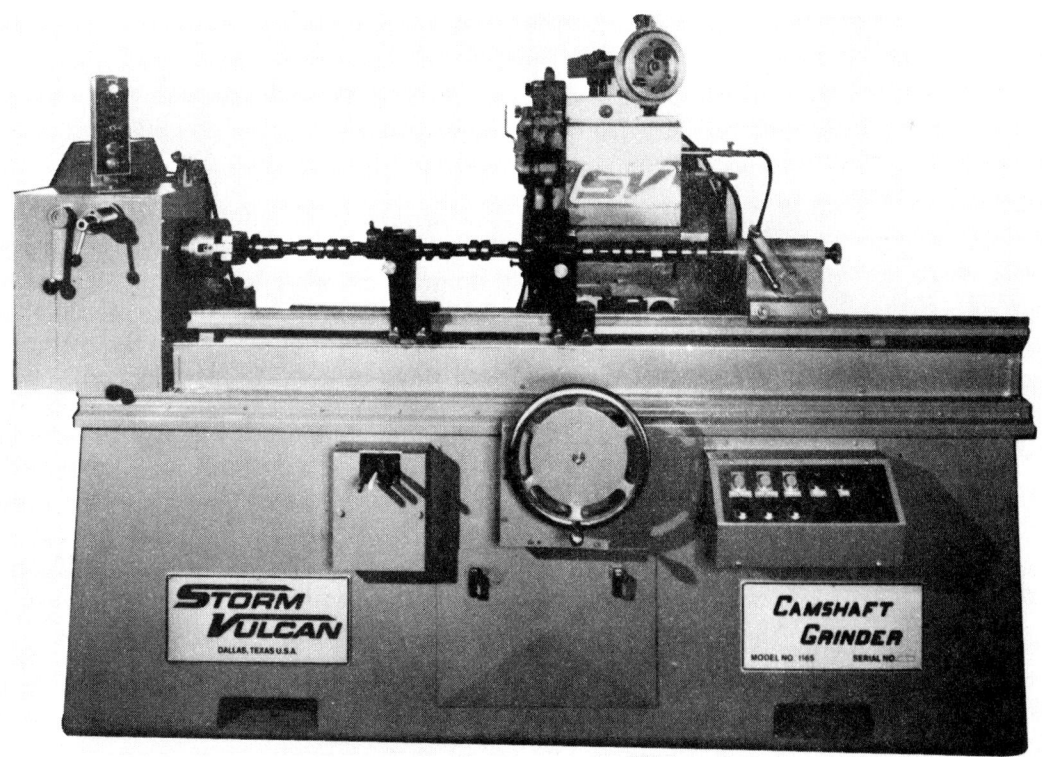

FIGURE 12-60   Camshaft grinder *(Courtesy of Storm Vulcan, Inc.)*

FIGURE 12-61 Grinder setup for large shafts *(Courtesy of Storm Vulcan, Inc.)*

align boring machine is set up as detailed in the owner's manual, the bearing can be cut. Remember that on many camshafts each journal is a different size.

After grinding, the camshaft surfaces should be sprayed or dipped (Figure 12-62) in a hard surface overlay coating such as molybdenum disulfide or manganese phosphate (often called lubrite or Parkerize treatment). Because this nonmetallic, oil-absorbing material gives an etched surface coating, it helps to give a rapid break-in without scuffing.

FIGURE 12-62 Typical lubrite unit *(Courtesy of Storm Vulcan, Inc.)*

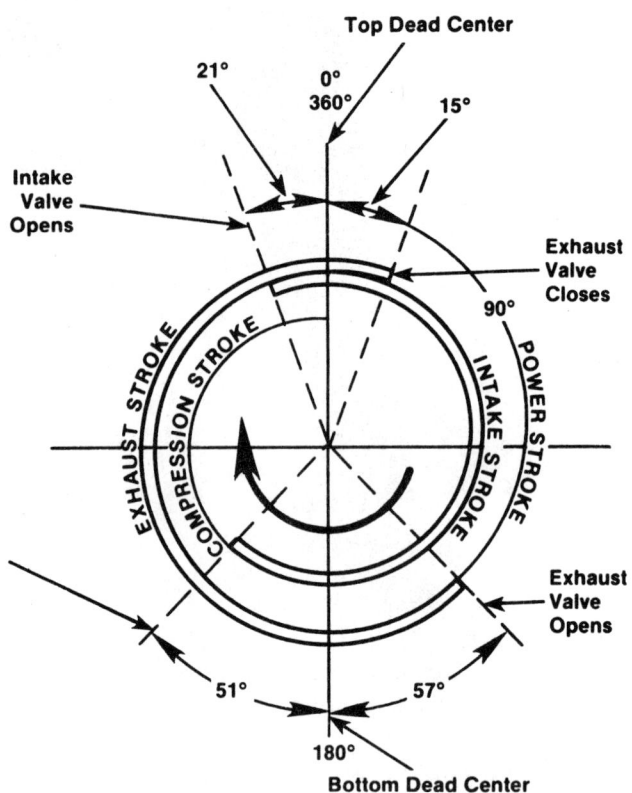

FIGURE 12-63 Typical valve timing diagram in relation to crankshaft rotation

CAMSHAFT/ CRANKSHAFT TIMING

For the valves to open and close in correct relation to the position of the crankshaft, the camshaft must be timed to the crankshaft. This means that the two shafts must be assembled so the lobes open the valves at a precise time in relation to the position of the piston and crankshaft. In the typical valve timing diagram shown in Figure 12-63, which shows the relation to crankshaft rotation, the intake valve starts to open at 21 degrees when the piston has reached top dead center (TDC) and remains open until it has traveled 51 degrees past bottom dead center (BDC). The number of degrees between the valve's opening and closing is called *intake valve duration time.* Figure 12-64 shows the intake valve duration time is 252 degrees of crankshaft rotation.

The exhaust stroke begins at 53 degrees before BDC and continues until 15 degrees after TDC, or a total exhaust valve duration time of 200 degrees of crankshaft rotation. Although both power and compression time are approximately the same, there is a period of time when both the exhaust and intake valves are open. This time period is known as the *valve overlap.*

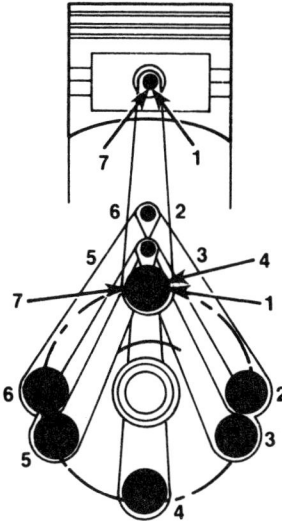

NOTE THAT PISTON TRAVEL IS GREATEST FROM LOCATING POINTS 1 TO 2 AND 6 TO 7.

Clockwise Crankshaft Rotation

Crank Throw Movement (Location Points)	Crank Rotation Between Locating Points (Degrees)	Total Crank Rotation (Degrees)	Piston Travel (Inches)
1–2	90	90	2-1/2
2–3	27	117	3/4
3–4	63	180	3/4
4–5	63	243	3/4
5–6	27	270	3/4
6–7	90	360	2-1/2

FIGURE 12-64 Piston travel as it relates to valve timing

Overlap is critical to exhaust gas scavenging and the development of proper cylinder pressure. A camshaft with a lot of overlap helps scavenge the cylinder at high engine speeds for improved efficiency. However, since both valves are open for a longer period of time, low rpm cylinder pressure tends to drop. Conversely, a camshaft with less overlap helps generate higher low rpm cylinder pressures, but does not scavenge the cylinders as effectively as high engine speeds. The amount of overlap, since it has an effect on cylinder pressure, would also be a factor in determining overall engine efficiency and exhaust emissions. Valve timing and overlap vary from one engine to another.

Lobe spread and lobe centers are two fairly new terms in camshaft work. Lobe centers describe where the maximum lift point of the intake and exhaust valve occur with respect to piston positioning during the intake and exhaust strokes. For example, suppose a camshaft has 117-degree exhaust and 107-degree intake lobe centers. This means that the maximum net lift point of the intake valve occurs 107

degrees after TDC during the intake cycle. The maximum lift point of the exhaust valve occurs 117 degrees before TDC during the exhaust cycle. Lobe spread is simply the average of the intake and exhaust lobe centers. In this case, the average of the 107-degree/ 117-degree camshaft would be 112 degrees. Hence, the cam has a 112-degree lobe spread.

To determine if the camshaft is advanced, retarded, or straight up (zero-degree advanced), one simply has to compare the lobe spread to the intake lobe center. If the intake lobe center is less than the lobe spread, the camshaft is advanced by the difference between the values. If the intake lobe center is greater than the lobe spread, the camshaft is retarded. If the values are the same, the camshaft is at zero degree timing or straight up. Since the intake lobe center is 107 degrees and lobe spread is 112 degrees, the camshaft is advanced 5 degrees.

It is possible for the rebuilder technician to advance or retard the camshaft by changing the lobe centering by using multiple keyway cam and crank sprockets, or offset keyway dowel pins (see Chapter 6). Advancing cam generally improves low-speed torque whereas retarding it usually improves high rpm power.

Timing adjustment marks can also be found on the vibration damper or crankcase pulley (Figure 12-65) on most production engines. A mark is usually a groove on the pulley to indicate the crankshaft angle in relation to a timing indicator. A magnetic timing probe allows a precise reading of crankshaft position for timing the engine. The TDC and BDC can be found when setting camshaft timing with a warp and alignment gauge (Figure 12-66).

With No. 1 at TDC at end of compression stroke, make a chalk mark at points 2 and 3 approximately 90° apart.

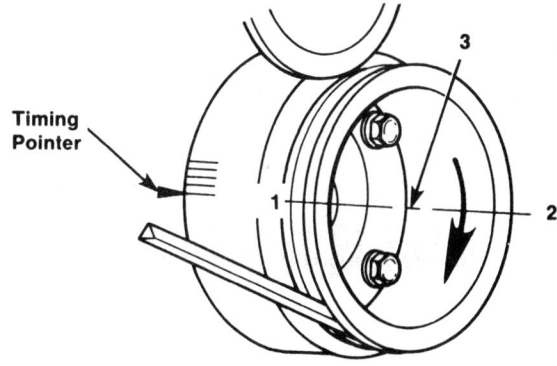

Position 1: No. 1 at TDC at end of compression stroke
Position 2: Rotate the crankshaft 180° (1/2 revolution) clockwise from position 1.
Position 3: Rotate the crankshaft 270° (3/4 revolution) clockwise from position 2.

FIGURE 12-65 The crankshaft must be in one of three positions to check valve clearances.

FIGURE 12-66 Warp and alignment gauge *(Courtesy of Goodson Shop Supplies)*

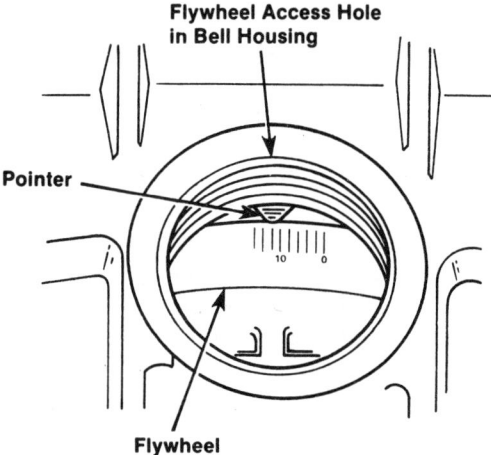

FIGURE 12-67 On some engines, the timing marks are located on the flywheel.

The flywheel can also carry the timing marks (Figure 12-67). Flywheel timing can be more accurate than timing with a smaller disc. With the transverse engine installation, the flywheel is more accessible for use in timing the engine.

Electronic timing signals for the engine computer can be taken from the flywheel. Advanced engines use a flywheel sensor to pick up magnetic pulses from the flywheel. The rpm sensor informs the computer of engine speed; the reference mark sensor reports on flywheel and crankshaft position in relation to TDC to time the firing of the spark plugs. In these engines, no distributor is needed. Engine timing at the flywheel will be used increasingly on engines. There are, however, millions of engines with timing at the front of the engine, usually on the vibration damper.

The rebuilder should keep in mind the following situations that can affect valve timing:

- Camshaft standardization is almost a must for the rebuilder. It is inconvenient to regrind ten different cams for a given engine, and it makes no sense to do so when the engine might be used in any one of several installations. So, most rebuilders use a standardized cam for a given family of engines within specific years. Although this has worked well in the past, it can lead to problems with today's computer-controlled engines. The differences in cam timing might be slight, but timing is part of a total system that works in harmony. Changing the cam can affect the whole system and alter fuel delivery, along with valve timing. Together, these will directly affect the combustion process.

- Along with cam changes, OEMs have changed the cam/crank sprockets to retard the cam. Using the early gears, as many rebuilders have traditionally done, might have made engines run better a few years ago, but the practice is suspect today. Again, it is the total system designed by the OEM that must remain in harmony if the computer is to work properly. Advancing the cam can have a detrimental effect on the system.

- Surfacing the blocks and heads on overhead cam engines affects cam timing. Because the distance between the cogs on the belt is fixed, when the head is moved downward toward the crank gear the cam is retarded to compensate for the change in distance. This advances the cam timing, again affecting system design.

- Valve train geometry can be affected by rebuilding. There are very few problems with hydraulic lifters as long as they are properly preset. However, if hydraulics are not set within their working range or mechanical lifters are improperly adjusted, the incorrect valve lash will affect both valve timing and engine performance. In the worst of cases, this affects combustion processes due to changes in timing and due to the valves themselves becoming overheated, causing preignition.

- Valves can cause abnormal combustion in other ways. If the rebuilder uses valves with thin margins or sinks them too deeply into the head, if the seats are too narrow or the springs are weak, or if the guides do not transfer heat properly, the valves will run hot. When this happens, the likelihood of detonation and preignition is greatly increased.

Anytime the valve timing is altered, the compression ratio is affected and compression heat varies. At the same time, the operational characteristics of the engine change, possibly beyond the coping range of the engine's control system. Compounding all of this are any valve train changes made by the rebuilder. The end result can be a significant increase in the possibilities for abnormal combustion.

Remember, proper engine valve timing is critical for good engine performance. Valve timing is always a compromise because the correct valve timing for any engine exists only at a given engine speed. The correct valve timing at any given rpm is different from any other. Fortunately, the variation in ideal valve timing, from idle to red line, is not so great as to preclude striking a compromise setting that is adequate for that entire range; especially if that rev range is narrow. However, several import vehicle manufacturers have or are planning to introduce variable valve timing, which should improve engine performance. As enough of these engines get into operation and wind up in rebuilders' shops, the technician and machinist will have to be able to deal with the new valve timing systems.

ENGINE BALANCING

In order to achieve the highest horsepower output, greatest fuel economy, longest component life, and smoothest operation, all rotating and reciprocating parts of an engine should be precision balanced. Performance engines have always been balanced. But now, ordinary units need it, too. Today's small high-revving designs are more sensitive to balance, yet carmakers' tolerances have not improved. That is why most engines benefit from balancing during a rebuild. In fact, balancing can be one of the most important steps in a successful rebuild.

As mentioned in Chapter 2, reciprocating engines fall into two general classifications. The first group is in-line engines, and these do not require bobweights. It is still necessary to match weight the pistons, rod, and so on. The in-line engines have a balanced crankshaft design, which is easy to identify in that the large counterweights on each end are in line, on the same side of the crankshaft. The second general classification of engines covers the V-type. These have an unbalanced crankshaft design where the large counterweights on each end are not in line and are on opposite sides of the crankshaft. This type of design requires bobweights. Some engines in this group are designed with part of the counterweight on the flywheel and dampener. It is most important to determine if the dampener and flywheel are counterweighted. If they are, they must be mounted when balancing the crankshaft.

The vehicle parts that must be weighed for precision balancing include:

- Pistons
- Piston rings
- Piston pins
- Connecting rods
- Connecting rod bearings and inserts

Engine components that should be balanced (preferably mounted on the balanced crankshaft) include:

- Flywheel
- Front pulley or torsional damper
- Clutch
- Clutch disc

Advanced electronic balance machines, such as the one shown in Figure 12-68, make the important calculations to balance a crankshaft. The operator's main task is weighing the piston rod assemblies and preparing bobweights for V-type engines.

FIGURE 12-68 Electronic balancer machine *(Courtesy of Stewart-Warner, Inc.)*

Computing the Bobweight. On the standard V-8 or V-6 engine, the bobweight total consists of 100 percent of the rotating weight and 50 percent of the reciprocating weight. For example, there are two rod and piston assemblies per throw on the standard V-8. The rotating weight would be the crank end of both rods and the bearing insert. The reciprocating weight would be the weight of one piston, one piston end of the rod, one set of rings, pin, and pin locks (if used). This means that 50 percent, or half of the actual weight for the reciprocating part, is the important factor. Keep track of the weight on a chart such as the one in Table 12-1.

By using a precision scale, the lightest piston is determined (Figure 12-69). Weigh all eight pistons, one at a time, find the lightest, and record this weight. Place the lightest piston on the scale and zero the scale. Match the remaining piston to within 1/2 gram.

If, after attempting to reduce the weight of the heavier piston, it is found that enough weight cannot be removed to equalize the set, either the heaviest or the lightest should be replaced so that a matched set can be attained. It is possible to add weight to the lightest piston by pressing an aluminum slug into the piston pin that goes with the light piston. However, this should only be done with the consent of the vehicle's owner. If it is done, the customer should be notified that this particular piston pin must always be placed with that piston; otherwise, the engine would be unbalanced.

FIGURE 12-69 Weighing pistons *(Courtesy of Stewart-Warner Corp.)*

Weight Removal from Pistons. Weight removal from pistons is usually accomplished in a lathe or, in some cases, a vertical mill (Figure 12-70). To prevent scoring of the ring grooves and lands, a piece of light-gauge strip steel should be wrapped around the piston. The width of this strip of steel should be such that it protects the piston from the chuck jaws. In those cases where the piston is balanced with the pin in place, the steel wrap should be long enough to extend over the pinhole so it will prevent the pin from coming out when the piston is rotated in the lathe. (Usually, the pin is removed from the piston when match weighing; however, in some cases it is necessary to match weigh the two as a unit.) When piston and pin are match weighed as a unit, caution must be taken to keep the pin and piston together, and the owner should be notified, indicating that the various pins and pistons are not interchangeable. Weight removal is accomplished on the inside of the piston skirt. Weight should be removed from the pistons in the same place the piston manufacturer removed weight—usually on the inside surface of the piston skirt.

TABLE 12-1: IMPORTANT BALANCING WEIGHTS

Rotating Weight (100 Percent)

Weight of crank end rod	_____
Weight of crank end rod	_____
Weight of set of bearing inserts	_____
Weight of set of bearing inserts	_____
Weight of locknuts (if separate)	_____
Weight of locknuts (if separate)	_____
Weight of oil (estimate)	_____

Reciprocating Weight (50 Percent)

Weight of piston	_____
Weight of piston pin	_____
Weight of piston pin lock (if used)	_____
Weight of one set of piston rings	_____
Weight of piston end of connecting rod	_____
Bobweight total	_____

FIGURE 12-70   Removing weight from pistons *(Courtesy of Stewart-Warner, Inc.)*

Care should be taken not to weaken or compromise the strength of the piston. If an oil control ring is located at the lower end of the skirt, extreme care should be taken in removing weight so as not to seriously weaken the piston at its thinnest section. The piston weight should be recorded not only for determining the weight of the bobweights but also for future reference in case a piston must be replaced at some future date. Remember that tolerance should be all piston +1/2 gram or –1/2 gram.

Match Weighing Connecting Rod.   To achieve a good balance, first the crankshaft ends and then the pin ends of the rods must be made equal in weight. This match weighing is accomplished by using a precision scale, plus a rod weighing device. When mounting the rod on the adapter, care should be taken so that free movement of the adapter is not hindered by the bearings contacting an oil hole or parting line.

A grinder, or preferably a belt sander, should be used to remove weight from rods. The manufacturer provides balance pads of additional material for easier weight removal. (If pads are not available, all the grinding should be made along the length of the rod, never across from side to side. Because decreasing thickness affects the strength directly, decreasing the depth affects the strength many times more.) Weigh all of the crank ends of the connecting rods as shown in Figure 12-71. When the weight of the crank ends of the rods have been found, the heavier crank ends should be matched by removing weight (Figure 12-72) so that they are made equal in weight with the lightest crank end of the connecting rods. This weight should then be recorded for future reference and for determining the weight of the bobweight.

The pin ends of the connecting rod should be matched weighed using the same procedure as was

FIGURE 12-71   Weighing crank ends of connecting rod *(Courtesy of Stewart-Warner, Inc.)*

used on the crank ends of the rods (Figure 12-73). The weight of the pin end of the rods should be recorded for future reference and for determining the weight of the bobweight.

Making Up the Bobweight.   Bobweights are devices attached to the throws of the crankshaft to simulate the affect of the rod and piston assembly. Bobweights consist of two halves held together on the throws by two calibrated nuts on

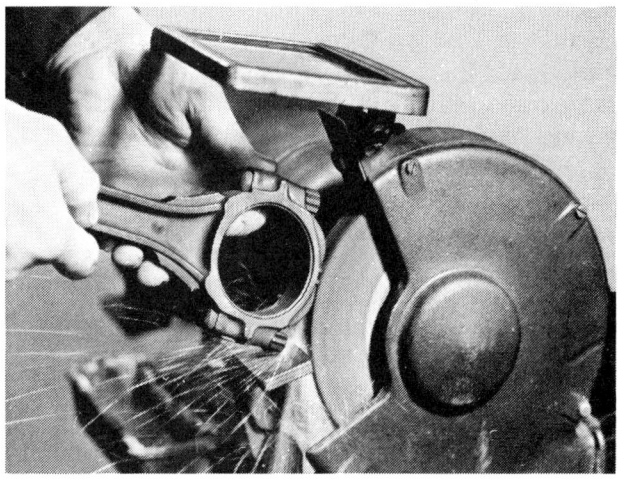

FIGURE 12-72   Removing weight from crank end of connecting rod *(Courtesy of Stewart-Warner, Inc.)*

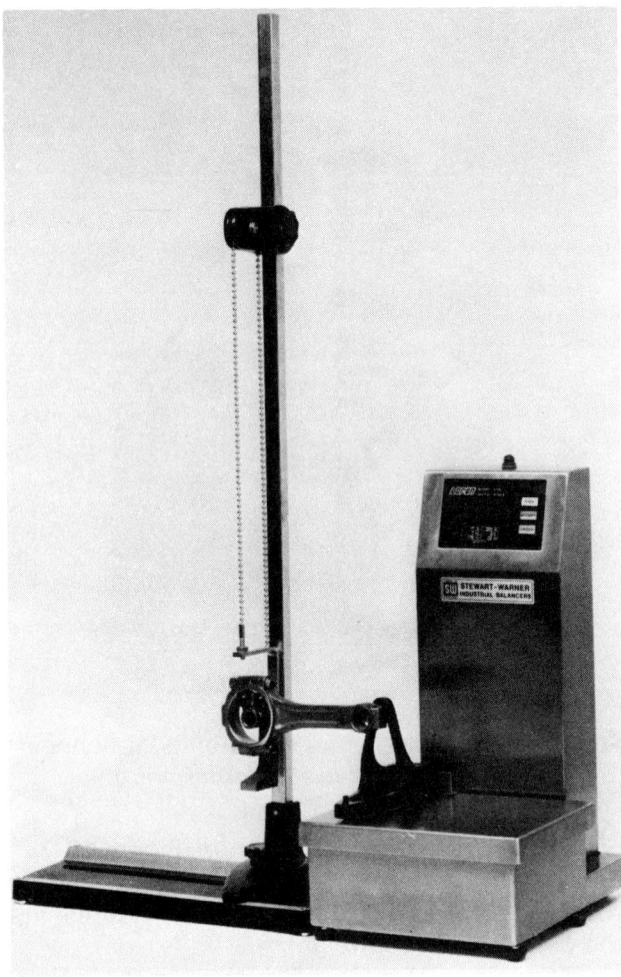

FIGURE 12-73 Weighing pin ends of connecting rods *(Courtesy of Stewart-Warner, Inc.)*

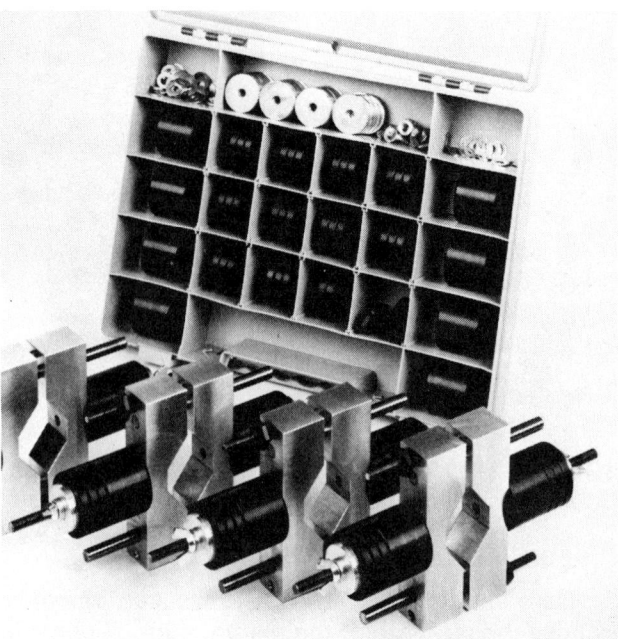

FIGURE 12-74 Assembling bobweights *(Courtesy of Hines Industries, Inc.)*

threaded studs (Figure 12-74). Calibrated nuts cannot be interchanged from bobweight to bobweight or stud to stud. (A letter on the calibrated nuts is matched to the correct half where a matching letter is stamped.) It is important that the calibrated nuts always be on their proper stud and bobweight half, or an unbalance can result. Each half has a threaded perpendicular stud and nut. The addition of matched weights to these studs increased the bobweight to its correct amount. Calibration of a bobweight can be checked by closing the bobweight and hand tightening the calibrated nuts. The calibrated nut reading should be the same on both nuts.

Place one complete bobweight on the scale and build two equal stacks of match weights until the scale reads the required total. Assemble the match-weight on the bobweight and replace the bobweight on the scale. Zero the scale and repeat the process for the remaining bobweights. Certain precautions

in assembling and mounting the bobweights are necessary. The matched weights should be placed symmetrically on the weight studs. The bobweight may be placed at any angle on the throw. However, for safety and ease of correction, they are usually aligned with the weight studs parallel to the nearest counterweight (Figure 12-75).

It is important that the match weights are centered through the throw. To obtain this when the halves are mounted on the throw, match the numbers on the calibrated nuts so that both nuts on the bobweight read the same. This automatically centers the match weights on the throw. The bob-

FIGURE 12-75 Bobweight mounted on crankshaft throw *(Courtesy of Stewart-Warner, Inc.)*

weights should be placed all to the right, left, or center. Any of the three ways is acceptable, but make sure that they are all in the same position.

Crankshaft Balancing Procedure. The crankshaft should be placed on the trunnion bearings and the bobweights installed. Adjust the crankshaft hold-down on the oil slinger so that the crankshaft will not shift and there is no drag. Follow the normal setup procedure described in the balancer machine's instruction manual. A typical procedure is as follows:

1. Once the crankshaft has been balanced to the correct tolerance, then the other components—flywheel, dampener, clutch pressure plate—are added, one at a time and balanced down to the correct tolerance. (Remember, if the flywheel and dampener are counterweighted, they must be on the crankshaft when it is balanced.)

2. Normally, corrections on the crankshaft are made on the counterweights using a 1/2-inch drill. However, there are circumstances when the counterweight will show light. Usually this is when modifications have been made to the engine such as stroking the crank or going from an external to an internal balance. If there are previous drill holes in the counterweight, add material such as lead weld or lead wool. If there are no holes in the proper position, drill (Figure 12–76) and add a heavier material. A very excellent material to add is a heavy metal, which is tungsten alloy. This material is millable and machinable and is approximately 1-1/2 times heavier than lead. There are several ways to add this material to the counterweights: drill and put a plug in the hole and weld a cap on, or drill and tap a hole and screw a plug into the counterweight. Probably the safest way is to drill a hole parallel to the axis of the crankshaft and press the plug in, then peen over the ends or pin it.

3. If the crankshaft is externally balanced, counterweighted flywheel and dampener, corrections can be made on the flywheel and dampener.

4. It is important that when balancing the flywheels and clutch pressure plate as an assembly on the crankshaft that they are indicated so that when the engine is assembled they will be in the same position as when they were balanced.

FIGURE 12-76   Drilling to add weight to the crankshaft *(Courtesy of Hines Industries, Inc.)*

5. Clutch pressure plates can be corrected by drilling on the studs on which the clutch springs are located or adding weight by welding or plugging. It is usually not necessary to balance the clutch disc, but if it is done it should be balanced on a mandrel and the correction made by grinding on the edge of the disc.

 SHOP TALK ————

Tolerance for performance and racing vehicles is 0.2 ounce inch per end, while street use cars are good at 0.5 inch/ounce per end. These same tolerances are used for each part added to the crankshaft when balancing and also when these parts are balanced by themselves.

In the last few years, the major automotive manufacturers have been concerned about annoying vibrations in small displacement engines. In six- and eight-cylinder engines the inertia forces of the reciprocating piston and rod assemblies can be offset as just described. This can be accomplished because of the closer crankshaft indexing. On four-cylinder engines this is not possible. To offset these forces some four-cylinder engines use one or two counterweighted balance shafts or gears turning in opposite directions and at twice the crankshaft speed. These counterweights provide equal but opposing

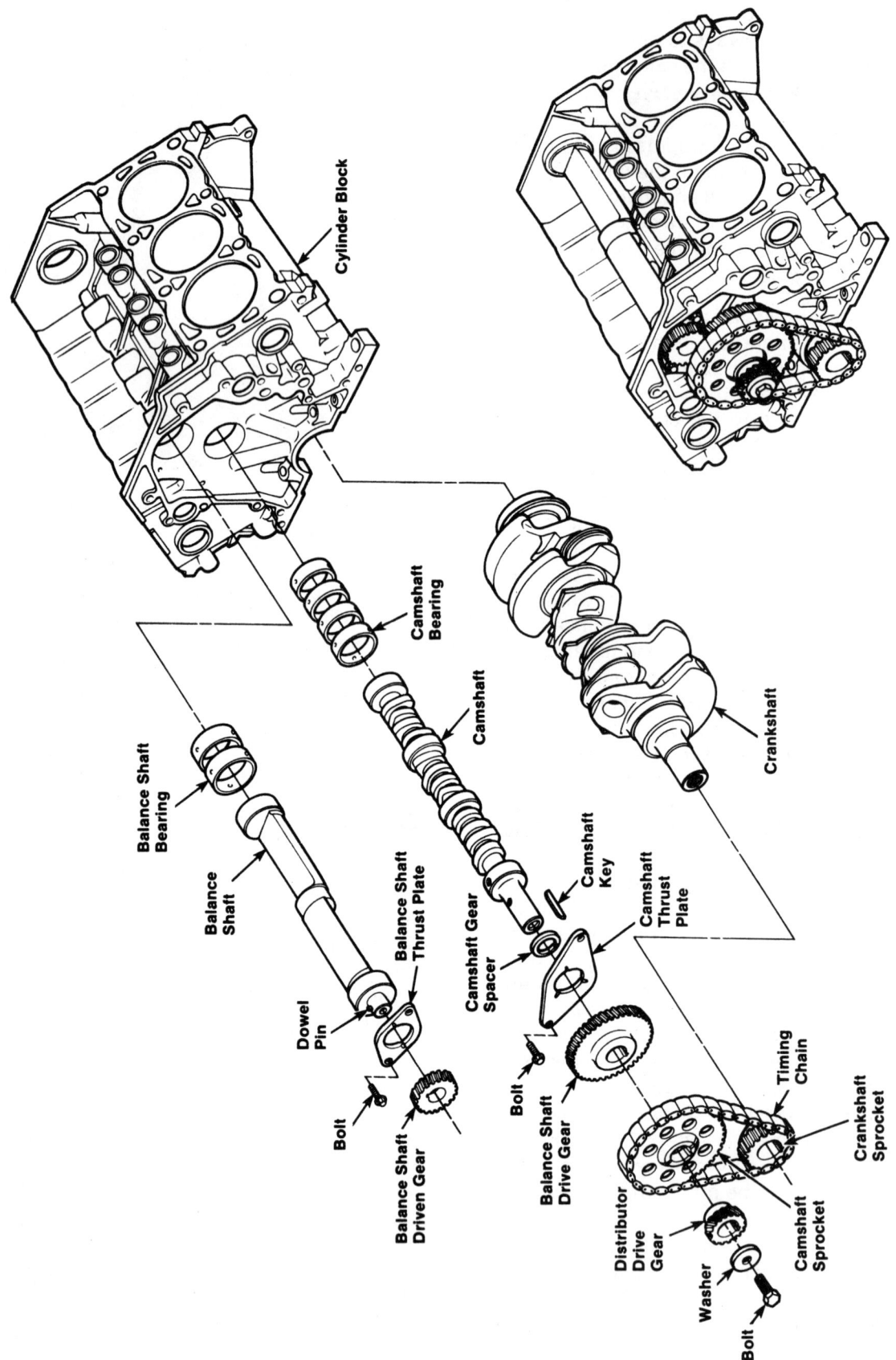

Cylinder Block

Camshaft Bearing

Camshaft

Crankshaft

Balance Shaft Bearing

Balance Shaft

Balance Shaft Thrust Plate

Camshaft Gear Spacer

Camshaft Key

Camshaft Thrust Plate

Dowel Pin

Bolt

Balance Shaft Driven Gear

Bolt

Balance Shaft Drive Gear

Timing Chain

Crankshaft Sprocket

Distributor Drive Gear

Washer

Camshaft Sprocket

Bolt

FIGURE 12-77 Single-balance shaft system

forces to engine vertical forces. The shafts or gears must be timed to crankshaft position.

As described and illustrated in Chapter 2, early balancing design contained two shafts. Turning at twice the engine speed, one of the shafts turned in the same direction as the crankshaft and the other turned in the opposite direction. The oil pump gears are used to drive the reverse turning shaft. Counterweights on the balance shafts are positioned to oppose the natural rolling action of the engine as well as the secondary vibrations caused by the piston and rod movements.

In the late 1980s, an engine with a single-balance shaft was introduced that fit neatly in the middle of the block above the camshaft (Figure 12-77). The balance shaft's moving mass helps to offset the reciprocating mass of the pistons and connecting rods to achieve the desired balancing effect. That is, the balance shaft is driven by the camshaft and crankshaft speed and is phased 180 degrees out of synchronization with the crankshaft so that the vibrations cancel each other out.

As already mentioned the flywheel and any other part attached to the crankshaft might require balancing. The method of balancing a flywheel varies according to the transmission. A heavy clutch plate can be drilled out at selected places to refine the balance. This reduces the heavy side. Often the torque converter drive plate of automatic transmission cars cannot be drilled out because it is too thin. Instead, flywheel balance clips are added to balance the flywheel by increasing the weight on the lighter side (Figure 12-78).

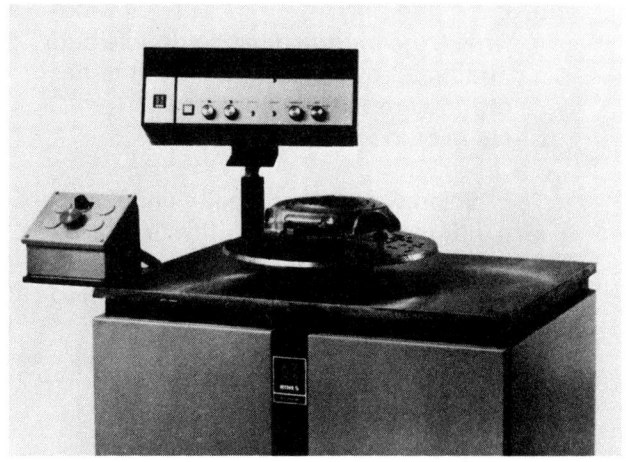

FIGURE 12-79 Balance arbor for flywheels and clutches

Some flywheels and clutches can be balanced independently on special balance arbors (Figure 12-79). But it should be kept in mind that some flywheels and harmonic balancers can be externally weighted as apart of a crankshaft balance and as such must be balanced as a part of the crankshaft assembly.

REVIEW QUESTIONS

1. Technician A says that crankshafts are generally classified by the number of main bearings. Technician B says that the shape of the crankshaft is not affected by the way the cylinders are numbered. Who is right?
    a. Technician A
    b. Technician B
    c. Both A and B
    d. Neither A nor B

2. Technician A says that acceptable alignment limits for a non-fillet-hardened crankshaft are usually twice as large as the limits for a fillet-hardened crankshaft. Technician B says the allowable bow for fillet-hardened crankshafts is 0.001 inch per foot of crankshaft length. Who is right?
    a. Technician A
    b. Technician B
    c. Both A and B
    d. Neither A nor B

3. Which of the following is false?
    a. Good grinding of a crankshaft requires a good coolant and its proper use.

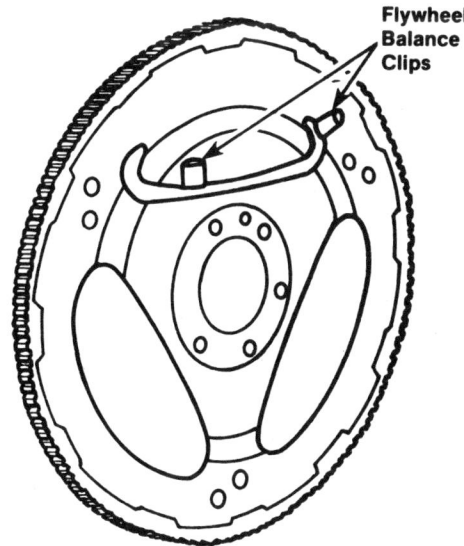

FIGURE 12-78 Balance clips are used to balance the thin style flywheel used with some automatic transmissions.

Flywheel Balance Clips

b. If the grinding wheel is not a full width wheel, the journal must be double cut.

c. The first areas of the crankshaft to be ground are the main bearings.

d. None of the above

4. Technician A says that the Melonite process is gaining favor over the Tufftriding process. Technician B says that both of these processes are types of the salt-bath nitriding process. Who is right?
   a. Technician A
   b. Technician B
   c. Both A and B
   d. Neither A nor B

5. Which of the following is true?
   a. A faulty vibration damper can only be replaced by an OEM unit.
   b. If a stone loads up and requires frequent dressing, it is probably too soft for the flywheel.
   c. Both A and B
   d. Neither A nor B

6. Which of the following is a cause of main bearing bore misalignment?
   a. warpage
   b. stretch
   c. spin
   d. all of the above

7. What is the general boring speed for large diesel blocks?
   a. 200 to 400 rpm
   b. 250 rpm minimum
   c. 300 rpm maximum
   d. 300 to 600 rpm

8. Camshafts are made of which of the following?
   a. low-carbon steel
   b. high-carbon steel
   c. iron
   d. all of the above

9. Technician A uses a ram press to straighten bent camshafts. Technician B uses the pressure from a blunted air chisel. Who is right?
   a. Technician A
   b. Technician B
   c. Both A and B
   d. Neither A nor B

10. Which of the following is false?
    a. A camshaft, like a crankshaft, can be built up by welding.
    b. A camshaft cannot be reground like a crankshaft.
    c. The first step in regrinding camshaft lobes is to make a master cam.
    d. None of the above

11. Which of the following is called valve overlap?
    a. The time when the exhaust valves of two consecutive cylinders are both open.
    b. The time when the intake valves of two consecutive cylinders are both open.
    c. Both a and b
    d. Neither a nor b

12. Technician A says that lobe spread is the average of the intake and exhaust lobe centers. Technician B compares the lobe spread to the intake lobe center to determine if the camshaft is advanced, retarded, or straight up. Who is right?
    a. Technician A
    b. Technician B
    c. Both A and B
    d. Neither A nor B

13. Which of the following is false?
    a. Altering the valve timing affects the compression ratio.
    b. Usually, the correct valve timing for any engine exists only at a given engine speed.
    c. Some engines have variable valve timing.
    d. None of the above

14. Which of the following requires bobweights?
    a. in-line engines
    b. V-type engines
    c. both a and b
    d. neither a nor b

15. Which of the following is true?
    a. If the flywheel and dampener are counterweighted, they must be on the crankshaft when it is balanced.
    b. The method of balancing a flywheel varies according to the transmission.
    c. Early crankshaft balancing designs contained two shafts.
    d. All of the above

CHAPTER THIRTEEN

LUBRICATING AND COOLING SYSTEMS

Objectives

After reading this chapter, you should be able to
- Name and describe the components of a typical lubrication system.
- Inspect and service an oil pump and engine bearings.
- Describe the crankcase ventilation system.
- Explain the oil service and viscosity ratings.
- List and describe the components of the cooling system.
- Describe the operation of the cooling system.
- Describe the function of the water pump, radiator, pressure radiator caps, and thermostats in the cooling system.
- Explain what an air cooling system is.

Proper operation of the engine's lubricating and cooling systems is important to its well-being. If a newly rebuilt engine does not supply oil or coolant to itself, all the work done by the rebuilder on the components of the engine will be quickly destroyed.

LUBRICATION SYSTEM

An engine's lubrication system (Figure 13–1) must perform several important functions:

1. Hold an adequate supply of oil
2. Remove contaminants from the oil
3. Deliver oil to all necessary areas of the engine

The main components of a typical lubrication system are:

- *Oil Pump.* The oil pump is the heart of the lubrication system. Just as the heart in the human body circulates blood through the body's veins, the oil pump circulates oil through the car's engine.

- *Oil Pump Pickup.* The oil pump pickup is a line from the oil pump to the oil stored in the oil pan (Figure 13–2). It usually contains a filter screen, which is submerged in the oil at all times. The screen serves to keep large particles from reaching the oil pump. This screen should be cleaned any time the oil pan is removed.

- *Oil Pan or Sump.* The oil pan attaches to the crankcase or block. It serves as the reservoir for the engine's lubricating oil. It is designed to hold all the oil necessary to lubricate the engine when it is running, plus a reserve. The oil pan helps to cool the oil through its contact with the outside air.

- *Pressure Relief Valve.* Since the oil pump is engine driven, an oil pressure relief valve (Figure 13–3) is included in the system to prevent excessively high system pressures from occurring as engine speed is increased. Once oil pressures exceed a preset limit, the spring-loaded pressure relief valve opens and allows the excess oil to bypass the rest of the system and return directly to the sump.

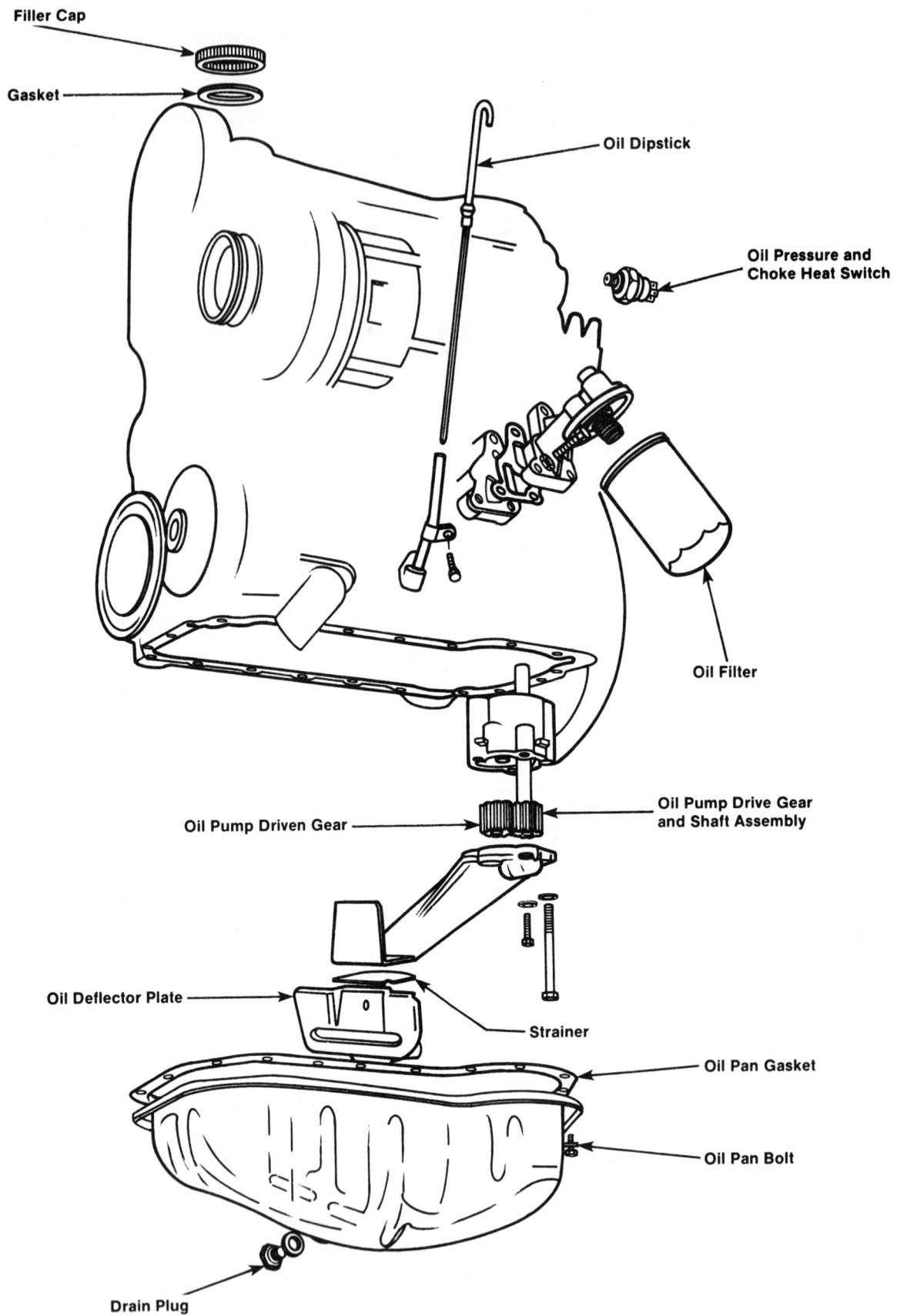

FIGURE 13-1 Components of a typical lubricating system

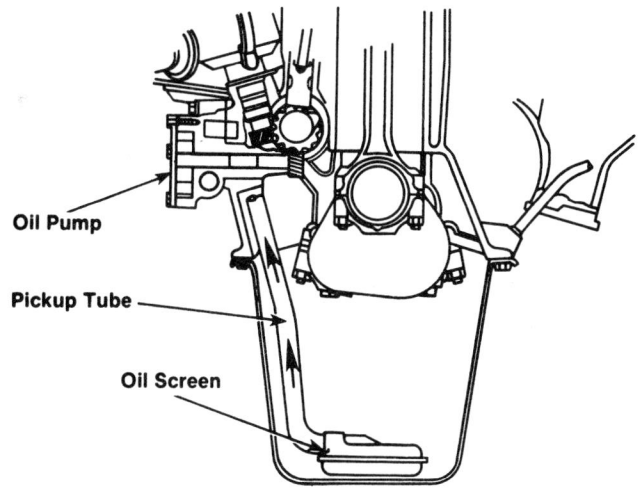

FIGURE 13-2 Oil pump pickup

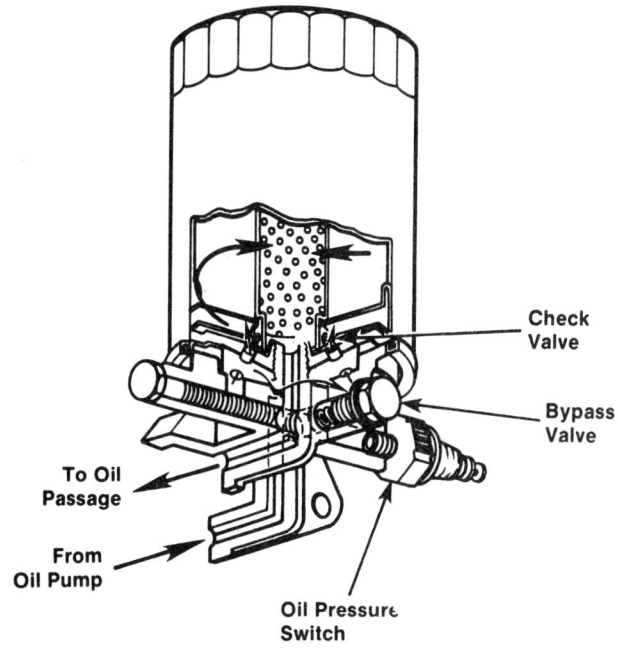

FIGURE 13-4 Typical oil filter circuit

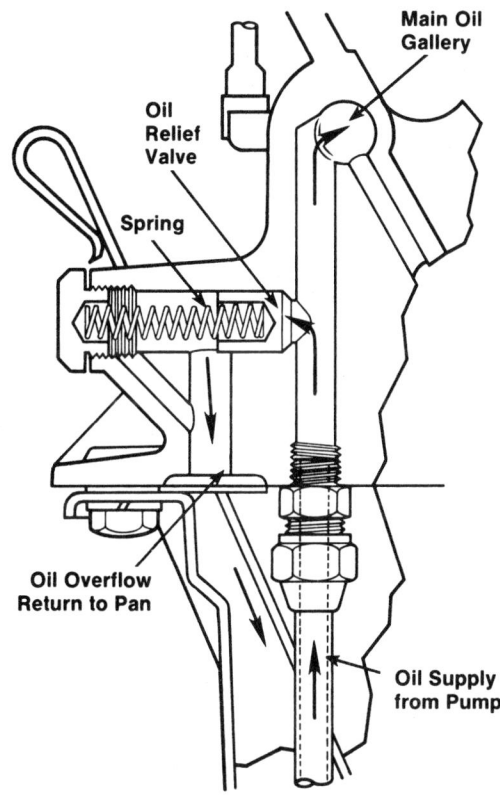

FIGURE 13-3 Oil pressure relief valve

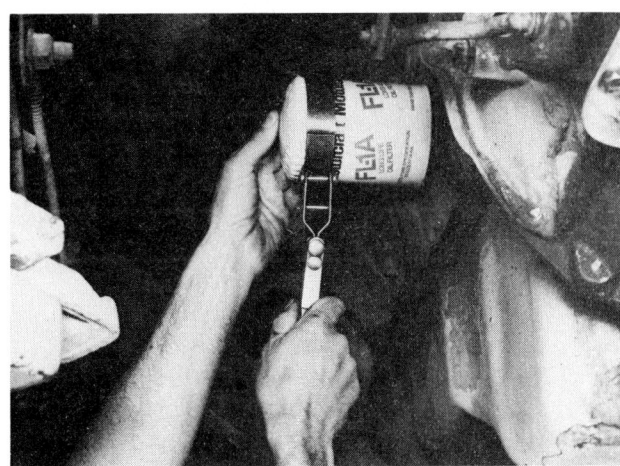

FIGURE 13-5 Typical method of installing an oil filter

• *Oil Filter.* Under pressure from the oil pump, oil flows through a filter (Figure 13-4) to remove any impurities that might have become suspended in the oil. This is necessary so impurities will not circulate through the engine and cause premature wear. Filtering also increases the usable life of the oil. Figure 13-5 illustrates the most common method of installing an oil filter.

• *Engine Oil Passages or Galleries.* From the filter, the oil then flows into the engine oil galleries (Figure 13-6). These galleries consist of interconnecting passages that have been drilled completely through the engine block during the manufacturing stage. The outside ends of the passages are then blocked off so that the oil can be routed through these galleries to various internal parts of the engine, including the engine

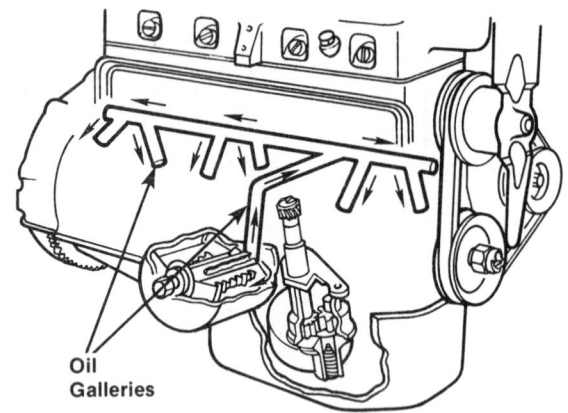

FIGURE 13-6 Oil galleries

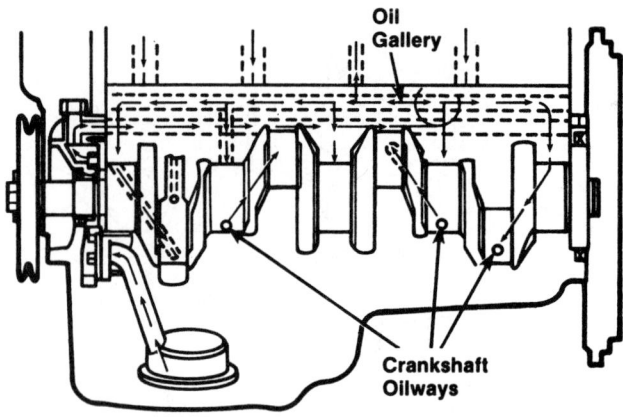

FIGURE 13-8 Oil throw-off

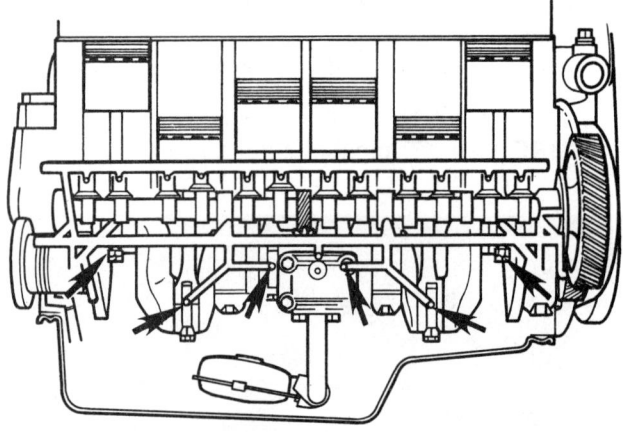

FIGURE 13-7 Oil paths in the engine

bearings. The crankshaft also contains oil passages (oilways) to route the oil from the main bearings to the connecting rod bearing surfaces (Figure 13-7).

- *Engine Bearings.* Since oil is delivered to the engine bearings by an oil gallery, an oil hole is machined in the bearing for alignment with the oil gallery in the engine block. In this manner, the bearing oil clearance space receives a constant supply of oil. In addition, as mentioned in Chapter 7, many engine bearings are manufactured with an oil groove to help distribute the oil over the surface of the bearing. Once the oil has been used by the bearing it flows out of the oil clearance space and is replenished by a fresh supply of oil under pressure from the engine oil pump. This oil is then thrown off the bearing surface by the spinning motion of the rotating engine part. This oil throw-off then lubricates many other parts of the engine, such as the cylinder walls and pistons (Figure 13-8).

- *Crankcase Ventilation.* Crankcase ventilation is necessary because of pressure that can build in the crankcase due to combustion pressure, which passes by the piston rings. Piston rings do not provide a complete positive seal of the combustion area and some combustion gases reach the oil pan. These gases can contaminate the oil and apply unwanted pressure to gaskets and seals.

- *Oil Pressure Indicator.* This system can be in the form of a gauge, which indicates the engine oil pressure at all times, or it can be a warning light that will come on whenever the engine is running with insufficient oil pressure. The warning light is the most common oil pressure indicator.

- *Oil Seals and Gaskets.* These are used throughout the engine to prevent both external and internal oil loss. The most common materials used for sealing are synthetic rubber, soft plastics, fiber, and cork. In critical areas these materials might be bonded to metal.

- *Dipstick.* The dipstick is used to measure the level of oil in the oil pan. The end of the stick is marked to indicate when the engine oil level is correct. It has a mark to indicate the need to add oil to the system.

- *Oil Coolers.* Some engines, mainly the diesel types, use an oil cooler. In passenger cars, it usually mounts on the cylinder block of the engine (Figure 13-9). Engine coolant is circulated through the oil cooler to remove excess heat from the oil passing through it on the way to the engine. Normal maximum engine oil temperature is considered to be 250 degrees Fahrenheit. Hot oil combining with oxygen will break down (oxidize) and form carbon and varnish. The higher the temperature, the faster these deposits build. An oil cooler helps keep the oil at its normal operating temperature.

CAUTION: If an engine equipped with an oil cooler experiences any internal metal failure, it is important that the cooler core be replaced when the engine is rebuilt.

OIL PUMPS

The oil pump is usually located in the oil pan. Its function is to supply lubricating oil to the various moving parts in the engine. To do this, two things are necessary: volume and pressure. The pump is designed to displace oil with more than adequate volume and pressure under all operating conditions. Both volume and pressure are directly related to the bleed-off through the clearances at the various parts of the engine. Engine bearing clearances have the greatest effect on oil flow. As the clearances increase so does the flow. However, when the flow increases beyond the capacity of the pump, then the pressure drops since there is no longer surplus capacity for pressure control by the regulating valve.

Most oil pumps incorporate a pressure-regulating or relief valve in the pump body, which limits the system pressure. A few engine designs have a relief valve in the block. The regulating valve consists of a ball or piston-type valve backed by a calibrated spring. This valve moves off its seat as oil pressure increases and uncovers a bypass port, which allows the excess oil to recirculate in the pump or dump

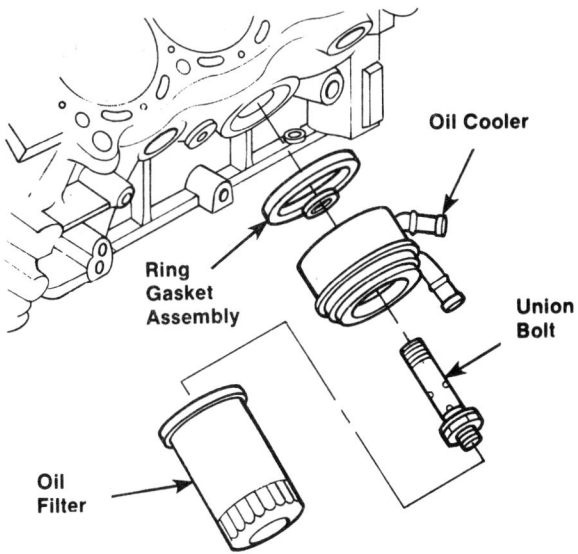

Ring
Gasket
Assembly

Oil Cooler

Union
Bolt

Oil
Filter

FIGURE 13-9   Location of typical oil cooler

into the pan. Valve opening is calibrated to maintain a designated level of oil pressure throughout the engine's normal operating range.

Oil under pressure is supplied to the main, connecting rod, and cam bearings and, in some engines, to the valve train. Also, in most engines, pressurized oil is routed to the hydraulic lifters and timing chain's sprocket or gears. Oil is put under pressure to ensure that a sufficient amount reaches the bearings and other pressure lubricated parts. Pressure in the lubrication system builds up because of resistance to the flow of oil caused by restrictions in the flow path from the pump through the part being lubricated. These restrictions are clearances between the engine bearings and journals as well as clearances in the valve train. For example, bearing oil clearances range from less than 0.001 inch to nearly 0.006 inch, depending on the engine. Since the pump delivers more than an adequate volume of oil to satisfy the engine's needs, it exerts pressure on the oil in the lubrication system's passages, which it attempts to compress. The pressure-regulating valve maintains pressure within the specified range by opening and bypassing a portion of the output to the inlet side of the pump or to the pan when pressure exceeds the range and by closing when it drops.

The primary cause of a drop in oil pressure is main and connecting rod bearing wear or damage, which increases the oil clearance. However, other parts mentioned, which are lubricated by pressurized oil, are also contributing factors as they wear. Increased clearances reduce the resistance to oil flow and, consequently, increase the volume of oil circulating through the engine while, at the same time, causing pressure throughout the operating range to decrease. The excess flow capacity of the pump is a safety measure to ensure lubrication of vital parts as the engine wears. Too much oil pressure is seldom a problem, but too little pressure can cause oil starvation in those pressure lubricated parts, which are more distant from the pump.

Engine oil pressure is also dependent upon oil viscosity, which is defined as the internal flow resistance of oil, and is temperature related. A high-viscosity oil has more flow resistance than a low-viscosity oil. Viscosity decreases as the temperature increases. (A more detailed look at viscosity is given later in this chapter.) For this reason, oil pressure is higher in a cold engine than it is when the engine is fully warmed up and has reached its normal operating temperature. The oil grade is related to its viscosity. For example, a 50W oil possesses a higher viscosity than a 10W oil. So it can be expected that an engine operating with 50W oil would have higher oil pressure than an engine operating with 10W oil.

It can be expected, then, that as vehicle mileage increases, oil pressure will decrease because engine wear causes an increase in operating clearances. Some United States automotive manufacturers' shop manuals specify oil pressure at a particular engine speed (for example, 2000 rpm). It will decrease as the engine speed drops to idle. As a suggested guideline, minimum safe oil pressure should be at least 10 psi per 1000 rpm. For example, minimum oil pressure at 2500 rpm would be 25 psi, while at a 700 rpm idle it would be 7 psi. Keep in mind that this is only a guide. Some engines can get by with less pressure; others might require more. Refer to a shop manual for more exact specifications.

Type of Oil Pumps

The most commonly used stock oil pumps are the rotor (gerotor) and gear type. Both are positive displacement pumps. That is, a fixed volume of oil passes through the pump with each revolution of its drive shaft. This is because the gears or rotors form a near perfect mechanical seal as they mesh, trapping fixed volumes of oil inside the pump and then pushing it out. Output volume is proportional to pump speed. So as engine rpm increases, pump output also increases. A pump is designed with an output that exceeds the engine's requirements so it will have sufficient capacity as the engine wears. Oil pumps usually rotate at camshaft speed, since in most engines the oil pump and distributor drive are connected.

A typical automotive rotor oil pump (Figure 13–10) consists of a four-lobe inner rotor, usually driven by the camshaft, and a five-lobe outer rotor, which is driven by the inner rotor. As the turning rotor lobes unmesh, oil is drawn in from the pan. Oil is trapped between the lobes, cover plate, and top of the pump cavity and moved to the outlet where the meshing lobes force the oil out. Output per revolution depends upon rotor diameter and thickness.

Gear pumps (Figure 13–11) consist of a drive gear connected to the input shaft and a driven gear. The drive gear turns the driven gear. Both gears trap oil between their teeth and the pump cavity wall, moving it from the inlet to the outlet. Output volume per revolution depends upon gear length and tooth depth. Another style of gear-type oil pump uses an idler gear with internal teeth that spins around the drive gear. In this style of pump, often called a *crescent* or *trochoidal* type, the gears are eccentric. That is, as the larger gear turns, it walks around the smaller, moving the oil in the space between.

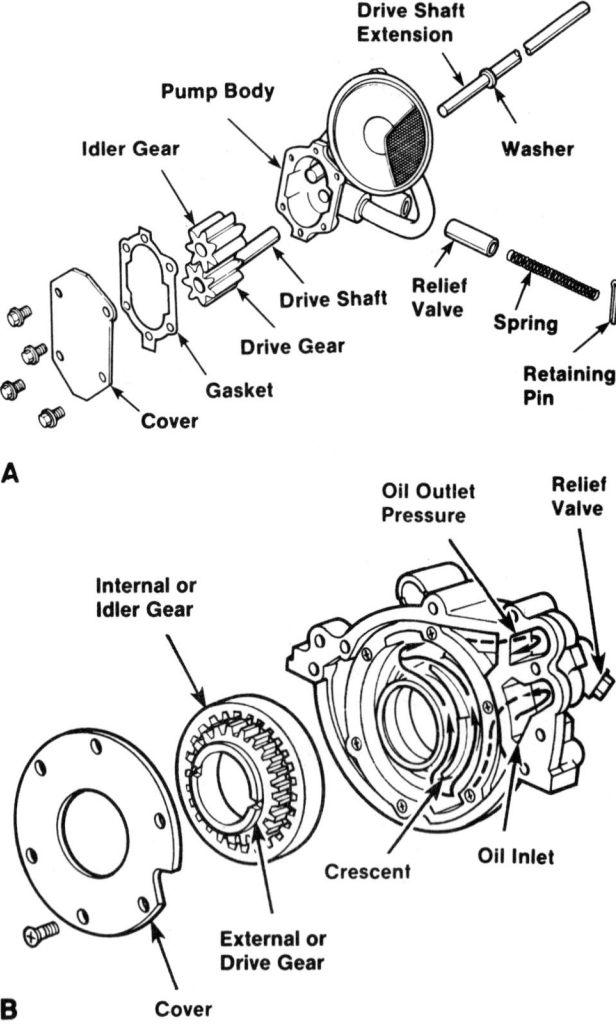

A

B

FIGURE 13–11 Two other popular types of gear-driven oil pumps: (A) external gear pump and (B) internal/external gear pump

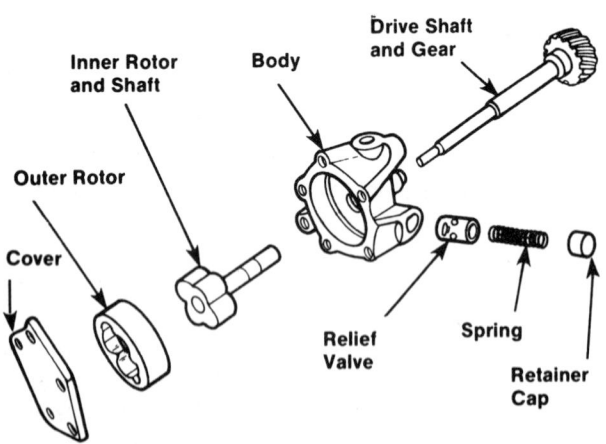

FIGURE 13–10 The rotor or gerotor oil pump

The rotor type pumps a greater volume of oil than a gear type because there is more room inside the open lobe of the outer rotor than there is room between the teeth of the gears of a gear-type pump. This means a greater volume of oil is sent to the engine.

The output volume of a high-volume oil pump is 20 to 25 percent more than a corresponding standard or stock replacement pump, depending on the application. This is accomplished by lengthening the pump gears or rotors, which increases the internal volume and, consequently, the pump output. Oil pressure will also be slightly higher. High-volume oil pumps are primarily used in engines where the bearing clearances have been increased. Increased bearing clearances allow more rapid oil bleed-off, which reduces oil pressure. Under these conditions, a high-volume pump maintains oil pressure at a safe level, while the increased oil flow also improves bearing cooling. It is usually recommended to increase the oil pan capacity by 1 or 2 quarts when using a high-volume oil pump. This assures an adequate oil supply at all times.

A high-pressure oil pump has a stiffer pressure relief valve spring than the stock pump. High-pressure pumps provide increased oil pressure but do not increase the volume of oil flowing through the engine. They are used in selected factory and modified performance engines.

Oil pumps are driven off the camshaft, some directly, others indirectly, by means of an intermediate shaft connected to the distributor (Figure 13–12). The oil pump is usually located in the oil pan. On a few engines the pump housing is an integral part of the timing gear cover but the gears are the same. Some pumps are mounted on the side of the engine; others are mounted internally.

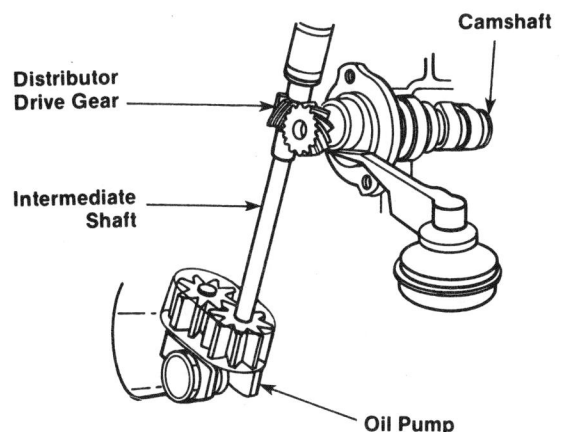

FIGURE 13–12   Gear on the end of this camshaft drives the distributor.

Pressure Regulation

As shown in Figure 13–10 and Figure 13–11, a pressure gauge and a valve have been added on the output side. The pressure gauge represents the oil pressure in the engine, and the valve is put there to represent the bearing clearances. As the valve is opened, the flow increases and the pressure drops. As the valve closes, the flow is reduced and the pressure increases.

Some means must also be provided to control the maximum oil pressure because the faster the pump turns the greater the pressure becomes. Therefore, a pressure-regulating valve is installed. Different types of designs are used; however, their function remains the same. The valve is loaded with a closely calibrated spring that allows oil to bleed off at a given pressure. If the engine manufacturer decides that a 50 psi oil pressure is desirable in the engine, then the pressure-regulating valve will not allow the pressure to go beyond 50 psi. When the pressure on the output side of the pump reaches this point, it presses against either a check valve, a ball, or a plunger, forcing it to unseat and allow oil to bypass and return to either the inlet side of the pump or to the crankcase.

Figure 13–13A is a cutaway of an oil pump showing the pressure-regulating valve in the *closed* position. Figure 13–13B is the same pump with the valve in the *open* position. Note how the oil can flow through the end of the plunger to the bypass and return to the inlet side of the pump. The plunger works within its bore with very little clearance; therefore, it can be seen that foreign material entering this area can jam or hinder the operation of the pressure-regulating valve.

On most engines made in recent years there is an oil filtering system somewhere on the output side of the pump. In some cases the filter is attached right to the pump; in other cases, it is attached in various locations depending on the manufacturer.

Oil Pressure Indicators

All automotive vehicles are equipped with either an oil pressure gauge or a low-pressure indicator light. An oil pressure indicator warns the driver of low oil pressure. The gauges are either mechanically or electrically operated.

In a mechanical system, oil travels up to the back of the gauge where a springy, flexible, hollow tube, called a *Bourdon tube,* uncoils as the pressure increases. A needle attached to the Bourdon tube moves over a scale to indicate the oil pressure.

Most pressure indicators found in vehicles today are electrically controlled (Figure 13–14). An oil

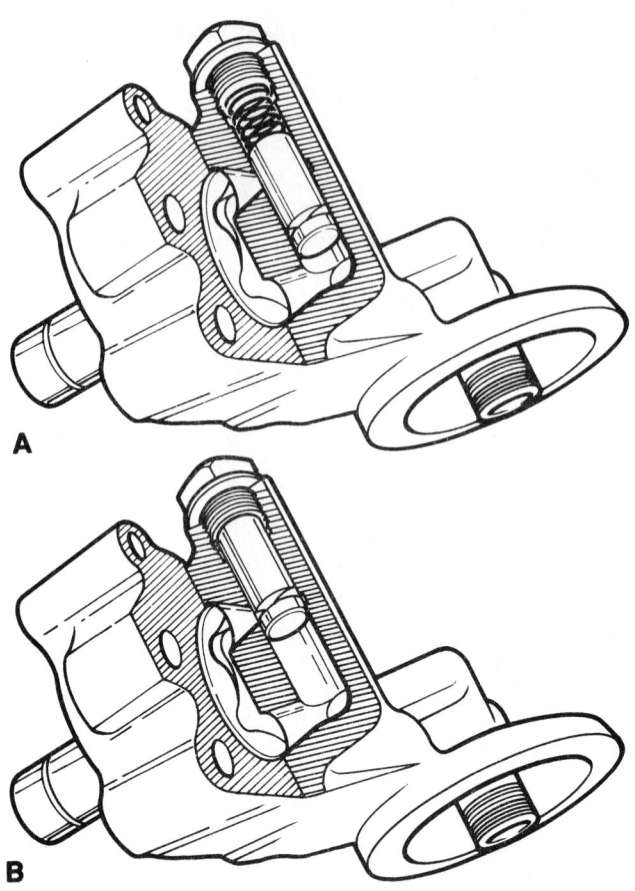

FIGURE 13-13 Cutaway of an oil pump showing the pressure-relief regulating valve in the (A) *closed* position and (B) in the *open* position. *(Courtesy of TRW, Inc.)*

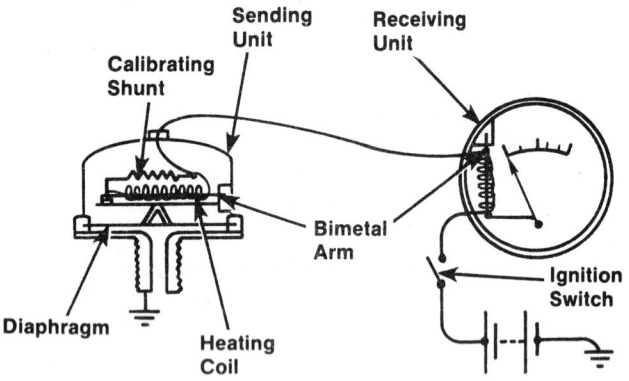

FIGURE 13-14 Schematic of an electric oil pressure indicator

pressure sensor is screwed into an oil gallery. As oil passes through an oil pressure sender (Figure 13-15), it moves a diaphragm, which is connected to a variable resistor. This resistor lowers the amount of current passing through an electrical circuit. A gauge on the dashboard reacts to the amount of

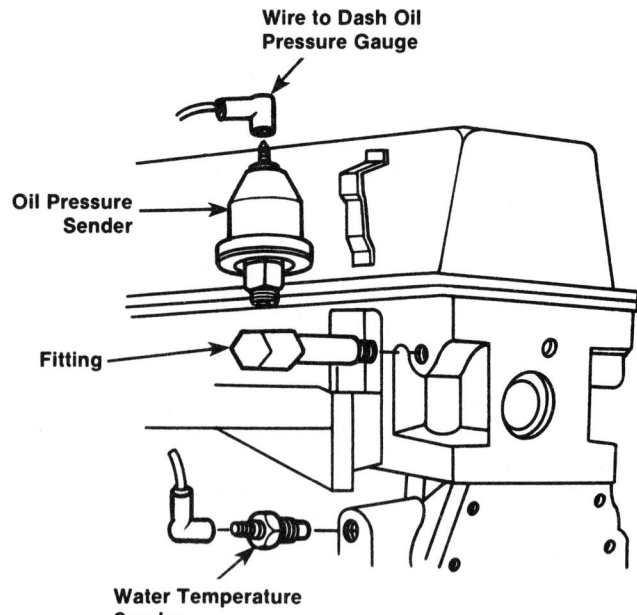

FIGURE 13-15 Oil pressure gauge sender

current passing through the electrical circuit and moves a needle over a scale to indicate the oil pressure.

In vehicles equipped with a warning light system, a diaphragm is connected to a sender switch or sensor. Under normal conditions, the sender switch is open. When oil pressure falls below the level necessary for safe operation, the lessening of pressure moves the diaphragm to close the sender switch (Figure 13-16), completing the electrical circuit. When this occurs, electricity flows and activates the warning light on the dashboard.

Some late-model vehicles are equipped with an electronic oil level indicator. It operates in the same manner as an electric light system except that a relay assembly is used in place of a diaphragm.

Oil Filtration System

Four types of oil filtration systems are commonly used.

- *Full Flow.* This type is used on most engines today (Figure 13-17). All of the oil going to the engine bearings goes through the filter first. However, should the filter become plugged, the relief valve contained in the filter will open and allow oil to bypass and go directly to the bearings, thus providing the bearings and the rest of the engine with necessary, though unfiltered, lubrication.
- *Bypass.* The bypass system is the least complex type of filtration. In Figure 13-18, note

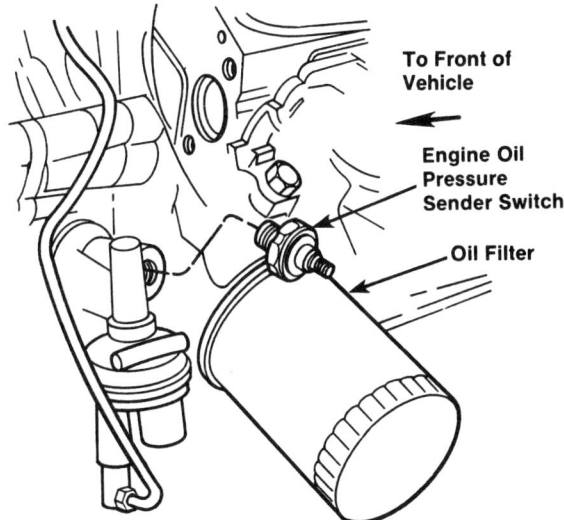

FIGURE 13-16 Oil pressure warning light sender

that on the output side of the pump the oil is fed directly to the engine bearings, the pressure-regulating valve, and the oil filter. The filtered oil is then returned to the oil pan. The oil is not filtered before it is sent to the bearings. Should the filter become plugged, no oil will flow through it, and the oil that is not used in the engine bearings will be bypassed through the pressure-regulating valve.

◉ SHOP TALK ⸻

In the bypass system, approximately 90 percent of the oil is pumped directly to the engine bearings and other moving parts, then drains back to the oil pan. The remaining 10 percent of the oil is diverted into the filter housing, where it is cleaned as it passes through the filter cartridge and out through the orifice to the sump.

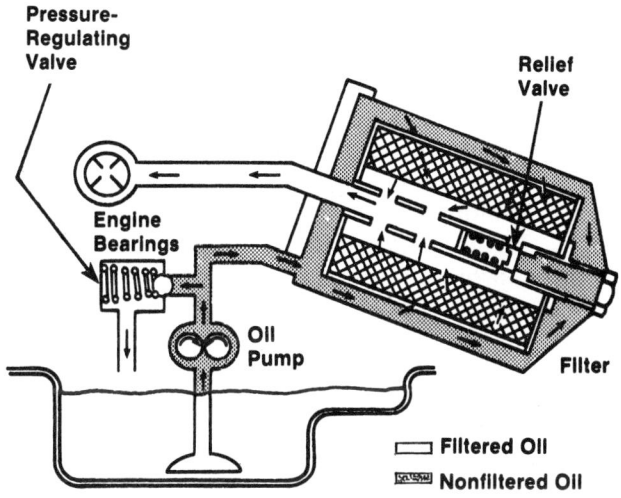

FIGURE 13-17 Full flow filtration system

- *Independent.* The independent system makes use of a separate filtration circuit, including its own auxiliary oil pump. A portion of the oil supply is pumped directly through the filter for cleansing and then returned to the crankcase.
- *Shunt.* The shunt-type system shown in Figure 13-19 is a little more sophisticated; however, it is only supplying partially filtered oil to the bearings. A mixture of filtered and unfiltered oil blends together at the end of a passageway or termination of the shunt.

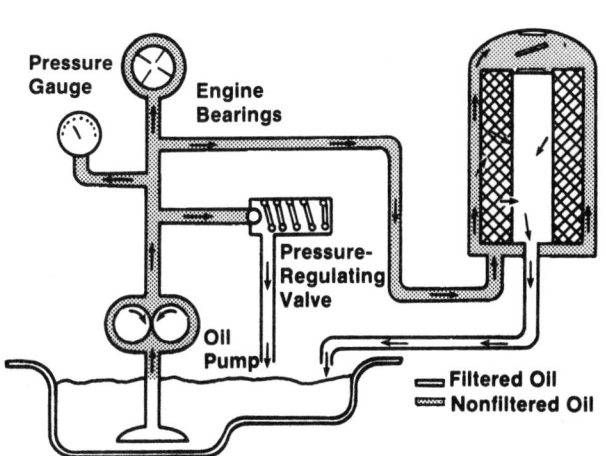

FIGURE 13-18 Bypass filtration system

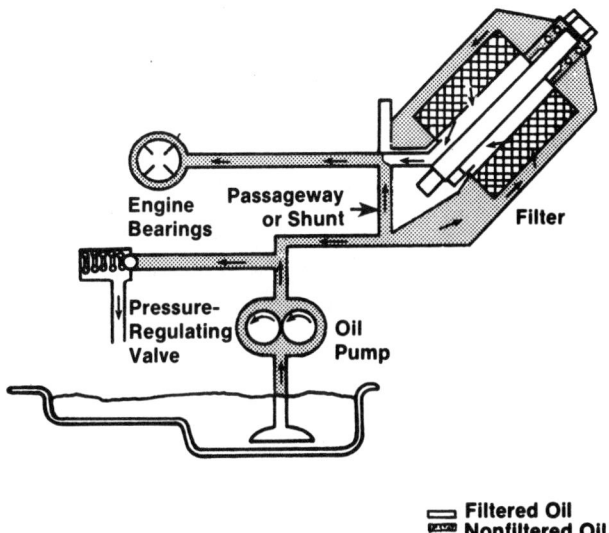

FIGURE 13-19 Shunt-type filtration system

Should this filter become plugged, oil will flow directly through the shunt passageway to the bearings.

 SHOP TALK _____

The full-flow system is almost identical to the shunt system. The only difference is the filter for the full-flow system must contain sufficient cartridges for a given unit to permit all the oil entering the filter to pass through the cartridges without bypassing through the relief valve under normal operating conditions. The bypass relief valve is designed to operate only in an emergency should the filter cartridges become plugged with contaminants.

Further information on oil filters is given later in the chapter.

Oil Pump Inspection and Service

Many engine rebuilders install a new or rebuilt oil pump (depending on the type) on each engine they rebuild. Integral pumps must be rebuilt. Bolt-on or nonintegral pumps can be rebuilt or replaced. If the old pump is to be reused, it should be carefully inspected for wear and thoroughly cleaned.

If the oil pressure tests low during diagnostic checks, any number of problems could be the cause, including:

- Worn or damaged oil pump
- Defective sending unit (Figure 13–20)
- Weak or damaged pressure relief valve spring
- Plugged oil pickup screen (Figure 13–21)
- Relief valve stuck in the *open* position

FIGURE 13-21 Plugged oil pickup screen

- Excessive oil dilution
- Buildup of sludge and dirt (Figure 13–22)
- Excess bearing clearances (crankshaft and camshaft)
- Air leak on the suction side of the oil pump (Figure 13–23). Since the air is compressible, it will cause the indicator to fluctuate and, in some cases, can cause the pressure-regulator valve to hammer back and forth. Under prolonged operation this can cause the valve to fail. It can also cause oil aeration, foaming, marginal lubrication, and failure of engine parts. Care should be exercised to make sure that all parts on the suction side fit tightly and that there is no place for air leakage. Air leakage often comes from cracked seams in the pickup tube.

Even though the oil pump is probably the best lubricated part of the engine, it is lubricated before the oil passes through the filter. As a result, it is subject to premature failure caused by foreign mate-

FIGURE 13-20 Defective sending unit

FIGURE 13-22 Sludge and dirt buildup

FIGURE 13-23   Air leak on the oil pump's suction side

rial entering the close tolerance area. Foreign particles can cause three kinds of trouble in a pump.

1. Fine abrasive particles gradually wear the surfaces, causing a reduction in efficiency.
2. Hard particles larger than the clearances can cause scoring and raising of metal as they pass through, finally resulting in seizure.
3. Large particles that cannot pass through will physically lock up the pump.

Of course, when the pump seizes or locks up, the intermediate or drive shaft is twisted off or sheared (Figure 13-24).

During normal operation the bypass valve seats on the cross strap (Figure 13-25), but when there is a demand for a large quantity of oil and the oil is cold and thick, the valve will unseat and allow the oil to bypass the screen and go directly into the pump. Of course, if the pump screen becomes plugged the bypass valve will be open during the majority of the

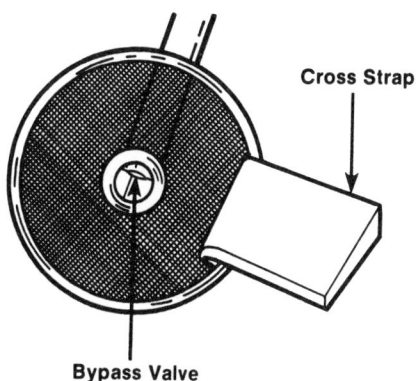

FIGURE 13-25   Pickup screen with cross strap bent out of position to show the bypass valve

time the engine is in operation. As the oil rushes through the bypass valve, it can create a vortex in the oil pan, which could draw up debris that is either floating in the oil or lying in the bottom of the pan. It should also be noted that much of this debris can be passed on through the pump.

To thoroughly inspect the oil pump, it must be disassembled. Carefully remove the pressure relief valve and note the direction in which it is pointing so that it can be reinstalled in its proper position (Figure 13-26). If the relief valve is installed backwards, the pump will not be able to build up pressure.

Before disassembling the pump, mark the gear teeth so that they can be reassembled with the same tooth indexing (Figure 13-27). Some pumps have the gears or rotors marked when they are manufactured. Once all the serviceable parts have been removed, clean them and dry them off with compressed air.

FIGURE 13-24   Sheared drive shaft

FIGURE 13-26   Remove pressure relief valve and note direction in which it is pointing.

FIGURE 13-27 Mark the gear teeth so they can be reassembled with the same indexing.

After the pump has been disassembled and cleaned, inspect the pump gears or rotors for chipping, galling, pitting, or signs of abnormal wear (Figure 13-28). Examine the housing bores for similar signs of wear. If any part of the housing is scored or noticeably worn, replace the pump as an entire assembly.

Check the mating surface of the pump cover for wear. If the cover mating surface is worn, scored, or grooved, replace the pump. Use a feeler gauge and straightedge to determine the cover flatness. The service manual gives the maximum and minimum acceptable feeler gauge thicknesses for the cover. If the cover is excessively worn, grooved, or scratched, it should be replaced.

Use an outside micrometer to measure the diameter and thickness of the outer rotor (Figure

13-29). The inner rotor's thickness should also be checked with an outside micrometer (Figure 13-30). If these dimensions are less than the specified amount given in the service manual, the rotors must be replaced.

With gerotor pumps, assemble the rotors back into the pump body, then use a feeler gauge to check the clearance between the outer rotor and pump body (Figure 13-31). If the manufacturer's specifications are not available, replace the pump or rotors if the measured clearance is greater than 0.012 inch.

After checking the outer rotor-to-pump housing clearance, position the inner and outer rotor lobes so that they face each other. Then measure the clearance between them with a feeler gauge (Figure 13-32). A clearance of more than 0.010 inch is unacceptable.

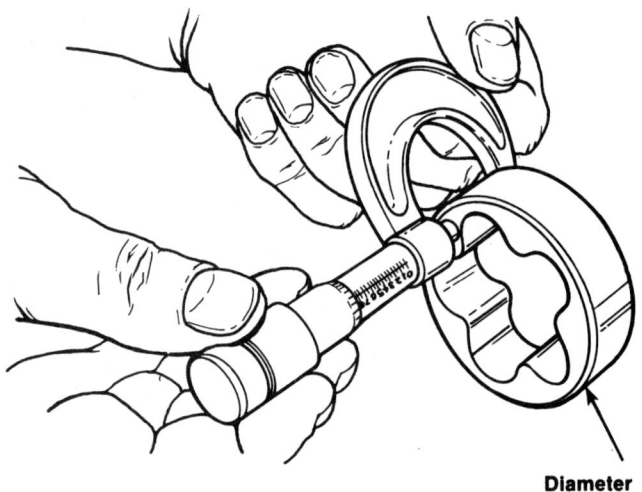

Diameter

FIGURE 13-29 Measuring the outer rotor

FIGURE 13-28 Inspect the pump gears or rotors for chipping or galling.

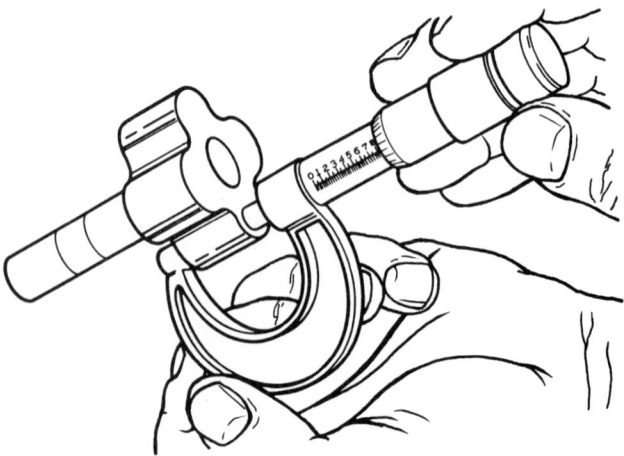

FIGURE 13-30 Measuring the thickness of the inner rotor

FIGURE 13-31 Checking clearance between the outer rotor and pump body

FIGURE 13-33 Taking several measurements around the housing

FIGURE 13-32 Measuring clearance between the inner and outer rotor lobes

FIGURE 13-34 Measuring clearance between a straightedge and gears

On a gear-type pump, it is important to measure the clearance between the gear teeth and pump housing. Take several measurements at various locations around the housing (Figure 13-33) and compare the readings. If the clearance at any point exceeds 0.005 inch, replace the pump as an assembly.

On both gear or rotor oil pumps, place a straightedge across the pump housing and measure the clearance between the straightedge and gears (Figure 13-34). To insure an accurate reading, make sure the housing surface is clean and free of residual gasket material and that the gears are bottomed in the bore. The desired end play clearance should not exceed 0.003 inch.

If the pump uses a hexagonal drive shaft, inspect the pump drive and shaft to make sure the

corners are not rounded (Figure 13-35). Check the drive shaft-to-housing bearing clearance by measuring the OD of the shaft and the ID of the housing bearing.

The gasket that is used to seal the end housing is also designed to provide the proper clearance between the gears and end plate. Consequently, do not substitute another gasket or make a gasket to replace the original one. If a precut gasket was not originally used, seal the end housing with a thin bead of anaerobic sealing material (Figure 13-36).

Inspect the relief valve spring for a collapsed or worn condition. Check the relief valve spring tension according to specifications for the specific engine. Also check the relief valve piston for scores and free operation in its bore.

The pickup screen and pump drive (Figure 13-37) should be replaced when an engine is rebuilt.

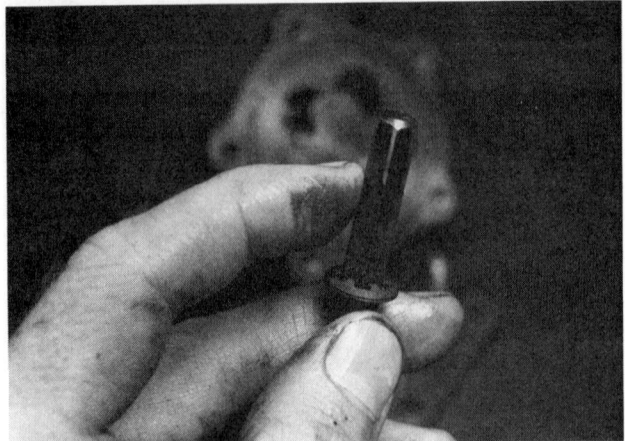

FIGURE 13-35 Inspecting pump and shaft for round corners

FIGURE 13-36 Sealing end housing with anaerobic sealing material

FIGURE 13-37 Pickup screen and pump drive

The screen and drive must be properly positioned. This is important to avoid oil pan interference and to ensure that the pickup is always submerged in oil. To make the oil pump pickup tube installation easier, there are several types of drivers available that are suitable for use with air-powered equipment or with a light mallet (Figure 13-38).

Check the pan for cracks, holes, damaged drain plug threads, and a loose baffle. Check the gasket surface for damage caused by overtightened bolts. Straighten the surface as required to restore original flatness. Replace the pan if repairs cannot be made.

On integral pumps, the timing case and gear thrust plate might also be worn. This will limit pump efficiency due to excess clearance. So, replace them as necessary. These oil pumps should be packed with petroleum jelly when they are rebuilt; otherwise, the pump will not be able to prime itself to begin pumping.

 SHOP TALK _____

Oil pump rebuilding kits are available. Figure 13-39 shows a standard application rebuild kit for gear-driven oil pumps in which the pump housing is part of the engine block or timing case. These kits contain a gear drive, two gears, an assortment of color-coded springs for the appropriate relief valve pressures, and complete installation instructions.

There are some other components that can be considered part of the lubrication system that help to assist in increasing engine performance. In newer engines, the baffle assembly (Figure 13-40) is one of these components. It is used to reduce windage in the oil sump.

FIGURE 13-38 Manual installation of pump pickup tube *(Courtesy of Goodson Shop Supplies)*

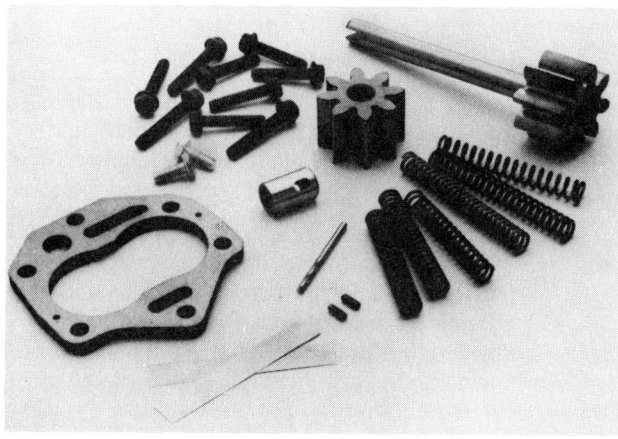

FIGURE 13-39 Parts of a rebuilding kit for gear-driven oil pump *(Courtesy of Perfect Circle/Dana)*

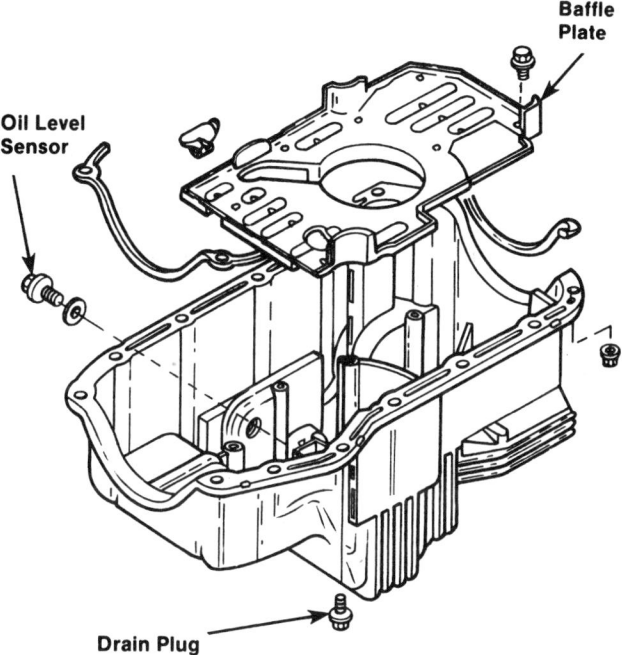

FIGURE 13-40 Oil pan assembly with baffle plate

 SHOP TALK

Windage is defined as the disturbance of air around a moving object, which, in this case, is the crankshaft. The purpose of the baffles is to lessen the windage disturbance by channeling oil draining from the crankshaft and upper portions of the engine quickly into the oil sump. The baffles also prevent oil in the sump from sloshing around and contacting the rotating crankshaft during hard cornering, braking, or acceleration.

As mentioned in Chapter 3, the PCV valve is also considered part of the lubrication system. Most engine rebuilding shops replace the PCV valve with a new one when rebuilding an engine. But remember that each engine type requires a different valve; the wrong PCV valve will affect engine performance.

OIL FILTERS

As stated previously, all oil leaving the oil pump is directed to the oil filter. The oil flows through the filter, then on to lubricate the engine. No unfiltered oil reaches the engine. This insures that very small particles of dirt and metal carried by the oil will not reach the close-fitting engine parts. The filter element and container are made as a unit, with a seal built in at the point the filter assembly contacts the block. The filter assembly threads directly on to the main oil gallery tube, eliminating external oil leaks and the possibility of oil leakage under pressure. The oil from the pump enters the filter can on the outside of the element, passes through the element to the center of the filter and into the main gallery.

The filter unit itself is a disposable metal container filled with a special type of treated paper or other filter substance (cotton, felt, and the like) that catches and holds the oil's impurities. When the engine is rebuilt, the oil filter is replaced. It is usually mounted on an adapter that bolts to the engine block (Figure 13-41).

Since there are several types of filter designs, always consult the vehicle manual for the one recommended for the engine being rebuilt.

OIL TYPES

The final component of the engine lubrication system to consider is the engine oil. Engine oils are carefully formulated to reduce the adverse effects of

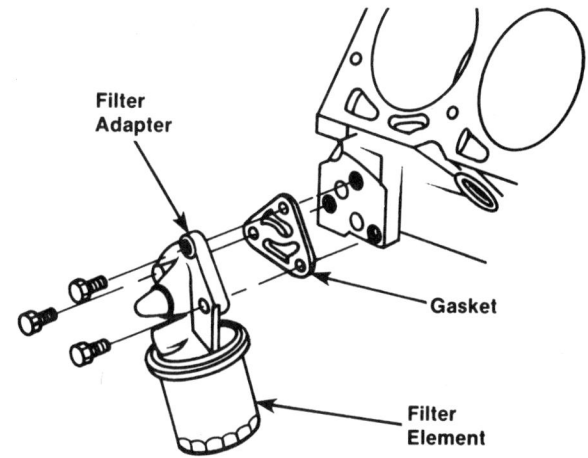

FIGURE 13-41 The oil filter is usually mounted on an adapter that bolts to the engine block.

engine operation. An effective engine oil should possess these important properties:

- Prompt circulation through the engine lubrication system
- Provide lubrication without foaming
- Reduce friction and wear
- Prevent rust and corrosion
- Prevent the formation of sludge and varnish deposits
- Provide cooling for engine parts
- Keep internal engine parts clean

To provide all these properties, modern engine oil contains many additives. Because of these additives, choosing the correct oil for each engine application can be a difficult task. However, the American Petroleum Institute (API) has developed service ratings that greatly simplify oil selection.

The American Petroleum Institute classifies engine oil based on the type of engine service it is suitable for. There are two basic classifications—a Standard or "S" class for passenger cars and light trucks and a Commercial or "C" class for heavy-duty commercial applications. Additionally, various grades of oil within each class are further classified alphabetically according to their ability to meet the engine manufacturers' warranty specifications (Tables 13–1 and 13–2).

 SHOP TALK ———————

*Among lubrication engineers, the designations S and C do not stand for **S**tandard and **C**ommercial vehicles respectively. The letters refer to the type of ignition that each engine utilizes. S is for gasoline engines that employ **S**park ignition; C is for the **C**ompression ignition system utilized by diesels.*

In addition to oil additives, oil viscosity is equally important in selecting an engine oil. The viscosity is affected by its temperature; for example, hot oil flows faster than cold. Since the rate of oil flow through the lubrication system is crucial for maintaining proper lubrication levels, viscosity is an important factor. Add to this the fact that the engine operates under a wide range of temperatures, and viscosity becomes even more important.

To standardize oil viscosity ratings, the Society of Automotive Engineers (SAE) has established an oil viscosity classification system that is accepted throughout the industry. This system is a numeric rating in which the higher viscosity, or heavier weight oils, receive the higher numbers. For exam-

	TABLE 13–1: API GASOLINE ENGINE DESIGNATION
SC	Service typical of gasoline engines in 1964 through 1967. Oil designed for this service provides control of high and low temperature deposits, wear, dust, and corrosion in gasoline engines.
SD	Service typical of gasoline engines in 1968 through 1970. Oils designed for this service provide more protection against high and low temperature deposits, wear, rust, and corrosion in gasoline engines. SD oil can be used in engines requiring SC oil.
SE	Service typical of gasoline engines in automobiles and some trucks beginning in 1972. Oil designed for this service provides more protection against oil oxidation, high temperature engine deposits, rust, and corrosion in gasoline engines. SE oil can be used in engines requiring SC or SD oil.
SF	Service typical of gasoline engines in automobiles and some trucks beginning with 1980. SF oils provide increased oxidation stability and improved antiwear performance over oils that meet API designation SE. It also provides protection against engine deposits, rust, and corrosion. SF oils can be used in engines requiring SC, SD, or SE oils.
SG	Service typical of gasoline automobiles and light-duty trucks, plus CC classification diesel engines beginning in late 1980. SG oils provide the best protection against engine wear, oxidation, engine deposits, rust, and corrosion. It can be used in engines requiring SC, SD, SE, or SF oils.

ple, an oil classified as an SAE 50 weight oil is heavier and flows slower than an SAE 10 weight oil. Heavy-weight oils are best suited for use in high-temperature regions; low-weight oils work best in low-temperature operations.

To meet the needs of the average motorist who might not want to change oils seasonally, oil manufacturers have developed multiviscosity oils. These oils carry a combined classification such as 10W-30. This classification means that the oil has a weight of 10 at ambient air temperature, but once the engine builds up heat, the oil will actually thicken to a weight of 30. An oil of this type would allow easy starting in cold weather and adequate protection at all operating temperatures.

The SAE classification (Table 13–3), like the API rating, is usually stamped on top of an oil can (Figure 13–42). Selecting oils that specifically meet or ex-

**TABLE 13-2: API DIESEL ENGINE
OIL DESIGNATION**

CA, CB	Oils with this designation are considered obsolete and should not be used unless specifically authorized by the engine manufacturer.
CC	Moderate-duty diesel and gasoline engine service. Service typical of certain naturally aspirated, turbocharged, or supercharged diesel engines operated in moderate- to severe-duty service and certain heavy-duty gasoline engines. Oils designed for this service provide protection from high-temperature deposits and bearing corrosion in these diesels and also from rust, corrosion, and low-temperature deposits in gasoline engines. These oils were introduced in 1961.
CD	Severe-duty diesel engine service. Service typical of certain naturally aspirated, turbocharged, or supercharged diesel engines where highly effective control of wear and deposits is vital or when using fuels of wide quality range including high sulfur fuels. Oils designed for this service were introduced in 1955 and provide protection from bearing corrosion and high-temperature deposits in these diesel engines.
CD-H	Severe-duty two-stroke cyclic diesel engine service. Service typical of two-stroke cycle diesel engines requiring highly effective control over wear and deposits. Oils designed for this service also meet all performance requirements of API service category CD.
CE	Severe-duty diesel engine service. Service typical of certain naturally aspirated, turbocharged, or supercharged heavy-duty diesel engines manufactured since 1983 and operated under both low-speed, high-load and high-speed, high-load conditions. Oils designed for this service must meet the requirements of API engine service category CC and CD.

ceed the manufacturer's recommendations and maintaining a regular oil change schedule is instrumental in obtaining maximum service life from an engine.

Lubricating oil used with depth or partial-flow filters is usually a heavy-duty, single-viscosity, detergent oil. Because of the wide variation in operating conditions, the specific lubricant is generally

determined by the user. Such judgments can be made based on recommendations by the engine manufacturer.

Two new terms have entered the oil field: energy-conserving and synthetic oil. Engine oils that are classified as energy-conserving, fuel-saving, or gas saving are designed to reduce friction, which in turn reduces fuel consumption. Friction modifiers and other additive changes are used to achieve this result. Energy-conserving oil is identified as such on the top of the oil can (Figure 13-42).

 SHOP TALK _____

Using the incorrect grade or type of oil in a rebuilt engine can cause a variety of problems. For example, the wrong viscosity oil can cause either an oil consumption problem or a low oil pressure problem, depending upon the weather and driving conditions. Using an oil with the incorrect service rating can result in inadequate protection for engine bearings and other moving engine parts. Always consult the engine manufacturer's recommendations to make certain that the engine is using the correct grade and type of oil. It is the responsibility of the rebuilder to inform the customer of the correct oil to use.

The introduction of synthetic motor oils dates back to World War II; it is often described as the "oil of the future." Synthetic oils are manufactured in a laboratory rather than pumped out of the ground and refined. They offer a variety of advantages over natural oils including better fuel economy, stability over a wide range of temperatures and operating conditions, and longevity. However, synthetic engine oil is not recommended by the Automotive Engine Rebuilders' Association for the break-in period.

FIGURE 13-42   API designation and SAE rating

TABLE 13-3: SAE GRADES OF MOTOR OIL

Lowest Atmospheric Temperature Expected	Single-Grade Oils	Multigrade Oils
32° F (0° C)	20, 20W, 30	10W-30, 10W-40, 15W-40, 20W-40, 20W-50
0° F (−18° C)	10W	5W-30, 10W-30, 10W-40, 15W-40
−15° F (−26° C)	10W	10W-30, 10W-40, 5W-30
Below −15° F (−26° C)	5W*	*5W-20, 5W-30

*SAE 5W and 5W-20 grade oils are not recommended for sustained high-speed driving.

Its outstanding ability to reduce wear by virtually eliminating friction between moving components is not desirable for a break-in oil. Certain predictable amounts of friction are required for the proper break-in of pistons and piston rings. AERA does not recommend the use of synthetic engine oils for the first 5,000 miles of service. Thereafter it is up to the vehicle owner to weigh the cost of more expensive synthetic motor oils, manufacturers' oil classification recommendations, and drain intervals.

COOLING SYSTEM

On most gasoline engines, the method of cooling is to circulate coolant through the cylinder block and cylinder head. Liquid cooling is preferable to air cooling because it is less noisy and better able to maintain a constant temperature at the cylinders. It also lets the engine operate more efficiently and makes a ready supply of hot coolant available to operate a heater for the passenger compartment.

The cooling system is made up of the following:

- *Pump.* Circulates the cooling liquid through the system. The liquid is a mixture of water and antifreeze; referred to as *coolant.*
- *Water Jackets.* Cored passages in the cylinder block and cylinder head that carry the coolant around the cylinders and combustion chambers.
- *Radiator.* Transfers heat in the coolant to the outside air as coolant flows through its tubes.
- *Fan.* A device that pulls cool outside air through the fins of the radiator to pick up heat. It can be driven by engine power or electricity.
- *Pressure Cap.* Maintains a pressure in the system to raise the boiling point of the coolant to a higher temperature. Also provides relief from excess pressure or vacuum.

- *Hoses.* Connect the components of the system to one another.
- *Thermostat.* Blocks off circulation in the system until a preset temperature is reached to speed engine warm-up. Also controls engine temperature at a predetermined level.
- *Temperature Indicator.* Warns the driver in case of overheating.
- *Belts and Pulleys.* Help to drive the water pump and the fan.
- *Fan Drive Clutch.* Allows the fan to cut out or run at lower speed when there is enough cold air available. This reduces the noise of the fan and conserves some of the power it uses up, especially at higher road speeds.
- *Coolant.* Usually a mixture of antifreeze and water. Antifreeze is a liquid, ethylene glycol, mixed with anticorrosion chemicals. Water alone will promote rust in the cooling system.
- *Temperature Warning System.* Alerts the driver of overheating. It consists of a temperature gauge and/or a light. A temperature sensor is screwed into a threaded hole in the water jacket (Figure 13-43).
- *Coolant Recovery System.* Consists of a special radiator cap with an upper sealing gasket, an overflow tube, and a recovery container.
- *Heater System.* Heated coolant flows through heater hoses to a small heater core, or radiator, located in a hollow container on either side of the fire wall. Air is directed or blown over the hot heater core, and the heated air flows into the passenger compartment (Figure 13-44).
- *Oil Cooler.* Some vehicles with automatic transmission have a sealed heat exchanger, or form of radiator, located in the coolant outlet tank. Metal or rubber hoses carry hot automatic transmission fluid to the heat ex-

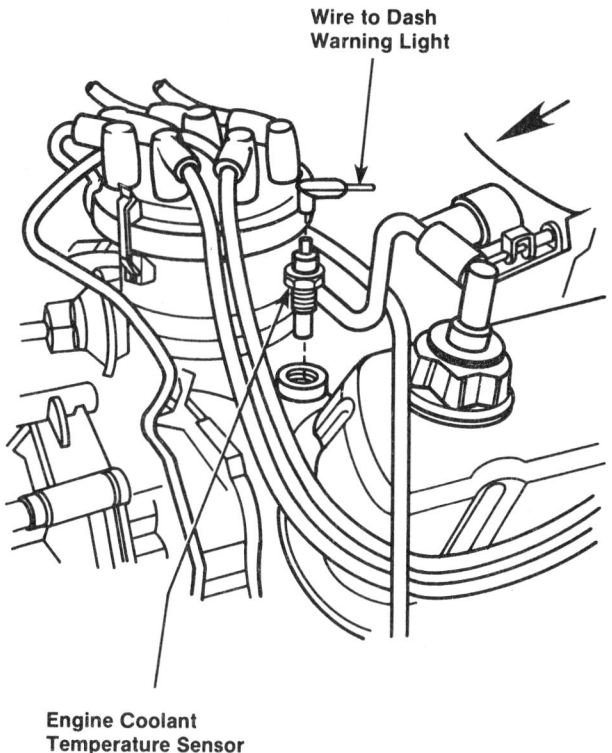

Wire to Dash Warning Light

Engine Coolant Temperature Sensor

FIGURE 13-43 Coolant temperature sensor

changer. The coolant passing over the sealed heat exchanger cools the fluid, which is then returned to the transmission (Figure 13-45). As mentioned earlier in this chapter, an oil cooler is used in some of the newer engines.

Only about 25 percent of the energy produced by internal combustion is actually transmitted to the drive wheels of the vehicle. The remaining 75 percent is wasted and must be dissipated or removed from the engine. If it is not, the engine will be destroyed by overheating. About half of this wasted heat goes out the exhaust or is lost to friction. The remaining half must be removed by the cooling system.

The cooling system removes heat at about the same rate it is generated to avoid severe engine damage due to extreme metal temperatures. At the same time, it must retain a certain amount of heat to enable the engine to warm up quickly and operate efficiently.

This delicate balance is accomplished by transferring the heat of the engine to the surrounding envelope of liquid coolant. The coolant is then pumped to a radiator where the heat is released into the atmosphere. This activity is influenced by many factors, the major ones being the following:

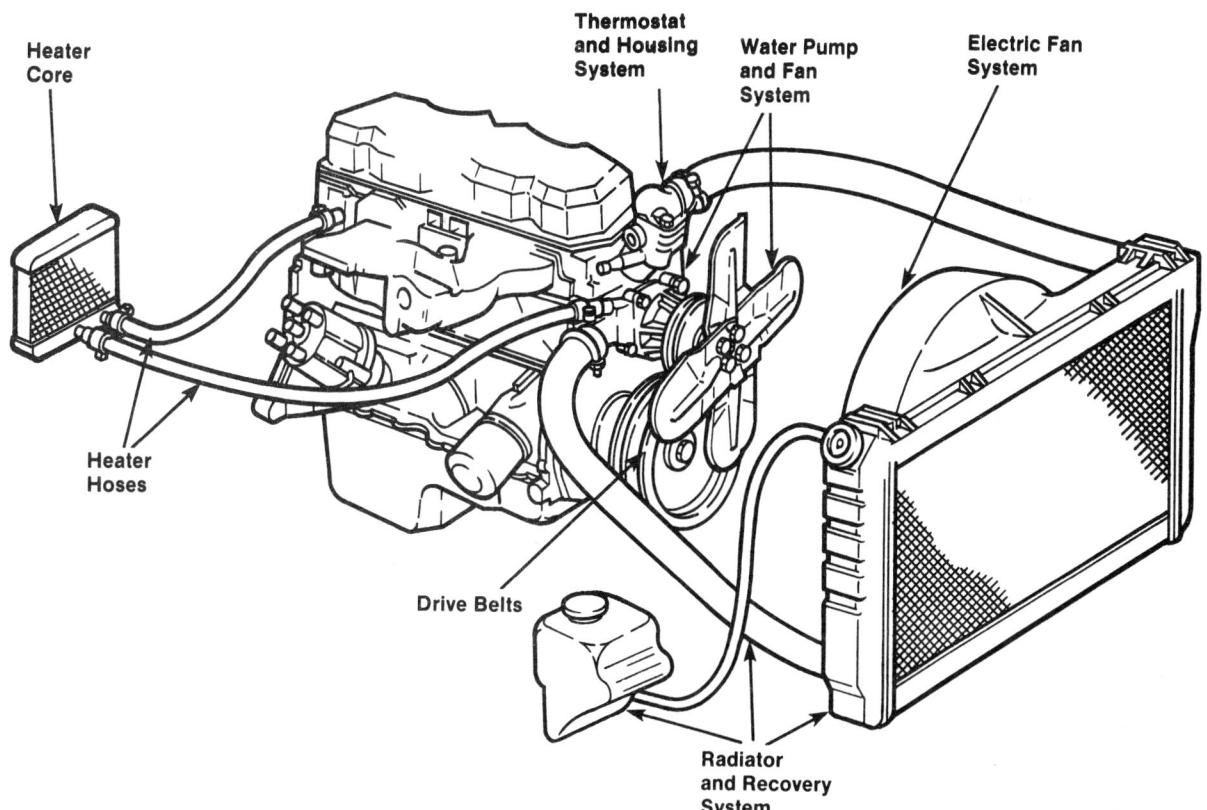

Heater Core

Thermostat and Housing System

Water Pump and Fan System

Electric Fan System

Heater Hoses

Drive Belts

Radiator and Recovery System

FIGURE 13-44 Components of a heater and cooler system

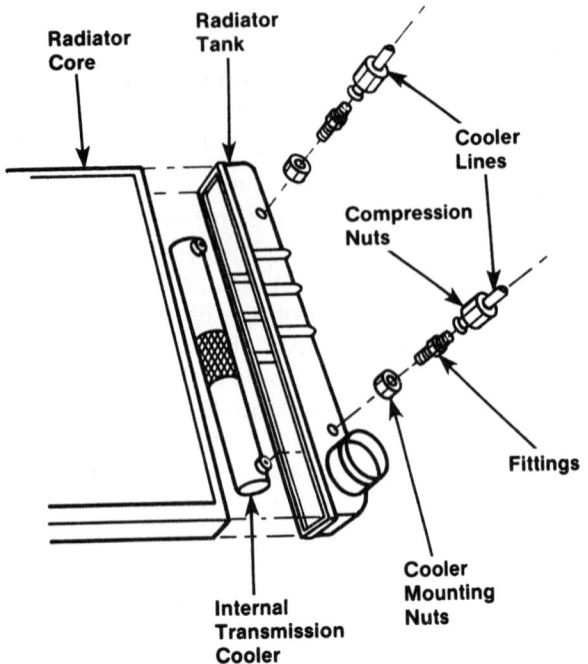

FIGURE 13-45 Components of a transmission oil cooler at the engine radiator

1. Cooling system capacity
2. Radiator size
3. Cooling pump rate
4. Engine size and the power it develops
5. Size and location of coolant passages within the engine

COOLING SYSTEM OPERATION

The water pump is belt driven by the engine crankshaft. The engine cooling fan can be mounted on the water pump or operated electrically. Coolant is pumped or circulated by the water pump from the bottom or side of the radiator through internal passages in the engine block and heads. The internal passages are often called the *water jacket*. The water pump then pumps the coolant through the thermostat to the top tank of the radiator. From the top tank the coolant flows slowly through a system of tiny tubes surrounded by cooling fins. These tubes and fins are called the *radiator core*. When the coolant reaches the opposite radiator tank the process begins again. Actually, from the point the thermostat opens, the coolant is constantly circulated through the engine and radiator. As the coolant travels through the engine block and heads, unwanted heat caused by combustion is transferred to the coolant. As the coolant flows through the radiator core, this heat is transferred to the air as the air passes through

the core and around the fins. The fan, usually mounted on the water pump, is used to move air through the radiator when the engine is idling or operating slowly, as in traffic. This insures sufficient air movement and good heat transfer at slow speeds. The cooling system's job is two-fold.

1. It carries away unwanted heat from the engine.
2. It must be able to maintain the engine temperature at the right point so that the engine can operate efficiently.

Water Pumps

Engine water pumps are generally the impeller type (Figure 13–46). An impeller is something that is designed to push, drive, or move something; in this case, engine coolant. The impeller or water pump is mounted on a shaft and designed like a rotating paddle. The shaft is mounted in the water pump housing and rotates on bearings. The pump contains a seal to keep the coolant from passing through it. At the drive end, the exposed end, a pulley is mounted to accept the belt, which is driven by the crankshaft. The pump housing usually includes the mounting point for the lower radiator hose.

When the engine is started, the crankshaft turns the water pump. The pump impeller pushes the water from its pumping cavity into the engine block. When the engine is cold the thermostat will be closed. This stops the coolant from reaching the top of the radiator. In order for the water pump to circulate the coolant through the engine during warm-up, a bypass passage is added below the thermostat, which leads back to the water pump. This passage must be kept free to eliminate hot spots in the engine during warm-up. It also allows the hot coolant to pass through the valve, which will open the thermostat.

Attached to the water pump at its drive pulley is the cooling or engine fan (Figure 13–47). The fan is

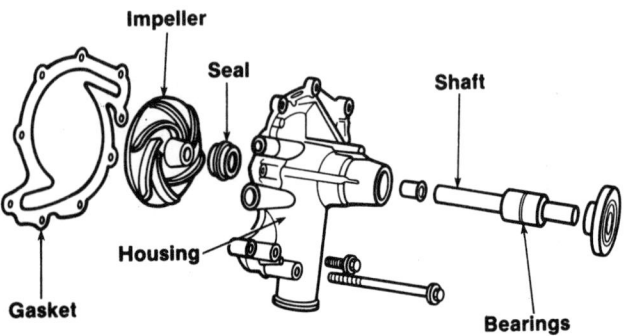

FIGURE 13-46 Impeller-type water pump

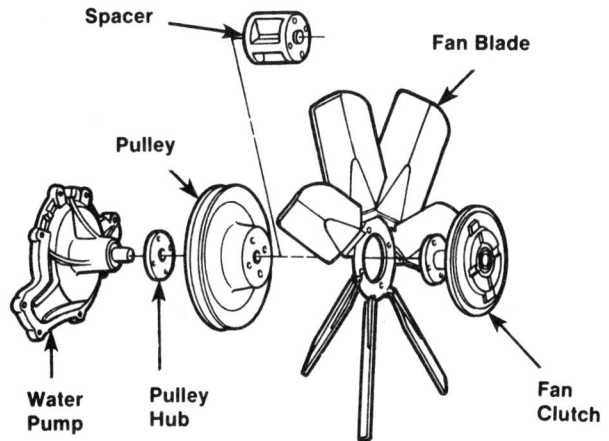

FIGURE 13-47   Engine fan is attached to the water pump

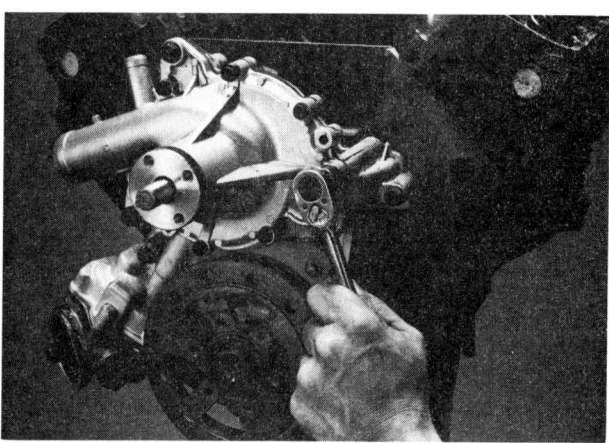

FIGURE 13-48   Installing a water pump *(Courtesy of Perfect Circle/Dana)*

driven or turned at approximately crankshaft speed. It is used to move air through the radiator at idle or low speed and has very little effect at high speed. Most modern fans will have four or more blades to supply good air movement. On many engines a fan shroud is used. The fan shroud is attached to the radiator and surrounds the fan. This concentrates the airflow to add to the system's cooling efficiency. Fan shrouds are found on most automobiles that are equipped with air conditioning.

Most rebuilders replace a water pump with a new one since it is usually easier and less expensive than attempting to repair one. When installing a water pump on an engine, apply a coating of good waterproof sealer to a new gasket and place it in position on the water pump. Coat the other side of the gasket with sealer and position the pump against the engine block until it is properly seated. Install the mounting bolts and tighten them evenly, in a staggered sequence, to specs with a torque wrench (Figure 13-48). Careless tightening could cause the pump housing to crack. Check the pump to make sure it rotates freely.

The remaining components of the cooling system are generally installed after the engine is placed back in the vehicle (see Chapter 15).

AIR COOLING SYSTEM

A few engines use a cooling system that employs air rather than liquid as the medium to transfer heat from the engine components to the atmosphere (Figure 13-49). Cylinders and heads have fins and are enclosed in a shroud to control airflow. Fins expose more of the surface area to airflow for better

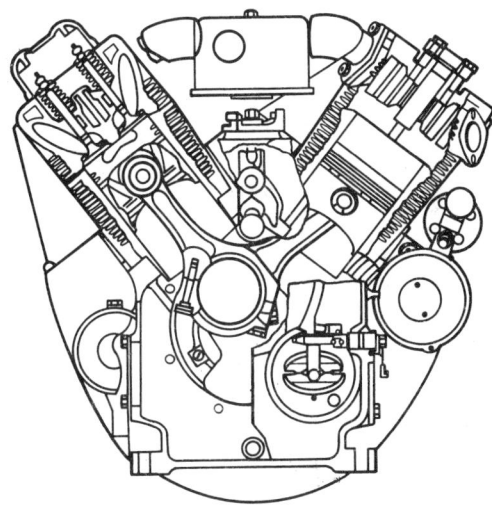

FIGURE 13-49   Cross section of a V-type air-cooled diesel engine

heat dissipation. Ducts and shrouds direct the airflow over the engine components, especially over the hotter cylinder head area. A belt-driven or electric blower provides the means for airflow. Fresh air is taken in and heated air expelled into the atmosphere. A thermostat connected to a control valve or door regulates airflow to control engine temperature.

In most engines, oil is cooled by the airflow across the oil pan. However, air-cooled engines run hotter and the oil requires more cooling. This additional cooling is provided by an oil cooler. Inside the air duct of the engine is an oil radiator. Air flowing from the blower carries heat away from the oil.

REVIEW QUESTIONS

1. Which of the following is a function of the engine's lubrication system?
   a. Hold an adequate supply of oil
   b. Remove contaminants from the oil
   c. Deliver oil to all necessary areas of the engine
   d. All of the above

2. Technician A says the primary cause of a drop in oil pressure is main and connecting rod bearing wear or damage. Technician B says that engine oil pressure is dependent upon oil viscosity. Who is right?
   a. Technician A
   b. Technician B
   c. Both A and B
   d. Neither A nor B

3. Technician A says that the gear pump pumps a greater volume of oil than a rotor pump. Technician B says that a high-pressure oil pump has a stiffer pressure relief valve spring than the stock pump. Who is right?
   a. Technician A
   b. Technician B
   c. Both A and B
   d. Neither A nor B

4. Technician A says that most pressure indicators found in vehicles today are electronically controlled. Technician B says that most engines made in recent years have an oil filtering system somewhere on the input side of the pump. Who is right?
   a. Technician A
   b. Technician B
   c. Both A and B
   d. Neither A nor B

5. Which type of oil filtration system is used on most engines today?
   a. bypass
   b. independent
   c. full flow
   d. shunt

6. Which of the following could cause low oil pressure during diagnostic checks?
   a. plugged oil pickup screen
   b. excessive oil dilution
   c. both a and b
   d. neither a nor b

7. Technician A uses a feeler gauge and straightedge to determine the pump cover flatness. Technician B uses an outside micrometer to measure the diameter and thickness of the outer rotor. Who is right?
   a. Technician A
   b. Technician B
   c. Both A and B
   d. Neither A nor B

8. Which of the following is used to reduce windage in the oil sump?
   a. oil coolers
   b. baffle assembly
   c. oil filter
   d. none of the above

9. Technician A says that the API classification "S" stands for standard passenger cars. Technician B says that the API classification "C" stands for vehicles with compression ignition. Who is right?
   a. Technician A
   b. Technician B
   c. Both A and B
   d. Neither A nor B

10. Technician A says that the American Petroleum Institute has established an oil viscosity classification system. Technician B says that higher viscosity oils receive the higher rating numbers. Who is right?
    a. Technician A
    b. Technician B
    c. Both A and B
    d. Neither A nor B

11. Which of the following is designed to reduce friction?
    a. SE oil
    b. CC oil
    c. energy-conserving oil
    d. synthetic oil

12. Which of the following is designed to offer stability over a wide range of temperatures and operating conditions?
    a. SF oil
    b. DD-H oil
    c. energy-conserving oil
    d. synthetic oil

CHAPTER FOURTEEN

SEALING THE ENGINE

Objectives

After reading this chapter, you should be able to
- Explain the principles and precautions of working with various fasteners.
- Determine how to select torque specifications for different size bolts.
- Explain how to make thread repairs.
- Explain the purpose of the various gaskets used to seal an engine.
- Identify the different types of gaskets and their use.
- Explain general gasket installation procedures.
- Determine the sealing requirements of a bimetal engine.
- Describe the methods of sealing the timing cover and rear main bearing.

Effectively sealing conventional gasoline and diesel engines means several things:

- Confine the extremely high combustion pressures.
- Keep the low-pressure liquids in the cooling system away from the cylinders and lubricating oil.
- Prevent both internal and external oil leaks.
- Suppress and muffle noise.

This chapter will discuss the sealing requirements of both conventional and performance engines from the standpoint of three crucial engine components: fasteners, gaskets, and seals. These components have a specific task, and only when they all are performing at optimum levels is the engine sealed perfectly.

FASTENERS

In recent years, adhesives, welding, and other assembly methods have eliminated the need for many automotive fasteners. Ironically, however, nearly as many new fastener applications have arisen for those that have been replaced. Nuts and bolts remain an excellent means of holding engine parts together. And for parts that must be disassembled and reassembled, there is no substitute for these fasteners. On the other hand, they are probably the

most neglected of all mechanical devices. Depending on the application, they are also potentially the most dangerous.

Many types and sizes of fasteners are used in the automotive industry. Each fastener is designed for a specific purpose and specific conditions that are encountered in vehicle operation. One of the most popular type of fastener is the threaded fastener. Threaded fasteners include bolts, nuts, screws, and similar items that allow the rebuilder to install or remove parts easily (Figure 14-1).

FIGURE 14-1 Many types of threaded fasteners are used on the automobile engine. *(Courtesy of Botts Auto Parts Company)*

When replacing fasteners the rebuilder technician should observe the following precautions:

- Always use the same number of fasteners as originally used by the OEM.
- Always use the same diameter, length, number of threads, and type of fasteners as used by the OEM.
- Always observe the OEM's recommendation given in the service manual for tightening sequence, tightening steps (increments), and torque valves.
- Always replace a used cotter pin.
- Always replace stretched fasteners or fasteners with damaged threads.
- Always use the correct washers and pins as specified by the OEM.

A number of terms have been used over the years to identify the various types of threads. Some of these have been replaced with new terms. The terms most commonly employed in the automotive trade—the United States Standard (USS), the American National Standard (ANS), and the Society of Automotive Engineers Standard (SAE)—have all been replaced by the Unified National Series. The Unified National Series consists of four basic classifications:

- Unified National Coarse (UNC or NC)
- Unified National Fine (UNF or NF) (SAE)
- Unified National Extrafine (UNEF or NEF)
- Unified National Pipe Thread (UNPT or NPT)

The two common metric threads are coarse and fine and can be identified by the letters SI (Système International d'Unités or International System of Units) or ISO (International Standards Organization).

BOLT USAGE

To identify the type of threads on a bolt, bolt terminology must be defined. The bolt has several parts (Figure 14–2).

- *Head.* The head is used to torque or tighten the bolt. A socket fits over the head, which enables the bolts to be tightened. Common U.S. Customary (USC) and metric sizes for bolt heads include those given in Table 14–1. The sizes are given in fractions of an inch and in millimeters. Some of the USC and metric sockets are very close in size. It is important not to use metric sizes for USC bolts or USC sizes for metric bolts. The bolt heads might be damaged.

- *Diameter.* Bolt diameter is the measurement across the major diameter of the threaded area or across the bolt shank.
- *Thread Pitch.* The thread pitch of a bolt in the English system is determined by the number of threads there are in one inch of threaded bolt length and is expressed in "number of threads per inch." The thread pitch in the metric system is determined by the distance in millimeters between two adjacent threads. To check the thread pitch of a bolt or stud, a thread pitch gauge (see Chapter 1) is used. Gauges are available in both English and metric dimensions.
- *Length.* Bolt length is the distance measured from the bottom of the head to the tip of the bolt.

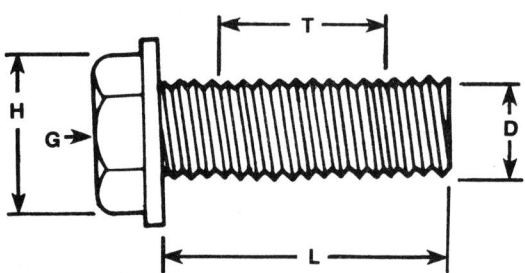

H—Head
G—Grade Marking (Bolt Strength)
L—Length (Inches)
T—Thread Pitch (Thread/Inch)
D—Nominal Diameter (Inches)

A

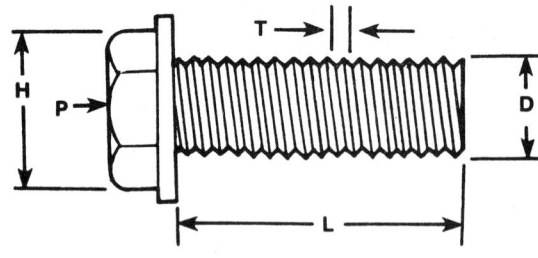

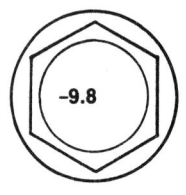

H—Head
P—Property Class (Bolt Strength)
L—Length (Millimeters)
T—Thread Pitch (Thread/Millimeter)
D—Nominal Diameter (Millimeter)

B

FIGURE 14–2 (A) English and (B) metric bolt measurement

- *Strength.* The bolt's tensile strength is the amount of stress or stretch it is able to withstand. The type of bolt material and the di-

TABLE 14-1: STANDARD BOLT HEAD SIZES

Common English (U.S. Customary) Head Sizes	Common Metric Head Sizes
Wrench Size (inches)	Wrench Size (millimeters)
3/8	9
7/16	10
1/2	11
9/16	12
5/8	13
11/16	14
3/4	15
13/16	16
7/8	17
15/16	18
1	19
1-1/16	20
1-1/8	21
1-3/16	22
1-1/4	23
1-5/16	24
1-3/8	26
7/16	27
1-1/2	29
	30
	32

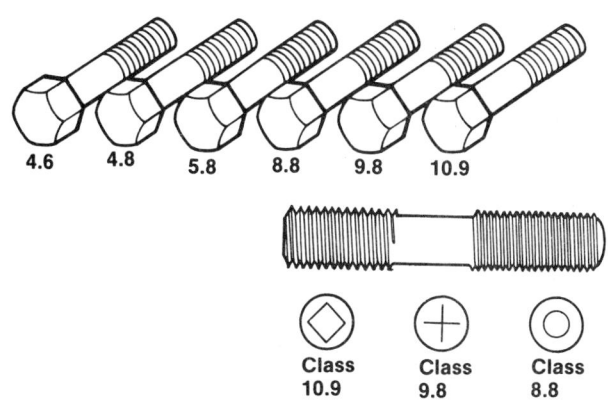

FIGURE 14-3 Metric bolts and studs are hardened by using the property class number.

ameter of the bolt determine its tensile strength. In the English system, the tensile strength of a bolt is identified by the number of radial lines (grade marks) on the bolt head. More lines mean higher tensile strength (Table 14-2). In the metric system, tensile strength of a bolt can be identified by a number (grade mark) on the bolt head (Figure 14-3). The higher the number, the greater the tensile strength.

It is very important to be familiar with the standard bolt indication measurements and grade markings. All bolts in the same connection must be of the same grade; otherwise, they will not perform equally. Likewise, nuts are graded to match their respective bolts (Table 14-3). For example, a Grade 8 nut must go with a Grade 8 bolt. If a Grade 5 nut was used instead, a Grade 5 connection would result.

TABLE 14-2: STANDARD BOLT STRENGTH MARKINGS

SAE GRADE MARKINGS	(no lines)	(3 lines)	(4 lines)	(5 lines)	(6 lines)
DEFINITION	No lines: unmarked indeterminate quality SAE Grades 0-1-2	3 Lines: common commercial quality Automotive & AN Bolts SAE Grade 5	4 Lines: medium commercial quality Automotive & AN Bolts SAE Grade 6	5 Lines: rarely used SAE Grade 7	6 Lines: best commercial quality NAS & Aircraft Screws SAE Grade 8
MATERIAL	Low Carbon Steel	Med. Carbon Steel Tempered	Med. Carbon Steel Quenched & Tempered	Med. Carbon Alloy Steel	Med. Carbon Alloy Steel Quenched & Tempered
TENSILE STRENGTH	65,000 psi	120,000 psi	140,000 psi	140,000 psi	150,000 psi

TABLE 14-3: STANDARD NUT STRENGTH MARKING

Inch System		Metric System	
Grade	Identification	Class	Identification
Hex Nut Grade 5	3 Dots	Hex Nut Property Class 9	Arabic 9
Hex Nut Grade 8	6 Dots	Hex Nut Property Class 10	Arabic 10
Increasing dots represent increasing strength		Can also have blue finish or paint dab on hex flat. Increasing numbers represent increasing strength.	

The Grade 5 nut cannot carry the loads expected of the Grade 8 bolt. Look for the nut markings which are usually located on one side. Grade 8 and critical applications require the use of fully hardened flat washers. They do not dish out like soft wrought washers that cause loss of clamp load.

Bolt heads can pop off because of fillet damage (Figure 14-4). The fillet is the smooth curve where the shank flows into the bolt head (Figure 14-5). Scratches in this area introduce stress to the bolt head, causing failure. It produces the same result as scratching a polished shaft. The bolt head can be protected by removing any burrs around the edges of holes. Also, place flat washers with their rounded, punched side against the bolt head and their sharp side to the work surface.

TORQUE PRINCIPLES

All metals are elastic, which means they can be stretched and compressed up to a certain point. This elastic, spring-like property is what provides the

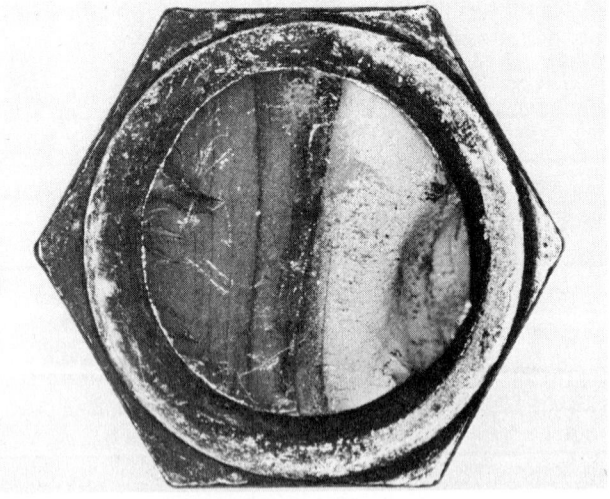

FIGURE 14-4 Shown above is the head of a bolt that failed due to fatigue as well as fillet damage. The fracture actually started at the 3 o'clock position. Each progressive impact created a benchmark. The bolt finally fractured when the crack had spread a little more than half the distance across the bolt.

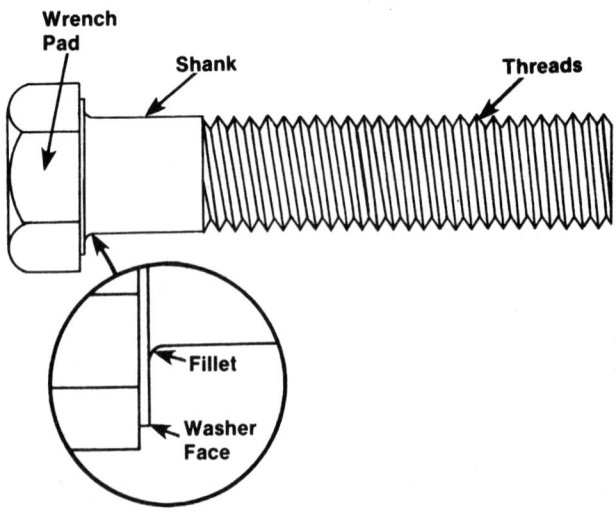

FIGURE 14-5 Details of the bolt fillet

clamping force when a bolt is threaded into a tapped hole or when a nut is tightened. As the bolt is stretched a few thousandths of an inch, clamping force or holding power is created due to the bolt's natural tendency to return to its normal, original length.

Like a spring, the more a bolt is stretched, the tighter it becomes. However, a bolt can be stretched too far. This is obvious when the grip on the wrench feels "mushy." At this point, the bolt can no longer safely carry the load it was designed to support.

Elasticity means that a bolt can be stretched a certain amount, and each time the stretching load is reduced, the bolt will return exactly back to its original, normal size. In other words, it is reusable. However, if the bolt is stretched into "yield," it takes a permanent set and never returns to normal (Figure 14-6); the bolt will continue to stretch more each time it is used, just like a piece of taffy that is stretched until it breaks.

Proper use of torque will avoid this yield condition. Torque values are calculated with a 25 percent safety factor below the yield point. There are some fasteners, however, that are torqued intentionally just barely into a yield condition, although not far enough to create the classic coke bottle shape of a necked out bolt. This type of fastener, known as a torque-to-yield (T-T-Y) bolt, will produce 100 percent of its intended strength, compared to 75 percent when torqued to normal values. These fasteners, however, should not be reused, unless otherwise specified. That is, some manufacturers caution against reusing T-T-Y bolts (Figure 14-7); others permit reuse. Consult the OEM's specifications and the gasket set instructions for bolt reuse. Some aftermarket gasket manufacturers make the decision easy for the technician by including new T-T-Y bolts.

Table 14-4 gives standard bolt and nut torque specifications. If the manufacturer's torque specifications are available, follow them precisely. If a torque-to-yield (T-T-Y) bolt is replaced with a new bolt of identical grade but torqued to a value found in

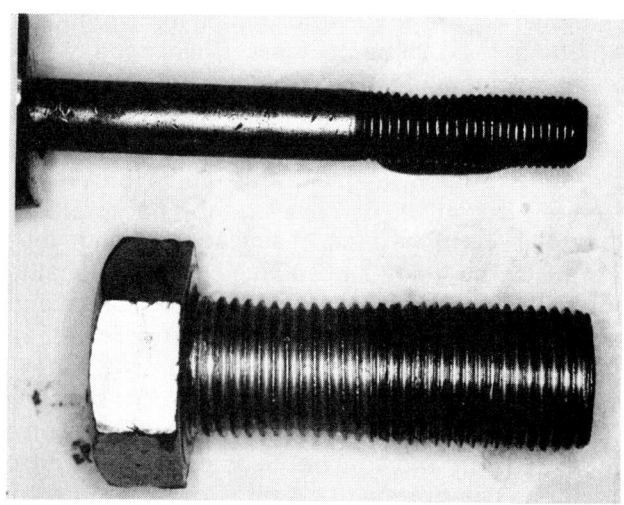

FIGURE 14-6   These bolts have been torqued past their yield points. Notice the classic coke bottle effect.

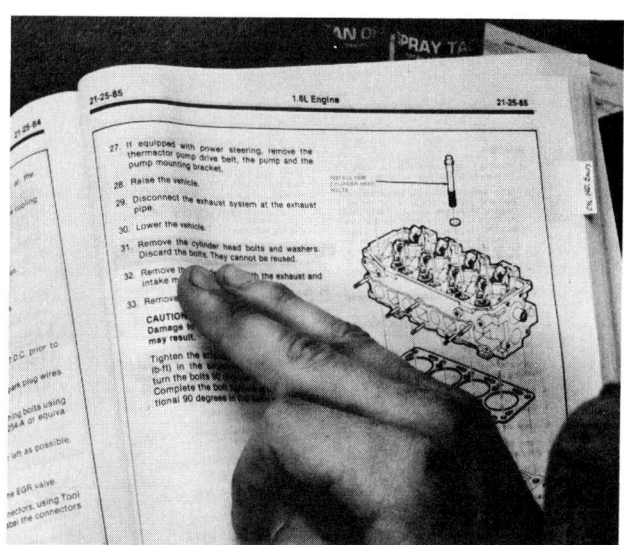

FIGURE 14-7   Some manufacturer's caution against reusing these bolts; consult the original equipment manufacturer's specifications and the gasket set instructions for bolt reuse.

TABLE 14-4:   STANDARD BOLT AND NUT TORQUE SPECIFICATIONS

Size Nut or Bolt	Torque (foot-pounds)	Size Nut or Bolt	Torque (foot-pounds)	Size Nut or Bolt	Torque (foot-pounds)
1/4–20	7–9	7/16–20	57–61	3/4–10	240–250
1/4–28	8–10	1/2–13	71–75	3/4–16	290–300
5/16–18	13–17	1/2–20	83–93	7/8–9	410–420
5/16–24	15–19	9/16–12	90–100	7/8–14	475–485
3/8–16	30–35	9/16–18	107–117	1–8	580–590
3/8–24	35–39	5/8–11	137–147	1–14	685–695
7/16–14	46–50	5/8–18	168–178		

a regular torque table, the clamping force produced will be at least 25 percent less. This is one reason many manufacturers are replacing OEM bolts with the next higher grade. As an example, a Grade 8 bolt torqued properly to its 25 percent safety factor will produce as much, if not more, clamping force than a Grade 5 bolt at 100 percent with no safety factor.

All bolts in the same or similar connection must be alike. The grade and surface condition, whether plated or nonplated, dry or lubricated, oil or anti-seize, cut threads or rolled, straight shank or reduced, will affect the torque/tension relationship and cause the performance of the connection to change. Because torque is actually a combination of both tension and torsion, it is also a function of friction. The bolt head or nut, whichever is being rotated, produces friction as it is turned, as do the threads when they gall together under the pressure of being in tension. Tests have proven that 90 percent of work energy is consumed by friction. Friction must first be overcome before any true work is done; for example, stretching the bolt. To compensate for surface variations, the following formula may be used to approximate the required torque:

$$T = FDC \div 12$$

where T = Torque in foot-pounds
F = Friction factor (torque coefficent)
D = Bolt diameter in inches
C = Bolt tension required in pounds

Table 14–5 gives the friction factor (F) for various surfaces and the percentage of torque required.

Nonplated bolts have a rougher surface than plated finishes. It therefore takes more torque to produce the same clamping force as on a plated bolt, even with one-third less friction. Add a lubricant and the torque might be as much as two-thirds lower.

Most printed torque values are for dry, plated bolts. But lubricants are beneficial when working with engines that have a great deal of oil around and in them. Lubricants provide smoother surfaces and more consistent and evenly loaded connections. They also help reduce thread galling.

Keep in mind that dry connections do not always turn out to be truly dry. They are often accidentally lubricated by touching the threads with dirty fingers. If a "dry" torque is used in this situation, the bolt might be torqued into yield or have its threads stripped because there might only be 75 percent friction and 25 percent energy to tension the bolt. This much energy can be too much, since only 10 percent of torque is needed to tighten the bolt. On the other hand, gritty dirt or nicked threads can increase friction. This results in decreased bolt tension and a joint that is not as tight as required.

Reusing a dry nut will produce a connection with decreasing clamp force each time it is used. Nuts should not be reused under these conditions. Nut threads are designed to collapse slightly to carry the bolt load. If dry nuts are reused, increased thread galling will result each time the nuts are reused at the same torque. Since torque is a function of friction, torque sees 100 percent use, or total output. It does not care if there is 90 percent friction and 10 percent work or 95 percent friction and only 5 percent work, as long as it totals 100 percent. It just means that not enough work is being produced to overcome the extra friction. In this situation, the torque wrench will say one thing, but the real output will actually be much less.

Lubrication of fasteners is recommended for consistency (Table 14–6). However, be sure to lubricate all the bolts with the same lubricant. Some lubricants are more slippery than others, which will affect torque values. Also, lubricate the bolt, never the hole; otherwise, the bolt will merely be tightening against the oil in the hole.

If a bolt with a reduced shank diameter (for example, a connecting rod bolt) is specified by the OEM, never replace it with a standard, straight shank bolt. A reduced shank diameter bolt looks "dog-boned." Its function is to reduce the stress on the threads by transferring it to the shank. A standard bolt under similar conditions would break very quickly at the threads.

Rolled threads are 30 percent stronger than cut threads (Figure 14–8). They also offer better fatigue resistance because there are no sharp notches to create stress points. Therefore, on any fastener placed in an application of high stress or cyclic mo-

TABLE 14–5: COMPENSATION FOR VARIOUS SURFACES

Surface	Friction Factor	Percentage of Torque Change Required
Dry, unplated steel	0.20	Use standard torque value shown.
Cadmium plating	0.15	Reduce standard torque 25%.
Zinc plating	0.21	Increase standard torque 10%.
Aluminum	0.15	Reduce standard torque 25%.
Stainless steel Supertanium Special alloy steel	0.30 0.20	Increase torque 50%.

TABLE 14-6: FRICTION FACTORS (F) AND TORQUE REDUCTIONS FOR LUBRICATED SURFACES ON ALLOY STEEL BOLTS

Lubricant	Friction Factor	Percentage of Torque Reduction Required
Collodial copper	0.11	Reduce standard torque 45%.
Never-seize	0.11	Reduce standard torque 45%.
Grease	0.12	Reduce standard torque 40%.
Moly-cote (molybdenum disulphite)	0.12	Reduce standard torque 40%.
Heavy oils	0.12	Reduce standard torque 40%.
Graphite	0.14	Reduce standard torque 30%.
White lead	0.15	Reduce standard torque 25%.

tion, never add any threads by cutting them onto existing rolled threads. Use bolts of proper length, threaded rod, or studs to achieve the correct thread length.

Studs are used to prevent continued damage to tapped holes by remaining in place at one end while the other end does the remainder of the work for future use. If a bolt or stud breaks in a tapped hole, the threads will become damaged from the shock of the breakage. The hole must then be retapped to reshape the damaged threads and assure even torque. Tapped holes in aluminum alloys should always be strengthened with thread inserts. Manufac-

turers do not add them during assembly because of costs and labor time. However, they are necessary for repeated maintenance.

Some manufacturers recommend a torque-turn method for head bolts. This is a very accurate approach, and it works for standard bolts and nuts as well. If the work thickness is no more than four times the bolt's diameter, snug the nut lightly against the work and rotate it one-third of a turn (120 degrees). If the work is between four to eight times the diameter, rotate the nut a half turn (180 degrees). This will produce a very reliable clamp load that is much higher than can be gotten just by torquing. There are no other variables to affect the results.

When installing during the sealing procedure, keep the following points in mind:

1. Visually inspect the bolts:
   - Threads must be clean (Figure 14-9) and undamaged. Discard any that are not acceptable.
   - Use liquid sealant or engine oil on the threads and seating face of the cylinder head bolts to prevent seizure from rust and corrosion. This is particularly important for an aluminum block or head. Use sealant on those bolts that hold in coolant or oil.
   - Install bolts in the proper holes. It is possible for an engine to have head bolts of different lengths.
   - Run a nut over the threads by hand. Discard if any binding occurs because the bolt is overstretched. If overstretched bolts are used with a resurfaced head and/or block or thinner gasket, the deformed threads will bind in the block threads. This gives a false torque yield point and causes early gasket failure.

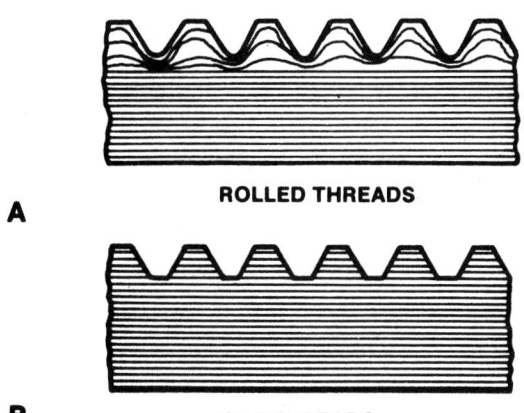

A ROLLED THREADS

B CUT THREADS

FIGURE 14-8 (A) Rolled threads are 30 percent stronger than (B) cut threads.

FIGURE 14-9 Cleaning bolt threads *(Courtesy of JP Industries, Inc.)*

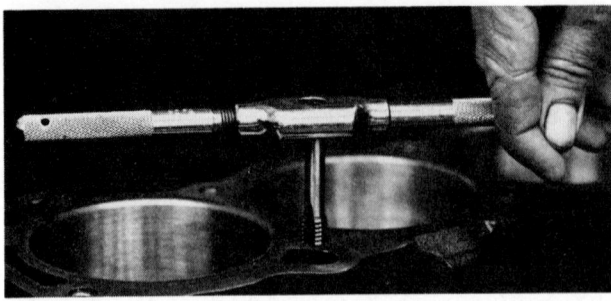

FIGURE 14-10 Cleaning bolt holes with a tap *(Courtesy of JP Industries, Inc.)*

- Clean bolt and cylinder block threads (Figure 14-10). Engine blocks with bolt holes entering a water jacket must be retapped.

2. Apply a light coat of 10W engine oil to threads and bottom face of bolt head (Figure 14-11). A sealer is required for a bolt that enters the water jacket (Figure 14-12). This will stop coolant seepage around the bolt threads. Seeping coolant could get in the oil or cause corrosion that might damage parts, resulting in engine failure.

3. Tighten head bolts in the recommended sequence. This is important to prevent a warp in the cylinder head.
   - Use an accurate torque wrench.
   - Tighten bolts to recommended torque in steps and proper sequence. (Refer to

manufacturer's cylinder head specification manual.)

4. Do not exceed manufacturer's specified torque.

5. If bolt heads are not tight against the cylinder head boss, the bolts should be removed and washers installed. Another new gasket must be installed.

6. Make sure that the bolt is the proper length and that it is not too long (Figure 14-13).

Bolt hole threads in the engine block will often pull up, leaving a raised edge around the hole (Figure 14-14). If the block has been resurfaced, the threads might run up to the surface. In either case, the bolt holes should be cleaned by chamfering (Figure 14-15) and the threads cleaned with an appropriate size bottoming tap. Always repair dam-

FIGURE 14-12 If the bolts enter the coolant passages, coat them with a flexible sealer. *(Courtesy of Fel-Pro Inc.)*

FIGURE 14-11 Prelubricating the bolt threads, underside of the bolt heads, and washers will minimize friction and help to ensure that the gasket receives the correct clamping force. *(Courtesy of Fel-Pro Inc.)*

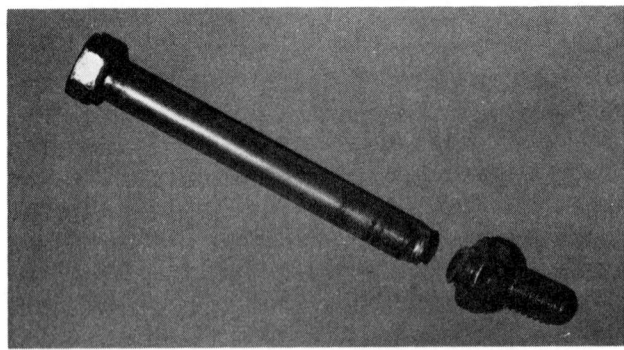

FIGURE 14-13 This bolt was too long. The two remaining threads above the nut could not absorb working load. The proper length bolt, that is, more threads above the nut, would have allowed for more torque. These threads, stressed during assembly, caused later fatigue.

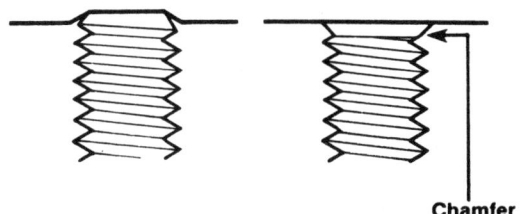

FIGURE 14-14 Bolt hole threads in the engine block will often pull up, leaving a raised edge. Also, if the bolt has been resurfaced, the threads might run up to the surface.

FIGURE 14-15 Chamfering bolt holes

aged threads as discussed later in the chapter to assure proper bolt performance.

Many different nuts are used by the automotive industry (Figure 14-16). But be wary of hexagon (hex) nut rotation on power wrenches. It is deceptively easy to place a bolt into a yield condition within seconds. Impact wrenches are the worst of-

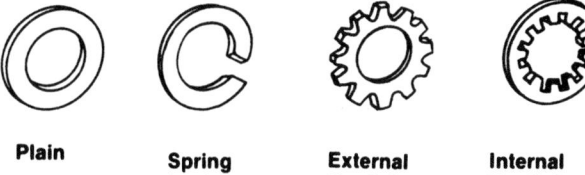

FIGURE 14-17 Washers are sometimes used to lock the bolts to the structure to keep them from coming loose and to prevent damage to soft metal parts.

fenders. Friction is needed to prevent the nut from spinning. If the nut is lubricated, there is no friction left to stop the impact wrench from hammering the nut past the bolt's yield point and/or stripping the threads.

Do not run the nut full speed onto the bolt. Instead, run it up slowly until it contacts the work, then mark the socket and watch how far it turns. Smaller air-powered speed wrenches do not produce the severe force of impact wrenches and are much safer to use. Follow this procedure with a torque wrench as well.

A rule of thumb about lock washers: If the connection did not come with one, do not add one. Lock washers are extremely hard and tend to break under severe pressure. Use locknuts and hard, flat washers. Properly torqued, this type of connection will never come loose—even if lubricated (Figure 14-17).

THREAD REPAIR

A common fastening problem is threads stripping inside an engine block, cylinder head, or other

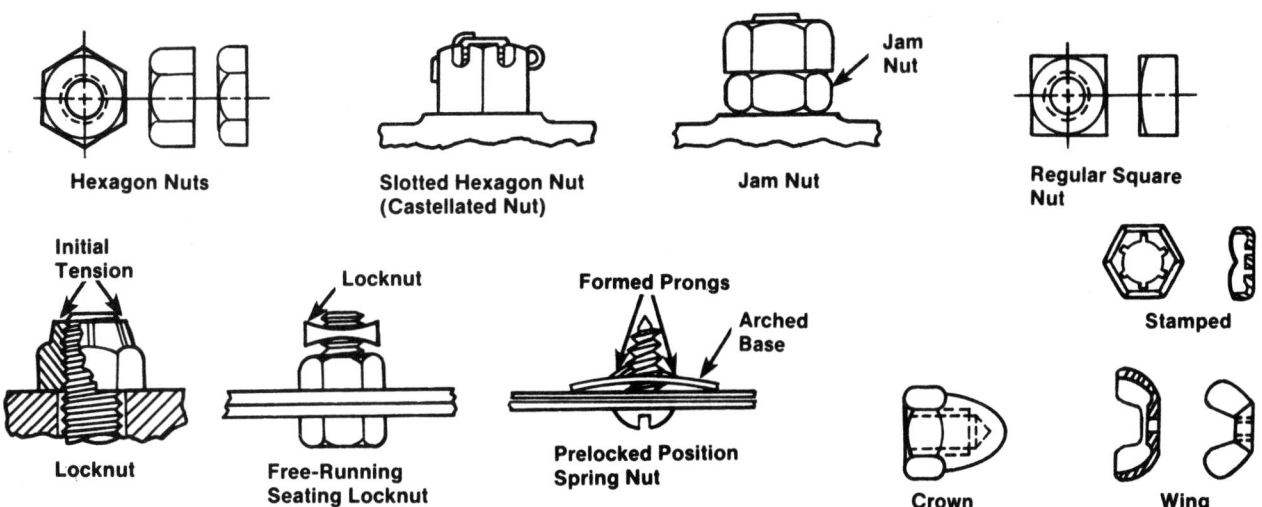

FIGURE 14-16 Many styles of nuts are used on the automobile. Each style has a specific purpose and application.

structure. This problem is usually caused by torque that is too high or by threading the bolt into the hole incorrectly. Rather than replacing the block or cylinder head, the threads can be replaced by the use of threaded inserts. Several threaded inserts are available—the helically coiled insert is the most popular (Figure 14–18). To install this and similar thread reconditioning inserts, proceed as follows:

1. Establish the size, pitch, and length of the thread required. Refer to the insert manufacturer's instructions for the proper size drill to use for the thread to be repaired.
2. Drill out the damaged threads with the specified drill. Clean out all metal chips from the hole.
3. Tap new threads in the hole using the specified tap. Lubricate the tap while threading the hole. Back out the tap every turn or two. When the hole is threaded to the proper depth, remove the tap and all metal chips from the hole.
4. Select the proper size insert and thread it onto the special installing mandrel or tool. Make sure the tool is engaged with the tang of the insert. Screw the insert in the hole by turning the installing tool clockwise. Lubricate the thread insert with motor oil if it is installed in cast iron (do not lubricate if installing in aluminum). Install the thread insert into the hole until it is flush with the surface or one turn below. Remove the installer. If the tang of the insert does not break off during mandrel

removal, break off the tang with a drift punch and remove it.
5. This completes the restoration of the threads and allows normal fastener use and torque.

GASKETS

Gaskets are used predominantly to prevent gas or liquid leakage between two parts that are bolted together (Figure 14–19). They are located in about fifteen to twenty different areas of the typical internal combustion engine. In addition to the sealing function mentioned earlier, gaskets also serve as spacers, wear insulators, and vibration dampers.

As a spacer, the gasket serves as a shim between two joined engine components, such as the fuel pump lever and cam. It also helps compensate for manufacturing or rebuilding tolerances in the cylinder head. In the wear insulator function, the gasket material is softer than the separated components. This allows for alternating expansion and contraction with less friction, thus preventing fretting or brinelling of machined surfaces. Keep in mind that the vibration damper function does not apply to liquid sealants that allow a metal-to-metal seal.

There are three basic classifications of engine gaskets:

- *Hard Gaskets.* These are steel, stainless steel, copper, or a combination metal enclosure with a compressible and heat-resistant clay/fiber compound sandwiched inside. Gaskets in this class are: cylinder head gaskets, exhaust manifold gaskets, and some intake manifold gaskets.
- *Soft Gaskets.* These consist of soft, flexible materials such as cork, rubber, a combination of cork/rubber, rubber-coated steel, paper, and other compressed fiber materials. Newer materials, sometimes called gaskets-in-a-tube, include silicone gaskets and anaerobics. Soft gaskets are used on valve covers, oil pans, rocker covers, pushrod covers, water pumps, thermostat housing, timing covers, and inspection plates.

  Silicone gasketing is available in two grades. The black is for general purpose, and the red is for high-temperature requirements. Most mechanics use silicone gasketing to aid gasket sealing in notches or dovetails.
- *Sealants.* Sealants and adhesives generally come in liquid form to be applied to all metal

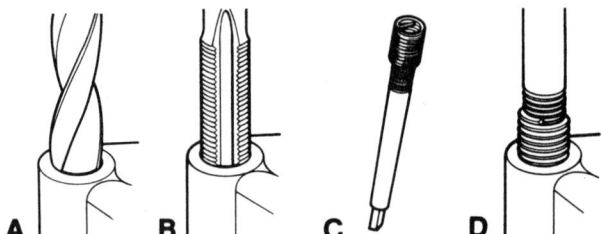

FIGURE 14–18 Steps in the installation of a helical screw repair coil: (A) Drill the damaged threads using the correct size drill bit. Clean all metal chips out of the hole. (B) Tap new threads in the hole, using the specified tap. The thread depth should exceed the length of the bolt. (C) Install the proper size coil insert on the mandrel provided in the installation kit. Bottom it against the tang. (D) Lubricate the insert with oil and thread it into the hole until flush with the surface. Use a punch or side-cutters to break off the tang.

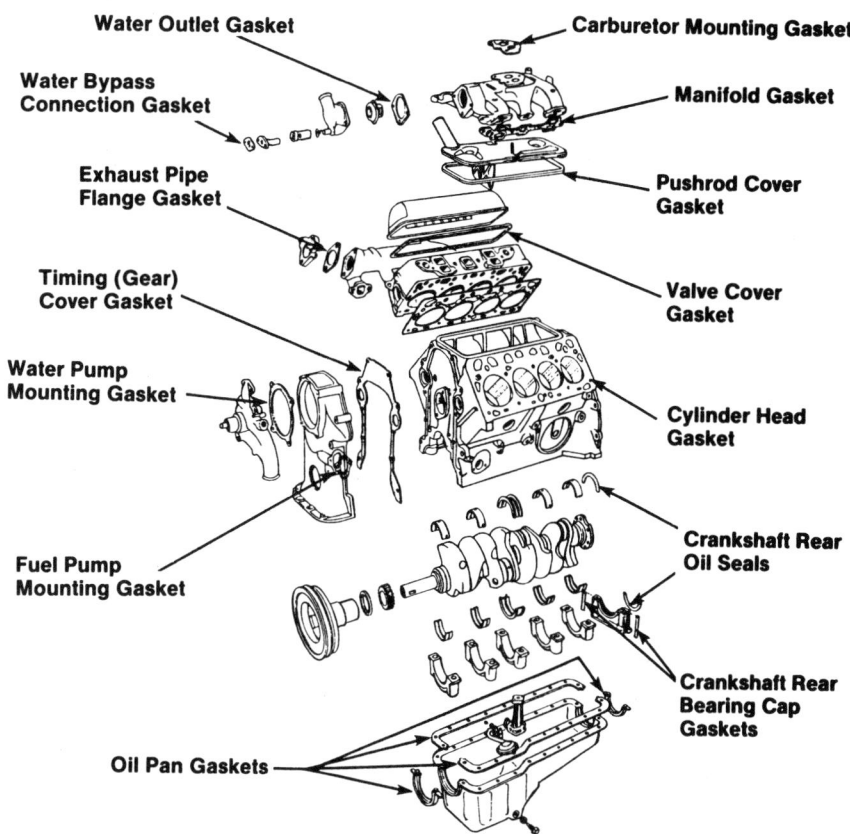

FIGURE 14-19 Engine gaskets

gaskets such as beaded steel-type cylinder head gaskets and intake manifold gaskets. It dries into a flexible, nonhardening sealant to help prevent leaks of oil, water, gasoline, air pressure, or vacuum. It is also useful as an antiseize for threads on bolts, studs, and fittings.

GASKET MATERIALS

Historically, with heavy-wall all-cast-iron engines, most gasket materials and constructions were similar, if not identical. Most engine applications were inherently forgiving, and gasketing was a low-applied technology business. One key to the auto industry's success in achieving late-model engine performance goals thought unlikely, impractical, or too costly just a few years ago has been the development of gaskets specifically designed to seal the new high-intake/exhaust engines. This includes exhaust-flange gaskets as well as exhaust rings—areas most subject to severe stresses of temperature and pressure. Make no mistake, these new sealing products have been highly engineered for virtually all critical areas of application in dozens of different engine designs, and inappropriate substitution of

non-OE-equivalent products in servicing or rebuilding them is not sufficient as it might have been in the past.

Another major change in gasket technology today is the elimination of asbestos. Although the automotive industry has voluntarily cut back the use of asbestos, the EPA has banned its use in all manufactured products after 1994. The irony is that asbestos is such an excellent gasket material that were it not for health liability issues, there would be no reason to replace it with anything else. It offers good strength and chemical resistance, and it is cheap compared to other materials. But the physical properties that make asbestos such a good gasket fiber also make it a hazardous substance to those who work with it.

Most late-model engines, including many small-block V-8 designs as well as the compact four- and six-cylinder engines, are equipped with premium gaskets made of space-age materials. Advanced gasketing technology incorporating flexible graphite cores, specialized surface coatings, new asbestos-free materials, elastomeric beading, reinforced cork products, wire-ring combustion seals, flat-plate hoop-strength constructions, and many other gasket design innovations all are different technological approaches by different OE suppliers to reach the same goals: long-term leak-free joints. The new gaskets have entirely different properties than traditional gasketing materials (asbestos, rubber, cork, and the like) that influence many critical design features including gasket thickness, compressibility, no-retorque characteristics, near-zero creep, greatly improved corrosion resistance, and superior thermal properties in sealing at temperatures exceeding 1000 degrees Fahrenheit.

Another area where gasket technology is continuing to evolve is in the use of molded silicone rubber valve cover and oil pan gaskets (Figure 14-20). Many domestic carmakers are following the European and Japanese lead by going to molded silicone rubber. Some of these gaskets are molded on a steel reinforcing carrier that makes installation easier, while preventing the gasket from shifting once it is in position.

The advantages of molded silicone rubber over cork/rubber cut gaskets include longevity, flexibility, conformability, and sealability. Oil makes silicone swell and produces an even tighter seal once the gasket is in use, but this also makes the gasket difficult to reuse. The primary disadvantage of silicone, however, is cost; silicone gaskets can cost several times as much as cork/rubber.

One thing rebuilders have to keep in mind about replacement gaskets is that the type of material used by the engine manufacturer may dictate what type of replacement gasket is required. In some instances, a

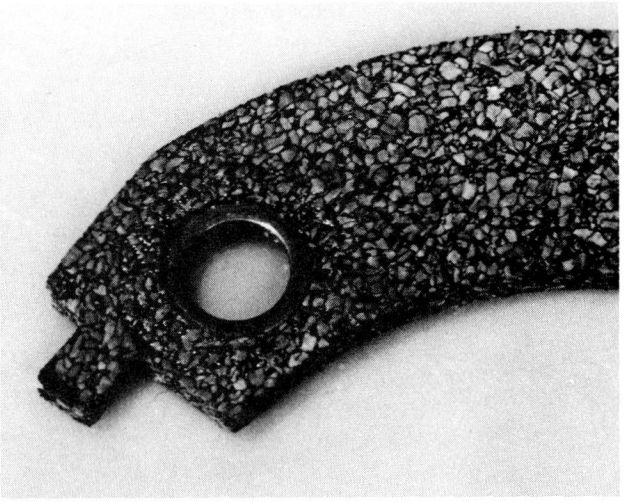

FIGURE 14-21 This oil pan gasket features metal grommets that help prevent gasket damage from overtightening the bolts.

molded silicone rubber gasket can be replaced by a cork/rubber cut gasket. But in other applications, the flange design on the valve cover or oil pan might make it impossible to use anything but a molded silicone replacement.

To help installers and rebuilders get the right amount of loading on these gaskets, many gasket manufacturers use small metal grommets in the gasket bolt holes so the gasket cannot be overtightened (Figure 14-21). The thickness of the grommet is such that the ideal amount of clamping force is achieved when the flange bottoms against the grommet.

Cork/rubber cut gaskets have also seen improvements in recent years. The addition of a metal carrier in the center of the gasket makes installation easier, increases the durability of the gasket, and helps to prevent shrinkage or movement once the gasket is installed.

Another change in gasket material has occurred on the intake side of the head in the form of injection molded and cut plastic gaskets with silicone sealing beads. These are finding their way into more and more engines these days because of lower temperatures in the intake manifold. Because of the change to multiport fuel injection, the exhaust crossover passage in the intake manifold has been eliminated, thus allowing for cooler incoming air temperatures. Some engines also lack an EGR valve, which is another source of heat in the intake manifold. This allows the use of lower temperature materials in intake manifold gaskets. Plastic intake manifolds are a real possibility in the near future, and some of them might have their own gaskets molded right in place.

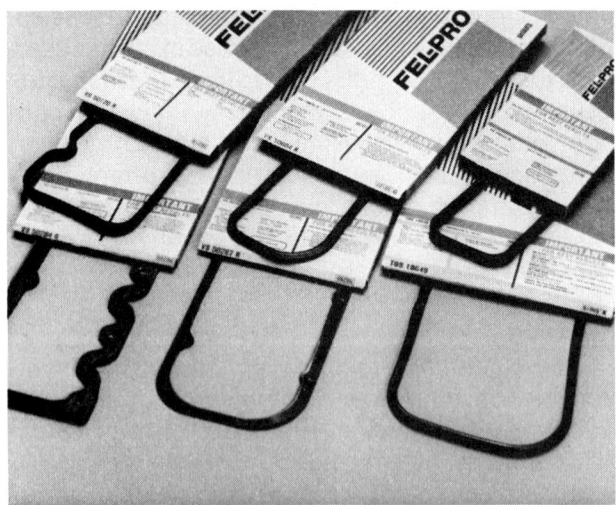

FIGURE 14-20 Some precision molded rubber gaskets are designed for a given vehicle. When replaced, the OEM type must be used. *(Courtesy of Fel-Pro Inc.)*

GENERAL GASKET INSTALLATION PROCEDURES

Although actual installation and assembly techniques of the various types of gaskets to different engine parts are covered later in this chapter and in Chapter 15, the following general instructions will serve as a helpful guide for installing gaskets. Because there are many different gasket materials and designs, it is impossible to list directions for every type of installation in this chapter. Remember to follow any special directions provided on the important instruction sheets packed with most gasket sets (Figure 14-22). These will give you additional tips on replacing the gaskets on the engine. The following are some general recommendations for installing gaskets:

1. *Never Reuse Old Gaskets.* Even if the old gasket appears to be in good condition, it will never seal as well as a new one. The old gasket has been exposed to high combustion temperatures and pressures, hot oil or coolant, and might have worn or damaged sealing surfaces. It also has been compressed by bolt torquing to fit the parts it has been sealing. It will not be able to re-form itself to the engine parts to create another good seat.

2. *Handle New Gaskets Carefully.* Be careful not to damage the new gaskets before placing them on the engine. Any bend or crease in the gasket material is a potential weak spot that might cause a leak after installation. This is especially true of the composition-type gaskets used on many cylinder heads and manifolds. If any attempt is made to straighten a bent or distorted gasket it could fracture the gasket and create a weak spot. Protect the new gaskets by keeping them in their package until installation.

3. *Proper Use of Sealants.* Use gasket sealants only when they are absolutely necessary. The hot, oily environment of an engine can cause some chemical sealants to react adversely with the binding compound in composition-type gaskets, making the gasket deteriorate and leak.

 SHOP TALK _____

Many installers tend to use too much sealant or grease on gaskets. Do not make this mistake. Because sealants and grease have less strength than gasket materials, they create weaker sealed joints. They also can prevent some gasket material from doing what it is supposed to do—soak up oil and swell to make a tight seal.

4. *Cleanliness Is Essential.* New gaskets seal best when used on clean surfaces. Thoroughly clean all mating surfaces of dirt, oil deposits, rust, old sealer, and gasket material (see Chapter 5). If any foreign substances remain, they can create a leak path. Scraping away the old gasket is not an easy job, but it is essential to assure a leak-free seal. When using a hand gasket scraper on aluminum parts, be very careful not to scratch the softer metal surfaces when removing the old gasket material (Figure 14-23).

5. *Using the Right Gasket in the Right Position.* Always compare the new gasket to the component mating surfaces to make sure that the right gasket is used. Comparing the new gasket with the original one is

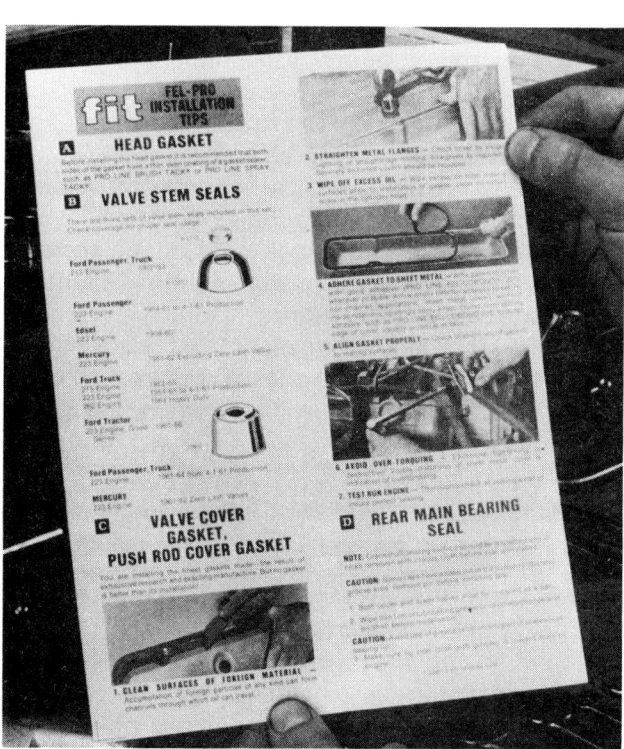

FIGURE 14-22 Typical manufacturer's installation sheet that comes with gasket sets. *(Courtesy of Fel-Pro Inc.)*

FIGURE 14-23 Scraping off an old gasket must be done carefully. *(Courtesy of Victor/Dana)*

another way to confirm that you are using the correct part (Figure 14-24). Check that all bolt holes, dowel holes, pushrod openings, coolant, and lubrication passages line up perfectly with the gasket. Some gaskets will have directions such as *top, front,* or *this side up* stamped on one surface (Figure 14-25). Remember to follow these instructions exactly when putting the gasket in place. An upside-down or reversed gasket can easily cause loss of oil pressure, overheating, and engine failure.

CYLINDER HEAD GASKETS

Cylinder head gaskets are the most sophisticated type of gasket. They also have the most demanding job to do. When first starting an engine in cold weather, parts near the combustion chamber might be subfreezing. Then, after only a few minutes of engine operation, these same parts might reach 400 degrees Fahrenheit. The inner edges of the cylinder head gasket are exposed to combustion flame

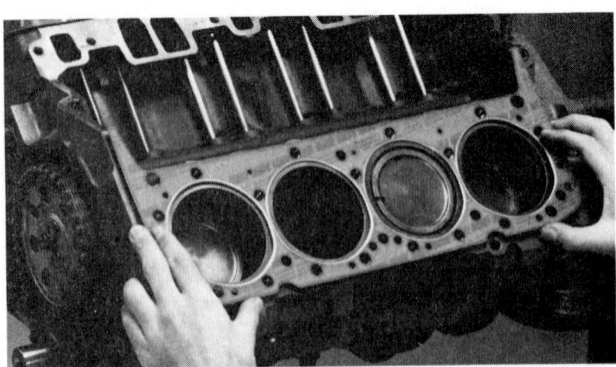

FIGURE 14-24 Checking head gasket for fit *(Courtesy of Fel-Pro Inc.)*

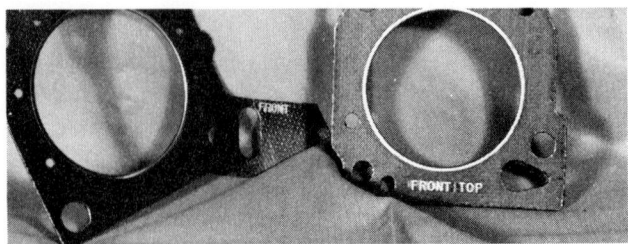

FIGURE 14-25 Check the directions stamped on the gaskets. *(Courtesy of JP Industries, Inc.)*

temperatures from 2000 to 3000 degrees Fahrenheit. Not only does this place a strain on the head gasket, it also affects the cylinder head and block, which compress the head gasket so it can do its job of sealing.

Pressures inside the combustion chamber also vary tremendously. On the intake stroke, a vacuum or negative pressure exists in the cylinder. Then, after combustion, pressure peaks of approximately 1,000 psi occur. This extreme change from suction to high pressure happens in a fraction of a second.

Cylinder head gaskets, under these conditions, must simultaneously:

- Seal intake stroke vacuum, combustion pressure, and the flame in the combustion chamber.
- Prevent coolant leakage, resist rust, corrosion and, in many cases, meter the coolant flow.
- Seal oil passages through the block and head while resisting chemical action.
- Allow for lateral and vertical head movement as engine heats and cools.
- Be flexible enough to seal minor surface warpage while being stiff enough to maintain adequate gasket compression.
- Fill small machining marks that could lead to serious gasket leakage and failure.
- Withstand forces produced by engine vibration.

Bimetal Engine Requirements

Another problem that head gaskets must face is the differing expansion rates of aluminum/cast-iron combination engines (Figure 14-26). Aluminum has a coefficient of thermal expansion two to three times greater than steel, depending on the alloy. This creates a lot of scrubbing action on a head gasket as the engine goes from cold to hot and back again. If the surface of the aluminum head has been roughly machined, it can grab and tear the head gasket fac-

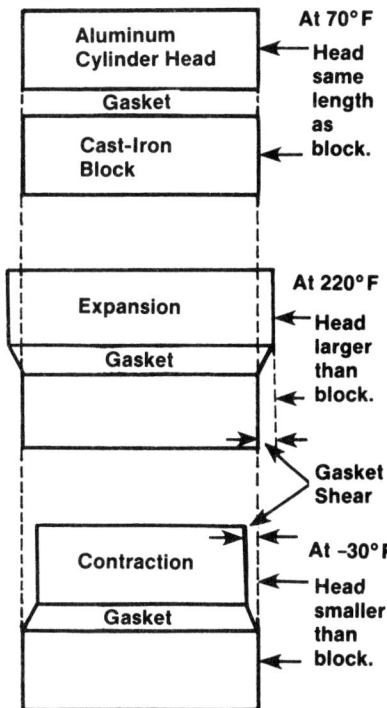

FIGURE 14-26 Bimetal engine thermal growth motions

ing material. To prevent this, slippery nonstick surface coatings and other materials are applied to the gaskets to create the needed lubrication between the head and gasket. Graphite is a natural lubricant, so graphite-faced gaskets are not coated.

Today's bimetal engines, as mentioned earlier, with thin-wall castings are not as rigid as their heavier cast-iron counterparts of a decade ago. There is more movement and flexing between the head, block, and cylinder bores, which means a more rigid gasket is often needed. This can be accomplished by using different facing materials, changing the design and loading of the combustion flanges, and/or by using silicone beading to increase sealing loads in critical areas.

As open deck engines become more common, the head gasket will play an increasingly important role in helping to stiffen the cylinder bores. The design and thickness of the bore flanges can affect the way in which the bores distort when the head is bolted down.

Cylinder Head Gasket Designs and Materials

Head gaskets have several designs. No single head gasket design is best for all applications. Gasket manufacturers engineer the correct head gasket design for each engine.

Some engines are assembled at the factory with the beaded steel-type head gasket that works well on clean and unwarped new cylinder heads. However, the beaded steel gaskets often do not seal the minute scratches and warp distortions common to used engines. Good sealing for service repair requires a perforated or nonperforated steel core covered with a special face-coating material. The coating acts as a microseal for surface imperfections such as minute machine marks.

Many new engines also require a composition-type gasket similar to the one used as a service gasket. This is because less rigid castings allow severe bending and distortion of the mating parts. Also, aluminum components are more likely to be damaged during assembly. Tougher performance requirements for new engines create a need for the composition gasket both on the assembly line and for builders.

The latest generation of aftermarket gaskets includes a bead sealant that increases clamping pressure around troublesome areas (Figure 14-27). Another common and desirable gasket design today is the no-retorque type. Older gasket designs, such as the steel-faced sandwich and perforated core, require the cylinder head of a builder engine to be retorqued after 300 to 500 miles. These gaskets take a set after initial engine operation and relax to the point where retorquing is needed to restore the clamping force (Figure 14-28). The extra labor and expense of retorquing a cylinder head gasket makes the repair more expensive than no-retorque designs in the long run. The no-retorque design retains a higher level of clamping force. Therefore, no retorquing is needed.

No-retorque gaskets typically combine the features of a dense, low compression core with a relatively thin facing material that can be compressed (to compensate for surface irregularities) without a significant loss of clamping force. Flexible graphite, multicoated steel, and various fiber facing products

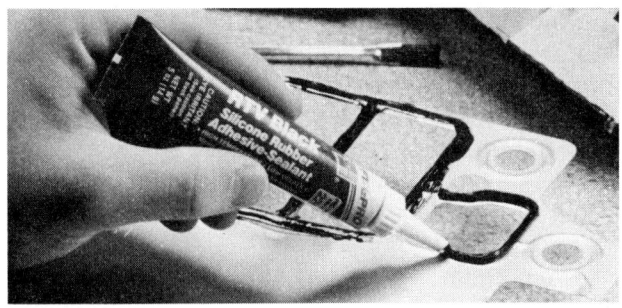

FIGURE 14-27 Clamping can be improved by using an adhesive/sealant *(Courtesy of Fel-Pro Inc.)*

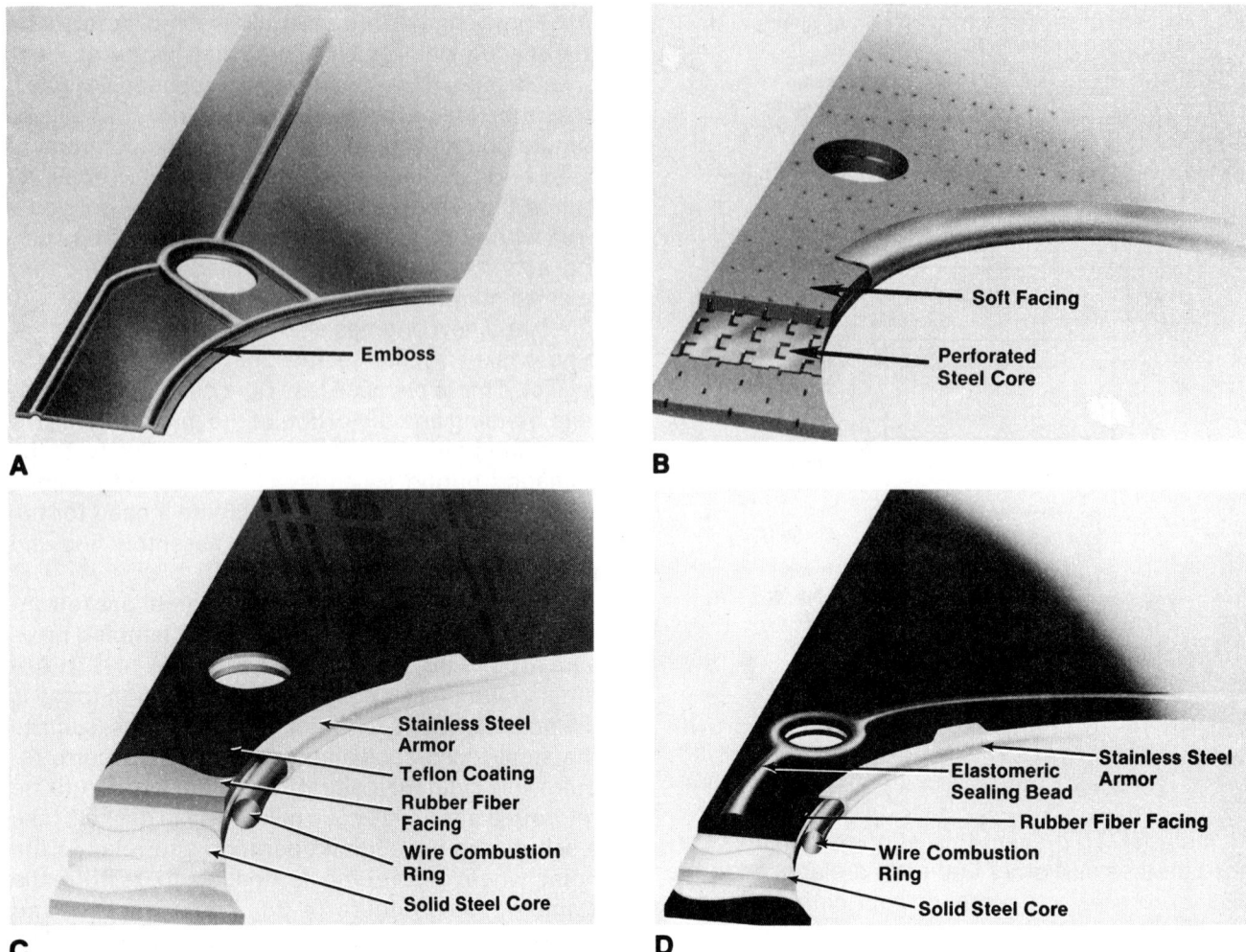

FIGURE 14-28 Types of gaskets: (A) Inexpensive to produce, used in older original equipment applications. Retains torque well. Must have supplementary sealer. Difficult to seal most aftermarket applications where warping and corrosion are found. (B) Original soft-surfaced design seals surface irregularities. No supplementary sealer is needed. Has moderate strength and moderate resistance to blowout. It is recommended that the head bolts should be retorqued after engine operation. (C) Wire ring combustion seal, encased in stainless steel armor added for increased strength in racing and heavy-duty applications. (D) Although the basic construction of this gasket is the same as the one shown in C, it is designed for added racing and heavy-duty applications. *(Courtesy of Fel-Pro Inc.)*

like Kevlar and fiberglass are examples of materials currently used to enhance the gasket's torque retention and sealing capabilities. Many of the new retorquing designs feature solid steel cores or wire combustion seals to help to resist blowouts. With such gaskets, the synthetic rubber sealing beads increase sealing forces at critical fluid openings (Figure 14-29). Usually no sealer is needed with the no-retorque type of gaskets.

To counter the effects of bimetal engine movement, some of the gasket coatings mentioned above also possess excellent antistick qualities. Substances like molybdenum disulfide (Teflon) and graphite provide a slippery surface that prevents the scrubbing action between the head and block from wearing out the gasket.

Installing Head Gaskets

In addition to the general gasket installation suggestions made earlier in this chapter, keep in mind that before installing the new head gasket, check it for fit on the head and block and against the old gasket. Combustion openings should be slightly larger than the engine bore and might have irregular shapes to accommodate valve pockets in the cylinder head. The coolant holes in the gasket might be fewer, smaller, or shaped differently than the coolant passages in the head or block. This is because head gaskets are designed to control or meter coolant flow (Figure 14–30). The gasket you purchase should conform to the most recent OEM recommendations. The installation of a gasket from a reputable manufacturer insures that the coolant holes are of proper size and configuration.

If the gasket surface is too smooth, the gasket cannot grip the surface to give a good combustion seal. If the surface is too rough, the gasket will not

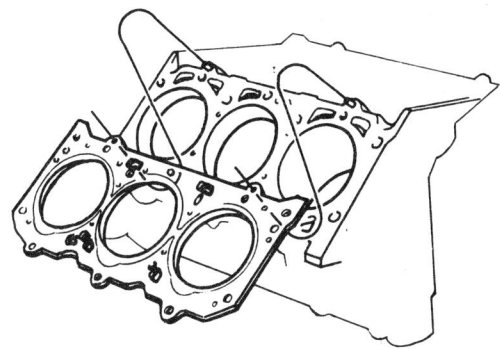

FIGURE 14–30   Coolant and oil flow directed from the engine, through a head gasket, to the cylinder block

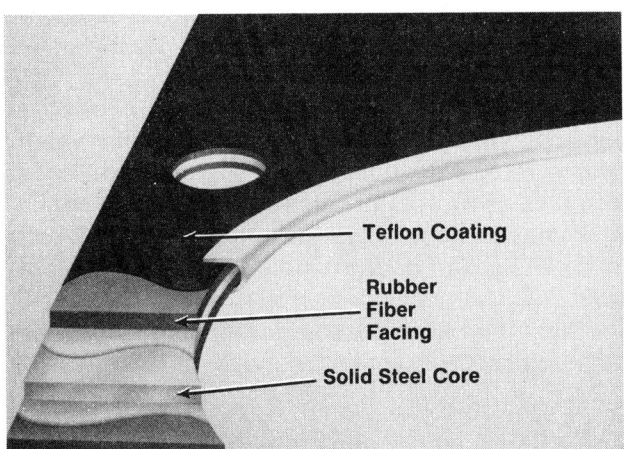

A

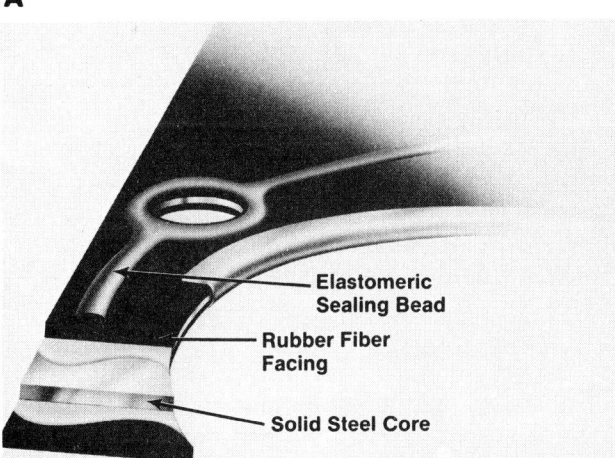

B

FIGURE 14–29   Types of no-retorque gaskets: (A) Solid steel core resists blowout. Rubber/fiber facings seal surface irregularities. Teflon coating creates initial seal, resists sticking, and seals small surface irregularities. Retains a high level of clamping force. No sealer needed. (B) Solid steel core resists blowouts. Rubber/fiber facings seal surface irregularities. Synthetic rubber sealing beads increase sealing forces at critical fluid openings. Retains a high level of clamping force. No sealer is needed *(Courtesy of Fel-Pro Inc.)*

Labels in Figure 14-29A: Teflon Coating, Rubber Fiber Facing, Solid Steel Core

Labels in Figure 14-29B: Elastomeric Sealing Bead, Rubber Fiber Facing, Solid Steel Core

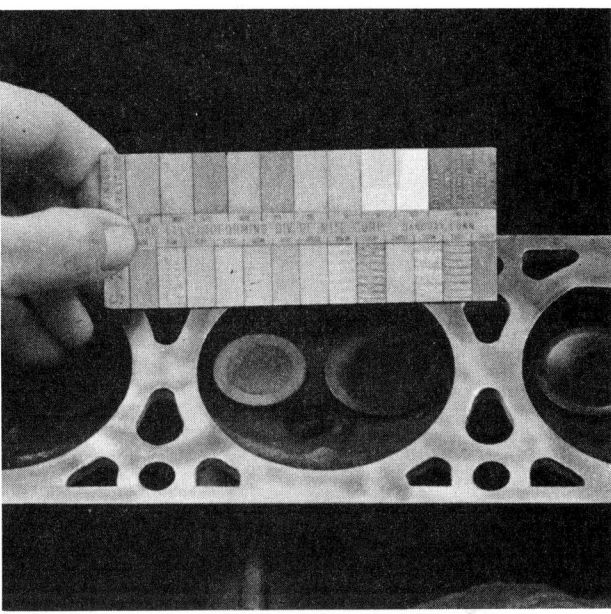

FIGURE 14–31   For optimum sealing, the surface finishes should measure between 90 and 110 microinches (RMS). *(Courtesy of Fel-Pro Inc.)*

conform to the surface, which causes fluid leaks. A finish of 90 to 110 microinches (RMS) is preferred (Figure 14–31). Many head gaskets will not seal when the surface is outside this range.

If the bolts have stretched or excessive machining has been done, the bolts can hit the bottom of blind holes. To check for this problem, put the correct bolt in the head. Put a pencil in the block hole. Mark the hole depth on the pencil. Then transfer the depth to the head bolt sticking out of the head. If it is determined that a bolt will hit the bottom of the hole, put a hardened steel washer under the thread of the bolt before assembly.

Impression Testing

The carbon impression paper technique is growing in popularity as a means of checking the effectiveness of the head gasket seal. Start by cutting a piece of carbon impression paper to about the size of the gasket and follow these steps:

1. Place the paper on top of the old gasket installed on the engine block.
2. Reinstall the cylinder head.
3. Tighten the head bolts to the recommended torque and in the proper sequence (Figure 14–32).
4. Remove the cylinder head.
5. Inspect the impression left on the paper (Figure 14–33). Dim or weak lines indicate

FIGURE 14–32 After carbon paper has been placed on the engine block and the cylinder head is reinstalled, tighten the head bolts to the recommended torque in the specified sequence. *(Courtesy of Victor/Dana)*

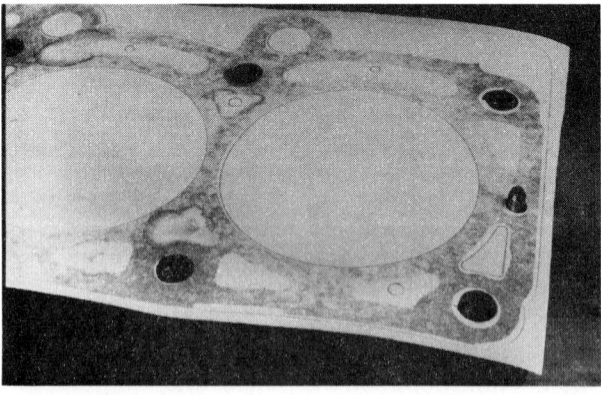

FIGURE 14–33 Inspect the impression left on the carbon paper. The dim or weak lines indicate poor gasket contact due to a warped head. *(Courtesy of Victor/Dana)*

poor gasket contact, probably due to a warped head.

DIESEL ENGINE HEAD GASKETS

Although the functions of cylinder head gaskets for conventional and diesel engines are basically the same, some differences do exist. Diesel engines boast higher combustion pressures, besides their basic design differences. In addition, some diesel engine manufacturers do not recommend cylinder head resurfacing more than 0.010 inch for their 350 cubic inch (5.7-liter) diesel engine. They recommend replacement of any cylinder head that is warped more than 0.010 inch.

Some diesel engine cylinder heads have pipe plugs covering the upper cylinder head bolts, which must be removed before the bolts. The plug threads must also be thoroughly cleaned and coated with a recommended sealer. Repeat this under the bolt heads. Failure to coat the bolts can cause coolant leaks.

Diesel engines have some other unique features related to head gasket installation:

- Dowels to position the cylinder head
- Camshaft key and crankshaft key position
- Specially designed cylinder head bolts for aluminum heads
- Reuse of cylinder head bolts is not recommended.

SHOP TALK

Some/all of the above diesel head features can be found in many gasoline engines.

Generally, the heavy-duty diesel engines in trucks and industrial equipment have replacement cylinder sleeves. These sleeves have a ridge (or bead) that protrudes above the cylinder block deck. The ridge imbeds into the head gasket to increase its ability to seal the extremely high combustion pressures. Some special diesel engine gaskets are made of steel plate (Figure 14–34) that can withstand up to 500,000 foot-pounds of torque load. Each engine make and model requires exacting installation dimensions. Obtain them from the engine manufacturer's service publications for a specific model.

The amount of protrusion is critical and must be measured with a dial indicator gauge while the sleeve is clamped into position (Figure 14–35). Gasket failure will quickly occur if the protrusion is

FIGURE 14-34 Typical solid steel gasket *(Courtesy of Victor/Dana)*

FIGURE 14-35 The amount of protrusion is critical and must be measured with a dial indicator gauge while the sleeve is clamped into position. *(Courtesy of Victor/Dana)*

incorrect. Major cylinder block repair and/or machining might be required if specifications are not met. The cylinder sleeve mounting shoulder in the block can also become worn, thus preventing proper protrusion of the sleeve. This condition must be corrected by replacing or repairing the block.

The gasket sets for some diesel engines include grommets that must be inserted into the water passage holes of each head gasket. Place the head gasket on a clean, level surface and press the grommets into position. Avoid grommet contact with fuel or lube oil. Diesel engines are identified as either single-corner studs or double-corner studs to select the head gasket. In addition, special cylinder head conversion plugs must be installed in selected water passages. Omission of these plugs will cause coolant leakage.

PERFORMANCE ENGINE HEAD GASKETS

Performance engines operate under conditions that demand a great deal from head gaskets. They produce very high horsepower and combustion pressures (frequently in the 1500 to 2000 psi range). Because of high operating speeds, performance engines are subject to high levels of vibration and motion, placing even greater stresses on the gasket's ability to seal.

Most performance head gaskets consist of fiber facings, which are laminated to a steel core. Typically, these gaskets are coated with a nonstick material, which seals minor surface imperfections and helps prevent the gasket from sticking when removed.

Another type of performance head gasket also uses laminated construction but, instead of a nonstick coating, includes sealing beads of elastomeric rubber to provide additional sealing in critical areas. Although these beads appear to be raised above the gasket surface, once compressed they become embedded into the gasket body, providing sealing on both sides of the gasket.

Both types of head gaskets include combustion seals consisting of metal O-rings (usually low-carbon steel or copper) contained within a stainless-steel gasket armor. This design provides highly concentrated sealing stress around the combustion chamber to withstand the high combustion forces.

Many of today's performance head gaskets are of a no-retorque design and retain torque better than conventional gaskets. However, all types of gaskets will exhibit some relaxation of torque, so it might be beneficial to retorque even no-retorque gaskets for maximum sealing and greater reliability under competitive conditions. The greatest torque loss occurs during the first hour of engine operation, so the retorquing should follow a 1-hour run-in and a complete cool down. Retorquing should never be performed while the engine is hot.

MANIFOLD GASKETS

There are three types of manifold gaskets:

- Intake manifold
- Exhaust manifold
- Combination intake and exhaust

Each type of manifold gasket has its own sealing characteristics and problems.

Intake Manifold Gaskets

Intake manifold gaskets come in two types: the valley cover type and the rail type (for V-design engines). The valley cover is known by many different names, including turtle back, turkey pan, dishpan, and bathtub. Intake manifold gaskets seal the joint between the intake manifold and the cylinder head. The gasket primarily seals the air/fuel mixture. Fuel-injected versions may seal air or air/fuel mixture with a vacuum. Some in-line designs also seal coolant that circulates around the passages of the intake manifold. Other intake manifold designs used especially on V-type engines must seal coolant and oil as well as exhaust crossover passages.

 SHOP TALK _____

If a cylinder head or block deck of a V-6 or V-8 engine is extensively machined, the intake manifold must also be machined (see Chapter 9). Before installing the intake manifold, make sure all necessary cleaning is done to assure proper alignment of all passages and bolt holes.

Most of today's intake manifolds are made of aluminum, with little surface area and few supports. Although this keeps weight to a minimum, it creates problems in sealing because aluminum deflects easily. When sealing the intake manifold to the cylinder head, the technician must take maximum advantage of the available bolt load. The load distribution of the bolts, manifold, and head may be uneven, but the gasket can be designed so that it concentrates the load where it is needed most.

Performance intake gaskets typically have been fabricated out of rubber-fiber sheets or compressed sheet materials. Although most rubber-fiber material is soft enough to accommodate warpage, it has a high torque loss and tends to absorb fuel. This reduces its effectiveness and strength and leads to the possibility that the gasket may be drawn into the intake ports. Although compressed gaskets provide higher strength and density and are less likely to absorb fuel, they do not accommodate warpage as easily.

As a solution, today's high-quality performance intake manifold gaskets feature a rubber-fiber sheet made of a rubber fiber and premium binder material that provides good fluid resistance. This process of elastomeric rubber sealing beads has been very successful at countering the nonuniform load distribution. The beads, which are located around each intake and water port, concentrate the available bolt load exactly where needed for optimum sealing. The elastomeric rubber material also provides superior resistance to gasoline and alcohol fuels.

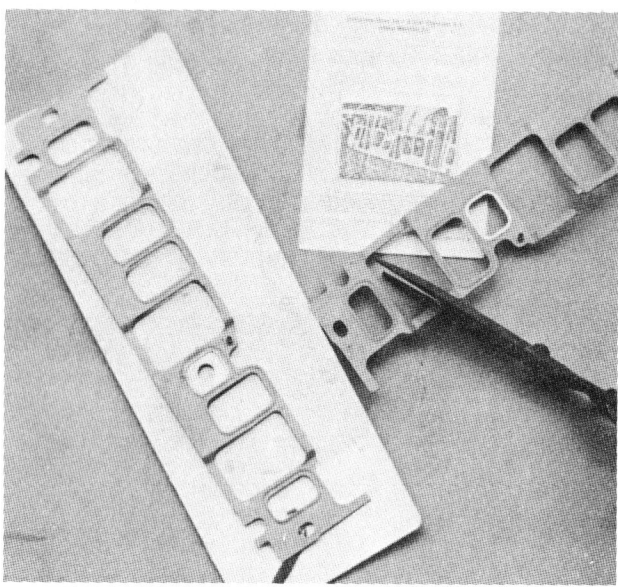

FIGURE 14-36 Cutting out a replacement gasket *(Courtesy of Victor/Dana)*

Most intake manifold replacement gasket sets include the special installation steps and precautions required by the new OE designs. A typical instruction sheet with a gasket shows how to properly cut the manifold gasket for an efficient and effective repair (Figure 14-36). Without the instructions a technician might unnecessarily remove pushrods or cut the gasket improperly.

 SHOP TALK _____

Typical manifold gasket kits are usually marked when required for the installation as follows:

- *R—gasket for the right-hand bank*
- *L—gasket for the left-hand bank*
- *Cut out—indicates proper place to cut*
- *This side up—indicates installation position when the design might not be clearly obvious*

To be sure the gasket will seal, check its fit (Figure 14-37). This will insure proper sealing of the manifold. Remember, only on steel shim-type gaskets is it a must to put a thin and even coat of a positioning sealer around vacuum port openings and a small bead of RTV silicone (described later in this chapter) around the coolant.

To secure the gasket in position, use a small amount of contact adhesive to hold the gasket in

FIGURE 14-37 Inspect the manifold for irregularities that might cause leaks, such as gouges, scratches, or cracks. Check for surface flatness with a straightedge and a feeler gauge and machine only enough to return surface to flat. Replace any parts that are cracked or badly warped. *(Courtesy of Fel-Pro Inc.)*

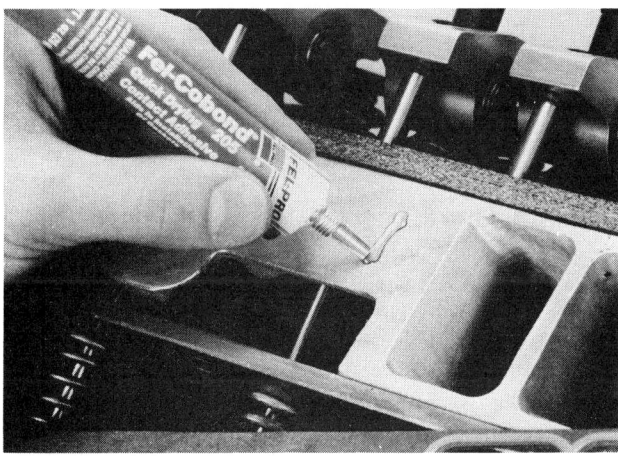

FIGURE 14-38 Applying contact adhesive to hold an intake manifold in place *(Courtesy of Fel-Pro Inc.)*

place (Figure 14-38). Align the gasket properly before the adhesive dries. Allow the adhesive to dry completely before proceeding with manifold installation. On some V-type engines that include end strip seals, the joint where the end strip seals meet the intake manifold gaskets should be sealed with a small bead of RTV silicone. This insures a complete seal around the manifold.

Exhaust Manifold Gaskets

The exhaust manifold gasket seals the joint between the head and exhaust manifold (Figure

FIGURE 14-39 Typical exhaust manifold gaskets *(Courtesy of Fel-Pro Inc.)*

14-39). Many new engines are assembled without exhaust manifold gaskets. This is possible because new manifolds are flat and fit tightly against the head without leaks. During use, the exhaust manifolds go through many heating/cooling cycles. This causes stress and corrosion in the exhaust manifold. Removing the manifold usually distorts the manifold slightly so it is no longer flat enough to seal without a gasket. Exhaust manifold gaskets are used to eliminate leaks when exhaust manifolds are reinstalled.

Some exhaust manifold gaskets have a perforated steel face on one side and a soft face on the other. The steel face is positioned toward the exhaust manifold on the engine because the exhaust manifold expands when it is heated during operation. As it expands, it moves on the head. This design seals the exhaust manifold while allowing it to move on the metal surface of the gasket, just like it does on the metal surface of the head in the absence of a gasket.

A new gasket design uses a ceramic-based facing on perforated steel. The ceramic-based facing does an excellent job of sealing high-temperature engines found on late-model vehicles.

Many heavy-duty engines use a separate embossed shim gasket at each port, usually with a single bead embossment. A double bead embossment is used in more critical sealing applications.

Combination Intake and Exhaust Manifold Gaskets

In-line engines that have the intake and exhaust manifold on the same side of the head often use a combination gasket (Figure 14-40). This gasket has a perforated steel core with the facing on one side at the exhaust ports and facings on both sides at the intake ports. Special steps must be taken when installing these gaskets to prevent intake or exhaust leaks. Instructions packaged with the gaskets describe the required installation steps.

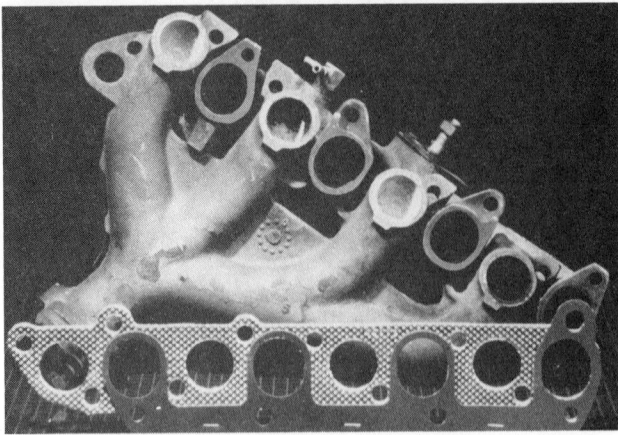

FIGURE 14-40 Typical combination intake/ exhaust manifold gasket *(Courtesy of Fel-Pro Inc.)*

VALVE COVER GASKETS

Valve cover gaskets have an increasingly difficult job in today's high-temperature engines. They must seal between a steel or aluminum (or molded plastic) stamping and a cylinder head surface that might not even be machined. Since the valve cover does not seal in pressure (the bolts that attach the cover to the cylinder head are usually widely spaced), it might appear that almost any gasket could do the job and, for the short run, almost any material can. But the gasket must also withstand high temperatures and the caustic action of acids in the oil.

To provide an effective seal, a performance valve cover gasket must conform to flange distortion (Figure 14-41). At the same time, it must provide good crush and split resistance to limit torque loss and chemical and heat resistance for long-term sealing. The most commonly used material is a blend of cork and rubber particles. The quality of the materials is crucial when using cork/rubber gaskets, because a low-quality cork/rubber mixture can develop leakage after a fairly short period of operation. Another popular valve cover material is made of synthetic rubber. But whether it is a cork/rubber or synthetic rubber gasket, it must fit properly (Figure 14-42).

When installing the gasket, apply a contact adhesive to the cover's sealing surfaces in small dabs. Mount the gasket on the cover and align it in position. If the gasket has mounting tabs, use them in tandem with the contact adhesive. Allow the adhesive to dry completely before mounting the cover on the head. Do not use a nonhardening sealer on the gasket; it will become a lubricant and might allow the gasket to slip or squeeze out of the joint. This is critical when installing rubber gaskets. High-quality

FIGURE 14-41 Examine the cover for cracks, flaws, and any distortion. Use a straightedge to check surface flatness. Tap distorted bolt holes with a hammer. Replace the cover if it is cracked or excessively warped. *(Courtesy of Fel-Pro Inc.)*

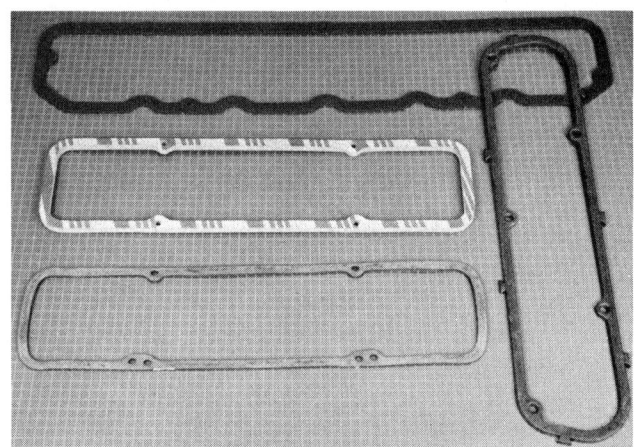

FIGURE 14-42 There are two primary valve cover gasket materials used for valve cover gaskets: synthetic rubber and cork/rubber compounds *(Courtesy of Fel-Pro Inc.)*

cover gaskets need no sealer because the gasket does the complete job.

OIL PAN GASKETS

The oil pan gasket seals the joint between the oil pan and the bottom of the block (Figure 14-43). In many cases, the oil pan gasket might also seal the bottom of the timing cover and the lower section of the rear main bearing cap.

It is difficult to locate oil pan gasket leaks. They might be in areas not immediately seen, making the oil pan gasket one of the most ignored gaskets on an

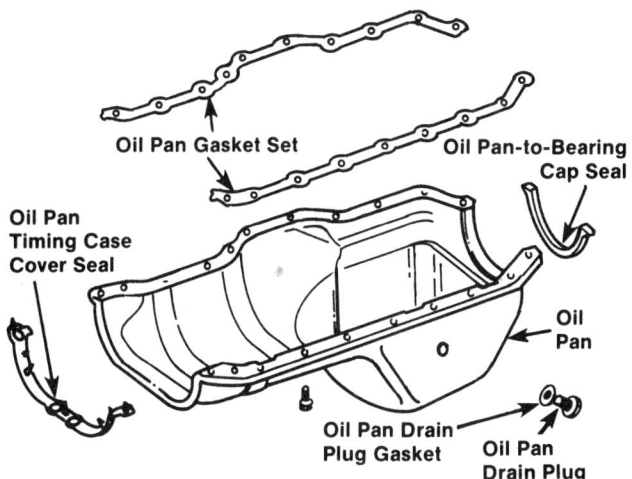

FIGURE 14-43 Typical oil pan assembly

Oil Pan Gasket Set

Oil Pan-to-Bearing Cap Seal

Oil Pan Timing Case Cover Seal

Oil Pan

Oil Pan Drain Plug Gasket

Oil Pan Drain Plug

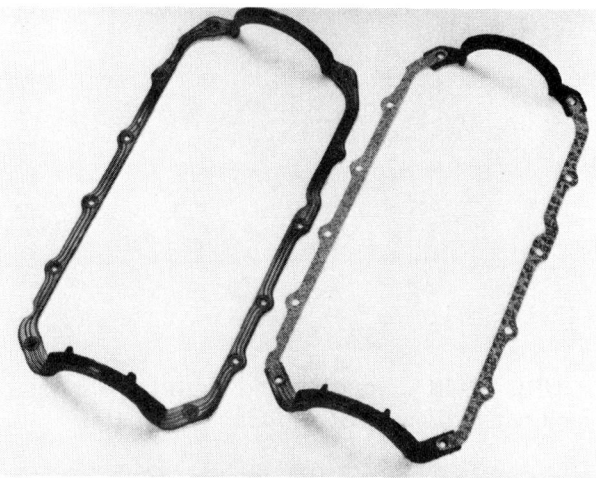

FIGURE 14-44 Typical one-piece oil pan gaskets *(Courtesy of Victor/Dana)*

engine. Oil leaks might be blamed on other gaskets or seals rather than the oil pan gasket.

Like valve cover gaskets, the oil pan gasket must resist hot, thin engine oil. Oil pans are usually made of stamped steel, but with a stronger gasket flange. Because of the added weight and splash of crankcase oil, the pan has more assembly bolts closely spaced and sometimes larger diameters than valve covers. As a result, the clamping force on the oil pan gasket is much greater, so the gasket is thinner and must resist crushing.

Oil pan gaskets, which are similar to valve covers, are made of several types of material. A popular material is synthetic rubber, known for its long-term sealing ability. It is tough, durable, and resists hot engine oil. Synthetic rubber gaskets are easy to remove, so the sealing surfaces need less cleanup.

Cork/rubber compounds are used for oil pan gaskets with many of the best properties of both cork and synthetic rubber. Cork/rubber resists high engine temperatures and crushing.

A third type of oil pan gasket has been developed for use in a harsh environment, such as racing. This oil pan gasket has a fiber core with a latex rubber coating. The fiber gives the gasket higher strength to resist crushing, blowout, heat, and over-torquing. The rubber coating aids in sealing fluid as well as in removal and cleanup.

Some oil pan gaskets are made in several segments for material and packaging economy. Others utilize single-piece designs. Single-piece oil pan gaskets are easier to install than the segmented type because there are no joints to be aligned (Figure 14-44). Molded rubber oil pan gaskets, especially the single-piece designs, are popular on import engines and are becoming more popular in new domestic engines because they do an excellent job of long-term sealing.

Before installing the oil pan gasket, check the flanges for warpage. Use a straightedge or lay the cover or pan, flange side down, on a flat surface with a flashlight underneath it to spot uneven edges (Figure 14-45). Check particularly around bolt holes. Minor distortions can be corrected with a hammer and block of wood (Figure 14-46). If the flanges are too bent to be repaired in this manner, the part should be replaced.

Carefully follow the chemical recommendations on the gasket package label and instruction forms. Read through all the installation steps before beginning. Take note of any of the original equipment manufacturer's recommendations in automotive manuals that could affect engine sealing. Check the assembly to determine where to mount the oil pan gasket(s). If the oil pan end seal has an extension that locates into the engine block (Figure 14-47),

FIGURE 14-45 Checking oil pan for flatness *(Courtesy of JP Industries, Inc.)*

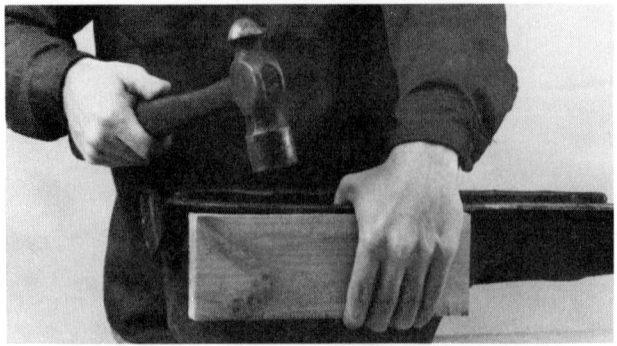

FIGURE 14-46 Straightening distorted flanges of an oil pan *(Courtesy of JP Industries, Inc.)*

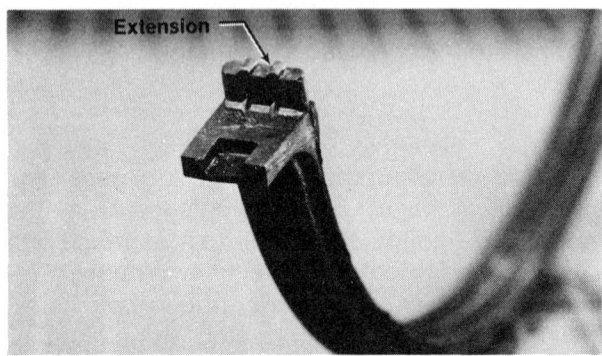

FIGURE 14-47 Oil pan extension *(Courtesy of Fel-Pro Inc.)*

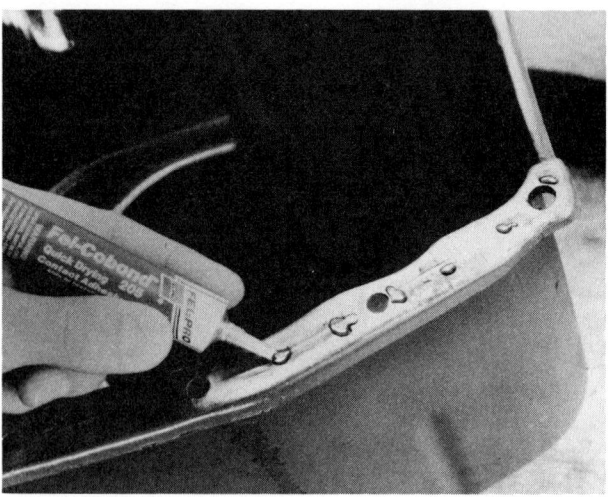

FIGURE 14-48 Securing the oil pan gasket in position. Determine where to mount the oil pan gasket(s) on the oil pan or on the block before applying the adhesive. *(Courtesy of Fel-Pro Inc.)*

rear main bearing cap, or the underside of the timing cover, the oil pan gasket(s) is attached to the engine block. Otherwise, the oil pan gasket(s) is attached to the oil pan before assembly of the oil pan to the block. The gasket should be mounted with a few dabs of quick-drying contact adhesive (Figure 14-48). Carefully align the gasket before the adhesive dries. Wait until the adhesive is dry before installing the pan.

ADHESIVES, SEALANTS, AND OTHER CHEMICAL SEALING MATERIALS

There are a number of chemicals that can be used to reduce labor and insure a good seal. Many gasket sets include a label with the proper chemical recommendation for use with that gasket set. Some even include sealers in the sets when the original equipment manufacturer used a sealer to replace a gasket and a gasket cannot be manufactured for that application. They also include sealers in some sets where gasket unions need a chemical sealant to assure a good seal.

Chemical adhesive and sealing compounds have long been used to enhance a gasket's performance, aid in installation, or facilitate removal. Today, however, the question of whether or not a supplementary sealing agent is needed is the subject of considerable, and inconclusive, discussion.

As previously stated, most high-quality precut gaskets feature a variety of special coatings that contribute to the overall sealing effectiveness of the gasket. For instance, soft synthetic coating materials like Teflon are designed to flow into surface imperfections, thereby making a chemical sealer unnecessary. Other high-tech coatings are designed to lubricate bimetal seams, improve torque retention, supply additional clamping load, or act as release agents during disassembly. Under these circumstances, the addition of a chemical sealer might impair or duplicate a feature that is already present.

 SHOP TALK _____

Chemical adhesives and sealants give added holding power and sealing ability where two parts are joined. Sealants usually are added to threads where fluid contact is frequent. Chemical thread retainers are either aerobic (cures in the presence of air) or anaerobic (cures in the absence of air). These chemical products can be used in place of lock washers.

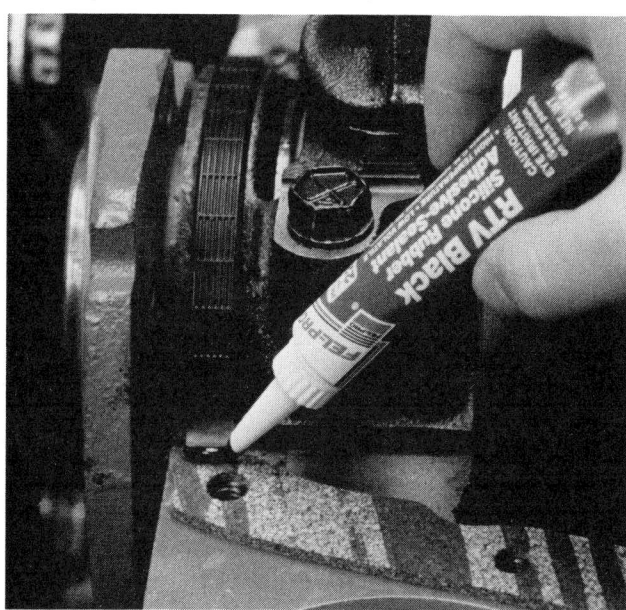

FIGURE 14-49 Applying a dab of silicone adhesive to the corner of a gasket *(Courtesy of Fel-Pro Inc.)*

A

B

FIGURE 14-50 Applying (A) a brush type and (B) an aerosal type of general-purpose sealant *(Courtesy of Fel-Pro Inc.)*

Of course, there are numerous locations in an engine where precut or premolded gaskets can benefit from the services of a chemical sealant. For example, to seal the intake manifold on a V-type engine, it is important to place a dab of silicone in the corners. The same can be said about the front cover-to-oil-pan joint, rear bearing cap seals, valley pan manifold installations, and so on (Figure 14-49).

ADHESIVES

Quick-drying contact adhesive is designed for bonding cork, rubber, fiber, and metal gaskets in place prior to assembly. Gasket adhesives form a tough bond when used on clean, dry surfaces. Adhesives do not aid the sealing ability of the gasket. They are meant only to hold gaskets in place during component assembly. Use small dabs; they will dry quicker for fast installation. Do not assemble components until the adhesive is completely dry. Most adhesives are ideal for use on gasket applications such as valve covers, pushrod covers, manifold and manifold end seals, and oil pan and oil pan end seals.

SEALANTS
General-Purpose Sealants

General-purpose sealers (sometimes called *chemical positioning agents*) come in liquid form and are available in a brush type (known as *brush*

tack) and an aerosal type (known as *spray tack*). These materials (Figure 14-50) form a tacky, flexible seal when applied in a thin, even coat that aids in gasket sealing by helping to position the gasket during assembly. The chemicals in a general-purpose sealant will not upset the designed performance of most mechanical gaskets. The possible exception to this is that sealant manufacturers do not recommend their use on rubber parts since these sealants are nonhardening and can cause rubber gaskets to slip.

CAUTION: Never use a hard-drying sealant (such as shellac) on gaskets. It will make future disassembly extremely difficult and might damage the gasket material.

Flexible Sealants

These sealants are most often used in rebuilding on threads of bolts that go into fluid passages (Figure 14-51). They are nonhardening sealers that fill voids, preventing the fluid from running up the threads. They resist the chemical attack of lubri-

FIGURE 14-51 When using RTV, assemble the joint as soon as possible before the sealant begins to cure. Be sure the bead of RTV is not disturbed when assembling the engine components. *(Courtesy of Fel-Pro Inc.)*

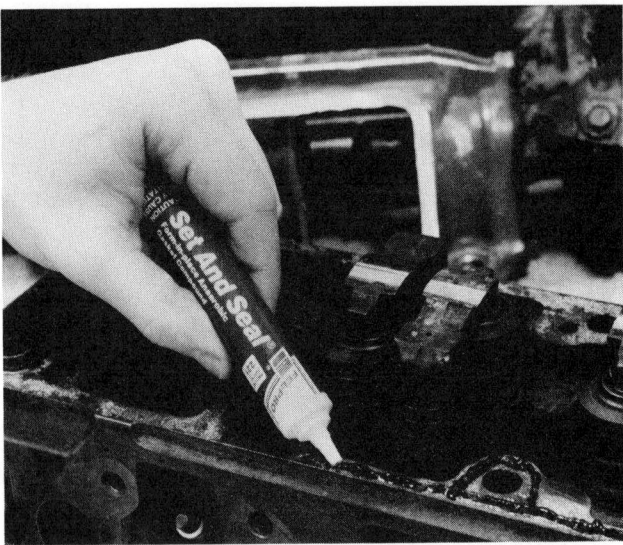

FIGURE 14-52 Using an anaerobic material as a formed-in-place gasket *(Courtesy of Fel-Pro Inc.)*

cants, synthetic oils, detergents, antifreeze, gasoline, and diesel fuel.

Silicone Formed-in-Place Sealants

Silicone gasketing can be used to replace conventional paper, cork, and cork/rubber gaskets. It is generally for use on oil pans, valve covers, thermostat housing, timing covers, water pumps, and other such installations. RTV silicone (room temperature vulcanizing) sealing products are the best known of the formed-in-place (FIP) gasket products. Although RTV silicone has been available in the automotive field since the early 1970s, the present formations alleviate most of the problems that plagued the earlier ones.

Today's RTV aerobic silicone formulations are impervious to most automotive fluids, extremely resistant to oil, oxygen-sensor safe, exhibit outstanding flexibility (a necessary feature on modern bimetal engines), and adhere well to a broad range of materials that include plastics, metal, and glass.

To use RTV silicone, proceed as follows:

1. Make sure that the mating surfaces are free of dirt, grease, and oil.
2. Apply a continuous 1/8-inch bead on one surface only (preferably the cover side), making sure to circle all bolt holes. Adjust the shape before a skin forms (in about 10 minutes).
3. Remove excess RTV silicone with a dry towel or paper towel.
4. Press the parts together. Do not slide the parts together—this will disturb the bead.
5. Tighten all retaining bolts to the manufacturer's specified torque.
6. Cure time is approximately 1 hour for metal-to-metal joints and can take up to 24 hours for 1/8-inch gaps.

WARNING: The uncured rubber contained in RTV silicone gasketing irritates the eyes. If any gets in your eyes, immediately flush with clean water or eyewash. If the irritation continues, see a doctor.

Anaerobic Formed-in-Place Sealants

These formed-in-place materials are used for thread locking as well as gasketing (Figure 14-52). As a retaining compound, they are mostly used to hold sleeves, bearings, and locking screw nuts in place where there is a high exposure to vibration.

CAUTION: Never use a sealant or formed-in-place gasket material for exhaust manifolds.

The major difference between aerobic and anaerobic sealants other than their method of curing is their gap-filling ability. Typically, 0.050 inch (3/64 inch) is the absolute limit of any anaerobic's gap-filling material (some are only designed to seal 0.005- to 0.010-inch gaps). Anaerobic sealers are intended to be used between the machined surfaces of rigid castings, not on flexible stampings (Figure 14-53).

 SHOP TALK _____

Once hardened, a good anaerobic bond is unbelievably tenacious and can withstand high temperatures. Therefore, care must be taken in their selection. They tend to be highly specialized and not readily interchangeable. For example, there are various levels of thread-locking products that range from medium-strength antivibration agents to high-strength, weld-like retaining compounds. The inadvertent use of the wrong product could make future disassembly an impossibility. Check the label to be certain that anaerobic material will suit the purpose of the application.

Hylomar

Hylomar, which stands for high temperature (hy), low cost (lo), Marston (mar) product, is neither an RTV nor an anaerobic. Chemically speaking, it is a combination of polyurethene paste and silica (not silicone) flakes mixed with methylene chloride solvent. When Hylomar is clamped in a joint, the silica flakes interlock and encapsulate the plastic paste, effectively shielding it from heat, liquids, and contaminants that might otherwise dissolve it. Because Hylomar never hardens or cures, the center remains soft and pliable—like an armor-plated sponge.

As a sealing supplement, Hylomar sticks to virtually any surface, it resists all fluids (including gasoline), and has a claimed temperature range of 60 to more than 600 degrees Fahrenheit. In addition, if a Hylomar-coated gasket is set down wrong, it can be peeled off and reseated without damage.

ANTISEIZE COMPOUNDS

These chemical-type materials are used on many fasteners, especially those used with aluminum parts. Its use prevents dissimilar metals from reacting with one another and seizing (Figure 14-54). Be sure to follow the manufacturer's recommendations on the use of this compound.

OIL SEALS

The job of seals is to keep oil and other vital fluids from escaping around a rotating shaft. They are pressed into stationary bores in the crankshaft and timing gear cover and generally consist of a metal casing to which a synthetic rubber sealing element is bonded.

FIGURE 14-53 Anaerobic compounds are often called thread locking and retaining compounds. *(Courtesy of Fel-Pro Inc.)*

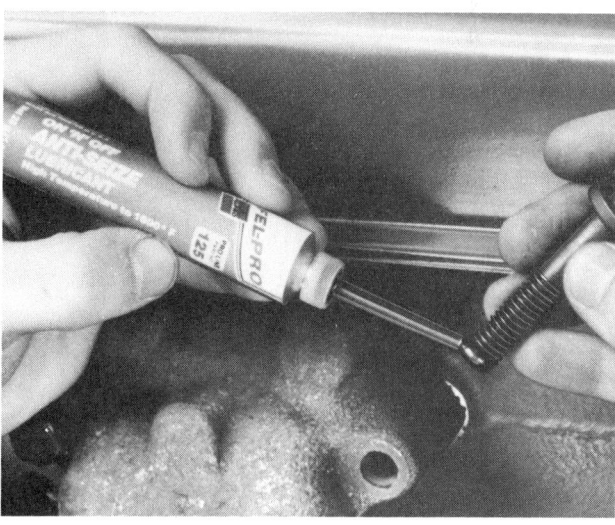

FIGURE 14-54 Using a high-temperature antiseize lubricant *(Courtesy of Fel-Pro Inc.)*

TIMING COVER OIL SEALS

An oil seal prevents oil from leaking out around the crankshaft at the front of the engine. The seal is usually located in the timing cover. It seals on the shaft hub of the harmonic balancer pulley hub or on the crankshaft (Figure 14-55). Sometimes this seal will wear a groove in the shaft or hub. A replacement seal will not be reliable if it is installed on the worn shaft. It will only seal properly on a smooth surface.

If the balancer shaft or pulley hub has a groove, it might have to be replaced or welded and ground to its original size and finish. To replace the sleeve, use a rebuilder replacement set that includes a sleeve, new front seal, anaerobic sealant, and any other gasket components required to seal the timing cover assembly. The sleeve will cover the groove in the balancer shaft or pulley hub and provide a new sealing surface for the front oil seal.

The hub or shaft of the harmonic balancer should be thoroughly cleaned. Apply a film of anaerobic sealant around the shaft. Start the sleeve on the shaft with the flared end. Carefully drive the sleeve onto the shaft with a press or a hammer and a clean block of wood until the sleeve is seated (Figure 14-56). Wipe the excess sealer from the shaft. If it is necessary to replace a sleeve, it can be loosened easily, using a dull cold chisel to indent the sleeve. This will stretch the sleeve enough so it will slide easily from the shaft.

When replacing the timing cover, remove the old gaskets and seals from the timing cover and

FIGURE 14-56 Driving the sleeve on the harmonic balancer pulley hub *(Courtesy of Fel-Pro Inc.)*

engine block. If the timing cover extends over the front lip of the oil pan, the front portion of the oil pan gasket will be exposed. With a sharp knife or razor blade, carefully cut off the front exposed portion of the oil pan gasket (Figure 14-57). Stubborn pieces of gasket debris can be softened and removed using a gasket remover, but avoid spraying any inside the engine. Use a nonmetallic scraper on aluminum and

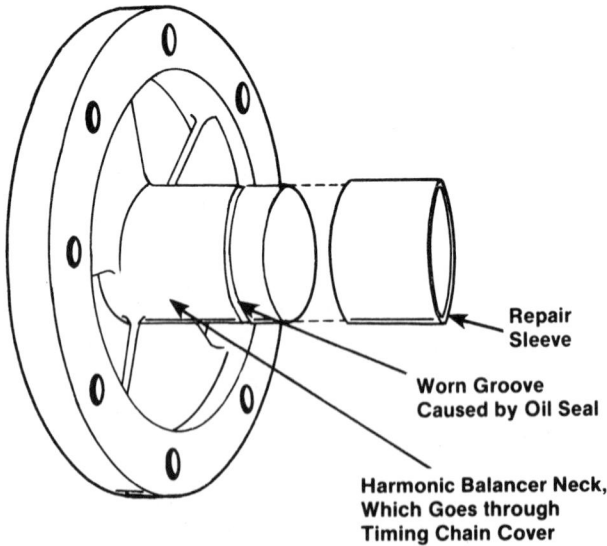

Repair Sleeve

Worn Groove Caused by Oil Seal

Harmonic Balancer Neck, Which Goes through Timing Chain Cover

FIGURE 14-55 Checking the harmonic balancer pulley hub. Replacement sleeves are available for worn valence necks (see Chapter 12).

FIGURE 14-57 Installing an oil pan gasket segment piece *(Courtesy of Fel-Pro Inc.)*

plastic oil pans and timing covers. Finish cleaning the surface with cleaner.

Install the new seal using a press, seal driver, or hammer and a clean block of wood (Figure 14-58). When installing the seal, be sure to support the cover underneath to prevent damage. Hold the new sections of the pan gasket and the molder seal, if supplied, in place with contact adhesive. Trim the ends of the new gasket segment at the block end. Then apply a small bead of RTV silicone sealant at the joint between the new segment and original pan rail gasket (Figure 14-59). Apply another bead of RTV at the joint between the new section of a pan rail gasket and molded rubber front oil pan seal, if so equipped.

FIGURE 14-58 Installing the new seal using a hammer and clean block of wood *(Courtesy of Fel-Pro Inc.)*

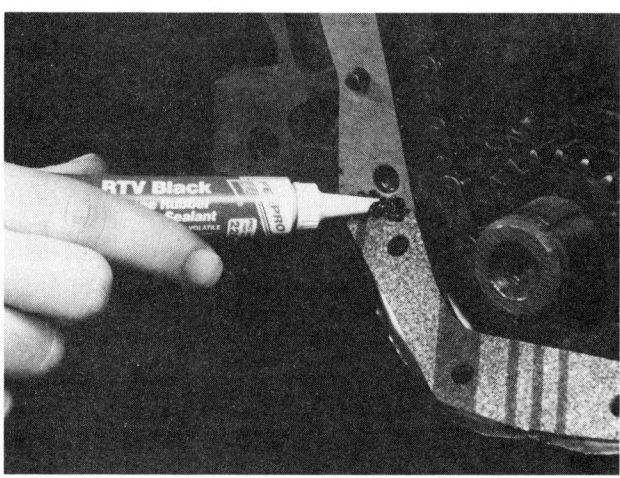

FIGURE 14-59 Apply a small bead of silicone sealant at the joint between the new segment and original pan rail gasket. *(Courtesy of Fel-Pro Inc.)*

REAR MAIN BEARING SEATS

Always make it a practice to replace oil seals on re-ring or overhaul jobs to avoid the chance of costly do-overs. This is a simple, inexpensive operation when the engine is torn down, but a costly and time-consuming job when seals start to leak after assembly.

Rear main bearing seals keep oil from leaking at the crankshaft around the rear main bearing. There are two basic types of constructions: wick- or rope-type packing and molded synthetic rubber.

Wick- or rope-type packings are common on many older engines. Molded synthetic rubber lip-type seals are used on many newer engines. They do a good job of sealing even when there is some eccentricity in the shaft, as long as the surface of the shaft is very smooth. Synthetic rubber seals may sometimes be retrofitted to some older engines that have wick seals, but only if the seals are offered as an option by the sealing manufacturer.

Three types of synthetic rubber are used for rear main bearing seals. Polyacrylate is commonly used because it is tough and abrasion resistant, with moderate temperature resistance to 350 degrees. Silicone synthetic rubber has a greater temperature range, but it has less resistance to abrasion and is more fragile than polyacrylate. Silicone seals must be handled carefully during installation to avoid damage. Viton (see Chapter 10) has the abrasion resistance of polyacrylate and the temperature range of silicone, but is the most expensive of the synthetic types.

No matter what the construction of the seal is, always check the shaft for smoothness. Shafts should be free of nicks and burrs to assure long oil seal life. Carefully remove any roughness with a very fine emery cloth and then clean the shaft thoroughly. The shaft should have a highly polished appearance and a smooth feel. Also, be sure to check and clean the oil slinger and oil return channel in the bearing cap.

Rubber Oil Seals

The split-type rubber formed seal is often used in rear main crankshaft applications. Remove the oil seals, including both the upper and lower split lip seals around the crankshaft, and the small side seals on the rear main bearing cap (Figure 14-60). The lower split seal and the side seals can be gently pried out of the bearing cap with a suitable tool. Be careful not to scratch or gouge the surface of the cap itself. The upper split seal can be tapped out gently by using a pin punch to start it moving, then pulling it

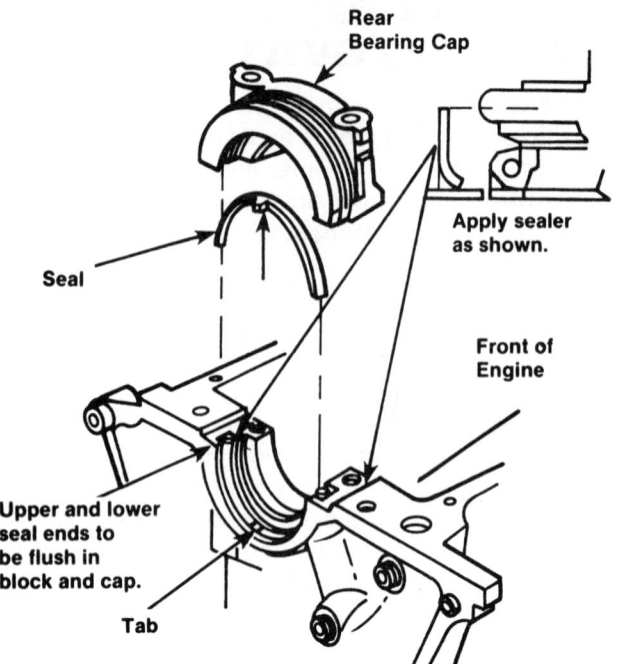

FIGURE 14-60 Two-piece rubber or neoprene seal

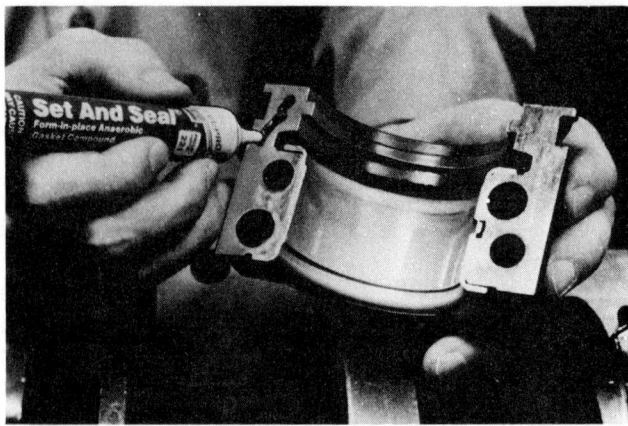

FIGURE 14-61 Applying anaerobic sealer to mating surfaces of the bearing cap *(Courtesy of Fel-Pro Inc.)*

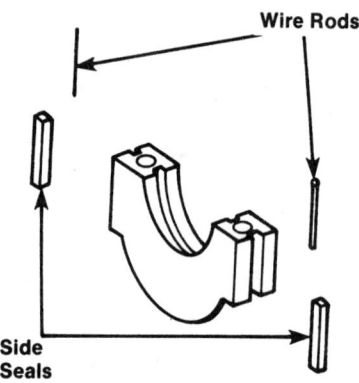

FIGURE 14-62 Some rear main engine bearings have side seals.

out with a pair of pliers. Thoroughly clean the crankshaft bearing cap and engine block sealing surfaces of oil deposits, oil sealer, and all foreign matter. Inspect all these areas for scratches, nicks, and gouges and remove imperfections with crocus cloth or fine emery paper.

Lubricate the lips of the new split seals with oil before installing. The upper split seal should be inserted into the groove in the engine block and rotated around the crankshaft, making sure that the sealing lip faces toward the front of the engine. In some cases engine repair manuals might suggest the use of a special tool to ease installation of this seal. Follow the same procedure to insert the lower split seal into the rear main bearing cap. Lubricate the sealing area on the crankshaft with a thin film of grease to prevent the seal from running dry. Apply a very thin coat of anaerobic sealer to the mating area of the bearing caps, being careful to avoid the ends of the rubber seals, and bolt the cap onto the engine block (Figure 14-61). Always torque all cap bolts to the engine manufacturer's specifications.

Some engines with a deep block skirt or rail have side seals on the rear main bearing cap or on a seal retainer. Dip each rear main bearing side seal in oil and insert seals into the cavities between the cap and the block, beveled end first. Gently tap in side seals until they are properly seated. Some side seals also have wire rods that should be tapped in place inside the seal to assure a secure fit (Figure 14-62).

Check the contents of the gasket set—if wire rods are provided, be sure to install them. Seals without rods will swell in place after the engine heats up to make a tight seal. Slight leakage might occur with the no-rod seals when the engine is first started, but it should stop after warm-up.

There are also one-piece rubber rear crankshaft oil seals (Figure 14-63A). To install this type, coat a new oil seal lip (Figure 14-63B) and the crankshaft oil seat casting (Figure 14-63C) with a light film of engine oil. Start the seal in the recess with the seal lip facing forward and install it with the rear oil seal replacer tool as shown in Figure 14-63D. Keep the tool straight with the centerline of the crankshaft and install the seal until the tool contacts the cylinder block surface. Remove the tool and inspect the seal to ensure it was not damaged during installation.

Rope- or Wick-Type Seals

To install rope-type oil seals, begin by placing the seal against the seal groove (Figure 14-64A).

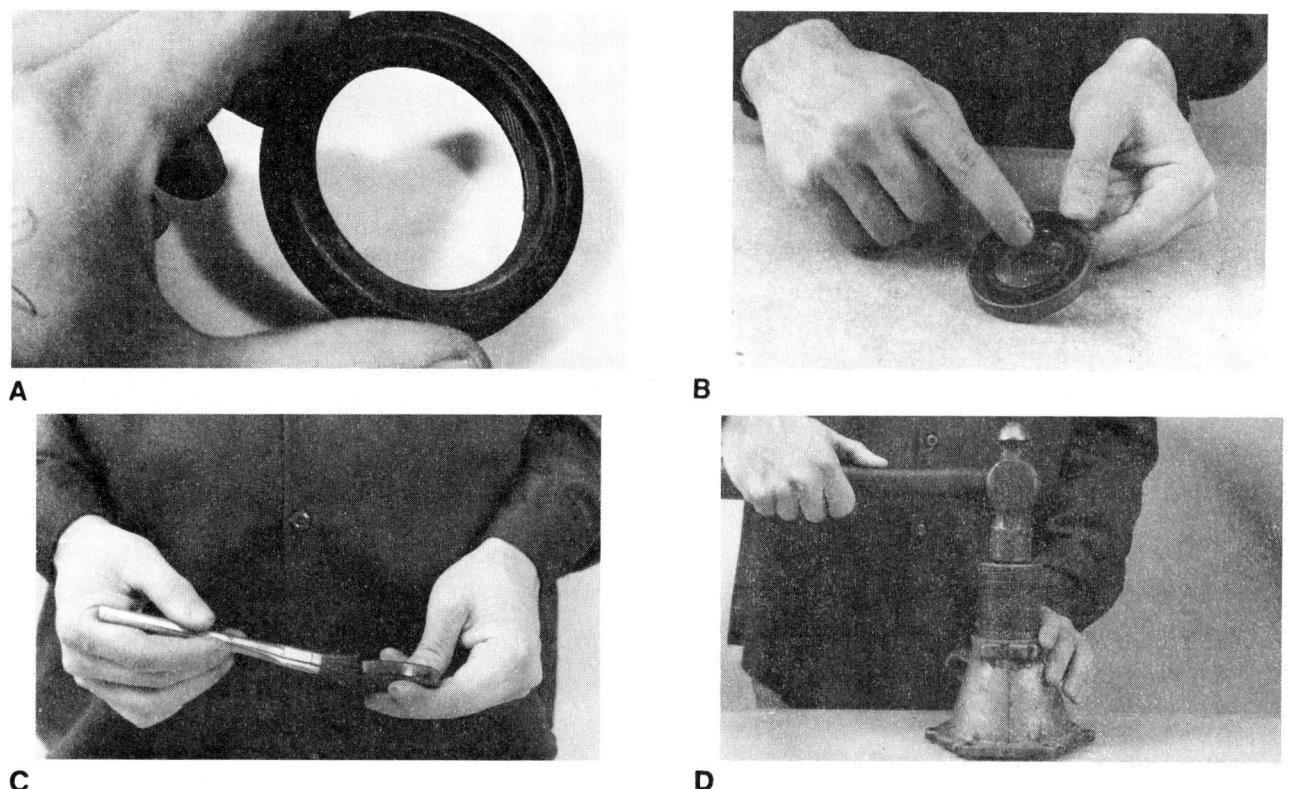

FIGURE 14-63   Steps for installing a one-piece rubber crankshaft oil seal *(Courtesy of JP Indusitres, Inc.)*

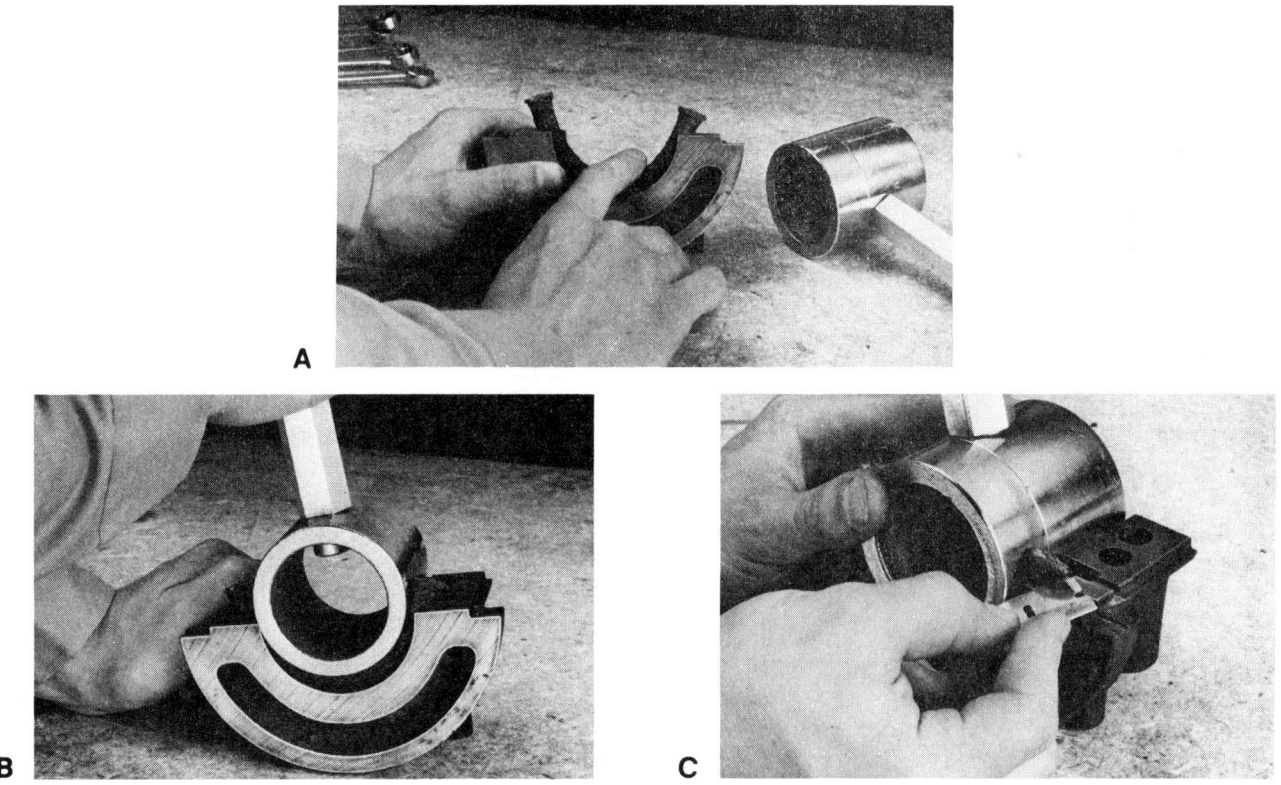

FIGURE 14-64   Steps for installing rope- or wick-type seals *(Courtesy of Fel-Pro Inc.)*

Using a rope packing installation tool or a smooth, round implement, such as a hammer handle or dowel, roll the seal into the groove until it bottoms (Figure 14-64B). Rolling the seal into place will prevent crushing of the rope fibers. When the seal has been rolled into place, trim off the excess with a sharp blade (Figure 14-64C). Liberally coat the seal with oil and allow time for it to soak in before installing the crankshaft.

CAUTION: Do not soak wick- or rope-type seals in oil *before* installation. To properly seal, the inner core threads of the rope must stay dry until completely installed. Soaking the rope can cause disintegration.

Miscellaneous Seals

Different parts of the engine have different sealing requirements. Some gaskets must seal pressure, others must seal vacuum. Some seal hot oil and others hot antifreeze. Some gaskets seal joints under flexible flanges, others seal joints between fairly rigid flanges held together with large mounting bolts. The construction of each gasket matches its sealing requirement. Choose the correct gasket or sealing material to insure that the proper sealing job is achieved.

Some overhead cam engines, for example, have solid, one-piece cam bearings and cam bearing supports. It is necessary to have a semicircular notch in the front and back of the head for machining the cam bearing bore or saddle at the factory (Figure 14-65). The notch is sealed with molded synthetic rubber semicircular plugs. These plugs can deteriorate in time, taking a set, and should be replaced anytime a cam cover gasket is replaced.

Most gasket sets for V-type engines include manifold end strip seals. They seal the joint between the manifold ends and the block (Figure 14-66). Molded rubber and cork/rubber are materials often used for end strip seals. Some have self-adhesive backings to aid in installation. The corner joints where the end strip seals meet the intake manifold gaskets should be sealed with a small amount of RTV silicone. This insures a complete seal all around the manifold.

The exhaust gas recirculation (EGR) valve recirculates some of the exhaust gases back into the engine, reducing NO$_x$ emissions by lowering the combustion temperature. The EGR valve gasket seal

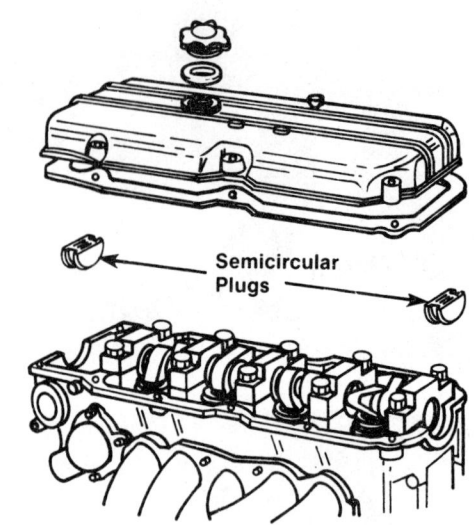

FIGURE 14-65 Typical OHC cover seals

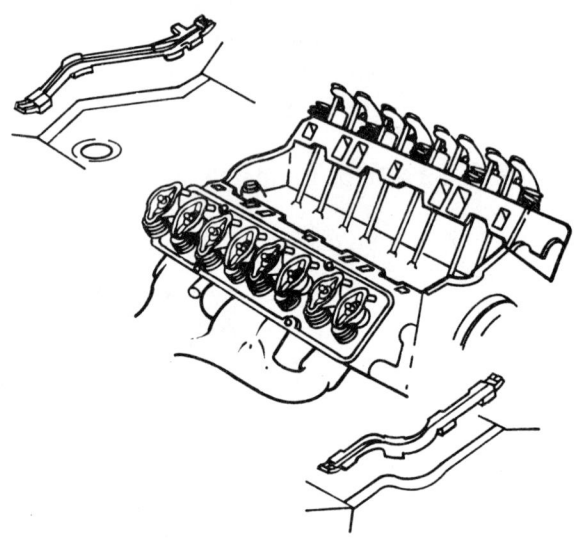

FIGURE 14-66 Typical end strip seals

(Figure 14-67) seals the valve and, in some applications, regulates the flow of exhaust by a specially sized hole in the gasket. Choosing the correct gasket for this application is critical because a "one-size-fits-all" gasket could change the exhaust gas flow and, as a result, change the performance of the engine.

The carburetor mounting gasket, located between the carburetor and the manifold, prevents vacuum and fuel leaks. Hotter running late-model engines require thicker carburetor mounting gaskets to help insulate the carburetor from engine heat. High-quality carburetor mounting gaskets use ferrules at the bolt holes (Figure 14-68), which help prevent carburetor base distortion that could dam-

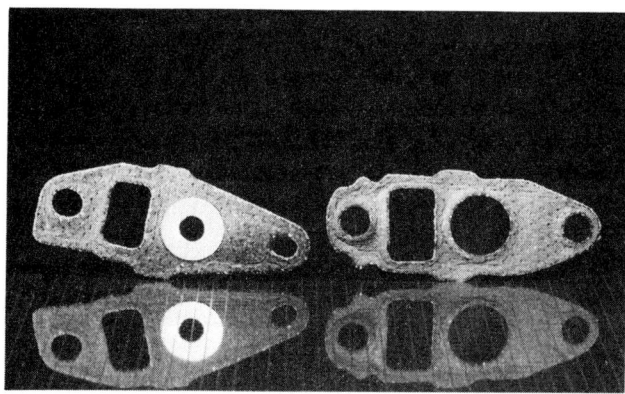

FIGURE 14-67 Typical EGR seals *(Courtesy of Fel-Pro Inc.)*

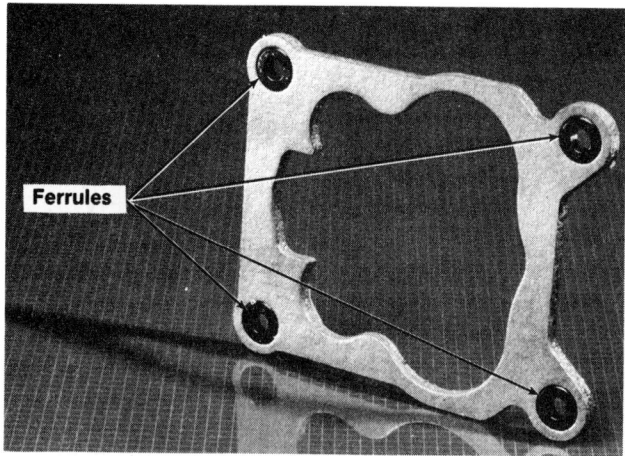

Ferrules

FIGURE 14-68 Carburetor engine mounting gasket that features ferrules *(Courtesy of Fel-Pro Inc.)*

age the casting, as well as binding throttle plates and fuel leaks. Spacers are used in some engine applications for emission control systems or heat insulation for the carburetor. The carburetor spacer is located between the carburetor and the manifold, with the space gasket mounted underneath the spacer.

Most transmissions require a gasket to seal the joint between the transmission pan and the case. Many pans use embossed sealing patterns to increase the sealing force on the gasket. Some pan gaskets have slightly smaller bolt holes to hold the bolts in place while the pan is being mounted to the case (Figure 14-69). This feature eliminates the need to use contact adhesive on the gasket, which could damage the transmission should an excessive amount of adhesive get inside. Transmission oil pan gasket installation procedures are similar to engine oil pan gasket procedures. Be sure that the bolts are in good condition and the flanges are straight and

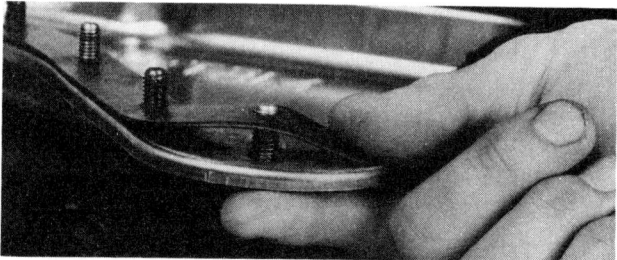

FIGURE 14-69 Transmission gasket seal bolts *(Courtesy of Fel-Pro Inc.)*

clean before the installation. Pans require replacement when they are severely warped or have been weakened after being straightened too many times.

Many overhead cam (OHC) import engine designs allow individual rebuilding of seals at the front of the engine. Figure 14-70 illustrates the typical timing sealing arrangement. Although it might save money to purchase only the parts that are needed, most rebuilders replace all of them.

SEALING DIESEL ENGINES

Sealing diesel engines is more difficult than sealing gasoline engines. The diesel engine has about 2-1/2 times the compression ratio of a gasoline engine. This results in much higher combustion pressure. The high combustion pressure requires the use of a head gasket with special sealing features.

Passenger vehicle diesel engines use two-stage combustion. Initial combustion takes place in a pre-

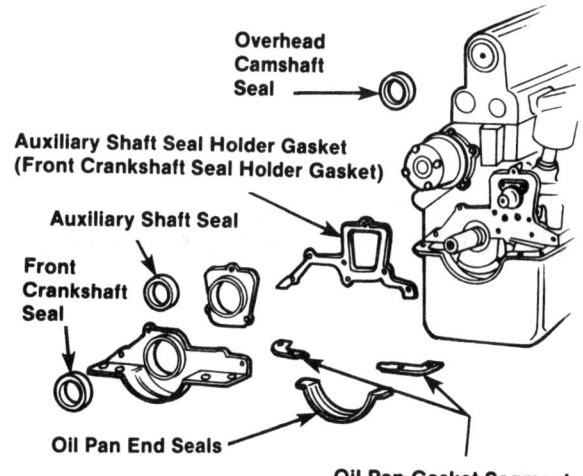

Overhead Camshaft Seal

Auxiliary Shaft Seal Holder Gasket (Front Crankshaft Seal Holder Gasket)

Auxiliary Shaft Seal

Front Crankshaft Seal

Oil Pan End Seals

Oil Pan Gasket Segments

FIGURE 14-70 Typical import vehicle's timing arrangement

combustion chamber. Combustion gases leave the precombustion chamber through a slot to complete the combustion process in the cylinder (Figure 14–71). Two-stage combustion produces less diesel combustion noise and smoother operation than single-stage direct injection combustion.

The precombustion chamber is located in the cylinder head, and a precup forms the bottom half of the chamber. The combustion chamber side of the precup should be smooth as well as flush with the head surface. However, there might be some production tolerances that allow the precup to stick out slightly from the head surface (proud) or to be slightly recessed. Head gasket sealing becomes more difficult when the precup is either proud or recessed (Figure 14–72).

Diesel engines are sensitive to changes in component dimensions from machining. Therefore, several different head gaskets for specific diesel engines are available. The gasket may have a standard size bore or an oversize bore, and the thickness of the gasket can be standard, or it can be thicker than

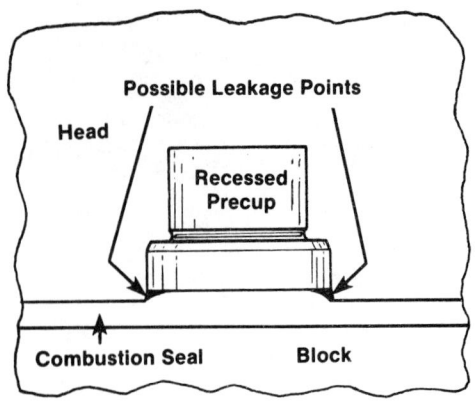

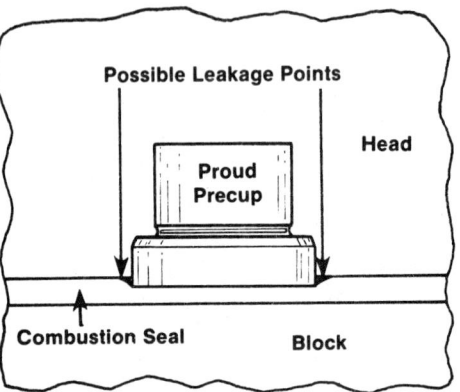

FIGURE 14–72 Diesel engine precombustion chamber

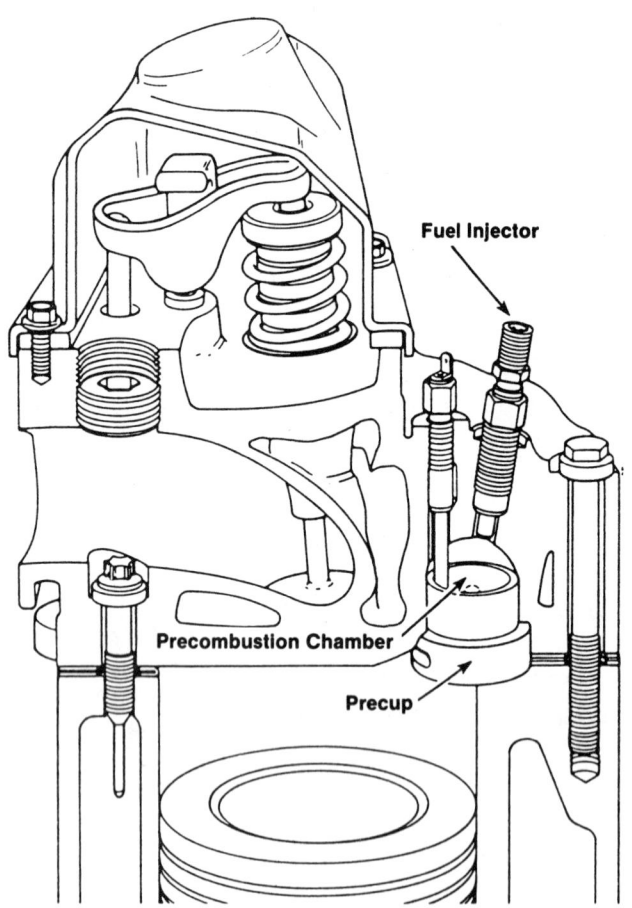

FIGURE 14–71 Diesel engine combustion chamber

standard. Because of this most aftermarket manufacturers do not package the head gaskets in the diesel engine head set or full set. The head gasket must be ordered separately, depending on what servicing or machining has been done on the head and/or block. Some of the unique features related to a diesel head gasket, for example, are (Figure 14–73):

• Dowels to position cylinder head
• Camshaft key and crankshaft key position
• Special designed cylinder head bolts for aluminum head
• Reuse of cylinder head bolts not recommended

Diesel engine cylinder head bolt torque procedure should be followed as specified by the manufacturer. The procedure varies with different models of engines and according to the grade or coating (cadmium or lubrite) of the bolts. The sequence can also be critical. Some procedures specify that each bolt should be tightened in steps.

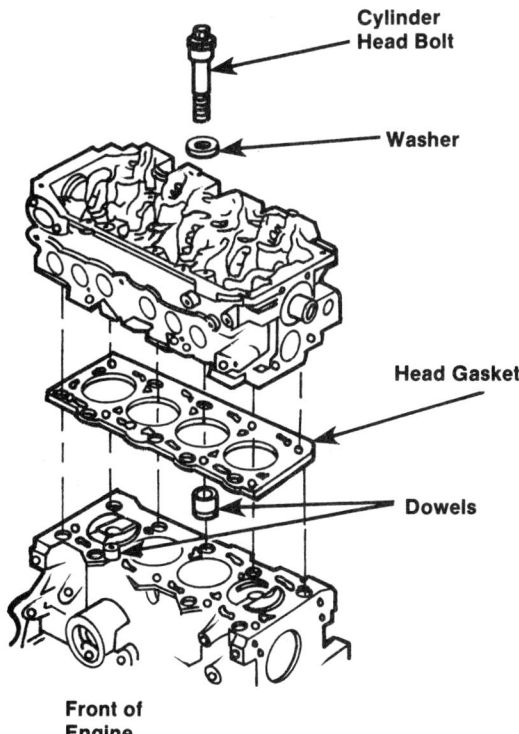

 SHOP TALK —————

Some diesel head gaskets are wax coated and packaged in airtight poly bags to prevent curing before installation to specified torque for operating thickness. Use no sealant of any kind.

FIGURE 14-73   Special diesel head consideration

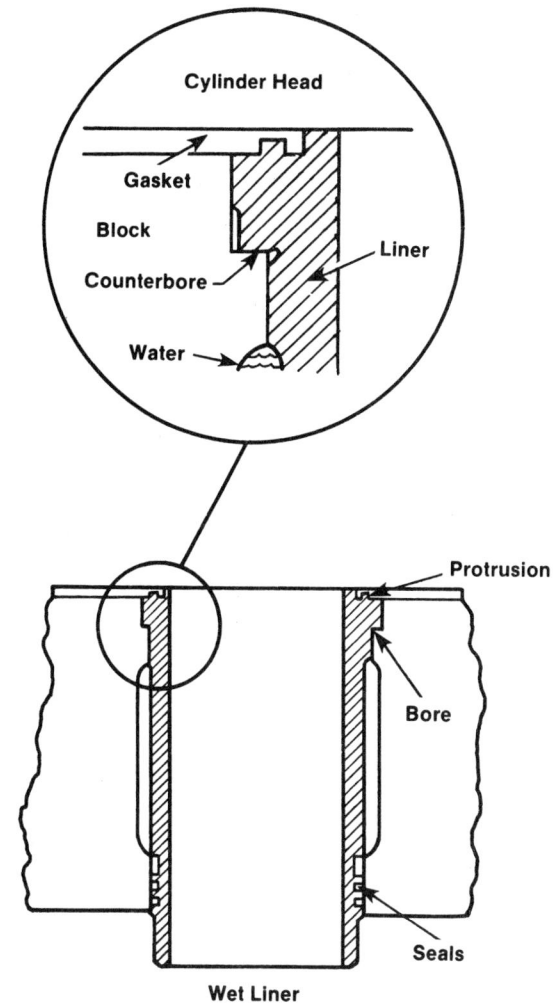

FIGURE 14-74   Typical diesel cylinder sleeve

Generally, the heavy-duty diesel engines in trucks have replacement cylinder sleeves (Figure 14-74). These sleeves have a ridge (or bead) that protrudes above the cylinder block deck. The ridge imbeds into the head gasket to increase ability to seal the extremely high combustion pressures. Each engine make and model requires close installation dimensions. Obtain them from the engine manufacturer's service publication for a specific model.

As mentioned, the amount of protrusion is critical and must be measured with a dial indicator gauge while the sleeve is clamped into position. Gasket failure will quickly occur if protrusion is incorrect. Major cylinder block repair and/or machining might be required if specifications are not met.

The cylinder sleeve mounting shoulder in the block can also be worn, which prevents proper protrusion of the sleeve. This condition must be corrected by replacing or reconditioning the block.

REVIEW QUESTIONS

1. Which of the following is a facet of engine sealing?
   a. suppresses and muffles noise
   b. confines high combustion pressures
   c. prevents internal and external oil leaks
   d. all of the above

2. In addition to their sealing function, gaskets also serve as _____ .
   a. vibration dampers
   b. spacers
   c. wear insulators
   d. all of the above

3. Technician A says that in the English system, the tensile strength of a bolt is identified by

the number of radial lines (grade marks) on the bolt head. Technician B says that the diameter of the bolt is a determining factor in its tensile strength. Who is right?
a. Technician A
b. Technician B
c. Both A and B
d. Neither A nor B

4. Which of the following is true?
a. T-T-Y bolts produce 100 percent of their intended strength when torqued just barely into a yield condition.
b. T-T-Y bolts can be reused unless otherwise specified.
c. Torque values are calculated with a 5 percent safety factor below the yield point.
d. None of the above

5. Which of the following is not classified as a soft gasket?
a. silicone
b. paper
c. rubber-coated steel
d. none of the above

6. Which of the following is the most sophisticated type of gasket?
a. intake manifold
b. combination intake and exhaust
c. cylinder head
d. none of the above

7. Which of the following is a type of intake manifold gasket?
a. rail type
b. valley cover type
c. turtle back type
d. all of the above

8. Which type of oil pan gasket is used for racing?
a. synthetic rubber
b. cork
c. cork/rubber compounds
d. fiber core with latex coating

9. How do aerobic sealants differ from anaerobic?
a. curing method
b. gap-filling ability

c. both a and b
d. neither a nor b

10. Technician A installs a new timing cover oil seal with a press. Technician B uses a hammer and a clean block of wood. Who is right?
a. Technician A
b. Technician B
c. Both A and B
d. Neither A nor B

11. Which rear main bearing seal is the most expensive?
a. polyacrylate
b. viton
c. silicone synthetic rubber
d. nylon

12. Which type of main bearing seal is commonly used because it is tough and abrasion resistant, with moderate temperature resistance to 350 degrees?
a. polyacrylate
b. viton
c. silicone synthetic rubber
d. nylon

13. Technician A soaks the rope-type seal in oil before installation. Technician B coats the seal with oil and allows time for it to soak in before installing the crankshaft. Who is right?
a. Technician A
b. Technician B
c. Both A and B
d. Neither A nor B

14. Most printed torque values are intended for what type of bolts?
a. dry, plated
b. wet, plated
c. dry, nonplated
d. wet, nonplated

15. Which of the following is a unique feature related to a diesel head gasket?
a. dowels to position cylinder head
b. camshaft key and crankshaft key position
c. reuse of cylinder head bolts not recommended
d. all of the above

CHAPTER FIFTEEN

ENGINE REASSEMBLY AND INSTALLATION

Objectives

After reading this chapter, you should be able to
- Explain the differences in installing cup-type and expansion-type core plugs.
- Install the cam bearings and check the camshaft end play.
- Install the main bearings and crankshaft and check the crankshaft end play.
- Install the pistons, connecting rods, timing components, and oil pump.
- Install the cylinder head and valve train.
- Install the valve, oil pan, and timing covers.
- Identify the methods of prelubricating an engine.
- Install a typical rebuilt engine and observe the correct starting and break-in procedures.

When reassembling an engine, the procedure recommended in Chapter 4 is reversed. That is, the basic steps for engine reassembly are as follows:

1. Reclean the block.
2. Install the core plugs.
3. Install the cam bearings and camshaft.
4. Install the main bearings and crankshaft.
5. Install the pistons and connecting rods.
6. Install the timing components.
7. Install the cylinder head and valve train.
8. Install the oil pump.
9. Install the timing cover, oil pan, and cylinder cover.

RECHECKING AND RECLEANING THE BLOCK

Before starting any engine rebuilding work, the engine block should be reinspected and recleaned. Throughout this book, it is obvious that there has been a variety of innovations over the past several years that have substantially affected cylinder head

and/or engine block machining and rebuilding. Smaller, lighter engines, higher revving engines, increased use of bimetallic engines, higher operating temperatures, and the near elimination of leaded gasoline have increased both the knowledge and the quality of rebuilder work required by today's shops.

As stated in the text several times, modern cylinder heads, both aluminum and cast iron, are using noticeably thinner castings than ever before. Careless machining of these castings can easily result in breaking through to water jackets, which turns a costly head into junk. For a variety of reasons, manufacturers are also limiting the amount of material that can be removed from a cylinder head during resurfacing. The measuring of critical wall thickness by using electronic (sonic) thickness testers (see Chapter 1), various cylinder thickness gauges (Figure 15-1), and micrometers and depth gauges (Figure 15-2) described in previous chapters must be done both before machining and reassembling an engine.

As heads have become lighter, thinner wall construction and higher operating temperatures have increased their susceptibility to cracking—often in

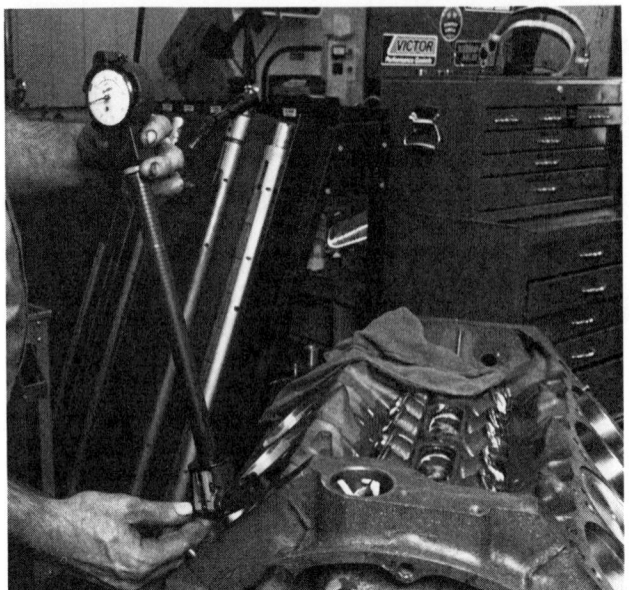

A

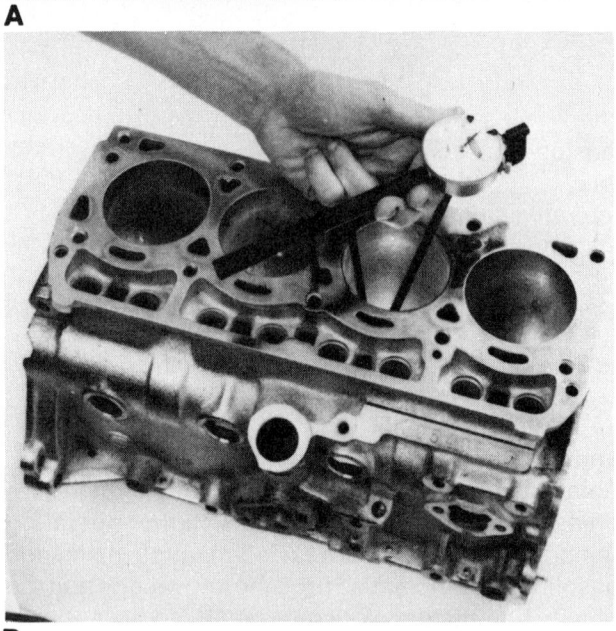

B

FIGURE 15-1 Measuring the cylinder with (A) bore gauge and (B) cylinder wall gauge. To use the latter, insert the probes in the cylinder and water jacket, or turn the block over to work from the bottom. *(Courtesy of Goodson Shop Supplies)*

areas that are not visible. Therefore, shops now should employ magnetic crack detection as well as pressure testing equipment (Figure 15-3).

It is important, too, that the block and its components are recleaned before their installation. A recleaning of the parts also gives the rebuilder a chance to catch any oversights. Dirt is the engine's number one enemy, and an engine rebuild is one of the most common ways for dirt to enter the engine.

FIGURE 15-2 Measuring cylinder head thickness with a depth gauge

FIGURE 15-3 Before reassembly, heads should be checked for leaks with magnetic particle detection and/or pressure testing equipment.

The block should be rewashed in the hot tank or in the jet spray washer. The cylinder walls should be washed with hot, soapy water (Figure 15-4). After the block is dry lube it with a light oil or paint (Figure 15-5). Run a brush through all oil galleries to clean them.

Many rebuilding shops have a special clean room for reassembly (Figure 15-6). The clean room is separated from all the machine operations. It might have special airlock doors and filters to prevent dust from floating in the air. To keep it clean, the room is mopped and wiped frequently.

If a clean room is not available, the rebuilder technician must take other precautions to insure that the engine is clean. The reassembly must be done as far away as possible from valve grinding and machine work. As each part is installed, it should be wiped carefully and inspected for dirt. Tools must be cleaned carefully before they are used. Whenever the engine is not being worked on, it should be covered; heavy plastic can be used—pull it over the block assembly and seal it with tape (Figure 15-7).

INSTALLING CORE PLUGS

The first step in reassembling the block is to install core or soft plugs. A nonhardening gasket sealer and the proper installation tool should be used to insert all freeze (frost) and oil galley plugs (Figure 15-8). Since most shops remove all plugs, it is important that the proper size and type are used (check a service manual for the proper replacement numbers). Brass core plugs are recommended because they will not rust through.

Prior to installing a core plug, the plug bore should be inspected for any damage that would interfere with the proper sealing of the plug. If the bore is damaged, it will be necessary to true the surface by boring for the next specified oversize plug. Oversize (OS) plugs are identified by the OS stamped in the flat located on the cup side of the plug.

Coat the plug and/or bore lightly with an oil-resistant (oil galley) or water-resistant (cooling jacket) sealer. The three basic core plugs are installed as follows:

1. *Disc- or Dished-Type.* This type fits in a recess in the engine casting with the dished side facing out (Figure 15-9A). The driver hammer hits the disc in the center of the crown and drives the plug into the bore until just the crown becomes flat. In this way the plug will expand properly and give a good tight fit.
2. *Cup-Type.* This type of plug is installed with the flanged edge outward (Figure 15-9B). The maximum diameter of this plug is located at the outer edge of the flange. The flange on cup-type plugs flares outward with the largest diameter at the outer (sealing) edge. The flanged (trailing) edge must be below the chamfered edge of the bore to effectively seal the plugged bore. If the core plug replacing tool has a depth seating surface, do not seat the tool against a nonmachined (casting) surface.
3. *Expansion-Type.* This type of plug is installed with the flanged edge inward (Figure 15-9C). The maximum diameter of this plug is located at the base of the flange with the flange flaring inward. When in-

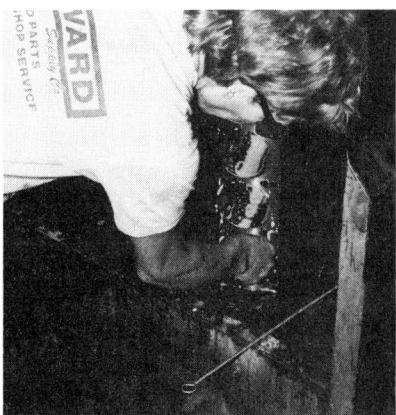

FIGURE 15-4 Washing a cylinder wall with hot soapy water *(Courtesy of Howard Supply Co.)*

FIGURE 15-5 After the block has been recleaned, give it a coat of light oil or paint. *(Courtesy of Perfect Circle/Dana)*

FIGURE 15-6 Engine assembly in many rebuilding shops is done in a special clean room.

FIGURE 15-7 Cover any engine that is not being worked on with heavy plastic.

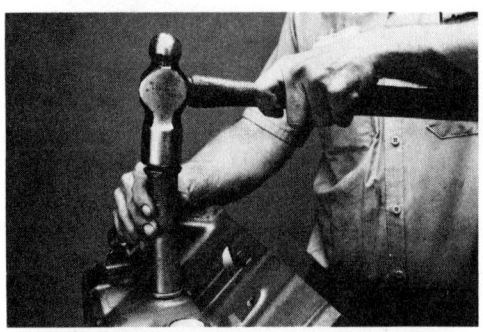

A

B

FIGURE 15-8 Freeze (frost) and oil galley plugs can be inserted by (A) hand or (B) power drivers. The latter is an air powered hammer. *(Courtesy of Goodson Shop Supplies)*

stalled, the trailing (maximum) diameter must be below the chamfered edge of the bore to effectively seal the plugged bore. As was the case with a cup type, if the core plug replacing tool has a depth seating surface, do not seat the tool against a nonmachined (casting) surface.

CAUTION: It is imperative to push or drive the core plug into the machined casting bore using a properly designed tool. As can be seen in Figure 15-9, all three types of plugs require a different installation tool. The use of the wrong tool could damage the plug and/or plug bore.

INSTALLING THE CAM BEARINGS AND CAMSHAFT

After the soft plugs are installed, the next step in the rebuilding process of the camshaft is usually installing the cam bearings. They should be replaced when the engine is rebuilt. Camshaft bear-

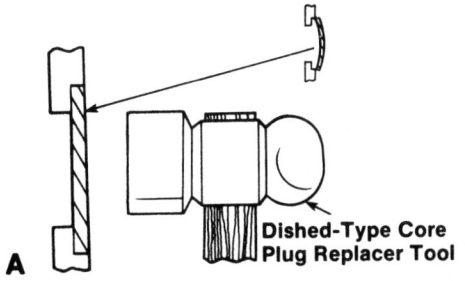

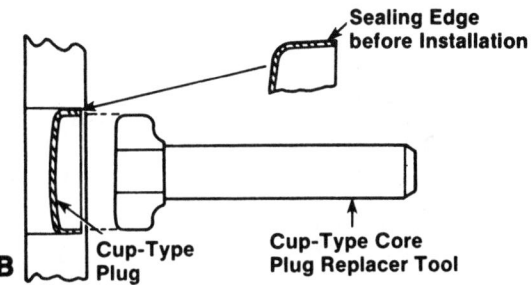

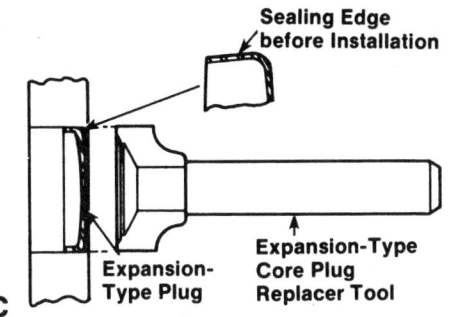

FIGURE 15-9   Typical core plugs and installation tools: (A) dished type; (B) cup; and (C) expansion type

ings should be sorted and laid out in order from rear to front. Then proceed as follows:

 1. Before the cam bearing installation, select either a short or long driving bar. Then select the mandrel that is closest to the cam bearing in size (Figure 15-10A).

2. Put the tool in the block, then place the bearing on the mandrel (it will fit loosely), and turn the bearing driver clockwise until the mandrel is snug in the bearing (Figure 15-10B). Start the bearing into the block by hand pressure.

3. Before driving the bearing into the block housing, be sure to back off on the driving bar 1/8 turn. This will allow the 0.003- to 0.006-inch press fit (Figure 15-10C).

4. Drive the bearing into the block by using sharp blows with a ball peen hammer (Figure 15-10D).

5. Generally, the bearing will be sealed correctly when the mandrel is flush with the face of the block (Figure 15-10E). Some cam bearings must be installed further in the bore so as to register the oil holes in the block. Refer to the engine manufacturer's specifications for the bearing location.

6. Remove the mandrel by withdrawing the driving bar (Figure 15-10F) and move on to the next bearing.

⚙ **SHOP TALK** _____

Regardless of the type of cam bearing used, a suitable tool is required to install the bearings. A worn or uneven driver could roll or swedge edges of the bearing, distorting it or not allowing the bearing to enter the block bore straight, gouging or burring the bearing. This creates cam installation problems and reduces actual bearing clearance. With adjustable mandrel installation tools, it is important to have the segmented separation slots in the expander properly aligned with the slots in the adjustable head (Figure 15-11). Failure to align the separation slots with the adjustable head can result in the bearing surface being burned or gouged. The rubber O-rings on the tool should all be in place for proper centering in the bearing.

Camshaft bearings, unlike rod and main bearings, must be pressed into place in cylindrical bearing bores with a recommended 0.003- to 0.006-inch interference to insure they remain in position. It is important to keep in mind that there are many variables that will affect cam bearing installation. These variables include

- Burrs in the housing bore or on the lead chamfer, which are the result of removing the old bearings

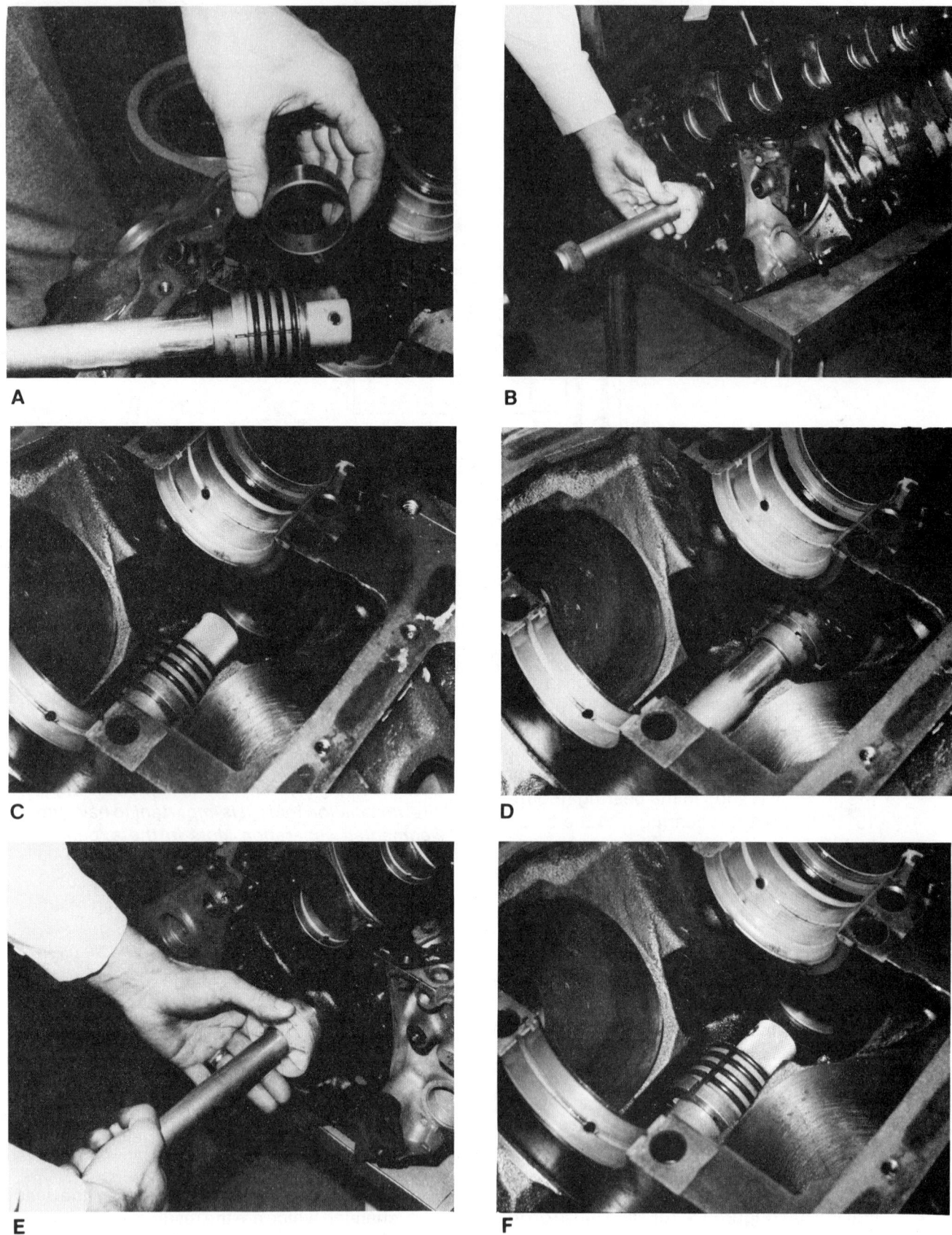

FIGURE 15-10 Steps in installing a cam bearing *(Courtesy of Sealed Power Corp.)*

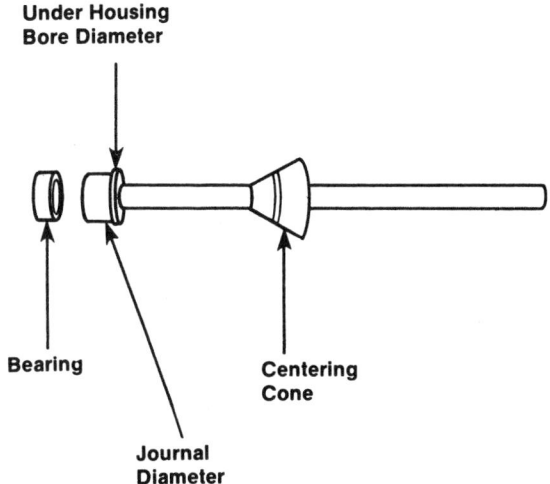

FIGURE 15-11   The smaller diameter of the solid bearing drivers should match the cam journal diameter.

- Inadequate lead chamfer when the block was initially machined by the original equipment manufacturer
- Slightly undersize housing hole diameter in the block (also the result of OEM discrepancies)
- Condition of the installation tool
- Overtightening of the adjustable head when using an expandable mandrel type of installation tool
- Misalignment of the cam bearing bores

Any of these conditions can cause the camshaft to bind or make it impossible to insert the shaft through all positions.

Some overhead camshafts do not have bearing inserts. In this case the bearing bores are usually split and the caps can be ground down and the bores honed to standard size. On others the camshaft bearings might be split two-piece inserts or the bushing type. While the procedure for installing the split-type bearings is basically the same as that for installing the crankcase main bearings, the bushing requires a special type of puller tool such as the one shown in Figure 15-12. Using this tool, it is possible to pull out the old bushing bearings and push in new ones. Be sure to follow the manufacturer's directions when using this tool.

CAUTION: The use of a standard bushing driver and hammer is not recommended because the aluminum bearing supports are very easily damaged or broken, requiring expensive head replacement.

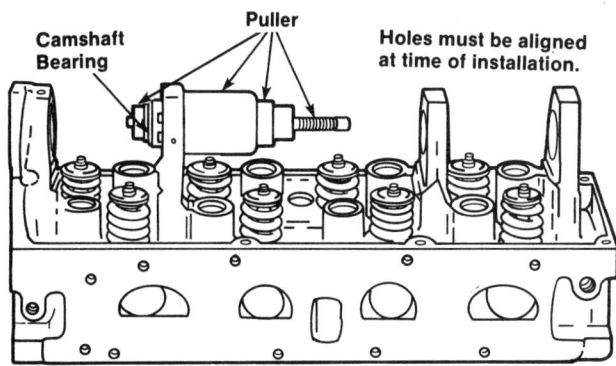

FIGURE 15-12   Replacing overhead bushing camshaft bearings. Some overhead camshaft designs do not have replaceable bearings.

After the cam bearings have been installed, the oil hole in the bearings should be aligned with those in the block. This will ensure correct lubrication and oil supply in vital engine areas. Proper alignment can be checked by inserting a wire through the holes (Figure 15-13) or by squirting oil into the holes. If the oil does not run out, then the holes are misaligned. This procedure should be repeated with each bearing.

INSTALLING THE CAMSHAFT

To install the camshaft, wipe off each cam bearing with a lint-free cloth, then spread clean oil on the bearings. Also coat the camshaft (and lifters) with an oil that meets or exceeds the engine manufacturer's specification (Figure 15-14). Current grades of oil that meet manufacturer's specifications are SF/SG

FIGURE 15-13   Aligning cam bearings *(Courtesy of TRW Inc.)*

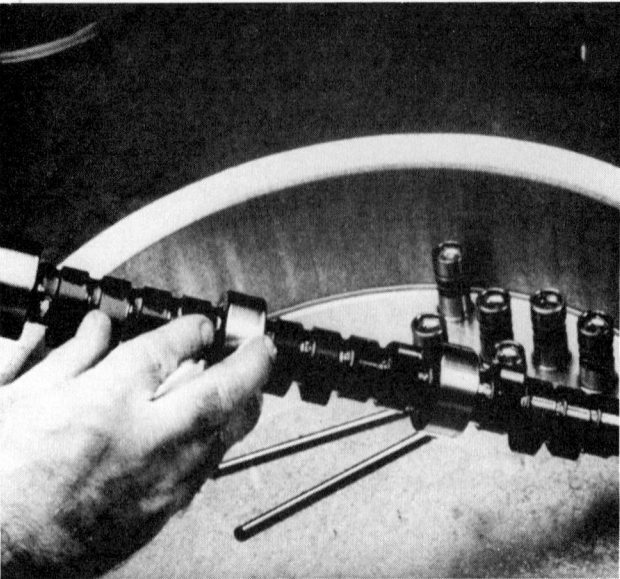

FIGURE 15-14 Lubricating the camshaft and lifters *(Courtesy of TRW Inc.)*

for gasoline engines and CC/CD for automotive diesels. Most premature cam wear develops within the first few minutes of operation. Prelubrication will help prevent this when the engine is started the first time. Special prelubricants could be used only if specifically recommended for camshaft and lifter break-in. Continued use of the above specified oil is generally recommended.

The camshaft should be carefully installed to avoid damaging the bearings with the edge of a cam lobe or journal. Be especially careful to keep it straight to prevent it from cutting or grooving the

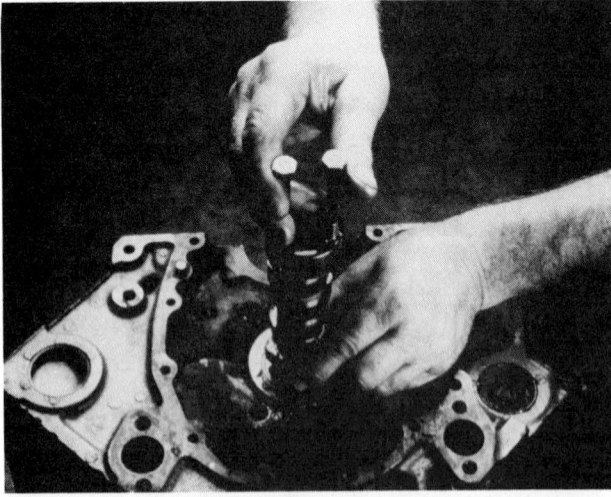

FIGURE 15-15 A threaded bolt in the front of the camshaft can be helpful in guiding the cam in place. *(Courtesy of TRW Inc.)*

bearings. A threaded bolt in the front of the camshaft can be helpful in guiding the cam in place. An alternative is to install the camshaft while the block rests on its end (Figure 15-15). When the camshaft is in place, install the thrust plate and/or the timing gear.

As mentioned in Chapter 12, some engines use a thrust plate to limit the end play of the camshaft. The thrust plate attaches to the front of the engine block behind the timing gear or sprocket (Figure 15-16). There might be a shoulder on the timing gear, sprocket, or camshaft, or there might be a spacer ring inside the thrust plate between the timing gear or sprocket and the front journal of the camshaft. The shoulder or spacer ring will be approximately 0.003 inch thicker than the thrust plate to allow for that much end clearance.

A camshaft timing gear can be pressed off and a replacement pressed on the camshaft (Figure 15-17). Be sure to align the thrust plate with the woodruff key during removal to prevent damage to the thrust plate. Both the thrust plate and timing gear must then be aligned with the woodruff key for assembly. Never hammer a gear or sprocket onto the shaft. Heat all metal and aluminum gears on a hot plate 200 to 300 degrees Fahrenheit. Be sure to install the gear while it is still hot to ensure ease of installation. The above step *does not* apply to fiber gears. These gears should be carefully installed. Press the camshaft into the gear and be sure to keep the gear square and aligned with the keyway at all times.

If the camshaft bearing journals were reground, (see Chapter 12), remember that the new bearing will have to be thicker to make up for the material machined from the bearing journals. These bearings are often described as undersize because they go with an undersize journal. Most engines use camshaft bearings that have the same outside diameter. In a few engines, the camshaft bearing housing sizes are different. It is a good idea to measure the inside diameter of each of the bearing bores. Use a telescoping gauge to do this, then compare the measurements with the outside diameters of each of the bearings.

Once the shaft is completely in the block the shaft should be able to be turned by hand. If the cam does not turn, binding might be the cause. Binding is the result of a damaged bearing, a nick on the cam's journal, or a slight misalignment of the block journals. When the problem is determined as well as which bearings are tight, hand scraping the journal might be necessary to clear up the problem (Figure 15-18). This can be done with an old camshaft where the end journal is cut for scraper grooves, a broken piston ring, or a bearing scraper.

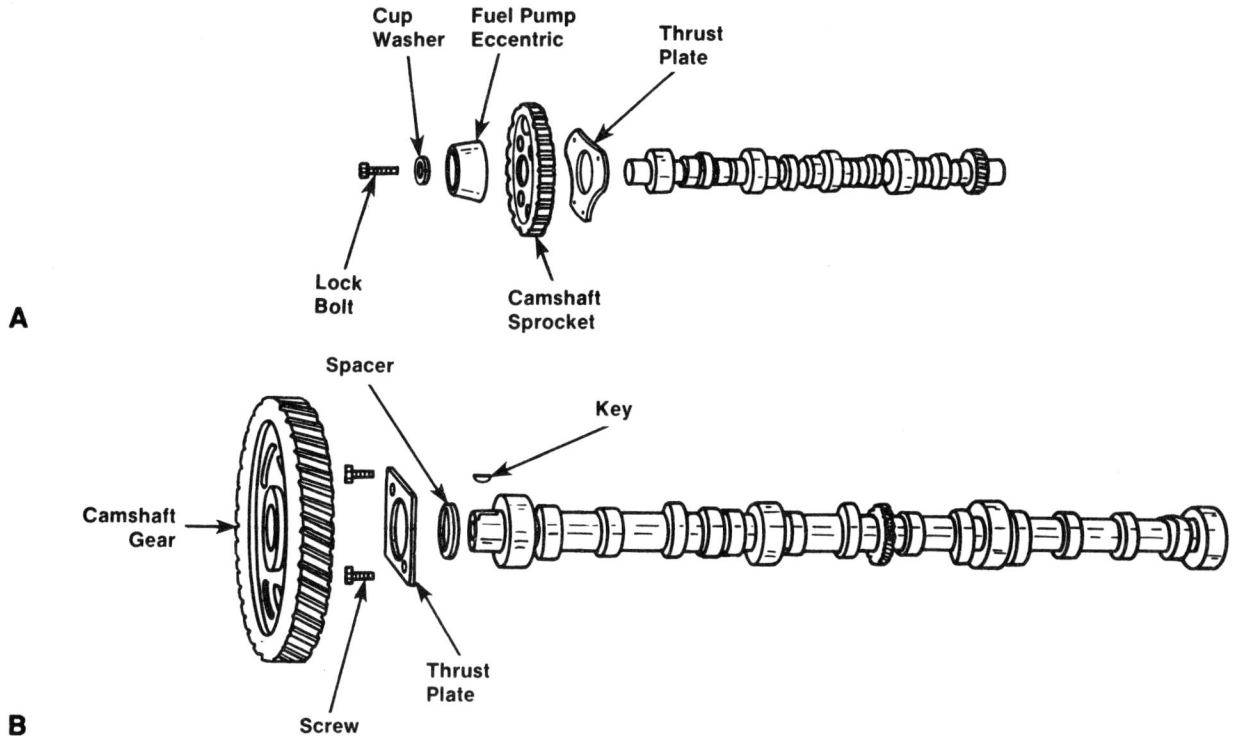

FIGURE 15-16 (A) The shoulder on the cam extends through the thrust plate to set the end play. (B) A spacer ring and thrust plate combination for control of the end play.

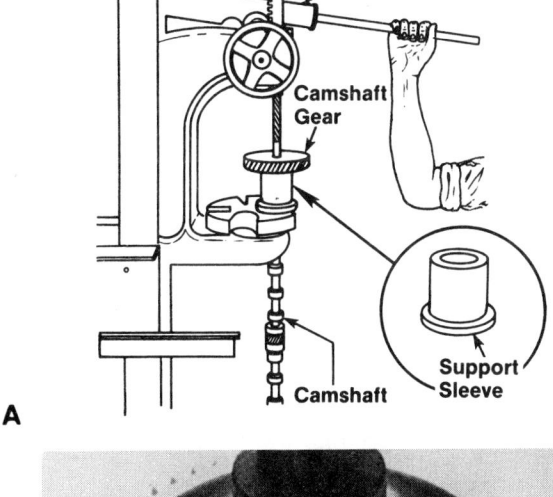

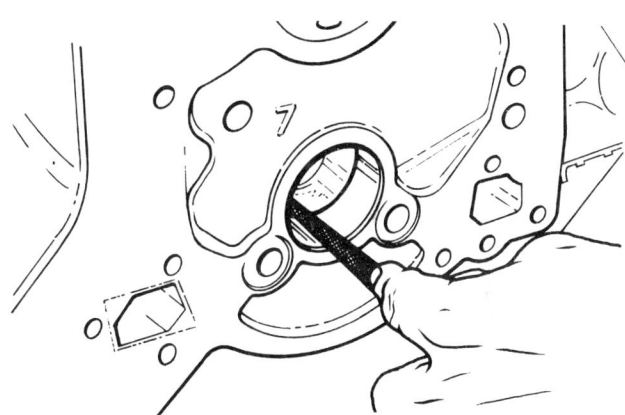

FIGURE 15-18 Hand scraping a journal *(Courtesy of Perfect Circle/Dana)*

FIGURE 15-17 (A) Using an arbor press to install a camshaft gear; (B) results of a pressed fit

CAMSHAFT END PLAY

Before proceeding to the next step in the reassembly, check to be sure that the clearance between the camshaft boss (or gear) and the backing plate is within manufacturer's specifications. Make this check with a feeler gauge. Install shims behind the thrust plate or reposition the camshaft gear and retest the end play. In some cases, adjustment is made by replacing the thrust plate.

To check end play, use a dial indicator setup as described in Chapter 7. Be sure the camshaft end play is not more than recommended in the service manual.

INSTALLING MAIN BEARINGS AND CRANKSHAFT

The main bearings and crankshaft are installed next. When selecting a new main bearing, remember that it must match the crankshaft diameter and main bearing housing size. If the crankshaft has been ground undersize, the main bearings will have to be thicker to make up the space; these thicker bearings are called undersize because they are used with an undersize crankshaft. Similarly, if the housing bores have been machined oversize by align boring, the bearings will have to take up this space. Bearing size is usually marked on the bearing box and on the back of the bearing (Figure 15-19).

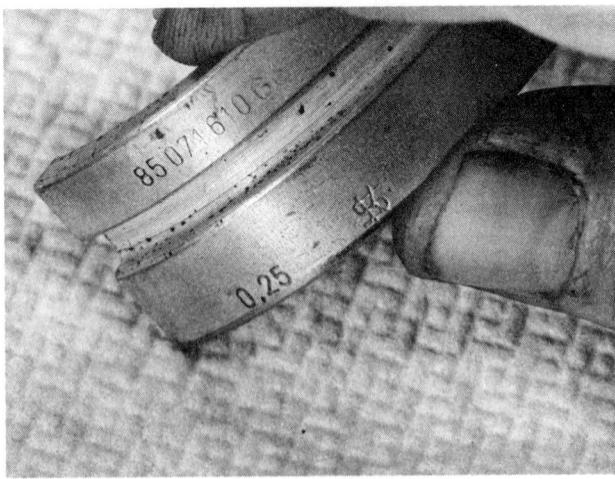

FIGURE 15-19 Methods of marking bearings on the back *(Courtesy of Federal Mogul Corp.)*

FIGURE 15-20 The main bearing saddles should be wiped clean with a lint-free cloth.

Consult the engine manufacturer's service literature for the exact engine buildup procedures. When the bearings are ready to be installed in the main bearing bores, make sure the bore is clean and dry before installing the bearing halves into place (Figure 15-20). Use a clean, lint-free cloth to wipe the bearing back and bore surface. Be certain that nothing is placed between the bearing back and the surface of the housing bore.

Put the new main bearing inserts into each of the main bearing caps and into the cylinder block housings. Be sure to wipe the caps and housings perfectly clean and dry first. Make sure all holes align (Figure 15-21). The backs of the main bearing inserts should never be oiled or greased. Place the crankshaft in the block on the new main bearing inserts and arrange the main bearing caps in the correct order and direction over the crankshaft. Follow the factory markings or use those made during disassembly.

The next step is to measure the oil clearance between the crankshaft and the main bearing in-

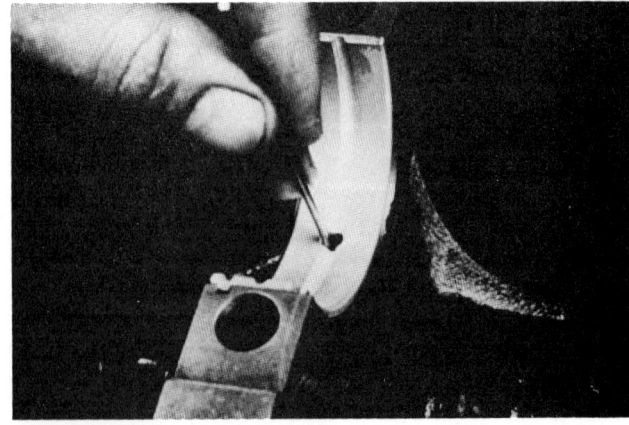

FIGURE 15-21 Checking oil hole alignment *(Courtesy of Federal Mogul Corp.)*

serts. Proper lubrication and cooling of the bearing depend on correct crankshaft oil clearances. Scored bearings, worn crankshaft, excessive cylinder wear, stuck piston rings, and worn pistons can result from too small an oil clearance. If the oil clearance is too great, the crankshaft might pound up and down, overheat, and weld itself to the insert bearings.

Plastigage (the fine plastic string mentioned in Chapter 7) is used to measure the oil clearance between the insert bearing and the crankshaft. The soft plastic string is relatively long and has a small diameter. Wipe the crankshaft and insert bearing free of oil. Cut a length of plastic string about 1/8 inch shorter than the bearing and set it on the bearing inside surface or on the crankshaft.

Set a main bearing cap in position over the crankshaft and tighten the bolts to the recommended torque wrench reading. Be careful not to turn the crankshaft. Remove the cap. The plastigage will have been squashed between the crankshaft and main bearing insert. To determine the clearance, use the width scale on the string package. Match the stripes on the package with the flattened string (Figure 15–22). Each stripe has a clearance measurement printed on it. When the correct stripe is found, read the clearance. This is the clearance between the main bearing insert and the crankshaft main bearing journal.

One side of the plastic-string package has clearance stripes for customary measurements; the other side has stripes for measuring clearance in metric measurements. The string can be purchased to measure different clearance ranges. Usually, only the smallest clearance range is necessary for reassembly work. The plastigage designed to measure wide clearances can be used for troubleshooting worn engines.

Repeat the measuring procedure for each of the main bearing inserts. If the clearance is not to specifications, the engine cannot be reassembled further until the cause has been determined—the new insert bearings might be the wrong size or an error might have been made in crankshaft grinding or align boring. As each bearing is installed, the locating lug must line up exactly with its recess and the bearing back must seat completely against the housing bore. Any flanged (thrust) bearings must be correctly positioned. Also check the alignment of any oil holes. At this time, make sure the rear crankshaft oil seal is installed as directed in Chapter 14.

Before the crankshaft is installed, wipe the bearing surfaces with a clean, lint-free cloth; then spread a good supply of oil on the bearing surface. The crankshaft should be gently and squarely put into place (Figure 15–23). Exercise extreme care at this time to avoid damage to the flange thrust surfaces. With the crankshaft in place, install the main bearing caps with the lower main bearing halves. Prior to assembly, wipe the bearing cap and bearing back with a clean, lint-free cloth and check the fit of the bearing (same as was done for the upper half). Spread a good supply of oil on the bearing surface before the cap is installed in place over the crankshaft. Make sure the main bearing caps are installed in the proper sequence, according to their position in the engine.

The threads of the main bearing cap screws must be clean and lightly lubricated. Once they are installed and finger tightened, check to make sure the flange thrust surfaces are properly aligned. This can be done by placing a large, flat-blade screwdriver between a crankshaft counterweight and the engine webbing and gently prying the crankshaft forward and backward in the engine (Figure 15–24). When doing this, avoid prying the crankshaft against the thrust surface itself.

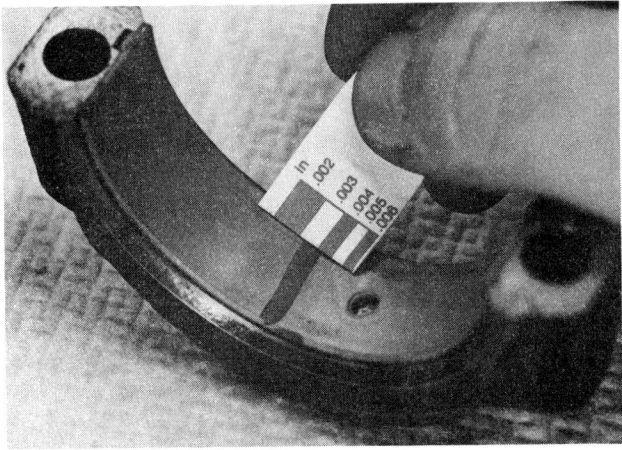

FIGURE 15-22   The plastigage oil clearance reading in the above illustration is 0.003 inch.

FIGURE 15-23   Crankshaft installation *(Courtesy of Federal Mogul Corp.)*

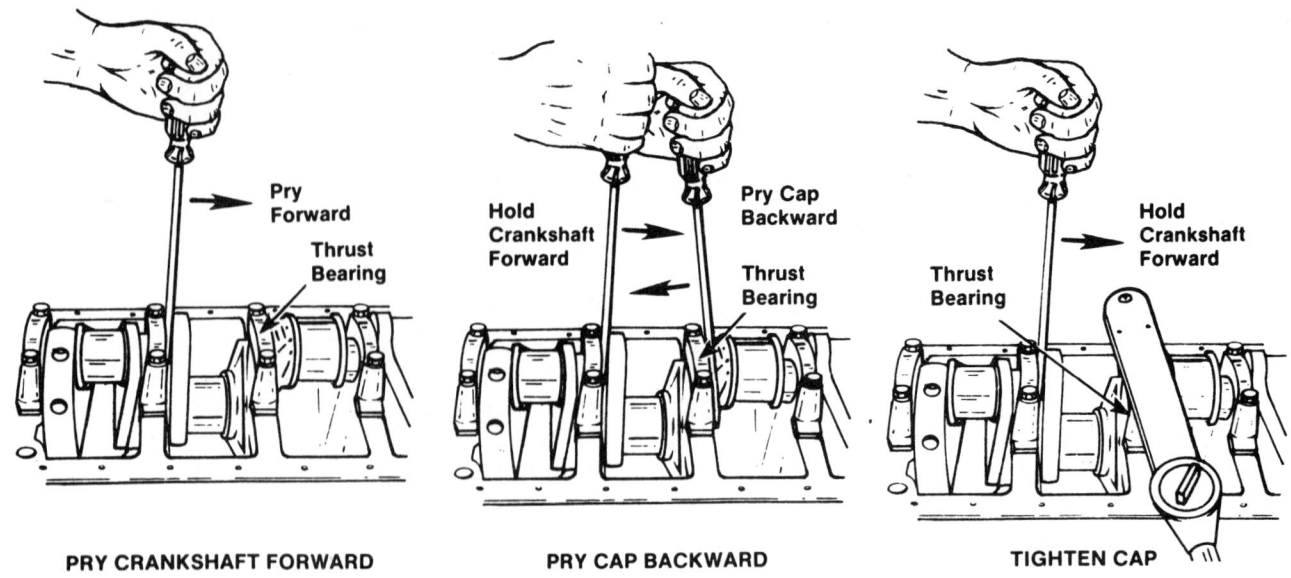

PRY CRANKSHAFT FORWARD   PRY CAP BACKWARD   TIGHTEN CAP

FIGURE 15-24 Method of aligning the thrust surfaces

 SHOP TALK ————

Stamp the crankshaft to indicate any reconditioning undersize information. Usually this information is located on the inside surface of the number one counterweight or bearing.

When the flange thrust surfaces are properly aligned, continue to tighten all of the cap screws alternately until they are fairly snug. Final tightening of the cap screws must be done with a torque wrench (Figure 15-25). Do not attempt to guess at torque tightening. Consult the engine manufacturer's specifications for the proper torque values and main cap tightening sequence (Figure 15-26).

FIGURE 15-25 Proper torquing *(Courtesy of Federal Mogul Corp.)*

After the cap screws have been torqued, rotate the crankshaft. If everything has been done properly, the crankshaft should rotate fairly easily. Also, no crankshaft installation is complete without measuring the crankshaft end play, the distance the crankshaft can move forward and backward in the engine. The crankshaft must have a slight amount of end play to turn freely. In many engines, end play is controlled by a thrust surface on one of the main bearing inserts, usually the center or rear main (see Chapter 7). Other engines use shims or thrust washers between the crankshaft and the cylinder block or flywheel.

Check the crankshaft end play by mounting a dial indicator stand on the front of the block with the dial indicator stem resting on the nose of the crankshaft, parallel to the crankshaft axis. Pry the crankshaft the extent of its travel rearward and set the indicator at zero. Pry the crankshaft forward to measure the end play.

It is also possible to measure the end play with a feeler gauge by prying the crankshaft rearward and measuring the clearance between the cap and the counterweight with a blade. Insert the feeler gauge at several locations around the rear thrust bearing face. If end play is uniform at all locations and the measurements are within specifications, the end play is correct (see Chapter 7).

If the end play is within the specified limits, the job is done. If the end play is too large or too small, the main bearing with the thrust surface must be exchanged for one with a thicker or thinner thrust surface. If the engine has thrust washers or shims, thicker or thinner washers or shims must be used.

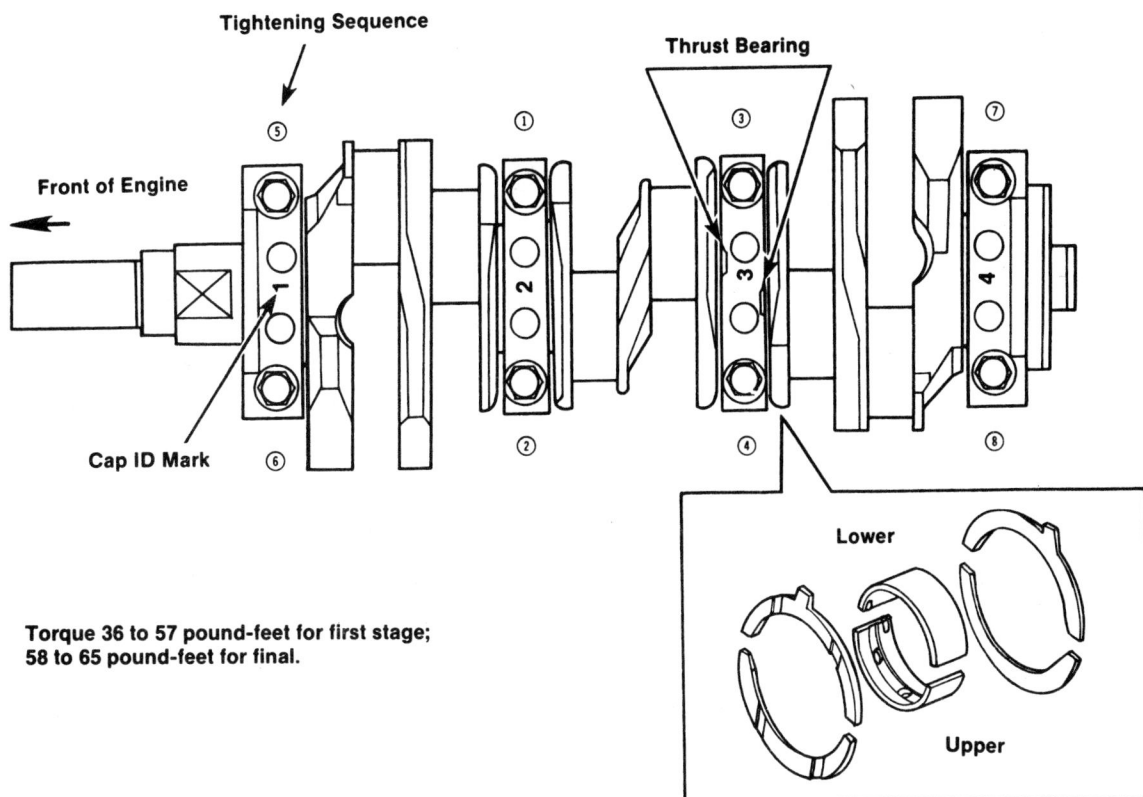

FIGURE 15-26   Typical service manual bearing and main cap installation instruction

INSTALLING PISTONS AND CONNECTING RODS

Once the crankshaft is in place, the next installation task is the piston and connecting rod assemblies. First, lay out the piston assemblies in order from the front to the rear of the engine. Double-check the marks on the connecting rod caps to make sure they are matched. Check the pistons to make sure they are installed on the connecting rods in the correct direction. Various markings are used to identify the positioning of pistons and connecting rods. Some of the more common markings are shown in Figure 15-27, but always check the manufacturer's service manual for exact component directions.

The insert bearings for the connecting rods, like those for the main bearings, must be the correct size. If the crankshaft has been machined undersize, matching rod bearing inserts must be installed. The size of the bearing inserts is printed on the box they come in and is stamped on the backs of the bearings, as shown in Figure 15-28.

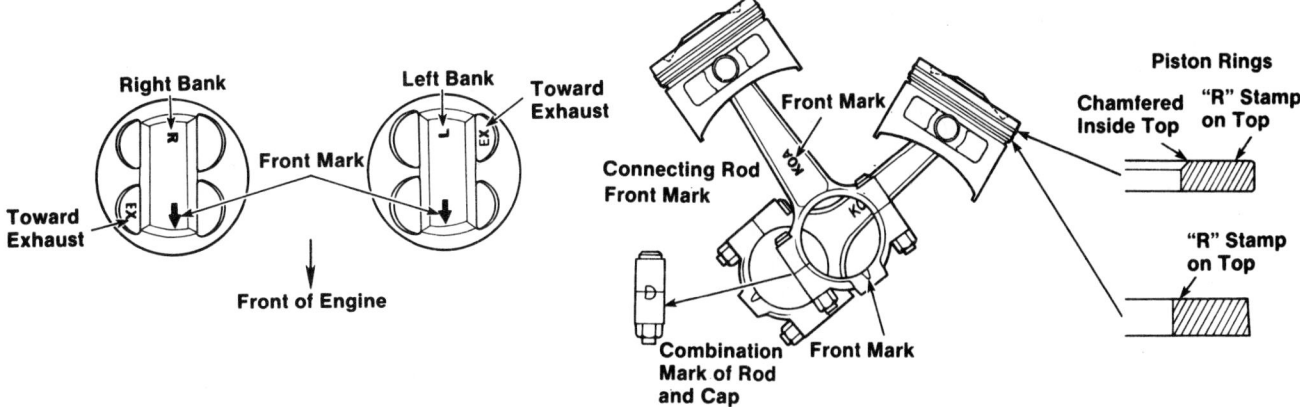

FIGURE 15-27   Common piston and connecting rod identification used with a typical DOHC engine

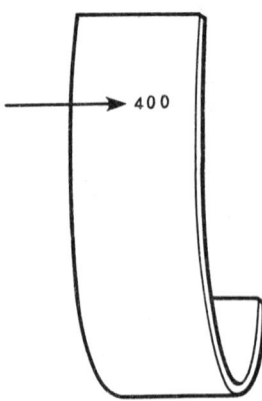

FIGURE 15-28 Connecting rod bearing size is marked on the back of the bearing.

Snap the new connecting rod bearing inserts into the connecting rods and rod caps. Make sure the tang on the bearing fits snugly into the matching notch (Figure 15-29). Also make certain the con-rod and piston move freely (Figure 15-30).

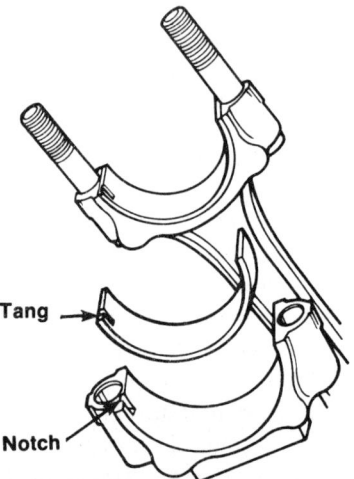

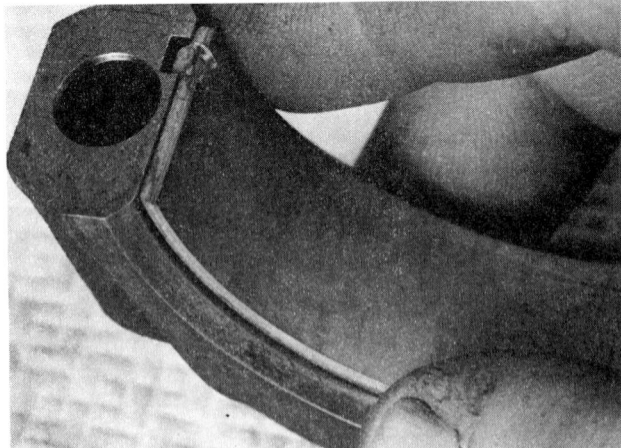

FIGURE 15-29 The new bearing is snapped into the rod and cap so that the tang engages the notch.

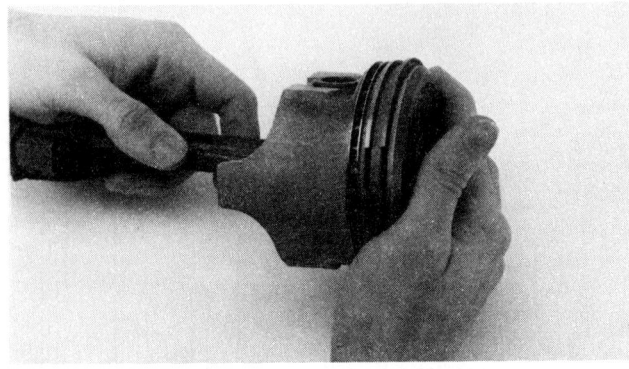

FIGURE 15-30 Make sure the piston moves freely.

The piston and rod can be assembled in the block in the following manner:

1. As mentioned in Chapter 11, use the head of the piston to position the ring so that it is square with the cylinder wall. In slightly worn cylinders that will not be resized, position the ring below the ring travel area. When the cylinders have been resized, the location of the ring is not critical. Next, use a feeler gauge to check the end gap (Figure 15-31A). Then check instructions on the piston ring box for correct installation.

2. Before attempting to install the piston/con-rod assembly in the cylinder bore, place rubber or aluminum protectors or boots over the threaded section of the rod bolts (Figure 15-31B). This will help to prevent bore and crankpin damage.

3. Lightly coat the piston, rings, rod bearings, cylinder wall, and crankpin with a light engine oil (Figure 15-31C). Some rebuilders submerge the piston in a large can of clean engine oil before it is installed. Also be sure to coat the cylinder wall with oil.

4. Fit the piston and ring into the cylinder bore by using a special compressor tool (Figure 15-31D). This tool is expanded to fit around the piston rings. The steps on

FIGURE 15–31   Installing a piston and connecting rod

the compressing tool are positioned downward. The compressing tool is tightened with an allen wrench to compress the piston ring. When the rings are fully compressed, the compressing tool will not compress any further. That is, the piston will fit snugly in the compressor, but not tightly.

5. Rotate the crankshaft until the crankpin is at its lowest level (BDC). Then place the piston/rod assembly into the cylinder bore until the steps on the compressing tool contact the cylinder block deck (Figure 15-31E). Make sure that the piston reference mark is in correct relation to the front of the engine. Also when placing the assembly, make certain that the rod threads do not touch or damage the crankpin. Then remove the protective covering from the rod bolts.

6. Lightly tap on the head of the piston with a mallet handle (Figure 15-31F) or block of wood until the piston enters the cylinder bore. Push the piston down the bore while making sure the connecting rod fits into place on the crankpin.

7. A drop of anaerobic thread-locking compound is good insurance against a loose rod cap nut and the resultant knock and thrown rod (Figure 15-31G). Position the matching connecting rod cap (Figure 15-31H) and finger tighten the rod nuts. Make sure the connecting rod blade and cap markings are on the same side. Gently tap each cap with a plastic mallet as it is being installed to properly position and seat it. Then torque the rod nuts to the specifications given in the service manual. When torquing the nuts, make sure that the socket does not interfere with the bearing cap rib to give a false reading. Lifting the socket slightly should alleviate this problem. Repeat the piston/rod assembly procedure for each assembly. Remember that connecting rods are numbered for easy identification and proper assembly (Figure 15-32).

When all the pistons and rods have been installed, the connecting rod side clearance can be measured (Figure 15-33). The side clearance is the amount of clearance between the crankshaft and the side of the connecting rod. On a V-type engine there are two connecting rods on each crankshaft journal. As mentioned earlier in this chapter, the side clear-

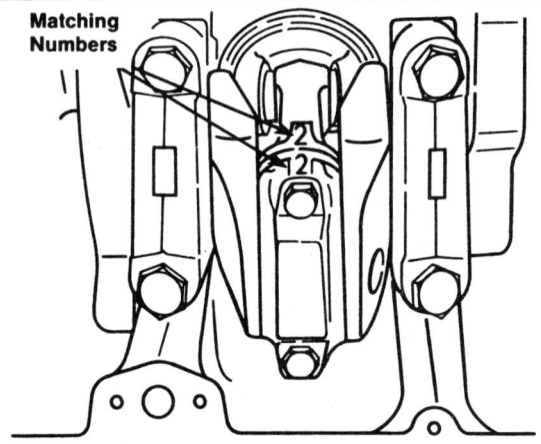

Matching Numbers

FIGURE 15-32 Connecting rods and caps are numbered in various ways for identification and proper assembly.

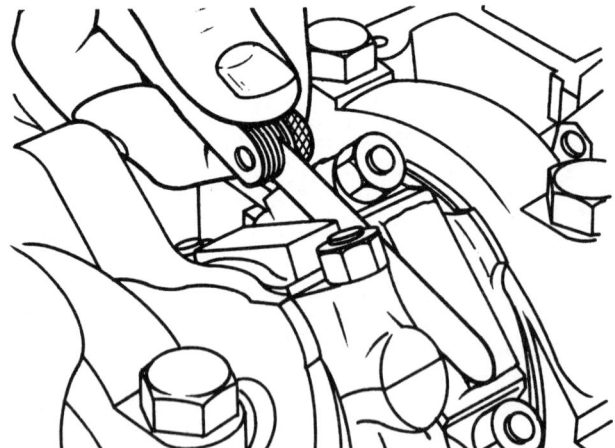

FIGURE 15-33 Measuring connecting rod side clearance

ance is measured by inserting a feeler gauge between the crankshaft and connecting rod side or between the two connecting rods. The clearance determines the amount of oil throw-off of the connecting rod bearings. The clearance can be increased by removing the connecting rod and grinding material off the sides. If the clearance is too great, the connecting rods might have to be replaced.

 SHOP TALK _____

Most expert rebuilders recommend that the bearing oil clearance of the connecting rod be checked against a manual's specifications. To do this, put a small strip of plastigage on the rod cap insert and hold it in place while positioning the insert on the connecting rod. Install the rod cap nuts and torque them to specifications. Remove the nuts and remove the cap. Compare the flattened plastic string to the plastigage package just as was done for the main bearings.

Be sure to coat the crankshaft assembly with engine oil (Figure 15-34).

 SHOP TALK _____

It is good practice to turn the crankshaft to check for any excessive increase in effort required to turn the engine each time a piston and rod assembly is installed. If excessive effort is required to turn the crankshaft, the reason for this should be determined and corrected. This may require the removal and inspection of the last piston and rod assembly installed.

FIGURE 15-34   Coating piston rings and cylinder walls with oil

INSTALLING THE TIMING COMPONENTS

During most engine rebuilds, a completely new timing assembly is usually installed. If wear exists on any component, replacement of the entire assembly is necessary. Wear in the chain, gears, or sprockets means a timing lag, which results in poor engine performance.

If the camshaft is mounted in the cylinder block, it can be installed after the piston assemblies have been put in place. Each camshaft lobe and each bearing journal is lubricated with assembly lubricant. The camshaft is then carefully pushed into the cylinder block, as discussed earlier in this chapter. Check it for fit in the cam bearings by turning it around several times. The camshaft should turn without any bind or drag; if it does not turn freely, as mentioned previously, the size of each cam bearing must be rechecked. The lifters are then lubricated and installed in their bores on top of the camshaft lobes.

With either a cylinder block or cylinder head mounted camshaft, the camshaft drive must be installed so that the camshaft and crankshaft are in time with each other. The camshaft sprocket is twice the diameter of the crankshaft sprocket (Figure 15-35). Both sprockets are held in position by a key or possibly a pin. There are factory timing marks on the crankshaft gear or sprocket and on the camshaft gear or sprocket (Figure 15-36). In some timing arrangements, the timing adjustment is on the crankshaft pulley damper (Figure 15-37). An overhead cam drive can have several idler gears or pulleys with timing marks (Figure 15-38). The timing marks

FIGURE 15-35   The camshaft sprocket (right) is twice the diameter of the crankshaft sprocket (left). *(Courtesy of Perfect Circle/Dana)*

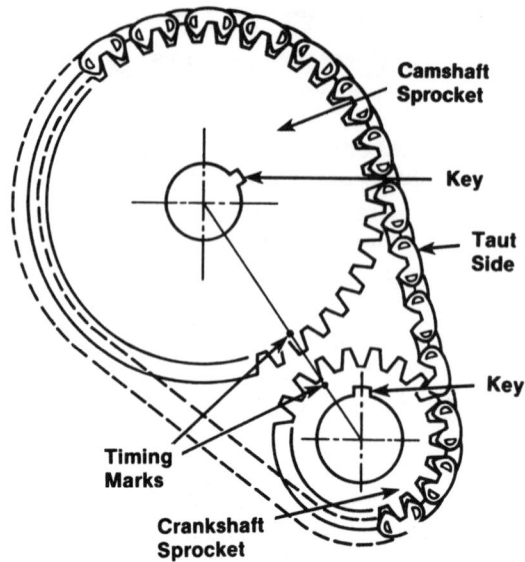

FIGURE 15-36 Factory timing marks on camshaft and crankshaft sprocket

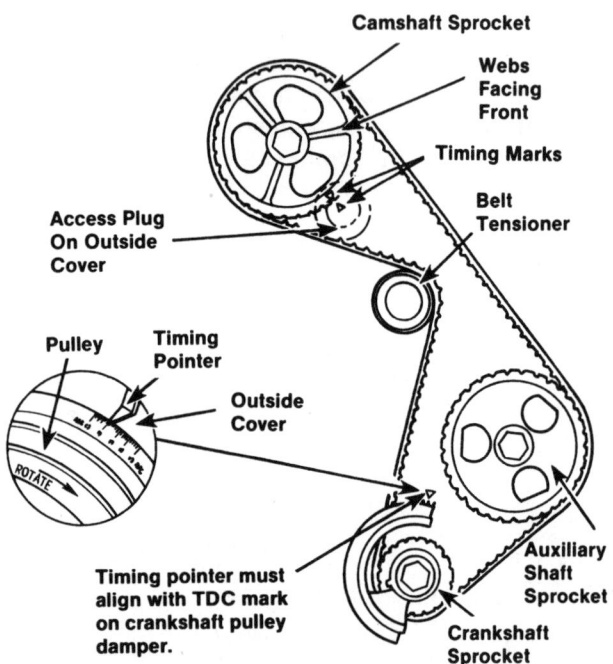

FIGURE 15-37 Many overhead camshaft engines now use cogged drive belts instead of chains.

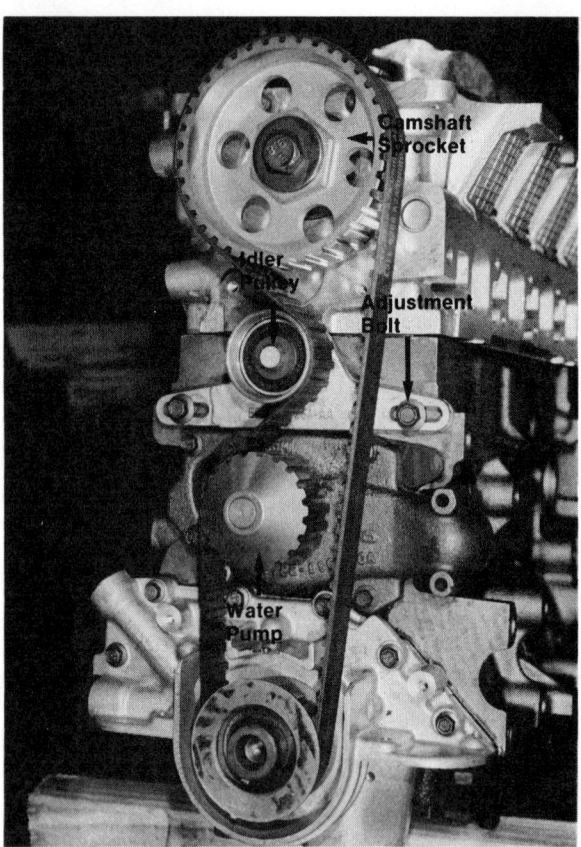

FIGURE 15-38 Typical idler pulley system

FIGURE 15-39 Installing a camshaft sprocket and chain. The sprocket must be aligned with a dowel pin or key in the camshaft. *(Courtesy of Perfect Circle/Dana)*

on all the gears must be positioned according to the manufacturer's instructions.

The chain or belt is generally installed with the gears in their correct positions. The chain (Figure 15-39) is installed on the crankshaft sprocket first, then around the camshaft sprocket. Never wind a chain onto the sprockets. Also, a screwdriver, pry bar, or hammer should never be used to force a chain into position. Prying or pounding on the chain will damage the links causing the chain to stretch and fail. Carefully place the chain on the sprockets and install the entire assembly as a unit by pressing both sprockets on evenly, keeping the keyways aligned; use a sleeve and tap gently into place keeping the sprockets parallel and aligned (Figure 15-40). When installing a timing belt, be sure to align the sprocket timing marks properly and adjust belt tension to the manufacturer's specifications (Figure 15-41). Certain guidelines should be followed to assure that a timing belt is not damaged. They are:

- Do not bend tightly, twist, or turn the belt inside out.
- Do not allow the belt to come into contact with oil, gasoline, water, or steam.

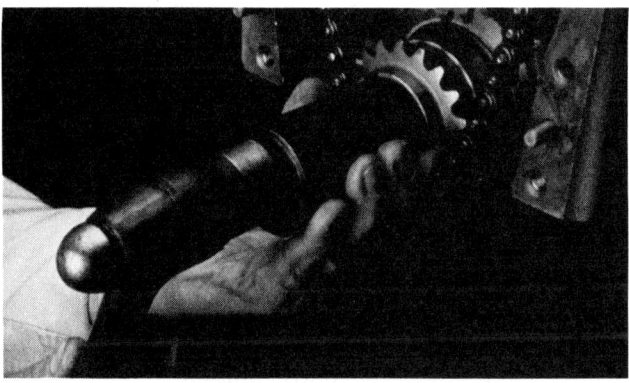

FIGURE 15-40 Installing a chain assembly onto the sprocket. Use care when installing. *(Courtesy of TRW, Inc.)*

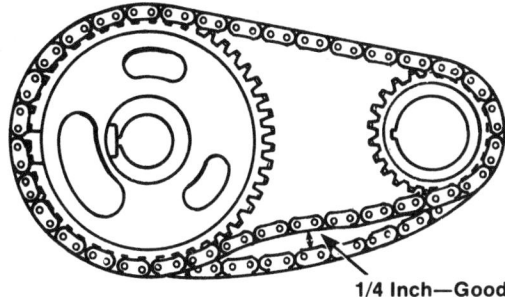

1/4 Inch—Good

FIGURE 15-41 The timing chain and sprockets should also be replaced on each overhaul or rebuild job.

- Keep the belt in a cool and dark room for long-term storage.
- Do not hit or squeeze the belt with a hammer or screwdriver during removal or installation.
- While handling the belt on a pulley, do not tighten or loosen the pulley set bolts without holding the shaft with a wrench or special jig. This prevents the timing from being altered.

Always check the backlash between the mating gear teeth with a feeler gauge or dial indicator as described in Chapter 7. The backlash should not be too loose (over 0.006 inch) or too tight (less than 0.003 inch) on the gear's pitch line. Clearance should always be checked on the same side of the tooth. For best accuracy, check the gear at four different meshings approximately 90 degrees apart.

To complete the installation, tighten the cam bolt (or bolts) in a staggered sequence (Figure 15-42). Use a torque wrench to tighten the bolts correctly to the manufacturer's recommended torque. Finally, be sure to recheck that all timing marks are correctly matched. The amount of play in the timing chain should be measured again (as explained in the unit on disassembly). Overhead camshaft drives usually use spring-loaded or hydraulic tensioners. The tensioners will require adjustment during assembly (Figure 15-43). The service manual for the engine will give the correct adjustment procedure (Figure 15-44).

The vibration damper and flywheel or converter flex plate are considered part of the crankshaft assembly. A bolt (or bolts) is used to retain the front pulley and the vibration damper. This goes on the end of the crankshaft. Most rebuilders do not install the crankshaft or flex plate assembly while the en-

FIGURE 15-42 Tighten the cam bolt (or bolts) in a staggered sequence *(Courtesy of TRW, Inc.)*

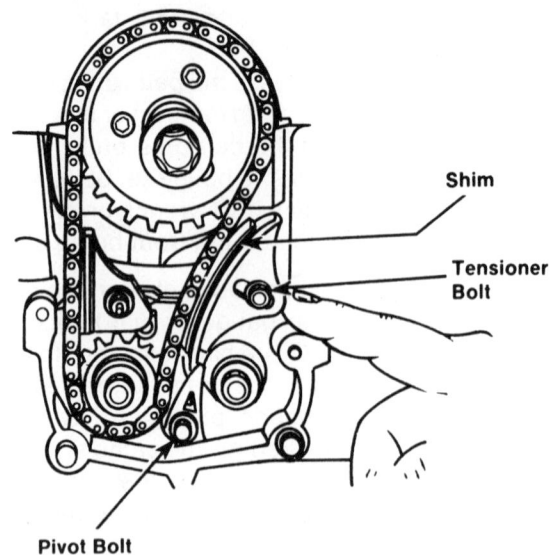

FIGURE 15-43   Adjusting chain tension

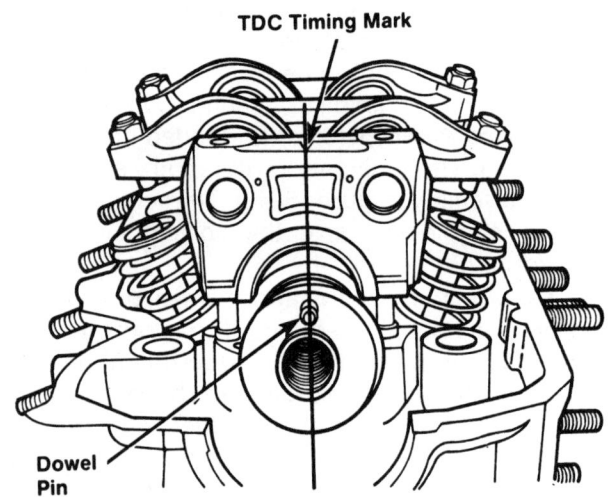

FIGURE 15-44   Position the cam correctly before installing the cylinder bead

gine is on the stand. Details on installing a flywheel or converter flex plate are given later in this chapter.

 SHOP TALK

As the valve train and timing system become more and more complex, the service manual becomes a must. This can be noted in Figure 15-45, which shows the crankshaft and camshaft timing belt installation for the dual overhead camshaft (DOHC) and four valves per cylinder design mentioned in Chapter 2.

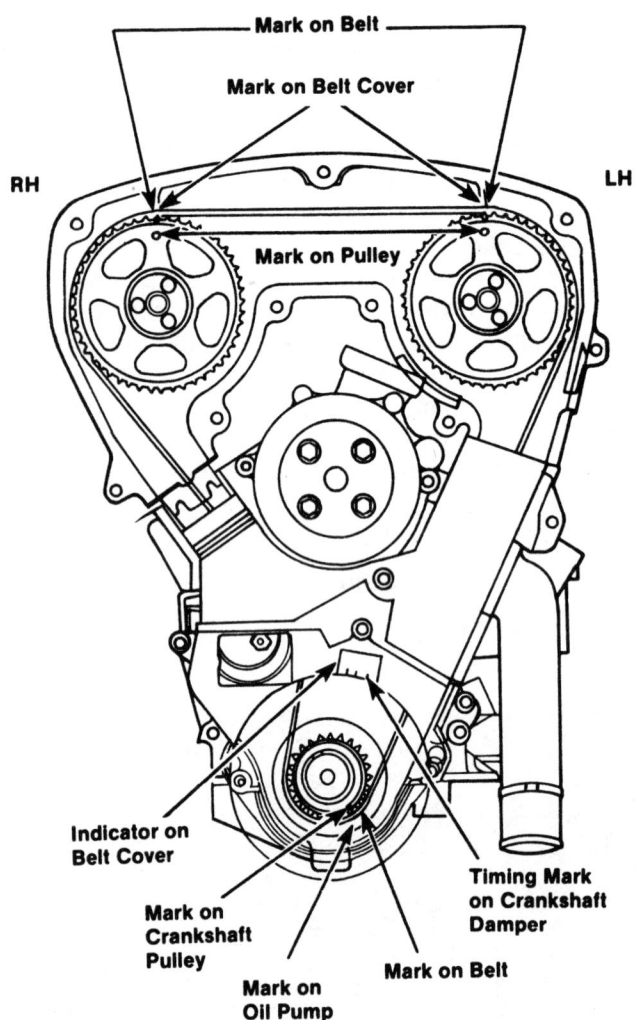

FIGURE 15-45   Crankshaft and camshaft timing belt installation for the DOHC engine

INSTALLING THE CYLINDER HEAD AND VALVE TRAIN

As mentioned several times, the valve train includes everything between the camshaft and valve. In the majority of domestic engines, it includes the lifter, pushrod, rocker arm, and rocker support. Once the basic assembly procedure described in Chapter 10 is completed, the lower and upper portions of the engine can be reassembled. The first step in doing this is to sort out the cylinder head bolts. Many engines use head bolts of different lengths (Figure 15-46). The appropriate service manual usually identifies where the long and short

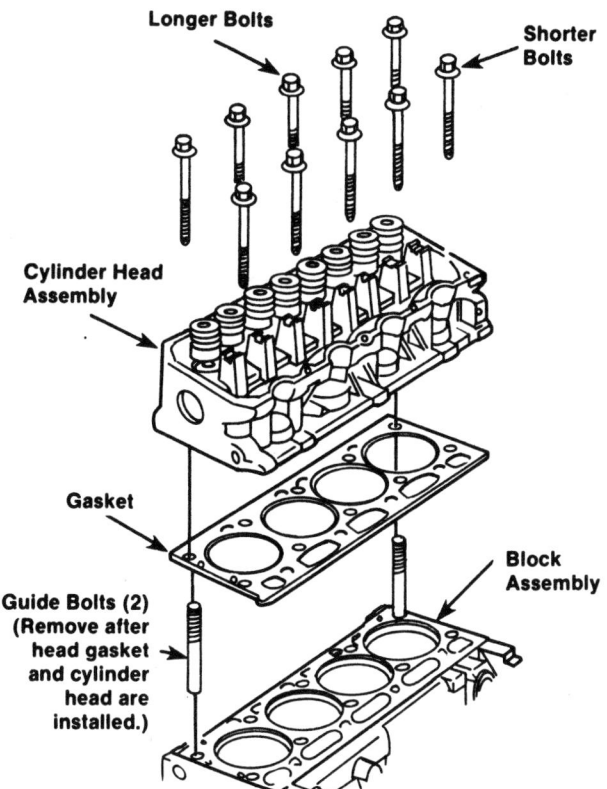

FIGURE 15-46 Head bolts are different lengths in some engines.

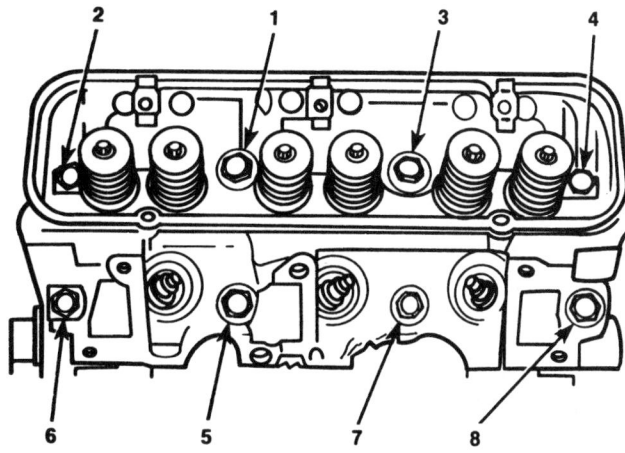

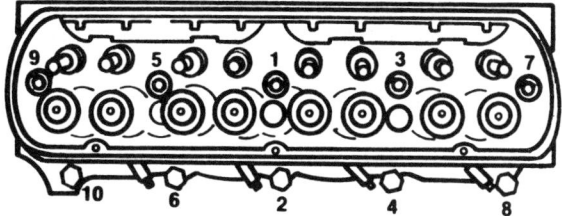

FIGURE 15-47 Examples of cylinder head bolt tightening sequences

bolts go. The threads of the bolts should be thoroughly cleaned and then lubricated with oil or anti-seize compound.

The head gasket should be positioned on the block and checked for fit as detailed in Chapter 14. If desired, the gasket may be coated with the proper type of sealant. The cylinder head is then placed in position on the block, and the bolts are inserted into the bolt holes.

Cylinder head bolts must be tightened in the correct order and to the proper amount of torque. A typical tightening sequence chart (Figure 15–47) is usually provided in the service manual. The chart shows which bolt should be tightened first, second, third, and so on. Most cylinder heads are tightened in a sequence that starts in the middle then moves out to the ends. The bolts are generally tightened in two or three stages. If the final torque is 100 foot-pounds, the bolts may first be tightened at 35 foot-pounds. Some manufacturers recommend that the head bolts be retorqued after the engine has been run and is hot.

When assembling the block and head, make sure the dowel pins are in place and the head and block are properly aligned. Threaded bolts entering the coolant jackets should be coated with nonhardening gasket sealant. Follow the manufacturer's specifications for torquing the head bolts (Figure 15–48). They should be torqued in three intervals.

When the camshaft in an OHC engine is installed, some clearance might be needed between

FIGURE 15-48 Torquing the head bolts

the cam and cam followers. Installed valve stem height (see Chapter 10) is critical to avoid incorrect valve train geometry and to correct valve lash.

Valve clearance is adjusted in one of three ways: with an adjustable rocker arm, with an adjustable tappet, or with adjustment discs or shims.

Adjustable rocker arms are found on overhead valve engines with the camshaft in the engine block. As noted earlier in this text, on some designs the entire rocker arm adjusts up and down on its support stud; other designs utilize an adjustment screw on the pushrod end of the rocker arm (Figure 15-49). The screw can be a self-locking capscrew or an adjusting screw with a locknut.

Before inserting the pushrods, apply a small drop of prelube to eliminate any metal-to-metal contact on start-up. Liberally coat the rocker arms with clean SF/SG grade oil (Figure 15-50). Then insert the rods and the rocker arms (Figure 15-51). Most modern domestic engines use positive stop rocker arm adjustments. This means when torquing the rocker arms to spec, the plunger is positioned into the lifter, giving correct lifter adjustment. This cannot happen if the lifter has been hand primed before installation. Therefore, it is not recommended that new lifters be primed before installation. This can damage the valve train components. Lifters are already preset.

 SHOP TALK

Some engines have rocker arms designated for right- or left-hand positions (Figure 15-52). They must be installed in the correct locations for proper pushrod alignment. Note the location of the pushrod socket to determine position.

On engines with adjustable rocker arms, position the plunger in the middle of the bore. Engines with nonadjustable rocker shaft assemblies might require longer or shorter pushrods to properly position the plunger in the middle of the lifter bore. The angular relationship of the top surface of the valve stem to the rocker arm, when the valve is closed, is important when assembling an engine (Figure 15-53A). Most OE engines use a 1/3:2/3 relationship. For camshafts with more than a 0.550-inch lift, a split setup is recommended (Figure 15-53B). This angular relationship is what determines rocker arm geometry.

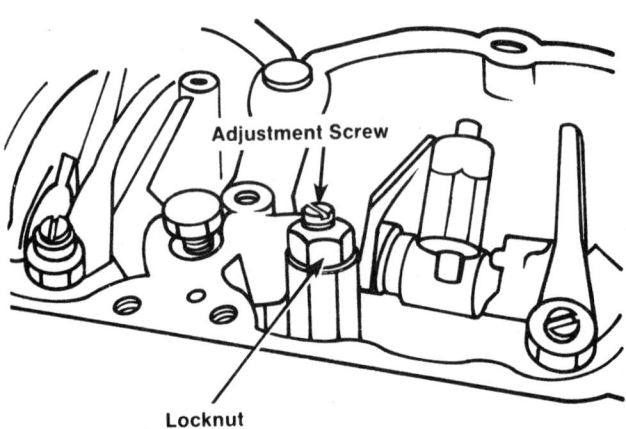

FIGURE 15-49 Adjustment screw and locknut on the pushrod end of a rocker arm

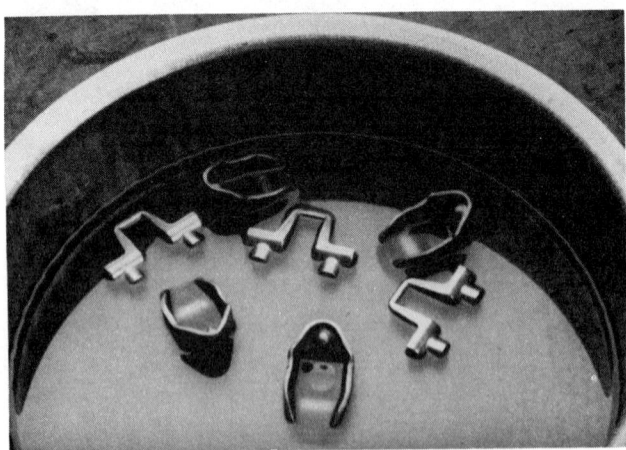

FIGURE 15-50 Rocker arm parts lubrication *(Courtesy of TRW, Inc.)*

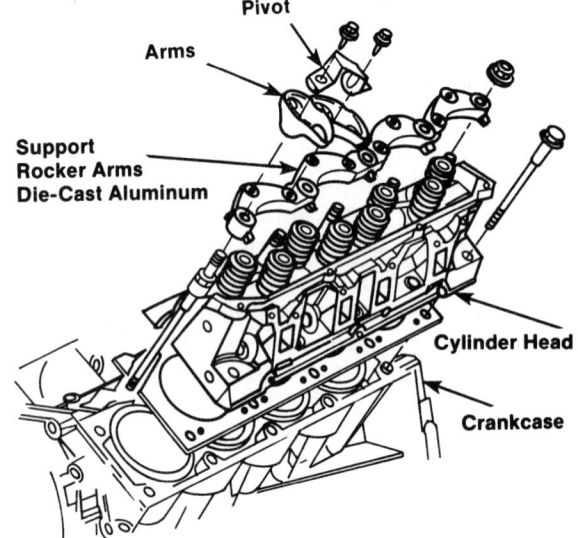

FIGURE 15-51 Rocker arm assembly

FIGURE 15-52   Correct rocker arm location *(Courtesy of TRW, Inc.)*

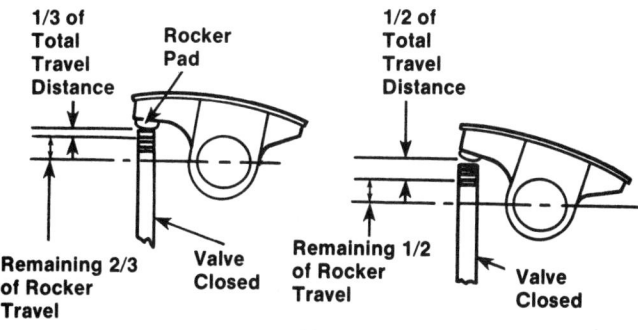

A   **B**

FIGURE 15-53   (A) Rocker arm moves one-third of the total travel distance before the rocker pad is centered on the valve end. (B) Rocker arm moves one-half of the total travel distance before the rocker pad is centered on the valve end.

If the metal was removed by machining from the gasket surface of either the block or cylinder heads, it will affect the valve train geometry. It lengthens the linkage. On engines with nonadjustable rocker arms (Figure 15-54), removal of more than 0.010 inch of material must be compensated for by shimming up the rocker arm assembly supports. Most rocker arm ratios are approximately 1.5:1. Therefore, the rocker arm supports should be shimmed 3/5 the amount removed from the gasket surface. For example, if 0.030 inch is removed from the gasket surface, then the rocker arm supports will require 0.018 inch shim. This is assuming that the valve tip protrusion was adjusted to specifications. If the stem protrusion is not correctly adjusted and/or the rocker arm supports are not shimmed as required, the hydraulic lifters can be totally collapsed and hold the valves open. When installing shims be sure not to block off oil passages to the rocker arms.

On engines so equipped, back off all tappet adjusting screws before tightening down the rocker arm assemblies. Never use a power speed wrench to tighten rocker arm assemblies because this can result in bent pushrods, broken rocker arms, or bent valves. After they have been tightened to the proper torque, adjust the tappets according to the vehicle manual specifications.

Adjustable tappets, or cam followers, have been on the automotive scene for years. An early design featured the tappet sitting on top of the cam lobe with a threaded adjusting device directly under the valve tip. Many of today's overhead cam engines utilize an adjustment screw; one side of the screw is flat and rests against the valve tip, and the other side

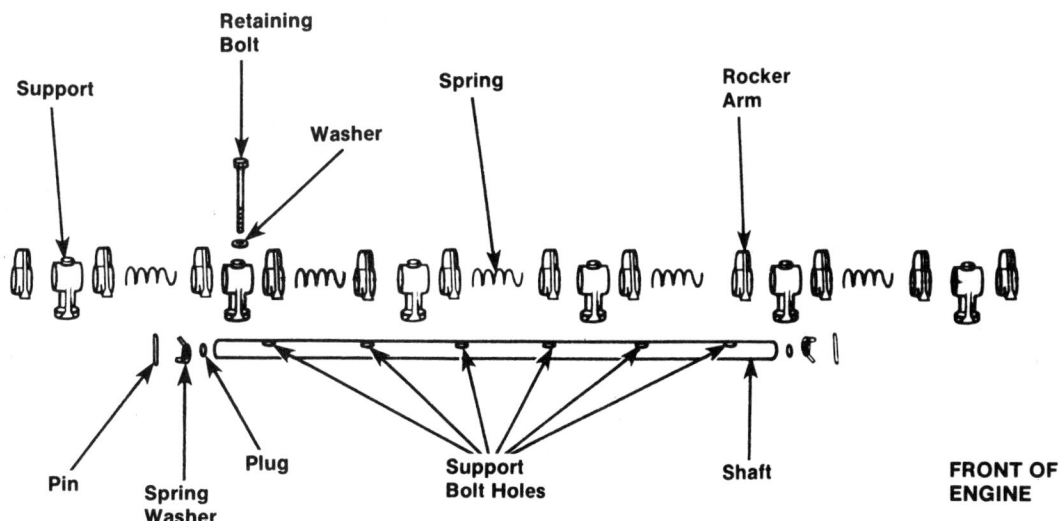

FIGURE 15-54   Rocker arm shaft assembly

is tapered and fits underneath the follower (Figure 15-55). By threading the screw in or out of the follower, the gap between the cam lobe surface and the top of the follower is made larger or smaller. The screw must be adjusted one complete turn at a time to keep the flat side against the valve tip; if necessary, install a different size adjustment screw to get proper clearance for turning.

When the camshaft in an OHC engine is installed, some clearance may be needed between the cam and cam followers (Figure 15-56). Installed

valve stem height is critical to avoid incorrect valve-train geometry and to correctly set valve lash (Figure 15-57).

Some overhead cam engines have an adjustment disc or shim between the cam lobe surface and the follower. To adjust clearance, a special tool and magnet must be used. To correct excessive clearance, a thicker disc or shim is added; if reduced clearance is the problem, a thinner disc or shim must be installed.

A static valve adjustment is required after an engine rebuilding to ensure proper engine starting and prevent damage to the valves from pistons hitting the valves. Normally, if the static valves are adjusted accurately, no further valve adjustment is required unless the cylinder heads are retightened. All valve adjustments should be made only within the limits prescribed in the manufacturer's vehicle shop manual (Figure 15-58). The shop manual is the best

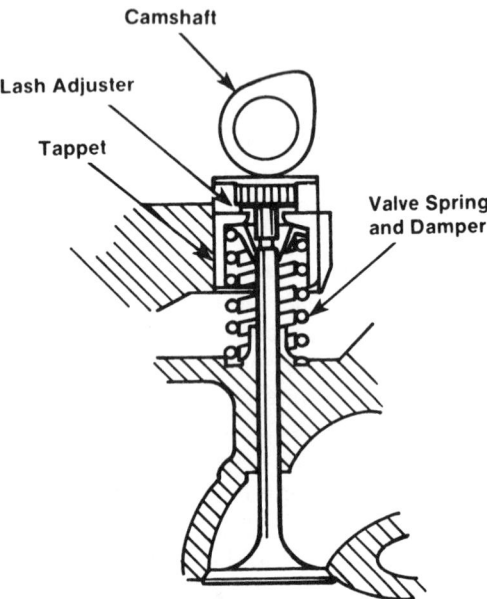

FIGURE 15-55 Some overhead cam engines feature a cam follower with an adjustment screw

FIGURE 15-57 Checking valve lash with a pry bar and feeler gauge.

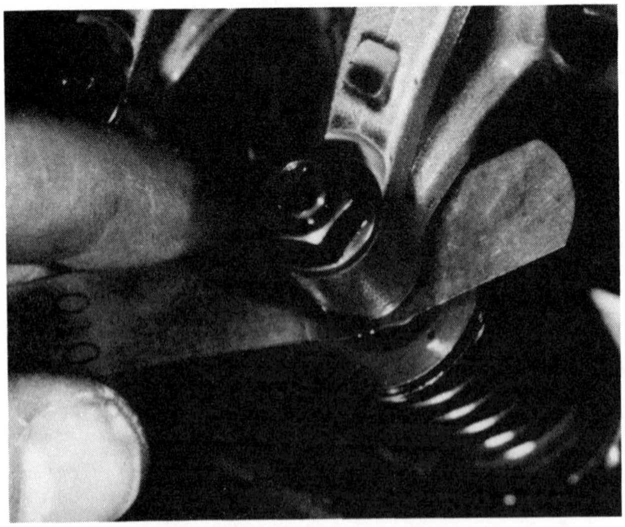

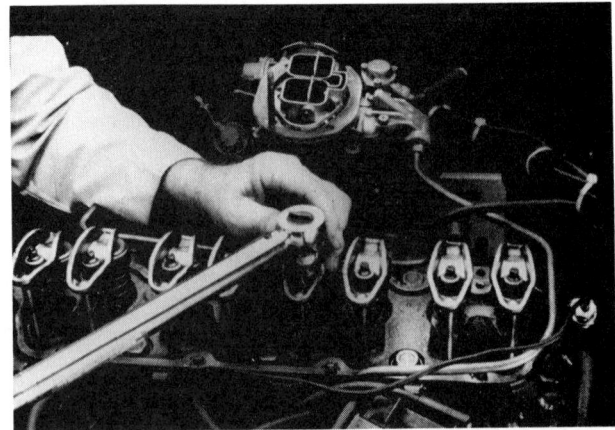

FIGURE 15-56 Checking valve clearance.

FIGURE 15-58 On valve trains with adjustment provisions, adjust the lifters according to the engine manufacturer's specifications. *(Courtesy of TRW, Inc.)*

source of information of the latest in cylinder head design (Figure 15-59).

INSTALLING THE OIL PUMP

The lubrication system of the newly reconditioned engine must be in perfect working order. If the supply of lubricating oil is inadequate, the new engine will be quickly destroyed. Many engine rebuilders install a new oil pump on each engine that is rebuilt. If the pump is not replaced, it should be inspected and rebuilt as described in Chapter 13.

Whether the pump is an integral or nonintegral type, it can be driven by a gear on the camshaft or by an extension shaft from the distributor. Care must be taken to make sure the gear or extension is engaged properly. Also, regardless of the type of pump, both should be primed before assembly. This can be done by submerging the pump in oil (Figure 15-60) or by packing the pump with lightweight assembly lube or oil. Rotate the pump shaft to distribute the oil within the pump body. In the case of a typical distributor-driven oil pump (Figure 15-61), the installation should be performed in the following manner:

1. Position the intermediate drive shaft into the distributor socket. With the shaft firmly

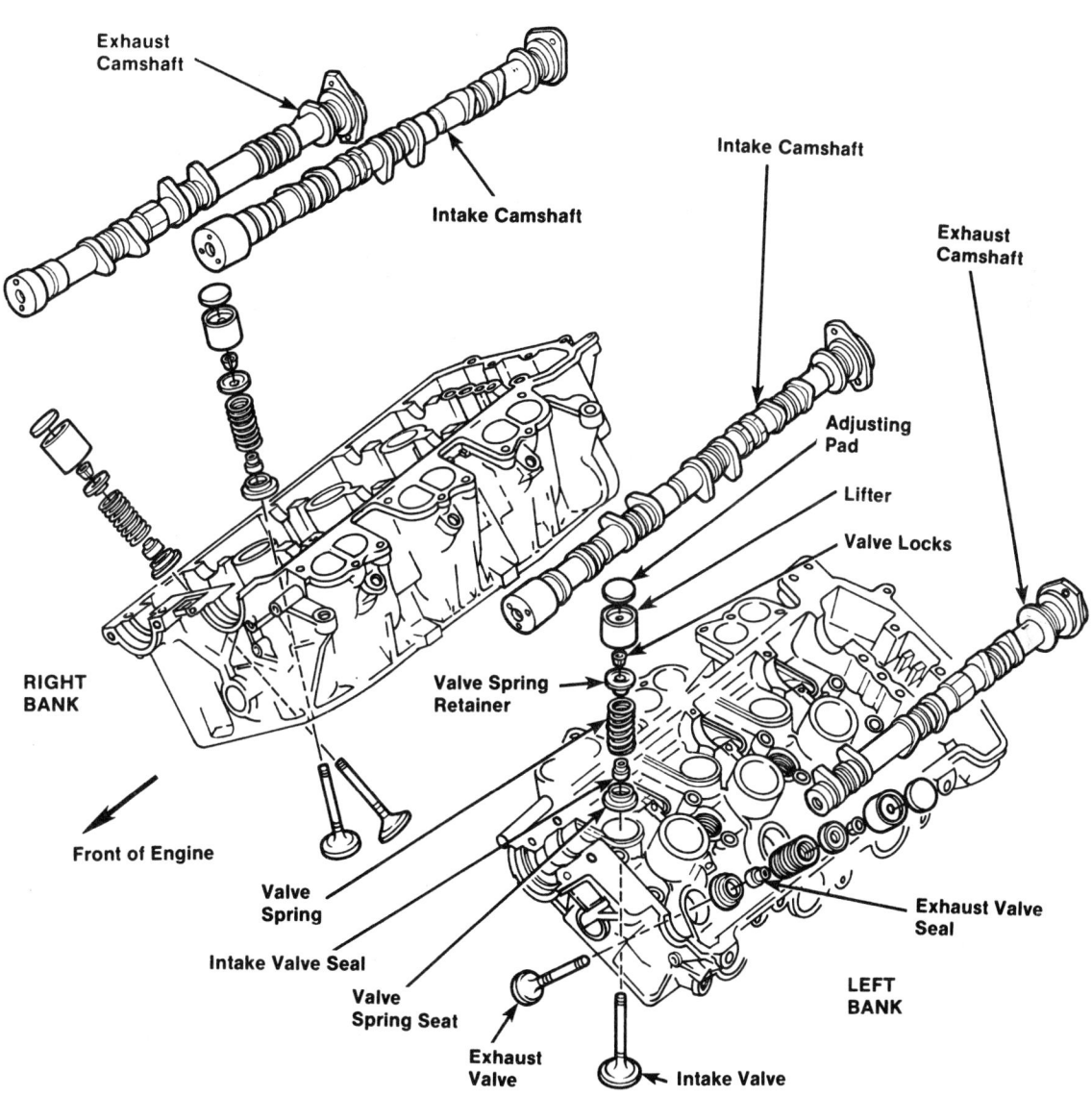

FIGURE 15-59   Cylinder head of a dual overhead camshaft and four-valve in-engine design

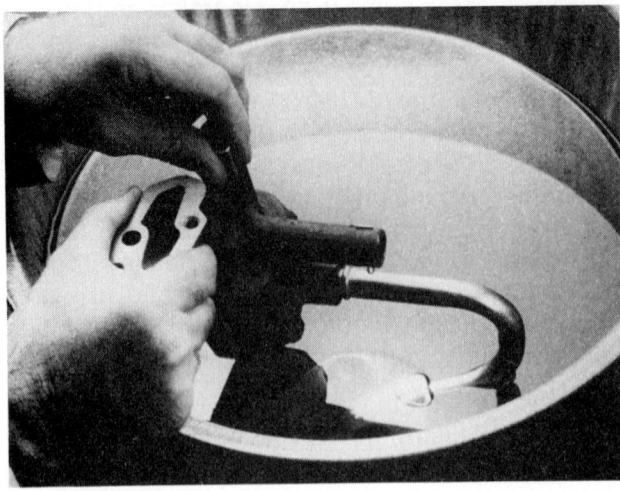

FIGURE 15-60 To prime an oil pump, submerge it in clear engine oil. *(Courtesy of TRW, Inc.)*

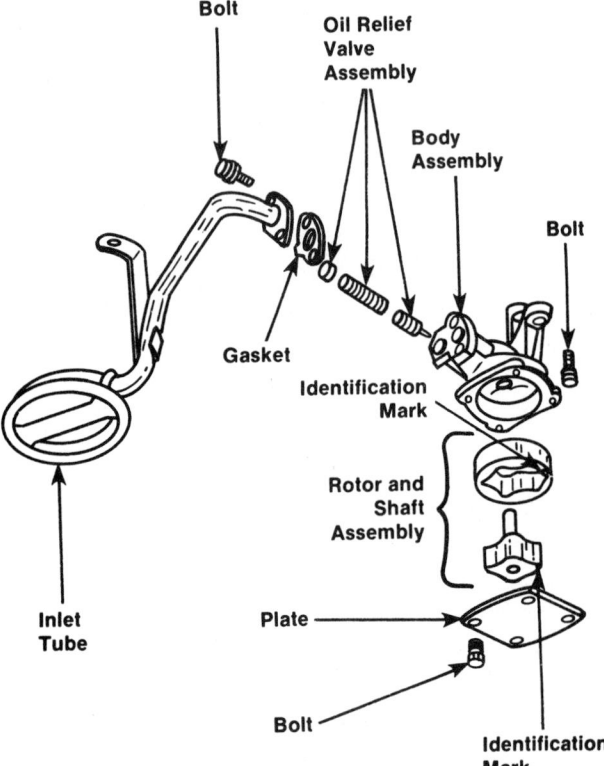

FIGURE 15-61 Oil pump and inlet tube

seated, the stop on the shaft should touch the roof of the crankcase. Remove the shaft and position the stop as necessary.

2. With the stop properly positioned, insert the intermediate drive shaft into the oil pump. Install the pump and shaft as an assembly. Do not attempt to force the

pump into position if it will not seat readily. The drive shaft hex might be misaligned with the distributor shaft. To align, rotate the intermediate drive shaft into a new position. Tighten the oil pump attaching screws to torque specifications.

3. Clean and install the oil pump inlet tube and screen assembly.

The installation of a typical camshaft-driven oil pump is as follows (Figure 15-62):

1. Prime the oil pump by filling the inlet opening with oil and rotating the pump shaft until oil emerges from the outlet opening.
2. Apply suitable sealant to the pump and to the block interface.
3. Install the pump to the full depth and rotate it back and forth slightly to ensure proper positioning and alignment through the full surface of the pump and the block machined interface surfaces (Figure 15-63).
4. Once installed, tighten the bolts or screws. The pump must be held in a fully seated position while installing bolts or screws (Figure 15-64).

 SHOP TALK _____

The instructions here for the installation of either type of pump are general. Specific directions can be found in the service manual.

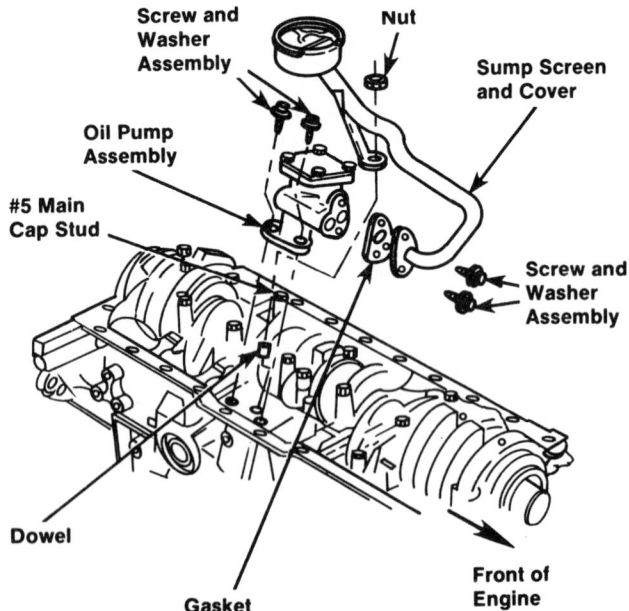

FIGURE 15-62 Camshaft-driven oil pump installed

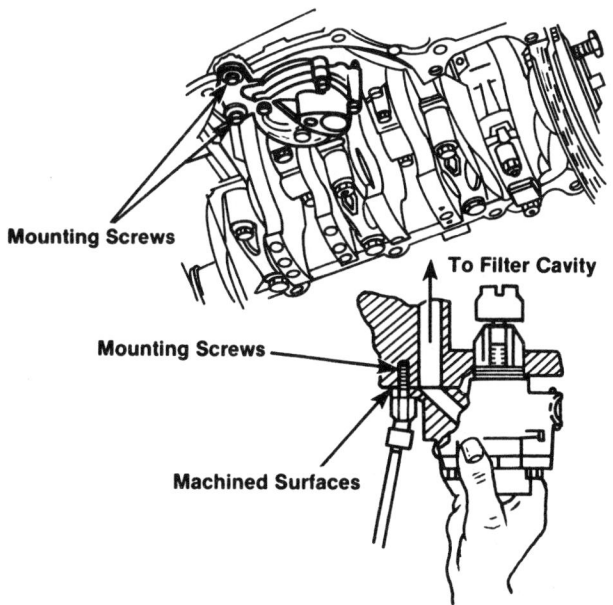

Mounting Screws

Mounting Screws

To Filter Cavity

Machined Surfaces

FIGURE 15-63   Oil pump replacement installed

FIGURE 15-64   Position and align the pump on the block. Use a new mounting gasket and tighten the mounting bolts uniformly. *(Courtesy of TRW, Inc.)*

INSTALLING VALVE, OIL PAN, AND TIMING FRONT COVERS

Engine covers—valve, oil pan, and front (timing)—are usually made of stamped steel, plastic, or die-cast aluminum. They cover otherwise exposed moving engine parts and help to seal the engine.

The installation of these covers is covered completely in Chapter 13.

 SHOP TALK _____

Many rebuilders install the water pump at this point or wait until the engine is back in the vehicle. When installing a water pump at anytime during the reassembly, apply a coating of good waterproof sealer to a new gasket and place it in position on the water pump. Coat the other side of the gasket with sealer and position the pump against the engine block until it is properly seated. Install the mounting bolts and tighten them evenly, in a staggered sequence, to specs with a torque wrench. Careless tightening could cause the pump housing to crack. Check the pump to make sure it rotates freely.

PRELUBRICATION CHECK

With the valve train, valve cover, and oil filter on the engine, prelubricate the engine with oil under pressure before installing the oil pan (Figure 15-65). Oil for the engine test is furnished by a prelubricator. It will take a brief period of time for the oil pump to supply adequate oil to a freshly overhauled engine when it is first started.

There are several ways to prelubricate an engine. One method is to drive the oil pump with an electric drill. With most engines, it is possible to make a drive that can be chucked in an electric drill motor to engage the drive on the oil pump. Insert the fabricated oil pump drive extension into the oil pump through the distributor drive hole. Drive the oil pump

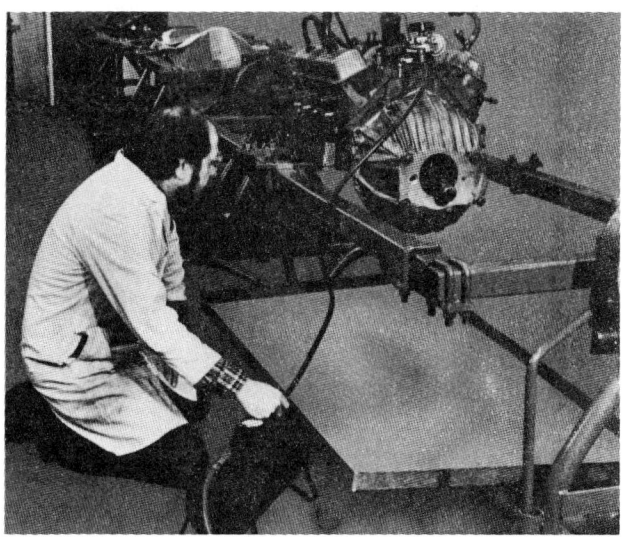

FIGURE 15-65   Priming the engine *(Courtesy of Federal Mogul Corp.)*

with the electric drill (Figure 15–66) and observe the rocker arms on the engine. Set the valve cover(s) on the engine to prevent oil splash. After running the oil pump for several minutes, remove the valve cover and see whether there is any oil flow to the rocker arms. If oil gets to the rocker arms, the lubrication system is full of oil and operating properly. If no oil is observed at the rocker arm area, there is a problem either with the pump, with an alignment of an oil hole in a bearing, or perhaps a plugged gallery.

CAUTION: A great deal of care must be taken when using an electric drill to prime an engine. An rpm that is too fast could damage the oil pump internally.

The best method of prelubricating an engine with oil under pressure without running it is to use a prelubricator. This device consists of an oil reservoir attached to a continuous air supply. When the reservoir is attached to the engine and the air pressure turned on, the prelubricator will supply the engine lubrication system with oil under pressure (Figure 15–67).

The engine prelubricator provides three important benefits.

- Prelubricates the engine after an overhaul to assure adequate lubrication from the moment the engine is turned over
- Helps determine the need for engine bearing replacement by showing the amount of oil leakage through the existing bearings
- Monitors the effectiveness of the repair job by showing the amount of leakage through the replacement bearings

FIGURE 15–67 Engine prelubricator *(Courtesy of Federal Mogul Corp.)*

Oil from the prelubricator is forced into the lines of the lubricating system under pressure equal to the normal road speed pressure. With the pan removed, the oil leakage at the bearings can be observed and relative oil clearance can be checked by the extent of the flow of oil from the bearing ends. Breaks or cracks in the oil lines are also easily located. Leakage of 20 to 120 drops per minute indicates that the bearing is in satisfactory condition. A more rapid flow means that the bearing probably should be replaced and that a positive check of its condition should be made. Before condemning a bearing, rotate the crankshaft one-half revolution to make sure that the rapid flow did not result from a registration of oil holes.

If the prelubricator check indicated that the overhaul was performed properly, install the oil pan (Figure 15–68), and fill the engine with the proper grade oil.

FIGURE 15–66 Priming the oil system for initial start-up using an electric drill *(Courtesy of Perfect Circle/Dana)*

FIGURE 15–68 Installing an oil pan

INSTALLING THE ENGINE

In many shops, after the engine is sprayed with an engine paint or rust inhibitor, the rebuilder's work is finished. The rebuilt engine is returned to the automotive repair shop where the repair shop mechanic will install the rebuilt engine back into the vehicle. But the last step in the reconditioning process is documenting the rebuilder's responsibilities as well as those of the customer. Do not jeopardize the product's reliability, the customer's satisfaction, or shop's reputation by omitting this part of the process. Mark the work order and indicate any special application or installation restrictions or instructions. Also include a copy of the product warranty with work receipts provided to the customer. Many shops install a heat tab for better warranty service (see Chapter 8). Another good way to record pertinent information such as bore, rod, and mains is to install an AERA metal engine tag (Figure 15-69). The metal engine tags can be ordered from AERA.

In shops that do their own installation, an auto mechanic usually does the work. In some instances the rebuilder is qualified to install the engine in the vehicle. It must be remembered that work on today's front-wheel-drive, computer-equipped, emission-controlled vehicle can be quite complex and requires special training and experience. Specialists in the various vehicle systems perform the various tasks involved in the installing of a rebuilt engine.

The following is a summary of the general steps necessary to install a typical rebuilt engine:

1. Attach the engine lifting sling or hoist and lift it slightly. Disconnect the engine from the engine stand.
2. Lower the engine carefully into the chassis. Make sure the dowels in the block engage the holes in the flywheel or converter housing. Follow the instructions given in the service manual for either automatic transmission or manual shift transmission reinstall.
3. Align the location mark on the crankshaft and flywheel or flex plate (the mark made when it was removed). Install with the new special bolts and torque to specifications. If used, remove the jack supporting the transmission.
4. Lower the engine until it rests on the engine support(s).
5. Tighten all engine bolts in a three-step process to specified torque. Remove the lifting sling or hoist.
6. Install the starter and connect the starter cable. Attach the automatic transmission fluid filler tube bracket, if so equipped.
   - On a vehicle with an automatic transmission, install the transmission oil cooler lines in the bracket at the cylinder block.
   - Connect the automatic transmission cable, then install the upper transmission-to-engine bolts.
   - Connect the transmission vacuum line at the junction.
7. Reconnect the vehicle exhaust system to the engine. Use new gaskets, where needed, to connect the exhaust manifolds to the exhaust pipes. Tighten the nuts to specifications.
   - Install the manifold heat shields. Certain engines use two side gaskets and two end seals to mount the intake manifold. In this case, to prevent leaks, it is critical to interlock the end seals and gaskets, to apply a bead of silicone sealer, and to prevent the intake manifold and gaskets from shifting when they are positioned and mounted on the heads. Always tighten the mounting bolts from the middle to the ends, alternating sides and tightening in stages until the correct torque is reached. Connect the transmission vacuum line to the manifold.
8. Connect the spark plug wires to the spark plugs. When installing spark plugs, apply a little antiseize lubricant to prevent electrolytic and other types of corrosion between metal parts (Figure 15-70).
9. Connect the crankcase vent hose to the inlet tube in the intake manifold. Install the PCV valve in the valve rocker arm cover.

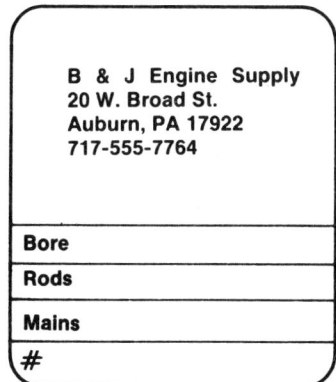

B & J Engine Supply
20 W. Broad St.
Auburn, PA 17922
717-555-7764

Bore

Rods

Mains

#

FIGURE 15-69 AERA engine tag

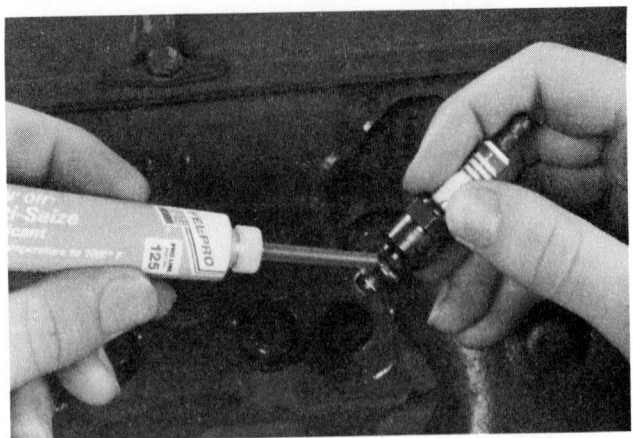

FIGURE 15-70 Applying an antiseize lubricant to a spark plug *(Courtesy of Fel-Pro Inc.)*

FIGURE 15-71 Checking fan belt tension *(Courtesy of TRW, Inc.)*

10. Connect the engine ground strap and the battery ground cable.

11. Connect the electronic engine control (EEC) harness to all sensors, if so equipped.
   - Connect the distributor and sender unit wires to the engine.

12. Install the accelerator cable and bracket assembly.

13. Connect the brake booster hose.

14. On a vehicle with an automatic transmission, connect the kickdown cable to the throttle body.
   - Connect the accelerator linkage to the carburetor and install the retracting spring. Connect the choke cable to the carburetor and hand throttle, if so equipped.
   - On a vehicle with power brakes, connect the brake vacuum line to the intake manifold.

15. Connect the coil primary wire, oil pressure, and coolant temperature sending unit wires, flexible fuel line, heater hoses, and the battery positive cable.

16. Install the alternator on the mounting bracket and connect the alternator wires.
   - On a vehicle with power steering, install the power steering pump on the mounting brackets.

17. Install the water pump pulley and belt (Figure 15-71), all hoses, the shroud, viscous fan drive, bolts on the fan clutch, the cooling fan, and drive belt. Tighten the fan bolts to specifications.
   - Adjust the tension of all drive belts to specification. Tighten the alternator,

power steering pump, and air compressor mounting bolts to specification.

18. Install the radiator. Connect the radiator lower hose to the water pump and the radiator upper hose to the coolant outlet housing. Connect the air compressor lines. If the air conditioner compressor and condenser were removed, install them at this point.
   - On a vehicle with an automatic transmission, connect the oil cooler lines.

19. On a vehicle with standard transmission, adjust the clutch pedal free travel.
   - On a vehicle with an automatic transmission, adjust the transmission control linkage. Check the fluid level and add as required to bring it to the proper level on the oil indicator.

20. Install the throttle body intake tubes.

21. Connect the fuel lines to the fuel pump and fuel injection system or carburetor.

22. Position the grille and lower gravel deflector.

23. Connect the starter wires and install the starter assembly.

24. Connect the speedometer cable.

25. Fill and bleed the cooling system. Connect the battery and fill the crankcase with the proper grade oil. If the oil filter was removed after the prelubrication test, reinstall the unit.

26. Check to be sure the emission controls are all in place (Figure 15-72). In fact, at this point, all equipment in the engine compartment should be reconnected.

27. Replace the hood, using the guide marks made during disassembly. Check the align-

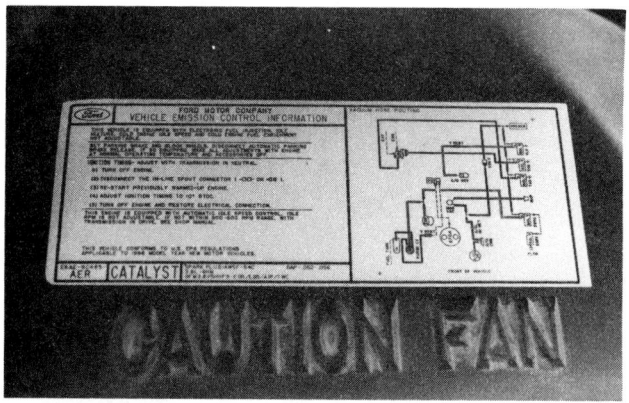

FIGURE 15-72 A vehicle emission control information (VECI) label contains vacuum circuits and other emission specifications.

ment by opening and closing the hood several times.

28. Make a final inspection of the engine compartment. Now the newly rebuilt engine is ready for the ultimate test: the starting and break-in procedures.

STARTING PROCEDURE

Set the valve, carburetor, and ignition timing as accurately as possible before starting the engine. The engine can be fine-tuned after it has been started and goes through the break-in test by using an engine analyzer or diagnostic tester (see Chapter 3). However, it is wise to time the ignition distributor before attempting to start the engine even though it is not the final adjustment.

WARNING: An engine that is not in correct time might backfire when it is first started. If this happens when there is a fuel leak, a fire might start and cause burns. Also be sure to connect the exhaust pipes ventilation system to prevent a dangerous buildup of carbon monoxide when the engine is running.

To adjust the ignition distributor, mark the crankshaft pulley notch or TDC mark and the timing cover pointer on the TDC mark with a dab of white paint. Turn the crankshaft until the number 1 piston, coming up on the compression stroke, aligns the TDC mark with the pulley or pointer (Figure 15–73). Watch the rocker arms on number 1 cylinder just after the intake arm closes and before the exhaust arm opens. The exhaust valve is TDC when the marks are aligned. Install the distributor with the

rotor pointing to the mark on the body that indicates number 1 position. It might be necessary to turn the rotor about 15 degrees away from the mark and wiggle the rotor slightly to engage the intermediate drive shaft or oil pump drive. Once the distributor base slides into solid contact with the block or manifold mounting, align the reference marks and tighten the mounting bolt slightly. You will have to check the ignition timing later, so do not completely tighten at this time.

When the engine is in time, fill the gasoline tank with several gallons of fresh gasoline. Have an associate crank the engine. It will take some time for fuel to be pumped from the tank to the engine. When the engine gets fuel it will try to start. Once it does start, set the throttle to an engine speed of approximately 25 mph (truck engines should be one-third throttle) until the engine coolant reaches normal operating temperature. Then shut down the engine and retorque the cylinder head cap screws and nuts, recheck carburetor or injector adjustments, ignition timing, and valve clearance. Look for any coolant or oil leaks. Run the engine at fast idle during the warm-up period to assure adequate initial lubrication for the piston rings, pistons, and cylinders.

BREAK-IN PROCEDURE

To prevent engine damage after it has been rebuilt and to insure good initial oil control and long engine life, the proper break-in procedure must be followed. That is, make a test run at 30 mph and accelerate at full throttle to 50 mph. Repeat the acceleration cycle from 30 to 50 mph at least ten times. No further break-in is necessary. If traffic conditions will not permit this procedure, accelerate the engine rapidly several times through the intermediate gears during the check run. The objective is to apply a load to the engine for short periods of time and in rapid succession soon after engine warm-up. This action thrusts the piston rings against the cylinder wall with increased pressure and results in accelerated ring seating.

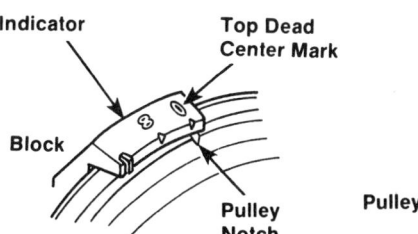

FIGURE 15-73 When the pulley and indicator are aligned, the number 1 piston is at top dead center.

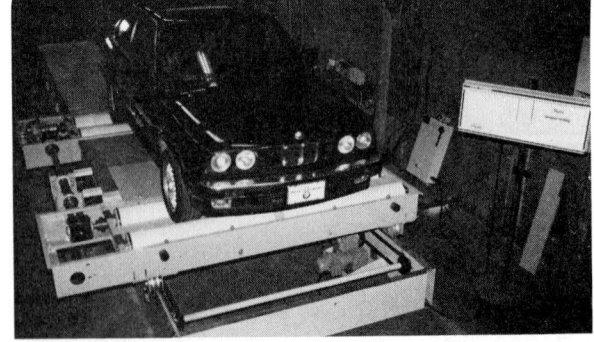

FIGURE 15-74 Engine dynamometer *(Courtesy of Clayton Dynamometer, Inc.)*

As mentioned in Chapter 3, an engine dynamometer is used to confirm and condition an engine before installation (Figure 15-74). New, rebuilt, or remanufactured engines can be tested and run-in prior to installation assuring that no car or truck receives a faulty engine. The dynamometer features repeatable testing with a "stay put" load setting. It will maintain the same speed and power curve after a full stop without readjustment of the controls. An automatic control system is available for constant speed and constant torque test modes (Figure 15-75). Preprogrammed test cycles based on speed, horsepower, and time can be conducted using the advanced system controller. Typical run-in procedures for both gasoline or diesel engines mounted on a dynamometer are given in Table 15-1.

Following the break-in procedure done either in the shop, on the road, or on a dynamometer, turn the vehicle over to the owner or operator with the following suggestions.

PASSENGER CARS OR LIGHT-DUTY TRUCKS

1. Drive the vehicle normally but avoid sustained high speed during the first 500 miles

FIGURE 15-75 Checking an engine installed in a vehicle *(Courtesy of Clayton Dynamometer, Inc.)*

TABLE 15-1: DYNAMOMETER RUN-IN PROCEDURES

Passenger Car and Light Truck Gasoline Engines

Step	Time	Manifold Vacuum (Inches of Hg)	RPM
1.	Warm-up period 10 minutes	No load	800
2.	10 minutes	15	1500
3.	15 minutes	10	2000
4.	15 minutes	10	2500
5.	15 minutes	6	3000
6.	5 minutes	Open throttle (full load)	3000

Truck and Other Heavy-Duty Gasoline Engines

Step	Time	Manifold Vacuum (Inches of Hg)	RPM
1.	Warm-up period 10 minutes	No load	600 to 800
2.	10 minutes	15	1500
3.	15 minutes	10	2000
4.	15 minutes	10	2500
5.	30 minutes	6	2500
6.	10 minutes	Open throttle (full load)	3000 or governed speed

Diesel Engines

Step	Time	Load	RPM
1.	Warm-up (approx. 15 minutes)	No load	600 to 800
2.	30 minutes	25% of rated HP	1200
3.	30 minutes	50% of rated HP	1600
4.	Retighten cylinder head studs and adjust tappet clearance and injectors.		
5.	30 minutes	75% of rated HP	1800
6.	60 minutes	75% of rated HP	Governed speed
7.	5 minutes	Power check (full load)	Governed speed

(break-in period).

2. Check the oil level frequently during the break-in period. It is not unusual to use 1 or 2 quarts of oil during this time.
3. The oil and oil filter should be changed at the end of the break-in time.
4. Retorque all cylinder head and intake manifold bolts.
5. Check the fuel injectors or carburetor adjustments.
6. Check the valve adjustments and distributor timing.

HEAVY-DUTY TRUCKS

If possible, place in light duty for the first 50 miles. At no time should the engine be lugged. Lug-ging is said to exist when the engine does not respond to further depression of the accelerator. Then follow steps 2 through 6 of the suggestions for driving passenger cars.

To keep engine rebuilding a viable profession, always listen to the customers' complaints or comments and make every effort to rectify all problems.

REVIEW QUESTIONS

1. What is the best way to clean the oil galleries?
   a. run a brush through them
   b. in a jet spray washer
   c. in a hot tank
   d. with light oil or rust penetrant

2. Which type of core plug is installed with its flanged edge facing outward?
   a. cup
   b. expansion
   c. both a and b
   d. neither a nor b

3. When installing a core plug, Technician A coats the plug with an oil-resistant sealer. Technician B coats the bore. Who is right?
   a. Technician A
   b. Technician B
   c. Both A and B
   d. Neither A nor B

4. What is used on some engines to limit the end play of the camshaft?
   a. camshaft boss
   b. thrust plate
   c. neoprene seal
   d. dial indicator setup

5. After installing the soft plugs, the next step is installing the _____ .
   a. camshaft
   b. cam bearings
   c. crankshaft
   d. main bearings

6. Camshaft binding is the result of which of the following?
   a. damaged bearing
   b. nick on the cam's journal
   c. slight misalignment of the block journals
   d. all of the above

7. Technician A checks to make sure the flange thrust surfaces are properly aligned after the main bearing cap screws are installed and finger tightened. Technician B checks for proper alignment by placing a large, flat-blade screwdriver between a crankshaft counterweight and the engine webbing and gently prying the crankshaft forward and backward in the engine. Who is right?
   a. Technician A
   b. Technician B
   c. Both A and B
   d. Neither A nor B

8. Technician A measures crankshaft end play with a dial indicator; Technician B uses a feeler gauge. Who is right?
   a. Technician A
   b. Technician B
   c. Both A and B
   d. Neither A nor B

9. What is the next installation task after installing the crankshaft?
   a. camshaft
   b. cam bearings
   c. pistons and connecting rods
   d. timing components

10. Technician A says that wear in the chain, gears, or sprockets of a timing assembly means a timing lag. Technician B says that if wear exists on any component in a timing assembly, a new part must be substituted into that assembly. Who is right?
    a. Technician A
    b. Technician B
    c. Both A and B
    d. Neither A nor B

11. Technician A winds a timing chain onto the sprockets. Technician B uses a screwdriver to pry a chain into position. Who is right?
    a. Technician A
    b. Technician B
    c. Both A and B
    d. Neither A nor B

12. In which of the following ways is the valve clearance adjusted?
    a. with an adjustable rocker arm
    b. with an adjustable tappet
    c. with adjustment discs
    d. all of the above

13. Which of the following is a benefit provided by an engine prelubricator?
    a. monitors the effectiveness of the repair job
    b. helps determine the need for engine bearing replacement
    c. both a and b
    d. neither a nor b

14. Which of the following statements concerning starting and break-in procedure is incorrect?
    a. Never time the ignition distributor before the engine is started.
    b. When the engine has started, set the throttle to approximately 25 mph until the engine coolant reaches normal operating temperatures.
    c. If traffic conditions will not permit the proper break-in procedure, accelerate the engine rapidly several times through the intermediate gears.
    d. The break-in procedure can be done on a dynamometer.

15. The proper break-in procedure is a test run at _____ .
    a. 50 mph and acceleration at full throttle to 60 mph
    b. 25 mph
    c. 30 mph and acceleration at full throttle to 50 mph
    d. 55 mph

GLOSSARY

Abrasion Wearing or rubbing away of a part.

Additive As used with reference to automotive oils, a material added to the oil to give it certain properties. For example, a material added to engine oil to lessen its tendency to congeal or thicken at low temperature.

Adhesion The property of lubricating oil that causes it to stick or cling to a bearing surface.

AERA Automotive Engine Rebuilder's Association.

Aftermarket Equipment sold to consumers after the vehicle has been manufactured.

Align Bore To bore, ream, or otherwise machine the crankshaft housing bore, or the assembled main bearings, creating centers all of which fall on a true centerline.

Alignment An adjustment to a line or to bring into a line.

Alloy A mixture of different metals such as solder, which is an alloy consisting of lead and tin.

Aluminum A metal noted for its lightness often alloyed with small quantities of other metals for automotive use.

Anticlockwise Rotation Rotating the opposite direction of the hands on the clock. The same as counterclockwise rotation.

Antifriction Bearing A bearing constructed with balls, rollers, or the like between the journal and the bearing surface to provide rolling instead of sliding friction.

ASME American Society of Mechanical Engineers.

Atmospheric Pressure The weight of the air at sea level; about 14.7 pounds per square inch; less at higher altitudes.

Backlash The clearance or "play" between two parts, such as meshed gears.

Balance (Dynamic) In balance, free from vibration while in motion.

Balance (Static) In equilibrium, or stationary balance.

Bearing The supporting part that reduces friction between a stationary and rotating part.

Bearing Back The outside area of the bearing that seats against the housing bore.

Bearing Bore The inside diameter of the assembled bearing.

Bearing Cap The removable half of the saddle that holds the bearing in place.

Bearing Lining The alloy adhered to the bearing back forming the bearing surface area.

Bearing Shell One member of a pair of insert type bearings.

Bearing Upper (Half Shell) The bearing half that is made for assembly in the engine block or connecting rod as opposed to the bearing cap.

Big End Large end of the connecting rod attached to the crankshaft.

Blowby The gas that leaks past the piston of an engine during the period of maximum compression in the combustion chamber.

Bore The diameter of the cylinder.

Boss A cast or forged part of a piston that can be machined for accurate balance.

Bourdon Tube A curved tube that straightens as the pressure inside it is increased. The tube is attached to a needle on a gauge, which senses the movement of the tube and transmits it as a pressure reading.

Break-in The process of wearing in to a desirable fit between the surfaces of two new or reconditioned parts.

Broach To finish the surface of metal by pushing or pulling a multiple-edge cutting tool over or through it.

Broach Relief The bearing wall at the parting line which is thinner than the bearing at the crown by an amount ranging from approximately 0.0005" to 0.0015". This metal is removed from the inside diameter of the bearing and blended smoothly into the surface approximately 1/4" to 3/8" from the parting line.

Burnish To smooth or polish with a sliding tool under pressure.

Bushing A removable liner for a bearing.

Calibrate To determine or adjust the graduation or scale of any instrument giving quantitative measurements.

Camshaft The shaft containing lobes or cams to operate the engine valves.

Camshaft Bearing A full round bearing used to hold the camshaft in position.

Carbon A common nonmetallic element that is an excellent conductor of electricity. It also forms in the combustion chamber of an engine during the burning of fuel and lubricating oil.

Carbonize The process of carbon formation within an engine, such as on the spark plugs and within the combustion chamber.

Carburetor A device for automatically mixing fuel in the proper proportion with air to produce a combustible gas.

Carrier An object that bears, cradles, moves, or transports some other object or objects.

Castellate Formed to resemble a castle battlement, such as in a castellated nut.

Center of Gravity The point at which a mass of matter balances. For example, the center of gravity of a wheel is the center of the wheel hub.

Centigrade A measurement of temperature used principally in foreign countries; zero on the centigrade scale is 32 degrees on the Fahrenheit scale.

Centrifugal Force A force that tends to move a body away from its center of rotation; for example: whirling a weight attached to a string.

Chamfer A bevel or taper at the edge of a hole or a gear tooth.

Chase To straighten up or repair damaged threads.

Check Valve A gate or valve that allows gas or fluid to pass in only one direction.

Cheek Crankshaft thrust collar.

Clearance The space allowed between two parts, such as between a journal and bearing.

Clockwise Rotation Rotation in the same direction as the hands of a clock.

Clutch A device for connecting and disconnecting the engine from the transmission or for a similar purpose in other units.

Combustion The process of burning.

Combustion Space or Chamber In automobile engines, the volume of the cylinder above the piston with the piston on top center.

Compound A mixture of two or more ingredients.

Compression The reduction in volume or the squeezing of a gas. As applied to metal, such as a coil spring, compression is the opposite of tension.

Compression Ratio A measure of how much the air has been compressed in a cylinder of an engine from TDC to BDC. Compression ratio will usually be from 8-1 to 25-1.

Compression Rings The upper rings on a piston designed to hold the compression in the cylinder.

Concentric Two or more circles having a common center.

Conformability The ability of the bearing lining material to move slightly and adjust itself to minute unevenness on the rotating shaft.

Connecting Rod The connecting link between the piston to the crankshaft.

Connecting Rod Bearing The bearing at the big end of the connecting rod in which the crankpin rotates.

Contraction A reduction of mass or dimension; the opposite of expansion.

Convection A transfer of heat by circulating heated air.

Coolant The liquid that circulates in an engine cooling system.

Core Plugs Plugs inserted into the block that allow the sand core to be removed during casting. Also, at times these plugs will pop out and protect the block if the coolant freezes.

Corrode To eat away gradually as if by gnawing, especially by chemical action such as rust.

Counterclockwise Rotation Rotating in the opposite direction as the hands of a clock. The same as anticlockwise rotation.

Counterbore To enlarge a hole to a given depth.

Countershaft An intermediate shaft that receives motion from a main shaft and transmits it to a working part; sometimes called a lay shaft.

Countersink To cut or form a depression to allow the head of a screw to go below the surface.

Counterweight Weight forged or cast into the crankshaft to reduce vibration.

Coupling A connecting means for transferring movement from one part to another; can be mechanical, hydraulic, or electrical.

Crankcase The area in the engine below the crankshaft. It contains oil and fumes from the combustion process.

Crankcase Breather A port or tube that vents fumes from the crankcase. An inlet breather allows fresh air into the crankcase.

Crankpin The machined cylindrical portion of the crank throw around which the connecting rod bearing is installed.

Crankshaft A mechanical device that converts reciprocating motion to vertical motion.

Crankshaft Counterbalance A series of weights attached to or forged integrally with the crankshaft placed to offset the reciprocating weight of each piston and rod assembly.

Crankshaft Counterweights Excess metal on the inner side of the crankshaft short arm; used to put the crankshaft into static and dynamic balance.

Crankshaft Long Arm The (usually) unmachined metal connector between two crankpins on a crankshaft.

Crankshaft Short Arm The (usually) unmachined metal connector between a main bearing journal and a crankpin; carries the crankshaft counterweights.

Crankshaft Thrust Collar A vertical disc-shaped machined area on the crankshaft against which the flange of the main thrust bearing rides.

Crank Throw The distance from the crankshaft main bearing centerline to the connect rod journal centerline. The stroke of any engine is the crank throw.

Crank Web The unmachined portion of a crankshaft that lies between two crankpins or between a crankpin and main bearing journal.

Crown The center area of a bearing half.

Crush The press fit allowance necessary to hold two bearing halves securely in the housing bore. In a bearing half, the amount of circumference in excess of a half circle.

Crush Relief Metal removed on the bearing surface at the parting faces extending the full length of the bearing.

Cylinder Internal holes in the cylinder block.

Cylinder Bore The internal diameter of a cylinder into which the piston is inserted.

Cylinder Head The top and cover for the cylinder, which houses the valves.

Cylinder Sleeve A round cylindrical tube that fits into the cylinder bore. Both wet and dry sleeves are used.

Dead Center The extreme upper or lower position of the crankshaft throw at which the piston is not moving in either direction.

Deceleration A decrease in speed.

Deflection Bending or movement away from normal due to loading.

Degree Abbreviated "deg." or indicated by a small "°" placed alongside a figure; can be used to designate temperature readings or angularity, 1 degree being 1/360 part of a circle.

Density Compactness; relative mass of matter in a given volume.

Detergent A compound of a soap-like nature used in engine oil to remove engine deposits and hold them in suspension in the oil.

Detonation As used in an automobile, indicates a too-rapid burning or explosion of the mixture in the engine cylinders. It becomes audible through a vibration of the combustion chamber walls and is sometimes confused with a "ping" or spark knock.

Diagnosis The use of instruments to determine the action or behavior of machine parts to locate the cause of failure.

Dial Gauge A type of micrometer wherein the readings are indicated on a dial rather than on a thimble.

Dial Indicator A measuring tool used to adjust small clearances up to 0.001 inch. The clearance is read on a dial.

Die One of a pair of hardened metal blocks for forming metal into a desired shape or a thread die for cutting external threads.

Die Casting An accurate and smooth casting made by pouring molten metal or composition into a metal mold or die under pressure.

Direct Drive Direct engagement between the engine and driveshaft where the engine crankshaft and the driveshaft turn at the same rpm.

Distortion A warpage or change in form from the original.

Distributor The mechanism in an ignition system that opens and closes the primary circuit through the distributor breaker points and directs the secondary high voltage to the spark plugs at the correct time for firing.

Dowel A pin extending from one part to fit into a hole in an attached part. Used for both location and retention.

Dowel Pin A pin inserted in matching holes in two parts to maintain those parts in fixed relation one to another.

Draw Filing A method of filing wherein the file is drawn across the work while held at a right angle to the length of the file.

Drill A tool for making a hole or to sink a hole with a pointed cutting tool rotated under pressure.

Drive-Fit Also known as a force-fit or press-fit. This term is used when the shaft is slightly larger than the hole and must be forced into place.

Drive Line The universal joints, drive shaft, and other parts connecting the transmission with the driving axles.

Drop Forging A piece of steel shaped between dies while hot.

Dynamic Balancing Equal distribution of weight on each side of a centerline of a wheel. Dynamic means moving or action, and dynamic balancing is done with the wheel moving or spinning.

Dynamometer A device used to brake or absorb power produced from an engine for testing purposes in a laboratory situation.

Eccentric One circle within another circle with neither having the same center; for example, a cam on a camshaft.

Eccentric Bearing A bearing in which the greatest wall thickness is at the crown and tapers to a few ten thousandths less toward the parting faces.

Eccentricity A physical characteristic designed into some bearings calling for an inside assembled vertical diameter that is slightly smaller than the horizontal diameter. Controlled by varying the wall thickness.

Efficiency A ratio of the amount of energy put into an engine as compared to the amount of energy coming out of the engine. Gas engines are about 28% efficient. A measure of the quality of how well a particular machine works.

Electromechanical Refers to a device that incorporates both electronic and mechanical principles in its operation.

Embedability The ability of the bearing lining material to absorb any dirt.

Enable A type of microcomputer decision that results in an automotive system being activated and permitted to operate.

End Clearance Play The extent of the possible forward and backward movement of the crankshaft, rod bearing on the crankpin, or the camshaft.

End Play The amount of axial or end-to-end movement in a shaft due to clearance in the bearings.

Engine The power source that propels the vehicle forward or in reverse. The engine can be of several designs, including the standard gasoline piston engine, diesel engine, and rotary engine.

Engine Block The main casting of an internal combustion engine.

Engine Displacement The sum of the displacement of all the engine cylinders.

Exhaust Pipe The pipe connecting the engine to the muffler to conduct the exhausted or spent gases away from the engine.

Expansion An increase in size. For example, when a metal rod is heated, it increases in length and perhaps in diameter; expansion is the opposite of contraction.

Fahrenheit A scale of temperature measurement ordinarily used in English-speaking countries. The boiling point of water is 212 degrees Fahrenheit as compared to 100 degrees Celsius.

Fatigue Deterioration of a bearing metal under excessive intermittent loads or prolonged operation.

Fatigue Strength Ability of a bearing to withstand the loads incurred during engine operation.

Feeler Gauge A metal strip or blade finished accurately with regard to thickness used for measuring the clearance between two parts. Such gauges ordinarily come in a set of different blades graduated in thickness by increments of 0.001 inch.

Ferrous Metal Metal that contains iron or steel and is therefore subject to rust.

File To finish or trim with a hardened rasp or file.

Fillets Small, rounded corners machined on the crankshaft for strength.

Filter (Oil, Water, Gasoline) A unit containing an element, such as a screen of varying degrees of fineness. The screen or filtering element is made of various materials depending on the size of the foreign particles to be eliminated from the fluid being filtered.

Fit The contact between two machined surfaces.

Flange A projecting rim or collar on an object for keeping it in place.

Flanged Bearing A main bearing constructed with vertical discs at the ends (flanges) that ride against a vertical machined face on the crankshaft for carrying thrust loads.

Flange Face Relief Same as broach relief, except that this undercut is on the parting lines of the flange face.

Flathead Engine Any engine with both intake and exhaust valve located in the cylinder block.

Floating Piston Pin A piston pin that is not locked in the connecting rod or the piston but is free to turn or oscillate in both the connecting rod and piston.

Flutter or Bounce As applied to engine valves, refers to a condition wherein the valve is not held tightly on its seat during the time the cam is not lifting it.

Flywheel A heavy circular device placed on the crankshaft. It keeps the crankshaft rotating when there is no power.

Foot-pound (ft-lb) A measure of the amount of energy or work required to lift 1 pound a distance of 1 foot.

Force-Fit Also known as a press-fit or drive-fit. This term is used when the shaft is slightly larger than the hole and must be forced in place.

Forge To shape metal by hammering while it is hot and plastic.

Forging Number A number that appears on the forged part (connecting rod or crankshaft) used to identify the engine model that the part belongs in.

Four-Cycle Engine Also known as otto cycle, wherein an explosion occurs every other revolution of the crankshaft; a cycle being considered as half a revolution of the crankshaft. These strokes are intake, compression, power, and exhaust.

Galling A displacement of metal, usually caused by a lack of lubrication or too loose fit.

Gasket A rubber, felt, cork, steel, copper, or asbestos material placed between two parts to eliminate leakage of gases, greases, and other fluids.

Gear A cylinder- or cone-shaped part with teeth on one surface that mate with and engage the teeth on another part that is not concentric with it.

Glaze As used to describe the surface of the cylinder, an extremely smooth or glossy surface, such as a cylinder wall, highly polished over a long period of time by the friction of piston rings.

Glaze Breaker A tool for removing the glossy surface finish in an engine cylinder.

Grind To finish or polish a surface with an abrasive wheel.

Ground The negatively charged side of a circuit. A ground can be a wire, the negative side of the battery, or even the vehicle chassis.

Gum In automotive fuels, this refers to oxidized petroleum products that accumulate in the fuel system, carburetor, or engine parts.

Harmonic Balancer A device to reduce the torsional or twisting vibration that occurs along the length of the crankshaft used in multiple cylinder engines. Also known as a vibration damper.

Heat Treated A process in which a metal is heated to a high temperature, then is quenched in a cool bath of water, oil, and brine (salt water). This process hardens the metal.

Heel The outside or larger half of the gear tooth.

Helicoil A device used to replace a set of damaged threads.

Hemispherical Combustion Chamber A type of combustion chamber that is shaped like a half of a circle. This combustion chamber has the valves on either side with the spark plug in the center.

Hg Chemical symbol for mercury. Also a reference to the amount of vacuum, that is, "inches of mercury."

Hone An abrasive tool for correcting small irregularities or differences in diameter in a cylinder, such as an engine cylinder or brake cylinder.

Housing Bore The machined surface into which the bearing is installed.

HP Abbreviation for horsepower, the amount of energy required to lift 550 pounds 1 foot in 1 second.

Hub The central part of a wheel or gear.

Hydraulic Pressure Pressure exerted through the medium of a liquid.

Hydrocarbon A chemical composition, made up of hydrogen and carbon, that is a component of exhaust emission.

ID Inside diameter.

Ignition In internal-combustion gasoline engines, the process of igniting the air/fuel mixture in the combustion chamber by an electrical spark from a spark plug.

I-Head A style of valve arrangement in an engine. I-head refers to the valves being placed directly above the piston in the cylinder head.

IHP Indicated horsepower developed by an engine and a measurement of the pressure of the explosion within the cylinder expressed in pounds per square inch.

Inertia Objects in motion tend to remain in motion. Objects at rest tend to remain at rest. Inertia is the force keeping these objects at rest or in motion.

Inhibitor A material to restrain or hinder some unwanted action, such as a rust inhibitor, which is a chemical added to cooling systems to retard the formation of rust.

Inlet or Intake Valve A valve that permits a fluid or gas to enter a chamber and seals against an exit.

In-Line Cylinders in an engine that are in one line or row, such as an in-line four or six cylinder. The cylinders are vertical as well.

Insert Guides Valve guides that are small cast cylinders pressed into the cylinder head.

Insert Type Bearing An interchangeable type of bearing as opposed to a cast-in type of bearing.

Integral Guides Valve guides that are manufactured and machined as part of the cylinder head.

Journal An inner bearing operated by a shaft.

Key A small block inserted between the shaft and hub to prevent circumferential movement.

Keyway or Key Seat A groove or slot cut to permit the insertion of a key.

Knock A general term used to describe various noises occurring in an engine; can be used to describe noises made by loose or worn mechanical parts, preignition, detonation, and so on.

Knurl To indent or roughen a finished surface.

Lapping The process of fitting one surface to another by rubbing them together with an abrasive material between the two surfaces.

Lay The direction or pattern of the predominant tool marks on a machined surface area.

L-Head A valve arrangement that has the valves located in the block and not in the head. Engines that have L-head designs are commonly called "flat head engines."

Liner Usually a thin section placed between two parts such as a replaceable cylinder line in an engine.

Limit A size or dimension either plus or minus the tolerance (high and low limit).

Lobe The part of the camshaft that raises the lifter.

Locating or Locking Lug A projection on a bearing back that nests in a slot in the bearing seat machined out to receive it. Used to locate the bearing in the housing bore and keep it from moving laterally.

Lubricant A substance capable of reducing friction between mating surfaces in relative motion through separation by oil film.

Lubricating System The subsystem on the engine that is used to keep all moving components lubricated.

Main Bearing A bearing that is usaed as a crankshaft support.

Main Bearing Clearance The clearane between the main bearing journal and the main bearings.

Main Bearing Journal A crankshaft journal that is supported by a main bearing.

Main Bearing Saddle Bore The housing that is machined to receive a main bearing.

Major and Minor Thrust The thrust forces applied to the piston on the compression and power strokes.

Manifold A pipe with multiple openings used to connect various cylinders to one inlet or outlet.

Mechanical Efficiency (Engine) The ratio between the indicated horsepower and the brake horsepower of an engine.

ally expressed as an arithmetical deviation from the mean or nominal surface line. Symbol (Y).

Micrometer A measuring instrument for either external or internal measurement in thou-

sandths and sometimes tenths of thousandths of inches.

Mill To cut or machine with rotating tooth cutters.

Millimeter One millimeter is the metric equivalent of 0.039370 of an inch or 1 inch being the equivalent of 25.4 mm.

Misalignment Bearings are not on the same centerline within good functional or working limits.

Misfiring Failure of an explosion to occur in one or more cylinders while the engine is running; can be continuous or intermittent failure.

OD Outside diameter.

Oil A viscous fluid, insoluble in water.

Oil Clearance The difference between the inside bearing diameter and the diameter of the journal.

Oil Film The thin layers of oil that protect the journal and bearing surfaces by separating them and preventing metal-to-metal contact while the engine is in operation.

Oil Gallery The main oil supply line in an engine block, often referred to as the *header*. Oil flows from this reservoir under pressure to the many parts that are to be lubricated.

Oil Groove A canal machined in the surface of a bearing to spread oil on a friction area or to permit the transfer of oil to another part.

Oil Pressure The pounds per square inch (psi) pressure as indicated by the oil gauge. It is the result of the oil pump delivery volume, limited by the oil clearance and modified by the pressure relief valve.

Oil Seal A device used to prevent oil seepage from the hole where the crankshaft extends through the crankcase.

Oil Starvation A condition whereby a friction area is deprived of adequate lubrication resulting in excessive wear or failure.

Out-of-Round An inside or outside diameter, designed to be perfectly round, having varying diameters when measured at different points across its diameter.

Overhead Cam Engine Any engine with the intake valves, exhaust valves, and camshaft located in the cylinder head.

Overhead Camshaft A camshaft located directly on top of the valves, used on I-head designs.

Overhead Valve Engine Any engine with the camshaft located in the block and the intake and exhaust valves located in the cylinder head.

Parting Edge The edge formed where the inside or outside surface of the bearing joins the parting face.

Parting Face   The surface that is in contact with the other bearing half when the bearing is assembled.

Parting Line   The theoretical line formed by the contacting parting faces.

Peen   To stretch or clinch over by pounding with the rounded end of a hammer.

Piston   A cylindrical part closed at one end attached to the crankshaft by the connecting rod. The force of explosion in the cylinder is exerted against the closed end of the piston causing the connecting rod to move the crankshaft.

Piston Collapse   A condition describing a collapse or reduction in diameter of the piston skirt due to heat or stress.

Piston Displacement   The volume of air moved or displaced by moving the piston from one end of its stroke to the other.

Piston Head   The part of the piston above the rings.

Piston Slap   The movement of the piston back and forth in the cylinder in a slapping motion.

Piston Stroke   The distance the piston travels from top dead center to bottom dead center.

Poppet Valve   A valve structure consisting of a circular head with an elongated stem attached in the center designed to open and close a circular hole or port.

Port   In engines, the openings in the cylinder block for valves, exhaust and inlet pipes, or water connections. In two-cycle engines, the openings for inlet and exhaust purposes.

Preload   A load within the bearing either purposely built in or resulting from adjustment.

Press-Fit   Also known as a force-fit or drive-fit. This term is used when the shaft is slightly larger than the hole and must be forced in place.

Psi   A measurement of pressure in pounds per square inch.

Pushrod   A connecting link in an operating mechanism, such as the rod interposed between the valve lifter and rocker arm on an overhead valve engine.

Quenching   A process of rapid cooling of hot metal by contact with liquid, gases, or solids.

Race   A channel in the inner or outer ring of an antifriction bearing in which the balls or rollers operate.

Race Cam   A type of camshaft for race cars that increases the lift of the valve, increases the speed of the valve opening and closing, increases the length of time the valve is held open, and so on.

Also known as full, three-quarter, or semirace cams, depending upon design.

Radial   Perpendicular to the shaft or bearing bore.

Radial Clearance (Radial Displacement)   Clearance within the bearing and between balls and races perpendicular to the shaft.

Ream   To finish a hole accurately with a rotating fluted tool.

Reciprocating   An up and down or back and forth motion.

Rocker Arm   In an automobile engine, a lever located on a fulcrum or shaft; one end of the level is on the valve stem, the other on the pushrod.

Rotary Engine   An engine construction wherein the crankshaft remains stationary and the cylinder spins around it as in a pinwheel.

Saddle Bore   The hole machined to receive the main bearings.

SAE   Society of Automotive Engineers.

Score   A scratch, ridge, or groove marring a finish surface.

Scuffing   Scraping and heavy wear from the piston on the cylinder walls.

Seal   A device used on rotating shafts to keep oil or other fluid on one side of the seal, thus eliminating leakage.

Seat   A surface, usually machined upon which another part rests or seats; for example, the surface upon which a valve face rests.

Seating   When two metals must seal gases and liquids, they must be worked together to make a good seal. This process of getting two metal surfaces to seal is called seating.

Seize   When one surface moving upon another scratches, it is said to seize. An example is a piston score or abrasion in a cylinder due to a lack of lubrication or over expansion.

Service Manual   A manual provided by the manufacturer or other publisher that describes service procedures, troubleshooting and diagnosis, and specifications.

Shim   Thin sheets, usually metal, which are used as spacers between two parts, such as the two halves of a journal bearing.

Shrink-Fit   Where the shaft or part is slightly larger than the hole in which it is to be inserted. The outer part is heated above its normal operating temperature or the inner part is chilled below its normal operating temperature or both and assembled in this condition. When cooled, an exceptionally tight fit is obtained.

Shrouding When a valve is placed close to the side of the combustion chamber, the air and fuel may be restricted by the side of the chamber. This restriction is referred to as shrouding.

Siamese Ports Intake or exhaust ports inside the cylinder head where two cylinders are feeding through the one port.

Slant An in-line cylinder arrangement that has been placed at a slant. This arrangement makes the engine have a lower profile for aerodynamic design.

Sleeve Valve A reciprocating sleeve or sleeves with ported openings placed between the piston and cylinders of an engine to serve as valves.

Sliding Fit Where sufficient clearance has been allowed between the shaft and journal to permit free running without overheating.

Slip-In Bearing A liner made to extremely accurate measurements that can be used for replacement purposes without additional fitting.

Slipper Skirt A piston that has a cutaway skirt so that the piston can come closer to the counterweights. This makes the overall size of the engine smaller.

Sludge In an automobile engine, it indicates a composition of oxidized petroleum products along with an emulsion formed by the mixture of oil and water. This forms a pasty substance that clogs oil lines and passages and interferes with engine lubrication.

Specifications Any technical data, numbers, clearances, and measurements used to diagnose and adjust automobile components. They are also called specs.

Spread The excess of diameter at the outside parting edges over the housing bore.

Stress The force or strain to which a material is subjected.

Stroke In an automobile engine, the distance moved by the piston.

Studs Rods with threads cut on both ends, such as a cylinder stud, that screws into the cylinder block on one end and has a nut placed on the other end to hold the cylinder head in place.

Sump The oil pan that is bolted to the engine block at the crankcase.

Tachometer A device for measuring and indicating the rotative speed of an engine.

Tap To cut threads in a hole with a tapered, fluted, threaded tool.

Tappets Another term for valve lifters.

Tension Effort that elongates or "stretches" a material.

Throw With reference to an automobile engine, usually the distance from the center of the crankshaft main bearing to the center of the connecting rod journal.

Throw-off The quantity of oil that escapes at the end of the bearings, and lubricates adjacent engine parts while engine is running.

Thrust Bearing An antifriction bearing designed to absorb any thrust along the axis of the rotating shaft.

Thrust Plate The plate used to bolt the camshaft to the block, which absorbs camshaft thrust.

Timing The process of identifying when air, fuel, and ignition occur in relation to the crankshaft rotation.

Timing Chain The chain that drives the camshaft and accessory shafts of an engine.

Timing Gears Any group of gears driven from the engine crankshaft to cause the valves, ignition, and other engine-driven apparatus to operate at the desired time during the engine cycle.

Tolerance A permissible variation between the two extremes of a specification or dimension.

Torque A twisting force applied to a shaft or bolt.

Torque Wrench A special wrench with a built-in indicator to measure the applied force.

Turbulence A disturbed or irregular motion of fluids or gases.

Undersize Either an inside or outside diameter that is less than standard size.

Valve A device used to open and close a port to let intake and exhaust gases in and out of the engine.

Valve Clearance The clearance or space between the valve and the rocker arm. As the parts heat up, the clearance is reduced because of expansion. This keeps the valves from remaining open when the engine is hot.

Valve (Exhaust) A poppet type valve consisting of a metal disc mounted on one end of a coaxial stem that opens and closes a circular hole leading to the exhaust manifold.

Valve Guide The part in the cylinder head that holds the stem of the valve.

Valve Lifter A pushrod or plunger placed between the cam and the valve on an engine; often adjustable to vary the length of the unit.

Valve Seat The matched surface upon which the valve face rests.

Valve Train The components of an engine necessary to convert camshaft movement to valve movement.

Varnish   A deposit in an engine lubrication system resulting from oxidation of the motor oil. Varnish is similar to, but softer than, lacquer.

Vibration Damper   A device to reduce the torsional or twisting vibration that occurs along the length of the crankshaft used in multiple cylinder engines; also known as a harmonic balancer.

Wear   The gradual change in dimension due to loss of surface metal caused by the friction and heat generated during the operation of the engine.

Wear (Excessive)   Wear caused by overloading or an out-of-balance condition of a factor affecting wear, resulting in lower-than-normal life expectancy of the part or parts being subjected to the adverse operating condition.

Wear (Normal)   The average expected wear when operating under normal conditions.

APPENDIX B

MEASUREMENT EQUIVALENTS

CONVERSION FACTORS		
Multiply	**By**	**To obtain**
Length		
Millimeters (mm)	0.03937	Inches
	0.1	Centimeters (cm)
Kilometers (km)	0.6214	Miles
	3281	Feet
Inches	25.4	Millimeters (mm)
Miles	1.6093	Kilometers (km)
Area		
Inches^2	645.16	Millimeters^2 (mm^2)
	6.452	Centimeters^2 (cm^2)
Feet^2	0.0929	Meters^2 (m^2)
	144	Inches^2
Volume		
Centimeters^3 (cc)	0.06102	Inches^3
	0.001	Liters (L)
Liters (L)	61.024	Inches^3
	0.2642	Gallons
	1.0567	Quarts
Inches^3	16.387	Centimeters^3 (cc)
Feet^3	1728	Inches^3
	7.48	Gallons
	28.32	Liters (L)
Fluid ounces (oz)	29.57	Milliliters (mL)
Mass		
Gram (g)	0.03527	Ounce
Kilograms (kg)	2.2046	Pounds
	35.274	Ounces
Force		
Ounce	0.278	Newton (N)
Pound	4.448	Newton (N)
Kilogram	9.807	Newton (N)
Torque		
Foot-pounds	1.3558	Newton-meters (Nm)
	0.1383	Kilogram/meter (kg/m)

CONVERSION FACTORS (CONTINUED)

Multiply	By	To obtain
Torque		
Inch-pounds	0.11298	Newton-meters (Nm)
	0.0833	Foot-pounds
Kilogram-meters (Kg/m)	7.23	Foot-pounds
	9.80665	Newton-meters (Nm)
Pressure		
Atmospheres	14.7	Pounds/square inch (psi)
	29.92	Inches of mercury (In. Hg)
Inches of mercury (In. Hg)	0.49116	Pounds/square inch (psi)
	13.1	Inches of water
	3.377	Kilopascals (kPa)
Bars	100	Kilopascals (kPa)
	14.5	Pounds/square inch (psi)
Kilogram/cu² (Kg/cm²)	14.22	Pounds/square inch (psi)
	98.07	Kilopascals (kPa)
Kilopascals (kPa)	0.145	Pounds/square inch (psi)
	0.2961	Inches of mercury (In. Hg)
Pascal	1	Newton/square meter (N/m²)
Fuel Performance		
Miles/gallon	0.4251	Kilometers/liter (km/L)
Velocity		
Miles/hours	1.467	Feet/second
	88	Feet/minute
	1.6093	Kilometers/hour (km/h)
Kilometers/hour	0.27778	Meters/second (m/s)

TORQUE CONVERSIONS
(FOOT-POUNDS, NEWTON-METERS, METER-KILOGRAMS)

ft/lb	0	1	2	3	4	5	6	7	8	9
0	0	1.35	2.70	4.05	5.40	6.75	8.10	9.45	10.8	12.1
10	13.5	14.9	16.2	17.6	18.9	20.3	21.6	22.9	24.3	25.6
20	27.0	28.3	29.7	31.0	32.5	33.7	35.1	36.4	37.8	39.1
30	40.5	41.8	43.2	44.5	45.9	47.2	48.6	49.9	51.3	52.6
40	54.0	55.3	56.7	58.0	59.4	60.7	62.1	63.4	64.8	66.1
50	67.5	68.8	70.2	71.5	72.9	74.2	75.6	76.9	78.3	79.6
60	81.0	82.3	83.7	85.0	86.4	87.7	89.1	90.4	91.8	93.1
70	94.5	95.8	97.2	98.5	99.9	101	102	103	105	106
80	108	109	110	112	113	114	116	117	118	120
90	121	122	124	125	126	128	129	130	132	133
100	135	136	137	139	140	141	143	144	145	147
110	148	149	151	152	153	155	156	157	159	160
120	162	163	164	166	167	168	170	171	172	174
130	175	176	178	179	180	182	183	184	186	187

Note: The following formulas can be used:

$$\text{ft-lb} \times 1.35 = \text{Nm}$$
$$\text{ft-lb} \div 7.23 = \text{mkg}$$

For meter-kilograms within 5%, divide Newton-meters by 10 (move the decimal point one place left).
For greater accuracy, divide by 9.81.

DECIMAL AND METRIC EQUIVALENTS

Fractions	Decimal (in.)	Metric (mm)	Fractions	Decimal (in.)	Metric (mm)
1/64	.015625	.397	33/64	.515625	13.097
1/32	.03125	.794	17/32	.53125	13.494
3/64	.046875	1.191	35/64	.546875	13.891
1/16	.0625	1.588	9/16	.5625	14.288
5/64	.078125	1.984	36/64	.578125	14.684
3/32	.09375	2.381	19/32	.59375	15.081
7/64	.109375	2.778	39/64	.609375	15.478
1/8	.125	3.175	5/8	.625	15.875
9/64	.140625	3.572	41/64	.640625	16.272
5/32	.15625	3.969	21/32	.65625	16.669
11/64	.171875	4.366	43/64	.671875	17.066
3/16	.1875	4.763	11/16	.6875	17.463
13/64	.203125	5.159	45/64	.703125	17.859
7/32	.21875	5.556	23/32	.71875	18.256
15/64	.234275	5.953	47/64	.734375	18.653
1/4	.250	6.35	3/4	.750	19.05
17/64	.265625	6.747	49/64	.765625	19.447
9/32	.28125	7.144	25/32	.78125	19.844
19/64	.296875	7.54	51/64	.796875	20.241
5/16	.3125	7.938	13/16	.8125	20.638
21/64	.328125	8.334	53/64	.828125	21.034
11/32	.34375	8.731	27/32	.84375	21.431
23/64	.359375	9.128	55/64	.859375	21.828
3/8	.375	9.525	7/8	.875	22.225
25/64	.390625	9.922	57/64	.890625	22.622
13/32	.40625	10.319	29/32	.90625	23.019
27/64	.421875	10.716	59/64	.921875	23.416
7/16	.4375	11.113	15/16	.9375	23.813
29/64	.453125	11.509	61/64	.953125	24.209
15/32	.46875	11.906	31/32	.96875	24.606
31/64	.484375	12.303	63/64	.984375	25.003
1/2	.500	12.7	1	1.00	25.4

TAP AND DRILL BIT DATA

	DECIMAL EQUIVALENTS AND TAP DRILL SIZES					
Drill Size	**Decimal**	**Tap Size**	**Drill Size**	**Decimal**	**Tap Size**
1/64	.0156		28	.1405	8–40
1/32	.0312		9/64	.1406	
60	.0400		27	.1440	
59	.0410		26	.1470	
58	.0420		25	.1495	10–24
57	.0430		24	.1520	
56	.0465		23	.1540	
3/64	.0469	0–80	5/32	.1562	
55	.0520		22	.1570	10–30
54	.0550	1–56	21	.1590	10–32
53	.0595	1–64, 72	20	.1610	
1/16	.0625		19	.1660	
52	.0635		18	.1695	
51	.0670		11/64	.1719	
50	.0700	2–56, 64	17	.1730	
49	.0730		16	.1770	12–24
48	.0760		15	.1800	
5/64	.0781		14	.1820	12–28
47	.0785	3–48	13	.1850	12–32
46	.0810		3/16	.1875	
45	.0820	3–56, 4–32	12	.1890	
44	.0860	4–36	11	.1910	
43	.0890	4–40	10	.1935	
42	.0935	4–48	9	.1960	
3/32	.0937		8	.1990	
41	.0960		7	.2010	1/4–20
40	.0980		13/64	.2031	
39	.0995		6	.2040	
38	.1015	5–40	5	.2055	
37	.1040	5–44	4	.2090	
36	.1065	6–32	3	.2130	1/4–28
7/64	.1093		7/32	.2187	
35	.1100		2	.2210	
34	.1110	6–36	1	.2280	
33	.1130	6–40	A	.2340	
32	.1160		15/64	.2344	
31	.1200		B	.2380	
1/8	.1250		C	.2420	
30	.1285		D	.2460	
29	.1360	8–32, 36	E, 1/4	.2500	

DECIMAL EQUIVALENTS AND TAP DRILL SIZES (CONTINUED)

Drill Size	Decimal	Tap Size	Drill Size	Decimal	Tap Size
F	.2570	5/16–18	31/64	.4844	9/16–12
G	.2610		1/2	.5000	
17/64	.2656		33/64	.5156	9/16–18
H	.2660		17/32	.5312	5/8–11
I	.2720	5/16–24	35/64	.5469	
J	.2770		9/16	.5625	
K	.2810		37/64	.5781	5/8–18
9/32	.2812		19/32	.5937	11/16–11
L	.2900		39/64	.6094	
M	.2950		5/8	.6250	11/16–16
19/64	.2968		41/64	.6406	
N	.3020		21/32	.6562	3/4–10
5/16	.3125	3/8–16	43/64	.6719	
O	.3160		11/16	.6875	3/4–16
P	.3230		45/64	.7031	
21/64	.3281		23/32	.7187	
Q	.3320	3/8–24	47/64	.7344	
R	.3390		3/4	.7500	
11/32	.3437		49/64	.7656	7/8–9
S	.3480		25/32	.7812	
T	.3580		51/64	.7969	
23/64	.3594		13/16	.8125	7/8–14
U	.3680	7/16–14	53–64	.8281	
3/8	.3750		27/32	.8437	
V	.3770		55/54	.8594	
W	.3860		7/8	.8750	1–8
25/64	.3906	7/16–20	57/64	.8906	
X	.3970		29/32	.9062	
Y	.4040		59/64	.9219	
13/32	.4062		15/16	.9375	1–12, 14
Z	.4130		61/64	.9531	
27/64	.4219	1/2–13	31/32	.9687	
7/16	.4375		63/64	.9844	
29/64	.4531	1/2–20	1	1.000	
15/32	.4687				

PIPE THREAD SIZES

Thread	Drill	Thread	Drill
1/8–27	R	1 1/2–11 1/2	1 47/64
1/4–18	7/16	2–11 1/2	2 7/32
3/8–18	37/64	2 1/2–8	2 5/8
1/2–14	23/32	3–8	3 1/4
3/4–14	59/64	3 1/2–8	3 3/4
1–11 1/2	1 5/32	4–8	4 1/4
1 1/4–11 1/2	1 1/2		

METRIC TAP DRILL SIZES

Diameter and Pitch	Metric Drill	Inch Drill
5 × .80	4.20	11/64
6 × 1.00	5.00	13/64
7 × 1.00	6.00	15/64
8 × 1.25	6.75	17/64
10 × 1.50	8.50	11/32
12 × 1.75	10.25	13/32

Note: Nominal outside diameter minus the pitch equals the tap drill size for 75% thread contact area.

DECIMAL EQUIVALENTS OF NUMBER SIZE DRILLS

No.	Size of Drill (Inches)	No.	Size of Drill (Inches)	No.	Size of Drill (Inches)	No.	Size of Drill (Inches)
1	.2280	21	.1590	41	.0960	61	.0390
2	.2210	22	.1570	42	.0935	62	.0380
3	.2130	23	.1540	43	.0890	63	.0370
4	.2090	24	.1520	44	.0860	64	.0360
5	.2055	25	.1495	45	.0820	65	.0350
6	.2040	26	.1470	46	.0810	66	.0330
7	.2010	27	.1440	47	.0785	67	.0320
8	.1990	28	.1405	48	.0760	68	.0310
9	.1960	29	.1360	49	.0730	69	.0292
10	.1935	30	.1285	50	.0700	70	.0280
11	.1910	31	.1200	51	.0670	71	.0260
12	.1890	32	.1160	52	.0635	72	.0250
13	.1850	33	.1130	53	.0595	73	.0240
14	.1820	34	.1110	54	.0550	74	.0225
15	.1800	35	.1100	55	.0520	75	.0210
16	.1770	36	.1065	56	.0465	76	.0200
17	.1730	37	.1040	57	.0430	77	.0180
18	.1695	38	.1015	58	.0420	78	.0160
19	.1660	39	.0995	59	.0410	79	.0145
20	.1610	40	.0980	60	.0400	80	.0135

DECIMAL EQUIVALENTS OF LETTER SIZE DRILLS

Letter	Size of Drill (Inches)	Letter	Size of Drill (Inches)
A	.234	N	.302
B	.238	O	.316
C	.242	P	.323
D	.246	Q	.332
E	.250	R	.339
F	.257	S	.348
G	.261	T	.358
H	.266	U	.368
I	.272	V	.377
J	.277	W	.386
K	.281	X	.397
L	.290	Y	.404
M	.295	Z	.413

HELICOIL TAP DRILL SIZES

Inches		Metric	
Thread	Tap Drill Size	Thread	Tap Drill Size
1/4–20	17/64	5 × 0.8	13/64
5/16–18	21/64	6 × 1.0	1/4
3/8–16	25/64	7 × 1.0	9/32
7/16–14	29/64	8 × 1.25	21/64
1/2–13	17/32	10 × 1.50	13/32
		12 × 1.75	31/64

APPENDIX D

REFERENCE TABLES

DRY TORQUE RECOMMENDATIONS

Diameter	Grade 5		Grade 6		Grade 8	
	Coarse	Fine	Coarse	Fine	Coarse	Fine
1/4	108[a]	120[a]	132[a]	156[a]	17	19
5/16	17	20	23	25	34	37
3/8	31	35	40	45	60	68
7/16	50	55	64	72	96	108
1/2	75	85	98	110	145	165
9/16	110	120	140	160	210	235
5/8	150	170	195	220	290	330

[a]Any torque less than 15-lb is given in in-lb.

LUBRICATED TORQUE RECOMMENDATIONS
(REDUCED 33 PERCENT)

Diameter	Grade 5		Grade 6		Grade 8	
	Coarse	Fine	Coarse	Fine	Coarse	Fine
1/4	72[a]	80[a]	88[a]	104[a]	136[a]	152[a]
5/16	135[a]	160[a]	15	17	22	25
3/8	21	23	26	30	40	45
7/16	33	37	43	48	64	72
1/2	50	57	65	73	96	110
9/16	73	80	93	107	140	157
5/8	100	113	130	147	193	220

[a]Any torque less than 15-lb is given in in-lb.

METRIC DRY TORQUE RECOMMENDATIONS

Diameter (mm)	Pitch	Grade 8.8	Grade 12.9
6	1.00	84[a]	132[a]
7	1.00	132[a]	20
8	1.25	18	29
8	1.00	20	32
10	1.50	33	58
10	1.25	35	61
10	1.00	38	64
12	1.75	59	100
12	1.50	65	110

[a]Any torque less than 15 ft-lb is given in in.-lb.

METRIC LUBRICATED TORQUE RECOMMENDATIONS
(REDUCED 33 PERCENT)

Diameter (mm)	Pitch	Grade 8.8	Grade 12.9
6	1.00	56[a]	88[a]
7	1.00	88[a]	160[a]
8	1.25	144[a]	19
8	1.00	160[a]	21
10	1.50	22	39
10	1.25	23	41
10	1.00	25	43
12	1.75	39	67
12	1.50	43	73

[a]Any torque less than 15 ft-lb is given in in.-lb.

TORQUE VALUES

The following charts give the standard torque values for bolts, nuts, and taperlock studs of SAE Grade 5 or better quality.

GENERAL TIGHTENING TORQUE FOR BOLTS, NUTS, AND TAPERLOCK STUDS			
Thread Diameter		Standard Torque	
Inches	Millimeters	ft.-lb.	N·m*
Standard Thread		Use these torques for bolts and nuts with standard threads (conversions are approximate).	
1/4	6.35	9 ± 3	12 ± 4
5/16	7.94	18 ± 5	25 ± 7
3/8	9.53	32 ± 5	45 ± 7
7/16	11.11	50 ± 10	70 ± 15
1/2	12.70	75 ± 10	100 ± 15
9/16	14.29	110 ± 15	150 ± 20
5/8	15.88	150 ± 20	200 ± 25
3/4	19.05	265 ± 35	360 ± 50
7/8	22.23	420 ± 60	570 ± 80
1	25.40	640 ± 80	875 ± 100
1-1/8	28.58	800 ± 100	1100 ± 150
1-1/4	31.75	1000 ± 120	1350 ± 175
1-3/8	34.93	1200 ± 150	1600 ± 200
1-1/2	38.10	1500 ± 200	2000 ± 275
		Use these torques for bolts and nuts on hydraulic valve bodies.	
5/16	7.94	13 ± 2	20 ± 3
3/8	9.53	24 ± 2	35 ± 3
7/16	11.11	39 ± 2	50 ± 2
1/2	12.70	60 ± 3	80 ± 4
5/8	15.88	118 ± 4	160 ± 6

GENERAL TIGHTENING TORQUE FOR BOLTS, NUTS, AND TAPERLOCK STUDS (CONTINUED)

Taperlock Stud   Use these torques for studs with taperlock threads.

1/4	6.35	5 ± 2	7 ± 3
5/16	7.94	10 ± 3	15 ± 5
3/8	9.53	20 ± 3	30 ± 5
7/16	11.11	30 ± 5	40 ± 10
1/2	12.70	40 ± 5	55 ± 10
9/16	14.29	60 ± 10	80 ± 15
5/8	15.88	75 ± 10	100 ± 15
3/4	19.05	110 ± 15	150 ± 20
7/8	22.23	170 ± 20	230 ± 30
1	25.40	260 ± 30	350 ± 40
1-1/8	28.58	320 ± 30	400 ± 40
1-1/4	31.75	400 ± 40	550 ± 50
1-3/8	34.93	480 ± 40	650 ± 50
1-1/2	38.10	550 ± 50	750 ± 70

*1 Newton-meter (N·m) is approximately the same as 0.1 mkg.

The torques shown in the charts that follow are to be used on the part of 37° flared, 45° flared and inverted flared fittings (when used with steel tubing), O-ring plugs, and O-ring fittings.

TORQUE FOR FLARED AND O-RING FITTINGS

Inverted 45° Flared   **37° Flared**

Tube Size (O.D.)	mm	3.18	4.78	6.35	7.92	9.52
	in.	.125	.188	.250	.312	.375
THREAD SIZE (in.)		5/16	3/8	7/16	1/2	9/16 5/8
TORQUE N·m		5 ± 1	11 ± 1	16 ± 2	20 ± 2	25 ± 3
TORQUE lb.-in.		45 ± 10	100 ± 10	145 ± 20	175 ± 20	225 ± 25

TORQUE FOR FLARED AND O-RING FITTINGS

		45° Flared		O-Ring Fitting—Plug			Swivel Nuts		
Tube Size (O.D.)	mm	12.70	15.88	19.05	22.22	25.40	31.75	38.10	50.80
	in.	.500	.625	.750	.875	1.000	1.250	1.500	2.000
THREAD SIZE (in.)		3/4	7/8	11/16	1-3/16 1-1/4	1-5/16	1-5/8	1-7/8	2-1/2
TORQUE N·m		50 ± 5	75 ± 5	100 ± 5	120 ± 5	135 ± 10	180 ± 10	225 ± 10	320 ± 15
TORQUE lb.-ft.		35 ± 4	55 ± 4	75 ± 4	90 ± 4	100 ± 7	135 ± 7	165 ± 7	235 ± 10

INDEX

Abrasive cleaners, 124, 136–39
  blaster, 137–39
    disadvantages, 139
    grit, 138
    shot materials, 137–38
    shot size, 138–39
  glass beads, 139
  how they work, 124, 136–37
  parts tumbler, 139
  vibrating tub, 139
Adhesives, 435
Aligning bar, 17–18
Aluminum heads, 243–47, 249–53
  straightening, 249–53
  welding, 243–47
Antiseize compounds, 437
ASE certification, 30–31

Balancer shaft, 70
Balancing, engine, 381–87
  bobweight, 382, 383–85
    computing, 382
    making up, 383–85
  components involved, 381
  connecting rod, match weighing, 383
  crankshaft, 385–87
    design, 385–87
    flywheel, 387
    procedure, 385
  pistons, weight removal, 382–83
Ball valve rotator, 164
Bearings, inspecting, 215–27
  common materials, 216
  crush, 217
  designs, 217
  distress, 220–25
  locating devices, 218
  oil clearance, 220
  oil grooves and holes, 218–19
  replacement, 225–27
    determining correct undersize, 225–27
    proper oil clearance, 225
  spread, 217
  types, 216
Belt-driven camshaft, 67
Belt surfacers, 254–58
Bolts, 412–14, 417–19
Bore and stroke, 40–41
Boring, cylinders, 317–21
Brake mean effective pressure, 44
Broaching machines, 261–63

Cam bearings, 450–53
Camshaft, 60–67, 167–71, 373–81
  basic operation, 60–61
  inspecting, 167–71
  overhead, 61
  pushrod guide plates, 64–65
  pushrods, 64
  reconditioning, 373–81
    basic materials, 373
    chilled cast iron, 376
    grinding, 376–78
    high-carbon steel, 375–76

low-carbon steel, 374–75
    malleable cast iron, 376
    spark test, 374
    straightening, 376
    timing, 378–81
  rocker arms, 65–66
  timing mechanisms, 66–67
    auxiliary functions, 67
    belt drive, 67
    chain drive, 66–67
    gear drive, 66
    timing belt, 67
    timing marks, 67
  valve lifters, 61–64
  hydraulic, 63–64
  solid, 61–63
Chain-driven camshaft, 66–67
Chemical cleaners, 124–32
  common agents, 125–26
    acid-based, 126
    alkaline-based, 125–26
    degreasers, 126
    metal inhibitors, 126
  hot spray tanks, 130–32
    advantages, 130–31
    maintenance, 132
    types, 131–32
  how they work, 124
  parts washers, 127–28
  soak tanks, 128–29
    cold, 128–29
    hot, 129
  steam, 127
  waste disposal, 124–25
Citrus chemical cleaning, 132–33
Cleaning methods, 123–42
  abrasives, 124, 136–39
    blaster, 137–39
    glass beads, 139
    how they work, 124, 136–37
    parts tumbler, 139
    vibrating tub, 139
  alternative methods, 132–33
  chemicals, 124–32
    common agents, 125–26
    hot spray tanks, 130–32
    parts washers, 127–28
    soak tanks, 128–29
    steam, 127
  manual, 139–41
  marking cleaned parts, 142
  thermal, 124, 133–36
    maintenance, 135–36
    procedure, 124, 134–35
    types of ovens, 133–34
  types of contaminants, 123–24
Combustion chamber, 51–54
  basic design, 51
    quenching, 51
    turbulence, 51
  chamber-in-piston, 53–54
  fast burn, 54
  hemispherical, 52–53

swirl, 53
wedge, 51–52
Compression gauge, 83–84
Compression ratio, 43, 267–69
Compression rings, 72
Compression testing, 81–86
Connecting rods, 72–73, 206–10, 335–40
    basic design 72–73
    common problems, 206–10
    diesel design, 210
    reconditioning bores, 337–40
        machinery, 337–38
        procedure, 338–40
    testing with an electronic gauge, 335–36
Cooling, engine, 406–9
    components, 406–7
    design, 407
    types, 408–9
        air, 409
        liquid, 408–9
Core plugs, 449–50
Cracks, 229–47
    causes, 229
    ceramic sealing, 238
    detecting, 230–34
        dye penetrant, 231–32
        magnetic, 230–31
        pressure testing, 232–34
    epoxy repair, 238–39
    pinning, 234–38
        external areas, 237
        installing a tapered plug, 234–35
        peening, 235–37
        plugs on both sides, 237–38
    shot peening, 239–40
        precautions, 240
        testing for maximum saturation, 240
        when to use, 239
    welding, 241–47
        aluminum head, 243–47
        heat or furnace, 243
        procedures, 240–42
        safety, 242–43
Crankshaft, inspecting, 67–70, 181–88, 337, 352–54
    basic design, 67–70
    clearance, 186
    end play, 186
    flywheel, 186–87
        automatic, 187
        manual shift, 186–87
    magnetic particle test, 354
    measuring journals, 182–83
    saddle alignment, 183–85
    specifications, 337
    straightness, 185–86
    vibration damper, 187–88
    visual exam, 181, 352–54
Crankshaft, reconditioning, 347–73
    align boring, 368–73
        causes of misalignment, 369–72
        checking alignment, 371
        mounting the block, 372–73
    building up, 365–66
        chromium plating, 365–66
        electro-welding, 366
    classifying, 348–51
        cylinder numbering, 348–50

journals, 350–51
grinding, 354–61, 362–65
    basic procedure, 358–61
    flywheel, 362–65
    setting up machinery, 356–58
journal hardening, 362
    heat treating, 362
    salt-bath nitriding, 362
polishing, 361
straightening, 366–68
vibration damper, 365
Cylinder blocks, 47–49, 55, 114–17, 210–15. *See also*
Cylinder heads, reconditioning
    aluminum, 49
    bores, inspecting, 212–15
        measuring, 213–15
        out-of-roundness, 213
        surface finish, 215
        taper, 213
    cast iron, 49
    core plugs, 55
    deck, inspecting, 210–12
        checking flatness, 210–11
        when to resurface, 210–12
    design, 48–49
    disassembly, 114–17
Cylinder bore dial gauge, 16–17
Cylinder head gaskets, 424–29
Cylinder heads, design, 49–51, 54
    aluminum, 50–51
    coolant passages, 54
    head gaskets, 49
    intake and exhaust ports, 54
    lightweight vs. heavy, 49–50
Cylinder heads, inspecting, 143–79
Cylinder heads, reconditioning, 110–14, 249–70
    aluminum, straightening, 249–53
    belt surfacers, 254–58
    broaching machines, 261–63
    clearance volume, 266–67
    compression ratio, 267–69
    disassembly, 110–14
    excessive surfacing problems, 265
    machining V-6 and V-8 engines, 269–70
    measuring surface finish, 253–54
    milling machines, 258–61
    surface grinders, 263–65
    when to resurface, 253
Cylinder leakage testing, 86–89
Cylinder power balance testing, 92–94
Cylinder sleeves, 55–56
Cylinders, reconditioning, 315–29
    boring, 317–21
    deglazing, 316–17
    dry sleeves, 326–27
    honing, 321–26
        finish problems, 325
        hone design, 321
        importance of, 321
        line type, 325–26
        machinery, 321
        procedure, 323–25
    importance of, 315–16
    wet sleeves, 328–29

Deglazing, cylinder, 316–17
Depth micrometer, 12–13

Diagnosis, of engine, 81–101
  checking oil consumption, 94–98
  compression test, 81–86
    analyzing results, 86
    compression gauge, 83–84
    dry, 84–85
    preparation, 84
    pressure specifications, 81–83
    remote starter switch, 84
    wet, 85–86
  cylinder leakage test, 86–89
    analyzing results, 88–89
    leakage tester, 86
    preparation, 87
    procedure, 87–88
  cylinder power balance test, 92–94
    precautions, 92–93
    procedure, 93–94
  diesel engine, 100–101
  evaluating engine condition, 94
  noise, 98–100
  common types, 99–100
  using a stethoscope, 98
  vacuum test, 89–92
    basic procedure, 89–90
    cranking, 90–91
    exhaust system, 91–92
    loss of compression, 92
    vacuum gauge, 89
Dial caliper, 8–9
Diesel engine, 76–79, 100–101, 443–45
  basic operation, 76–77
  combustion chambers, 77–78, 443–44
  diagnosing common problems, 77–78
  fuel injection, 78–79
  sealing, 443–45
    cylinder sleeves, 445
    gaskets, 444
Disassembly, of engine, 103–17
  cylinder block, 114–17
    freeze plugs, 116–17
    main bearings, 116
    oil gallery plugs, 117
    piston and rod assembly, 116
    ridge, 115
    timing components, 114–15
  cylinder head, 110–14
    camshaft, 112–14
    rocker assembly, 110
    valve cover, 110
    valve train, 114
  lifting out, 108–10
    equipment, 108
    procedure, 109–10
  vehicle preparation, 104–8
Displacement, 42–43
Dry compression test, 84–85
Dye penetrant crack detection, 231–32
Dynamometer, 81

Electronic gauges, 19–21
  comparing dimensions, 20
  maintenance, 21
  reading, 19
  setting tolerances, 20
  switching scales, 19–20
Engine analyzer, 81, 92
Engine efficiency, measuring, 43

Epoxy repair, 238–39
Exhaust manifold, 47
Exhaust valves, 56–57
Expansion plugs, 55
Fasteners, 411–20
  bolts, 412–14, 417–19
    component parts, 412–13
    fillet damage, 415
    installing, 417–19
  thread repair, 419–20
  torque, 414–17
    elasticity, 414–15
    specifications, 415–17
  types and sizes, 411–12
Feeler gauge, 14
Flywheel, 70, 186–87, 362–65, 387
Followers, 304–5
Four-gas analyzer, 81
Four-stroke engine, 34–38
  basic operation, 34–36
  combustion cycle, 36–38
    compression stroke, 36
    exhaust stroke, 36–38
    intake stroke, 36
    power stroke, 36

Gaskets, 73, 420–34
  basic function, 73, 420
  basic types, 420–21
  cylinder head, 424–29
    basic operation, 424
    bimetal engines, 424–25
    design and materials, 425–26
    diesel engines, 428–29
    impression testing, 428
    installing, 426–27
    performance engine, 429
  installation precautions, 423–24
  manifold, 429–31
    combination, 431
    exhaust, 431
    intake, 429–31
  materials, 421–22
  oil pan, 433–34
  valve cover, 432–33
Gauges, 13–17, 19–21
  cylinder bore dial, 16–17
  electronic, 19–21
    advantages, 19–20
    maintenance, 21
  feeler, 14
  out-of-roundness, 17
  radius, 14–15
  screw pitch, 15
  small hole, 13–14
  telescoping, 13
  valve seat runout, 16
Gear-driven camshaft, 66
Grinding, 271–76, 292–97, 354–61, 363–65
  crankshaft, 354–61
  flywheel, 362–65
  valves, 271–76
  valve seats, 292–97

Hand tools, 5
Hazardous waste, 28–30
  characteristics of, 29
  employer responsibility, 28
  testing for, 30

Honing, cylinder, 321–26
Hydraulic valve lifters, 63–64, 172–73

In-line engine, 38
Inside micrometer, 12
Installing the engine, 475–79
Intake manifold, 45–47
Intake valves, 56–57

Journals, 68–69, 182–83, 350–51, 359–61, 362

Knurling, 279–82, 330–31
    pistons, 330–31
    valve guides, 279–82

Leakage, of oil, 94–98
Leakage tester, cylinder, 86
Lubrication, 389–406. *See also* Oil, checking
consumption
    filters, 391, 403
    filtration, 396–98
        bypass, 396–97
        full flow, 396
        independent, 397
        shunt, 397–98
    oil pressure problems, 393–94
    other components, 389–93
    pressure indicators, 395–96
    pressure regulation, 395
        gauges, 395
        valves, 395
    pumps, 389, 393, 394–95, 398–403
        design, 389, 393
        gear, 394–95
        inspecting, 398–99
        rotor, 394
        servicing, 400–403
    types of oil, 403–6

Machinist's rule, 8
Magnetic crack detection, 230–31
Main bearings, 456–57
Manifold gaskets, 429–31
Manifolds, 45–47
    exhaust, 47
    intake, 45–47
Mechanical valve lifters, 61–63, 171–72
Micrometer, 9–13
    common components, 9
    depth, 12–13
    inside, 12
    outside, 9–11
        inch-graduated, 9–10
        metric, 10–11
        vernier scale, 10
Milling machine, 258–61
Mobile crane, 108
Multivalve engine, 57–60

Noise, diagnosing, 98–100
    abnormal combustion, 100
    damper or flywheel, 100
    main or thrust bearings, 99–100
    piston pin knock, 99
    piston slap, 99
    ridge, 99
    rings, 99
    rod bearings, 99
    tappets, 100
    using a stethoscope, 98

Oil, checking consumption, 94–98. *See also* Lubrication
    external leaks, 94–95
        abnormal crankcase pressure, 95
        normal crankcase pressure, 94–95
    high temperature thinning, 95
    leakage indicators, 96–98
        exhaust pipe, 96–97
        spark plugs, 97
        valve guides, 97–98
    testing for leaks, 96
        fluorescent tracer dye, 96
        high-pressure, 96
        low-pressure, 96
        unburned fuel dilution, 96
Oil control rings, 72
Oil gallery plugs, removing, 117
Oil pan gaskets, 433–34
Oil seals, 73
Opposed cylinder engine, 38
O-rings, valve stem, 303
Out-of-roundness gauge, 17
Outside mircometer, 9–11
Overhead camshaft, 61

Piston pins, 204–5, 207–10, 340–45
    assuring good fit, 340
    causes of failure, 207–10
    design, 204–5
    reconditioning, 341–45
        full-floating pin, 341, 342–43
        heat method, 345
        oscillating pin, 341, 343
        press method, 343–44
Piston rings, 71–72, 200–204, 205–6, 332–35
    causes of failure, 205–6
    checking clearance, 202–3, 332
    coatings, 201–2
    compression rings, 200–201
    design, 71–72
    diesel, 203–4
    installing, 332–35
    leakage, 332
    low-tension, 202
Pistons, 70–71, 116, 191–200, 203–4, 329–31
    inspecting, 70–71, 191–200, 203–4
        causes of failure, 194–200
        design, 70–71, 192–93
        diesel, 203–4
        measuring, 193–94
    reconditioning, 329–31
        feeler strip and scale, 330–31
        knurling, 330
        measuring clearance, 329–30
    removing, 116
Piston skirt, 71
Plunge grinding, 360–61
Pneumatic tools, 5–6
Pressure testing for cracks, 232–34
Pushrods, 64–65, 175–76, 305

Radius gauge, 14–15
Reaming, valve guides, 282–83
Reassembling, the engine, 447–79
    cam bearings, 450–53
        common problems, 451–53
        procedure, 450–51
    camshaft, 453–56
        end play, 455–56
        procedure, 453–54

core plugs, 449–50
  cup, 449
  disc, 449
  expansion, 449–50
crankshaft, 457–58
cylinder head, 466–68
  final installation, 475–79
  engine break-in, 477–78
  engine start-up, 477
  precautions, for owner, 478–79
  procedure, 475–77
main bearings, 456–57
oil pump, 471–72
pistons and connecting rod, 459–63
prelubrication check, 473–74
  benefits, 474
  methods, 473–74
rechecking the block, 447–48
recleaning the block, 448–49
valve train, 468–71
Remote starter switch, 84
Removal, of engine. *See* Disassembly
Replacement parts, 118–21
  complete engine assembly, 120
  cores, 120–21
    credit guarantee, 120–21
    import vehicles, 121
    suppliers, 121
  engine parts kit, 119
  lower end engine kit, 118–19
  ordering, 118
  short block assembly, 120
  super or master kit, 119–20
Rocker arms, 65–66, 110, 176–79, 305–8

Safety, 24–30, 242–43
  hazardous wastes, 28–30
  personal, 24–25
  tools and machinery, 26–28
  welding, 242–43
  work area, 25–26
Salt bath cleaning, 133
Screw pitch gauge, 15
Sealing materials, 435–43
  adhesives, 435
  antiseize compounds, 437
  other types of seals, 442–43
  rear main bearing, 439
  rope- or wick-type, 440–42
  rubber oil, 439–40
  sealants, 435–37
  timing cover oil, 438–39
Service manuals, types of, 21–22
Siamese ports, 54
Slant cylinder engine, 38
Small hole gauge, 13–14
Spring-loaded valve rotator, 164
Surface grinders, 263–65
Sweep grinding, 360

Telescoping gauge, 13
Testing, of engine. *See* Diagnosis
Thermal cleaning, 124, 133–36
Timing, camshaft/crankshaft, 378–81
Torque, 414–17
Torque indicating wrench, 18–19
Two-stroke engine, 74–75

Ultrasonic cleaning, 132
Vacuum gauge, 89
Vacuum testing, 89–92
Valve cover gaskets, 432–33
Valve designs 40
Valve seat runout gauge, 16
Valves, inspecting, 143–67
  breakage, 153–56
  burning, 146–53
  guides, 158–59
  keeper grooves, 166–67
  locks, 166
  rotators, 163–65
  seats, 159–60
  springs, 161–63
  stem seals, 165–66
  worn stems and tips, 156–57
Valves, intake and exhaust, 56–57
Valves, reconditioning, 271–303
  grinding, 271–76
  guides, 276–89
  seats, 290–300
  stem seals, 300–303
Valve timing drive assembly, inspecting, 188–90
  chain and sprocket, 189
  rubber timing belt, 190
  timing chain deflection, 188–89
  timing gear backlash, 190
Valve train, inspecting, 167–79
  camshaft, 167–71
  lifters, 171–75
  pushrods, 175–76
  rocker arms, 176–79
  static engine analysis, 167
Valve train, reconditioning, 303–13
  followers, 304–5
  installed spring height, 309–13
  installed stem height, 308–9
  pushrods, 305
  rocker arms, 305–8
V-blocks, 18
Vehicle identification number, 3
Vernier scale micrometer, 10
Vibration damper, 70, 187–88, 365

Water pumps, 408–9
Welding, 241–47
  aluminum heads, 243–47
  heat or furnace, 243
  procedures, 240–42
  safety, 242–43
Wet compression test, 85–86
Workshop procedures, 1–3, 6, 22–28
  engine overhaul, 1–3
  engine tune-up, 1
  keeping records, 22–24
    billing, 24
    dispatch sheet, 23
    labor charges, 24
    parts requisition, 22–23
    work order, 22
  safety, 24–28
    personal, 24–25
    tools and machinery, 26–28
    work area, 25–26
  setting up a productive work space, 6